AF334020

Carlos Cotta, Simeon Reich, Robert Schaefer and Antoni Ligęza (Eds.)

Knowledge-Driven Computing

Carlos Cotta, Simeon Reich, Robert Schaefer and Antoni Ligęza (Eds.)

Knowledge-Driven Computing

Studies in Computational Intelligence, Volume 102

Editor-in-chief
Prof. Janusz Kacprzyk
Systems Research Institute
Polish Academy of Sciences
ul. Newelska 6
01-447 Warsaw
Poland
E-mail: kacprzyk@ibspan.waw.pl

Further volumes of this series can be found on our
homepage: springer.com

Vol. 78. Costin Badica and Marcin Paprzycki (Eds.)
Intelligent and Distributed Computing, 2008
ISBN 978-3-540-74929-5

Vol. 79. Xing Cai and T.-C. Jim Yeh (Eds.)
*Quantitative Information Fusion for Hydrological
Sciences*, 2008
ISBN 978-3-540-75383-4

Vol. 80. Joachim Diederich
Rule Extraction from Support Vector Machines, 2008
ISBN 978-3-540-75389-6

Vol. 81. K. Sridharan
Robotic Exploration and Landmark Determination, 2008
ISBN 978-3-540-75393-3

Vol. 82. Ajith Abraham, Crina Grosan and Witold
Pedrycz (Eds.)
Engineering Evolutionary Intelligent Systems, 2008
ISBN 978-3-540-75395-7

Vol. 83. Bhanu Prasad and S.R.M. Prasanna (Eds.)
*Speech, Audio, Image and Biomedical Signal Processing
using Neural Networks*, 2008
ISBN 978-3-540-75397-1

Vol. 84. Marek R. Ogiela and Ryszard Tadeusiewicz
*Modern Computational Intelligence Methods
for the Interpretation of Medical Images*, 2008
ISBN 978-3-540-75399-5

Vol. 85. Arpad Kelemen, Ajith Abraham and Yulan Liang
(Eds.)
Computational Intelligence in Medical Informatics, 2008
ISBN 978-3-540-75766-5

Vol. 86. Zbigniew Les and Mogdalena Les
Shape Understanding Systems, 2008
ISBN 978-3-540-75768-9

Vol. 87. Yuri Avramenko and Andrzej Kraslawski
Case Based Design, 2008
ISBN 978-3-540-75705-4

Vol. 88. Tina Yu, David Davis, Cem Baydar and Rajkumar
Roy (Eds.)
Evolutionary Computation in Practice, 2008
ISBN 978-3-540-75770-2

Vol. 89. Ito Takayuki, Hattori Hiromitsu, Zhang Minjie
and Matsuo Tokuro (Eds.)
Rational, Robust, Secure, 2008
ISBN 978-3-540-76281-2

Vol. 90. Simone Marinai and Hiromichi Fujisawa (Eds.)
*Machine Learning in Document Analysis
and Recognition*, 2008
ISBN 978-3-540-76279-9

Vol. 91. Horst Bunke, Kandel Abraham and Last Mark (Eds.)
Applied Pattern Recognition, 2008
ISBN 978-3-540-76830-2

Vol. 92. Ang Yang, Yin Shan and Lam Thu Bui (Eds.)
Success in Evolutionary Computation, 2008
ISBN 978-3-540-76285-0

Vol. 93. Manolis Wallace, Marios Angelides and Phivos
Mylonas (Eds.)
*Advances in Semantic Media Adaptation and
Personalization*, 2008
ISBN 978-3-540-76359-8

Vol. 94. Arpad Kelemen, Ajith Abraham and Yuehui Chen
(Eds.)
Computational Intelligence in Bioinformatics, 2008
ISBN 978-3-540-76802-9

Vol. 95. Radu Dogaru
*Systematic Design for Emergence in Cellular Nonlinear
Networks*, 2008
ISBN 978-3-540-76800-5

Vol. 96. Aboul-Ella Hassanien, Ajith Abraham and Janusz
Kacprzyk (Eds.)
*Computational Intelligence in Multimedia Processing:
Recent Advances*, 2008
ISBN 978-3-540-76826-5

Vol. 97. Gloria Phillips-Wren, Nikhil Ichalkaranje and
Lakhmi C. Jain (Eds.)
Intelligent Decision Making: An AI-Based Approach, 2008
ISBN 978-3-540-76829-9

Vol. 98. Ashish Ghosh, Satchidananda Dehuri and Susmita
Ghosh (Eds.)
*Multi-Objective Evolutionary Algorithms for Knowledge
Discovery from Databases*, 2008
ISBN 978-3-540-77466-2

Vol. 99. George Meghabghab and Abraham Kandel
Search Engines, Link Analysis, and User's Web Behavior,
2008
ISBN 978-3-540-77468-6

Vol. 100. Anthony Brabazon and Michael O'Neill (Eds.)
Natural Computing in Computational Finance, 2008
ISBN 978-3-540-77476-1

Vol. 101. Michael Granitzer, Mathias Lux and Marc Spaniol
(Eds.)
Multimedia Semantics - The Role of Metadata, 2008
ISBN 978-3-540-77472-3

Vol. 102. Carlos Cotta, Simeon Reich, Robert Schaefer and
Antoni Ligęza (Eds.)
Knowledge-Driven Computing, 2008
ISBN 978-3-540-77474-7

Carlos Cotta
Simeon Reich
Robert Schaefer
Antoni Ligęza (Eds.)

Knowledge-Driven Computing

Knowledge Engineering and Intelligent Computations

With 107 Figures and 43 Tables

 Springer

Prof. Carlos Cotta
ETSI Informática (3.2.49)
UMA, Campus de Teatinos
29071 Málaga
Spain
ccottap@lcc.uma.es

Prof. Simeon Reich
Department of Mathematics
The Technion – Israel Institute of Technology
32000 Haifa
Israel
sreich@tx.technion.ac.il

Prof. Robert Schaefer
Institute of Informatics
AGH – University of Science and Technology
al. Mickiewicza 30
30-059 Kraków
Poland
schaefer@agh.edu.pl

Prof. Antoni Ligęza
Institute of Automatics
AGH – University of Science and Technology
al. Mickiewicza 30
30-059 Kraków
Poland
ligeza@agh.edu.pl

ISBN 978-3-540-77474-7   e-ISBN 978-3-540-77475-4

Studies in Computational Intelligence ISSN 1860-949X

Library of Congress Control Number: 2008920257

Cover design: Deblik, Berlin, Germany

Printed on acid-free paper

9 8 7 6 5 4 3 2 1

springer.com

Preface

Computers are probably the most sophisticated tools invented by humans throughout the history of mankind. They are also the most versatile – the range of their applications seems to be unlimited. In contrast to most other machines, the decisive factor leading to the breakthrough in power and success of computers consists in separating the knowledge about how they should act (and therefore the algorithms driving their behavior) from the physical substrate carrying the information. This allows fast, flexible, far-going modifications of the knowledge component, and hence boosts their development as increasingly useful and powerful tools.

During the relatively short recent history of computer development, they have gained more and more domains of application. Starting from the domain of binary operations, through purely mathematical, numerical calculations, the focus moved towards databases, text processing, image processing and pattern recognition, and finally artificial intelligence and nature-inspired computing. In this sense, the last years have witnessed intensive research and search for new computational models in numerous application domains.

Related to these previous considerations, it has been observed for numerous challenging problems of interest that classical mathematical approaches are simply insufficient, or that potential models are becoming too complex and thus computationally unmanageable. It is certainly a widespread phenomenon observed throughout numerous areas of research involving modeling, control, and optimization of complex systems, in which classical mathematical methods have reached some limits of applicability. New mathematical approaches are required, incorporating a significant component of Knowledge responsible for Driving the Computational Process.

Knowledge-Driven Computing constitutes an emerging area of intensive research located at the intersection of Computational Intelligence and Knowledge Engineering with strong mathematical foundations. It embraces methods and approaches coming from diverse computational paradigms, such as evolutionary computation and nature-inspired algorithms, logic programming and constraint programming, rule-based systems, fuzzy sets and many others. The use of various knowledge representation formalisms and knowledge processing and computing paradigms is oriented towards the efficient resolution of computationally complex and difficult problems. The use of various forms of knowledge – from simple rules to meta-heuristic techniques – to control the computational processes constitutes a common core for different knowledge-driven computing paradigms.

The domain of Knowledge-Driven Computing is far from being a uniform, well-established branch of science. It is rather an emerging, diverse area of knowledge concerning the answer to current computational challenges, and covering both numerical and symbolic computing. The main focus of research is on building computationally efficient models which can provide useful solutions with reasonably simple tools. The knowledge component may refer to knowledge representation, search strategy, computational paradigm, etc. Although far from exhaustive, the following list of techniques illustrates the technologies and paradigms typically used in this area:

- Genetic algorithms,
- Evolutionary programming,
- Evolution strategies,
- Genetic programming,
- Memetic algorithms,
- Scatter search,
- Estimation of distribution algorithms,
- Ant colony optimization,
- Particle swarm optimization,
- Multi-agent systems,
- Innovation strategies,
- Knowledge representation,
- Knowledge processing,
- Rule-based systems,
- Ontologies, description logics, XML,
- Soft, fuzzy, temporal and spatial issues.

The common denominator for research in Knowledge-Driven Computing and related areas is that reasonable efficiency is obtained with simple and intuitive models, usually following some biological, social, or human related phenomena. The knowledge level of the model plays an important role in the overall success of these approaches.

The role of Knowledge in Knowledge Driven Computing is at least three-fold. First, Knowledge Representation (KR) plays an important role in developing an efficient representation of the domain of interest, its models and characteristics. Second, Knowledge Processing (KP) paradigms, such as inference and computation (both numeric and symbolic) are crucial for the transformation of the input knowledge and generation of problem solutions. Finally, Knowledge Control (KC) – i.e., the use of strategies, rules, heuristics and constraints to deal with computationally hard problems in an efficient way – seems to be an intrinsic factor for successful applications.

These three factors are observable features of the collected material: Modern, advanced KR formalisms are key issues in several chapters of this volume. In the chapter entitled *Processing and Querying Description Logic Ontologies Using Cartographic Approach* by Krzysztof Goczyła et al., Description Logic is used as a KR formalism and a new inference system incorporating a novel Knowledge Cartography approach is proposed. Also the chapter *A Parallel Deduction for Description Logic with ALC Language* by Adam Meissner and Grażyna Brzykcy investigates advanced parallel inference issues with Description Logic as a tool. The chapter *XML Schema Mappings Using Schema Constraints and Skolem Functions* by Tadeusz Pankowski investigates the problem of transforming knowledge representation schemes using XML as a KR language. Investigation of exploration and interpretation of association rules obtained through data mining is presented in chapter *Query-Driven Exploration of Discovered Association Rules* by K. Świder et al.; a special Predictive Model Markup Language, based on XML, is also developed for analysis of complex mining models and knowledge extraction. XML is also used in the chapter *Handling the Dynamics of Norms – A Knowledge-Based Approach* by Jolanta Cybulka and Jacek Martinek, where the main focus is on capturing the changes of legal acts. Knowledge management in time is based on relatively simple concepts of static and dynamic facts, events and dates with rules encoded in Prolog.

The chapter *Temporal Specifications with XTUS. A Hierarchical Algebraic Approach* by Antoni Ligęza and Maroua Bouzid presents an extended, hierarchical formalism for efficient specification of temporal knowledge, and the chapter *Temporal Specifications with FuXTUS. A Hierarchical Fuzzy Approach* by Maroua Bouzid and Antoni Ligęza, outlines a fuzzy version of XTUS for dealing with imprecise temporal knowledge specifications.

In two related chapters (*Design and Analysis of Rule-Based Systems with Adder Designer* by Marcin Szpyrka and *Methodologies and Technologies for Rule-Based Systems Design and Implementation. Towards Hybrid Knowledge Engineering* by Grzegorz Jacek Nalepa) the issue of efficient design of rule-based systems is investigated. The former uses generalized decision tables for knowledge specification and verifies their properties with a new kind of Petri Nets (the so-called Real-Time Colored Petri Nets), while the latter puts forward a novel design procedure incorporating visual tools, and integrating the logical design with the verification stage. Rule-based systems as knowledge representation tools and their development in the context of medical knowledge are investigated in the chapter *How to Acquire and Structuralize Knowledge for Medical Rule-Based Systems?* by Beata Jankowska and Magdalena Szymkowiak. The main focus of the paper is on providing an algorithmic approach for the organization of knowledge.

The chapter *Outline of Modification Changes* by Josep Lluís de la Rosa et al. analyzes the conditions under which the current implementation of a system (with its hardware realization and software knowledge) becomes insufficient for achieving more complex goals, and thus a new system, evolving from the old one by introducing structural changes is necessary. Their concept of Modification Systems emerging from Automatic Control, Multi-Agent Systems and Artificial Intelligence is an interesting study of the philosophy of innovation where new needs and new challenges call for new solutions and new knowledge developed to overcome existing limitations.

In the chapter *A Universal Tool for Multirobot System Simulation* by Wojciech Turek et al. an advanced distributed simulation environment for modeling robots is presented. It enables three-dimensional simulation of kinematics and dynamics as well as control algorithms development.

The objective of the chapter *Bond Rating with π Grammatical Evolution* by Anthony Brabazon and Michael O'Neill is to introduce a variant of grammatical evolution showing a capability to discriminate between investment and junk rating classifications. The models thus developed are highly competitive with MLP models based on the same datasets. In a related chapter entitled *Experiments with Grammatical Evolution in Java* by Loukas Georgiou and William J. Teachan, the optimal governing of grammatical evolutionary computation in a distributed environment is discussed.

The analysis of the asymptotic behavior of a dynamical system generated by an evolutionary process is analyzed in chapter *On Use of Unstable Behavior of a Dynamical System Generated by Phenotypic Evolution* by Iwona Karcz-Dulęba. The knowledge obtained from this study can be exploited in tuning the genetic process applied to optimization tasks, or to identify parameters of an unknown fitness function in the case of *black-box* tasks.

The chapter *Application of Genetic Algorithms in Realistic Wind Field Simulations* by Rafael Montenegro et al. performs the parameter adjustment of a three dimensional mass-consistent numerical model of the atmosphere movement over a complex, mountainous region. The knowledge-based genetic global optimization strategy allows the authors to overcome the multimodality and weak regularity of the complicated objective function formulated in this problem. An interesting hybrid approach to multi-objective optimization is presented in the chapter *Improving Multi-Objective Evolutionary Algorithms by Using Rough Sets* by Alfredo G. Hernández-Díaz et al. The authors consider a multi-objective version of a differential evolution algorithm, which is run for a low number of function evaluations, and whose output is then enhanced via the use of rough-set theory. They show how this approach can successfully compete with other state-of-the-art approaches such as the conspicuous NSGA-II.

Ramón Sagarna and José Antonio Lozano approach an interesting problem in software engineering, namely test data generation, via estimation of distribution algorithms (EDAs) in the chapter *Software Metrics Mining to Predict the Performance of Estimation of Distribution Algorithms in Test Data Generation*. They add an interesting twist to this line of research by studying the performance of EDAs when applied to this problem, and building performance predictors using machine learning techniques. This work paves the way for the use of more sophisticated Data Mining techniques on this domain.

Finally, Francisco Fernández de Vega and Gustavo Olague propose a new nature-inspired algorithm with application to image processing in the chapter *Advancing Dense Stereo Correspondence with the Infection Algorithm*. Their approach is termed Infection Algorithm, and blends ideas from epidemic algorithms and cellular automata. The usefulness of this approach is validated by a real-world application to stereo matching: computing the correspondence between pixels in different images.

As can be seen in the list of articles outlined before, the main aim of this volume has been to gather together a selection of recent papers providing new ideas and solutions for a wide spectrum of Knowledge-Driven Computing approaches. More precisely, the ultimate goal has been to collect new knowledge representation, processing and computing paradigms which could be useful to practitioners involved in the area of discussion. To this end, contributions covering both theoretical aspects and practical solutions, and dealing with topics of interest for a wide audience, and/or cross-disciplinary research were preferred. The main source of inspiration for this volume was a series of international conferences on Computer Systems and Methods held in Cracow, Poland, starting in 1997. Some of the contributions included here are actually based on selected papers presented at these conferences.

The editors would like to cordially thank all the people who made possible the completion of this volume. First of all, thanks are due to all the authors who contributed to the scientific quality of this book. Thanks also to all the referees who contributed to the selection and improvement of the contents of this volume. We also aknowledge the work done by Jarosław Warzecha who managed the technical edition of this volume. Last, but not least, thanks are due to Prof. Janusz Kacprzyk for his support during the development of this volume. To all of them, we extend our gratitude and sincere acknowledgement that without their help and support, this volume would have never come into existence.

Carlos C. Cotta
Simeon Reich
Robert Schaefer
Antoni Ligęza

Summer 2007

List of Referees

Marian Adamski
University of Zielona Góra, Poland

Zbigniew Banaszak
University of Zielona Góra, Poland

Joachim Baumeister
University of Würzburg, Germany

Anthony Brabazon
University College Dublin, Ireland

Janez Brest
University of Maribor, Slovenia

Krzysztof Cetnarowicz
AGH – University of Science and Technology, Cracow, Poland

Carlos Cotta
University of Málaga, Spain

Diana Cukierman
Simon Fraser University, Surrey, Canada

Antonio J. Fernández
University of Málaga, Spain

Ewa Grabska
Jagiellonian University, Cracow, Poland

Elżbieta Hajnicz
Institute of Computer Science, Polish Academy of Science, Warsaw, Poland

Francisco Herrera
University of Granada, Spain

Zdzisław Hippe
University of Information Technology and Management, Rzeszów, Poland

Ian Horrocks
University of Manchester, UK

Radosław Klimek
AGH – University of Science and Technology, Cracow, Poland

Rainer Knauf
Technische Universität Ilmenau, Ilmenau, Germany

Witold Kosiński
Polish Japanese Institute of Information Technology, Warsaw, Poland

Krzysztof Kozłowski
Poznań University of Technology, Poland

William B. Langdon
University College London, UK

Andrzej Łachwa
Jagiellonian University, Cracow, Poland

Bing Liu
University of Illinois at Chicago, USA

Lawrence Mandow
University of Málaga, Spain

Robert Marcjan
AGH University of Science and Technology, Cracow, Poland

Zygmunt Mazur
Wrocław University of Technology, Poland

Zbigniew Michalewicz
University of Adelaide, Australia

Wojciech Moczulski
Silesian University of Technology, Gliwice, Poland

Abdel-Illah Mouaddib
University of Caen, France

Malek Mouhoub
University of Regina, Canada

Mieczysław Muraszkiewicz
Warsaw University of Technology, Poland

Piotr Orantek
Silesian University of Technology, Gliwice, Poland

Gregor Papa
Jožef Stefan Institute, Ljubljana, Slovenia

Jaroslav Pokorny
Charles University, Praha, Czech Republic

Lech Polkowski
Polish Japanese Institute of Information Technology, Warsaw, Poland

Jacek Ruszkowski
Department of Medical Informatics and Biomathematics, Medical Centre of
Postgraduate Education, Warsaw, Poland

Simonas Šaltenis
Aalborg University, Denmark

Rob Saunders
University of Sydney, Australia

Bernhard Seeger
Philipps-University Marburg, Germany

Patrick Siarry
Université Paris XII, France

Vilem Srovnal
VSB Technical University of Ostrava, Czech Republic

Zbigniew Suraj
Rzeszów University, Poland

Tadeusz Szuba
AGH – University of Science and Technology, Cracow, Poland

Piotr Szwed
AGH – University of Science and Technology, Cracow, Poland

Halina Ślusarczyk
Jagiellonian University, Cracow, Poland

Bartłomiej Śnieżyński
AGH – University of Science and Technology, Cracow, Poland

Alicja Wakulicz-Deja
University of Silesia, Katowice, Poland

Marek Wojciechowski
Poznań University of Technology, Poland

Contents

List of Contributors

Maroua Bouzid
University of Caen
France

Anthony Brabazon
University College Dublin
Ireland

Grażyna Brzykcy
Poznań University of Technology
Poland

Rafael Caballero
University of Malága
Spain

Krzysztof Cetnarowicz
AGH – University of Science and
Technology
Cracow, Poland

Carlos A. Coello Coello
CINVESTAW-IPN
Mexico

Jolanta Cybulka
Poznań University of Technology
Poland

Josep Lluís de la Rosa
University of Girona
Spain

J.M. Escobar
University of Las Palmas de Gran
Canaria
Spain

Santiago Esteva
University of Girona
Spain

Francisco Fernández de Vega
University of Extremadura
Spain

Alberto Figueras
University of Girona
Spain

Loukas Georgiou
University of Wales
Bangor, United Kingdom

Krzysztof Goczyła
Gdańsk University of Technology
Poland

J.M. González-Yuste
University of Las Palmas de Gran
Canaria
Spain

Alfredo G. Hernández-Díaz
Pablo de Olavide University
Seville, Spain

Salvador Ibarra
University of Girona
Spain

Bartosz Jędrzejec
Rzeszów University of Technology
Poland

Iwona Karcz-Dulęba
Wrocław University of Technology
Poland

Antoni Ligęza
AGH – University of Science and
Technology
Cracow, Poland

Jose A. Lozano
University of the Basque Country
San Sebastian, Spain

Evelyne Lutton
INRIA Rocquencourt
France

Robert Marcjan
AGH – University of Science and
Technology
Cracow, Poland

Jacek Martinek
Poznań University of Technology
Poland

Adam Meissner
Poznań University of Technology
Poland

Julian Molina
University of Malága
Spain

Rafael Montenegro
University of Las Palmas de Gran
Canaria
Spain

G. Montero
University of Las Palmas de Gran
Canaria
Spain

Grzegorz Jacek Nalepa
AGH – University of Science and
Technology
Cracow, Poland

Gustavo Olague
CICESE Research Center
Mexico

Michael O'Neill
University of Limerick
Ireland

Tadeusz Pankowski
Poznań University of Technology
Poland

Cynthia B. Perez
CICESE Research Center
Mexico

Beata Puchałka-Jankowska
Poznań University of Technology
Poland

Christian Quintero
University of Girona
Spain

Josep Antoni Ramon
University of Girona
Spain

E. Rodríguez
University of Las Palmas de Gran
Canaria
Spain

Ramon Sagarna
University of the Basque Country
San Sebastian, Spain

Louis V. Santana-Quintero
CINVESTAV-IPN
Mexico

Marcin Szpyrka
AGH – University of Science and
Technology
Cracow, Poland

Magdalena Szymkowiak
Poznań University of Technology
Poland

Krzysztof Świder
Rzeszów University of Technology
Poland

William J. Teachan
University of Wales
Bangor, United Kingdom

Wojciech Turek
AGH – University of Science and
Technology
Cracow, Poland

Wojciech Waloszek
Gdańsk University of Technology
Poland

Marian Wysocki
Rzeszów University of Technology
Poland

Teresa Zawadzka
Gdańsk University of Technology
Poland

Michał Zawadzki
Gdańsk University of Technology
Poland

Temporal Specifications with FuXTUS.
A Hierarchical Fuzzy Approach

Maroua Bouzid[1] and Antoni Ligęza[2]

[1] GREYC, Campus II Sciences 3, BD Maréchal Juin, 14032 Caen Cedex,
   `bouzid@info.unicaen.fr`
[2] AGH – University of Science and Technology, al. Mickiewicza 30, 30-059 Kraków
   Poland, `ligeza@agh.edu.pl`

Summary. Specification and efficient handling of imprecise temporal knowledge is
an important issue in design and implementation of contemporary information systems. In domains such as natural language processing, modern databases and data
warehouses, knowledge-based systems or decision support systems qualitative and
imprecise temporal information is often in use at various levels of abstraction. This
paper explores an approach based on TUS, the Time Unit System, and its extended
version called XTUS, both providing algebraic tools for constructing simple yet powerful crisp temporal specifications of hierarchical nature. The main contribution of
this paper consists in extending the TUS/XTUS approach by means of elements of
fuzzy set theory. In particular, a fuzzy extended version of TUS, called FuXTUS,
is introduced and its basic operations and properties are shown. It is argued that
this simple and consistent with natural language and natural calendar way of building temporal specifications is capable of efficient dealing with imprecise temporal
specifications. Numerous simple examples illustrate the presented ideas.

1 Introduction

Representation of temporal knowledge and particularly imprecise temporal
knowledge constitutes a core issue in the development and use of numerous
complex information systems. In many applications, with natural language
processing and knowledge-based systems communicating with man in the first
place, efficient specification and processing of imprecise temporal knowledge
is of primary interest. Being able to define and deal with such imprecise temporal representations at different grain levels is an important research theme
in the area of Artificial Intelligence and Natural Language Processing as well
as in the domains of Database Systems, Data Warehouses and Decision Support Systems. Furthermore, the inherent abstraction power of the granularity
concept has been successfully exploited in several application domains, including temporal and spatial reasoning, hierarchical planning, natural language

M. Bouzid and A. Ligęza: *Temporal Specifications with FuXTUS. A Hierarchical Fuzzy
Approach*, Studies in Computational Intelligence (SCI) **102**, 1–16 (2008)
`www.springerlink.com`   © Springer-Verlag Berlin Heidelberg 2008

understanding, temporal database design, medical informatics, image processing, real-time system modeling, design, analysis and verification [3, 8, 20].

In this paper, we outline the basic research issue involving join representation of *imprecise temporal specifications* and several levels of *temporal granularity*. Any temporal granularity can be viewed as a partitioning of the temporal domain into groups of elements, where each element is perceived as an indivisible unit (a granule). The description of a fact, an action, or an event can use these granules to provide them with a temporal qualification at the appropriate abstraction level. Examples of standard time granularities are *days, weeks, months*, while user-defined granularities may include *business-weeks, trading-days, working-shifts, school-terms* are often used with imprecise qualifications such as *at the beginning of, in the middle of, the last days/hours of*, or *most of the time/period*, etc.

The presented approach is based on [14, 15] which presents a formalization of a system of granularities in the context of interval calculus. Granules are formed in a hierarchical way from the so-called called Time Units and are defined as finite sequences of integers. They are organized in a linear hierarchy *(year, month, day, hour, minute, second)*. This paper refers to an algebraic version of extended TUS, called XTUS; for details refer to [19]. The main focus of this work is on representing imprecise temporal statements – we present a further extension to XTUS towards incorporating fuzzy temporal specifications.

The structure of the paper is as follows. In Section 2 a basic motivation for fuzzy temporal knowledge specification is provided. Section 3 presents flat fuzzy intervals and Section 4 – basic fuzzy functions applied in temporal qualifications. Fuzzy eXtended TUS (FuXTUS) is presented in Section 5. Section 6 refers to practical applications and examples. Section 7 provides a short discussion of related work. Finally, Section 8 concludes the paper.

2 Motivation for Fuzzy Temporal Specifications

Temporal specifications used in natural language are often imprecise, qualitative and rough. Many people still prefer a classical, analogue watch to a digital one. In some, even technologically advanced countries, the social feeling and respect of time is far from being exact.

Even when trying to formally describe certain phenomena and their mutual temporal relationship one often uses qualitative linguistic descriptions. Some most typical examples are of the form:

- at the *beginning of* some period, e.g. at the *beginning of this month*,
- in the *middle of* some period, e.g. in the *middle of this year*,
- at the *end of* some period, e.g. at the *end of* this week.

More complicated examples may include also specification of imprecisely defined *repeated events* and semi-logical combinations based on conjunction, disjunction, negation or exclusive-or. Moreover, qualitative quantifiers like *most of the time, frequently within a certain period of time* or *about half of the weekends* referring to imprecise temporal specification of integral nature often appear in natural language formulations. Such linguistically specified imprecise denotations of time are easily interpreted and dealt with by human. On the other hand, on the man-machine frontier only precise specifications are accepted, with SQL and its temporal built-in capabilities being the most practical example at hand.

A further problem concerns data analysis. For example, large databases of banks or stores contain very precise data of every transaction. However, for higher-order analysis of such data some abstraction must be carried out so as to discover and describe some medium and long term trends, etc. Discovering interesting, human-understandable relations requires translating temporal descriptions into qualitative, less precise but more transparent simplified language [18].

3 Flat Fuzzy Intervals

Below some basic ideas concerning elements of *FuXTUS – Fuzzy eXtended Time Unit System* are introduced.

Let us define first a *basic flat fuzzy term* for representing fuzzy temporal intervals. A basic flat fuzzy term corresponds to a convex interval, but the fuzzy degree of belonging to an interval may be less than or equal to 1. So the definition is a generalization of a flat convex crisp interval of XTUS.

Let T denote the flat, lowest level universe of time composed from all the Basic Time Units (BTU) [14, 15]. Note that assuming that BTU is the lowest level (and simultaneously most precise) description of time interval, we introduce the finest granularization which induces *discrete time space* for further considerations. Assuming that BTU is defined as one second for example, one cannot ask about more precise representation of time. The question of beginning, middle or end of the BTU period are to be excluded, since there is no way to express a more precise time instant[1].

Consider first the level of BTUs. Let μ denote a global function of the form:

$$\mu\colon T \to [0, 1]$$

with the obvious meaning of a degree to which a BTU from T satisfies some condition (a fuzzy degree). In fact μ defines a fuzzy set over T.

Definition 1 *A Basic Time Unit fuzzy term t at the level of BTUs is a pair*

$$t = (z, \mu(z)),$$

---

[1] Obviously unless a next, more detailed level of granularization is introduced; however, for clarity of the discussion we assume that once established, the scheme of terms is kept over.

where $z \in T$ is a BTU in the assumed time space and $\mu(z) \in [0,1]$ is its fuzzy degree; the traditional notation coming from fuzzy sets theory of the form $t = z/\mu(z)$ will also be used.

For simplicity we shall also write μ_z instead of $\mu(z)$.

The meaning of z/μ_z is that z satisfies the discussed specification with the degree of μ_z. For example, consider that BTUs are days, and we say something about holidays in August. The *last days of the holidays* may be defined as: 31/1, 30/0.9, 29/0.8, etc.

At the higher level of abstraction one can define *fuzzy intervals* specified with a fuzzy membership function, as well as fuzzy sets composed of discrete elements. Let **Z** denote the set of integers.

Definition 2 *Let $Z = [z^-, z^+]$ denote a convex and crisp (classical) interval. Here z^- and z^+ can be both elements of BTUs or they may correspond to a higher level granules of time. Further, let*

$$\mu_Z \colon \mathbf{Z} \to [0,1]$$

be a function such that

$$\mu_Z(z) = 0 \text{ for } z < z^-$$
$$\mu_Z(z) \geq 0 \text{ for } z^- \leq z \leq z^+$$
$$\mu_Z(z) = 0 \text{ for } z > z^+$$

We call $t = (Z, \mu_Z)$ a basic flat fuzzy term. The interval Z is the support of the fuzzy term and function μ defines the fuzzy membership degree.

In case Z is composed of BTUs the definition of μ_Z provides a sequence of pairs of the form $z_1/\mu_Z(z_1), z_2/\mu_Z(z_2), \ldots, z_m/\mu_Z(z_m)$, where $z^- = z_1$ and $z^+ = z_m$. All the values of the μ_Z function greater than 0 are assigned to indivisible BTUs and located within interval Z.

In case Z is specified at a higher abstraction level, its elements (including z^- and z^+) are no longer BTUs, i.e. they denote time some higher order units composed of BTUs. However, to any $z \in Z$ μ_Z assigns a single number exactly as before. One can imagine application of the flattening function f to Z (see [19]), in order to find d^- and d^+ being BTUs, such that $f(Z) = [d^-, d^+]$; hence $f(Z)$ is the interval equivalent to Z but expressed with BTUs. The fuzzy degree function can be evaluated for any BTU in $[d^-, d^+]$. For all BTUs belonging to the same $z \in Z$ it remains a constant function what is the consequence of the assumption of discrete time and Def. 2. Perceived in the domain of BTUs $d \in T$, $\mu_Z(d)$ is a piecewise constant function (a stepwise function).

An important consequence of the above definition is that if Z is expressed by a single integer z, which simultaneously it is no longer a BTU, it can be assigned a fuzzy degree as a constant number only. In order to be able to assign it a function (not only a single number) one must go to a lower level of hierarchy. For example, if $z = 17$ is a day, the passing to the level of hours one can define the fuzzy function of the form $\mu\colon \{0, 1, 2, 3, \ldots, 23\} \to [0, 1]$. Further, if minute is the BTU, then at the lowest level the μ function can be defined over the set $\{0, 1, 2, 3, \ldots, 59\}$.

By analogy, let us introduce also a fuzzy set composed of discrete elements.

Definition 3 *A finite discrete fuzzy set is a pair* (Z, μ_Z)*, where* $Z = \{z_1, z_2, \ldots, z_k\}$ *is a set of integers denoting time units at some level of granularity (days, months, hours,...) and* μ_Z *is a function defining the fuzzy membership degree of the form:*

$$\mu_Z\colon Z \to [0, 1].$$

A finite discrete fuzzy set is usually represented as a set of pairs of the form $\{z_1/\mu_Z(z_1), z_2/\mu_Z(z_2), \ldots, z_k/\mu_Z(z_k)\}$ where all $\mu_Z(z_i)$ are single numbers such that $\mu_Z(z_i) \in [0, 1]$, $i = 1, 2, \ldots, k$.

Definition 3 is analogous in construction to Definition 2; the main difference is that in case of Def. 2 we have the domain in the form of a sequence of integers while in the case of Def. 3 it can be any arbitrary set of integers.

Note also that the classical definition of a finite, discrete fuzzy set, e.g. Def. 3 allows to assign only *constant numbers* to elements of z. In case it is necessary to define the fuzzy degree in a more detailed way one must go a step downward the hierarchy, and assign a new new fuzzy function to the units of the lower level.

Below some elements of the Fuzzy eXtended Time Unit System (FuXTUS) are introduced.

Definition 4 (Flat fuzzy term) *A flat fuzzy term* t *defining fuzzy time interval or a set of intervals at a single level of hierarchy is:*

- *a pair of the form* $z/\mu(z)$ *where* z *is a constant integer of BTUs or any higher level units and* $\mu(z) \in [0, 1]$ *is the fuzzy degree of* z *(a singleton function),*
- *a pair of the form* (Z, μ_Z)*, where* $Z = [z^-, z^+]$ *is a range of integers and* μ_Z *is a fuzzy degree function of the form* $\mu_Z\colon [z^-, z^+] \to [0, 1]$*,*
- *a finite discrete fuzzy set represented as a set of pairs of the form* $\{z_1/\mu_Z(z_1), z_2/\mu_Z(z_2), \ldots, z_k/\mu_Z(z_k)\}$ *where all* $\mu_Z(z_i)$ *are single numbers such that* $\mu_Z(z_i) \in [0, 1]$*,* $i = 1, 2, \ldots, k$
- *any union of the above.*

For simplicity, a single element set (or interval) will be considered equivalent to its unique element. For intuition, a single pair composed of an integer number such as 7 and a fuzzy degree coefficient such as 0.7, i.e. the pair $7/0.7$ denotes a certain interval of time and its fuzzy degree; depending on the assigned interpretation it can be a month (July), a day (24 hours), an hour (60 minutes), etc. In fact, in order to assign meaning to a flat term one must specify its *type* (unit) [19]. This can be denoted as a pair *type*: *term*, e.g. *month*: 12 or *day*: 29. If the type is known, its specification will be omitted.

A set of integers specify several intervals of the same type (months, days, hours) which are not necessarily adjacent. For example $[1,3,5]$ may denote Monday, Wednesday and Friday. To any day one can assign a constant fuzzy coefficient or, in general, a fuzzy degree function defined at the level of BTUs.

In case the integers are subsequent ones, instead of writing $[z^-, z^- + 1, z^- + 2, \ldots, z^+]$ one simply writes $[z^-, z^+]$, i.e. specifies the range of integers; for example, $[9-16]$ is the equivalent for $[9, 10, 11, 12, 13, 14, 15, 16]$, and when speaking about hours it denotes the interval beginning at 9:00 and ending at 16:59 using one minute as a BTU. For the overall interval a single fuzzy function defined at the level of BTUs can be assigned.

4 Basic Fuzzy Functions

Below, some examples of fuzzy functions are presented for intuition. In what follows, any flat or hierarchical crisp term will be called a temporal term. A single, general function defining the fuzzy degree is introduced. The function covers several well-known cases of particular membership functions.

Definition 5 (The π function (fuzzy)) *Let s be a convex temporal term and let $f(s) = [f(s)^-, f(s)^+]$ after the flattening operation; for simplicity we put $f(s)^- = s^-$ and $f(s)^+ = s^+$. Further, let s^1, s^2 be two integers, such that $s^- \leq s^1 \leq s^2 \leq s^+$. We define the $\pi^{\langle s^1, s^2 \rangle}$ function as follows:*

$$\pi^{\langle s^1, s^2 \rangle}(s^-) = 0 \text{ for } s^- < s^1$$
$$\pi^{\langle s^1, s^2 \rangle}(s^-) = 1 \text{ for } s^- = s^1$$
$$\pi^{\langle s^1, s^2 \rangle}(s^1) = 1$$
$$\pi^{\langle s^1, s^2 \rangle}(s^2) = 1$$
$$\pi^{\langle s^1, s^2 \rangle}(s^+) = 1 \text{ for } s^+ = s^2$$
$$\pi^{\langle s^1, s^2 \rangle}(s^+) = 0 \text{ for } s^+ > s^2$$

and is constructed by piecewise linear interpolation among these points. We distinguish the following special cases of function π:

1. $\Delta^{s^1} = \pi^{\langle s^1, s^1 \rangle}$ obtained for $s^- < s^1 = s^2 < s^+$;
2. $\Gamma = \pi^{\langle s^+, s^+ \rangle}$ obtained for $s^- < s^1 < s^2 = s^+$;
3. $\gamma = \pi^{\langle s^-, s^- \rangle}$ obtained for $s^- = s^1 < s^2 < s^+$.

In fact, the presented definition is a bit informal – these are the most typical shapes used in practice.[2]

A schematic presentation of the proposed function is presented in Fig. 1. Figures 2, 3, and 4 may serve as examples of a γ function, Γ function and a Δ function, respectively.

---

[2] One can ask for more formal requirements, e.g. exactly one region where the function is greater than 0, unimodality or one or a limited number of discontinuity points. This discussion seems however to be beyond the scope of this paper.

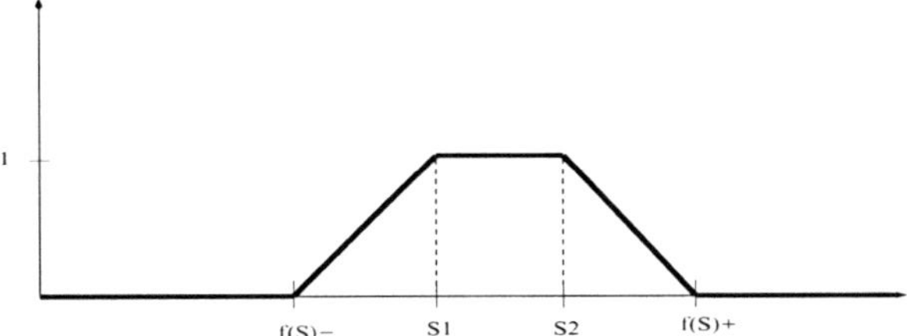

Fig. 1. The overall scheme of the π function

Note also, that the function is defined for a domain located at the level of BTUs; in fact, it can be located at any arbitrary level, depending on the required level of detail.

5 Fuzzy Extended TUS

Having defined the necessary elements such as various forms of flat fuzzy terms we can present the concept of fuzzy terms of FuXTUS.

Recall that the original TUS, as introduced by Ladkin [15], allows only for constant terms expressed as sequences of integers. This constitutes a strong limitation with respect to the expressive power of TUS – although in fact arbitrary convex intervals can be specified using the *convexify*$(.,.)$ operation, there seems to be other straightforward possibilities of extending the notation in a simple and transparent way.

XTUS, as presented in the accompanying paper [19], possess higher expressive power. The main extension concerned the flat terms placed at some i-th position in TUS terms – we allowed for use of intervals (ranges) and sets of integers. Further, an anonymous variable denoted with '_' was introduced which allows for partial specification and specification of repeated events (cycles). Below a further extension towards fuzzy time specification is proposed.

Definition 6 (Hierarchical fuzzy term) *A hierarchical fuzzy term (or a fuzzy term for short) specifying fuzzy interval of time and using k levels of hierarchy is any sequence of the form*

$$[t_1, t_2, \ldots, t_k]$$

where any $t_i \in \{t_1, t_2, \ldots, t_{k-1}\}$ are crisp flat terms or an anonymous variable denoted with _ and t_k is a flat fuzzy term defined as in Def. 4.

For intuition, the $k-1$ levels define or allocate the fuzzy interval defined with the last, most precise k-th term in a fuzzy way. For example, if the scheme of the terms is specified as [*year*, *month*, *day*], then the expression *the last days of June 2005* can be expressed as:

$$[2005, 6, \{20/0, 21/0.1, 22/0.2, 23/0.3,$$
$$24/0.4, 25/0.5, 26/0.6, 27/0.7, 28/0.8, 29/0.9, 30/1\}]$$

or using some piecewise linear Γ functions as

$$[2005, 6, ([20, 30], \Gamma)],$$

where $\Gamma(20) = 0$, $\Gamma(30) = 1$.

Note that the flattening operation allows to flatten the fuzzy term and represent it as a pair $([d^-, d^+], \mu_t)$, where $f(t) = [d^-, d^+]$ and d^-, d^+ are BTUs.

Below we introduce the idea of an extended hierarchical fuzzy term. This time we allow for fuzziness at any level of hierarchical specification.

Definition 7 *An* extended hierarchical fuzzy term *(or an* extended fuzzy term *for short) specifying composed fuzzy interval of time and fuzzy specifications at k levels of hierarchy is any sequence of the form*

$$[t_1, t_2, \ldots, t_k],$$

where any $t_i \in \{t_1, t_2, \ldots, t_k\}$ are flat fuzzy terms defined as in Def. 4 or an anonymous variable denoted with _.

Note that with respect to the above definition we arrive at a real possibility of using fuzzy specifications at *all* the levels of granularities.

Referring to the recent example, a specification like *the last days of June during beginning of the XXI century* can now be given as:

$$[\{2001/1, 2002/0.9, 2003/0.8, 2004/0.7, 2005/0.6,$$
$$2006/0.5, 2007/0.4, 2008/0.3, 2009/0.2, 2010/0.1, 2011/0\},$$
$$6, \{20/0, 21/0.1, 22/0.2, 23/0.3, 24/0.4,$$
$$25/0.5, 26/0.6, 27/0.7, 28/0.8, 29/0.9, 30/1\}]$$

or equivalently as

$$[([2001, 2005], \gamma), 6, ([20, 30], \Gamma)]$$

with appropriately defined piecewise linear functions γ and Γ.

5.1 Calculating the Fuzzy Degree

Let $\odot$ denote any triangular norm, such as *min* or *times* $(\cdot)$. Let

$$t = [t_1, t_2, \ldots, t_k]$$

be a specification of an extended hierarchical fuzzy term defining some fuzzy constraints and let

$$s = [s_1, s_2, \ldots, s_k]$$

be another hierarchical fuzzy term (or precise classical crisp term of TUS; in the other case we put 1 as the default μ function for any flat term at any level). The degree to which s satisfies the specification of t can be calculated as

$$(t_1 \odot s_1) \odot (t_2 \odot s_2) \odot \ldots \odot (t_k \odot s_k).$$

In case t_i is an anonymous variable the results of $t_i \odot s_i$ equals to s_i (and vice versa) for $i = 1, 2, \ldots, k$.

6 A Fuzzy Approach for Qualitative Temporal References

In this section, we develop a method to deal with qualitative temporal references. By temporal references, we mean a temporal location of an information given some temporal reference marks, where as such reference marks we can consider *start, end, middle* as examples. The temporal information we consider are expressed in XTUS and we need to assess the degree of satisfaction of this temporal information according to one of these temporal references. Indeed, in many real-world environments we need to temporally locate, often in an approximate way, a temporal interval using temporal references.

For example, let us consider the following sentence *At the end of the year, I am on holidays.* The temporal reference *end of the year* could be the end of November, the beginning of December, etc. In other words, the question is to define to what degree the end of November or the first week of December satisfy the reference *end of the year*. To develop a reasoning scheme on such temporal propositions, we need a graceful representation expressive enough.

In the following, we restrict ourselves to the temporal references such as

$$R_t = \{start, middle, end\}$$

Consider two positions k and $k^{'}$ in a hierarchical term of XTUS and an interval x. Our objective consists of determining to what degree x satisfies a temporal reference on unit of position k. To do that, we introduce the function *number-of-unit*$(k, k^{'})$ which allows us to determine the number of units of position $k^{'}$ composing a unit of position k. Roughly speaking:

$$number\text{-}of\text{-}unit(1, 1) = 1$$

$$number\text{-}of\text{-}unit(1, 2) = 12$$

$$number\text{-}of\text{-}unit(2, 3) \in \{28, 29, 30, 31\}$$

$$number\text{-}of\text{-}unit(3, 4) = 24$$

$$number\text{-}of\text{-}unit(1, 3) \in \{365, 366\}$$

These examples illustrate the meaning of this function using universal time units (year, month, day, hour, minute, second), but we can easily extend it to other time units like week, semester or biennale.

6.1 Temporal Reference Satisfaction Using Fuzzy Functions

We consider a set of temporal references $R = \{start, middle, end\}$. Let k and k' be two scales in the hierarchical time unit such that $k < k'$. Let x be a single integer number such as 7. It denotes a certain interval of time; depending on the assigned interpretation it can be a month (July), a day (Sunday), an hour (7h), etc.

We define a function $start_k^{k'}$ defined from $\{1, 2, \ldots, number\text{-}of\text{-}unit(k, k')\}$ to the interval $[0, 1]$:

$$start_k^{k'} : \{1, 2, \ldots, number\text{-}of\text{-}unit(k, k')\} \rightarrow [0, 1]$$

such that $start_k^{k'}$ computes for each $x \in \{1, 2, \ldots, number\text{-}of\text{-}unit(k, k')\}$ the degree α ($0 \leq \alpha \leq 1$) to which x satisfies the temporal reference $start$ on the unit of position k. As depicted in Figure 2, the more one moves away from the origin (which represents the beginning of the unit of position k expressed in units of position k'), the more the satisfaction degree of the interval for the temporal reference on k decreases. We consider that the first third of $number\text{-}of\text{-}unit(k, k')$ ($[1, \frac{number-of-unit(k,k')}{3}]$) satisfies the reference with degree 1, that middle of $number\text{-}of\text{-}unit(k, k')$ ($\frac{number\text{-}of\text{-}unit(k,k')}{2}$) satisfies the reference $middle_k^{k'}$ with degree 1, and that the third third of $number\text{-}of\text{-}unit(k, k')$ ($[2 \cdot \frac{number-of-unit(k,k')}{3}, number\text{-}of\text{-}unit(k, k')]$) satisfies the reference $end_k^{k'}$ with degree 1. From these intuitions, we define three fuzzy functions to represent respectively the three temporal references as below.

The function $start_k^{k'}$ is thus calculated on three different intervals such as it is illustrated in Figure 2. The value of this function on these three sub-intervals is as follows:

$$start_k^{k'}(x) = \begin{cases} 1 \\ \qquad \text{if } x \in [1, \dfrac{number\text{-}of\text{-}unit(k,k')}{3}]; \\[2em] \dfrac{-3}{number\text{-}of\text{-}unit(k,k'))} \cdot x + 2 \\ \qquad \text{if } x \in]\dfrac{number\text{-}of\text{-}unit(k,k')}{3}, \dfrac{2 \cdot number\text{-}of\text{-}unit(k,k')}{3}[; \\[2em] 0 \\ \qquad \text{if } x \in [\dfrac{2 \cdot number\text{-}of\text{-}unit(k,k'))}{3}, number\text{-}of\text{-}unit(k, k'))] \end{cases}$$

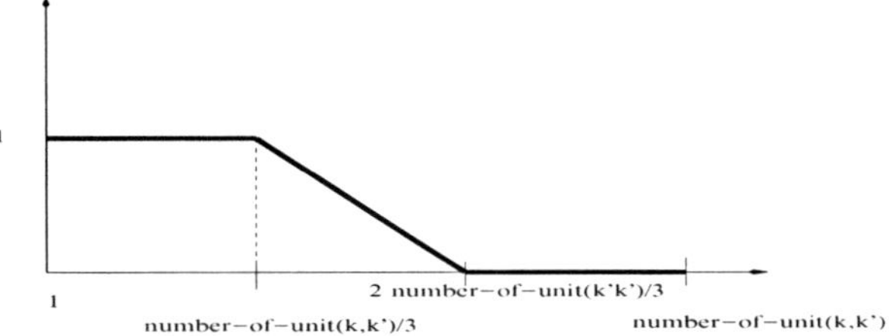

Fig. 2. The curve of the function *start*

As an example consider the expression "*The last days of june 2007*" is represented as follow:

$$[2007, 6, ([1 - 30], start_2^3(x))]$$

Thus, for each day of June, we can calculate its satisfaction degree for the reference beginning of the month thanks to the function *start*.

In the same way, the function $end_k^{k'}$ is calculated on three sub-intervals, such as illustrated in Figure 3 by respecting the intuitions mentioned above and which gives the following function:

$$end_k^{k'}(x) = \begin{cases} 0 \\ \quad \text{if } x \in [1, \dfrac{number\text{-}of\text{-}unit(k,k')}{3}]; \\[2em] \dfrac{3}{number\text{-}of\text{-}unit(k,k'))} \cdot x - 1 \\ \quad \text{if } x \in]\dfrac{number\text{-}of\text{-}unit(k,k'))}{3}, \dfrac{2 \cdot number\text{-}of\text{-}unit(k,k'))}{3}[; \\[2em] 1 \\ \quad \text{if } x \in [\dfrac{2 \cdot number\text{-}of\text{-}unit(k,k'))}{3}, number\text{-}of\text{-}unit(k, k'))] \end{cases}$$

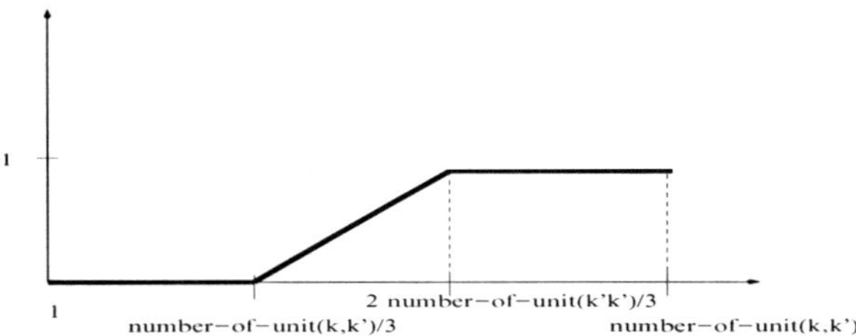

Fig. 3. The curve of the function *end*

Finally, the function $middle_k^{k'}$ from the set $\{1, 2, \ldots, number\text{-}of\text{-}unit(k, k')\}$ to $[0, 1]$ is defined and it assigns to each $x \in \{1, 2, \ldots, number\text{-}of\text{-}unit(k, k')\}$ a degree α $(0 \leq \alpha \leq 1)$ to which the temporal reference $middle$ satisfies a unit at position k. On the same intuitive bases as the other fuzzy functions, this function (see Figure 4) takes value 1 at one point only and it is calculated as follows:

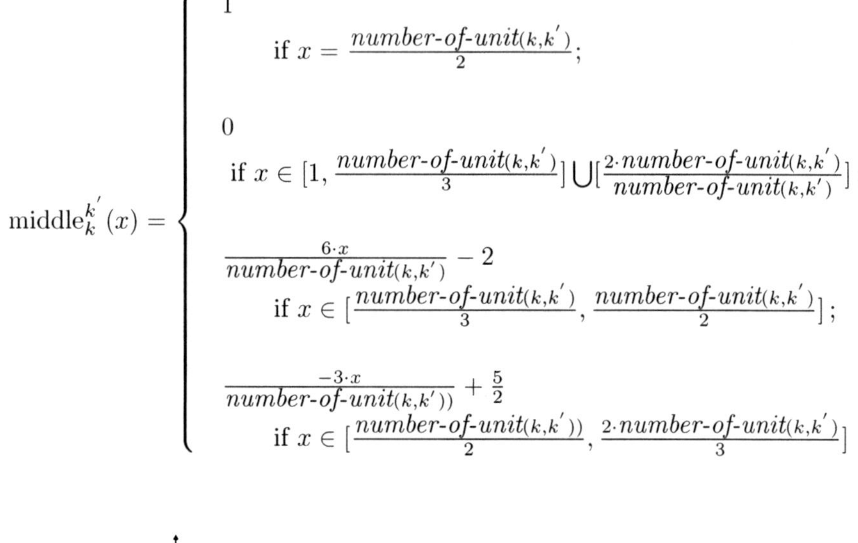

$$
middle_k^{k'}(x) =
\begin{cases}
1 \\
\quad \text{if } x = \dfrac{number\text{-}of\text{-}unit(k,k')}{2}; \\[2ex]
0 \\
\quad \text{if } x \in [1, \dfrac{number\text{-}of\text{-}unit(k,k')}{3}] \cup [\dfrac{2 \cdot number\text{-}of\text{-}unit(k,k')}{number\text{-}of\text{-}unit(k,k')}]; \\[2ex]
\dfrac{6 \cdot x}{number\text{-}of\text{-}unit(k,k')} - 2 \\
\quad \text{if } x \in [\dfrac{number\text{-}of\text{-}unit(k,k')}{3}, \dfrac{number\text{-}of\text{-}unit(k,k')}{2}]; \\[2ex]
\dfrac{-3 \cdot x}{number\text{-}of\text{-}unit(k,k'))} + \dfrac{5}{2} \\
\quad \text{if } x \in [\dfrac{number\text{-}of\text{-}unit(k,k'))}{2}, \dfrac{2 \cdot number\text{-}of\text{-}unit(k,k')}{3}]
\end{cases}
$$

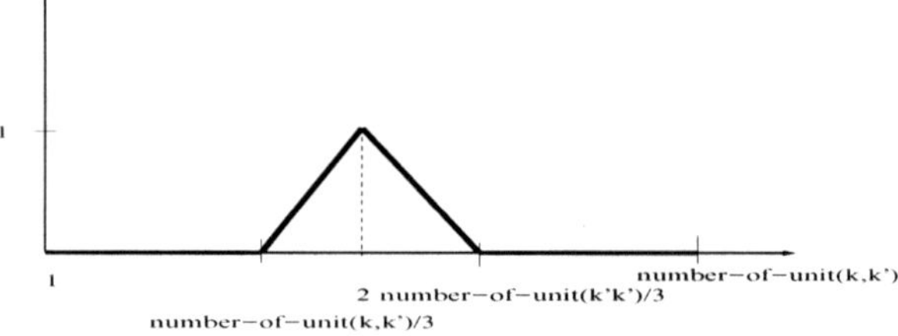

Fig. 4. The curve of the function $middle$

6.2 Composition of Fuzzy Functions Representing a Mixture of Temporal References

The objective consists in evaluating the degree to which an interval satisfies a composition of temporal references expressed on units at different positions in a hierarchical term. For example, let us consider the expression *At the beginning of months of the end of the year, I have a training* and let us consider as the interval the third day of December. To determine the degree of satisfaction of this interval one should first determine the degree to which the interval satisfies the temporal description of start of a month. Then, we combine this degree with satisfaction of the end of the year of reference. To do that, we use the operator $\odot$ of the accepted t-norm which allows us to compose fuzzy functions such that:

$$R1_{k_1}^{k_2}(x_1) \odot R2_{k_2}^{k_3}(x_2) \odot \ldots \odot Ri_{k_i}^{k_j}(x_i)$$

where Ri are temporal references and x_i are intervals.

Let us consider for example the reference *beginning of the end of the year*, one would like to determine the satisfaction degree of the interval November 15 for the reference *the beginning of the end of the year*. For this purpose, we first apply the fuzzy function *end* on the interval *year*, then, over the period where this function is positive we apply the function *start* finally, we apply the operator fuzzy *min* on these two functions as described in the Figure 5.

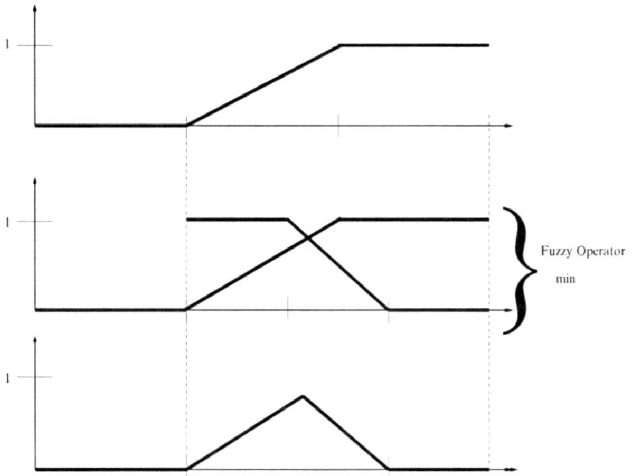

Fig. 5. The composition of fuzzy functions

7 Related Work

Mechanizing temporal knowledge[3] has been an active research domain in Artificial Intelligence for quite a long period [8]. In [13] an attempt was made at developing an approach for assisting knowledge-based system in problem solving. The paper focused on how time specialists formulate and organize statements involving temporal references and time relative knowledge, check consistency of such knowledge and use the knowledge in question answering.

---

[3] This is in fact the exact title of the paper by Kahn and Gorry [13].

From that time numerous approaches to formalize temporal logics were developed [8, 11]. In the companion paper on XTUS and a crisp, algebraic approach to defining temporal specifications we provide a short review of several papers loosely related to our work, including issues such as cycles, periodicity and infinite data [2, 7, 12, 21], granularity of time specifications [3, 20], absolute and relative temporal specifications [9, 10], and unions of intervals [5, 17].

The main foundations of this paper, however, are constituted by the ideas of Ladkin concerning his Time Unit System, i.e. TUS [14–16]. The proposed hierarchical time specifications were further developed in [4] and in the accompanying paper [19]. The main contribution of this paper is the extension of the XTUS formalism towards incorporating fuzzy temporal specifications at all levels of granularity in the hierarchical specifications; perhaps this is the first work concerning a combined, hierarchical and fuzzy approach to development of temporal specifications of imprecise nature.

On the other hand, it should be noticed that application of fuzzy set theory to modelling imprecise temporal knowledge even at single level of flat representations is not sufficiently studied in Artificial Intelligence. Symptomatically, the recent handbook [8] does not seem to perceive this issue. Our early paper [6] introduced some simple and intuitive solution based on the idea of characteristic functions. Similar ideas are explored in modern literature, however the papers dealing with fuzzy temporal approaches are relatively rare.

In [22] the concept of *fuzzy temporal interval* is introduced and explored in some details. The basic idea is that a fuzzy function is applied over a flat interval to denote the fuzzy degree. The intervals are flat, i.e. no hierarchy is introduced. In [18] the concept of a *fuzzy calendar* is introduced. A calendar, understood as a structured collection of time intervals is associated with a fuzzy membership function. The concept is applied for discovering fuzzy association rules.

8 Conclusions

This paper develops XTUS [19] being an attractive and powerful algebraic tool for constructing crisp temporal specifications and performing some operations on them toward imprecise temporal specification by means of fuzzy sets theory. In particular, an extended version of XTUS, called FuXTUS is introduced and its basic operations and properties are shown. One of the main features of FuXTUS consists in incorporating imprecise temporal specifications simultaneously at different levels of temporal specification hierarchy. In such a way quite complex but imprecise natural language specifications can be easily expressed and handled.

The paper was aimed at presenting a new proposal for fuzzy algebraic temporal specifications of time constraints. Special attention is paid to easy definition of single-level and multi-level fuzzy temporal algebraic specifications of varying granularity. An important, intrinsic feature of the formalism consists in its extendibility – the precision of representation can easily be handled by simple scheme modification within the same framework.

The most important part of the paper refers to FuXTUS, a Fuzzy eXtended TUS, which is an original proposal of this paper. Through incorporating fuzzy specification of elements of XTUS and introducing some mathematical operations the formalism is extended towards handling imprecise time specifications and operations over them. This is especially important for dealing with natural language specifications and understanding human sense of time. The proposed approach can perhaps be applied in knowledge-based systems dealing with temporal knowledge and cooperating with human.

References

1. Allen JF (1983) Maintaining knowledge about temporal intervals. Communications of the ACM, 26:832–843
2. Bettini C, Wang X, Ferrari E, Samarati P (1998) An access control model supporting periodicity constraints and temporal reasoning. ACM Transactions on Database Systems, 23:231–285
3. Bettini C, Wang X, Jajodia S (1998) A general framework for time granularity and its application to temporal reasoning. Annal of Mathematics and Artificial Intelligence, 22:29–58
4. Bouzid M, Ligęza A (2005) Algebraic temporal specifications with extend TUS: Hierarchical terms and their applications. Proceedings of the 17th IEEE International Conference on Tools with Artificial Intelligence, 249–253
5. Bouzid M, Ladkin P (2002) Simple reasoning with time-dependent propositions. International Journal of Interest Group in Pure and Applied Logic (IGPL), 10:379–399
6. Bouzid M, Ligęza A (1995) Temporal logic based on characteristic functions. KI-95: Advances in Artificial Intelligence, 19th Annual German Conference on Artificial Intelligence, Bielefeld,Germany, Springer, Lecture Notes in Computer Science 981:221–232
7. Cukierman D, Delgrande J (1998) Expressing time intervals and repetition within a formalisation of calendars. Computational Intelligence, 14:563–597
8. Fischer M, Gabbay D, Vila L (eds) (2005) Handbook of Temporal Reasoning in Artificial Intelligence. Elsevier, Foundations of Artificial Intelligence, Amsterdam
9. Hajnicz E (1989) Absolute dates and relative dates in an inferential system on temporal dependencies between events. International Journal of Man-Machine Studies, 30:537–549
10. Hajnicz E (1991) A formalization of absolute and relative dates based on the point calculus. International Journal of Man-Machine Studies, 34:717–730
11. Hajnicz E (1996) Time Structures. Formal Description and Algorithmic Representation, volume 1047 of Lecture Notes in Artificial Intelligence. Springer, Berlin, Heidelberg
12. Kabanza F, Stévenne J-M, Wolper P (1995) Handling infinite temporal data. Journal of Computer and System Sciences, 51:3–17

13. Kahn K, Gorry GA (1977) Mechanizing temporal logic. Artificial Intelligence, 9:87–108
14. Ladkin PB (1986) Primitives and units for time specification. In: Proceedings of the 5th National Conference on AI, AAAI'86, Morgan Kaufmann, 354–359
15. Ladkin PB (1987) The Logic of Time Representation. PhD thesis, University of California at Berkeley
16. Ladkin PB (1986) Time representation: A taxonomy of interval relations. In: Proceedings of the 5th National Conference on AI, Morgan Kaufmann, AAAI'86, 360–366
17. Leban B, Mcdonald D, Foster D (1986) A representation for collection of temporal intervals. In: Proceedings of the 5th National Conference on AI, AAAI'86, 354–359
18. Lee W-J, Lee S-J (2004) Fuzzy Calendar Algebra and Its Applications to Data Mining. 11th International Symposium on Temporal Representation and Reasoning TIME 2004, Combi C, Ligozat G (eds), Los Alamitos, California, IEEE Computer Society, 71–78
19. Ligęza A, Bouzid M (2007) Temporal specifications with XTUS. A hierarchical algebraic approach. In: Knowledge-Driven Computing, Studies in Computational Intelligence, Springer-Verlag (this volume), Berlin, Heidelberg
20. Montanari A (1996) Metric and layered temporal logic for time granularity. Technical report, ILLC Dissertation, University of Amsterdam
21. Niézette M, Stévenne J-M, Leban B, Mcdonald D, Foster D (1992) An efficient symbolic representation of periodic time. In: Proceedings of International Conference on Information and Knowledge Management, 161–168
22. Ohlbach HJ (2004) Relations Between Fuzzy Time Intervals. 11th International Symposium on Temporal Representation and Reasoning TIME 2004, Combi C, Ligozat G (eds), Los Alamitos, California, IEEE Computer Society, 44–51

Bond Rating with πGrammatical Evolution

Anthony Brabazon[1] and Michael O'Neill[1]

Natural Computing Research and Applications Group,
University College Dublin, Ireland.
anthony.brabazon@ucd.ie; m.oneill@ucd.ie

1 Introduction

Most large firms use both share and debt capital to provide long-term finance for their operations. The debt capital may be raised from a bank loan, or may be obtained by selling bonds directly to investors. As an example of the scale of US bond markets, the value of new bonds issued in 2004 totaled $5.48 trillion, and the total value of outstanding marketable bond debt at 31 December 2004 was $23.6 trillion [1]. In comparison, the total global market capitalisation of all companies quoted on the New York Stock Exchange (NYSE) at 31/12/04 was $19.8 trillion [2]. Hence, although company stocks attract most attention in the business press, bond markets are actually substantially larger.

When a company issues traded debt (e.g. bonds), it must obtain a credit rating for the issue from at least one recognised rating agency (Standard and Poor's (S&P), Moody's and Fitches'). The credit rating represents an agency's opinion, at a specific date, of the creditworthiness of a borrower in general (a bond-issuer credit-rating), or in respect of a specific debt issue (a bond credit rating). These ratings impact on the borrowing cost, and the marketability of issued bonds. Although several studies have examined the potential of both statistical and machine-learning methodologies for credit rating prediction [3–6], many of these studies used relatively small sample sizes, making it difficult to generalise strongly from their findings. This study by contrast, uses a large dataset of 791 firms, and introduces πGE to this domain.

In common with the related corporate failure prediction problem [7], a feature of the bond-rating problem is that there is no clear theoretical framework for guiding the choice of explanatory variables, or model form. Rating agencies assert that their credit rating process involves consideration of both financial and non-financial information about the firm and its industry, but the precise factors, and the related weighting of these factors, are not publicly disclosed. In the absence of an underlying theory, most published work on credit rating prediction employs a data-inductive modelling approach, using firm-specific financial data as explanatory variables, in an attempt to 'recover' the model

A. Brabazon and M. O'Neill: *Bond Rating with πGrammatical Evolution*, Studies in Computational Intelligence (SCI) **102**, 17–30 (2008)
www.springerlink.com

used by the rating agencies. This produces a high-dimensional combinatorial problem, as the modeller is attempting to uncover a 'good' set of model inputs, and model form, giving rise to particular potential for evolutionary automatic programming methodologies such as GE.

1.1 Structure of Chapter

The next section provides a concise overview the bond rating process, followed by a sections which introduce Grammatical Evolution and its variant πGE. Next, a description of the data set and methodology adopted is provided. The remaining sections provide the results of the experiments followed by a number of conclusions.

2 Background

Several categories of individuals would be interested in a model that could produce accurate estimates of bond ratings. Such a model would be of interest to firms that are considering issuing debt as it would enable them to estimate the likely return investors would require if the debt was issued, thereby providing information for pricing the bonds. The model could also be used to assess the creditworthiness of firms that have not issued debt and hence do not already have a published bond rating. This information would be useful to bankers or other companies that are considering whether they should extend credit to that firm.

2.1 Notation for Credit Ratings

Although the precise notation used to denote the creditworthiness of a bond or issuer varies between rating agencies, the credit status is generally denoted by means of a discrete, mutually exclusive, letter rating. Taking the rating structure of S&P as an example, the ratings are broken down into 10 broad classes. The highest rating is denoted AAA, and the ratings then decrease in the following order, AA, A, BBB, BB, B, CCC, CC, C, D. Ratings between AAA and BBB (inclusive) are deemed to represent *investment grade*, with lower quality ratings deemed to represent debt issues with significant speculative characteristics (also called *junk bonds*). A 'C' grade represents a case where a bankruptcy petition has been filed, and a 'D' rating represents a case where the borrower is currently in default on their financial obligations. As would be expected, the probability of default depends strongly on the initial rating which a bond receives (Table 1). Ratings from AAA to CCC can be modified by the addition of a + or a - to indicate at which end of the rating category the bond rating falls.

Table 1. Rate of default by initial rating category (1987-2002) (from Standard & Poor's, 2002)

Initial Rating	Default Rate (%)
AAA	0.52
AA	1.31
A	2.32
BBB	6.64
BB	19.52
B	35.76
CCC	54.38

2.2 Rating Process

Rating agencies earn fees from bond issuers for evaluating the credit status of new issuers and bonds, and for maintaining credit rating coverage of these firms and bonds. A company obtains a credit rating for a debt issue by contacting a rating agency and requesting that an issue rating be assigned to the new debt to be issued, or that an issuer rating be assigned to the company as a whole. As part of the process of obtaining a rating, the firm submits documentation to the rating agency including recent financial statements, a prospectus for the debt issue, and other non-financial information. Discussions take place between the rating agency and management of the firm and a rating report is then prepared by the analysts examining the firm. This rating report is considered by a rating committee in the rating agency which decides the credit rating to be assigned to the debt issue/issuer.

Rating agencies emphasise that the credit rating process involves consideration of financial as well as non-financial information about the firm, and also considers industry and market-level factors. The precise factors and related weighting of these factors used in determining a bond's rating are not publicly disclosed by the rating agencies. Subsequent to their initial rating, a bond may be re-rated upwards (upgrade) or downwards (downgrade) if company or environmental circumstances change. A re-rating of a bond below investment grade to junk bond status (such bonds are colourfully termed fallen angels) may trigger a significant sell-off as many institutional investors are only allowed, by external or self-imposed regulation, to hold bonds of investment grade.

3 Grammatical Evolution

Grammatical Evolution (GE) is an evolutionary algorithm that can evolve computer programs in any language [8–11], and can be considered a form of grammar-based genetic programming. Rather than representing the programs as parse trees, as in GP [12], a linear genome representation is used. A genotype-phenotype mapping is employed such that each individual's variable length binary string, contains in its codons (groups of 8 bits) the information to select production rules from a Backus Naur Form (BNF) grammar. The grammar allows the generation of programs in an arbitrary language that are guaranteed to be syntactically correct, and as such it is used as a generative grammar, as opposed to the classical use of grammars in compilers to check syntactic correctness of sentences. The user can tailor the grammar to produce solutions that are purely syntactically constrained, or they may incorporate domain knowledge by biasing the grammar to produce very specific forms of sentences. BNF is a notation that represents a language in the form of production rules. It is comprised of a set of non-terminals that can be mapped to elements of the set of terminals (the primitive symbols that can be used to construct the output program or sentence(s)), according to the production rules. A simple example BNF grammar is given below, where <expr> is the start symbol from which all programs are generated. The grammar states that <expr> can be replaced with either <expr><op><expr> or <var>. An <op> can become either +, -, or *, and a <var> can become either x, or y.

```
<expr>  ::= <expr><op><expr>  (0)
          | <var>             (1)

   <op>  ::= +                (0)
          | -                 (1)
          | *                 (2)

  <var>  ::= x                (0)
          | y                 (1)
```

The grammar is used in a developmental process to construct a program by applying production rules, selected by the genome, beginning from the start symbol of the grammar. In order to select a production rule in GE, the next codon value on the genome is read, interpreted, and placed in the following formula:

$$Rule = Codon\ Value\ Mod\ Num.\ Rules$$

where Mod represents the modulus operator. Given the example individual's genome (where each 8-bit codon has been represented as an integer for ease of reading) in Fig. 3, the first codon integer value is 220, and given that we have 2 rules to select from for <expr> as in the above example, we get $220\ Mod\ 2 = 0$. <expr> will therefore be replaced with <expr><op><expr>.

Beginning from the left hand side of the genome codon integer values are generated and used to select appropriate rules for the left-most non-terminal

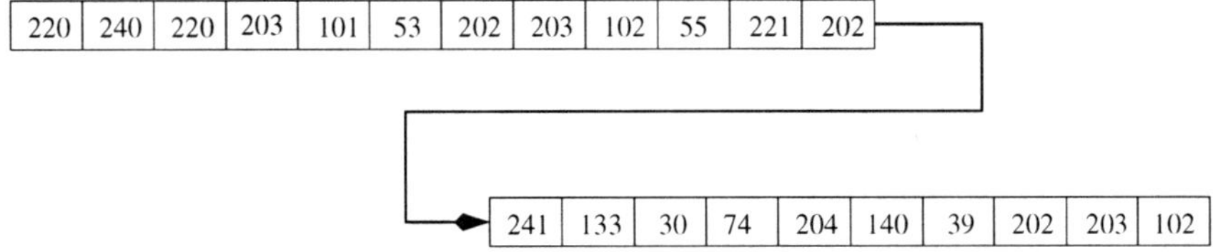

Fig. 1. An example GE individual's genome represented as integers for ease of reading.

in the developing program from the BNF grammar, until one of the following situations arise:

- A complete program is generated. This occurs when all the non-terminals in the expression being mapped are transformed into elements from the terminal set of the BNF grammar.
- The end of the genome is reached, in which case the *wrapping* operator is invoked. This results in the return of the genome reading frame to the left hand side of the genome once again. The reading of codons will then continue unless an upper threshold representing the maximum number of wrapping events has occurred during this individual's mapping process.
- In the event that a threshold on the number of wrapping events has occurred and the individual is still incompletely mapped, the mapping process is halted, and the individual assigned the lowest possible fitness value.

Returning to the example individual, the left-most `<expr>` in `<expr><op>` `<expr>` is mapped by reading the next codon integer value 240 and used in 240 *Mod* 2 $=$ 0 to become another `<expr><op><expr>`. The developing program now looks like `<expr><op><expr><op><expr>`. Continuing to read subsequent codons and always mapping the left-most non-terminal the individual finally generates the expression `y*x-x-x+x`, leaving a number of unused codons at the end of the individual, which are deemed to be introns and simply ignored. A full description of GE can be found in [8].

4 πGrammatical Evolution

The GE mapping process can be divided into a number of sub-components including the transcription and translation processes as outlined in the previous section. The πGE variant of GE replaces the translation process to allow evolution to specify the order in which production rules are mapped as opposed to the strict depth-first, left to right, mapping of the standard GE algorithm. In πGE we use the genotype to dictate which non-terminal from those present to expand next, before deciding which production rule to apply to the selected non-terminal. The genome of an individual in πGE is different in that there are two components to each codon. That is, each codon corresponds to the pair of values (*nont*, *rule*).

In the first derivation step of the example mapping presented earlier, `<expr>` is replaced with `<expr><op><expr>`. Then in the standard GE genotype-phenotype mapping process, the left-most non-terminal (the first `<expr>`) in the developing program is always expanded first. The πGE mapping process differs in an individual's ability to determine and adapt the order in which non-terminals will be expanded [13]. To this end, a πGE codon corresponds to the pair (*nont*, *rule*), where *nont* and *rule* are represented by N bits each (N=8 in this study), and a chromosome, then, consists of a vector of these pairs. In πGE, we analyse the state of the developing program before each derivation step, counting the number of non-terminals present. If there is more than one non-terminal present in the developing program the next codon's *nont* value is read to pick which non-terminal will be mapped next according to the following mapping function:

$$Non - terminal = Codon\ nont\ Value\ Mod\ Number of non - terminals$$

In the above example, there are 3 non-terminals ($<expr>_0<op>_1<expr>_2$) after application of the first production rule. To decide which non-terminal will be expanded next we use Number of non-terminals $= 9\ \%\ 3 = 0$, i.e., $<expr>_0$ is expanded. The mapping rule for selecting the appropriate rule to apply to the current non-terminal is given in the normal GE fashion:

$$Rule = Codon\ rule\ Value\ Mod\ Number of Rules$$

In this approach, evolution can result in a derivation subsequence being moved to a different context as when counting the number of non-terminals present we do not pay attention to the type of non-terminals (e.g. `<expr>` versus `<op>`).

An example of the application of πGE is provided in Fig. 4. In the top derivation tree, 9 Mod 3=0 (this derivation step is labelled **b**), hence the left-most non-terminal is expanded first. In the bottom derivation tree a mutation event transforms the second codon's *nont* value from 9 to 8, giving 8 Mod 3=2 (step **b**), hence the right-most non-terminal is expanded instead. The three subsequent subtrees (derivation steps labelled **c & d**, **e**, and **f & g**) that are produced are redistributed amongst other non-terminals.

In this instance, the single mutation is acting in a similar fashion to a multiple sub-tree exchange or crossover within the individual. The top derivation tree corresponds to the expression (x * x) - y, whereas the bottom tree gives x * (x - y).

We note that πGE could be implemented in more than one way. An alternative approach would be to respect non-terminal types and only allow choices to be made between non-terminals of the same type, thus preserving the semantics of the following derivation subsequence, and simply changing the position in which it appears in the developing program.

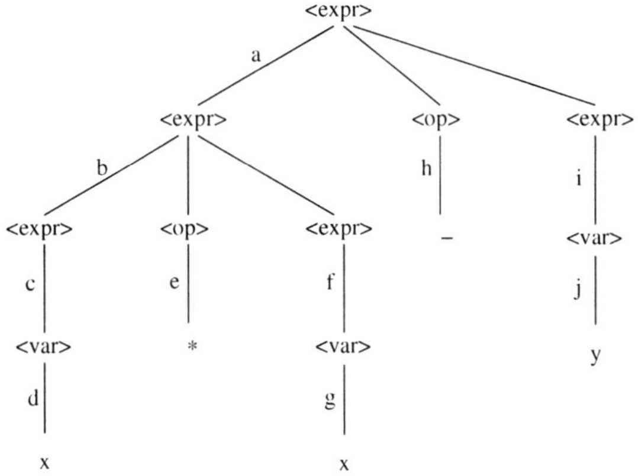

(23, 88), (9, 102), (20, 11), (5, 18), (16, 8), (27, 3), (12, 4), (4, 4), (3, 7), (6, 9)........
a   b   c   d   e   f   g   h   i   j

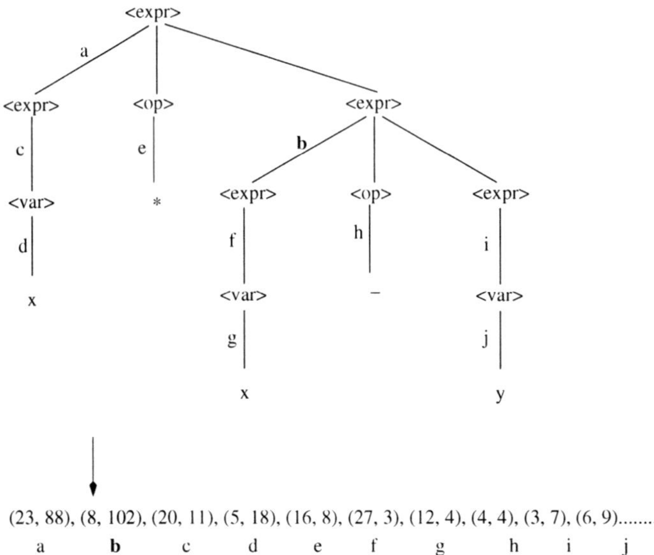

(23, 88), (8, 102), (20, 11), (5, 18), (16, 8), (27, 3), (12, 4), (4, 4), (3, 7), (6, 9)........
a   b   c   d   e   f   g   h   i   j

Fig. 2. An example of πGE, illustrating a single mutation event in the *nont* position in the second codon.

5 Experimental Approach

The dataset consists of financial data of 791 industrial and service US companies, along with their associated bond-issuer credit-rating, drawn from the S&P Compustat database. Of these companies, 57% have an investment-grade rating (AAA, AA, A, or BBB), and 43% have a junk rating. To allow time for the preparation of year-end financial statements, the filing of these statements with the Securities and Exchange Commission (S.E.C), and the development of a bond rating opinion by Standard and Poor rating agency, the bond rating of the company as at 30 April 2000, is matched with financial information drawn from their financial statements as at 31 December 1999. A subset of 600 firms was randomly sampled from the total of 791 firms, to produce two groups of 300 investment grade and 300 junk rated firms. The 600 firms were randomly allocated to the training set (420) or the hold-out sample (180), ensuring that each set was equally balanced between investment and non-investment grade ratings.

Five groupings of explanatory variables, drawn from financial statements, are given prominence in prior literature as being the prime determinants of bond issue quality and default risk:

   i. Liquidity
  ii. Debt
 iii. Profitability
 iv. Activity / Efficiency
   v. Size

Liquidity refers to the availability of cash resources to meet short-term cash requirements. Debt measures focus on the relative mix of funding provided by shareholders and lenders. Profitability considers the rate of return generated by a firm, in relation to its size, as measured by sales revenue and/or asset base. Activity measures consider the operational efficiency of the firm in collecting cash, managing stocks and controlling its production or service process. Firm size provides information on both the sales revenue and asset scale of the firm and also provides a proxy metric on firm history. The groupings of potential explanatory variables can be represented by a wide range of individual financial ratios, each with slightly differing information content. The groupings themselves are interconnected, as weak (or strong) financial performance in one area will impact on another. For example, a firm with a high level of debt, may have lower profitability due to high interest costs.

Following the examination of a series of financial ratios under each of these headings, a total of eight financial variables was selected for inclusion in this study. The selection of these variables was guided both by prior literature in bankruptcy prediction [14–16] and literature on bond rating prediction [17–20]. These ratios were then further filtered using statistical analysis. The ratios selected were as follows:

   i. Current ratio

  ii. Retained earnings to total assets

 iii. Interest coverage

 iv. Debt ratio

  v. Net margin

 vi. Market to book value

 vii. Log (Total assets)

viii. Return on total assets

Table 2. Means of input ratios for investment and junk bond groups of companies

	Investment grade	Junk grade
Current ratio	1.354	1.93
Retained earnings/Total assets	0.22	-0.12
Interest coverage	7.08	1.21
Debt ratio	0.32	0.53
Net margin	0.07	-0.44
Market to book value	18.52	4.02
Total assets	10083	1876
Return on total assets	0.10	0.04

The objective in selecting a set of proto-explanatory variables is to choose financial variables that vary between companies in different bond rating classes, and where information overlaps between the variables are minimised (the financial ratios chosen during the selection process are listed at the end of this section). Comparing the means of the chosen ratios (see Table 2) for the two groups of ratings, reveals a statistically significant difference at the 1% level, and as expected, the financial ratios in each case, for the investment ratings are stronger than those for the junk ratings. The only exception is the current ratio, which is stronger for the junk rated companies, possibly indicating a preference for these companies to hoard short-term liquidity, as their access to long-term capital markets is weak. A correlation analysis between the selected ratios indicates that most of the cross-correlations are less than | 0.20 |, with the exception of the debt ratio and (Retained Earnings/Total Assets) ratio pairing, which has a correlation of -0.64. The grammar adopted is as follows:

```
<lc> ::= if( <expr> <relop> <expr> )

        class=''Junk'';
    else
        class=''Investment Grade'';

<expr> ::= ( <expr> ) + ( <expr> )

        | <coeff> * <var>

<var> ::= var3[index] | var4[index]
```

```
          | var5[index] | var6[index]

          | var7[index] | var8[index]

          | var9[index] |var10[index]

          | var11[index]

  <coeff> ::=  ( <coeff> ) <op> ( <coeff> )

            | <float>

  <op> ::= + | - | *

  <float> ::= 9 | 8 | 7 | 6 | 5 | 4
            | 3 | 2 | 1 | -1 | .1

  <relop> ::= <=
```

where **var3** = Current Ratio, **var4** = Retained Earnings to total assets, **var5**= Interest Coverage, **var6** = Debt Ratio, **var7** = Net Margin, **var8** = Market to book value, **var9** = Total Assets, **var10** = ln (Total Assets), **var11** = Return on total assets.

6 Results

The results from our experiments are now provided. Each of the πGE experiments is run for 100 generations, with variable-length, one-point crossover at a probability of 0.9, one point bit mutation at a probability of 0.01, roulette selection, and steady-state replacement. To assess the stability of the results across different randomisations of the dataset between training and test data, we recut the dataset five times, maintaining an equal balance of investment and non-investment grade ratings in the resulting training and test datasets. In our experiments, fitness is defined as the number of correct classifications obtained by an evolved discriminant rule. The results for the best individual of each cut of the dataset, where 30 independent runs were performed for each cut, averaged over all five randomisations of the dataset, for a population size of 500 is given in Table 3, and Figure 6 displays the evolution of the mean average and mean best results over time.

Table 3. Average performance for the five recuts of the best evolved rules on their in and out-sample datasets.

	Fitness	TP	TN	FP	FN
In-sample	0.8450	182.8	172.1	37.9	27.2
Out-sample	0.8500	77.9	75.1	14.9	12.1

To assess the overall hit-ratio of the developed models (out-of-sample), Press's Q statistic [21] was calculated for each model. In all cases, the null

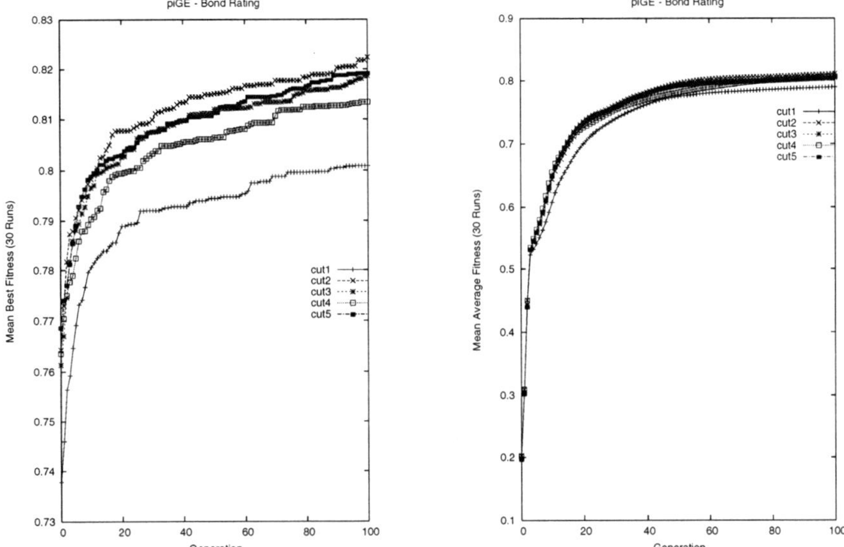

Fig. 3. Mean average (left) and mean best (right) over 30 runs, across all 5 recuts.

hypothesis, that the out-of sample classification accuracies are not significantly better than those that could occur by chance alone, was rejected at the 1% level. A t-test of the hit-ratios also rejected a null hypothesis that the classification accuracies were no better than chance at the 1% level. Across all the data recuts, the best individual achieved classification accuracy of 86% in-sample and 87% out-of-sample.

When applying any model induction approach, it is important to reduce the possibility of overfitting. A number of practical steps to reduce the chance of overfitting include the collection of a sufficiently large dataset relative to the number of explanatory variables included in the model, and the testing of the developed model on a sizeable out-of-sample dataset. In this study we have trained the models using 420 data vectors, tested the evolved models using a sizeable out-of-sample dataset (180 data vectors), and have restricted the evolved models to use a maximum of eight explanatory variables. As noted above, the in-sample and out-of-sample classification accuracies are very similar, indicating that overfitting does not seem to have been a problem. Given that the evolved models were restricted to use a maximum of eight explanatory variables we have not implemented a regularisation term in the error function.

Examining the structure of one of the best individuals shows that the evolved discriminant function had the following form:

$$\text{IF } (0 \leq -2 + \textit{Debt Ratio} - \textit{Total Assets} - 5*\frac{\textit{Retained Earnings}}{\textit{Total Assets}})$$
$$\text{THEN 'Junk' ELSE 'Investment Grade'}$$

Examining the signs of the coefficients of the evolved rules does not suggest that they conflict with common financial intuition. The rules indicate that low/negative retained earnings, low/negative total assets or high levels of debt finance are symptomatic of a firm that has a junk rating. It is noted that similar risk factors have been identified in predictive models of corporate failure which utilise financial ratios as explanatory inputs [7, 22]. Conversely, low levels of debt, a history of successful profitable trading, and high levels of total assets are symptomatic of firms that have an investment grade rating.

6.1 Comparison of Results

To provide a benchmark for the results obtained by πGE we compare them with the results obtained on the same recuts of the dataset, using a fully-connected, feedforward multi-layer perceptron (MLP), trained using the back-propagation algorithm. The developed networks utilised all the explanatory variables. The optimal number of hidden-layer nodes was found following experimentation on each separate data recut, and varied between two and four nodes. The classification accuracies for the networks, averaged over all five recuts is provided in Table 4.

Table 4. Performance of the MLPs on the training and out-of-sample datasets, averaged over all five recuts of the dataset.

	Fitness	TP	TN	FP	FN
In sample	0.869	181.8	183.2	26.8	28.2
Out-sample	0.850	75.8	77.2	12.8	14.2

The levels of classification accuracy obtained with the MLP are competitive with earlier research, with for example [17] obtaining an out-of-sample classification accuracy of approximately 83.3%, although it is noted that the size of the dataset in their study was small. Comparing the results from the MLP with those of πGE on the initial fitness function (Table 3) suggests that πGE has proven competitive with an MLP methodology, in terms of producing a similar classification accuracy. Benchmark results were also obtained using an LDA methodology. Utilising the same dataset recuts as πGE, LDA produced results (averaged across all five recuts) of 82.74% in-sample, and 85.22% out-of-sample. Again, πGE is competitive against these results in terms of classification accuracy. Comparing the results obtained by the linear classifiers (LDA and πGE) against those of an MLP, suggests that strong non-linearities between the explanatory variables and the dependent variable are not present.

7 Conclusions & Future Work

The objective of this chapter was to introduces a novel classification system based on a variant of Grammatical Evolution, πGE, and to assess the utility of this methodology using information drawn from the financial statements of bond-issuing firms. Despite using data drawn from companies in a variety of industrial sectors, the developed models showed an impressive capability to discriminate between investment and junk rating classifications. The πGE developed models also proved highly competitive with a series of MLP models developed on the same datasets. Several extensions of the methodology in this study are indicated for future work. One route is the inclusion of non-financial company and industry-level information as input variables. A related possibility would be to concentrate on building rating models for individual industrial sectors. Another avenue of research would be to extend the grammar used by πGE in this study to encompass multi-class bond rating predictions.

References

1. Bond Market Association – Research Quarterly (2005), `http://www.bondmarkets.com`
2. NYSE (2005) Market Statistics, `http://www.nyse.com`
3. Ederington H (1985) Classification models and bond ratings. Financial Review 20(4):237–262
4. Gentry J, Whitford D, Newbold P (1988) Predicting industrial bond ratings with a probit model and funds flow components. Financial Review 23(3):269–286
5. Huang Z, Chen H, Hsu C, Chen W, Wu S (2004) Credit rating analysis with support vector machines and neural networks: a market comparative study. Decision Support Systems 37(4):543–558
6. Shin K, Han I (2001) A case-based approach using inductive indexing for corporate bond rating. Decision Support Systems 32:41–52
7. Brabazon A, O'Neill M, Matthews R, Ryan C (2002) Grammatical Evolution and Corporate Failure Prediction. In: Spector L et al. (eds) Proceedings of the Genetic and Evolutionary Computation Conference (GECCO 2002). Morgan Kaufmann, New York, pp. 1011–1019
8. O'Neill M, Ryan C (2003) Grammatical Evolution: Evolutionary Automatic Programming in an Arbitrary Language. Kluwer Academic Publishers, Boston
9. O'Neill M (2001) Automatic Programming in an Arbitrary Language: Evolving Programs in Grammatical Evolution. PhD thesis, University of Limerick
10. O'Neill M, Ryan C (2001) Grammatical Evolution. IEEE Trans. Evolutionary Computation 5(4):349–358
11. Ryan C, Collins JJ, O'Neill M (1998) Grammatical Evolution: Evolving Programs for an Arbitrary Language. In: Proceedings of the First European Workshop on Genetic Programming. Springer-Verlag, Berlin, pp. 83–95
12. Koza J (1992) Genetic Programming. MIT Press, Boston

13. O'Neill M, Brabazon A, Nicolau M, Mc Garraghy S, Keenan P (2004) πGramm-atical Evolution. In: Deb K et al. (eds) Proceedings of GECCO 2004. Springer-Verlag, Berlin, pp 617–629
14. Altman E (1993) Corporate Financial Distress and Bankruptcy. John Wiley and Sons Inc, New York
15. Morris R (1997) Early Warning Indicators of Corporate Failure: A critical review of previous research and further empirical evidence. Ashgate Publishing Limited, London
16. Altman E (1998) The importance and subtlety of credit rating migration. Journal of Banking & Finance 22:1231–1247
17. Dutta S, Shekhar S (1988) Bond rating: a non-conservative application of neural networks. In: Proceedings of IEEE International Conference on Neural Networks. IEEE Press, Piscataway, New Jersey, pp 443–450
18. Kamstra M, Kennedy P, Suan TK (2001) Combining Bond Rating Forecasts Using Logit. The Financial Review 37:75–96
19. Singleton J, Surkan A (1991) Modeling the Judgment of Bond Rating Agencies: Artificial Intelligence Applied to Finance. Journal of the Midwest Finance Association 20:72–80
20. Brabazon A, O'Neill M (2006) Biologically Inspired Algorithms for Financial Modelling. Springer, Berlin
21. Hair J, Anderson R, Tatham R, Black W (1998) Multivariate Data Analysis. Prentice Hall, Upper Saddle River, New Jersey
22. Brabazon A, O'Neill M (2003) Anticipating Bankruptcy Reorganisation from Raw Financial Data using Grammatical Evolution. In: Raidl G et al. (eds) Proceedings of EvoIASP 2003. Springer-Verlag, Berlin, pp 368–378

Handling the Dynamics of Norms – A Knowledge-Based Approach

Jolanta Cybulka[1] and Jacek Martinek[2]

[1] Institute of Control and Information Engineering, Poznań University of Technology, pl. M. Skłodowskiej-Curie 5, 60-965 Poznań, Poland
`jolanta.cybulka@put.poznan.pl`
[2] Institute of Control and Information Engineering, Poznań University of Technology, pl. M. Skłodowskiej-Curie 5, 60-965 Poznań, Poland
`jacek.martinek@put.poznan.pl`

1 Model of Norms Dynamics

The widely observable frequent changes of norms in many normative systems existent in different countries stimulated the research regarding the invention of models used to represent the dynamics of norms. The models concern norms expressed in the form of legal provisions. The application of the event calculus appeared to be a suitable way to model the considered dynamics ([1–3]). According to this approach the authors designed a model ([4]) on the basis of which the changes of legal provisions in time can be handled in a relevant knowledge base. The main purpose of this knowledge base is to support the task of searching for the current version of the provisions.

The proposed legal provisions dynamics model (*LPDModel*) brings to light the layered structure of any legal system that is represented by means of the set of legal statutory documents ([5]). It is assumed that the documents contain the textual representation of the legal provisions. According to the layered structure, provisions are divided into:

1) *substantial provisions*, concerned with some subject matters,
2) *meta-provisions*, having the control nature and
3) *supra-provisions* used to formulate the legislation.

In Polish legislation, the supra-provisions are contained in the regulation called "the rules of legislative technique" ([6]), which controls the production of the legal provisions and, at the same time, serves as the basis of the considered model. The regulation constitutes the three types of statutes: a *main*, an *amending* and an *introductory* one. The types reflect one of dynamic facets of the legal system: one may claim that only the main statutes change in time due to their introduction or amendments, while the others remain unchanged.

J. Cybulka and J. Martinek: *Handling the Dynamics of Norms – A Knowledge-Based Approach*,
Studies in Computational Intelligence (SCI) **102**, 31–43 (2008)
`www.springerlink.com`   © Springer-Verlag Berlin Heidelberg 2008

Also, we take for granted that in statutes, the legal provisions are considered to be interrelated pieces of a legal text.

The supra-provisions specify features and categorisations of all the legal provisions, what enables to distinguish the layers of the earlier mentioned *substantial* and the *meta-provisions*. The substantial provisions are concerned with the subject matters of the regulated domain. Some of them are regarded as *basic* provisions, which may be either specialised by some *specialising* provisions or generalised by the *generalising* ones. The specialisation and the generalisation are both examples of *structural relations* holding between the provisions. These relations undergo changes in the process of the statute amendments. As to the meta-provisions, they act on the other provisions in that they enable to enact, repeal, amend or interpret the substantial provisions. Therefore, they constitute the core rules handling the legal provisions dynamics.

The *LPDModel* is semantically grounded both on the supra-provisions and on the modelling ideas taken from the event calculus. The event calculus engine manages the specifications of the three knowledge sources, namely the description of the incoming events (the first source), which influence the properties of the states of affairs (fluents, the second source) and the propositions concerning the fluents at a given point of time (the third source). Following this, *LPDModel* handles:

1) the ordered list of events of the legislative nature (i.e. the statute promulgation or enactment),
2) the control knowledge expressed in meta-provisions that contains the "frozen" actions influencing the states of affairs and
3) the current status of legal provisions in points of time (i.e. the textual form and the structural relations).

The *LPDModel* enables to generate the output in the form of 3) on the basis of 1) and 2) given as the input.

Let us consider in more detail the form of the control knowledge. The meaning of a meta-provision is expressed by something what we call "meta-norms". Every meta-norm depends on the occurrence of some mentioned above event and defines an *action* which should be executed in order to introduce, repeal or amend the existing provisions. We classify actions as *instantaneous* and *durative*. Among the durative actions we distinguish the *temporary provision suspension*, the *temporary provision prolongation* or the *vacatio legis enactment*. Every durative action is ended (or initiated) by a special *limitary event* which, in turn, conditions the meta-norm enforcing the execution of an indicated instantaneous action. For example, the enactment of the meta-provision, which enforces the suspension of the indicated provision in some period, may be modelled via the meta-norm, which imposes the execution of the "suspension" action during this period. When the limitary event occurs, which ends the execution of the suspension action, the instantaneous action (the suspended provision enactment) should be executed. In the special case of the enactment of a retroactive provision, the limitary event begins the execution of a durative action in some point of time in the past.

As to the events handling procedure, it is necessary that the knowledge base relevant to the *LPDModel* should contain the data concerned with all the events, both these, which are passed and also the chronologically ordered list of the forthcoming ones. The list of incoming events must contain the limitary events of the durative actions which are under execution. Also, this list holds the events concerning the promulgation of new or amending statutes.

It is worth noticing that the process of the retrieval of the current version of a provision can be done in an efficient way due to the special index, which was introduced to the *LPDModel*-based knowledge base. It contains the mnemonic names for the basic provisions and enables their browsing. The current textual version of the required basic provisions is retrieved accompanied by the provisions that are actually structurally related to it.

The aim of the next section is to describe the structure of the *LPDModel*-based knowledge base.

2 Knowledge Base Structure

The prototype knowledge base is implemented in SWI-Prolog extended by the object-oriented XPCE module [7]. The crucial point of the control knowledge representation is the specification of the meta-norms, which are connected with the meta-provisions. It is assumed that meta-norms are also attached to the subject knowledge pieces (the substantial provisions) - every such provision is accompanied by the default meta-norm which describes its enactment method, indicates its current textual version and specifies the "structural facts", which define the structural relations that hold between the considered provision and the other ones.

The *LPDModel* in Prolog (*LPDKB*) uses "facts", either time invariable (the constant facts) or facts which start or terminate at some point of time (indicated by a date). The constant facts are expressed in the knowledge base by means of the Prolog clauses, while the non-constant facts are expressed by terms, initiated or terminated by some events, which in turn have dates of their occurrence attached to them. This is a standard description method used in the event calculus. The statute as a whole has a *LPDKB* knowledge base representation in the form of a clause having three arguments (the constant data): the number *Nr* of the statute, the term *Kind* (which is an element of the set {*main, amending, introductory*}) and the term *Name* of the statute. These constitute the following expression:

$$statute(Nr, Kind, Name).$$

Also, the structural relations between the amending (or the introductory) and the main statutes should be given by means of the clauses:

$$amends(Nr, MainStatuteNr),$$

$$introduces(Nr, MainStatuteNr).$$

Each substantial provision is described by a term of the form:

$$p(StatuteNr, ArticleNr, SectionNr, ProvisionNr)$$

in which *StatuteNr* is always the main statute number disregarding the provision origin (i.e. the main or the amending statute). The term *ArticleNr* is used to represent the article's number (the textual contents of the article is the "source" of the provision), *SectionNr* constitutes the section number or zero if the article is not divided into sections, and the last argument, *ProvisionNr* is the number of the provision. The numbers are assigned to the provisions during the pre-processing phase of the statute text. The provisions are recognised and marked by a person who describes the statute in order to model it.

The meta-provision is encoded via a term of the form:

$$p(StatuteNr, ArticleNr, SectionNr, ProvisionNr).$$

The *StatuteNr* is the number of the statute containing the provision, and the other arguments have the same meaning as the arguments of the provision term above. The important attribute of the provision is its textual contents. It is assigned as *Text* to the provision identifier *ProvisionId* via the term:

$$text(ProvisionId, Text).$$

The recorded events are identified by the relevant identifiers, which are built of a name and some specifying arguments. Here are, for example, the events of the statute promulgation and its coming into force:

$$statute\_promulgation(StatuteNr),$$
$$statute\_enactment(StatuteNr).$$

The occurrence of an event *EventId* from the events list in a given point of *Time* (a day on which the event happened), is expressed by the clause:

$$happens(EventId, Time).$$

The result of such an activity is a change concerning the states of affairs, being the effect either of the initiation or of the termination of some facts (fluents) listed in *FactList*:

$$initiates(EventId, FactList)$$
$$terminates(EventId, FactList).$$

It is worth noticing that one event may cause the beginning or the termination of many facts. Meta-norms assigned to a provision (identified by the *ProvisionId*) are specified by clauses of the form:

$$mn(ProvisionId, EventId, InstantaneousAction),$$
$$mn(ProvisionId, EventId, DurativeAction,$$
$$LimitaryEventId, LimitaryEventTime).$$

Here *EventId* is the conditioning event and the third argument defines the action to be executed. In the case of a durative action the limitary event identifier and the limitary event time are also specified in the clause.

It appeared that the wide spectrum of legal states of affairs properties that occur after the legislative events could be modelled by the results of the execution of eight instantaneous and five durative actions of the meta-norms. The detailed description of the set of elementary operations that constitute the proposed actions is given in [4]. For example, the following term represents the instantaneous action of the enactment of the indicated provision (*ProvisionId*) with the textual contents (*Text*), which brings about some structural relations (*NewStructFacts*) between provisions:

$$provision\_enactment(ProvisionId, Text, NewStructFacts).$$

The data concerning the mentioned structural relations (the non-constant facts) between provisions are recorded into the *LPDKB* with the use of four types of terms:

basic(ProvisionId),
specifies(ProvisionId1,ProvisionId2),
generalises(ProvisionId1,ProvisionId2).

Sometimes a statute is in force temporarily only, up to the happening of some event (the limitary event *statute_temporal_force_ends*) that terminates its being in force. Such a situation can be modelled via a durative *temporary_holding_in_force* action accompanied by information about its limitary event:

temporary_holding_in_force(StatuteNr),
LimitaryEventId = statute_temporal_force_ends(StatuteNr).

To explore the current status of legal provisions in points of time (searching for non-constant facts), one has to define the Prolog clauses, which solve the problem. We formulate them on the grounds of the solution given in [1]. In the clauses the data gathered in earlier described predicates *happens*, *initiates* and *terminates* are used. The clauses under consideration are as follows:

holds_at(Fact, Time) :-
     happens(Event, TimeI),
     TimeI =< Time,
     initiates_fact(Event, Fact),
     not(broken(Fact, TimeI, Time)).

initiates_fact(Event, Fact) :-
     initiates(Event, FactList),
     member(Fact, FactList).

broken(Fact, TimeI, Time) :-
     happens(Event, TimeE),

$$TimeI < TimeE, TimeE =< Time,$$
$$terminates\_fact(Event, Fact).$$

$$terminates\_fact(Event, Fact) :-$$
$$terminates(Event, FactList),$$
$$member(Fact, FactList).$$

3 Knowledge Base User Interface

The user of the *LPDKB* knowledge base is equipped in the GUI interface consisting of the menu bar (of operations) and the two dialog windows. The upper one serves as a text presentation tool for provisions, while in the lower window the messages of the system-user interaction are shown. There are four main menu operations that result in:

1) *the presentation of the incoming events* – the list of such events is shown to the user,
2) *the registration of the event* – the event indicated by the user is enrolled on the incoming events list; the event is given its name and the date of its occurrence; the chronologically ordered current list of the incoming events appears in the lower window,
3) *the occurrence of the event* – the chronologically first event of the incoming events list is both selected to the occurrence and deleted form the list; the data about the occurred event are recorded into the knowledge base; all the actions which are conditioned by the event are executed; the message about the event occurrence is signalled in the lower dialog window,
4) *the presentation of provisions relevant to the selected date* – the user inputs the date to the system via an auxiliary window; then the scrolled list of the names of basic provisions relevant to the date is presented; the user may indicate one of them and then its current textual version is shown in the upper dialog window; the basic provision is accompanied by the relevant structurally related provisions.

The considered operations handled by the user enable to simulate the real situations that happen in time in legal systems. Briefly speaking, the situations (legal states of affairs) are controlled by the legal events and the consequences of their incoming. When an event occurs, all the actions, which are conditioned by the event, should be executed. But first, the search for all the meta-norms, which are conditioned by the considered event, must be done. It is only when the actions frozen in the meta-norms are executed. It concerns both the instantaneous and the durative actions. The execution of an instantaneous action results in the insertion into the knowledge base of the appropriate *initiates* and *terminates* formulae which correspond to the changes caused by the action. These formulae are associated with the considered event. Sometimes the action imposes the limitary event time change, in which case the list of incoming events is modified in an appropriate way. The execution of a durative action starts with the enrollment of its limitary event on the list of the incoming events. After that, the starting operations of this action are executed. They are performed in a way similar to the execution of an instantaneous action (see [4] for more details).

4 Provision Description Example

Let us now consider a short example illustrating how the provisions may be represented in the *LPDKB*. The example is based on the Polish statute on social security, to which we assign the number 1. The second section of the article 2a of the statute contains a deeming provision, which defines a farmer's monthly income from one hectare. By means of the auxiliary predicate index the name '*Income from one hectare*' is attached to the generated provision identified by *p(1,'2a',2,1)*:

$$index(p(1,\text{'}2a\text{'},2,1),\text{'}Income\ from\ one\ hectare\text{'}).$$

Due to some technical reasons the aiding text identifier *t(1,1,'2a',2,1)* is introduced together with the term representing the source textual contents of the considered provision:

$$full\_text(t(1,1,\text{'}2a\text{'},2,1),\text{''}Art.2a.2.\ In\ the\ context\ of\ this\ statute\ it\ is$$
$$assumed\ that\ the\ monthly\ income\ from\ one\ re\text{-}calculated$$
$$hectare\ amounts\ to\ 204\ zl.\text{''}).$$

The provision is both the substantial and the basic one. There are no special indications in the statute as to the enactment method of this provision, so it should be enacted when the whole statute is enacted (which is the default enactment procedure). The following meta-norm gathers all the data:

$$mn(p(1,\text{'}2a\text{'},2,1),\ statute\_enactment(1),\ provision\_enactment(p(1,\text{'}2a\text{'},2,1),$$
$$t(1,1,\text{'}2a\text{'},2,1),\ [basic(p(1,\text{'}2a\text{'},2,1))]))).$$

The *LPDKB* is encoded in Prolog, but an additional interface to it was also created, which enables the specification of provisions in the XML-based language. The specification language and its translator to Prolog are described in the next sections.

5 XML-Based Specification of Provisions

Nowadays, many information systems, especially Web services, take the advantage of using the XML language as their data-encoding format. It is easy to define the structure of an XML specification (in the form of a Document Type Definition DTD or an XML schema specification in the XML Schema Definition language – XSD) and this structure can be easily modified and translated to other useful forms. Such a situation concerns also the legal texts, in which case the legal authorities publish the current versions of statutes, resolutions and other acts encoded in the XML. In this paper the authors propose the XML structure of the document which is relevant to describe the legal provisions that are contained in the statutes. The structure is defined by means of the DTD which formats the input data concerning provisions processed in *LPDModel.* Now, the XML document type definition will be presented and described.

In *LPDModel* the legal provisions are contained in the texts of statutes, that is why the whole DTD represents a `<STATUTE>` and its types: a `<MAIN_STATUTE>`, an `<AMENDING_STATUTE>` or an `<INTRODUCTORY_STATUTE>`. The contents model of each of the statute types consists of the nonempty sequence of items called `<ELEMENT>` and each statute tag has the attributes denoting the number and the name of the statute. The amending (the introductory) statute tag has the additional attribute `AMENDS (INTRODUCES)`, which states the structural relation between the amending (the introductory) and the main statutes. The `<ELEMENT>` item represents a provision that is described by a sequence of the three elements: a `<PROVISION_TEXT>`, the optional `<PROVISION_NAME>` and the specification of `<META_NORMS>` connected with the provision. A provision text is given a short identifier and the provision name is connected with the provision identifier (via the relevant attributes). The meta-norms are distinguished according to the action type, namely there exist different tags for meta-norms with instantaneous action (`<IA_META_NORM>`) and with the durative one (`<DA_META_NORM>`). These meta-norms tags have attributes defined via the `ATTRS` entity mechanism – they are the provision identifier `PROVISION_ID`, the specification of the `EVENT` and the `CONDITIONS` of the action. The `<DA_META_NORM>` tag has two more attributes, which are the specification of the `LIMITARY_EVENT` of the action and some description of the event `LIM_EVENT_DESCRIPTION`.

The set of instantaneous actions tags contains eight elements:

- `<PROVISION_ENACTMENT>`
- `<SUSPENDED_PROVISION_ENACTMENT>`
- `<PROVISION_CHANGE>`
- `<STATUTE_REPEALING>`
- `<PROVISION_REPEALING>`
- `<PROLONGATED_PROVISION_REPEALING>`
- `<RETROACTIVITY_RECORDING>`
- `<LIMIT_EVENT_TIME_CHANGE>`

Each of them has an empty model of contents and some dedicated values assigned to the actions by means of the attributes (see the DTD below). There exist five durative actions tags, which are:

- `<TEMPORARY_HOLDING_IN_FORCE>`
- `<VACATIO_LEGIS>`

- `<PROVISION_PROLONGATION>`
- `<PROVISION_SUSPENSION>`
- `<RETROACTIVE_PROVISION_ENACTMENT>`

They also have the empty contents model and the values assigned by means of the attributes. The whole definition of the DTD is given below. Of course, it is possible to define an XML schema relevant to the considered DTD, but such a definition is more verbose and less readable for the human reader.

```
<?xml version = "1.0" standalone="yes" encoding="UTF-16" ?>
<!DOCTYPE STATUTE [
<!ELEMENT STATUTE (MAIN_STATUTE | AMENDING_STATUTE |
                   INTRODUCTORY_STATUTE)>
<!ELEMENT MAIN_STATUTE (ELEMENT +) >
  <!ATTLIST MAIN_STATUTE
      NR NMTOKEN #REQUIRED
      NAME CDATA #REQUIRED >
<!ELEMENT AMENDING_STATUTE (ELEMENT +)>
  <!ATTLIST AMENDING_STATUTE
      NR NMTOKEN #REQUIRED
      NAME CDATA #REQUIRED
      AMENDS NMTOKEN #REQUIRED >
<!ELEMENT INTRODUCTORY_STATUTE (ELEMENT +) >
  <!ATTLIST INTRODUCTORY_STATUTE
      NR NMTOKEN #REQUIRED
      NAME CDATA #REQUIRED
      INTRODUCES NMTOKEN #REQUIRED >
<!ELEMENT ELEMENT (PROVISION_TEXT, PROVISION_NAME?, META_NORMS)>
<!ELEMENT PROVISION_TEXT (#PCDATA)>
  <!ATTLIST PROVISION_TEXT
      TEXT_ID CDATA #REQUIRED >
<!ELEMENT PROVISION_NAME (#PCDATA)>
  <!ATTLIST PROVISION_NAME
      PROVISION_ID CDATA #REQUIRED >
<!ELEMENT META NORMS ((IA_META_NORM | DA_META_NORM) +) >
<!ELEMENT IA_META_NORM (INSTANTANEOUS_ACTION ) >
<!ELEMENT DA_META_NORM (DURATIVE_ACTION ) >
<!ENTITY % ATRS "PROVISION_ID CDATA #REQUIRED
                 EVENT CDATA #REQUIRED
                 CONDITIONS CDATA REQUIRED" >
  <!ATTLIST IA_META NORM %ATRS; >
  <!ATTLIST DA_META NORM %ATRS;
      LIMITARY_EVENT CDATA #REQUIRED
      LIM_EVENT_DESCRIPTION CDATA #REQUIRED >
<!ELEMENT INSTANTANEOUS ACTION   ( PROVISION_ENACTMENT |
             SUSPENDED_PROVISION_ENACTMENT |
             PROVISION_CHANGE | STATUTE_REPEALING|
             PROVISION_REPEALING |
             PROLONGATED_PROVISION_REPEALING |
             RETROACTIVITY_RECORDING |
```

```
                          LIMIT_EVENT_TIME_CHANGE) >
   <!ELEMENT PROVISION_ENACTMENT (EMPTY) >
   <!ENTITY % ATRS1 "PROVISION_ID CDATA #REQUIRED
                     TEXT CDATA #REQUIRED
                     NEW_FACTS CDATA #REQUIRED" >
     <!ATTLIST PROVISION_ENACTMENT %ATRS1; >
   <!ELEMENT SUSPENDED_PROVISION_ENACTMENT (EMPTY) >
     <!ATTLIST SUSPENDED_PROVISION_ENACTMENT %ATRS1; >
   <!ELEMENT PROVISION_CHANGE (EMPTY) >
     <!ATTLIST PROVISION_CHANGE %ATTRS2;
         TEXT CDATA #REQUIRED >
   <!ELEMENT STATUTE_REPEALING (EMPTY) >
     <!ATTLIST STATUTE_REPEALING
         STATUTE_NR NMTOKEN #REQUIRED >
   <!ELEMENT PROVISION_REPEALING (EMPTY) >
   <!ENTITY % ATTRS2 "PROVISION_ID CDATA #REQUIRED" >
     <!ATTLIST PROVISION_REPEALING %ATTRS2; >
   <!ELEMENT PROLONGATED_PROVISION_REPEALING (EMPTY) >
     <!ATTLIST PROLONGATED_PROVISION_REPEALING %ATTRS2; >
   <!ELEMENT RETROACTIVITY_RECORDING (EMPTY) >
     <!ATTLIST RETROACTIVITY_RECORDING %ATTRS2; >
   <!ELEMENT LIMIT_EVENT_TIME_CHANGE (EMPTY) >
     <!ATTLIST LIMIT_EVENT_TIME_CHANGE
         LIMITARY_EVENT CDATA #REQUIRED
         NEW_TIME CDATA #REQUIRED >
   <!ELEMENT DURATIVE_ACTION (TEMPORARY_HOLDING_IN_FORCE |
               VACATIO_LEGIS | PROVISION_PROLONGATION |
               PROVISION_SUSPENSION |
               RETROACTIVE_PROVISION_ENACTMENT) >
   <!ELEMENT TEMPORARY_HOLDING_IN_FORCE (EMPTY) >
     <!ATTLIST TEMPORARY_HOLDING_IN_FORCE
         STATUTE_NR NMTOKEN #REQUIRED >
   <!ELEMENT VACATIO_LEGIS (EMPTY) >
     <!ATTLIST VACATIO_LEGIS
         STATUTE NR NMTOKEN #REQUIRED >
   <!ELEMENT PROVISION_PROLONGATION (EMPTY) >
     <!ATTLIST PROVISION_PROLONGATION %ATTRS2; >
   <!ELEMENT PROVISION_SUSPENSION (EMPTY) >
     <!ATTLIST PROVISION_SUSPENSION %ATTRS2; >
   <!ELEMENT RETROACTIVE_PROVISION_ENACTMENT (EMPTY) >
     <!ATTLIST RETROACTIVE_PROVISION_ENACTMENT %ATTRS2; > ]>
```

Now the XML encoding of the short example described in Section 4 will
be presented. The provision comes from the main statute. The provision description is a valid (according to the DTD) XML document.

```
<STATUTE>
   <MAIN_STATUTE
       NR = "1"
       NAME = "The statute of January 20, 1990 on social security">
```

```
<ELEMENT>
  <PROVISION_TEXT    TEXT_ID = "t(1,1,'2a',2,1)">
                     Art. 2a. 2. In the context of this statute
                     it is assumed that the monthly income from
                     one re-calculated hectare amounts to 204 zl.
  </PROVISION_TEXT>
  <PROVISION_NAME  PROVISION_ID = "p(1,'2a',2,1)">
                     Income from one hectare
  </PROVISION_NAME>
  <META_NORMS>
     <IA_METANORM PROVISION_ID = "p(1,'2a',2,1)"
                  EVENT = "statute_enactment(1)"
                  CONDITIONS = "true">
                  <INSTANTANEOUS_ACTION>
                    <PROVISION_ENACTMENT
                        PROVISION_ID = "p(1,'2a',2,1)"
                        TEXT = "t(1,1,'2a',2,1)"
                        NEW_FACTS = "[basic(p(1,'2a',2,1))]"/>
                  </INSTANTANEOUS_ACTION>
     </IA_META_NORM>
  </META_NORMS>
</ELEMENT>
</MAIN_STATUTE>
</STATUTE>
```

6 Translation Rules From XML Into the Internal Database Representation

The translator implements the syntax-directed translation controlled by the
DTD definition given in the previous section. It was assumed that the specifications of provisions might be very long. Therefore the program reads, recog-
nises and translates the text in portions of the following types:

- the initial portion from the beginning up to the first "ELEMENT" component, which ends with the `</ELEMENT>` tag,
- the portion containing one "ELEMENT" component (except the first one),
  which also ends with the `</ELEMENT>` tag,
- the final portion, i.e. the text after the last `</ELEMENT>` tag.

The syntax analysis of the specification and its simultaneous translation
into Prolog is performed by means of the DCG grammar mechanism. The
translation result is a sequence of Prolog clauses (and terms). They may be
used as the input into the *LPDKB* to implement a provisions retrieval system. It is worth to note some technical details of the translator. During the
translation, the conditional clauses may be generated. When the XML specifications of the clause head and its body contain occurrences of the same

variable, then also the Prolog encoding of this clause should contain relevant occurrences of the same Prolog variable. To this end, during the translation, at first the string of characters is formed to represent the clause and then this string is transformed into the Prolog term by means of the following *tra_term* predicate:

$$tra\_term(String, Term) :-$$
$$string\_to\_atom(String, A),$$
$$term\_to\_atom(Term, A).$$

The predicate is defined by means of two system predicates, which are used to transform a string into an atom and then the atom into a term. The second of these transformations may generate an exception signal if the atom characters do not form a syntactically correct term. However, the exception signal may be caught by means of the system *catch(Goal, Signal, Recovery)* predicate and the recovery mechanism may send a message to the user that the analysed text portion contains a syntax error. Two versions of the translator exist. One of them accepts the correct specifications written in English and the second one concerns specifications written in Polish.

7 Conclusion

The presented knowledge base may be applied to represent the dynamics of the legal provisions contained in a wide spectrum of the legal statutes. The authors made the prototype version of the system publicly available on `http://mica.ai-kari.put.poznan.pl/~jcyb/LPDKB.html`. The system was positively tested on short-sized examples in the Windows XP/SWI-Prolog/XPCE environment. In the future, the challenging task to do will be the testing of the system performance, based on the real set of provisions. Moreover, it seems reasonable to work out some methodological hints for the users of how to separate and describe the provisions, using the proposed method and tools. It is also possible and practically valuable to formally specify the *LPDModel* conceptualisation in terms of a relevant ontology to make it accessible in the Semantic Web environment.

The research was supported by Poznań University of Technology, the grant 45-087/06-BW and in part by Polish Ministry of Science and Higher Education research and development grant, 2006-2009.

References

1. Kesim FN, Sergot M (1996) A Logic Programming Framework for Modelling Temporal Objects. IEEE Transactions on Knowledge and Data Engineering, 8(5):724–741
2. Kowalski RA, Sergot M (1986) A Logic-Based Calculus of Events. New Generation Computing, 4:67–95

3. Marin RH, Sartor G (1999) Time and norms: a formalisation in the event-calculus. In: Proc. of the 7th International Conference on Artificial Intelligence and Law (ICAIL 99). ACM Press, New York
4. Martinek J, Cybulka J (2005) Dynamics of Legal Provisions and its Representation. In: Proc. of the 10th International Conference on Artificial Intelligence and Law (ICAIL 05). ACM Press, New York
5. Palmirani M (2005) Time Model in Normative Information Systems. In: Palmirani M, Engers van T, Traunmuller R (eds), The Role of Legal Knowledge in e-Government. Wolf Legal Publishers, Tilburg
6. Uchwała Nr 147 Rady Ministrów z dnia 5 listopada 1991 r. w sprawie zasad techniki prawodawczej. Monitor Polski z 16 grudnia 1991 r., M.P.91.44.3. (in Polish)
7. Wielemaker J, Anjewierden A (1999–2001) Programming in XPCE/Prolog. University of Amsterdam, `http://www.swi-prolog.org/packages/xpce/`

Experiments with Grammatical Evolution in Java

Loukas Georgiou[1] and William J. Teahan[1]

School of Informatics, University of Wales, Bangor,
Dean Street, Bangor, Gwynedd LL57 1UT, U.K.
loukas@informatics.bangor.ac.uk, wjt@informatics.bangor.ac.uk

Keywords

Grammatical Evolution, Genetic Algorithms, Evolutionary Computation, jGE, libGE, GP

Summary. Grammatical Evolution (GE) is a novel evolutionary algorithm that uses a genotype-to-phenotype mapping process where variable-length binary strings govern which production rules of a Backus Naur Form grammar are used to generate programs. This paper describes the Java GE project (jGE), which is an implementation of GE in the Java language, as well as some proof-of-concept experiments. The main idea behind the jGE Library is to create a framework for evolutionary algorithms which can be extended to any specific implementation such as Genetic Algorithms, Genetic Programming and Grammatical Evolution.

1 Motivation

The main goal of the Java GE project at Bangor (jGE v0.1, 2006) is the implementation of an Evolutionary Algorithms (EA) framework which will facilitate further research into Evolutionary Algorithms (and especially Grammatical Evolution). Grammatical Evolution (O'Neill & Ryan 1999) was chosen as the main Evolutionary Algorithm of the jGE Project because it facilitates, due the use of a BNF Grammar, the evolution of arbitrary structures and programming languages.

The objectives of the jGE project as follows:

- To provide an open and extendable framework for the experimentation with Evolutionary Algorithms;
- To create an agent-oriented evolutionary system (using an agent-based framework);

L. Georgiou and W.J. Teahan: *Experiments with Grammatical Evolution in Java*, Studies in Computational Intelligence (SCI) **102**, 45–62 (2008)
www.springerlink.com

- To bootstrap further research on the application of the principles of the Evolutionary Synthesis theory (Mayr, 2002) in machines;
- To provide integration and interoperability with other projects such as evolutionary algorithms with knowledge sharing (Teahan *et al.*, 2005);
- To provide integration with other open source and free Java projects like Robocode (e.g. evolution of simulated robots using GE).

The main purpose of this paper is to describe the jGE Library by demonstrating its use in proof-of-concept experiments. Part of the work in this paper has been published as an extended abstract in (Georgiou & Teahan, 2006). However, the work here provides greater detail, and includes a full description of the experimental results.

2 The Grammatical Evolution System

Grammatical Evolution (O'Neill & Ryan, 2001) is an evolutionary algorithm that can evolve complete programs in an arbitrary language using a variable-length binary string. The binary string (genome) determines which production rules in a Backus Naur Form (BNF) grammar definition are used in a genotype-to-phenotype mapping process to generate a program. Namely, a chosen Evolutionary Algorithm (typically a variable-length genetic algorithm) creates a population of individuals. Then, for each individual:

1. The genotype (a variable-length binary string) is used to map the start symbol of the BNF Grammar into terminals. (The grammar is used to specify the legal phenotypes.)
2. The GE algorithm reads "codons" of 8 bits and generates the corresponding integer (RNA) each time a non-terminal to be translated has more than one production rule.
3. The selected production rule is calculated with the formula: $rule = codon \bmod rules$ where $codon$ is the codon integer value, and $rules$ is the number of rules for the current non-terminal.

After the mapping process (i.e. the creation of the phenotype), the fitness score is calculated and assigned to each individual (phenotype) according to the given problem specification. These fitness scores are sent back to the evolutionary algorithm which uses them to evolve a new population of individuals.

O'Neill and Ryan (2001) take inspiration from nature and claim that Grammatical Evolution embraces the developmental approach and draws upon principles that allow an abstract representation of a program to be evolved. This abstraction enables GE to do the following things: it separates out the search and solution spaces; it allows evolution of programs in an arbitrary language; it enables the existence of degenerate genetic code; and it adopts a wrapping operation that allows the reuse of the genetic material.

According to O'Neill and Ryan (2001), Grammatical Evolution is based on the principles of Evolutionary Automatic Programming, Molecular Biology, and Grammars. Although Grammatical Evolution is a form of Genetic Programming, it differs from traditional GP in three ways: it employs linear genomes; it performs an ontogenetic mapping from genotype to phenotype; and it uses a grammar to dictate legal structures in the phenotypic space.

Instead of trying to evolve computer programs directly, which is the case in Genetic Programming (Koza 1992; 1994), Grammatical Evolution uses a variable length linear "genome" which governs how a Backus Naur Form grammar definition is mapped to an executable computer program. Regarding the use of grammars, O'Neill and Ryan (2001) state that they provide a simple, yet powerful, mechanism that can be used for the description of any complex structure such as languages, graphs, neural networks, mathematical expressions, molecules compounds. This was the main reason why GE was chosen as the main and default Evolutionary Algorithm of the jGE Library.

Grammatical Evolution takes the approach that a Genotype must be mapped to a Phenotype, like some other former approaches (Genetic Algorithm for Developing Software, Paterson & Livesey, 1997), but it does not use a one-to-one mapping, and moreover it evolves individuals that contain no introns. It uses a Genetic Algorithm to control what production rules are fired when there are more than one choice for a Backus Naur Form non-terminal symbol (Ryan *et al.*, 1998a).

In natural biology, there is no direct mapping between the genetic code and its physical expression. Instead, genes guide the creation of proteins which affect the physical traits either independently or in conjunction with other proteins (Ryan *et al.*, 1998a). Grammatical Evolution treats each transition as a "protein" which cannot generate a physical trait on its own. Instead, each one protein can result in a different physical trait depending on its position in the genotype and consequently, the previous proteins that have been generated.

Grammatical Evolution uses all the standard operators of Genetic Algorithms, plus two new operators: Prune and Duplicate. The gene duplication is analogous to the production of more copies of a gene or genes, in order to increase the presence of a protein or proteins in the cell of a biological organism. The gene pruning reduces the number of introns in the genotype and according to Ryan *et al.* (1998a) it results in dramatically faster and better crossovers (Later research questions the usefulness of this operator because of the important role of introns (O'Neill *et al.* 2001).

The main advantages of Grammatical Evolution according to Ryan *et al.* (1998a) are the following:

- It can evolve programs in any language;
- theoretically, it can generate arbitrary complex functions; and
- it has closer biological analogies to nature than Genetic Programming.

But Grammatical Evolution is, like Genetic Programming, subject to problems of dependencies (Ryan *et al.*, 1998a). For example, the further a gene is from the root of the genome, the more likely it will be affected by the previous genes. Ryan *et al.* (1998a) suggest the biasing of individuals to a shorter length and the progressive generation of longer genomes.

3 libGE vs. jGE

The libGE library is an implementation of the Grammatical Evolution system written in the C++ language. A recent version is the 0.26 beta 1, 3 March 2006 (Nicolau, 2006). Characteristics of libGE are presented in (Nicolau, 2006) and (O'Neill & Ryan, 2001). libGE implements the Grammatical Evolution mapping process. It can be used by an evolutionary computation algorithm in order to map the genotype (the result of the search algorithm) to the phenotype (the program to be evaluated). As Nicolau (2006) says in the documentation "On its default implementation, it maps a string provided by a variable-length genetic algorithm onto a syntactically-correct program, whose language is specified by a BNF (Backus-Naur Form) context-free grammar."

Our implementation of the Grammatical Evolution system, the jGE Library, uses the Java programming language. The main difference between jGE and libGE is that jGE incorporates the functionality of libGE as a component and provides implementation of the Search Engine as well as the Evaluator. Namely, as will be shown below, the jGE is a more general framework for the execution of Evolutionary Algorithms. Indeed, it still provides, like libGE, the feature of using any other Search Engine and Evaluator beyond that already provided by default in jGE. Individual components of the jGE, such as the GE Mapping Mechanism, the BNF Parser, and the Mathematical Functions classes, may also be used separately for special purpose projects.

Another main difference between jGE and libGE is the goal of each project. libGE provides an implementation of the Grammatical Evolution mapping process, whereas the goal of the jGE Project is the development of a general Evolutionary Algorithms framework which facilitates the incorporation and evaluation of Evolutionary techniques; and the incorporation of agent-oriented principles to develop implementations for parallel distributed systems.

Java was chosen as the implementation language for the jGE project mainly to fit in with other artificial intelligence projects being developed at Bangor. However, Ghanea-Hercock (2003) also lists several advantages of using Java for the development of Evolutionary Algorithms applications: automatic memory management, pure object-oriented design, high-level data constructs (e.g. dynamically resizable arrays); platform independent code; and the availability of several complete EA libraries for EA systems. On the other hand, he mentions that the main price we have to pay in using Java is the significant increase in execution time of interpreted Java programs compared with compiled languages like C and C++. But he adds that the recent work of Sun and other companies has resulted in "Just in Time" compilers which significantly improve the execution speed of the Java programs. Also, he mentions that future developments in computer languages may lead to better alternatives to Java with for example improved speed. Such an example is the release of C# from Microsoft (Ghanea-Hercock, 2003). Of course, in spite of the last argument of Ghana-Hercock, there is a huge dispute of whether Java is slower or faster than C# (or even than C and C++). And to put it in the correct context: whether an implementation in JVM (Java Virtual Machine) is slower or faster than one in the CLR (Common Language Runtime) of Microsoft.

4 Overview of jGE (v0.1)

This section briefly describes the overall design of jGE v0.1 and further details of one of its components, the Genetic Operations package. The main idea behind the development of the jGE Library—it can be downloaded at (jGE v0.1, 2006)—is to create a framework for evolutionary algorithms which can be extended to any specific implementation such as Genetic Algorithms, Genetic Programming and Grammatical Evolution. This means that instead of using a mapper-centric approach like libGE, jGE uses a GA-oriented approach. Namely, instead of being just the implementation of the mapping mechanisms between the Search Engine and the Evaluation Engine as for libGE, it provides libraries for both of these components. This means that someone using jGE is able to specify the core strategy of the evolutionary process by selecting the following parameters (see Fig. 1a):

- the desired implementations of the genetic operators (selection, crossover, mutation, etc.);
- the genotype to phenotype mapping mechanism;
- the evaluation mechanism;
- the initial population;
- the initial environment (although currently not yet implemented).

The purpose of the last parameter is to allow the developer to specify an environment in which the population is living, and to influence the creation of new generations as well as the phenotype of the individuals, their growth process, and finally their own genotypes before they reproduce new offspring.

The Mapper component is responsible for the genotype-to-phenotype mapping. It provides an interface which accepts as an argument an individualȨćs genotype and returns its corresponding phenotype. An implementation of any mapping process can be added and used in the jGE system as long as it satisfies the required interface. Currently, two mapping processes are implemented and supported: No-Mapping (Classical GA) and BNF-Based Mapping (Grammatical Evolution).

The Evaluator component assigns a fitness value to the phenotype of an individual. It defines a standard interface and any implementation of a problem

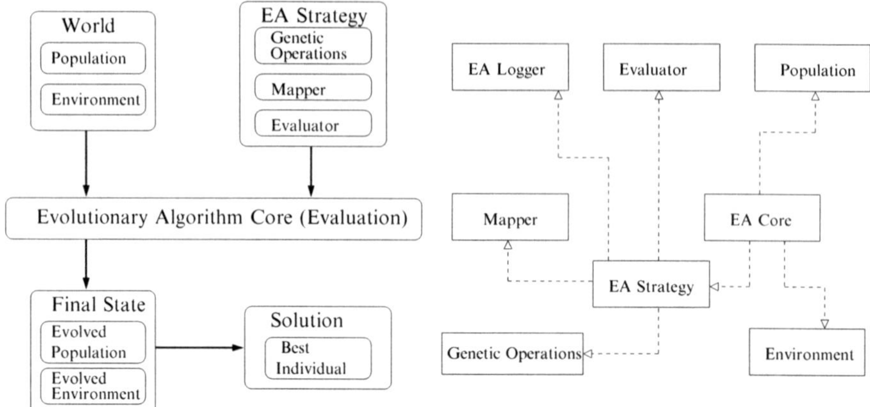

Fig. 1. (a) jGE Architecture; (b) Component diagram for jGE EA framework

specification must implement this interface. In the current version, two problem specifications (and their corresponding evaluators) are available: Hamming Distance and Symbolic Regression. The Evaluator implementation is the only component of the system that has to be created for any new category of problems which will be tackled by the jGE system.

The problem specification and evolutionary strategy can be created in an external XML file which is loaded by the core mechanism of the jGE framework. The core mechanism is then responsible for allocating and executing the appropriate actions and directives, and to produce the final results.

A summary of the main components of the jGE Library (see Fig. 1b) and its packages can be found at (Georgiou & Teahan, 2006). A detailed description of each class and its services can be found in the Java Documentation of the library (jGE v0.1, 2006). Even though jGE is focused on the implementation of the Grammatical Evolution system, it contains all the necessary functionality for the execution and construction of other Evolutionary Algorithms as well. Currently, as well as GE, two other EC algorithms are implemented: Standard GA and Steady-State GA.

jGE decomposes and implements some services which are required by EC algorithms and provides functionalities for the ad-hoc implementations of other evolutionary based systems. In version v0.1, the library is concentrated on Grammatical Evolution. But this library can also be used by any other Java System for the creation of evolutionary algorithms as well as for other functionalities such as the parsing and representation of BNF Grammar definitions, the compilation and execution of Java programs, and the generation of random numbers in specific ranges.

4.1 Genetic Operations Component

One of the most useful components of jGE is the Genetic Operations component. Its classes implement various versions and types of the genetic operators as static methods. In this way, ad-hoc implementations of evolutionary algorithms can easily access the various genetic operations and use them in different combinations. Currently the following operators are implemented:

- **Genesis**: random creation of an initial pool of binary string genotypes; and random creation of an initial population of individuals.
- **Selection**: roulette wheel selection; rank selection; N best and M worst selection.
- **Crossover**: standard one-point crossover for fixed-length genotypes; standard one-point crossover for variable-length genotypes.
- **Mutation**: standard one-point mutation.
- **Duplication**: standard duplication.
- **Pruning**: standard pruning.

An abstract class EvolutionaryAlgorithm defines common properties and behaviours for evolutionary algorithms like Genetic Algorithms, Genetic Programming, and Grammatical Evolution. An Evolutionary Algorithm simulates the biological process of evolution. The evolution unit of this process is the population as Darwinism argues (Mayr, 2002). The basic strategy of an Evolutionary Algorithm is the following:

Set current population $P = N$ individuals
For *Generation = 1 to MaxGenerations*
  Competition: Evaluate the individuals of P
  Selection: Select from P the individuals to mate
  Variation: Apply Crossover, Mutation, etc. to the selected individuals
  Reproduction: Create the new population P' and set $P = P'$
End For

The subclasses of this class must implement the concrete steps of the above strategy in order to provide specific versions of Evolutionary Algorithms. Further, two evolutionary algorithms have been implemented: the Standard Genetic Algorithm and a version of a Steady-State Genetic Algorithm. For the former, the following process is implemented:

Set current population $P = N$ individuals
Perform fitness evaluation of the individuals in P
While *(solution not found* **and** *MaxGenerations not exceeded)*
  Create a new empty population, P'
  Repeat *until P' is full*
    Select two individuals from P to mate using Roulette Wheel Selection
    Produce two offspring using standard one-point crossover with probability P_c

Perform Point Mutation with probability P_m on the two offspring
Perform Duplication with probability P_d on the two offspring
Perform Pruning with probability P_p on the two offspring
Add the two offspring into P'
End Repeat
Replace P with P'
Perform fitness evaluation of individuals in P
End While
Return the best individual, S (the solution) in current population, P

For the Steady-State Genetic Algorithm (SSGA), the main idea is that a portion of the population P survives in the new population P' and that only the worst individuals are replaced. Namely, a few good individuals will mate and their offspring will replace the worst individuals. The rest of the population will survive. The portion of the population P that will be replaced in P' is known as the Generation Gap and is a fraction in the interval $(0,1)$. The default implementation of SSGA uses a fraction, $G = 2/N$ (where N the size of the population). Namely, two individuals will mate and their offspring will replace the two worst individuals. In general, the number of the individuals which will be replaced in each generation is $G \times N$. In the case where $G \times N$ is not an even integer, then the larger even integer less than $G \times N$ and larger than 0 will be used.

The SSGA process implemented by this class is the following:

Set current population $P = N$ individuals
Set $G = $ the generation gap
Perform fitness evaluation of the individuals in P
While *(solution not found **and** max generations not exceeded)*
Create a new empty population, P'
Repeat *until (new offspring $= G \times N$)*
Select two individuals from P to mate using Roulette Wheel Selection
Produce two offspring using standard one-point crossover
with probability P_c
Perform Point Mutation with probability P_m on the two offspring
Perform Duplication with probability P_d on the two offspring
Perform Pruning with probability P_p on the two offspring
Add the two offspring into P'
End Repeat
Add the best $(N - G \times N)$ individuals of P into P'.
Replace P with P'
Perform fitness evaluation of individuals in P
End While
Return the best individual, S (the solution) in current population, P

The Grammatical Evolution class implements the default version of GE (with a minor exception regarding the steady state replacement mechanism as mentioned below). The default implementation as described by O'Neill and Ryan, uses a Steady-State replacement mechanism such that two parents produce two children, the best of which replace the worst individual in the population only if the child has a greater fitness than the individual to be replaced. Our implementation uses a slightly different replacement mechanism which is described above in the SSGA process. Also, there is the option to use a Generational replacement mechanism like in Standard GA.

Regarding the configuration of a Grammatical Evolution run, O'Neill and Ryan suggest the following: a typical wrapping threshold is 10; the size of the codon is 8-bits; and typical probabilities are: crossover–0.9; mutation–0.01; duplication–0.01; pruning–0.01. This configuration is the default of the GrammaticalEvolution class. Further, this implementation uses the following default values: max. generations: 10; searching mechanism: Steady-State GA; Generational Gap of the Steady-State GA, (n = the population size).

The next section describes some proof-of-concept experiments performed with the jGE Library.

5 Experiments

Experiments in (Ryan *et al.*, 1998; 1998b) ashow that Grammatical Evolution is able to tackle Symbolic Regression, Trigonometric Identity, and Symbolic Integration problems well. The adoption of the Steady State approach (Ryan & O'Neill, 1998) dramatically improves the performance of the Grammatical Evolution algorithm, making it as efficient in the mentioned problems types as the Genetic Programming algorithm. Also, Nicolau (2006) demonstrated the generation of multi-line code in the classical Santa Fe Ant Trail problem. Indeed, the last experiment showed that Grammatical Evolution outperforms Genetic Programming (Koza 1992; 1994) in this specific problem when GP does not use the solution length fitness measure (Nicolau 2006).

Three different proof-of-concept experiments with jGE were performed — Hamming Distance, Symbolic Regression and Trigonometric Identity (see sections 5.3 to 5.4 below). For the first problem, two further evolutionary algorithms were tried for comparison—Standard GA, and Steady State GA. The second and third problems are based on the experiments which have been performed by Michael O'Neill and Conor Ryan (1998, 2001, and 2003). The objective of these experiments is to demonstrate the applicability and effectiveness of the jGE v0.1 library for the execution of EA experiments.

Before the results are described, however, the next section provides a brief discussion on some of the Java issues encountered during the experiments, and this is followed by some sample Java source code to illustrate the ease with which these experiments were set up using the jGE library.

5.1 Java Issues

The first version of the Evaluation component used the JavaCompiler class to evaluate the Java programs (phenotypes). This compiles (using the javac.exe compiler), and executes (using the java.exe runtime), once in each generation of a run, the dynamically created java source code which are the phenotypes of all the individuals of the population. This is an extremely time consuming task and for problems such as Symbolic Regression, this is the most important factor which effects the execution speed. In each Symbolic Regression experiment, the compilation/execution takes place once when a new run starts (for the creation of the initial population) and once in each generation (during the evaluation of the individuals of the population).

Although the time complexity with respect to the compilation/execution of Java code is linear ($O(N)$, where N is the number of generations of a run), it is a significantly time consuming task which can significantly degrade overall performance. Moreover, other problems will have a higher rate of growth of execution time if they need to frequently use the source code compilation and bytecode execution tasks.

For the above reasons, alternative methods for the compilation and execution of Java code were investigated. The experimental evidence (see jGE v0.1, 2006) leads to the conclusion that a much better solution than using javac.exe and java.exe is the following setup: a) Use of the Jikes compiler for the compilation of the java source code (Jikes, 2004); b) Utilization of the Dynamic Class Loading and Introspection features of the Java Virtual Machine (ClassLoader class, and the Reflection API).

Jikes is an open source Java compiler written in the C++ language and translates Java source files into the bytecode instructions set and binary format defined in the Java Virtual Machine Specification. Jikes has the following advantages as noted in the Jikes official web site: open source; strictly Java compatible; high performance; dependency analysis; constructive assistance.

The Java ClassLoader is an important component of the Java Virtual Machine which is responsible for finding and loading classes at runtime. It loads classes on demand into memory during the execution of a Java program. Furthermore, it is written in the Java Language and can be extended in order to load Java classes from every possible source (local or network file system, network resources, etc.). Using both the ClassLoader and the Reflection API, it is possible to perform the loading of Java bytecode and its execution from inside of any Java program using the same instance (process) of JVM. In the current version (v0.1), the jGE Library provides the option of using either the Sun JVM or the IBM Jikes for the execution and compilation of Java code.

The experiments described below are the first experiments with jGE v0.1 using real data based on the suggested configurations provided by O'Neill and Ryan (2003). Because this was the first time a large amount of data was used (e.g. populations of 500 individuals, sample of 50 data points etc.) an unexpected problem arose. During the evaluation of a Grammatical Evolution run on Symbolic Regression, the Java compiler (both Sun JDK 1.5 and IBM Jikes) threw an error during the compilation of the produced Java class which was responsible for calculating the Raw Fitness of all the individuals of a population. The error message in Sun JDK 1.5 was the following: "Code too large". The reason for this error was tracked down to an undocumented limitation of the Java compiler which cannot compile a method with bytecode size larger than 64Kb. This problem forced the re-factoring of the Symbolic Regression class in order for the compilation of the class which runs and evaluate the Java code (phenotype) of the individuals to be possible. The whole code which was placed in the main method of the temporary class has been broken into many smaller methods instead (one for each individual of the population).

5.2 Sample Java Source Using jGE

This section provides a sample of the Java source code used for these experiments. The source code used for all three problems is essentially the same—except for a small amount of variation to specify the problem itself and the evolutionary algorithm used. Fig. 2 lists the Java source code for the Hamming Distance problem. The line labelled by (1) in the figure provides the problem specification (this will vary for the three types of problems). Lines labelled by (3) set the parameters to be used by the evolutionary algorithm. Line (4) executes the algorithm and returns the solution.

```
/* This method shows the use of jGE in a Hamming Distance problem.
 * @return The solution of the Hamming Distance experiment. */

  public Individual<String, String> hdExperiment() {
      Individual<String, String> solution = null;
      String target = "11100011100010101010101010101010";
(1) HammingDistance hd = new HammingDistance(target);
(2) // Insert EA specification here (see Fig. 3)
(3) ea.setCrossoverRate(0.9);
(3) ea.setMutationRate(0.01);
(3) ea.setDuplicationRate(0.01);
(3) ea.setPruningRate(0.01);
(3) ea.setMaxGenerations(100);
(3) ea.setLogger(log);
(4) solution = ea.run();
      // This method returns the following: Number of generations =
      // ea.lastRunGenerations(); Fitness = solution.rawFitness();
      // Solution = solution.getPhenotype().value().
      return solution;
  }
```

Fig. 2. Java source code for the Hamming Distance problem.

Fig. 3 lists the alterations needed to the source code in Fig. 2 to configure for the different evolutionary algorithms. That is, line (2) should be replaced by the code shown in Fig. 3 depending on the algorithm that is chosen. Additionally, an extra line needs to be inserted before (3) but only for the Steady State Genetic Algorithm. Any of the variations of the static method shown in Fig. 2 performs a simple run of the corresponding experiment. An external application is needed to call the method using a loop in order to execute a full experiment with many runs.

Standard GA:
Replace line (2) in Fig. 2 with:

```
StandardGA ea = new StandardGA(50, 1, 30, 30, hd);
```

Steady-State GA:
Replace line (2) Fig. 2 with:

```
SteadyStateGA ea = new SteadyStateGA(50, 1, 30, 30, hd);
```

Insert before line (3) above:

```
ea.setFixedSizeGenome(true);
```

Grammatical Evolution:
Replace line (2) Fig. 2 with:

```
GE: BNFGrammar bnf = null;
bnf = new BNFGrammar("BinaryGrammar.bnf");
GrammaticalEvolution ea =
   new GrammaticalEvolution(bnf, hd, 50, 8, 20, 40);
```

Fig. 3. Source code alterations to Fig. 2 required for the different Evolutionary Algorithms in the Hamming Distance problem.

The next three subsections describe the experimental results for the three problems investigated. Each sub-section provides a Grammatical Evolution Tableau (Ryan & O'Neill, 1998) description of the problem, the BNF grammar used by the Grammatical Evolution system, and the experimental results.

5.3 Hamming Distance Experiments

The Hamming Distance problem involves the finding of a given binary string. The target string was: 1110001110001010101010101010101010. For this problem, Grammatical Evolution, Standard GA, and Steady-State GA were compared.

Table 2 compares the number of generations and raw fitness that resulted for the three types of evolutionary algorithm experimented with. The results show that Grammatical Evolution (or more precisely, the jGE implementation of it) outperforms the Standard Genetic Algorithm with a significantly higher average raw fitness from less number of generations on average. It is also competitive with the Steady State Genetic Algorithm with comparable average raw fitness, but costing a greater number of generations on average.

```
<phenotype> ::=
    <binary><binary><binary><binary><binary><binary><binary><binary>
    <binary><binary><binary><binary><binary><binary><binary><binary>
    <binary><binary><binary><binary><binary><binary><binary><binary>
    <binary><binary><binary><binary><binary><binary>
<binary> ::= 0|1
```

Fig. 4. BNF Grammar used for Hamming Distance problem.

Table 1. Hamming Distance GE Tableau.

Objective: Find the target binary string
Terminal Operands: 0 and 1
Terminal Operators: none
Fitness cases: The target string
Raw Fitness: Target String Length—Hamming Distance
Standardised Fitness: Same as raw fitness
Wrapper: None
Parameters: Population Size (M) = 50; Maximum Generations (N) = 100; Prob. Mutation (P_m) = 0.01; Prob. Crossover (P_c) = 0.9; Prob. Duplication (P_d) = 0.01; Prob. Pruning (P_p) = 0.01; Selection Mechanism = Steady State GA with Generation Gap (G) = 0.9; Codon Size = 8

Table 2. Results for Hamming Distance problem.

Run	Standard GA Number of Generations	Raw Fitness	Steady-State GA Number of Generations	Raw Fitness	Grammatical Evolution Number of Generations	Raw Fitness
1	100	27	43	30	100	26
2	100	2	30	30	39	30
3	100	28	41	30	32	30
4	100	2	40	30	100	26
5	100	26	19	30	100	26
6	100	26	37	30	4	30
7	100	27	24	30	46	30
8	100	27	33	30	48	30
9	100	29	27	30	100	26
10	64	30	26	30	100	26
Average	96.4	22.4	32	30	66.9	28

5.4 Symbolic Regression Experiments

Symbolic Regression problems are problems of finding some mathematical expression in symbolic form that matches a given set of input and output pairs. The particular function experimented with was the following: $f(x) = x^4 + x^3 + x^2 + x$. Two different BNF grammars were tried (Figs. 5 and 6).

```
<expr> ::=  <expr> <op> <expr> | ( <expr> <op> <expr> ) |
            <pre-op> ( <expr> ) | <var>
<op> ::=  + | - | / | *
<pre-op> ::=  Math.sin | Math.cos | Math.log
<var> ::=  x | 1.0
```

Fig. 5. BNF Grammar (A) used for Symbolic Regression problem.

```
<expr> ::= <expr><op><expr>|<var>
<op> ::=  + | - | / | *
<var> ::=  x
```

Fig. 6. BNF Grammar (B) used for Symbolic Regression problem.

Results show that the BNF Grammar (B) in Fig. 6 is a much better grammar to use for tackling this problem (as evidenced by the raw fitness values in Table 5 being close to 1.0 in 50% of the runs as compared to 0% of the runs for BNF Grammar (A) in Fig. 5.

5.5 Trigonometric Identity Experiments

The particular function experimented with was $cos(x)$, and the desired trigonometric identity $1 - 2sin^2 x$ (for this reason the Java unary operator `Math.cos` was not included in the BNF Grammar of this problem). The objective of these experiments was to find a mathematical expression identical to the function. The results in Table 7 show that the raw fitness values are above 0.5 in 50% of the runs.

```
<expr> ::=  <expr> <op> <expr> | ( <expr> <op> <expr> ) |
            <pre-op> ( <expr> ) | <var>
<op> ::=  + | - | / | *
<pre-op> ::=  Math.sin
<var> ::=  x | 1.0
```

Fig. 7. BNF Grammar used for Trigonometric Identity problem.

Table 3. Symbolic Regression GE Tableau.

Objective: Find a function of one independent variable and one dependent variable, in symbolic form that fits a given sample of 20 (x_i, y_i) data points, where the target function is the quadratic polynomial $x^4 + x^3 + x^2 + x$.
Terminal Operands: x (the independent variable), the constant 1.0.
Terminal Operators: The binary operators +, -, /, *, and -. The unary operators `Math.sin`, `Math.cos`, and `Math.log`.
Fitness cases: The given sample of the pairs (x_i, y_i) of 20 data points in the interval $[-1, +1]$. The input data points (x_i) are randomly created and their corresponding output points (y_i) are automatically created by the expression $x^4 + x^3 + x^2 + x$.
Raw Fitness: The sum, of the absolute values of errors taken over the fitness cases (x_i, y_i). With the above Raw Fitness, the best individuals have lower values, so an Adjusted Fitness is used and assigned to each individual. This is typically defined for an individual i as following: $Fa(i) = 1/(1 + Fs(i))$ where Fs the Standardised Fitness of i. In this case the Adjusted Fitness of an individual i is calculated as following: $Fa(i) = 1/(1 + Fr(i))$ where Fr the Raw Fitness of i. The fitness value varies from 0 to 1 and Invalid individuals will have Raw Fitness Value 0.
Standardised Fitness: Same as raw fitness.
Wrapper: Standard productions to generate a Java Class with a `main()` method which prints the fitness values in the standard output.
Parameters: Population Size $(M) = 500$; Maximum Generations $(N) = 50$; Prob. Mutation $(P_m) = 0.01$; Prob. Crossover $(P_c) = 0.9$; Prob. Duplication $(P_d) = 0.01$; Prob. Pruning $(P_p) = 0.01$; Selection Mechanism $=$ Steady State GA with Generation Gap $(G) = 0.9$; Codon Size $= 8$.

Table 4. Results for Symbolic Regression problem using BNF Grammar (A).

Run	N	Phenotype	Raw Fitness
1	50	((1.0 + x)* x)/Math.sin (Math.cos(x))	0.4694745127276528
2	50	(1.0 + x) * Math.sin (x/Math.sin (Math.cos ((Math.log (1.0) + x)- x * Math.log (Math.cos (Math.log (Math.sin (Math.cos (1.0)))))))))	0.6421964929997449
3	50	x * x + (Math.sin (x)+ x)+ x * Math.sin (x)	0.2013364798843769
4	50	x +(1.0 *((x * x)*(x + Math.sin (1.0)+ 1.0)))	0.35628132431978915
5	50	(x *(1.0 +(x + x *(x * 1.0) + x)))	0.27187284673050927
6	50	x *(1.0 +(Math.sin (x)+ x + Math.cos (1.0)))	0.23027579709471763
7	50	(x *(x +(1.0 / Math.sin (1.0))))	0.20312605083167595
8	50	1.0 + x - Math.cos ((x + Math.cos (x - 1.0)))/1.0	0.12454074453273598
9	50	(x / 1.0 *(x + (1.0 + (Math.cos (1.0)+ x)* x * 1.0)))	0.36138848058931
10	50	1.0 + Math.sin (x) + Math.sin (x - Math.cos (x) + Math.sin (Math.sin (x - Math.cos (x)* 1.0)))	0.16294511118210622

Table 5. Results for Symbolic Regression problem using BNF Grammar (B).

Run	N	Phenotype	Raw Fitness
1	50	x + x * x	0.28564566880148
2	50	x * x + x * x * x + x	0.2551146843044831
3	40	x * x * x + x * x * x / x * x * x / x * x / x * x * x / x + x * x + x / x * x	1.0
4	50	x * x + x + x * x + x / x * x * x * x	0.2697923828640847
5	50	x * x * x * x + x * x + x + x * x * x	0.9999999999999991
6	50	x - x + x + x + x * x - x * x / x + x * x * x * x - x * x + x * x - x + x + x * x * x	0.9999999999999991
7	50	x * x * x + x + x * x + x - x + x + x * x - x * x + x + x * x - x - x	0.2467044521737591
8	50	x + x * x + x * x * x - x + x * x * x * x + x /x * x	0.9999999999999998
9	50	x * x + x	0.1996232819981532
10	50	x * x * x + x * x * x * x + x + x * x	0.9999999999999998

Table 6. Trigonometric Identity GE Tableau.

Objective: Find a new mathematical expression, in symbolic form that equals a given mathematical expression, for all values of its independent variables. Examined Function: `Math.cos(2 * x)`.
Desired Trigonometric Identity: $1 - 2Sin^2 x$.
Terminal Operands: x, the constant 1.0.
Terminal Operators: The binary operators $+$, $-$, $/$, $*$, and $-$. The unary operator `Math.sin`.
Fitness cases: The given sample of the pairs (x_i, y_i) of 20 data points in the interval $[0, 2\pi]$. The input data points (x_i) are randomly created and their corresponding output points (y_i) are automatically created by the expression `Math.cos(2 * x)`.
Raw Fitness: The sum, of the absolute values of errors taken over the fitness cases (x_i, y_i). With the above Raw Fitness, the best individuals have lower values so an Adjusted Fitness will be used and assigned to each individual. This is typically defined for an individual i as following: $Fa(i) = 1/(1 + Fs(i))$ where Fs the Standardised Fitness of i. In this case the Adjusted Fitness of an individual i is calculated as following: $Fa(i) = 1/(1 + Fr(i))$ where Fr the Raw Fitness of i. The fitness value varies from 0 to 1 and Invalid individuals will have Raw Fitness Value 0.
Standardised Fitness: Same as raw fitness.
Wrapper: Standard productions to generate a Java Class with a `main()` method which prints the fitness values in the standard output.
Parameters: Population Size $(M) = 500$; Maximum Generations $(N) = 50$; Prob. Mutation $(P_m) = 0.01$; Prob. Crossover $(P_c) = 0.9$; Prob. Duplication $(P_d) = 0.01$; Prob. Pruning $(P_p) = 0.01$; Selection Mechanism = Steady State GA with Generation Gap $(G) = 0.9$; Codon Size = 8

Table 7. Results for Trigonometric Identity problem.

Run	N	Phenotype	Raw Fitness
1	50	Math.sin (x+ (x + 1.0))	0.10793425584713574
2	50	Math.sin (x+ x + 1.0 + Math.sin (Math.sin(1.0)))	0.31125330920116906
3	50	Math.sin (Math.sin ((x+ x) + 1.0 + Math.sin (Math.sin (1.0))) * 1.0)	0.25691103931194736
4	50	Math.sin ((1.0 - x) + (1.0 - x))	0.1466328694322308
5	50	Math.sin ((x + (Math.sin (1.0 * Math.sin (1.0)/1.0)+ Math.sin (1.0))+ x/ 1.0))	0.84065701811658
6	50	Math.sin ((1.0 - (x - Math.sin (Math.sin (Math.sin (Math.sin (Math.sin (Math.sin (1.0)))))) + x)))	0.8182130706092126
7	50	Math.sin (1.0 + (x+ Math.sin (Math.sin (1.0))) + x)	0.3567880989976856
8	50	Math.sin ((x + (1.0/1.0 + (Math.sin (Math.sin (Math.sin (Math.sin (1.0)))) + x))))	0.5933482508093464
9	50	Math.sin ((x + x + (Math.sin (1.0) + ((1.0+ x + x + Math.sin(Math.sin (1.0))) - 1.0 - x) - x)))	0.8178615390653478
10	50	Math.sin (Math.sin (Math.sin (1.0)) + x + Math.sin (Math.sin(1.0))+ x)	0.5348593720523398

6 Discussion

The results of the above experiments with jGE confirmed two expected findings. First, that jGE using Grammatical Evolution is able to produce useful solutions even though these are not the best possible. Secondly, that different set-ups and configurations of the searching and evaluation mechanisms have a significant impact on the quality (degree of correctness) of the solution.

The above findings will guide the next steps in this project. Namely, the next version of jGE will provide implementations of more genetic operators which will facilitate experiments in a larger range of possible configurations. Also, the need for more than a standard PC's processing power is prominent in order to be improved the time-scale of new experiments with jGE. It is well known (Ghanea-Hercock, 2003) that evolutionary algorithms are processing-power demanding algorithms and that they can take advantage of parallel processing architectures. This need for more processing power, in order to reduce the time-scale of the new experiments, will be tackled by incorporating into jGE a parallel distributed processing framework making it possible to execute the problems in question transparently on many machines (i.e. if there are m machines and n individuals then each one machine will execute genotype-to-phenotype mapping and assign fitness values to n/m individuals). Finally, the previously mentioned improvements will facilitate future research into incorporating knowledge-sharing within the process of an evolutionary algorithm as a precursor to tackling some of the issues concerning thought, knowledge, evolution and search that were raised in (Teahan et al., 2005).

References

1. Georgiou L, Teahan WJ (2006) jGE—A Java implementation of Grammatical Evolution. 10th WSEAS Int. Conf. on Systems, Athens, July
2. Ghanea-Hercock R (2003) Applied Evolutionary Algorithms in Java. New York, NY: Springer
3. jGE v0.1. (2006), Java GE (jGE) Official Web Site. School of Informatics, Univ. Wales, Bangor, U.K. http://www.informatics.bangor.ac.uk/~loukas/jge
4. Jikes 1.22. IBM Corp. (2004), USA: NY. http://jikes.sourceforge.net
5. Koza JR (1992) Genetic Programming: On the Programming of Computers by the Means of Natural Selection. Cambridge, MA: MIT Press
6. Koza JR (1994) Genetic Programming II: Automatic Discovery of Reusable Programs. Camb., MA: MIT Press
7. Mayr E (2002) What Evolution Is. London: Phoenix
8. Nicolau M (2006), libGE: Grammatical Evolution Library for version 0.26beta1, 3 March 2006. http://waldo.csisdmz.ul.ie/libGE/libGE.pdf
9. O'Neill M, Ryan C (1999) Evolving Multi-line Compilable C Programs. In: Proc. of the 2nd European Workshop on Genetic Prog., 1999, pp. 83–92
10. O'Neill, M, Ryan C (2001) Grammatical Evolution. IEEE Transactions on Evolutionary Computation 5(4), 349–358
11. O'Neill M, Ryan C (2003) Grammatical Evolution: Evolutionary Automatic Programming in an Arbitrary Language. USA: Kluwer
12. O'Neill M, Ryan C, Nicolau M (2001) Grammar Defined Introns: An Investigation into Grammars, Introns, and Bias in Grammatical Evolution. In Proceedings of GECCO 2001
13. Paterson N, Livesey M (1997) Evolving caching algorithms in C by GP. In Genetic Programming 1997, pp. 262–267. MIT Press
14. Ryan C, Collins JJ, and O'Neill M (1998a) Grammatical Evolution: Evolving Programs for an Arbitrary Language. Lecture Notes in Comp. Sci. 1391. First European Workshop on Genetic Programming 1998
15. Ryan C, O'Neill M (1998) Grammatical Evolution: A Steady State Approach. In Proceedings of the 2nd Int. Workshop on Frontiers in Evolutionary Algorithms, 1998, pp. 419–423
16. Ryan C, O'Neill M, Collins JJ (1998b) Grammatical Evolution: Solving Trigonometric Identities. In Proceedings of Mendel 1998: 4th Int. Mendel Conf. on Genetic Algorithms, Optimisation Problems, Fuzzy Logic, Neural Networks, Rough Sets held in Brno, Czech Republic June 24-26 1998, pp. 111–119
17. Teahan WJ, Al-Dmour N, Tuff PG (2005) On thought, knowledge, evolution and search. In Proceedings of Computer Methods and Systems CMS'05 Conference held in Krakow, Poland, 14–16 November 2005

Processing and Querying Description Logic Ontologies Using Cartographic Approach

Krzysztof Goczyła, Wojciech Waloszek, Teresa Zawadzka, and Michał Zawadzki

Gdańsk University of Technology, Department of Software Engineering,
ul. Gabriela Narutowicza 11/12, 80-952 Gdańsk, Poland
{kris,wowal,tegra,michawa}@eti.pg.gda.pl

Summary. Description Logic (DL) is a formalism for knowledge representation that recently has gained widespread recognition among knowledge engineers. After a brief introduction to DL, the paper presents a DL reasoner developed at Gdańsk University of Technology (GUT). The reasoner, called KASEA, is based on an original idea called Knowledge Cartography. The paper presents basics of Knowledge Cartography, its potentials and limitations, and compares the solution with other DL reasoners.

1 Introduction

Description Logic (DL) is a formalism for knowledge representation that is based on the first-order logic (FOL) [1]. DL has recently gained widespread popularity among knowledge engineers, mainly due to the fact that OWL-DL [18], the language for Semantic Web ontologies [4, 20] promoted by W3C, is based on a Description Logic dialect. During the past several years a number of DL reasoners have been developed, the most prominent being Racer Pro [14], FaCT++ [7], Pellet [16] and InstanceStore [3]. In this chapter we present another DL reasoner KASEA, developed at GUT in the course of a EU 6th FP project PIPS [19]. The approach we applied in our reasoner is quite different from the approaches in other DL reasoners. In our approach (primarily proposed by W. Waloszek, see [9]), dubbed as Knowledge Cartography, we treat a universe as a map of overlapping concepts. Each concept consists of a number of atomic regions in the map. Each region is assigned a unique bit position in a string of bits called a *signature* of a concept. In that way, we reduce inference problems to appropriate Boolean operations on signatures.

The rest of the paper is organized as follows: In Section 2 we give a brief introduction to DL and related inference problems. In Section 3 we introduce ideas of Knowledge Cartography. In Section 4 we discuss implementation of

K. Goczyła et al.: *Processing and Querying Description Logic Ontologies Using Cartographic Approach*, Studies in Computational Intelligence (SCI) **102**, 63–80 (2008)
www.springerlink.com

the approach in KASEA. In Section 5 some experimental results are presented. Section 6 concludes the chapter with further work perspectives.

2 Ontologies and Description Logics

This section is devoted for those readers who are not familiar with ontologies and Description Logic. We give a short introduction "by example"; for a comprehensive study an interested reader is referred to [1].

2.1 Ontologies

Informally, an ontology is a description of terms that are important for a given domain of interest. An ontology can be viewed as an encyclopedia or a dictionary where we can find knowledge on some subject. Being a bit more formal, an ontology is an explicit (i.e. expressed in a formal language) specification of conceptualization [13]. The "conceptualization" is meant as a model of some part of the world, expressed in terms of concepts, their properties and relations between the concepts. In this definition, the term "concept" is of crucial importance for understanding ontologies.

Description Logic (DL) is a formalism (or rather – a family of formalisms) based on the first-order logic. DL has both precise formalism and easy to understand set-theory interpretation. For these reasons it has recently become a commonly accepted basis for knowledge representation and management systems. Actually, the OWL-DL language, promoted by W3C within the Semantic Web initiative [20], is based on a DL dialect.

The approach to representing knowledge about the world as a DL ontology is based on the following three natural and simple assumptions:

1. There is a *universe* (domain of interest) to be described as an ontology.
2. The universe consists of *individuals* that are instances of *concepts*.
3. Concepts are related to each other by binary relations called *roles*.

According to the above, a DL ontology consists of two parts: a *terminological* part (TBox) and an *assertional* part (ABox). TBox contains concepts, roles and axioms that define constraints on concepts and roles. ABox contains instances of concepts (in the form of unary assertions) and instances of roles (in the form of binary assertions). An ontology does not have to have both parts. Very often, ontologies contain only a TBox, although one can also imagine an ontology with ABox only, with implicit concepts and roles.

DL ontologies are data for DL *knowledge bases* (KB). For ontologies to be processed and reasoned over, a KB must be equipped with a reasoner, or inference engine. There must also exist a language to express the ontology.

Any language from a rich family of DL languages has basic components:

- atomic concepts, with the universal concept $\top$ (Top) representing the universe, and the empty concept $\bot$ (Bottom) that cannot have any instances;

- atomic roles;
- constructors that are used to create complex concepts and roles.

In this paper we assume that ontologies are specified in $\mathcal{ALC}$—a DL dialect that is simple, yet powerful enough to define non-trivial ontologies. Table 1 presents constructors of the language.

2.2 Inference Problems

In the following, an exemplary ontology of family relations will be used (see Table 2). The ontology contains the concepts Person, Man, Woman, Parent, Father and Mother, and the role hasChild. TBox of the ontology contains also axioms expressed with two operators: concept equivalence ($\equiv$) and concept inclusion ($\sqsubseteq$), with obvious meaning.

The axioms of the ontology state that a Man and a Woman are Persons, and no one can be simultaneously a Man and a Woman; a Parent is a Person that has at least one human child; a Father (or a Mother) is a Parent that is a Man (or a Woman). ABox contains some known facts about Ann, Jane, and Charles who are members of the family (e.g. Jane and Charles are children of Ann). We will use the above ontology to explain basic DL inference problems.

Subsumption: Concept C *subsumes* concept D (denotation: $D \sqsubseteq C$) if the set of instances of D is always a subset of the set of instances of C. In our example, some subsumptions are given explicitly as axioms. Other can be inferred, e.g. Father $\sqsubseteq$ Person and Mother $\sqsubseteq$ Person.

Table 1. Constructors of $\mathcal{ALC}$

$\neg C$	Negation of a concept
$C \sqcap D$	Intersection of two concepts
$C \sqcup D$	Union of two concepts
$\exists R.C$	Existential quantification: a set of individuals (role subjects) that are at least once related by role R to an individual (role filler) that is an instance of concept C
$\forall R.C$	General quantification: a set of individuals (role subjects) for which all (possibly zero) relations by role R are with individuals (role fillers) that are all instances of concept C

Table 2. An exemplary ontology

TBox	ABox
Man $\sqsubseteq$ Person	Woman (Ann)
Woman $\sqsubseteq$ Person	Woman (Jane)
Woman $\sqcap$ Man $\equiv \perp$	Man (Charles)
Parent $\equiv$ Person $\sqcap$ $\exists$.hasChild.Person	hasChild (Ann, Jane)
Father $\equiv$ Man $\sqcap$ Parent	hasChild (Ann, Charles)
Mother $\equiv$ Woman $\sqcap$ Parent	

Satisfiability: Concept C is *satisfiable* if it can have instances. In other words, concept C is *unsatisfiable* if $C \equiv \bot$. In our example, all the concepts are satisfiable. If we, however, add to TBox a new concept Fosterer defined as Fosterer $\equiv$ Mother $\sqcap$ Father, then Fosterer is unsatisfiable.

Disjointness: Concepts C and D are *disjoint* if sets of their instances are always disjoint ($C \sqcap D \equiv \bot$). In our example, pairs of concepts: (Woman, Man) and (Mother, Father) are disjoint.

Equivalence: Concepts C and D are *equivalent* ($C \equiv D$) if sets of their instances are always equal.

The above inference problems all refer to TBox. It is easy to show that they are not independent of each other. For instance, checking if two concepts are disjoint is equivalent to checking satisfiability of their intersection. Actually, all the above problems can be reduced to subsumption [1].

Existence of ABox leads to other basic inference problems that refer to the whole ontology. Below we use a self-explanatory functional notation of DIG interface [2].

Instance retrieval: Retrieve from a knowledge base all individuals that are instances of a given concept (denotation: *instances*(C)). For example, a response to *instances* (Mother) is a singleton set {Ann}.

Instance check: Check if a given individual is an instance of a given concept (denotation: *instance*(x, C)). For example, the response to *instance* (Ann, Mother) is true.

Realization: For a given individual x find most specific (i.e. minimal in the sense of partial ordering) named concepts C_i such that *instance*(x, C_i) holds for each i.

Consistency check: A knowledge base K is said to be *consistent* (denotation: *consistent*(K)) if all the named concepts from TBox are satisfiable and ABox does not contain any false individual (an individual is said to be false if it cannot be an instance of any satisfiable concept). In other words, an inconsistent KB has no FOL model.

Our exemplary knowledge base is consistent. But let us assume that we add to ABox two new assertions: Woman(Mary), Man(Mary). As a result, the knowledge base has lost its consistency because Mary must be an instance of the empty concept $\bot$, which is absurd. (Note that a KB consisting of only two above assertions, without the TBox part, is consistent.)

Now we extend our exemplary ontology by the role hasSon (with its natural meaning). It is clear that having a son implies having a child, so we can formulate the following axiom: hasSon $\sqsubseteq$ hasChild. But how to express that a son (the right-hand side filler of hasSon) must be a Man? In other words, how to define the *range* of role hasSon? We do it in axiom $\exists$hasSon.$\neg$ Man $\equiv \bot$ that can be interpreted as follows: if any individual has a son who is not a Man, then this individual is false (i.e. it does not exist). Similarly, we can state that only Parents may have sons (i.e. we define the *domain* of a role): $\exists$hasSon.$\top \sqsubseteq$ Parent, which means that if any individual has a son, then this individual is a Parent.

Having defined the domain and the range of hasSon, we add to ABox a new fact: hasSon (Charles, John). Now let us issue against our knowledge base two queries: *types*(John) and *types*(Charles). (A *types*(x) query asks a KB for all named concepts from TBox that x is an instance of.) Because Charles is not a false individual, John must be a Man (a member of the range of hasSon), and also a Person (any Man is a Person). Hence, the response to *types*(John) is {Man, Person}. The response to *types*(Charles) is {Parent, Man, Person}, which is derived from the axiom defining the domain of hasSon and from explicitly given contents of ABox and TBox.

Traditional approaches to solving the basic inference problems exploit different kinds of structural analysis of terminology (*structural subsumption*) or so-called tableau algorithms [1]. In general, the inference problems are of exponential time/space complexity, so it is of crucial importance for real-life applications to optimize algorithms used in order to obtain acceptable performance of an inference engine. These issues will be discussed in Section 4.

2.3 Applications

DL ontologies, usually specified in OWL-DL, are presently vigorously developed within the Semantic Web activity framework. They are used as a means of interchange of information, or knowledge, among web-enabled information systems, to attain interoperability between them. One of such systems is PIPS (*Personalized Information Platform for Life and Health Services* [19]), currently under development within a 6th EU Framework Programme, Priority "e-health" (Contract No 507019). One of its elements is a knowledge base that processes OWL-DL ontologies in areas of medicine, nutrition and healthy lifestyle. The PIPS KB is managed by KASEA, a system for processing and reasoning over DL ontologies, based on a novel idea of Knowledge Cartography. The rest of the paper is a presentation of Knowledge Cartography.

3 Knowledge Cartography

The idea of Knowledge Cartography is based on the following assumptions:

1. TBox component of the knowledge base is modified so rarely that in practice it may be considered constant.
2. A knowledge base is queried much more often than it is updated (by update we mean here addition of new assertions into ABox). Therefore, performance of retrieval is crucial, while performance of updates is not.
3. A knowledge base should be able to store and efficiently process numerous individuals.

The Knowledge Cartography and algorithms behind it aim at storing in a knowledge base as many conclusions (inferred facts) about concepts and individuals as possible. The conclusions are derived during ontology loading

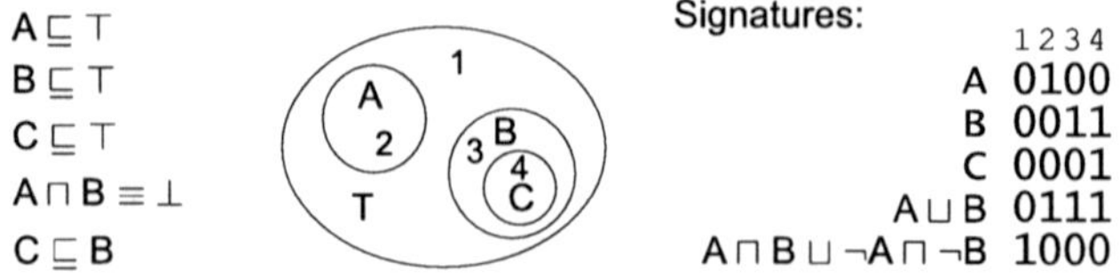

Fig. 1. An example of a map of concepts with signatures of atomic and complex concepts

and updating. Any conclusion can be quickly retrieved in the process of query answering and remains always valid due to the fact that the terminology is not changed. By appropriate organization of a knowledge base identical conclusions can be applied to many individuals, which facilitates efficient information retrieval and reduces the size of a database that is used to store the KB.

3.1 The General Idea

Knowledge Cartography takes its name after a notion of "a map of concepts". A map of concepts is basically a (graphical and symbolic) description of relationships that hold between concepts in a terminology treated as sets of individuals. The map is created during the knowledge base creation. The map of concepts can be graphically represented as a Venn diagram (see Fig. 1). Each atomic area of the map (i.e. an area that does not contain any other area; called henceforth a *region*) represents a single valid intersection of concepts. By valid we mean an intersection that is satisfiable with respect to a given terminology. Unsatisfiable regions (not allowed by terminological axioms) are excluded from the map (as in Fig. 2, where two axioms: d) and e) excluded four regions from the map). The algorithm for processing ontologies, called Cartographer, calculates a number of valid regions n and assigns each region a subsequent integer from the range $[1, n]$ (shown in Figs. 1 and 2 as ordinals inside regions). Because any area in the map consists of some number of regions, any area can be represented by a string of binary digits (bits) of length n with "1"s at positions corresponding to contained regions and "0"s elsewhere.

According to this rule, any concept in a terminology is assigned a *signature*—a string of bits representing the area covered by the concept in the map. It is important that in this way we can represent any combination of complement, union and intersection of concepts by simply performing Boolean negation, disjunction and conjunction.

Formally speaking, we define a function s from concepts in TBox to elements of a Boolean algebra $\mathbf{B}^n = \{0, 1\}^n$ (the set of bit strings of length n). It can be shown that the only requirement to be met by function s is:

$$s(C) \leq s(D) \iff C \sqsubseteq D \tag{1}$$

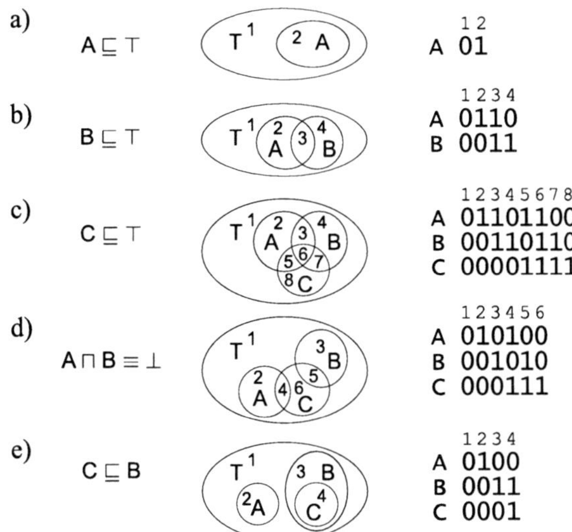

Fig. 2. Steps of building a map of concepts for the terminology from Fig. 1

The $\leq$ operator is understood in terms of a Boolean algebra (i.e. a bit string that includes all "1"s of another string at the same positions is greater than or equal to the other string). In such a function regions are mapped to atoms of $\mathbf{B}^n$, i.e. strings of "0"s with a single "1". Moreover, for any concepts C and D the following equalities hold:

$$s(\neg C) = \neg s(C) \tag{2}$$

$$s(C \sqcap D) = s(C) \wedge s(D) \tag{3}$$

$$s(C \sqcup D) = s(C) \vee s(D) \tag{4}$$

Having determined s, any basic TBox reasoning problem can be solved by appropriate operations on signatures. Indeed:

- A query for equivalence of concepts C and D can be performed by checking whether $s(C) = s(D)$.
- A query for subsumption of concepts C and D can be performed by checking whether $s(C) \leq s(D)$.
- A query for disjointness of concepts C and D can be performed by checking whether $s(C) \wedge s(D) = \{0\}^n$.

Also advanced reasoning problems, like *least common subsumer* problem or *most specific concept* problem [1], can be solved by simple operations on signatures.

From (1) we deduce that the order of the range of function s should be equal to the number of terminologically unequivalent concepts that can be expressed in the terminology. With introduction of $\exists R.C$ and $\forall R.C$ constructs this number reaches infinity (namely $\aleph_0$), because existential and general quantifiers can be nested to theoretically unlimited level (for instance, a concept in TBox can be defined as $\exists R.\exists R.\exists R.\exists R....\exists R.C$).

Due to this fact, we have made an important decision restricting the use of $\exists R.C$ and $\forall R.C$ constructs in queries. The only concepts of the form $\exists R.C$ and $\forall R.C$ that can be used in queries are those explicitly defined in the terminology. In this way we limit the number of unequivalent concepts, making it possible to fix the length of signatures. This restriction limits expressiveness of ad hoc queries accepted by the system, but our experiences gained from using KaSeA in a real-life (although still experimental) environment show that this limitation is not severe for knowledge base users.

Analogical techniques as for TBox can be applied to reasoning over ABox. We assign each individual a in ABox a signature of the most specific concept the individual is an instance of (we denote this concept as C_a; this concept need not be defined explicitly in TBox). After determination of signatures for all individuals in the KB we can reduce all ABox reasoning problems to TBox reasoning problems, which in turn can be solved by operations on signatures. For example, checking if individual a is an instance of concept C is reduced to checking whether concept C_a is subsumed by concept C. And checking consistency of a knowledge base consists in simply checking if there exists in the KB any concept or any individual whose signature is equal to $\{0\}^n$.

3.2 The Map Creation Algorithm

The main algorithmic problem in Knowledge Cartography is determination of function s, i.e. creation of the map of concepts. We can formulate this problem as follows: Given an $\mathcal{ALC}$ terminology $\mathcal{T}$, for each atomic concept and each concept of the form $\exists R.C$ (called jointly *mapped concepts*) generate its signature. (Note that concepts of the form $\forall R.C$ do not have to be considered because they can be converted to the equivalent form of $\neg \exists R.\neg C$.) The problem is not polynomial, however, some optimization techniques can make the process of map (and signatures) creation efficient for real-life ontologies.

The process of creation of signatures is based on the fact that regions of the map are mapped to the atoms of $\mathbf{B}^n$ that constitute the range of function s. Regions can be viewed simply as valid intersections of all mapped concepts in $\mathcal{T}$ or their complements, i.e. all possible complex concepts of the form:

$$L_1 \sqcap L_2 \sqcap \ldots \sqcap L_k \tag{5}$$

where k is the number of mapped concepts, and L_i is a placeholder for i-th mapped concept or its complement. Using this approach we see a terminology as a set of first-order logic formulae, mapped concepts being variables, and reduce the problem of a map creation to finding a truth table for the terminology. Each satisfiable combination of variable values constitute a region.

From among many techniques available we applied Ordered Binary Decision Diagrams (OBDD), originally proposed in [5]. Cartographer systematically builds an OBDD tree for the whole terminology by combining (using the logical AND operation) a current tree with trees generated for formulae corresponding to consecutive axioms. The initial tree is a tree with a single node (OBDD for a tautology).

Axioms are converted into first-order logic formulae as proposed in [5], but the method is somehow simplified because concepts of the form $\exists R.C$ (or $\forall R.C$) are represented as one variable. Each new mapped concept is assigned a new variable name. For example, the axiom:

$$\text{Momo} \equiv \text{Person} \sqcap \forall \text{ hasChild.Man} \sqcap \exists \text{ hasChild.Man}$$

is converted to the equivalent form:

$$\text{Momo} \equiv \text{Person} \sqcap \neg\exists \text{ hasChild.}\neg \text{ Man} \sqcap \exists \text{ hasChild.Man}$$

and subsequently to the formula:

$$c_1 \longleftrightarrow c_2 \wedge \neg e_1 \wedge e_2$$

An outline of the algorithm is presented below:

Input: A terminology $\mathcal{T}$ expressed in $\mathcal{ALC}$.
Output: An OBDD tree T for terminology $\mathcal{T}$.
    Initialize T to OBDD of any tautology.
    For each axiom A from $\mathcal{T}$:
        Convert A to the formula F in the way described above.
        Generate the OBDD U for the formula F.
        $T := T \wedge U$ (where $\wedge$ denotes conjunction of two OBDD
            trees in the sense of [5])
    Next

The detailed example of subsequent steps of the algorithm would require us to introduce the notion of OBDDs in detail, which is out of scope of this paper. However, as OBDDs track satisfiability of logical formulas for various combinations of variable values, the steps of the algorithm are in accordance with the example given in Fig. 2.

Direct application of the above algorithm to some terminologies may lead to generation of spurious (unsatisfiable) regions, as shown in Fig. 3a. According to the axioms in the terminology, an individual cannot belong to $\exists R.B$ and not belong to $\exists R.A$ (because, according to the inclusion axiom, each member of B is a member of A). In order to exclude the spurious regions we perform post-processing that produces a tree T' on the basis of T. The post-processing consists in checking whether for each combination of values of variables e_i that concern the role R and are satisfiable with respect to T it is possible to create a set of individuals that would satisfy the combination.

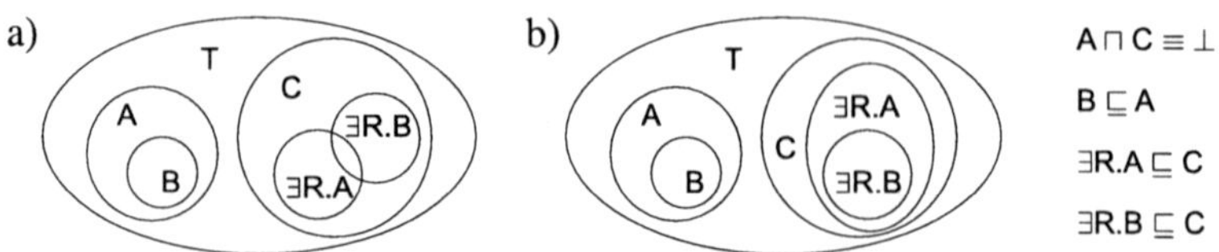

Fig. 3. A map of concepts before post-processing (a) and after post-processing (b)

For instance, in Fig. 3a the following atomic intersection is satisfiable with respect to T:

$$\neg A \sqcap \neg B \sqcap C \sqcap \neg \exists R.A \sqcap \exists R.B$$

This region (represented in Fig. 3a as the part of $\exists R.B$ that is not contained in $\exists R.A$) should be, however, excluded from T' during post-processing because the following family of concepts:

$$\neg A \sqcap B \sqcap \ldots$$

is not satisfiable with respect to T (because there cannot exist any individual that belongs to B and does not belong to A).

Details of post-processing are outside the scope of this paper. It can be shown that the post-processing is sound and complete, i.e. all intersections it excludes are invalid and there is no invalid intersection it does not exclude.

4 The KaSeA System

Cartographic Approach has been successfully applied in a prototype of the PIPS system. The Knowledge Inference Engine (KIE), a vital component of KASEA, uses Cartographer to load ontologies and to infer and store inferred facts. KIE allows for processing a terminology (TBox) and assertions (ABox). Data are stored in a relational database (Oracle 9i), which allows for using built-in optimization techniques like indexing or query optimization.

4.1 Query-Processing

Knowledge base updates (*tells*) and queries (*asks*) are formulated in DIGUT [6], an interface based on DIG/1.1 especially modified for querying KASEA. *Tells* that can be handled by KASEA are concept assertions of the form $C(a)$ and role assertions of the form $R(a, b)$. Due to assumptions stated at the beginning of Sec. 3, KASEA does not handle ad hoc *tells* that concern terminological axioms.

For processing a concept assertion $C(a)$, C has to be an expression built of concepts used in the terminology (constructs of the form $\exists R.A$ are allowed only if a signature for this construct has been determined during loading the terminology into the KB). The course of actions consists of the following two steps: (1) calculate a signature $s(C)$, and then (2a) if a is not in the database, add it to the database and assign it the signature $s(C)$, (2b) otherwise combine the signature of C_a with $s(C)$ using logical AND operation and if C_a has been changed update the neighborhood (see below) of a. In processing a role assertion $R(a, b)$ only neighborhood update is performed.

Necessity of updating neighborhood is a consequence of the fact that changing our knowledge about membership of an individual a may change our knowledge about individuals related to a. In the current version of KASEA simple mechanisms of *positive role checking* and *negative role checking* have been applied. In the positive role checking every pair (a, b) related with role R is checked against all mapped concepts of the form $\exists R.C$. If b is a member of the concept C, the signature of C_a is combined with $s(\exists R.C)$ using logical AND operation (because a has to be a member of the concept $\exists R.C$). In the *negative role checking* every pair (a, b) related with R is also checked against all mapped concepts of the form $\exists R.C$. If a is a member of the concept $\neg \exists R.C$, the signature of C_b is combined with $s(\neg C)$ (because b may not be a member of C). An example of the positive (Step 3) and negative (Step 4) role checking is shown in Fig. 4.

This process is recursively repeated if the signature of any individual has been changed. The process eventually stops after a finite number of steps because the number of individuals in ABox is finite and in each update the number of "1"s in signatures of individuals being processed may only decrease. Potentially, the process of neighborhood update is prone to combinatorial explosion, but in real-life ontologies there is little chance for that.

Most *asks* can be brought down to subsumption checking and, consequently, to simple operations on signatures. For example, the instance retrieval problem for a concept C can be performed as finding all individuals a such that C_a is subsumed by C. During subsumption checking some preliminary tests are used to quickly exclude existence of the subsumption relation (e.g. counting "1"s in segments of signatures). Then bitwise Boolean operations are performed in order to check whether two signatures stand in the $\leq$ relation.

4.2 Related Work

A similar approach of mapping $\mathcal{ALC}$ terminologies into Boolean algebra can be found in [17]. However their work substantially differs from the one presented here in terms of motivation and application. In [17] the mapping was a tool to support use of information contained in TBox in a system of algebraic (in)equalities. This allowed for solving optimization problems for numerical properties associated with DL concepts (an example of such property might be the number of individuals being members of a specific concept). In this paper we deal with performance problems and we use the mapping to create *signatures* that can be stored in the database and associated with both concepts and individuals ([17] focused only on TBox) and used to improve

a)

Patient	111100000
PatientUnderThreat	001100000
∃hasBPResult.BPResultHigh	001000000
∃hasRecord.ClearRecord	010000000
ClinicalRecord	000011000
ClearRecord	000001000
BPResult	000000111
BPResultHigh	000000001

b)

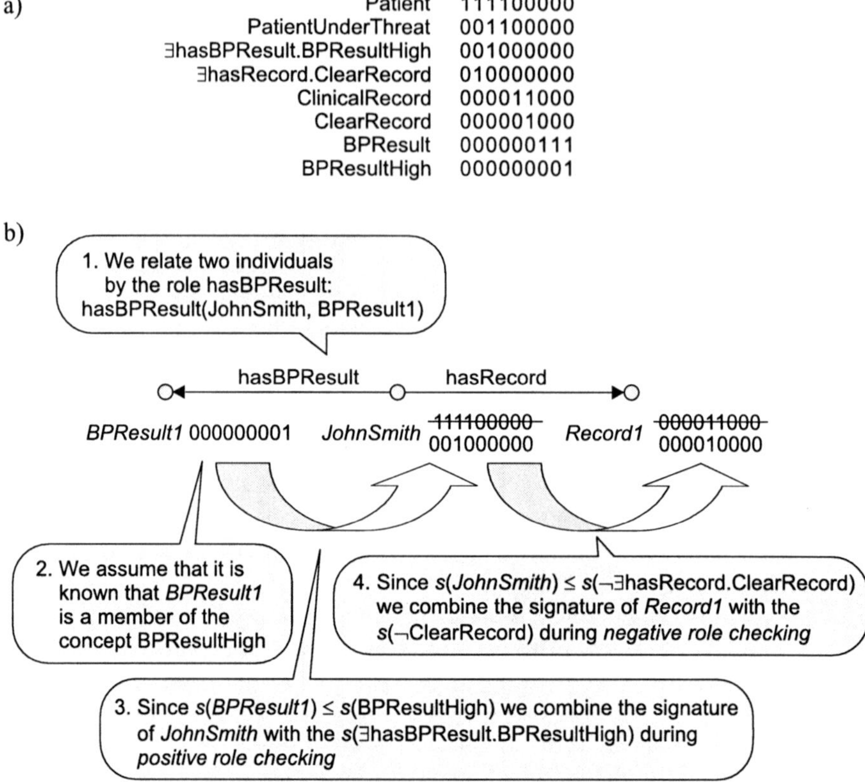

Fig. 4. a) Concepts from an exemplary terminology along with their signatures and b) An example of positive and negative role checking; the changes may further propagate through role instances not shown in the figure

performance of query answering. The notion of *signatures* in a similar meaning appears in papers on optimizing object oriented databases and is used there for answering queries about set-valued attributes (see e.g. [21]).

The signatures allow for storing inference results in a relational database, from where they can be quickly retrieved in the process of query answering. A similar idea stays behind *InstanceStore* [3]. However *InstanceStore* cannot make inferences on its own and needs another reasoner to evaluate relationships among concepts and individuals. Additionally it can only handle role-free ABoxes. Due to these features it substantially differs from KaSeA, intented to be a lightweight stand-alone reasoner holding all TBox and ABox data needed for making inferences in a relational database.

ABox reasoning optimization problems are also discussed in [8]. In their approach, ABox is also kept in a relational database. The kind of reasoning they are focused on is limited to checking KB consistency, while we strive to optimize a broad range of ABox queries execution.

Notion of region calculus was also exploited by Gotts et al. in [12], however in their work mainly spatial properties of regions were taken into consideration, while in our work such properties are inadequate.

5 Experimental Comparison with Other Reasoners

Performance of KASEA has been tested and compared with several freely available tools, especially with Racer Pro [14] and Pellet [16]. There was also a trial to compare KASEA with Kaon2 [15]. However, we did not succeed in testing Kaon2—each time we were sending a DIG request we received server error ("immature end of file"). The tests were performed on a PC with Pentium 4 3GHz and 2GB RAM.

Our tests revealed that Racer Pro 1.9.0 and Pellet 1.3 have some troublesome features:

- They are not suitable for using them in environment where there are often changes in ABox.
- They use main memory as a storage for ontologies.
- Therefore they are not scalable with respect to number of individuals.

We loaded to Racer Pro an ontology created for the PIPS project. The ontology concerns a few related domains: food, drugs, persons and personal clinical information. The ontology consists of about 600 concepts. Two kinds of tests were performed. In the first one various sets of individuals were loaded to KB. The sets contained respectively 20, 20, 2000 and 20000 individuals. Then tests for two DIG queries (*instance* and *instances* for complex concept $Person \sqcup ClinicalRecord \sqcup Anamnesis$) have been performed. Results are presented in Table 3 (all times are in milliseconds).

Two groups results are shown in the table. The first is the time of answering a query just after creating a new KB and inserting assertions to this KB. The second result for *instance* query is the time of answering the same query again. This time is very short, presumably because a cached response is returned. The great advantage of Racer Pro is that a consecutive query, with ABox unchanged, is responded in a very short time. For example, when we ask about instance of concept *Anamnesis* just after creating a new KB, the response time is equal to approximately 10 seconds. Then we ask for instances of *Person* and the query response time is just about 750 ms. However, in real-life applications ABox seems to change quite often and in such cases the time of response is very long.

In the second row of Table 3, the result (marked with 2)) is the time of response to the *instances* query after executing the *instance* query. For both queries the complex concept we ask for is exactly the same. The second time is shorter than the first time but both results are unsatisfactory in comparison with KASEA (see two last rows of Table 3). These tests show that for queries

Table 3. Results of experiments for *instance* and *instances* queries for various sets of individuals for Racer Pro, Pellet and KASEA

No. of individuals	20	200	2000	20000
RacerPro – *instance*	1) 10579 2) 16	1) 11437 2) 15	1) 27156 2) 16	1) 298094 2) 16
RacerPro – *instances*	1) 7610 2) 485	1) 55938 2) 1250	1) 59953 2) 15250	1) 300032 2) 115000
Pellet – *instance*	1297	1157	1485	4204
Pellet – *instances*	1172	1031	error	error
KASEA – *instance*	390	328	359	735
KASEA – *instances*	609	593	828	1516

about instances of a concept the response time grows quasi-linearly with the number of individuals, which for large ABoxes is unacceptable.

We have performed exactly the same tests for Pellet reasoner. The Pellet web site states that the reasoner has worse response times in comparison to Racer Pro, however in our tests results were rather in favor of Pellet, although we were not able to perform tests for *instances* query for 2000 or more individuals. The tests were carried out using DIG Client for Racer Pro, but the error appeared on the server side. The results in Table 3 contain times of answering queries after creating a new KB.

The second suite of tests was focused on scalability of reasoners with respect to number of individuals and complexity of concepts in the queries. The charts in Fig. 5 (note the logarithmic scale on both axes) present results of executing *instances* query for an atomic concept A, for an unsatisfiable concept B and for two complex concepts C and D defined as follows:

$$B \equiv Anamnesis \sqcap Person \sqcap FemalePerson \sqcap MalePerson \tag{6}$$

$$C \equiv (Anamnesis \sqcap Person \sqcap FemalePerson \sqcap MalePerson) \sqcup Aux1 \tag{7}$$

$$D \equiv (((Anamnesis \sqcap Person \sqcap FemalePerson \sqcap MalePerson) \sqcup Aux1) \sqcap$$
$$\sqcap Aux2) \sqcup$$
$$(\neg(FruitVegetables \sqcup Numbers \sqcup (IngredientRation \sqcap Pathology))) \sqcup$$
$$\sqcup (\exists isGTNumberOf.\top) \tag{8}$$

Firstly, we notice that Racer Pro and Pellet, in contrast to KASEA, are not able to response the query with 60000 individuals loaded to KB and—for the more complex concept D—even for 6000 individuals. Moreover, Pellet—which responds quickly to *instances* query for atomic concepts—is inefficient for complex concepts. In contrast, for KASEA the response time mainly depends on the number of individuals belonging to the concept specified, not on the number of all individuals loaded to the KB, as it is for Racer Pro and Pellet.

The chart in Fig. 6 presents the response times for *instance* query. For all the reasoners these times do not depend on the complexity of the con-

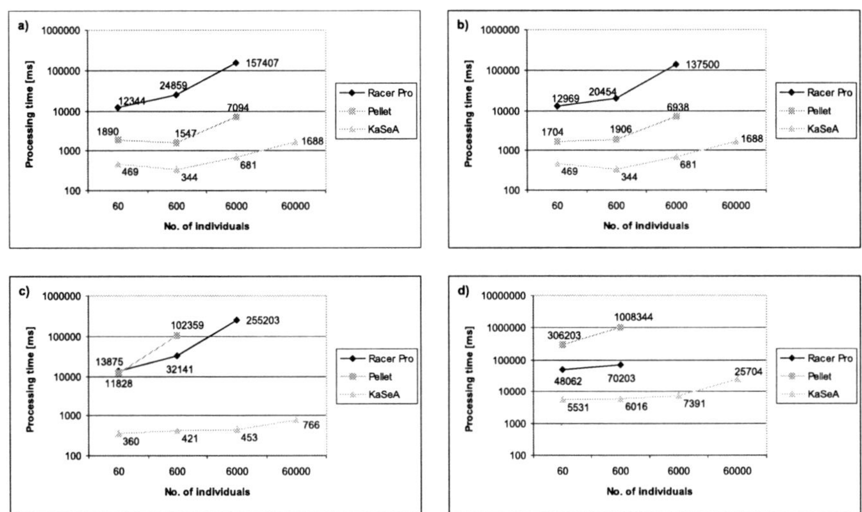

Fig. 5. Response times for *instances* queries for concepts a) A, b) B, c) C, d) D

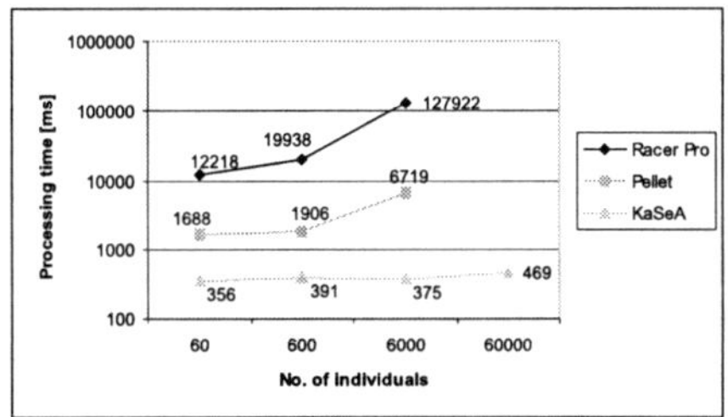

Fig. 6. Response times for *instance* query

cept specified in the query. For KASEA the time grows insignificantly with increasing number of individuals, in contrast to Racer Pro and Pellet.

To sum up: The experiments (also other ones not presented here due to shortage of space) showed that KASEA reveals much better scalability with respect to number of individuals in KB, and is much less sensitive to complexity of concepts appearing in *instance* and *instances* queries.

6 Further Development

Our present work focuses on overcoming some limitations of KASEA. The most important ones are: restriction of use of $\exists R.C$ concepts in queries and inefficiency of neighborhood update. Although the first limitation may seem inherent for Cartographic Approach, it may be overcome by using signatures with variable length. The second limitation stems from complexity of the neighborhood update mechanism and from the fact that it is not fully OWA-compliant.

Moreover, we are gradually extending KaSea capabilities in order to support such constructs as cardinality constraints and symmetric, transitive and functional roles, so that the system could cover whole $\mathcal{SHION}$ (a DL dialect that OWL-DL is based on). We have recently introduced reasoning over concrete domains [11]. We are also extending signature analysis on roles to support role hierarchies analogously to concept hierarchies. Other research topics focus on integration of knowledge from different ontologies and on reasoning over ontologies that are not fully trustworthy (e.g. they come from knowledge sources that cannot be fully trusted).

References

1. Baader FA, McGuiness DL, Nardi D, Patel-Schneider PF (2003) The Description Logic Handbook: Theory, implementation, and applications. Cambridge University Press
2. Bechhofer S (2003) The DIG Description Logic Interface: DIG/1.1. Univ. of Manchester
3. Bechhofer S, Horrocks I, Turi D Implementing the Instance Store. CEUR `http://sunsite.informatik.rwth-aachen.de/Publications/CEUR-WS/Vol-115`
4. Berners-Lee T, Hendler J, Lassila O (2001) The Semantic Web. Scientific American, May 2001
5. Bryant RE (1986) Graph-based algorithms for Boolean function manipulation. IEEE Transaction on Computers
6. (2005) DIGUT Interface, V1.3. KMG@GUT, 2005. `http://km.pg.gda.pl/km/DIGUT_Interface_1.3.pdf`
7. FaCT++. `http://owl.man.ac.uk/factplusplus/`
8. Fokoue A, Kershenbaum A, Ma L, Schonberg E, Srinivas K, Williams R (2006) SHIN ABox Reduction. In: Proceeding of DL Workshop 2006.
9. Goczyła K, Grabowska T, Waloszek W, Zawadzki M (2005) The Cartographer Algorithm for Processing and Querying Description Logics Ontologies. LNAI, Vol. 3528, pp. 163–169
10. Goczyła K, Grabowska T, Waloszek W, Zawadzki M (2006) The Knowledge Cartography — A new approach to reasoning over Description Logics ontologies. LNCS, Vol. 3831, pp. 293–302
11. Goczyła K, Waloszek A, Waloszek W (2006) Concrete-domain Reasoning Techniques in Knowledge Cartography. Semantics 2006 (to be published)
12. Gotts NM, Gooday JM, Cohn AG (1995) A connection based approach to commonsense topological description and reasoning. The Monist: An International Journal of General Philosophical Inquiry, 79(1)
13. Gruber TR (1993) A translation approach to portable ontologies. Knowledge Acquisition, 5(2):199–220
14. Haarslev V, Möller R (2005) RacerPro Reference Manual. August 12, 2005, `http://www.racer-systems.com/products/racerpro/reference-manual-1-9.pdf`
15. KAON2. `http://kaon2.semanticweb.org/`

16. MINDSWAP. (2003) Maryland Information and Network Dynamics Lab Semantic Web Agents Project, Pellet, `http://www.mindswap.org/2003/pellet/`
17. Ohlbach HJ (1999) Set Description Languages and Reasoning about Numerical Features of Sets. International Workshop on Description Logics, 1999
18. OWL – Web Ontology Language Guide. (2004) W3C, `www.w3.org/2004/OWL`
19. Personal Information Platform for Life and Health Services. `http://www.pips.eu.org`
20. Semantic Web Activity. `http://www.w3.org/2001/sw`

21. Tousidou E, Bozanis P, Manolopoulos Y (2002) Signature-based structures for objects with set-valued attributes. Inf. Systems Vol. 27, pp. 93–121, Elsevier 2002.

Rough Sets Theory for Multi-Objective Optimization Problems

Alfredo G. Hernández-Díaz[1], Luis V. Santana-Quintero[2],
Carlos A. Coello Coello[2], Rafael Caballero[3], and Julián Molina[3]

[1] Pablo de Olavide University, Department of Quantitative Methods, Seville, Spain
   `agarher@upo.es`
[2] CINVESTAV-IPN, Computer Science Section, México
   `lvspenny@hotmail.com, ccoello@cs.cinvestav.mx`
[3] University of Málaga, Department of Applied Economics (Mathematics), Spain
   `rafael.caballero@uma.es, julian.molina@uma.es`

Summary. In this chapter, we propose the use of rough sets to improve the approximation provided by a multi-objective evolutionary algorithm. The main idea is to use this sort of hybrid approach to approximate the Pareto front of a multi-objective optimization problem with a low computational cost (only 3000 fitness function evaluations). The hybrid operates in two stages: in the first one, a multi-objective version of differential evolution is used as our search engine in order to generate a good approximation of the true Pareto front. Then, in the second stage, rough sets theory is adopted in order to improve the spread of the solutions found so far. To assess our proposed hybrid approach, we adopt a set of standard test functions and metrics taken from the specialized literature. Our results are compared with respect to the NSGA-II, which is an approach representative of the state-of-the-art in the area.

1 Introduction

Multi-Objective Programming (MOP) is a research field that has raised great interest over the last thirty years, mainly because of the many real-world problems which naturally have several (often conflicting) criteria to be simultaneously optimized [7,17].

In recent years, a wide variety of multi-objective evolutionary algorithms (MOEAs) have been proposed in the specialized literature [3,4]. However, the study of hybrids of MOEAs with other types of techniques is still relatively scarce. This chapter presents a study of the use of rough sets theory as a local search explorer able to improve the spread of the solutions produced by a MOEA. Our main motivation for such a hybrid approach is to reduce the overall number of fitness function evaluations performed to approximate the true Pareto front of a problem. Our proposed hybrid is able to produce

A.G. Hernández-Díaz et al.: *Rough Sets Theory for Multi-Objective Optimization Problems*,
Studies in Computational Intelligence (SCI) **102**, 81–98 (2008)
`www.springerlink.com`   © Springer-Verlag Berlin Heidelberg 2008

reasonably good approximations of the Pareto front of a variety of problems of different complexity with only 3000 fitness function evaluations.

The organization of the rest of the chapter is the following. Section 2 provides some basic concepts required to understand the rest of the chapter. An introduction to rough sets theory is provided in Section 3. In Section 4, we introduce differential evolution, which is the approach adopted as our search engine. Section 5 describes the relaxed form of Pareto dominance adopted for our secondary population (called Pareto-adaptive ϵ-dominance). Our proposed hybrid is described in Section 6. The experimental setup adopted to validate our approach and the corresponding discussion of results are provided in Section 7. Finally, our conclusions and some possible paths for future research are provided in Section 8.

2 Basic Concepts

We are interested in solving problems of the type[1]:

$$\text{Minimize } \mathbf{f}(\mathbf{x}) := [f_1(\mathbf{x}), f_2(\mathbf{x}), \dots, f_k(\mathbf{x})] \tag{1}$$

subject to:

$$g_i(\mathbf{x}) \leq 0 \quad i = 1, 2, \dots, m \tag{2}$$

$$h_i(\mathbf{x}) = 0 \quad i = 1, 2, \dots, p \tag{3}$$

where $\mathbf{x} = [x_1, x_2, \dots, x_n]^T$ is the vector of decision variables, $f_i : \mathbb{R}^n \to \mathbb{R}$, $i = 1, \dots, k$ are the objective functions and $g_i, h_j : \mathbb{R}^n \to \mathbb{R}$, $i = 1, \dots, m$, $j = 1, \dots, p$ are the constraint functions of the problem. To describe the concept of optimality in which we are interested, we will introduce next a few definitions.

Definition 1. Given two vectors $\mathbf{x}, \mathbf{y} \in \mathbb{R}^k$, we say that $\mathbf{x} \leq \mathbf{y}$ if $x_i \leq y_i$ for $i = 1, \dots, k$, and that $\mathbf{x}$ **dominates** $\mathbf{y}$ (denoted by $\mathbf{x} \prec \mathbf{y}$) if $\mathbf{x} \leq \mathbf{y}$ and $\mathbf{x} \neq \mathbf{y}$.

Definition 2. We say that a vector of decision variables $\mathbf{x} \in \mathcal{X} \subset \mathbb{R}^n$ is **nondominated** with respect to $\mathcal{X}$, if there does not exist another $\mathbf{x}' \in \mathcal{X}$ such that $\mathbf{f}(\mathbf{x}') \prec \mathbf{f}(\mathbf{x})$.

Definition 3. We say that a vector of decision variables $\mathbf{x}^* \in \mathcal{F} \subset \mathbb{R}^n$ ($\mathcal{F}$ is the feasible region) is **Pareto-optimal** if it is nondominated with respect to $\mathcal{F}$.

Definition 4. The **Pareto Optimal Set** $\mathcal{P}^*$ is defined by:

$$\mathcal{P}^* = \{\mathbf{x} \in \mathcal{F} | \mathbf{x} \text{ is Pareto-optimal}\}$$

---

[1] Without loss of generality, we will assume only minimization problems.

Definition 5. The **Pareto Front** $\mathcal{PF}^*$ is defined by:

$$\mathcal{PF}^* = \{\mathbf{f}(\mathbf{x}) \in \mathbb{R}^k | \mathbf{x} \in \mathcal{P}^*\}$$

We thus wish to determine the Pareto optimal set from the set $\mathcal{F}$ of all the decision variable vectors that satisfy (2) and (3).

3 Rough Sets Theory

Rough Sets theory is a new mathematical approach to imperfect knowledge. The problem of imperfect knowledge has been tackled for a long time by philosophers, logicians and mathematicians. Recently, it also became a crucial issue for computer scientists, particularly in the area of artificial intelligence (AI). There are many approaches to the problem of how to understand and manipulate imperfect knowledge. The most used one is the fuzzy set theory proposed by Lotfi Zadeh [26]. Rough sets theory was proposed by Pawlak [19], and presents another attempt to this problem. Rough sets theory has been used by many researchers and practitioners all over the world and has been adopted in many interesting applications. The rough sets approach seems to be of fundamental importance to AI and cognitive sciences, especially in the areas of machine learning, knowledge acquisition, decision analysis, knowledge discovery from databases, expert systems, inductive reasoning and pattern recognition. Basic ideas of rough set theory and its extensions, as well as many interesting applications, can be found in books (see [20]), special issues of journals (see [15]), proceedings of international conferences, and in the internet (see `www.roughsets.org`).

Let's assume that we are given a set of objects U called the *universe* and an indiscernibility relation $R \subseteq U \times U$, representing our lack of knowledge about elements of U (in our case, R is simply an equivalence relation based on a grid over the feasible set; this is, just a division of the feasible set in (hyper)-rectangles). Let X be a subset of U. We want to characterize the set X with respect to R. The way rough sets theory expresses vagueness is employing a boundary region of the set X built once we know points both inside X and outside X. If the boundary region of a set is empty it means that the set is *crisp*; otherwise, the set is *rough* (inexact). A nonempty boundary region of a set means that our knowledge about the set is not enough to define the set precisely (see Figure 1).

Then, each element in U is classified as *certainly* inside X if it belongs to the lower approximation or *partially* (*probably*) inside X if it belongs to the upper approximation (see Figure 1). The *boundary* is the difference of these two sets, and the bigger the boundary the worse the knowledge we have of set X. On the other hand, the more precise is the grid implicity used to define the indiscernibility relation R, the smaller the boundary regions are. But, the more precise is the grid, the bigger the number of elements in U, and then,

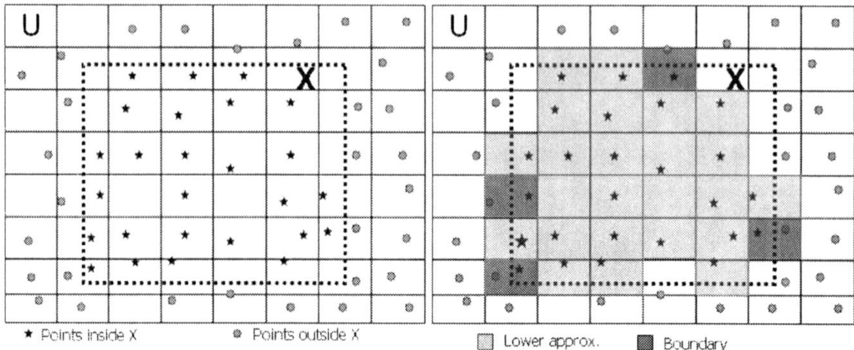

Fig. 1. Rough sets approximation

the more complex the problem becomes. Then, the less elements in U the better to manage the grid, but the more elements in U the better precision we obtain. Consequently, the goal is obtaining "small" grids with the maximum precision possible. These two aspects are called **Density** and **Quality** of the grid. If q is the number of criteria (in our case, the number of objectives), Q_i is the i-th criteria, b_j^i is the j-th value of the i-th criteria (we assume these value are ordered increasingly), then:

$$Density(G) = \sum_{i=1}^{q} \sum_{j=1}^{|Q_i|} x_j^i$$

$$Quality(G) = \frac{|Low(X)|}{|X|}$$

where x_j^i is 1 if b_j^i is active in the grid and $|Low(X)|$ is the cardinality of the lower approximation of X.

3.1 Use of Rough Sets in Multi-Objective Optimization

For our MOP problems we will try to approximate the Pareto front using a Rough Sets grid. To do this, we will use an initial approximation of the Pareto front (provided by any other method) and will implement a grid in order to get more information about the front that will let us improve this initial approximation. Then, at this point we have to face the following problem: the more precise the grid is, the higher the computational cost required to manage it, and the less precise the grid is, the less knowledge we get about the Pareto front. Thus, we need to design a grid that balances these two aspects. In other words, a grid that is not so expensive (computationally speaking) but that offers a reasonably good knowledge about the Pareto front to be used to improve the initial approximation. To this aim, we must design a grid and decide which elements of U (that we will call *atoms* and will be just rectangular portions of decision variable space) are inside the Pareto optimal set and which are not. Once we have the *efficient atoms*, we could easily intensify the search over these atoms as they are built in decision variable space.

To create this grid, as an input we will have N feasible points divided in two sets: the nondominated points (ES) and the dominated ones (DS). Using these two sets we want to create a grid to describe the set ES in order to intensify the search on it. This is, we want to describe the Pareto front in decision variable space because then we could easily use this information to generate more efficient points and then improve this initial approximation. Figure 2 shows how information in objective function space can be translated into information in decision variable space through the use of a grid.

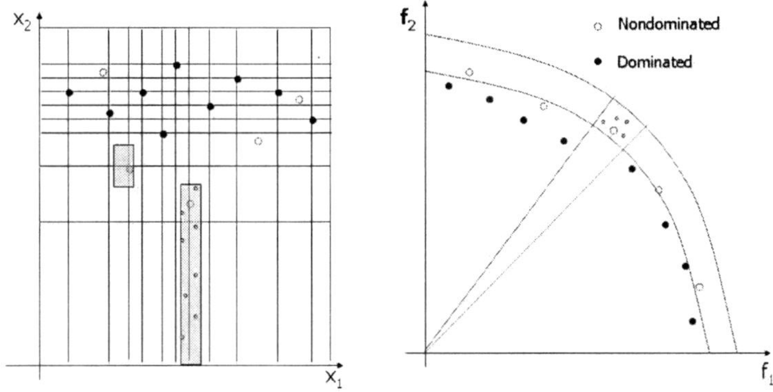

Fig. 2. Decision variable space (left) and objective function space (right)

We must note the importance of the DS set as in a rough sets method the information comes from the description of the boundary of the two sets. Then, the more efficient points provided the better. However, it is also required to provide dominated points, since we need to estimate the boundary between being dominated and being nondominated. Once this information is computed, we can simply generate more points in the "efficient side". The way in which these atoms are computed is described in Section 6.

Since the computational cost of managing the grid increases with the number of points used to create it, we will try to use just a few points. However, such points must be as far from each other as possible, because the better the distribution the points have in the initial approximation the less points we need to build a reliable grid. On the other hand, in order to diversify the search we build several grids using different (and disjoint) sets DS and ES coming from the initial approximation. To ensure these sets are really disjoint we will mark each point as explored or non-explored (if it has been used or not to compute a grid) and we will not allow repetitions. Algorithm 1 describes a Rough Sets iteration.

Algorithm 1 Rough Sets Iteration

1: Choose $NumEff$ non-explored points of ES.
2: Choose $NumDom$ non-explored points of DS.
3: Generate $NumEff$ efficient atoms.
4: **for** $i = 0$ to $NumEff$ **do**
5:   **for** $j = 0$ to $Offspring$ **do**
6:     Generate (randomly) a point new in atom i and send to ES
7:     **if** new is efficient **then**
8:       Include in ES
9:     **end if**
10:     **if** A point old in ES is dominated by new **then**
11:       Send old to DS
12:     **end if**
13:     **if** new is dominated by a point in ES **then**
14:       Remove new
15:     **end if**
16:   **end for**
17: **end for**

---

4 Differential Evolution

Differential Evolution (DE) [21, 24] is a relatively recent heuristic designed to optimize problems over continuous domains. DE has been shown to be not only very effective as a global optimizer, but also very robust producing in many cases a minimum variability of results from one run to another. DE has been extended to solve multi-objective problems by several researchers (see for example [1, 2, 11, 13, 16, 18, 22, 25]). However, in such extensions, DE has been found to be very good at converging close to the true Pareto front (i.e., for coarse-grained optimization), but not so efficient for actually reaching the front (i.e., for fine-grained optimization). Thus, we will show how these features can be exploited by our hybrid, which uses rough sets theory as a local optimizer in order to improve the spread of the nondominated solutions obtained by the MOEA adopted (which is based on differential evolution in our case).

In DE, each decision variable is represented in the chromosome by a real number. As in any other evolutionary algorithm, the initial population of DE is randomly generated, and then evaluated. After that, the selection process takes place. During the selection stage, three parents are chosen and they generate a single offspring which competes with a parent to determine who passes to the following generation. DE generates a single offspring (instead of two as a genetic algorithm) by adding the weighted difference vector between two parents to a third parent. In the context of single-objective optimization, if the resulting vector yields a lower objective function value than a predetermined population member, the newly generated vector replaces the vector with respect to which it was compared. In addition, the best parameter vector $X_{best,G}$ is evaluated for every generation G in order to keep track of the progress that is made during the minimization process. More formally, the process is described as follows:

For each vector $\overrightarrow{x_{i,G}}; i = 0, 1, 2, \ldots, N - 1.$, a trial vector $\overrightarrow{v}$ is generated using:

$$\overrightarrow{v} = \overrightarrow{x_{r1,G}} + F \cdot (\overrightarrow{x_{r2,G}} - \overrightarrow{x_{r3,G}})$$

with $r_1, r_2, r_3 \in [0, N - 1]$, integer and mutually different, and $F > 0$.

The integers r_1, r_2 and r_3 are randomly chosen from the interval $[0, N - 1]$ and are different from i. F is a real and constant factor which controls the amplification of the differential variation $(\overrightarrow{x_{r2,G}} - \overrightarrow{x_{r3,G}})$.

5 Pareto-adaptive ϵ-dominance

One of the concepts that has raised more interest within evolutionary multi-objective optimization in the last few years is, with no doubt, the use of relaxed forms of Pareto dominance that allow us to control the convergence of a MOEA. From such relaxed forms of dominance, ϵ-dominance [14] is certainly the most popular. ϵ-dominance has been mainly used as an archiving strategy in which one can regulate the resolution at which our approximation of the Pareto front will be generated. This allows us to accelerate convergence (if a very coarse resolution is sufficient) or to improve the quality of our approximation (if we can afford the extra computational cost). However, ϵ-dominance has certain drawbacks and limitations. For example: (1) we can lose a high number of nondominated solutions if the decision maker does not take into account (or does not know) the geometrical characteristics of the true Pareto front, (2) the extrema of the Pareto front are normally lost and (3) the upper bound for the number of points allowed by a grid is not easy to achieve in practice.

In order to overcome some of these limitations, the concept of $pa\epsilon$-dominance was proposed in [10]. Briefly, the main difference is that in $pa\epsilon$-dominance the hyper-grid generated adapts the sizes of the boxes to certain geometrical characteristics of the Pareto front (e.g., almost horizontal or vertical portions of the Pareto front) as to increase the number of solutions retained in the grid. This scheme maintains the good properties of ϵ-dominance but improves on its main weaknesses. In order to do this, it considers not only a different ϵ for each objective but also the vector $\epsilon = (\epsilon_1, \epsilon_2, \ldots, \epsilon_m)$ associated to each $f = (f_1, f_2, \ldots, f_m) \in R^m$ depending on the geometrical characteristics of the Pareto front. This is, the scheme considers different intensities of dominance for each objective according to the position of each point along the Pareto front. Then, the size of the boxes is adapted depending on the portion of the Pareto front that is being covered. Namely, the boxes are, for example, smaller at the extrema of the Pareto front (since these regions are normally more difficult to cover), and they become larger towards the middle portions of the front.

In [10], it is empirically shown that the advantages of $pa\epsilon$-dominance over ϵ-dominance make it a more suitable choice to be incorporated into a MOEA and therefore our decision of adopting this scheme for the work reported in this chapter.

6 The Hybrid Method: DEMORS

Our proposed approach, called DEMORS (Differential Evolution for Multi-objective Optimization with Rough Sets) [9], is divided in two different phases, and each of them consumes a fixed number of fitness function evaluations. During Phase I, our DE-based MOEA is applied for 2000 fitness function evaluations. During Phase II, a local search procedure based on rough sets theory is applied for 1000 fitness function evaluations, in order to improve the solutions produced at the previous phase. Each of these two phases is described next in more detail.

6.1 Phase I: Use of Differential Evolution

The pseudo-code of our proposed DE-based MOEA is shown in Algorithm 2 [23]. Our approach keeps three populations: the main population (which is used to select the parents), a secondary (external) population, which is used to retain the nondominated solutions found and a third population that retains dominated solutions removed from the second population.

First, we randomly generate 25 individuals, and use them to generate 25 offspring. Phase I has two selection mechanisms that are activated based on the total number of generations and a parameter called $sel_2 \in [0, 1]$, which regulates the selection pressure. For example, if $sel_2 = 0.6$ and the total number of generations is $G_{max} = 200$, this means that during the first 120 generations (60% of G_{max}), a random selection will be adopted, and during the last 80 generations an elitist selection will be adopted. In both selections (random and elitist), a single parent is selected as reference. This parent is used to compare the offspring generated by the three different parents. This mechanism guarantees that all the parents of the main population will be reference parents for only one time during the generating process. Both types of selection and recombination operators are described in [23].

Differential evolution does not use an specific mutation operator, since such operator is somehow embedded within its recombination operator. However, in multi-objective optimization problems, we found it necessary to provide

---
Algorithm 2 Phase I pseudo-code

1: Initialize vectors of the population P
2: Evaluate the cost of each vector
3: **for** $i = 0$ to G **do**
4:    **repeat**
5:       Select (randomly) three different vectors
6:       Perform crossover using DE scheme
7:       Perform mutation
8:       Evaluate objective values
9:       **if** offspring is better than main parent **then**
10:          replace it on population
11:       **end if**
12:    **until** population is completed
13:    Identify nondominated solutions in population
14:    Add nondominated solutions into secondary population
15:    Add dominated solutions into third population
16: **end for**

---

an additional mutation operator in order to allow a better exploration of the search space. We adopted uniform mutation for that sake [8].

As indicated before, our proposed approach uses an external archive (also called secondary population). In order to include a solution into this archive, it is compared with respect to each member already contained in the archive using the $pa\epsilon$-dominance grid [10]. Any member that is removed from the secondary population is included in the third population. The $pa\epsilon$-dominance grid is created once we obtain 100 nondominated solutions. If Phase I is not able to find at least 100 nondominated solutions, then the grid is created until Phase II (if during this second phase it is possible to find at least 100 nondominated solutions). The minimum number of nondominated solutions needed to create the grid is critical in several aspects:

- If we create the grid with just a few points, then the performance of the grid may significantly degrade.
- Once we create the grid, the number of points in this second population significantly decreases, and we have to ensure a minimum number of points that will be used by the Phase II.
- The behavior of the Phase II is a lot better if the grid was created during Phase I, since this ensures that the secondary population has a good distribution of solutions.

An exhaustive set of experiments undertaken by the authors indicated that 100 points was a good compromise to cover the three aspects indicated above.

The third population stores the dominated points needed for the Phase II. Every removed point from the secondary population is included in the third population. If this third population reaches a size of 100 points, a

$pa\epsilon$-dominance grid will be created in order to manage them and thus ensure a good distribution of points.

6.2 Phase II: Local Search Using Rough Sets

Upon termination of Phase I, we start Phase II, which departs from the non-dominated set generated in Phase I (ES). This set is contained within the secondary population. We also have the dominated set (DS), which is contained within the third population. It is worth remarking that ES can simply be a list of solutions or a $pa\epsilon$-dominance grid, depending on the moment at which the grid is created (if Phase I generated more than 100 nondominated solutions, then the grid will be built during that phase). This, however, does not imply any difference in the way in which the Phase II works.

Algorithm 3 Phase II pseudo-code

1: $ES \leftarrow$ nondominated set generated by Phase I
2: $DS \leftarrow$ dominated set generated by Phase I
3: eval $\leftarrow 0$
4: **repeat**
5:     $Items \leftarrow NumEff$ points $\in$ ES $\& NumDom$ points $\in$ DS
6:     Range Initialization
7:     Compute Atoms
8:     **for** i $\leftarrow 0$, Offspring **do**
9:        eval $\leftarrow$ eval $+ 1$
10:      ES $\leftarrow Offspring$ generated
11:      Add Offspring into ES set
12:    **end for**
13: **until** $1000 <$ eval

From the set ES we choose $NumEff$ points previously unselected. If we do not have enough unselected points, we choose the rest randomly from the set ES. Next, we choose from the set DS $NumDom$ points previously unselected (and in the same way if we do not have enough unselected points, we complete them in a random fashion). These points will be used to approximate the boundary between the Pareto front and the rest of the feasible set in decision variable space. What we want to do now is to intensify the search in the area where the nondominated points reside, and refuse finding more points in the area where the dominated points reside. For this purpose, we store these points in the set $Items$ and perform a rough sets iteration:

1. **Range Initialization:** For each decision variable i, we compute and sort (from the smallest to the highest) the different values it takes in the set $Items$. Then, for each decision variable i, we have a set of $Range_i$ values, and combining all these sets we have a (non-uniform) grid in decision variable space.

2. **Compute Atoms:** We compute $NumEff$ rectangular atoms centered in the $NumEff$ efficient points selected. To build a rectangular atom associated to a nondominated point $x^e \in Items$ we compute the following upper and lower bounds for each decision variable i:

   - Lower Bound i: Middle point between x_i^e and the previous value in the set $Range_i$.
   - Upper Bound i: Middle point between x_i^e and the following value in the set $Range_i$.

   In both cases, if there are no previous or subsequent values in $Range_i$, we consider the absolute lower or upper bound of variable i. This setting lets the method to explore close to the feasible set boundaries.

3. **Generate Offspring:** Inside each atom we randomly generate $Offspring$ new points. Each of these points is sent to the set ES (that, as mentioned, can be a $pa\epsilon$-dominance grid) to check if it must be included as a new nondominated point. If any point in ES is dominated by this new point, it is sent to the set DS.

7 Computational Experiments

In order to validate our proposed approach, our results are compared with respect to those generated by the NSGA-II [5], which is a MOEA representative of the state-of-the-art in the area.

The first phase of our approach uses three parameters: crossover probability (Pc), elitism (sel_2) and population size (Pop). On the other hand, the second phase uses three more parameters: number of points randomly generated inside each atom ($Offspring$), number of atoms per generations ($NumEff$) and the number of dominated points considered to generate the atoms ($NumDom$). Finally, the minimum number of nondominated points needed to generate the $pa\epsilon$-dominance grid is set to 100 for all problems.

Our approach was validated using 27 test problems, but due to space constraints, only 9 were included in this chapter: 5 from the **ZDT** set [27] and 4 from the **DTLZ** set [6]. In all cases, the parameters of our approach were set as follows: $Pc = 0.3$, $sel_2 = 0.1$, $Pop = 25$, $Offspring = 1$, $NumEff = 2$ and $NumDom = 10$. The NSGA-II was used with the following parameters: crossover rate $= 0.9$, mutation rate $= 1/\text{num\_var}$ (num_var $=$ number of decision variables), $\eta_c = 15$, $\eta_m = 20$, population size $= 100$ and maximum number of generations $= 30$. The population size of the NSGA-II is the same as the size of the grid of our approach, in order to allow a fair comparison of results, and both approaches adopted real-numbers encoding and performed 3000 fitness function evaluations per run.

In order to allow a quantitative comparison of results, we adopted the three following performance measures:

Size of the space covered (SSC): This metric was proposed by Zitzler and Thiele [28], and it measures the hypervolume of the portion of the objective space that is dominated by the set, which is to be maximized. In other words, SSC measures the volume of the dominated points. Hence, the larger the SSC value, the better.

Unary additive epsilon indicator ($I^1_{\varepsilon+}$): The epsilon indicator family has been introduced by Zitzler et al. [29] and comprises a multiplicative and additive version. Due to the fact that the additive version of ϵ-dominance has been implemented in the hybrid algorithm, we decided to use the unary additive epsilon indicator ($I^1_{\varepsilon+}$) as well. The unary additive epsilon indicator of an approximation set A ($I^1_{\varepsilon+}(A)$) gives the minimum factor ϵ by which each point in the real front R can be add such that the resulting transformed approximation set is dominated by A:

$$I^1_{\varepsilon+}(A) = inf_{\epsilon \in \mathbf{R}}\{\forall z^2 \in R \backslash \exists z^1 \in A : z^2_i \leq z^1_i + \epsilon \; \forall i\}.$$

$I^1_{\varepsilon+}(A)$ is to be minimized and a value smaller than 0 implies that A strictly dominates the real front R.

Standard Deviation of Crowding Distances (SDC): In order to measure the spread of the approximation set A, we compute the standard deviation of the crowding distance of each point in A:

$$SDC = \sqrt{\frac{1}{|A|} \sum_{i=1}^{|A|}(d_i - \overline{d_i})^2}$$

where d_i is the crowding distance of the $i-th$ point in A (see [4] for more details of this distance) and $\overline{d_i}$ is the mean value of all d_i. Nevertheless, other types of measures could be use for d_i. Now, $0 \leq SDC \leq \infty$ and the lower the value of SDC, the better the distribution of vectors in A. A perfect distribution, that is $SDC = 0$, means that d_i is constant for all i.

7.1 Discussion of Results

Table 1 shows a summary of our results. For each test problem, we performed 30 independent runs per algorithm. The results reported in Table 1 are the mean values for each of the three performance measures and the standard deviation of the 30 runs performed. The best mean values in each case are shown in **boldface** in Table 1.

It can be clearly seen in Table 1 that our DEMORS produced the best mean values in all cases. The graphical results shown in Figures 3 and 4 serve to reinforce our argument of the superiority of the results obtained by our DEMORS. These plots correspond to the run in the mean value with respect to the unary additive epsilon indicator. In all the bi-objective optimization problems, the true Pareto front (obtained by enumeration) is shown with a continuous line and the approximation obtained by each algorithm is shown with black circles. In Figures 3 and 4, we can clearly see that in the ZDT problems, the NSGA-II is very far from the true Pareto front, whereas our

DEMORS has already converged to the true Pareto front after only 3000 fitness function evaluations. The spread of solutions of our DEMORS is evidently not the best possible, but we argue that this is a good trade-off (and the performance measures back up this statement) if we consider the low computational cost achieved. Evidently, the quality of the spread of solutions is sacrificed at the expense of reducing the computational cost required to obtain a good approximation of the Pareto front.

Our results indicate that the NSGA-II, despite being a highly competitive MOEA is not able to converge to the true Pareto front in most of the test problems adopted when performing only 3000 fitness function evaluations. If allowed a higher number of evaluations, the NSGA-II would certainly produce a very good (and well-distributed) approximation of the Pareto front.

Table 1. Comparison of results between our DEMORS and the NSGA-II for the ZDT and DTLZ problems adopted. σ refers to the standard deviation over the 30 runs performed.

Function	SSC				$I^1_{\varepsilon+}$				SDC			
	DEMORS		NSGA-II		DEMORS		NSGA-II		DEMORS		NSGA-II	
	Mean	σ	Mean	σ	Mean	σ	Mean	σ	Mean	σ	Mean	σ
ZDT1	**0.852**	0.001	0.635	0.021	**0.006**	0.001	0.193	0.022	**0.008**	0.004	0.051	0.010
ZDT2	**0.794**	0.014	0.555	0.032	**0.031**	0.036	0.342	0.053	**0.033**	0.026	0.159	0.041
ZDT3	**0.788**	0.002	0.647	0.025	**0.017**	0.006	0.154	0.020	0.091	0.016	**0.073**	0.005
ZDT4	**0.993**	0.002	0.866	0.029	**0.002**	0.001	0.137	0.030	**0.011**	0.012	0.128	0.070
ZDT6	**0.899**	0.002	0.333	0.042	**0.004**	0.002	0.572	0.054	**0.016**	0.030	0.211	0.089
DTLZ1	**0.997**	0.0007	0.996	0.002	**0.023**	0.007	0.046	0.009	0.096	0.013	**0.040**	0.018
DTLZ2	**0.941**	0.0017	0.930	0.004	**0.067**	0.008	0.079	0.015	0.026	0.011	**0.007**	0.007
DTLZ3	**0.996**	0.0006	0.996	0.004	**0.042**	0.018	0.060	0.014	0.110	0.036	**0.043**	0.016
DTLZ4	0.821	0.115	**0.890**	0.032	0.352	0.078	**0.245**	0.038	0.136	0.050	**0.039**	0.010

7.2 Evaluating the Importance of Using Rough Sets

A natural question to ask regarding the use of rough sets in this case is if they really provide an aggregated value to the MOEA adopted. Some may think that the multi-objective extension of differential evolution that we adopted for the first stage of our approach is powerful enough as to converge to the Pareto front of the problems that we studied without any further help. We have argued that this is not the case, but some numerical results may be a more convincing argument. For that sake, we conducted a small experimental study in which we evaluated the outcome produced when applying only the first stage of the algorithm, and then we compared such results with respect to those generated upon applying the second stage. Table 2 shows this comparison of results. The values in **boldface** are the best mean results. By looking at Table 2, one can clearly appreciate that in most cases, and with respect to the three performance measures adopted, the use of rough sets improved (on average) the performance of the algorithm (mainly with respect to the unary additive epsilon indicator metric).

Table 2. Comparison of results between our DEMORS and the Phase 1 of our algorithm for the ZDT and DTLZ problems adopted. σ refers to the standard deviation over the 30 runs performed.

	SSC		$I^1_{\varepsilon+}$		SDC	
Function	DEMORS	Phase-I	DEMORS	Phase-I	DEMORS	Phase-I
	Mean $\quad\sigma$	Mean $\quad\sigma$	Mean $\quad\sigma$	Mean $\quad\sigma$	Mean $\quad\sigma$	Mean $\quad\sigma$
ZDT1	**0.852** 0.001	0.849 0.002	**0.006** 0.001	0.020 0.008	**0.008** 0.004	0.031 0.023
ZDT2	0.794 0.014	**0.796** 0.010	0.031 0.036	**0.016** 0.012	**0.033** 0.026	0.035 0.026
ZDT3	**0.788** 0.002	**0.788** 0.002	**0.017** 0.006	0.023 0.007	0.091 0.016	**0.073** 0.009
ZDT4	**0.993** 0.002	0.992 0.002	**0.002** 0.001	0.004 0.003	**0.011** 0.012	0.024 0.020
ZDT6	**0.899** 0.002	0.896 0.002	**0.004** 0.002	0.017 0.005	**0.016** 0.030	0.062 0.036
DTLZ1	**0.997** 0.0007	0.996 0.007	**0.023** 0.007	**0.023** 0.008	0.096 0.013	**0.019** 0.014
DTLZ2	**0.941** 0.0017	0.933 0.002	**0.067** 0.008	0.077 0.015	**0.026** 0.011	0.029 0.015
DTLZ3	**0.996** 0.0006	0.995 0.008	**0.042** 0.018	**0.042** 0.014	0.110 0.036	**0.024** 0.016
DTLZ4	**0.821** 0.115	0.7570 0.108	**0.352** 0.078	0.394 0.038	**0.136** 0.050	0.185 0.056

8 Conclusions and Future Work

We have presented a new technique to improve the results of a MOEA based on a local search mechanism inspired on Rough sets theory. The proposed approach was found to provide very competitive results in a variety of test problems, despite the fact that it performed only 3000 fitness function evaluations. Within this number of evaluations, NSGA-II, a highly competitive MOEA, is not able to converge to the true Pareto front in most of the test problems adopted. This led us to conclude that Rough Sets is a suitable tool to be hybridized with a MOEA in order to improve the local exploration around the nondominated solutions found so far. If the search engine adopted to produce a coarse-grained approximation of the Pareto front is efficient (as in our case), then a good approximation of the true Pareto front can be achieved with a low computational cost.

As part of our future work, we are interested in coupling the local search mechanisms described in this chapter to different search engines. Particularly, we are interested in exploring a hybridization with particle swarm optimization [12], which has also been found to be a very effective search engine in multiobjective optimization.

Acknowledgement. The second author acknowledges support from CONACyT through a scholarship to pursue graduate studies at the Computer Science Section of the Electrical Engineering Department at CINVESTAV-IPN. The third author acknowledges support from CONACyT project number 42435-Y.

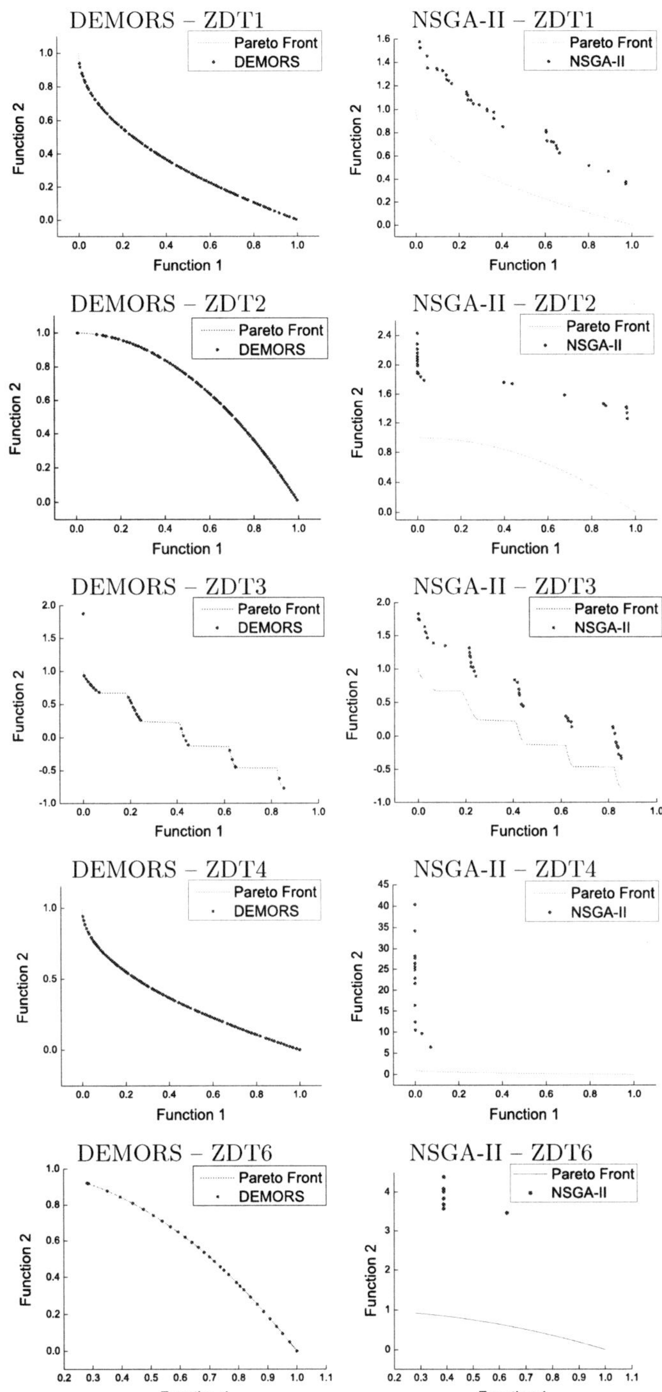

Fig. 3. Pareto fronts generated by DEMORS (left) and NSGA-II (right) for ZDT1, ZDT2, ZDT3 , ZDT4 and ZDT6

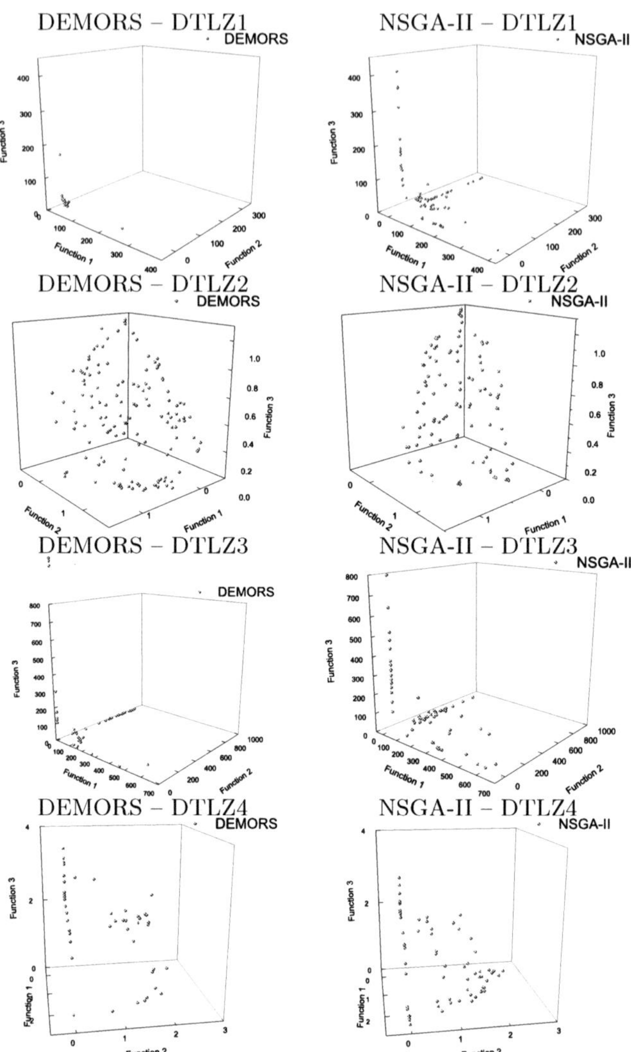

Fig. 4. Pareto fronts generated by DEMORS (left) and NSGA-II (right) for DTLZ1, DTLZ2, DTLZ3 and DTLZ4

References

1. Abbass HA (2002) The Self-Adaptive Pareto Differential Evolution Algorithm. In: Congress on Evolutionary Computation (CEC'2002), volume 1, pp. 831–836, Piscataway, New Jersey, May 2002. IEEE Service Center
2. Babu BV, Jehan MML (2003) Differential Evolution for Multi-Objective Optimization. In: Proceedings of the 2003 Congress on Evolutionary Computation

(CEC'2003), volume 4, pp. 2696–2703, Canberra, Australia, December 2003. IEEE Press

3. Coello Coello CA, Van Veldhuizen DA, Lamont GB (2002) Evolutionary Algorithms for Solving Multi-Objective Problems. Kluwer Academic Publishers, New York, May 2002 ISBN 0-3064-6762-3

4. Deb K (2001) Multi-Objective Optimization using Evolutionary Algorithms. John Wiley & Sons ISBN 0-471-87339-X

5. Deb K, Pratap A, Agarwal S, Meyarivan T (2002) A Fast and Elitist Multiobjective Genetic Algorithm: NSGA–II. IEEE Transactions on Evolutionary Computation, 6(2):182–197, April 2002

6. Deb K, Thiele L, Laumanns M, Zitzler E (2005) Scalable Test Problems for Evolutionary Multiobjective Optimization. In: Abraham A, Jain L, Goldberg R (eds), Evolutionary Multiobjective Optimization. Theoretical Advances and Applications. pp. 105–145, Springer, USA

7. Ehrgott M (2005) Multicriteria Optimization. Springer, Berlin, second edition. ISBN 3-540-21398-8

8. Goldberg DE (1989) Genetic Algorithms in Search, Optimization and Machine Learning. Addison-Wesley Publishing Co., Reading, Massachusetts, USA

9. Hernández-Díaz AG, Santana-Quintero LV, Coello Coello C, Caballero R, Molina J (2006) A new proposal for multi-objective optimization using differential evolution and rough sets theory. In: 2006 Genetic and Evolutionary Computation Conference (GECCO'2006), Seattle, Washington, USA, July 2006. ACM Press. (accepted)

10. Hernández-Díaz AG, Santana-Quintero LV, Coello Coello CA, Molina J (2006) Pareto adaptive - ϵ-dominance. Technical Report EVOCINV-02-2006, Evolutionary Computation Group at CINVESTAV, México, March 2006

11. Iorio AW, Li X (2004) Solving rotated multi-objective optimization problems using differential evolution. In: AI 2004: Advances in Artificial Intelligence, Proceedings, pp. 861–872. Springer-Verlag, Lecture Notes in Artificial Intelligence Vol. 3339

12. Kennedy J, Eberhart RC (2001) Swarm Intelligence. Morgan Kaufmann Publishers, California, USA

13. Kukkonen S, Lampinen J (2004) An Extension of Generalized Differential Evolution for Multi-objective Optimization with Constraints. In: Parallel Problem Solving from Nature - PPSN VIII, pp. 752–761, Birmingham, UK, September 2004. Springer-Verlag. Lecture Notes in Computer Science Vol. 3242

14. Laumanns M, Thiele L, Deb K, Zitzler E (2002) Combining convergence and diversity in evolutionary multi-objective optimization. Evolutionary Computation, 10(3):263–282, Fall 2002

15. Lin TY (1996) Special issue on rough sets. Journal of the Intelligent Automation and Soft Computing, 2(2):*, Fall 1996

16. Madavan NK (2002) Multiobjective Optimization Using a Pareto Differential Evolution Approach. In: Congress on Evolutionary Computation (CEC'2002), volume 2, pp. 1145–1150, Piscataway, New Jersey, May 2002. IEEE Service Center

17. Miettinen KM (1999) Nonlinear Multiobjective Optimization. Kluwer Academic Publishers, Boston, Massachusetts

18. Parsopoulos KE, Taoulis DK, Pavlidis NG, Plagianakos VP, Vrahatis MN (2004) Vector Evaluated Differential Evolution for Multiobjective Optimization. In:

2004 Congress on Evolutionary Computation (CEC'2004), volume 1, pp. 204–211, Portland, Oregon, USA, June 2004. IEEE Service Center

19. Pawlak Z (1982) Rough sets. International Journal of Computer and Information Sciences, 11(1):341–356, Summer 1982

20. Pawlak Z (1991) Rough Sets: Theoretical Aspects of Reasoning about Data. Kluwer Academic Publishers, Dordrecht, The Netherlands ISBN 0-471-87339-X

21. Price KV, Storn RM, Lampinen JA (2005) Differential Evolution. A Practical Approach to Global Optimization. Springer, Berlin, Germany ISBN 3-540-29859-6

22. Robič T, Filipič B (2005) DEMO: Differential Evolution for Multiobjective Optimization. In: Coello Coello CA, Hernández Aguirre A, Zitzler E (eds), Evolutionary Multi-Criterion Optimization. Third International Conference, EMO 2005, pp. 520–533, Guanajuato, México, March 2005. Springer. Lecture Notes in Computer Science Vol. 3410

23. Santana-Quintero LV, Coello Coello CA (2005) An algorithm based on differential evolution for multi-objective problems. International Journal of Computational Intelligence Research, 1(2):151–169

24. Storn R, Price K (1997) Differential Evolution - A Fast and Efficient Heuristic for Global Optimization over Continuous Spaces. Journal of Global Optimization, 11:341–359

25. Xue F, Sanderson AC, Graves RJ (2003) Pareto-based Multi-Objective Differential Evolution. In: Proceedings of the 2003 Congress on Evolutionary Computation (CEC'2003), volume 2, pp. 862–869, Canberra, Australia, December 2003. IEEE Press

26. Zadeh LA (1965) Fuzzy sets. Information and Control, 8(1):338–353, Fall 1965

27. Zitzler E, Deb K, Thiele L (2000) Comparison of Multiobjective Evolutionary Algorithms: Empirical Results. Evolutionary Computation, 8(2):173–195, Summer 2000

28. Zitzler E, Thiele L (1999) Multiobjective Evolutionary Algorithms: A Comparative Case Study and the Strength Pareto Approach. IEEE Transactions on Evolutionary Computation, 3(4):257–271, November 1999

29. Zitzler E, Thiele L, Laumanns M, Fonseca CM, da Fonseca VG (2003) Performance assessment of multiobjective optimizers: an analysis and review. IEEE Transactions on Evolutionary Computation, 7(2):117–132, Summer 2003

How to Acquire and Structuralize Knowledge
for Medical Rule-Based Systems?

Beata Jankowska[1] and Magdalena Szymkowiak[2]

[1] Poznań University of Technology, Institute of Control and Information
Engineering, Pl. Sklodowskiej Curie 5, 30-965 Poznań,
`beata.jankowska@put.poznan.pl`
[2] Institute of Mathematics, ul. Piotrowo 3a, 60-965 Poznań, `mszymkow@amu.edu.pl`

Summary. The intention of a medical expert system is to help doctors make right
diagnostic and therapeutic decisions concerning, sometimes not very well-known to
them, diseases. This expert system needs a high quality knowledge base. In order to
design such a base one has to reach sources containing knowledge that is current,
rich and based on reliable medical experiments. At the same time, due to various
formats of this knowledge storing, its acquisition and structuralization to the form
required by expert systems is not an easy task. Focusing our attention on medical
rule-based systems, we propose the algorithms and tools that will be useful while
designing such a knowledge base.

1 Introduction

The number of modern diseases is a great one. Furthermore, the disease's
process is different depending on region, season, population specificity and a
patient himself. General practitioner (GP) is this physician whom the patient
visits at the beginning. That is why, GP has an important and not easy role
to play. He or she is required to make initial diagnosis, recommend some
pharmacotherapy and direct the patient to a specialist doctor. Obviously, any
specialized and effective medical expert system would be helpful in GPs' work.

However, the following questions arise:

– is it possible to structuralize heterogeneous medical knowledge so as to
  obtain some homogeneous, independent of its acquisition source, form?
– is it possible to encode medical knowledge, without any loss, in a compact
  and clear form – also from non-specialist's point of view?
– can medical knowledge be expressed in form of production rules that make
  it possible to reason automatically?

A lot of old [12, 15, 19] and modern [16, 20] medical expert systems have
proved that the above questions can be answered "yes". In this paper, we

B. Jankowska and M. Szymkowiak: *How to Acquire and Structuralize Knowledge for Medical
Rule-Based Systems?*, Studies in Computational Intelligence (SCI) **102**, 99–116 (2008)
`www.springerlink.com`

propose a systematic method of medical knowledge transformation into the form required by rule-based systems. The outline of this method has been already presented in [9].

2 Knowledge Base of Medical Expert System

A medical expert system is primarily expected to give essential and reliable opinions. The reaction time is only of secondary importance. Therefore, to guarantee good medical system performance, we need a high quality knowledge base. Designing such a base is not an easy task, mainly – because of knowledge uncertainty. The most important design decision concerns the method of knowledge representation and – associated with it – the method of approximate reasoning. While selecting these methods, the designers should take account of the domain of system application.

The most willingly used methods of uncertain knowledge representation are rules with uncertainty and various non-symbolic methods, such as belief networks and neural networks. Non-symbolic methods, especially belief networks, are very suitable for specifying diagnostic procedure. This procedure resolves itself into a series of simple tasks (medical interviewing, medical examination, interpreting laboratory tests). Their results remain in known, empirically confirmed numerical relations. Understanding these relations, a doctor is able to make reliable diagnosis, also when the diagnostic information is incomplete.

For a change, it seems that facts and rules are the most convenient in order to express the cause-effect relations between a diagnosed disease, its symptoms and pharmacotherapy versus expected effects of its application. Establishing of these relations is being done on the grounds of medical evidence, especially – this one comprising results of clinical research.

In our considerations we deal with the expert system aiding both diagnostic and treatment decisions regarding a class of diseases. So, we set our mind on representing medical knowledge by means of facts and rules with uncertainty. In consequence, the knowledge base of our system will consist of:

– temporary facts, specifying information about the state of health of a patient,
– permanent rules, representing medical knowledge that concerns diagnostics and/or treatment of various diseases and that was acquired experimentally.

To illustrate the form and meaning of knowledge stored in this knowledge base, let us give a simple example. We shall introduce into the base a series of facts which characterize the present state of health of a patient John Smith. Each fact is given a probability mass PM-f that reflects its reliability:

f_1: (temperature 39.0) PM-f 1.0
f_2: (generally feeling unwell) PM-f 0.8
f_3: (lack of appetite) PM-f 1.0

```
f₄: (bad headache) PM-f 1.0
f₅: (abdominal pain) PM-f 0.8
f₆: (strong catarrh) PM-f 1.0
```

Similarly, each rule's conclusion is given a probability mass PM-c and each rule is given a global probability mass g-PM that reflects its reliability as a whole. Let us assume that the knowledge base contains the following diagnostic rule at the moment:

```
r₁: it happens g-PM 0.75 :
       if      (temperature ≥ 38.5) and
               (generally feeling unwell) and
               (bad headache) and
               (strong catarrh)
       then    (infection of upper respiratory tracts) PM-c 0.6
               (inflammation of sinuses) PM-c 0.2
```

Let us assume that the final probability p_{ij} of the j-th conclusion of the rule r_i is calculated using the formula:

$p_{ij} = (\text{g-PM}_i) \cdot \min\{(\text{PM-f}_{ik}) : f_{ik}$ is the premise (fact) from the rule $r_i\} \cdot (\text{PM-c}_{ij})$.

Then for the rule r_1 we can advance, with final probability $p_{11} = 0.75 \cdot 0.8 \cdot 0.6 = 0.36$, a hypothesis of "infection of upper respiratory tracts" in John Smith and, with final probability $p_{12} = 0.75 \cdot 0.8 \cdot 0.2 = 0.12$, a hypothesis of "inflammation of sinuses".

Similarly, if the knowledge base contains the treatment rule r_2 illustrating effects of paracetamol:

```
r₂: it happens g-PM 0.6 :
       if      (37.5 < temperature ≤ 39.0) and
               (bad headache) and
               (catarrh) and
               (paracetamol in a due dosage for two days)
       then    (temperature ≤ 37.5) PM-c 0.8
               (reduction in headache severity) PM-c 0.9
```

then, with final probability $p_{21} = 0.6 \cdot 1.0 \cdot 0.8 = 0.48$, we can advance a hypothesis of "temperature ≤ 37.5°C" and, with final probability $p_{22} = 0.6 \cdot 1.0 \cdot 0.9 = 0.54$, a hypothesis of "reduction in headache severity" as a result of taking the medicine for two days.

Each of the above rules consists of:

- premises, specifying constrains of its using (e.g. patient's age or sex, results of his or her laboratory tests, intensity of symptoms of the disease in the patient, as well as detailed dosage of the medicine taken),
- conclusions, formulating hypotheses about the present (diagnostic rule) or future (treatment rule) patient's state of health (e.g. kind and/or intensity level of disease, risk of death, necessity of hospitalization,possibility of

replacing a toxic medicine with a less aggressive one, or change of the disease's course from persistent to intermittent one),
- probability mass of conclusions PM-c, specifying reliability of conclusions (under the constraint that rule's premises are fulfilled strictly),
- a global probability mass of the rule g-PM, meaning the rule's importance compared with the other rules enclosed in the knowledge base.

Farther on we mention the most important sources of medical knowledge and analyze their usefulness for designing rules with uncertainty discussed above.

3 Sources for Medical Knowledge Acquisition

The basic sources of medical knowledge are: traditional – textbooks, medical journals and individual patients' files from consulting rooms, as well as more modern ones – electronic patients' files from hospitals and clinics and electronic registers of clinical trials.

The textbooks are reliable source of basic knowledge. However, because of their wide application, they rarely cover specialized knowledge, that is indispensable from diagnostic and therapeutic points of view. Moreover, the progress of medical sciences is so quick that laborious publishing procedures make textbook knowledge outdated in a short time.

The current, specialized knowledge is presented in medical journals. Some electronic bases of journals, for example *Best Evidence* and *Medline*, are kept and available via the Internet. Searching such bases is quick, also, knowledge they cover is of huge cognitive value. The only, nevertheless essential problem is that articles, written in natural language, cannot be processed automatically with the use of well known syntax analysis methods.

Let us consider in more details the other sources of medical knowledge that we mentioned above.

3.1 Individual Patients' Files from Consulting Rooms

An individual patient's file includes information about a patient and a course of his or her illness over many years, and not just a few weeks.

During this period, the doctor, taking permanent care of the patient, has possibility to observe not only short-term but also long-term effects of various pharmacotherapies. Unfortunately, due to a specific for each doctor – not subjected to any norms – way of recording the observations, the use of this source of knowledge, despite a big amount of information it contains, seems to be very difficult.

Table 1. Individual patient's file from consulting room

<table>
<tr><td colspan="2" align="center">Consultation Report</td></tr>
<tr><td>Dictated by:</td><td>A.M., GP</td></tr>
<tr><td>Date:</td><td>March 29, 2006</td></tr>
<tr><td>Patient:</td><td>Jack Novak</td></tr>
<tr><td>Sex:</td><td>Male</td></tr>
<tr><td>Birthday:</td><td>February 1, 2000</td></tr>
<tr><td colspan="2">History of Disease
Hospitalizations: three times in 2000 (March, April, July) – airway obstruction, bronchitis, once in 2002 (May) dyspnoea, once in 2003 (March) pneumonia.
Medicines: antibiotic, short-acting theophylline preparation, short-acting beta2-agonist, inhaled anticholinergic, inhaled glucocorticosteroid (GCS), intravenous GCS, inhaled cromone.</td></tr>
<tr><td colspan="2">Physical Exam
   – lungs: recurring episodes of wheezing, intensive cough,
     prolonged expiration (symptoms are more intensive at night),
   – skin: erythematous rash, palm surface – eczema.</td></tr>
<tr><td colspan="2">Assessment
   – bronchial asthma,
   – food allergy.</td></tr>
<tr><td colspan="2">Plan
To keep control over the course of asthma – administer permanently:
   – inhaled GCS – Budesonid mitte (2 puffs twice a day),
   – inhaled cromone – Cropoz (2 puffs 3 times a day).
To diminish the severity of nagging symptoms – administer:
   – short-acting beta2-agonist – Berodual (2 puffs twice a day),
   – anticholinergic – Atrovent (2 puffs 3 times a day).
To avoid food allergy – administer:
   – Nutramigen (special anti-allergic milk).
 Signed by: A.M., GP on March 29, 2006 </td></tr>
</table>

A piece of consultation report is exemplified in Table 1. It gives us information about a child (Jack Novak) suffering from the persistent asthma disease.

3.2 Electronic Patients' Files from Hospitals and Polyclinics

Knowledge enclosed in individual patients' files is substantial. However, due to heterogeneous format of its representation, also this knowledge is not suitable for automatic processing with the use of well known syntax analysis methods. Such a possibility appears together with electronic standard records that are written in some universal format. The proposal of such a format was put forward by the Health Level Seven (HL7) organization. The general aim of standardization is to facilitate exchange, management and integration of electronic health care information [1].

HL7 works with the XML technology [3]. OBX (observation/result segment) is the smallest indivisible unit of HL7 report. It specifies results of physical exams, laboratory tests, X-ray examinations, tomography and the like. Having connected OBX segments, we obtain a protocol of the patient's course of treatment and/or evaluation. Next, combining numerous protocols altogether, we acquire knowledge about the compliance, safety, and clinically essential outcomes of the treatment under consideration. As a consequence, we can estimate the treatment effectiveness and compare one treatment approach with another.

A detailed form of HL7 standard records can be found, for example, on the website [6]. The mentioned below Table 2 contains nothing but an exemplary series of segments with: message header (segment MSH), personal data of our Jack Novak (segment PID), a description of his stay in hospital in March 2003 (segment PV1), an order to undergo some laboratory test during this hospitalization (segments ORC and OBR) and an analytical result of this test (segments OBX) [7].

If the whole (most of) medical documentation was recorded in HL7 stan-

Table 2. HL7 report – a single OBX segment

```
MSH| ^~ \&|LAB1||DESTINATION||20030305083000||ORU^R03|LAB1003929
PID|||3131313||NOVAK^JACK|||
PV1||I|
ORC|RE
OBR||A485388^OE|H29847^LAB1|BLOOD CULTURE|||
OBX||FT|SDES^SOURCE||BLOOD-RAPID||
OBX||FT|EXAM^MICROSCOPIC||GRAM POSITIVE COCCI IN GROUPS||
```

dard, it could be processed statistically and – in automatic or semi-automatic way – transformed to the form of rules with uncertainty. To achieve this, two essential difficulties should be overcome. Firstly, the standard HL7 should become general. To achieve this purpose, a specialized world medical network should be created. However, it means as well high expenditure of time, money and effort as resistance of most of hospital manager staff.

The second difficulty results from personal data protection. In order to have insight into personal medical evidence, a written assent of the Commission on Medical Ethics is needed. It is common knowledge that the Commission gives such an assent unwillingly. Meanwhile, obtaining the rules on the basis of small medical documentation generates doubts about rules' reliability.

3.3 Registers of Clinical Trials

Electronic, available via the Internet, registers of clinical trials seem to constitute this valuable source of knowledge that is helpful while establishing the therapy. The registers include results of intentional clinical research, made to evaluate effectiveness of various pharmacotherapies. In order to standardize this evaluation, clinically essential outcomes are appointed. They reflect (directly or indirectly) state of health and general feeling of patients. It is easy to search these electronic registers out, and the usefulness of information they cover can be never overrated.

Each register is a protocol of an experiment carried out on two groups of patients chosen at random:

- a treatment group (TG), that is given an intentional medical intervention,
- a control group (CG), that takes a regular cure.

Table 3. Protocol of clinical trials – experiment constraints

Parti-cipants	N = 271
	AGE – 2-18 years
	BASELINE SEVERITY – PEF$<0.5 \cdot$PEF$_{max}$ or asthma score $=12$–15
Inter-ventions	**Protocol**
	– Fixed: 60 minutes
	– Observation period: up to 248 minutes
	– Multiple doses
	Treatment Group (TG)
	–Albuterol(Beta2-Agonist)2.5mg($<$20kg)/5.0mg($\geq$20kg)q20minutesx3&
	IB (Anticholinergic) 500 mcg at 20 and 40 minutes
	Control Group (CG)
	–Albuterol 2.5mg ($<$ 20kg)/ 5.0mg ($\geq$20kg) q20 minutes x3&
	Saline (placebo) at 20 and 40 minutes
	Co-intervention (other medicines used during the study)
	– Systemic GCS – all received oral steroids at 20 minutes
	Device – nebulizer

The basic protocol's components are the following ones:

- collective characteristics of participants of the experiment,
- description of medical constrains put on the experiment course (treatment duration, names of drugs, drug dosing methods, number of daily doses and dose amount),
- numerical results of the experiment.

The 6-year old patient Jack Novak, we analyzed before, could take part in the experiment to study the influence of administering modern anticholinergics to children with persistent bronchial asthma in a case of rapid extracerbation, on the decrease in number of necessary hospitalizations. The components of the protocol of this experiment are shown in Tables 3 and 4, including: Table 3 – description of experiment constrains and Table 4 – experiment results.

The superiority of clinical trials registers over the other sources of knowledge results from the fact that they store the real evidence comprising not

Table 4. Protocol of clinical trials – experiment results (the meaning of parameters shown will be explained in Section 5)

Comparison:	Anticholinergic (multiple doses) + Beta2-Agonist vs Beta2-Agonist Alone		
Outcome:	Hospital admission		
Co-intervention:	GCS during study		

Study	Treatment Group $p_1 = A_1/N_1$	Control Group $p_0 = A_0/N_0$	Relative Risk	95% CI
Qureshi 1997	9/36	14/31	0.55	(0.28; 1.10)
Qureshi 1998 (mod)	8/79	9/84	0.95	(0.38; 2.33)
Qureshi 1998 (sev)	51/136	71/135	0.71	(0.54; 0.93)
Zorc 1999 (mod)	18/98	25/96	0.71	(0.41; 1.21)
Zorc 1999 (sev)	7/22	12/29	0.77	(0.36; 1.63)
Zorc JJ 1999 (mil)	6/60	4/57	1.43	(0.42; 4.79)
Total	99/431	135/432	0.73	(0.60; 0.91)

single but numerous patients. Such evidence can be transformed into the form of rules with uncertainty. Assuming that the intentional experiments were carried out honestly (regardless of commercial interests), we can obtain rules of high reliability.

Moreover, evidence from clinical trials registers is not subjected to any special protection. So, its availability is much bigger than this one from individual patients' files from hospitals or consulting rooms.

In the light of the above considerations, we can grant clinical trials registers the highest value in the information quality scale. Farther on, we will focus our attention on the problems of acquisition and structuralization of knowledge coming from specifically this source. We will concentrate on using knowledge that is necessary for therapeutic purposes.

4 Preliminary Knowledge Structuralization

The first step to transform the protocol to the rule with uncertainty is about elaborating the constraints put on the experiment course. The basic constraint refers to the medicines used – their names, dosing method (inhaled, oral – in tablets or syrup, intramuscular, intravenous), as well as the number of daily doses and dose amount. The next constraints refer to the participants of the experiment. They usually concern: sex and age, the disease diagnosed and the stage of its progress, as well as intensity of some symptoms of the disease before starting the therapy. While transforming, the constrains should be given a form of simple formulae, and next they should be connected by means of conjunction connectives to obtain a formula-premise.

Next, analogously, we elaborate the description of clinically essential outcomes that have been tested during the experiment. The outcomes – risk of death, necessity of hospitalization, increase or recession levels of the symptoms of the disease after the pharmacotherapy – reflect indirectly the influence of the medicine taken on the state of health and general feeling of patients. Each outcome description should be transformed to a simple formula. The obtained formulae-conclusions, together with the previously obtained formula-premise, can be now used to specify a few simple rules with uncertainty (recommended solution) or, after connecting by means of conjunction connectives – to specify one, complex rule with uncertainty.

If two (or more) different protocols report on similar experiments (the experiments carried out in order to test the main and side effects of the same pharmacotherapy), then such protocols can be joined to form a whole. This resultant protocol reports on a virtual experiment covering a large group of people. Without detailed considerations, we shall only reveal that the constraints of the resultant protocol are formed by adding up the corresponding constraints from all the component experiments, whereas the outcomes – by adding up the corresponding outcomes tested during these experiments. It is an usual case that the component experiments are carried out under the constraints that are slight different, one from the other. So, it might happen that the formula-premise will be a full-domain specification. Also, essential discrepancies between the component experiments results could, after adding up, give full-domain outcomes. In consequence, in the rule with uncertainty obtained, the premise or some conclusions could be the truth values. In an extreme case, we could obtain a rule which is an usual fact, or which is worthless.

By the way we shall notice that whereas the process of transforming protocols into rules with uncertainty should be carried out manually, the pre-selection of electronic materials can be performed automatically, using the well known syntax analysis methods [2].

5 Advanced Knowledge Structuralization

The main protocol component is numerical juxtaposition of the experiment results. The most popular form of their specification is a contingency table [5] (see Table 5). The rows of the table contain the results obtained in groups of patients, respectively: the treatment group (TG) and the control group (CG).

Table 5. Contingency Table

	Outcome		
	Yes	No	Totals
TG	A_1	B_1	N_1
CG	A_0	B_0	N_0

Alternatively, the results – having done further transformation – can be also presented by means of well-known statistical parameters (see Table 4). Let us analyze the semantics of the most frequently used parameters and investigate the relationships between these parameters and probability mass of conclusions of the rule with uncertainty designed.

5.1 Relative Risk and Other Statistical Parameters

In the protocols of clinical trials, the most frequently used statistical parameters are the following ones [4, 8]:

- risk of symptoms in the treatment group: $p_1 = \frac{A_1}{N_1}$;
- risk of symptoms in the control group: $p_0 = \frac{A_0}{N_0}$;
- Absolute Risk Reduction: $ARR = |p_0 - p_1|$;
- Relative Risk: $RR = \frac{p_1}{p_0}$;
- Relative Risk Reduction: $RRR = 1 - RR = \frac{p_0 - p_1}{p_0}$;
- Number Needed to Treat – number of patients who have to undergo the treatment in order to avoid (or provoke) one, tested outcome: $NNT = \frac{1}{ARR}$.

Discovering the meaning of evidence stored in protocols of clinical trials is not an easy task. Most often, we should consider the value of not just one, but of a few statistical parameters. We should realize the relationships that hold between them. To this end, we need some knowledge as well from medicine, as from statistics.

It seems that from all the mentioned statistical parameters, the most important is Relative Risk RR (see Table 4). This parameter informs us, how many times the risk of symptom presence is smaller ($RR < 1.0$) or bigger ($RR > 1.0$) in the treatment group, than in the control group. The notion of "risk" is not always understood literary. The symptoms studied can be of negative character (e.g. increase in frequency of necessary hospitalizations), or of positive one (e.g. increase in number of days at work). $RR = 1.0$ means that the risk of symptoms presence is the same in both experimental groups. If so, the medical intervention does not have any influence on the symptoms.

However, let us remark that Relative Risk RR (analogously to the other statistical parameters) is determined on the basis of tests within a small, representative group of people. Therefore, the question is whether this result can be extrapolated on the whole population of people who conform the experiment's criteria. Statistics answers this question with "yes" and offers two different methods of such an extrapolation: the hypothesis test method and the confidence interval method. The new, referring to a wide population, relative risk is denoted with rr.

Let us focus our attention on the estimation method that calculates Confidence Interval CI, including – with given, high probability $1 - \alpha$ – a value of the tested parameter rr. In statistical research we usuallyassume $1 - \alpha = 0.95$;

then Confidence Interval is called 95% CI. From an analytical point of view, both position of CI and its span are important. If 1.0 belongs to CI for rr, then the result is not statistically significant. The span s of CI shows the precision of the parameter value: the shorter CI (smaller s) the higher precision.

The following Katz's formula [4] can be used to estimate CI including – with probability $1 - \alpha$ – the $\ln rr$ value:

$$\ln RR + u_{\frac{\alpha}{2}} \cdot se_{\ln rr} < \ln rr < \ln RR + u_{1-\frac{\alpha}{2}} \cdot se_{\ln RR}$$

(the standard error $se_{\ln RR}$ stands here for $\sqrt{\frac{1-p_1}{N_1 \cdot p_1} + \frac{1-p_0}{N_0 \cdot p_0}}$, and $u_{\frac{\alpha}{2}}$, $u_{1-\frac{\alpha}{2}}$ – for critical values of normal distribution $N(0,1)$ of respective orders).

As we can see, the span s of CI depends on the standard error value $se_{\ln RR}$ – the smaller error the shorter interval. Then the standard error value depends on the sample size – the number of people in the treatment and control groups. In consequence, to obtain a short CI, both tested groups of patients should be large enough.

5.2 Determining Reliability of Conclusions

As it was said before, none of the statistical parameters mentioned is complete enough to give us precise information about the effectiveness of medical intervention considered with reference to a wide population of people who conform to the experiment criteria. Thus, none of the parameters is complete enough to calculate the probability mass of conclusions of the rule with uncertainty designed. To eliminate this difficulty, we put forward yet another statistical parameter and give it a name of the expected effectiveness $ee_{1-\alpha}$. This parameter is to reflect in one number both: the known effectiveness of medical intervention carried out on a small group of patients, and the belief that the next such intervention will have – with probability $1 - \alpha$ – the same effectiveness as this one considered. All in all, the expected effectiveness $ee_{1-\alpha}$ must depend on:

- Relative Risk RR, calculated with reference to the group of participants of the experiment,
- the span s of CI including – with probability $1 - \alpha$ – relative risk rr for the whole population of people conforming to the experiment constrains.

Calculating the $ee_{1-\alpha}$ value is rational only when CI does not include 1.0 (the result must be statistically significant). Precisely, in a case of $RR < 1.0$ (equivalent to decrease in frequency of symptoms tested during the experiment), the expected effectiveness $ee_{1-\alpha}(-)$ of avoiding the symptoms can be determined by the formula:

$$ee_{1-\alpha}(-) = \frac{1 - RR}{1 + \frac{s}{2}}(1 - \alpha).$$

In a case of RR > 1.0 (equivalent to increase in frequency of symptoms tested during the experiment), the expected effectiveness $ee_{1-\alpha}(+)$ of multiplying the symptoms can be determined by the formula:

$$ee_{1-\alpha}(+) = \frac{1 - \frac{1}{RR}}{1 + \frac{s}{2}}(1 - \alpha).$$

The parameter proposed can be used directly as the probability mass of conclusions PM-c of the rule with uncertainty designed.

5.3 Determining Reliability of Rule as a Whole

While designing a rule with uncertainty, we should determine not only probability mass of its conclusions, but also probability mass of the rule as a whole (the global probability mass g-PM). This factor declares the rule's reliability and, consequently, the rule's priority. Therefore, it can influence the reasoning process – including its final result [10, 11]. The considered factor depends on many parameters of the medical experiment. The most important of them are [14]:

- the experiment methodology (the best one is double-blind test),
- the number of participants of the experiment (it should not be smaller than 300 individuals),
- the number of constraints to qualify patients for the experiment (age, sex, weight, past diseases, medicines taken, etc.),
- the constraints imposed on the course of recommended pharmacotherapy (the definite medicine versus a medicine from the list, the rigorously determined dose amount versus a dose from some range, etc.).

While estimating reliability of the rule as a whole, other parameters of the experiment may also be worth considering:

- the number of hospitals in which the experiment was carried out,
- duration of the experiment (desirably not shorter than 12 months),
- differentiation between patients according to their age,
- extra-medical information about patients (family background, level of education, type of work, etc.),
- necessity for carrying out the experiment.

If the above information is accessible, the global probability mass can be calculated by means of a neural network that has been previously designed and learned.

6 Designing the Knowledge Base of System RiAD

Let us illustrate our considerations by means of an actual application. It refers to the bronchial asthma disease – one of the most common and onerous diseases of the XXI century.

The quoted examples consider the piece of real medical evidence found in electronic, available via the Internet, bases. The evidence is first processed statistically, and then it is transformed into the form of rules with uncertainty. The obtained rules can be included in the knowledge base of the rule-based expert system RiAD (Relief in Asthma Disease), that is being designed by a group of students interested in applications of artificial intelligence in medicine [13].

6.1 Bronchial Asthma Disease and Its Treatment

Bronchial asthma is a chronic inflammatory disorder of airways, caused by the immune system dysfunction [21]. The immune system of an asthmatic person identifies some air components as dangerous ones (saprophytes, mould, pollen). Then, numerous antibodies are produced to prevent the imaginary threat. Acute and frequent inflammatory states are the reasons for bronchial hyperresponsiveness and obstruction. Next, the obstruction causes a limited flow of air through the airways and leads to recurring episodes of wheezing, shortness of breath, chest tightness, and cough, all of which usually occur at night and in the early morning, as well as after strenuous physical exertion.

Asthma is recognized on the basis of subjective symptoms mentioned above. Also, in the laboratory conditions, we can measure a quantitative limitation of lungs' functioning. This can be expressed, among others, with the intensive Forced Expiratory Volume in 1st second FEV_1 and Peak Expiratory Flow PEF. There are four degrees of asthma severity: intermittent, mild persistent, moderate persistent and severe persistent.

Periodical asthma exacerbations cause more frequent visits of patients in consulting rooms, more frequent hospitalizations and a bigger risk of death. In consequence, the quality of patient's life deteriorates. Moreover, treatment and social costs (e.g. due to absences at work) increase. Thus, one of the most important goals of asthma treatment is preventing its exacerbations. Many years of experience have shown that, although asthma is incurable, in most cases we can gain and keep control over its course. Then chronic symptoms are minimized, among others, disease exacerbations are rare, there is a smaller demand for drugs relieving of cough and wheezing, and there is a little PEF and FEV_1 variability in 24-hours (PEF_{var} and $FEV_{1,var}$ are slight).

The most effective drugs reducing the frequency of asthma exacerbations are inhaled glucocorticosteroids (GCS). Other anti-asthmatic drugs, administered instead of or together with GCS, are: long-acting beta2-agonists, antileukotrienes, anti-bodies IgE, long-acting theophylline preparations and cromones. Beside regular administering drugs that control the course of the disease, in an emergency, to diminish severity of nagging symptoms, more drugs have to be used, such as: short-acting beta2-agonists, inhaled anticholinergics, or short-acting theophylline preparations.

The drugs mentioned above can be taken both singly or in combinations. Their chemical constitution and doses are matched with the degree of asthma severity, as well as with the actual patient's complaints.

6.2 Designing of Therapeutic Rules – 1st Example

Now, we shall have a closer look at the contents of a protocol of clinical trials which can be found in the electronic base of *The Cochrane Library*. The register, prepared by Plotnick and Ducharme [17], includes research results concerning influence of inhaled anticholinergics combined with beta2-agonists and GCS in children suffering from persistent asthma. In both the treatment and control groups, children were administered beta2-agonists and GCS, and, additionally, in the treatment group – anticholinergics.

The experiment was carried out on six independent groups of patients. The results covered several outcomes, including the one that is interesting for us – hospitalization necessity factor. The evidence comprising the factor studied is presented in Table 4.

Let us quote once again the values of Relative Risk RR and 95%CI for rr obtained in six component experiments (the last column in Table 4):

1. $RR = 0.55$,   95%CI for rr – (0.28; 1.10);
2. $RR = 0.95$,   95%CI for rr – (0.38; 2.33);
3. $RR = 0.71$,   95%CI for rr – (0.54; 0.93);
4. $RR = 0.71$,   95%CI for rr – (0.41; 1.21);
5. $RR = 0.77$,   95%CI for rr – (0.36; 1.63);
6. $RR = 1.43$,   95%CI for rr – (0.42; 4.79).

As we shall see, all the experiments, except the last one, show that administration of anticholinergics results in decreased hospitalization necessity ($RR < 1.0$). However, we should notice that from all the results obtained, the one only, referring to the 3rd component experiment, is statistically significant (CI for rr does not include 1.0).

After having added the partial results up, we get the following total one (the last row in Table 4): $RR = 0.73$ and 95%CI for rr equal to (0.60; 0.91). Note that the obtained result is statistically significant, and the total CI is shorter in comparison to each component CI.

Having done further calculations, we can finally conclude that the influence of anticholinergics on reducing hospitalization necessity reaches the expected effectiveness $ee_{0.95}(-) = 0.22$.

In order to transform the considered protocol into a rule with uncertainty, we should first elaborate the constraints put on the course of the experiment. Their informal specification for the 3rd component experiment is presented in Table 3. Having elaborated the constraints from all of six component experiments, we add them up and obtain the following formula-premise of the rule designed:

(1 $\leq$ age $\leq$ 18) and

```
(persistent asthma disease) and
(PEF < 0.7·PEFmax) and
(inhaled anticholinergics <multiple doses> + beta2-agonists +
 inhaled GCS)
```

Analogously, we should elaborate the specification of the outcome considered. As a result we obtain the following formula-conclusion:

```
(no necessity of hospitalization)
```

Finally, we should determine global probability mass g-PM of the rule considered. We must remember that g-PM decides about rule's priority, and so, about its activation while reasoning. A low g-PM value can even prevent putting the rule into the agenda.

The global probability mass g-PM was conclusively calculated by means of a feedforward neural network [13]. The network, of the architecture 10-8-6-2-1 (three hidden layers), was designed with help of 7SPHINX 4.1 environment [22]. It was trained on 250 vectors, of the form consulted with doctors and statistics experts.

Having determined the value g-PM = 0.62 (and earlier – the value PM-c equal to $ee_{0.95}(-) = 0.22$), now we can design, for the Plotnick and Ducharme register presented before, the following rule with uncertainty:

```
r3: it happens g-PM 0.62:
      if     (1 ≤ age ≤ 18) and
             (persistent asthma disease) and
             (PEF < 0.7·PEFmax) and
             (inhaled anticholinergics <multiple doses> +
             beta2-agonists + inhaled GCS)
      then   (no necessity of hospitalization) PM-c 0.22
```

Knowing that the rule's premise is strictly fulfilled (PM-f = 1), we can calculate a final probability of the hypothesis of "no necessity of hospitalization" as equal to $p_{3\,1} = 0.62 \cdot 1.0 \cdot 0.22 = 0.14$.

6.3 Designing of Therapeutic Rules – 2nd Example

At last – yet another example of transforming medical evidence into a rule with uncertainty. A new protocol of clinical trials comes from the electronic journal *JAMA*, which is available via the Internet. For a change, we shall consider the protocol presenting results of the research on asthma prevention in adults [18]. The experiment considered was carried out for minimum three months, on adults suffering from persistent asthma. One of the outcomes tested in the experiment was the risk of asthma exacerbations while taking inhaled glucocorticosteroids (GCS). The control group was instead administered placebo or short-acting beta2-agonists.

For the outcome mentioned above, the authors calculated Relative Risk RR = 0.46 and 95% CI for rr equal to (0.34; 0.62). The result is statistically significant and the expected effectiveness of reducing asthma exacerbations is $ee_{0.95}(-) = 0.45$. The rule with uncertainty takes the following form:

```
r₄: it happens g-PM 0.85:
   if     (age > 19) and
          (persistent asthma disease <symptoms once a week>) and
          (night sleep and day activity disorders) and
          ((FEV₁ < 0.8·FEV₁,due) or
           (PEF < 0.8·PEFmax) or
           (PEVvar > 0.2) or
           (FEV₁,var > 0.2)) and
          (inhaled GCS <daily ⇔ 500μg beclomethasone>
           <period ≥ 3 months>)
   then   (no necessity of hospitalization) PM-c 0.45
          (no necessity of unplanned visits to doctors) PM-c 0.45
          (no necessity of intravenous GCS) PM-c 0.45
          (PEFmorning ≥ 0.75·PEFmax) PM-c 0.45
```

Analogously as in the case of the rule r_3, we can now advance hypotheses of: "no necessity of hospitalization", "no necessity of unplanned visits to doctors", "no necessity of intravenous GCS" and "$PEF_{morning} \geq 0.75 \cdot PEF_{max}$" with final probabilities $p_{41} = p_{42} = p_{43} = p_{44} = 0.85 \cdot 1.0 \cdot 0.45 = 0.38$.

Let us notice, that the probability mass of the conclusion of "no necessity of hospitalization", PM-c $= 0.45$, is in the case of r_4 essentially higher than the probability mass of the same conclusion from r_3, PM-c $= 0.22$. Although the dispersion of results was similar in both cases, the results of the last experiment were more radical. Let us further notice that both rules differ also in the global probability mass g-PM ($0.85 > 0.62$). The difference, in favour of the rule r_4, is caused by a big number and rigorousness of rule's premises (concerning patients and the pharmacotherapy applied), as well as by rigorous approach to the experiment itself (treating in the same way patients from both, the treatment and the control, groups).

7 Final Remarks

Knowing that the evidence stored in registers of clinical trials can be transformed into strict form of rules with uncertainty, we have future prospects of development of expert systems aiding GPs as well in diagnostics as in treatment decisions. Since the results of clinical research are commonly available (via the Internet) and their reliability is certified by expert-reviewers, then, no matter which disease we mean, the knowledge about its effective treating is growing more and more.

Although the registers of clinical trials are generally of high quality, we can easily differentiate between the most reliable and the remaining ones. Having a neural network (or any system) that is able to estimate the register's reliability, soon after this, we can fix priorities of the rules with uncertainty obtained. In this way, we can influence a course of reasoning, and – in system that is not locally confluent – a final conclusion itself.

By the way, one can bring it in question that, in the rule-based system proposed, final probabilities p_{ij} of the conclusions are low indeed. However, we should remember that the remaining probabilities $(1 - p_{ij})$ are reserved not for the opposite conclusions but for the ignorance range instead. On the other hand, the formula for calculating the parameter of expected effectiveness $ee_{1-\alpha}$ can be still improved.

Let us remark that the most valuable rules with uncertainty are being obtained from protocols of those experiments that were carried out on large groups of participants. Such rules could be constructed by using an additional semantic model (an ontology) of medical knowledge processed in the system. Then, it would be possible to join the results coming from various protocols. The problem mentioned is the subject of our current studies.

Finally, let us notice that if the evidence was specified more formally, keeping the assumed worldwide standards (HL7), we could automate the whole process of its transformation. It could be done with the use of well known syntax analysis methods.

References

1. About HL7, `http://www.hl7.org/about/hl7about.htm`
2. Cybulka J, Jankowska B, Nawrocki JR (2002) Automatyczne przetwarzanie tekstów. AWK, Lex i YACC (Automatic Processing of Texts. AWK, Lex and YACC), NAKOM, Poznań
3. Dolin RH, Alschuler L, Boyer S, Beebe C, Kona Editorial Group (2000) An Update on HL7's XML-based Document Representation Standards. In: Proceedings of AMIA Symp:190–194
4. Gerstman BB (2003) StatPrimer, `http://www.sjsu.edu/faculty/gerstman/StatPrimer`
5. GraphPad Prism, Version 4.03 (2005), `http://www.graphpad.com`
6. HL7 version 2.3, `http://www.medinfo.rochester.edu/hl7/v2.3/httoc.htm`
7. Interface HL7 pomiędzy szpitalnym systemem informatycznym (HIS) a specjalizowanym modułem diagnostycznym, wersja 1.2 (The HL7 Interface between the Hospital Information System and a Specialized Diagnostic Module, ver.1.2) (2004) Report published within the Project HL7PL, `http://hl7pl.uhc.com.pl`, Lublin
8. Jaeschke R, Cook D, Guyatt G (1998) Ocena artykułów o leczeniu lub zapobieganiu – cz. I–IV (The Evaluation of Articles about Treating and Preventing – part I–IV). In: Medycyna Praktyczna 02–05
9. Jankowska B, Szymkowiak M (2005) Knowledge Acquisition for Medical Expert Systems. In: Proceedings of the 5th Conference on Computer Methods and Systems 1:319–328, Kraków
10. Jankowska B (2006) Fast Reasoning in a Rule-based System with Uncertainty. A paper submitted to the Foundations of Computing and Decisions Sciences

11. Jankowska B (2004) How to speed up reasoning in a system with uncertainty?. In: Innovations in Applied Artificial Intelligence, LNAI 3029. Springer-Verlag:817–826
12. Lucas PJF, Segarr RW, Janssens AR, (1989) HEPAR: an expert system for diagnosis and disorders of the liver and biliary tract. In: Liver 9:266–275
13. Michalak M et al. (2006) The project designed by students of the subject: Applications of Artificial Intelligence in Medicine, Poznań University of Technology, Poznań
14. Mruczkiewicz J (2004) Podstawy Evidence Based Medicine, czyli o sztuce podejmowania trafnych decyzji w opiece nad pacjentami (EBM, or the Art of Taking Right Decisions about Nursing a Patient). In: Medycyna Praktyczna 06
15. Musen M et al. (1987) OPAL: Use a Domain Model to Drive an Interactive Knowledge-Editing Tool. In: International Journal of Man-Machine Studies 26:105–121
16. Oniśko A, Druzdzel MJ, Wasyluk H (2000) Extension of the HEPAR II model to multiple-disorder diagnosis. In: Intelligent Information Systems. Advances in Soft Computing Series, Physica-Verlag Heidelberg:303–313
17. Plotnick LH, Ducharme FM (2005) Combined inhaled anticholinergics and beta2-agonists for initial treatment of acute asthma in children. In: The Cochrane Library
18. Sin DD, Man J, Sharpe H, Qi Gan W, Man SFP (2004) Pharmacological management to reduce exacerbations in adults with asthma: a systematic review and meta-analysis. In: JAMA 292:367–376
19. Shortliffe E (1976) Computer-Based Medical Consultations: MYCIN, American Elsevier
20. Suermondt HJ, Cooper GF, Heckerman D (1991) A combination of cutset conditioning with clique-tree propagation in the Pathfinder system. In: UAI '90, Proceedings of the Sixth Annual Conference on Uncertainty in Artificial Intelligence, Elsevier:245–254
21. Światowa strategia rozpoznawania, leczenia i prewencji astmy. Raport NHLBI/WHO (Worldwide Strategy of Asthma Diagnosing, Treatment and Prevention. Report NHLBI/WHO) (2002). In: Medycyna Praktyczna 6
22. 7SPHINX 4.1 – Pakiet Sztucznej Inteligencji (Package of Artificial Intelligence Software) (2005) Artificial Intelligence Laboratory, Katowice

On Use of Unstable Behavior of a Dynamical System Generated by Phenotypic Evolution

Iwona Karcz-Dulęba[1]

Institute of Computer Engineering, Control and Robotics, Wrocław University of Technology, Wyb. Wyspianskiego 27, 50-362 Wrocław, Poland, `iwona.duleba@pwr.wroc.pl`

Summary. A dynamical system model of the simple evolutionary process is considered. Evolution running with proportional selection and normally distributed mutation is regarded on phenotypic level. Fixed points of the system and their stability for various fitness functions: symmetrical and asymmetrical, uni- and bimodal, are provided. The system demonstrates an unstable behavior for certain parameters of the process. The knowledge obtained from the study can be exploited to set parameters of the process applied to optimization tasks or to identify parameters of an unknown fitness function in the case of "black-box" tasks.

1 Introduction

Although evolutionary algorithms are easy to implement and apply in various domains, they are analytically tractable only in some simple cases. Still, there is an urgent need for theoretical foundations of the methods. The theory of dynamical systems has been recently used as one among a few methods facilitating the analysis of evolutionary methods. Vose and coworkers [14, 15] proposed dynamical system models of genetic algorithms, described the expected behavior of the algorithms for infinite populations. The dynamical system model of phenotypic evolution applied to small populations and a real-valued infinite search space was proposed in [2, 4, 6]. The phenotypic evolution was characterized by proportional selection and normally distributed mutation. The populations are studied in a space of population states. The analysis of the process is performed for the case of very small, two-element populations and one-dimensional search space. Expected values of the population states generated a dynamical system. The asymptotic behavior of the evolutionary process is determined by calculating fixed points of the system and determining their stability. The number and localization of fixed points depends on an evaluation function and the only parameter of evolution - standard deviation of mutation σ. In general, fixed points are located near optima and saddles of the evaluation function. Usually, fixed points are stable for small values of

I. Karcz-Dulęba: *On Use of Unstable Behavior of a Dynamical System Generated by Phenotypic Evolution*, Studies in Computational Intelligence (SCI) **102**, 117–131 (2008)
`www.springerlink.com`   © Springer-Verlag Berlin Heidelberg 2008

the standard deviation of mutation. When σ is increased, fixed points became unstable, periodical orbits arise and, for some evaluation functions, a period doubling road to chaos is observed. In this paper we concentrate on exploiting unstable behavior of the dynamical system to gain some knowledge about the system itself and parameters that impact the system. The critical value of the standard deviation of mutation σ_c, for which a fixed point loses its stability, may help to set parameters of the evolutionary process applied for optimization problems. The unstable behavior could also be used to gain the knowledge about an unknown evaluation function in the case of "black-box" tasks. Some researchers suggested that chaotic behavior could also be used for restoring diversity in the population trapped at the local optimum [16].

The paper is organized as follows. In Section 2 the model of phenotypic evolution for two-element population is presented. The dynamical system generated by the evolutionary process is described in Section 3. In Section 4 fitness functions are introduced, fixed points and their stability for the functions are examined. In Section 5 some hints to set values of the process' parameter are put forward and remarks how to gain information about a fitness function based on evolution of the system are provided. Conclusions are collected in Section 6.

2 The Model of a Two-Element Population

The model of Darwinian phenotypic evolution is of interest [3]. A population P of m individuals evolves in continuous and unbound search space $T = R^n$. Individual $\boldsymbol{x}$ is characterized by its type, i.e. n-dimensional vector of traits $\boldsymbol{x} = (x_1, x_2, \ldots, x_n)$, $\boldsymbol{x} \in R^n$, and its quality, described by the value of fitness function $q(\boldsymbol{x})$. In this paper an evaluation function is identified with a fitness function. The evaluation function is usually derived from the problem being solved and gives a performance measure. The fitness is an internal quality measure assigned to individuals. Often, also in the paper, the functions are not distinguished. However, sometimes it is necessary to transform a real evaluation function into a non-negative fitness function.

Reproduction of the population begins with proportional selection followed by normally distributed mutation. Evolution in the next generation is ruled by probability distribution function $f_{\boldsymbol{x}}^{i+1}(\boldsymbol{x}|P^i)$ which describes the probability of individual $\boldsymbol{x}$ to be the member of the $(i+1)$st generation when the population in the ith generation is P^i

$$f_{\boldsymbol{x}}^{i+1}(\boldsymbol{x}|P^i) = \sum_{k=1}^{m} \alpha(\boldsymbol{x}_k^i) \cdot g(\boldsymbol{x}, \boldsymbol{x}_k^i) = \sum_{k=1}^{m} \frac{q(\boldsymbol{x}_k^i)}{\sum_{j=1}^{m} q(\boldsymbol{x}_j^i)} \cdot g(\boldsymbol{x}, \boldsymbol{x}_k^i), \qquad (1)$$

where
P^i - population in the ith generation,
$\boldsymbol{x}_k^i \in R^n$ - type of the kth individual in the ith generation,

$\alpha(\boldsymbol{x}_k^i)$ - probability of the individual $\boldsymbol{x}_k^i$ selection,

$q(\boldsymbol{x}_k^i)$ - fitness of the individual $\boldsymbol{x}_k^i$,

$g(\boldsymbol{x}, \boldsymbol{x}_k^i)$ - distribution of mutation of the kth individual.

The introduced evolutionary process can be viewed as a discrete-time Markov process with infinite number of states. Because a new population depends only on state of the current population (in a probabilistic sense), a population can be regarded as a Markov process state. Thus, evolution is analyzed in the space of population states S where every point denotes the whole population [2, 4, 6]. The structure of the state space is more complicated than the structure of search space T. The dimensionality of S is equal to $dim(S) = n \cdot m$. Furthermore, the state should not depend on ordering of individuals within the population. Therefore the equivalence relation U that identifies all points corresponding to permutations of individuals within the population is to be defined on S. Consequently, the space S is transformed into the factor space $S_U = R^{n \cdot m}/U$. The fitness function in space S must represent the whole population, so it is defined as the average fitness of all individuals within the population.

The advantage to examine evolving populations in more complex space S appears when two-element populations and individuals with only one trait ($m = 2$ and $n = 1$) are considered. The whole population can be described as a point (x_1, x_2) and visualized easily on S plane. The following equivalence relation, ordering the individuals in population, is assumed:

$$R^2 = S \xrightarrow{\;U\;} S_U \subset R^2 : \quad (x_1^i, x_2^i) \rightarrow \begin{cases} (x_1^i, x_2^i) & \text{for } x_1^i \geq x_2^i \\ (x_2^i, x_1^i) & \text{for } x_1^i < x_2^i \end{cases}.$$

The factor space S_U covers the right half-plane bounded by the line $x_1 = x_2$, called the *identity axis*. The fitness function for the population is the average fitness from both individuals $q_S = (q(x_1) + q(x_2))/2$.

The analysis of the evolution in the state space is more convenient after counterclockwise rotation of $X_1 X_2$ coordinate frame with the angle $\gamma = \pi/4$ around the origin. In the new coordinate frame WZ, coordinates of point (x_1, x_2) are transformed to $w = (x_1 + x_2)/\sqrt{2}$ and $z = (x_2 - x_1)/\sqrt{2}$. The population P is mapped to the state $\boldsymbol{s} = (w, z)$ placed in the right half-plane ($w \geq 0$), bounded by the identity axis which becomes Z axis.

The new coordinates admit interesting interpretation. Coordinate w describes the distance of population state from the identity axis and it can be considered as a measure of the population diversity. Coordinate z locates the state along the identity axis and can be viewed as the measure of the population mean. Evolution of two-element population in the state space S_U (the WZ coordinate frame) according to rules of proportional selection and normally distributed mutation can be easily simulated and visualized. Two exemplary runs of the evolution are depicted in Fig. 1.

Simulations and analytical results [2, 4, 7] showed that the evolution appears to be a two-speed process: the population quickly reaches states located

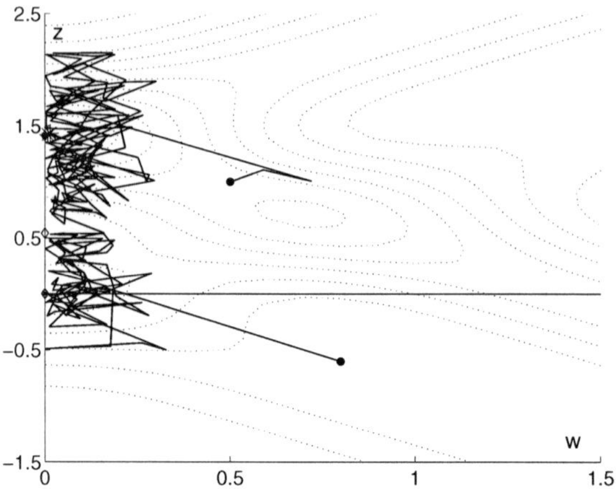

Fig. 1. Evolution of two-element population in the space of population states WZ. Two initial states $\boldsymbol{s}^0 = (0.8, -0.6)$ and $\boldsymbol{s}^0 = (0.5, 1.0)$ are marked with dots. Contour dotted lines represent average fitness of bimodal Gaussian fitness function (14); $a_1 = a_2 = 5$, $h = 2$, $\sigma = 0.1$

near the identity axis and then slowly drifts to the optimum (optima) of the fitness function (Fig. 1). The fast phase indicates unification of the population. When an initial population is widely diversified (large coordinate w) it becomes almost homogeneous (small coordinate w) just after a few generations. In the second phase coordinate w of the population state practically does not change whereas coordinate z changes slightly, with steps depending on the standard deviation of mutation.

Since the distributions (1) are independent for each individual, the distribution of the population state in the factor space is given by

$$\tilde{f}_{S_U}^{i+1}(\boldsymbol{s}|\boldsymbol{s}^i) = m! \prod_{j=1}^{m} f_x^{i+1}(\boldsymbol{x}_j|\boldsymbol{s}^i) = m! \prod_{j=1}^{m} \sum_{k=1}^{m} \alpha(\boldsymbol{x}_k^i) \cdot g(\boldsymbol{x}_j, \boldsymbol{x}_k^i). \qquad (2)$$

For two-element population, Eq. (2) takes the form:

$$\tilde{f}_{S_U}^{i+1}(x_1, x_2|\boldsymbol{s}^i) = 2f_x^{i+1}(x_1|\boldsymbol{s}^i) \cdot f_x^{i+1}(x_2|\boldsymbol{s}^i). \qquad (3)$$

When transforming the distribution (3) to coordinate frame WZ, the expected value of the population state can be calculated. Coordinates w and z of the expected value are equal to

$$E^{i+1}\left[w|\boldsymbol{s}^i\right] = \sqrt{\frac{2}{\pi}}\sigma + (1 - \Psi^{i2})\sigma \cdot \theta(\frac{w^i}{\sigma}), \qquad (4)$$

$$E^{i+1}\left[z|\boldsymbol{s}^i\right] = z^i + \Psi^i \cdot w^i, \qquad (5)$$

where

$$q_1 = q((w + z)/\sqrt{2}), \quad q_2 = q((z - w)/\sqrt{2}),$$

$$\Psi(w, z) = \frac{q_1 - q_2}{q_1 + q_2}, \quad \Psi^i = \Psi(w^i, z^i), \quad \theta(\xi) = \phi_0(\xi) + \xi \cdot \Phi_0(\xi),$$

$$\phi_0(\xi) = \frac{1}{\sqrt{2\pi}}(\exp\left(-\xi^2/2\right) - 1), \quad \Phi_0(\xi) = \frac{1}{\sqrt{2\pi}} \int_0^\xi \exp(-t^2/2)dt.$$

The first element of expected value $E[w|\boldsymbol{s}]$ sum depends only on the standard deviation of mutation σ while the second depends also on the current value of w and on the fitness function via coefficient Ψ. The value of $E[w|\boldsymbol{s}]$ is bounded [12]

$$\sqrt{2/\pi}\sigma \leq E^{i+1}[w|\boldsymbol{s}^i] \leq \sqrt{2/\pi}\sigma + w^i/2. \tag{6}$$

The second element in Eq. (4) takes the value close to zero if the difference in fitness of two individuals are significant. Thus, the expected value $E[w|\boldsymbol{s}]$ becomes constant and approaches its lower bound. When the value of w is small, the second element influences the expected value. The expected value $E[z|\boldsymbol{s}]$ depends on the current population state and on the coefficient Ψ.

In Fig. 2 expected values of population states for different initial states are displayed. The expected trajectories resemble trajectories of evolution (cf. Fig. 1). The two phases are also observed. The fast phase indicates rapid decrease of $E[w|\boldsymbol{s}]$ until it attains its lower bound $w_l = \sqrt{2/\pi}\sigma$. Then, the slow phase appears. It is characterized by small state transitions towards the optimum of the fitness function.

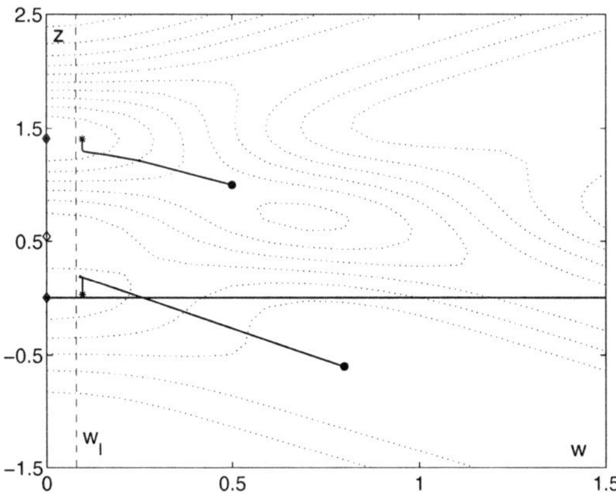

Fig. 2. Trajectories of expected values of population states for initial states $\boldsymbol{s}^0 = (0.8, -0.6)$ and $\boldsymbol{s}^0 = (0.5, 1.0)$. Contour dotted lines represent average fitness of bimodal Gaussian fitness function (14); $a_1 = a_2 = 5$, $h = 2$, $\sigma = 0.1$

3 Dynamical System

The analysis of evolution considered as the stochastic process described by the population distribution (1) or (2) is difficult. Its deterministic approximation using the expected values of the population state (4)-(5) for two-element populations facilitates the study. The expected values generate the discrete dynamical system in the space S_U

$$\begin{cases} w^{i+1} = & \sqrt{\frac{2}{\pi}}\sigma + (1 - \Psi^{i^2}) \cdot \sigma \cdot \theta(\frac{w^i}{\sigma}) \\ z^{i+1} = & z^i + \Psi^i \cdot w^i. \end{cases} \tag{7}$$

Fixed points (w^s, z^s) of the system and their stability can be determined. The fixed points are characterized by the conditions

$$w^s \simeq 0.97\sigma, \tag{8}$$

$$\Psi(w^s, z^s) = 0. \tag{9}$$

Coordinate w of the fixed points (8) depends on the standard deviation of mutation only. Because w indicates population diversity, in the equilibrium state the population does not consist of copies of the optimal individual but individuals' trials vary about σ one from another. Coordinate z^s depends on the fitness function and it is derived from the equality $q_1 = q_2$ checked at the fixed point

$$q((z^s + w^s)/\sqrt{2}) = q((z^s - w^s)/\sqrt{2}). \tag{10}$$

Solutions of (10) are intersection points of two shifted fitness functions q_1 and q_2. The number of the fixed points (intersection points) is determined by the fitness function modality and the standard deviation of mutation. In general, for a fitness function with k optima the system has no more than $(2k + 1)$ fixed points. The points are located in the vicinity of optima and saddles of the fitness.

To characterize the asymptotic behavior of the dynamical system (7), the stability of the fixed points are determined. The Jacobi matrix of linear approximation is diagonal [7, 12] and its eigenvalues are equal to $\lambda_1 = \Phi_0(w^s)$, $\lambda_2 = 1 + w^s \frac{\partial \Psi(w,z)}{\partial z}|_{(w^s,z^s)}$. Because $\lambda_1 < 1$, the fixed points stability is determined by the second eigenvalue λ_2. The fixed point (w^s, z^s) is stable if inequality

$$-2 \leq w^s \frac{\partial \Psi(w, z)}{\partial z}|_{(w^s,z^s)} \leq 0 \tag{11}$$

is satisfied. Thus, the stability depends on the parameter of evolutionary process σ and on the fitness function.

4 Fixed Points and Their Stability

The asymptotic behavior of the system (7) depends on properties of fitness functions: the number of optima and symmetry. Later on the following types of fitness functions are examined.
Unimodal tent functions (Fig. 4.a)

$$q(x) = \begin{cases} x/A + 1 & \text{for } x \in [-A, 0), \ \ A > 0 \\ -x/B + 1 & \text{for } x \in [0, B], \ \ B \geq A \\ 0 & \text{otherwise,} \end{cases} \tag{12}$$

when $A = B$, the function is symmetrical.
Unimodal Gaussian functions (Fig. 4.b)

$$q(x) = \begin{cases} \exp(-a_1 x^2) & \text{for } x \leq 0 \\ \exp(-a_2 x^2) & \text{for } x > 0. \end{cases} \tag{13}$$

For $a = a_1 = a_2$, the function is symmetrical.
Bimodal Gaussian functions (Fig. 4.c,d)

$$q(x) = \exp(-a_1 x^2) + h \exp(-a_2(x - 1)^2). \tag{14}$$

($h > 0$, a_1, $a_2 > 0$). For $a_1 = a_2$ and $h = 1$, the function is symmetrical.
Below, a summary of obtained results $[6\text{--}9, 11, 12]$ is presented.

4.1 Symmetrical Fitness Functions

Unimodal fitness functions
Unimodal fitness functions with the optimum at point $(0, 0)$ (i.e. symmetrical functions (12) and (13)) have at most one fixed point. The fixed point $(0.97\sigma, 0)$ is placed on the symmetry axis ($z^s = 0$) at the distance $w^s \simeq 0.97\sigma$ from the optimum. Thus the population in the stable state consists of two individuals with the same quality but located on the opposite slopes of the fitness function ($x_1^s \simeq 0.69\sigma$ and $x_2^s \simeq -0.69\sigma$). The smaller σ, the better localization of the optimum by the evolutionary process. When parameters of the fitness function are determined, the fixed point's stability depends on the standard deviation of mutation. For small values of σ, the fixed point is stable and loses its stability if σ exceeds its critical value σ_c obtained from condition (11). To simplify calculation of σ_c the critical value of w-coordinate $w_c \simeq 0.97\sigma_c$ is used instead. For symmetrical tent function (12) with $A = B$, the critical value $w_c = 2/3A$ and for unimodal Gaussian function (13) with $a = a_1 = a_2$, the critical value $w_c = \sqrt{2/a}$. For mutations with $\sigma > 1.03 w_c$, the fixed point loses its stability and a periodic orbit arises (Fig. 4).
Bimodal fitness functions
The bimodal fitness function (14) is symmetrical for $h = 1$ and $a_1 = a_2$. In this

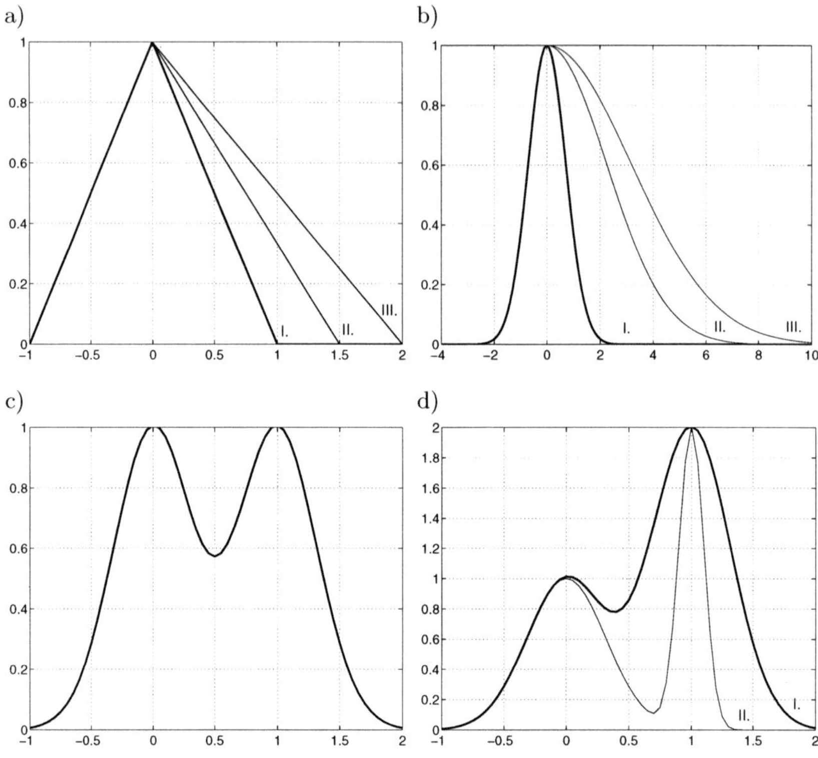

Fig. 3. Fitness functions a) unimodal tent function (12), $A = 1$, $B = 1; 1.5; 2$, I/II/III; b) unimodal Gaussian function (13), $a_1 = 1$, $a_2 = 1; 0.1; 0.05$, I/II/III; c) bimodal symmetrical Gaussian function (14), $h = 1$, $a_1 = a_2 = 5$; d) bimodal asymmetrical Gaussian function (14), $h = 2$, $a_1 = 5$, $a_2 = 5; 50$, I/II.

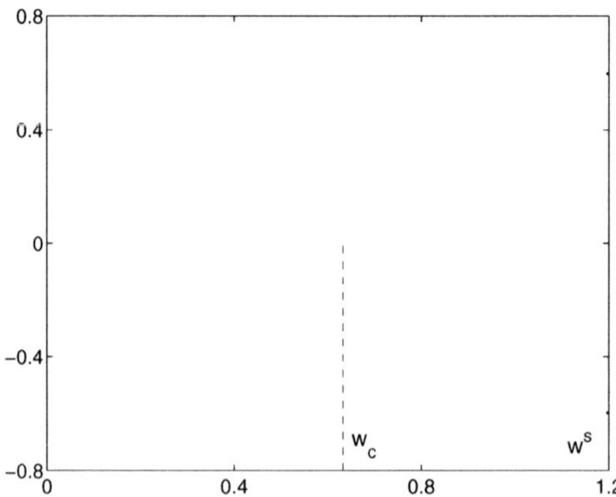

Fig. 4. Bifurcation diagram for symmetrical unimodal Gaussian function (13), $a = 5$

case, the dynamical system (7) displays one or three fixed points depending on σ (Fig. 5). Three fixed points, two symmetrical optima and one saddle point, appear when the standard deviation of mutation is small. When the parameter is increased, two optima fixed points come closer and closer to each other and, finally, collapse at one fixed point located in the vicinity of the saddle. Both optima fixed points are stable. The saddle fixed point is unstable for small σ. When all fixed points become a single fixed point, the point is stable. Finally, for large σ, an orbit of period 2 is observed. Generally, for a multi-modal fitness function with k optima, the number of fixed points is odd and vary from one to $2k + 1$. One fixed point is always located on the symmetry axis, while the others, are symmetrically paired. With the increase of the standard deviation of mutation the number of the points decreases and they lose their stability.

4.2 Asymmetrical Fitness Functions

Unimodal fitness functions
The asymmetry of the fitness function influences the z-coordinate of the fixed point. Using Eq. (10) formulas describing coordinate z^s were obtained for functions (12) and (13). For the asymmetrical tent function, z^s and the critical value w_c are equal to

$$z^s = w^s \frac{B - A}{B + A}, \qquad w_c = \frac{4AB(A + B)}{8AB + (A + B)^2}.$$

For the asymmetrical Gaussian function (13), z-coordinate of the fixed point is given by

$$z^s = w^s \cdot \frac{a_1 + a_2 - 2\sqrt{a_1 \cdot a_2}}{a_2 - a_1}.$$

The critical value of w_c is equal to

$$w_c = \sqrt{2/\sqrt{a_1 a_2}}.$$

For mutations with $\sigma > 1.03 w_c$, the fixed point loses its stability and a new unstable structure arises. In the case of asymmetrical functions not only the orbit of period 2 but also period doubling road to chaos was observed (Fig. 6). Series of period doubling bifurcations and chaos appeared when the function asymmetry is evident i.e. $B \geq 1.5$ for the function (12) and $a_2 < 0.1$ for the function (13) [9].

Bimodal fitness functions
The are three or one fixed points for analyzed examples of bimodal asymmetrical fitness function (14) with different parameters. One fixed point is located near to the local optimum, the second one close to the saddle and the third one near to the global optimum. As the standard deviation of mutation increases, fixed points that correspond to the local optimum and the saddle get closer and, eventually, disappear leaving only one equilibrium point.

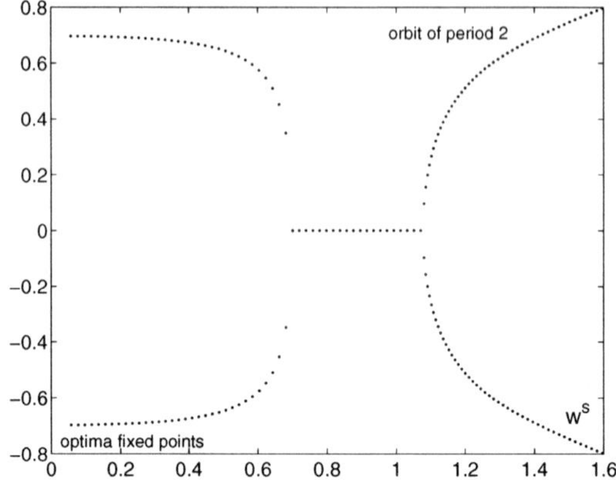

Fig. 5. Equilibrium points of the dynamical system (7) for symmetrical bimodal Gaussian fitness function (14), for two initial states $s^0 = (0.8, 0.6)$ and $s^0 = (0.8, -0.6)$; $h = 1.0$, $a = 5$

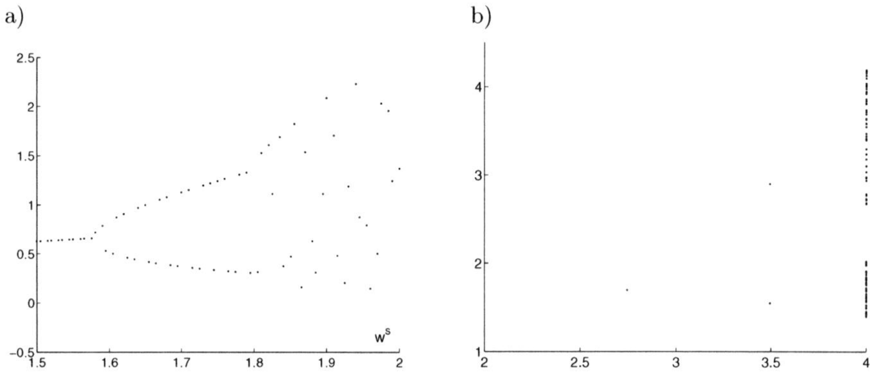

Fig. 6. Equilibrium points of the dynamical system (7) a) asymmetrical unimodal tent fitness function (12); $a_1 = 1.0$, $a_2 = 0.05$ b) asymmetrical unimodal Gaussian fitness function (13); $a_1 = 1.0$, $a_2 = 0.05$

Different aspects of asymmetry can be regarded while considering fixed points stability for bimodal functions. When asymmetry is defined as the difference in hills' heights but not in hills' widths, there is no qualitative difference between this case and symmetrical functions with equi-height peaks. Local fixed points are usually stable but quickly disappear with the increase of the standard deviation of mutation. The global fixed point becomes unstable for large values of parameter σ and period 2 orbits appeared (Fig. 7). If the hills are narrow, the dynamics of the system resembles that of a separate unimodal hill. Additional bifurcations appear for global fixed points (also for local fixed points when the saddles are deep enough) for the critical value $w_c = \sqrt{2/a}$ (Fig. 7.b,c).

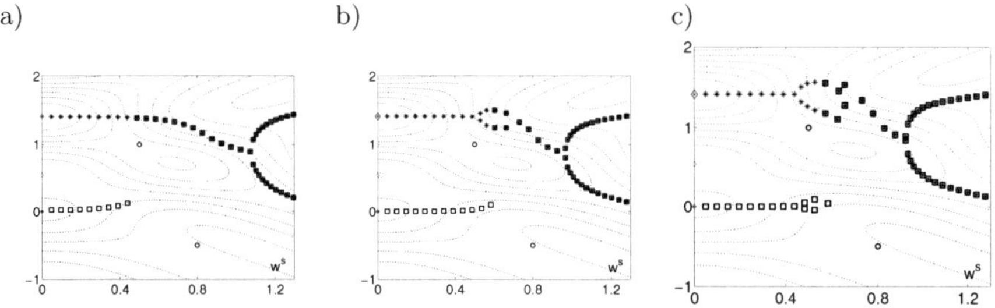

Fig. 7. Equilibrium points of the dynamical system (7) for bimodal Gaussian fitness function (14) with equi-width hills, for two initial states $s^0 = (0.8, -0.5)$ and $s^0 = (0.5, 1.0)$ represented by, respectively, squares and stars. $h = 2.0$; a) $a = 5$; b) $a = 8$; c) $a = 10$.

When fitness hills have different widths, the behavior of the dynamical system (7) is more complex (Fig. 8). If the width of the global optimum is small, i.e. its basin of attraction is smaller than the basis of the local one, the global fixed point disappeared faster with the increase of σ and the local fixed point remains. Moreover, the global fixed point becomes unstable very fast and not only the orbit of period two but also period doubling orbits leading to chaos for very narrow global peak are observed. The local fixed point is stable for relatively large σ but if it loses its stability, the system becomes chaotic through period doubling bifurcations. In simulations orbits of the period of 3 were observed what, according to the Sarkovsky theorem, indicates the presence of orbits of all other periods as well. The existence of bifurcations of different periods and chaos was confirmed by Lyapunov exponents [13].

5 Settings Parameters of Evolutionary Process and Fitness Functions

Properties, or at least some hints, obtained from studying evolutionary processes facilitate applying the evolutionary methods to optimization problems. The analysis and simulation results presented above may help to set parameters of the process and/or to get information about unknown fitness functions.

Selection of the optimal parameters' values is a hard problem for appliers of optimization methods. In the case of analyzed evolutionary process, which can also serve as an evolutionary optimization method, the task is much easier as there is only one parameter to play with, the standard deviation of mutation σ.

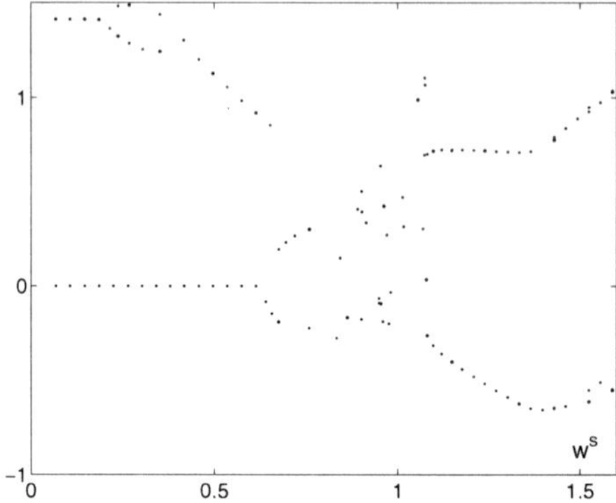

Fig. 8. Equilibrium points of the dynamical system (7) for bimodal Gaussian fitness function (14), for two initial states $s^0 = (0.8, -0.6)$ and $s^0 = (0.4, 1.0)$, $a_1 = 5$, $a_2 = 50$, $h = 2.0$.

Yet, the mutation rate is important parameter for majority of evolutionary methods, often responsible for their efficiency. When observing average time to reach the fixed point (orbit) [10], the fastest stabilization of the system is observed for σ a little smaller than σ_c. For very small σ the time is increasing. Still, a value of the parameter cannot be too large, as the optima will be located very inaccurately and unstable behavior may appear easily. It cannot be also too small, as computational costs to approach the optimum increase significantly. It seems that the reasonable heuristics are to set the value of σ 5 to 10 times smaller than the critical value σ_c, for which the fixed point loses its stability.

Evolutionary algorithms are often applied to solve optimization tasks with an unknown fitness function. We consider a parameter optimization problem where a set of control values has to be optimized to fulfill some requirements (in this paper - to maximize the fitness function). Such a problem might be regarded as a 'black-box' task with input parameters to be adjusted and output indicating how good the parameters are. For example, in many industrial processes an object controled may output the fitness function values, however the analytical model of the object is unknown (difficult to derive or unreliable). Some other forms of a fitness uncertainty, such as noisy, robust, approximated or dynamic functions, are also considered in the scope of evolutionary computations [1, 5].

Information obtained from studying the dynamical system behavior provides some basic information about the shape of the fitness function. The number of fixed points is tightly correlated with the number of optima. Unstable behavior designates the symmetry of the function: an orbit of period two appeared for symmetrical functions, a chaotic behavior displayed functions with considerable asymmetry. In Fig. 9 the critical values of the function parameter (A or a) for unimodal fitness (12), (13) in symmetrical and asymmetrical versions are presented. When the value of w_c (for which the system becomes unstable) is known, the fitness parameter can be read from the diagrams.

In the case of bimodal Gaussian functions (14) with different hills widths, the optima fixed points lost their stability for $w_c = \sqrt{2/a}$, where a is a decline of a given hill. When the peaks are equi-width, the global fixed point becomes unstable for $w_c \simeq 2 \cdot \sqrt{2/a}$. Additional bifurcations for smaller σ may appeared when the hills are narrow and the saddle is deep. The local fixed point vanished earlier when heights of the hills were different, and earlier when the difference in heights was bigger. The optima fixed points disappeared when $\sigma_d > 0.73d$, where d denotes the distance between optima for the Gaussian function (14). When the condition is fulfilled, the functions q_1 and q_2 have only one intersection point (10). Thus, based on the value of the standard deviation of mutation for which only one fixed point remains, one can judge about the distance between optima for Gaussian functions. More complicated behavior of the dynamical system: losing stability by the global fixed point for relatively small σ, regaining stability again for some range of σ, orbits with different periods (cf. Fig. 6) indicates that the global hill is narrow and its influence on the population is weak.

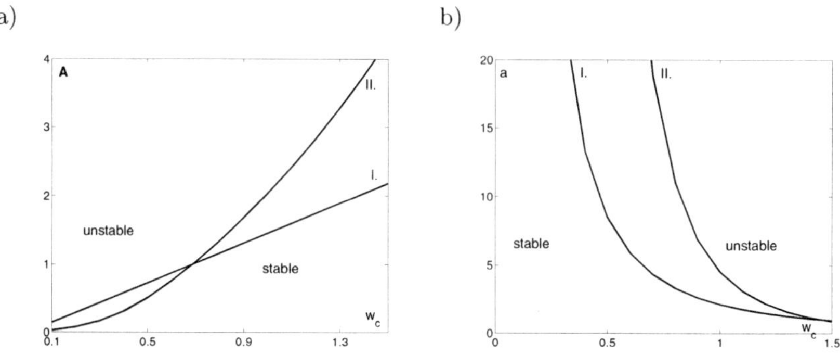

Fig. 9. Ranges of parameters a/A and w_c, for which the unimodal quality function is stable: a) tent unimodal function (12): I. symmetrical, II. asymmetrical, $B = 1.0$; b) Gaussian unimodal function (13): I. symmetrical, II. asymmetrical

6 Conclusion

The analysis of asymptotic behavior of the dynamical system generated by evolutionary process can be useful in setting parameters of the process and determining certain features of the fitness function in the case of "black-box" tasks. The value of the standard deviation of mutation for which the fixed point loses its stability, may prompt a reasonable range of effective values of the parameter, when the process is applied to optimization tasks. The number of fixed points suggested modality of the fitness. Chaotic behavior observed for increasing value of the standard deviation of mutation may indicate that the fitness function is asymmetrical. It is expected that the presented approach could be extended for other unimodal (bimodal) fitness functions as numerous functions can be approximated with the linear segments of the tent or with slopes of the Gaussian functions.

References

1. Branke J, Jin Y (eds) (2006) Special Issue on evolutionary computation in the presence of uncertainty. IEEE Trans. Evol. Comp., 10(4)
2. Chorążyczewski A, Galar R, Karcz-Dulęba I (2000) Considering Phenotypic Evolution in the Space of Population States. In: Rutkowski L, Tadeusiewicz R (eds) Proc. Conf. Neural Networks and Soft Computing. Zakopane, 615–620
3. Galar R (1985) Handicapped individua in evolutionary processes. Biol.Cybern., 51, 1–9
4. Galar R, Karcz-Dulęba I (1994) The evolution of two: An Example of Space of States Approach. In: Sebald AV, Fogel LJ (eds) Proc. 3rd Annual Conf. on Evolutionary Programming. San Diego CA, World Scientific, 261–268
5. Jin Y, Branke J (2005) Evolutionary optimization in uncertain environments - A survey. IEEE Trans. Evol. Comp., 9(3), 303–317
6. Karcz-Dulęba I (2000) Dynamics of two element populations in the space of population states: the case of symmetrical quality functions. Proc. 7 Nat. Conf. Evol. Comp. and Global Opt., Lądek Zdrój, 115–122, (in polish)
7. Karcz-Dulęba I (2002) Evolution of Two-element Population in the Space of Population States. Equilibrium States for Asymmetrical Fitness Functions. In: Arabas J (ed) Evolutionary Algorithms and Global Optimization Warsaw University of Technology Press, Warsaw, 35–46
8. Karcz-Dulęba I (2004) Asymptotic Behavior of Discrete Dynamical System Generated by Simple Evolutionary Process. Journ. of Applied Math. and Comp. Sc., 14(1), 79–90
9. Karcz-Dulęba I (2004) Period-doubling bifurcations in discrete dynamical system generated by evolutionary process. Proc. 7 Nat. Conf. Evol. Comp. and Global Opt., Kazimierz Dolny, Poland, 83–88
10. Karcz-Dulęba I (2004) Time to the convergence of evolution in the space of population states. Journ. of Applied Math. and Comp. Sc., 14(3), 279–287
11. Karcz-Dulęba I (2005) Bifurcations and chaos in phenotypic evolution for unimodal fitness functions. In: Greblicki W, Smutnicki C (eds) Control Systems. WKiL, 41–50

12. Karcz-Dulęba I (2006) Dynamics of two-element populations in the space of
    population states. IEEE Evol. Comp., 10(2), 199–209
13. Karcz-Dulęba I (2006) Chaos detection with Lyapunov exponents in dynamical
    system generated by evolutionary process. LNCS, Springer Verlag, (to appear)
14. Vose MD, Wright AH (1994) Simple genetic algorithms with linear fitness. Evo-
    lutionary Computation, 4(2), 347–368
15. Vose MD (1999) The Simple Genetic Algorithm. Foundations and Theory, MIT
    Press
16. Wright AH, Agapie A (2001) Cyclic and Chaotic Behavior in Genetic Algo-
    rithms. In: Proc. Genetic and Evolutionary Computation Conf. GECCO-2001.
    Morgan Kaufmann, San Francisco, 718–724

Temporal Specifications with XTUS.
A Hierarchical Algebraic Approach

Antoni Ligęza[1] and Maroua Bouzid[2]

[1] AGH – University of Science and Technology, al. Mickiewicza 30, 30-059 Kraków
  Poland, `ligeza@agh.edu.pl`
[2] GREYC, Campus II Sciences 3, BD Maréchal Juin, 14032 Caen Cedex,
  `bouzid@info.unicaen.fr`

Summary. Representation of temporal knowledge and efficient handling of temporal specifications is an important issue in design and implementation of contemporary information systems, such as databases, data warehouses, knowledge based-systems or decision support systems. This paper explores and further develops the ideas of TUS, the Time Unit System being an algebraic tool for constructing simple yet powerful temporal specifications. In particular, an extended version of TUS, called XTUS is presented in details and its basic operations and properties are shown. It is argued that this simple and consistent with natural calendar way of building temporal specifications is capable of dealing with phenomena such as nested cycles. A somewhat extended example is used to illustrate the ideas and application.

1 Introduction

Representation of temporal knowledge constitutes a core issue in the development and use of almost any complex information systems. Starting from spreadsheet applications, through relational databases, and up to data warehouses and knowledge-based real-time systems, specification of temporal dimension is an important part of data analysis and inference.

Although specification and efficient handling of temporal knowledge is an important issue in design and implementation of contemporary information systems, efficient dealing with temporal knowledge is far from being definitely solved. Numerous theory-oriented *Temporal Logics* [7] introduce variety of notations and operations frequently far from needs generated by realistic applications and unacceptably complex from the engineering point of view. Due to sophisticated and philosophical nature, majority of the proposals cannot be accepted by practitioners. Moreover, most of the proposals are based on flat (either point-based or interval-based) representations which do not fit well into systems using natural calendar.

Contrary to numerous complex, theory-oriented approaches based on temporal logics [7, 10] the presented formalism is based on simple algebraic

A. Ligęza and M. Bouzid: *Temporal Specifications with XTUS. A Hierarchical Algebraic Approach*, Studies in Computational Intelligence (SCI) **102**, 133–148 (2008)
`www.springerlink.com`

approach introduced in [12, 13]. Moreover, it is a hierarchical approach incorporating the notion of *granularity*. Any temporal granularity can be viewed as partitioning of temporal domain into groups of elements, where each element is perceived as an indivisible unit (a granule). The description of a fact, an action, or an event can use these granules to provide the temporal qualification specified at the appropriate abstraction level. Examples of standard time granularities are *days, weeks, months*, while user-defined granularities may include *business-weeks, trading-days, working-shifts* or *school-terms*.

The foundations for this work are found in [12, 13] which present a formalization of a system of granularities in the context of the interval calculus. Granules (called Time Units) are defined as finite sequences of integers. They are organized in a linear hierarchy of the form *(year, month, day, hour, minute, second)*. Depending on the needs the basic structure can be shortened and hence operating at the level of bigger granules of time. TUS, the Time Unit System presented in [13], follows the line of interval time specification and introduces a hierarchical, granular temporal specifications based on natural calendar and time. Its specifications refer to absolute time defined with sequences of integers.

In this paper we present details of XTUS, an eXtended Time Unit System, initially introduced in [4]. It is a generalized version of TUS, the Time Unit System introduced by Ladkin [12, 13]. The proposed extensions improve the expressive power of TUS so that floating time intervals can also be specified and hence repeated (cyclic) specifications can be constructed in a straightforward way. It is argued that this simple and consistent with natural calendar way of building logical specification is capable of dealing with phenomena such as nested cycles. A somewhat extended example is used to illustrate the ideas and application.

The structure of the paper is as follows. In Section 2 basic notions of the TUS system are recalled in brief and a short outline of the classical TUS is presented. In Section 3 we present the bases of the eXtended Time Unit System called XTUS. Section 4 describes some basic operations in XTUS. Section 5 is devoted to the presentation of an extensive example illustrating the presented ideas and the expressive power of XTUS. In Section 6 related work is referred to. Concluding remarks are given in Section 7.

2 TUS – Time Unit System

The Time Unit System (TUS, for short) was introduced by Ladkin in his early work [12] and further developed in his Ph.D. thesis [13]. TUS provides a way for concise specification of time instants and time intervals based on the common way of expressing dates. In fact, instead of direct linear notation it uses granular, hierarchical specification.

2.1 Basic Ideas of TUS

Let us recall the basic ideas of TUS. TUS introduces the notion of *Basic Time Units* (BTU, for short), such as *seconds*, *hours* or even *days*. The selection of specific BTU depends on the required precision and induces the finest level of *granularity* into the considered time model.

As TUS refers to absolute time, the basic time units are defined by providing a complete specification of the date; typically it includes *year*, *month*, *day*, *hour*, *minute* and *second*. Although the system is extendible to arbitrarily fine BTUs, in [13] it is assumed that up to the six levels of granularity given above will be used. Hence, assuming that our basic time unit is a second, any such BTU can be specified in a unique way as a sequence of integers i having the following form:

$$i = [year, month, day, hour, minute, second]. \tag{1}$$

For example, $i = [2005, 6, 8, 11, 55, 0]$ denotes the first second of the 11:55 hour on the 8-th day of June (actually: Wednesday) of the year 2005.

Sequences shorter than six integers specify intervals of time interpreted in a natural way. Some examples can be as follows:

- $[2005, 6, 8, 11, 55]$ represents the one-minute long interval starting at 11:55 on the 8-th of June, 2005; obviously, the interval consists of 60 BTUs,
- $[2005, 6, 8, 11]$ represents the one-hour long interval starting at 11:00 on the 8-th of June, 2005; the interval consists of 3600 BTUs,
- $[2005, 6, 8]$ represents the one-day long interval starting at midnight June 7, 2005; it consists of 86,400 BTU-s,
- $[2005, 6]$ represents the one-month long interval starting at midnight May 31, 2005; since June is 30 day long, it consists of 2,592,000 BTUs,
- $[2005]$ represents the one-year long interval starting at midnight December 31, 2004.

Note that the shorter the sequence is, the more coarse the level of granularity is in use. Finer specification of time intervals requires longer sequences of TUS. Obviously, TUS is human-oriented – the granular, hierarchical specification is easily readable and interpreted, but from a technical point of view it can be always recalculated to a flat integer expressed in basic time units.

2.2 Convex Intervals

Since the sequences of TUS refer to a fixed clock units of time, they do not allow for specification of arbitrary convex intervals. In [13] a special operator named *convexify* and defining a minimal cover of any two intervals is introduced. For any two intervals i and j, $convexify(i, j)$ is the smallest interval of time containing both i and j. The result of $convexify(i, j)$ is defined in a unique way. Examples of application of *convexify* are as follows:

- *convexify*([2001], [2005]) denotes the five-year long interval composed of the first five years of the XXI Century,
- *convexify*([2005, 6, 11], [2005, 6, 12]) denotes a specific weekend in June 2005 (Saturday and Sunday),
- *convexify*([2005, 6, 10, 17, 30], [2005, 6, 13, 9]) denotes the effective weekend time, i.e. time one may spend out of work starting on Friday, June 10 at 17:30 and ending on Monday, June 13, at 9:00. The sequences do not need to be of the same length.

It may also happen that the result of *convexify* may be expressed in a simpler way, e.g. *convexify*([2005, 1], [2005, 12]) = [2005].

2.3 A Note on the Domains of TUS

Consider all the sequences i of six positive integers of the form (1); obviously, not all of them specify legal dates, e.g. [2005, 2, 29]. Clearly, the range of admissible integers at position j may depend on preceding values in the particular sequence. Further considerations are restricted to feasible sequences only, i.e. ones representing legal dates.

After [13] let us introduce the function $max_j([a_1, a_2, \ldots, a_{j-1}])$ defining the maximal legal value appearing at the j-th position in the sequence. In TUS, referring to the calendar in use, only the number of days depends on the year and the month. Hence we have:

- $max_3([a_1, 2]) = 28$, provided that a_1 is not divisible by 4 or it is divisible by 100, but not by 400,
- $max_3([a_1, 2]) = 29$ in other cases,
- $max_3([a_1, a_2]) = 31$ for arbitrary a_1 when $a_2 \in \{1, 3, 5, 7, 8, 10, 12\}$,
- $max_3([a_1, a_2]) = 30$ for arbitrary a_1 when $a_2 \in \{4, 6, 9, 11\}$.

Further specification of the max functions is straightforward: $max_2([a_1]) = 12$, $max_4([a_1, a_2, a_3]) = 23$, $max_5([a_1, a_2, a_3, a_4]) = 59$, and, finally, in case of seconds $max_6([a_1, a_2, a_3, a_4, a_5]) = 59$ for any values of a_1, a_2, a_3, a_4, a_5. The maximal value of the year is undefined – it can be any integer.

Year is assumed to take integer values (positive or negative, without 0) [13]. Since the minimal value for any other component of the sequence is well defined[1] it is straightforward to define the domain for any a_i, $i \in \{1, 2, 3, 4, 5, 6\}$. We shall denote such a domain as $\mathcal{T}_i$. Clearly, any domain is a countable ($\mathcal{T}_1$) or finite ($\mathcal{T}_2 - \mathcal{T}_6$) set of integers; moreover, $\mathcal{T}_3$ depends on the current values of a_1 and a_2. The minimal element in any domain will be denoted symbolically as min_i and we put $min_1 = -\infty$; obviously, min_i depends only on i (the position) and not on the current values in the preceding sequence.

---

[1] It is 1 for the month and day, and 0 for hour, minute and second.

3 Extended Time Unit System

The original TUS, as introduced by Ladkin [13], allows only for constant terms expressed as sequences of integers, such as in pattern (1). This constitutes a strong limitation with respect to the expressive power of TUS – although in fact arbitrary convex intervals can be specified using the *convexify*(., .) operation, there seems to be other straightforward possibilities of extending the notation in a simple and transparent way.

3.1 Introduction to Extended TUS

Let us introduce the concept of time domain. A time domain (*TD*) is any convex (finite or infinite) subset of positive integers denoting subsequent basic time units. Depending on the current needs one can allocate the discussion within a specific interval of time, either absolute (referring to some calendar and clock time) or relative. In simple words, a *TD* is a window of time to which the discussion is restricted.

Consider a particular *TD* selected for current discussion.

Definition 8 (Flat term) *A* flat term *defining time interval at a single level of hierarchy is:*

- *any constant integer $z \in TD$,*
- *any range of integers $[z_1 - z_2]$, such that $z_1, z_1 + 1, z_1 + 2, \ldots, z_2 \in TD$,*
- *any union of flat terms.*

For simplicity, a single integer is considered equivalent to a single-element set containing it. A single integer number such as 7 denotes a certain interval of time; depending on the assigned interpretation it can be a month (July), a day (24 hours), an hour (60 minutes), etc. In fact, in order to assign meaning to a flat term one must specify its *type* (unit). This can be denoted as a pair *type* : *term*, e.g. *month* : 12 or *day* : 29. If the type is known, specification of it will be omitted. Every time we refer to a flat term it should be situated within a well-defined time domain *TD*, either explicit or implicit one.

A set of integers may be used to specify several intervals of the same type (months, days, hours) which are not necessarily adjacent. For example $[1, 3, 5]$ may denote Monday, Wednesday and Friday. In case the integers are subsequent ones, instead of writing $[z, z + 1, z + 2, \ldots, z + j]$ one simple writes $[z - (z + j)]$, i.e. specifies the range of integers; for example, $[9 - 13]$ is the equivalent for $[9, 10, 11, 12, 13]$, and when speaking about hours it denotes the interval beginning at 9:00 and ending at 13:59 using one minute as a BTU.

Definition 9 (Hierarchical term) *A* hierarchical term *(or a term for short) specifying interval of time at k levels of hierarchy is any sequence of the form*

$$[t_1, t_2, \ldots, t_k],$$

where t_1 is a flat term and any $t_i \in \{t_2, t_2, \ldots, t_k\}$ is a flat term or an anonymous variable denoted with $\_$.

In order to assign meaning to terms, one must define its *scheme* as

$$[type_1, type_2, \ldots, type_k]$$

or alternatively use typed flat terms, i.e. $[type_1 : t_1, type_2 : t_2, \ldots, type_k : t_k]$. For example, one can define a term like $[year : 2005, month : 5, day : \_, weekday : [1-5]]$, or simply write $[2005, 5, \_, [1-5]]$ if the type specification is known.

A *temporal term* is either a *flat term* (see Definition 8) or a *hierarchical term* (see Definition 9).

The underscore '$\_$' denoting *any* flat term (as the anonymous variable in PROLOG) will also be used. For example, $[2005, 5, \_]$ denotes any day of May 2005. The ε symbol will be used to denote an empty (or impossible) term.

Consider a hierarchical term $t = [t_1, t_2, \ldots, t_k]$. A basic time unit (at level k) $b \in TD$ can be *covered* by t if it is consistent with the specification of t or can be staying out of t if it does not fall into the scope of BTUs defined by t. For example, if day is BTU and we restrict the TD to be the 2007 year then 77 is covered by $[2007, 3, [16, 17, 18]]$, while 78 is not.

Consider a scheme given by $[year, month, day, weekday]$ which is redundant. Note that, the specification of time with hierarchical terms can be:

- ambiguous – the same time period can be often specified in different ways; for example $[2005, 5, [10, 12, 14], [1-5]] = [2005, 5, [10, 12], [2, 4]] = [2005, 5, [10, 12], \_]$,
- redundant – since days of the month and weekdays may be used together, specifications such as $[2005, 5, [10, 12, 14], [2, 4, 6]]$ is redundant, one does not need the weekdays,
- inconsistent – this is a consequence of redundancy, e.g. $[2005, 5, 10, 7]$ is inconsistent and as such empty (equal to ε). Even in case of using the $\_$ symbol inconsistency may occur, e.g. $[2005, \_, 31, 5]$ is inconsistent since 31 was never on Friday in 2005.

There is no simple way to deal with these issues; on the other hand they seem to be no harmful for majority of potential applications. They are intrinsic features of any more rich formal or informal language for knowledge specification, including logic. Since XTUS is aimed at achieving high expressive power (close but still inferior to natural language), all the three features constitute a natural consequence of this assumption.

The expressive power of a *single* hierarchical term is obviously insufficient to cover *any* natural language specification. For example, *the last day of February and July 2005* cannot be expressed with a single term. Such a specification will require more complex construction with the union operator (see section 4).

3.2 Handling Specification of Cycles

Observe that the introduction of the underscore denoting unspecified flat term allows to represent *periodical* specifications, since the definition of the interval may be not anchored at a precise instant of time. For example, $[2005, \_, \_, [1-5]]$ is the default specification of working days in 2005; since the month and specific days are absent, it defines a repetition over the subsequent weeks.

Similarly, the schooldays in France in 2005 can be defined as $[2005, [1-6] \cup [9-12], \_, [1, 2, 4, 5]]$, i.e. every year, during all months apart from July and August, the children go to school on Monday, Tuesday, Thursday and Friday. This example specification is illustrated below. Note that this time a cycle

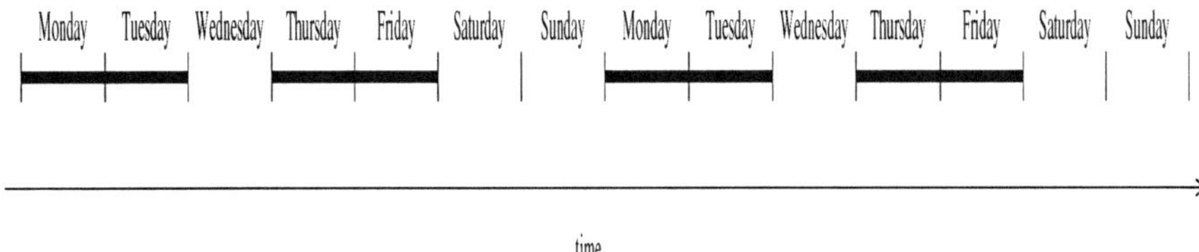

Fig. 1. An example specification of cyclic repetition.

within a cycle was defined in fact: the school weeks repeat over the given months.

Definition 9 requires that the underscore cannot be used at the first position in specification; this would denote in fact an infinite cycle. Although one may imagine and perhaps deal with such specifications for certain purposes of theoretical considerations, this would make the discussion unnecessarily complex.

Obviously, not all the cycles can be specified with a single hierarchical term. Those referring to a period such as *every second year* or *every third month* are not expressible within the simple language. On the other hand, the specific cycles mentioned can be handled:

- either by introducing new types of units (not a very elegant solution), or
- by defining new constraints over values (e.g. even or odd year, month number divisible by three, etc.), or
- by defining new restricted domains for the components (e.g. day $\in 1, 7, 14, ...$ or month $\in 3,6,9,12$.

An elegant solution may perhaps consist in defining a language of higher expressive power with symbolic constraints used instead of simple numeric values.

4 Algebraic Operations in XTUS

Since the XTUS specification of time intervals can be always mapped into extensional definition of flat, non-convex intervals, some typical algebraic operations can be specified.

4.1 Hierarchical versus Flat Specifications

First let us define two basic operations for *flattening* (f) and reconstructing the hierarchical structure (hierarchising) (h) of time specification.

Definition 10 *If t is a hierarchical term of XTUS, then $f(t) = [a_1, b_1] \cup [a_2, b_2] \cup \ldots [a_m, b_m]$, where any $[a_i, b_i]$ is a convex flat interval expressed in basic time units (BTUs) for $i = 1, 2, \ldots, m$ and where $b_i < a_{i+1}$ for $i = 1, 2, \ldots, m - 1$, covering t and minimal, i.e. any $b \in TD$ covered by t satisfies $b \in [a_i, b_i]$ for some $i \in \{1, 2, \ldots, m\}$ and $f(t)$ is minimal, i.e. any $b \in f(t)$ is covered by t; a_1 is the starting point of t and will be denoted also as $f(t)^-$ while b_m is the end point of t and will be denoted also as $f(t)^+$.*

Obviously, if t is convex (it denotes a single, convex interval), then $f(t) = [a, b]$ is also convex. Let for example restrict the time domain to be the 2007 year, and let $t = [2007, [3-4], [3, 5, 7]]$ with days used as BTUs. Then $f(t) = [62, 97]$, since the 3rd of March 2007 was the 62nd day of the year and the 7th of April was the 97th day of the year.

Definition 11 *If $[a_1, b_1] \cup [a_2, b_2] \cup \ldots [a_m, b_m]$ is a sum of convex flat intervals expressed in BTUs, then $h([a_1, b_1] \cup [a_2, b_2] \cup \ldots [a_m, b_m])$ is the equivalent hierarchical term (if exists) or an equivalent set of XTUS terms.*

Observe that this operation may lead to either a single term or a set (sum) of terms necessary to specify the equivalent intervals (see below). For example, taking the year 2007 as the domain and days as BTUs, an interval $[33, 51]$ may be transformed to $[2007, 2, [2-20]]$. For interval $[57, 64]$ the equivalent can be expressed as $[2007, 2, [26-28]] \cup [2007, 3, [1-5]]$. Also in case of several intervals one can find some equivalent construction of hierarchical terms. In this way $h([a_1, b_1], [a_2, b_2], \ldots, [a_k, b_k])$ may be constructed as a more complex expression of terms composed by algebraic operators (see below).

Having defined the flattening operation it is straightforward to define the *diameter* of a constant term of XTUS.

Definition 12 *Let t be a constant term of XTUS and let f denote the flattening operation. The diameter $\delta(t)$ of t is defined as*

$$\delta(t) = f(t)^+ - f(t)^-.$$

Obviously, $\delta(t)$ is the length of an interval expressed in the basic time units. In case of convex term t, $\delta(t)$ is also the *length* of t, and we have $f(t) = [f(t)^-, f(t)^+]$.

4.2 Basic Algebraic Operations

Let r and s denote two terms specifying intervals of XTUS. The following operations have natural, straightforward interpretation:

- $r \cup s$ – the union of intervals,
- $r \cap s$ – the intersection of intervals,
- $r \setminus s$ – the difference of intervals.

XTUS terms may represent convex and non-convex intervals (unions of convex intervals). Further, even if the initial XTUS terms are convex, the result of union and difference operations are not necessarily single convex intervals. In general, unions-of-convex-intervals are necessary to express some specifications [13]; the expressive power of such language significantly increases. In such a case we shall write $r_1 \cup r_2 \cup \ldots \cup r_m$ or $\{r_1, r_2, \ldots, r_m\}$, where $r_1, r_2, \ldots, r_m$ are convex intervals. The above operations are extended over non-convex intervals in a straightforward way.

In certain cases, the sum of intervals (convex or non-convex) can be combined into a single term, e.g. $[2005, [6-7], [16-18], \_] \cup [2005, [6, 7], [21-23], \_]$ $= [2005, [6-7], [16, 17, 18, 21, 22, 23], \_]$, where the resulting term is a non-convex one. On the other hand, there may exist convex specifications which are not expressible with a single XTUS term, e.g. $[2004, [10-12], \_]$ $\cup$ $[2005, [1-3], \_]$ which is convex. This seems to be a straightforward consequence of the assumption of staying close to the natural language specification of time which admits such practical constructions.

Note that the intersection operation can be performed directly at the internal levels of hierarchical terms. We have the following observation.

observation 1 *Let* $r = [r_1, r_2, \ldots, r_k]$ *and* $s = [s_1, s_2, \ldots, s_k]$ *are two terms. The intersection of hierachical terms can be performed as*

$$r \cap s = [r_1 \cap s_1, r_2 \cap s_2, \ldots, r_k \cap s_k],$$

where $r_i \cap \_ = r_i$ *and* $\_ \cap s_i = s_i$ *for* $i = 1, 2, \ldots, k$, *provided that* $r_i \cap s_i \neq \emptyset$ *for any* $i \in \{1, 2, \ldots, k\}$ *and* $r \cap s = \varepsilon$ *otherwise.*

Observe that the result of intersection may be inconsistent without a visible empty intersection occurring at some position $i \in \{1, 2, \ldots, k\}$, e.g. $[2007, [1, 2], [1 - 31]] \cap [2007, [2, 3], [29 - 31]] = [2007, 2, [29 - 31]$ which is inconsistent. An intersection of hierarchical terms will be also referred to as composition.

5 Example: Temporal Specification for a Thermostat System

In this section an example specification of temporal requirements for a thermostat system are presented. The example comes from the handbook [18] and it constitutes a perfect example for investigating rule-based systems [16]. The goal is to define the set-point of a thermostat system depending on current time, where the time specification refers to season, weekday and hours. The original specification covers the following 18 rules.

```
Rule 1
if    the day is Monday or the day is Tuesday or the day is Wednesday
      or the day is Thursday or the day is Friday
then  today is a workday

Rule 2
if    the day is Saturday or the day is Sunday
then  today is the weekend

Rule 3
if    today is workday and the time is 'between 9 am and 5 pm'
then  operation is 'during business hours'

Rule 4
if    today is workday and the time is 'before 9 am'
then  operation is 'not during business hours'

Rule 5
if    today is workday and the time is 'after 5 pm'
then  operation is 'not during business hours'

Rule 6
if    today is weekend
then  operation is 'not during business hours'

Rule 7
if   the month is January or the month is February or the month is December
then  the season is summer

Rule 8
if    the month is March the month is April or the month is May
then  the season is autumn

Rule 9
if    the month is June or the month is July or the month is August
then  the season is winter

Rule 10
if   the month is September or the month is October or the month is November
then  the season is spring

Rule 11
if    the season is spring and operation is 'during business hours'
then  thermostat_setting is '20 degrees'

Rule 12
if    the season is spring and operation is 'not during business hours'
then  thermostat_setting is '15 degrees'
```

```
Rule 13
if    the season is summer and operation is 'during business hours'
then  thermostat_setting is '24 degrees'

Rule 14
if    the season is summer and operation is 'not during business hours'
then  thermostat_setting is '27 degrees'

Rule 15
if    the season is autumn and operation is 'during business hours'
then  thermostat_setting is '20 degrees'

Rule 16
if    the season is autumn and operation is 'not during business hours'
then  thermostat_setting is '16 degrees'

Rule 17
if    the season is winter and operation is 'during business hours'
then  thermostat_setting is '18 degrees'

Rule 18
if    the season is winter and operation is 'not during business hours'
then  thermostat_setting is '14 degrees'
```

Observe that preconditions of the rules form a perfect example of temporal specifications. In [16] we have developed a logical specification of these rules using attributive logic. In this paper the goal is to show a possible application of XTUS and its expressive power.

In order to develop the necessary temporal specifications let us first establish the correct scheme for representing time intervals. For covering the original specification we shall use five-component sequences of the form:

$$[season, \ month, \ day, \ weekday, \ hour]$$

allocated within the time domain equal to one year, where:

- *season* is the season of the year with the following arbitrary enumeration: Spring = 1, Summer = 2, Autumn = 3, and Winter = 4,
- *month* is the month of the year with the following enumeration: January = 1, February = 2, ..., December = 12,
- *day* is the day of the month,
- *weekday* is the day of the week with the following arbitrary enumeration: Monday = 1, Tuesday = 2, ..., Sunday = 7,
- *hour* is the hour (from 0 to 23).

We allocate the scheme within one year long window of time, and for shortening the notation the year position is omitted. Note that it is not necessary to represent the year in an explicit way – the specification of thermostat is year-independent; this means that one can apply it to the current year or the next one, etc. In fact, one can consider the specification as one covering finite (if finite TD is specified) or even infinite number of cycles. As a straightforward consequence of this simplification the universal variable symbol '_' can occur at the first position of hierarchical temporal terms.

Further, the specification is obviously redundant: the month determines the season (see rules 7-10). In fact we also make no use of the day of the month; it is kept only to keep consistency with former discussion. The finest level of granularity is hour, so there is no need to represent minutes or seconds.

Consider now the first two rules. Let wd denote a workday and wk – weekend. These intervals can be specified as follows:

$$wd = [\_, \_, \_, [1-5], \_], \tag{2}$$

$$wk = [\_, \_, \_, [6-7], \_]. \tag{3}$$

Recall that $[\_, \_, \_, [1-5], \_]$ is a shorthand for $[\_, \_, \_, [1, 2, 3, 4, 5], \_]$.

The above specifications are obviously cyclic ones: every season, month, and independently on the day number, the working days are from Monday to Friday, while weekend days are Saturday and Sunday.

Next, consider time specification. Let bh denote business hours and nbh not business hours. Extracting the information from rules 3-5 we have

$$bh = [\_, \_, \_, \_, [9-16]], \tag{4}$$

$$nbh = [\_, \_, \_, \_, \{[0-8], [17-23]\}]. \tag{5}$$

Recall that $\{[0-8], [17-23]\}$ is a non-convex interval and hence nbh given by (5) stays for the sum $[\_, \_, \_, \_, [0-8]] \cup [\_, \_, \_, \_, [17-23]]$.

Now we can construct the time specification covering the knowledge of rules 3-6. Let dbh denote 'during business hours', and $ndbh$ denote 'not during business hours'. What we need is the composition of specifications given by (2) – (5). So we have:

$$dbh = [\_, \_, \_, [1-5], \_] \cap [\_, \_, \_, \_, [9-16]], \tag{6}$$

$$ndbh = ([\_, \_, \_, [1-5], \_] \cap [\_, \_, \_, \_, \{[0-8], [17-23]\}]) \cup [\_, \_, \_, [6-7], \_]. \tag{7}$$

The first specification can be simplified to

$$dbh = [\_, \_, \_, [1-5], [9-16]], \tag{8}$$

and it constitutes a repetition of a convex interval. The second specification can be simplified to

$$ndbh = [\_, \_, \_, [1-5], \{[0-8], [17-23]\}] \cup [\_, \_, \_, [6-7], \_], \tag{9}$$

and it constitutes a repetition of a non-convex interval. For intuition, it says that 'not during business hours' means before 9:00 or after 17:00 on working days and any time during weekend. Next, let us pass to rules 7-10. Let *spr* denote Spring, *sum* denote Summer, *aut* denote Autumn and *win* denote

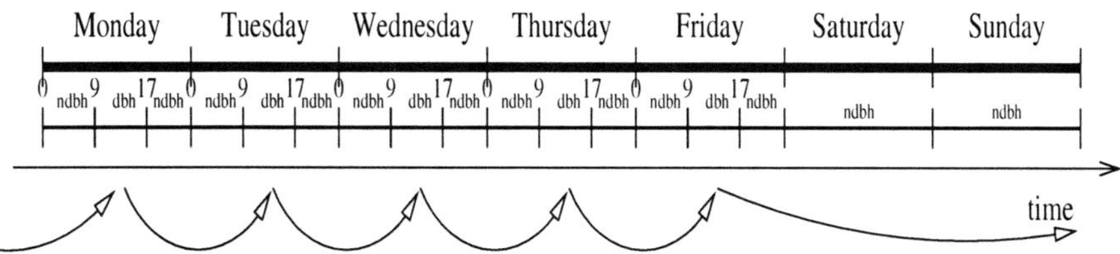

Fig. 2. An example specification of overlying cycles.

Winter. We have the four following straightforward specifications equivalent to rules 7-10:

$$spr = [1, [9{-}11], \_, \_, \_], \tag{10}$$

$$sum = [2, \{[1, 2, 12]\}, \_, \_, \_], \tag{11}$$

$$aut = [3, [3{-}5], \_, \_, \_], \tag{12}$$

$$win = [4, [6{-}8], \_, \_, \_]. \tag{13}$$

Finally, by appropriate composition of specifications of *dbh* (8) and *ndbh* (9) with specifications of season given by (10) – (13) we arrive at the temporal specifications of preconditions of rules 11-18; for simplicity they are named *r11* to *r18* in order to keep easy track of reference; so we have:

$$r11 = spr \cap dbh = [1, [9{-}11], \_, [1{-}5], [9{-}16]], \tag{14}$$

$$r12 = spr \cap ndbh =$$
$$[1, [9{-}11], \_, [1{-}5], \{[0{-}8], [17{-}23]\}] \cup [1, [9{-}11], \_, [6{-}7], \_], \tag{15}$$

$$r13 = sum \cap dbh = [2, [1, 2, 12], \_, [1{-}5], [9{-}16]], \tag{16}$$

$$r14 = sum \cap ndbh =$$
$$[2, [1, 2, 12], \_, [1{-}5], \{[0{-}8], [17{-}23]\}] \cup [2, [1, 2, 12], \_, [6{-}7], \_], \tag{17}$$

$$r15 = aut \cap dbh = [3, [3{-}5], \_, [1{-}5], [9{-}16]], \tag{18}$$

$$r16 = aut \cap ndbh =$$
$$[3, [3{-}5], \_, [1{-}5], \{[0{-}8], [17{-}23]\}] \cup [3, [3{-}5], \_, [6{-}7], \_], \tag{19}$$

$$r17 = win \cap dbh = [4, [6{-}8], \_, [1{-}5], [9{-}16]], \tag{20}$$

$$r18 = win \cap ndbh =$$
$$[4, [6{-}8], \_, [1{-}5], \{[0{-}8], [17{-}23]\}] \cup [4, [6{-}8], \_, [6{-}7], \_]. \tag{21}$$

Note that in fact the obtained specifications can be further simplified. For example, since the specification of months defines in fact the season, the specification of it is redundant. Moreover, since the day was not used one can further simplify the scheme to [*month, weekday, hour*]. Hence the specification of the intervals from the last two rules for example can take form as simple as $r17 = [[6{-}8], [1{-}5], [9{-}16]]$ and $r18 = [[6{-}8], [1{-}5], \{[0{-}8], [17{-}23]\}] \cup [[6{-}8], [6{-}7], \_]$.

6 Related Work

The main source of inspiration for the presented work were the papers by Ladkin. In [12, 13] granularities are defined as a particular set of sequences (e.g., YEARS is the set of sequences of length 1) or through the transitive closure of a *meet* operator that allows to define periodical repetitions of granules (e.g., Mondays as the "repetition" every 7 days of the Monday identified by *[2005,06,20]*. A limitation of this formalism is that the position in sequence is associated with a specific granularity. For example, one may want to define academic years and semesters and identify *[2005,1]* as the first semester of the academic year, while in [12] its interpretation is fixed to January 2005.

Another relevant work on *union-of-convex-intervals* and repetition is [14], but the emphasis here is more on reasoning with qualitative relations than on calendar expression representation. In [3], time units were used to represent the *union-of-convex-intervals* during which we evaluate the irreflexive temporal propositions. In this last work also, the emphasis is more reasoning with temporally qualified proposition than calendar expression. However, the work in [6] can be seen as an extension of the time unit system proposed in [12]. Similarly to integer sequences, calendar expression in [12] can identify the n-th granule of granularity within a granule of coarser one, but the granularity identifier is explicitly written and not implied by the position. The formalism in this work includes existential and universal quantification.

Several works are loosely related to presented proposal. Formalization of absolute and relative temporal specifications incorporating the terms introduced by Ladkin [12,14] knowledge was investigated in [8] and [9]. An appropriate axiomatization and logical inference rules for absolute and relative dates based on the point calculus were given in [8]. In [9] the notions of absolute and relative dates are explored in the context of inferring temporal dependencies between events. A comprehensive in-depth study and wide presentation of temporal knowledge specification an temporal inference in Artificial Intelligence is provided in [10].

Handling infinite temporal data[2] was the main focus of [11]. A framework extending the classical relational database approach by introducing generalized tuples referring to repeating points and simple constraints over them was presented there.

Time granularity has been extensively studied in the last few years. Among other, the work [2] and other papers by the same authors propose a general framework for the mathematical characterization of time granularities and investigate its applications in several AI and database (DB) areas. Logical aspects are deeply investigated in [17] proposing a multi-sorted temporal logic framework for time granularity. In particular, applications to real time systems specification and reasoning in the event calculus are studied. A symbolic formalism based on collections of temporal intervals was proposed in [15] with

---

[2] This is in fact the exact title of the paper by Kabanza et al. [11].

the aim to represent temporal expression occurring in natural language. Another issue is the slice formalism which was introduced in [19] as an alternative to the collection formalism in order to have a simple underlying evaluation procedure for the symbolic expression. Considering extensions of the slice formalism, in [1] slice expressions are used to represent granularities in the specification of temporal authorizations for database access control.

7 Conclusions and Future Work

This paper re-explores TUS being an attractive and powerful yet simple algebraic tool for constructing temporal specifications and performing some operations on them. In particular, an extended version of TUS, called XTUS is presented and its basic operations and properties are shown. One of the main features of XTUS consists in relaxing the constraint that time terms must be anchored with respect to the calendar. This is achieved through simple introduction of the unspecified flat term denoted with the underscore. It is argued that this simple and consistent with natural calendar way of building logical specification is capable of dealing with phenomena such as nested cycles. A further extension is that set and interval values are allowed in the specifications at any level of hierarchy. The expressive power of such terms is hence higher than of the ones based on constant integers only. A somewhat extended example is used to illustrate the ideas and application.

Contrary to most of the papers listed in the section devoted to related work, the paper is aimed at presenting a new proposal for algebraic temporal specifications and their handling, and not at presenting yet another set of temporal inference rules over temporal relations. Special attention is paid to easy definition of convex intervals and unions-of-convex-intervals as well as repeated (cyclic) events and nested cycles. Some basic operations on the algebraic time expressions of XTUS are defined. As the ultimate result we arrived at a simple yet powerful tool for hierarchical, algebraic temporal specifications of variable granularity. An important, intrinsic feature of the formalism consists in its extendibility – the precision of representation can easily be handled by simple scheme modification within the same framework.

Future work will be focused on introducing variables and defining constraints over them; this should still improve the expressive power of the formalism. Algebraic operations as well as their properties and implementation are also to be studied in more details. A separate paper devoted to fuzzy extensions of XTUS, the so-called FuXTUS [5] is presented in this volume.

References

1. Bettini C, Wang X, Ferrari E, Samarati P (1998) An access control model supporting periodicity constraints and temporal reasoning. ACM Transactions on Database Systems, 23:231–285

2. Bettini C, Wang X, Jajodia S (1998) A general framework for time granularity and its application to temporal reasoning. Annal of Mathematics and Artificial Intelligence, 22:29–58
3. Bouzid M, Ladkin P (2002) Simple reasoning with time-dependent propositions. International Journal of Interest Group in Pure and Applied Logic (IGPL), 10:379–399
4. Bouzid M, Ligęza A (2005) Algebraic temporal specifications with extend TUS: Hierarchical terms and their applications. In: Proceedings of the 17th IEEE International Conference on Tools with Artificial Intelligence, pp. 249–253
5. Bouzid M, Ligęza A (2007) Temporal Specifications with FuXTUS. A Hierarchical Fuzzy Approach. Studies in Computational Intelligence. Springer-Verlag, Berlin, Heidelberg (this volume)
6. Cukierman D, Delgrande J (1998) Expressing time intervals and repetion within a formalisation of calendars. Computational Intelligence 14:563–597
7. Fischer M, Gabbay D, Vila L (eds) (2005) Handbook of Temporal Reasoning in Artificial Intelligence, vol. 1. Elsevier, Amsterdam
8. Hajnicz E (1989) Absolute dates and relative dates in an inferential system on temporal dependencies between events. International Journal of Man-Machine Studies, 30:537–549
9. Hajnicz E (1991) A formalization of absolute and relative dates based on the point calculus. International Journal of Man-Machine Studies, 34:717–730
10. Hajnicz E (1996) Time Structures. Formal Description and Algorithmic Representation. Lecture Notes in Artificial Intelligence, vol. 1047. Springer, Berlin, Heidelberg
11. Kabanza F, Stévenne J-M, Wolper P (1995) Handling infinite temporal data. Journal of Computer and System Sciences, 51:3–17
12. Ladkin PB (1986) Primitives and units for time specification. In: Proceedings of the 5th National Conference on AI, AAAI'86, pp.354–359. Morgan Kaufmann
13. Ladkin PB (1987) The Logic of Time Representation. PhD thesis, University of California at Berkeley
14. Ladkin PB (1986) Time representation: A taxonomy of interval relations. In: Proceedings of the 5th National Conference on AI, AAAI'86, pp.360–366. Morgan Kaufmann
15. Leban B, Mcdonald D, Foster D (1986) A representation for collection of temporal intervals. In: Proceedings of the 5th National Conference on AI, AAAI'86, pp.354–359 Morgan Kaufmann
16. Ligęza A (2006) Logical Foundations for Rule-Based Systems. Studies in Computational Intelligence, vol. 11. Springer-Verlag, Berlin, Heidelberg
17. Montanari A (1996) Metric and layed temporal logic for time granularity. Technical report, ILLC Dissertation, University of Amsterdam
18. Negnevitsky M (2002) Artificial Intelligence. A Guide to Intelligent Systems. Addison-Wesley, Harlow, England; London; New York
19. Niezette N, Stevenne J, Leban B, Mcdonald D, Foster D (1992) An efficient symbolic representation of periodic time. In: Proceedings of International Conference on Information and Knowledge Management, pp.161–168

A Parallel Deduction for Description Logics with $\mathcal{ALC}$ Language

Adam Meissner and Grażyna Brzykcy

Institute of Control and Information Engineering, Poznań University
of Technology, pl. M. Skłodowskiej-Curie 5, 60-965 Poznań, Poland
`Adam.Meissner@put.poznan.pl`, `Grazyna.Brzykcy@put.poznan.pl`

1 Introduction

The term Description Logics (DLs) is commonly accepted to indicate a certain class of formal logics for representing knowledge and reasoning about it in information systems. These logics are descendants of a formal calculus that was proposed by Brachman in the KL-ONE system [5]. To represent the important entities of a given domain in DLs one can use atomic concepts, roles and individuals (instances of the concepts). Additionally, a set of constructors for denoting complex concepts and roles is defined to obtain the adequate expressivity of this formalism. Description Logics can be classified by the languages they support; one of the basic languages in this area is called $\mathcal{ALC}$ (the acronym stands for *Attribute Concept Description Language with Complements*).

Most of DLs are decidable subsets of the first order predicate calculus with efficient inference algorithms. The computational tractability together with the expressivity of DLs are the good reasons to use them as a knowledge representation formalism in various information systems. These systems are successively applied in different areas, e.g. software engineering, object data bases, medical expert systems, control in manufacturing and action planning in robotics. During the second half of nineties an interest in DLs increased significantly due to the emergence of the Semantic Web idea [14].

Reasoning systems for DLs provide a user with a basic service that is query answering. The queries are regarded as hypotheses, which are to be automatically inferred from the knowledge gathered in a system as a set of logical formulas. One of standard inference methods in the domain of DLs is the tableau-based algorithm ([2]) originating from the semantic tableau calculus [3]. This algorithm was implemented in many reasoning systems (also called inference systems or automated deduction systems) for DLs, such as FaCT, Racer or DLP ([2]). The efficiency of automated deduction systems can be generally increased by introducing distributed and parallel computations to them. This approach has been developed since early nineties and it

A. Meissner and G. Brzykcy: *A Parallel Deduction for Description Logics with $\mathcal{ALC}$ Language*,
Studies in Computational Intelligence (SCI) **102**, 149–164 (2008)
`www.springerlink.com`

yields many theoretical results [4]. However, bringing these ideas into practice is quite difficult. The main reason of problems seems to lie in the fact, that a method of distributing computations is a part of a program execution strategy. The strategy, in turn, is either defined in a program (as for example, in imperative programming) or it is a fixed part of a runtime environment (as for instance, in the Prolog language). In consequence, a constructor of an inference system usually has to implement also a method of parallelization and distribution of computations, which makes the whole construction more complicated and, therefore, it increases a probability of error. This solution may also cause problems with scalability, that is, with adapting the system to changes in a computational environment.

The matter looks different in the Mozart system [12], which is a runtime environment for the Oz programming language. The language supports constraint programming methodology [1] and it enables a program to be executed according to various strategies defined in the Mozart system. In particular, one of them implements parallel computations on distributed machines. We find the tableau-based inference algorithm very similar to the way the computations are performed in the constraint programming model. Therefore, in this paper we propose, how to take advantage on these similarities in order to parallelize the reasoning procedures for DLs. The one main inference problem in the logics is checking the unsatisfiability of a concept. The algorithm constructs a tableau, which represents all possible interpretations of the tested concept. As any branch of the tree is built independently of the others, the whole tree can be constructed in parallel and, consequently, an inference process might be shortened. We performed some computational experiments for combinatorial problems on the multi-agent architecture [11] of Mozart system and observed nearly linear processing time improvement with the growing number of agents in the search engine.

A possible real life application of this approach is the Semantic Web domain. To realize this new Web vision the community of knowledge engineers searches for efficient methods of knowledge representation and reasoning. Open standards are built for the representation of Web-based knowledge in a machine readable manner (RDF, OWL) and different tools for supporting reasoning in the Semantic Web are proposed (with SWRL submitted by W3C). A significant part of the tools is based on DLs, thus finding efficient inference methods in this area is a particularly important task.

The organization of this paper is as follows. The basic formalism of DLs is presented in section 2. In section 3, the essential constraint programming terms are depicted. A parallelization of an inference process for DL, performed in the constraint programming system Mozart, is defined in section 4. Section 5 contains some final remarks. An earlier version of this article was published as [6].

2 Description Logics with $\mathcal{ALC}$ Language

Every DL system contains two components called *concepts* and *roles*. A concept is a set of individuals, which are called its *instances*. If we consider such notions as *carbon, sodium* and *hydrogen*, we may treat them as instances of the *chemical element* concept. A role is a binary relation, which holds between two concepts. For instance, the role *sells* that intuitively relates a merchant to an article, is a subset of the set *merchant* $\times$ *article*, where terms *merchant* and *article* denote concepts. In DL there exist special expressions being names of concepts (called *concept descriptions*) and names of roles (called *role descriptions*). In the sequel, if it does not lead to misunderstanding, we will usually identify descriptions with their meanings.

In the applied notation we assume that concept instances and role descriptions start with lower-case letters, e.g. *carbon, sells*. As to the concept descriptions, they may be *atomic descriptions* or *complex descriptions*. At this point we assume that an atomic description is any alphanumeric string which starts with an upper-case letter, e.g. *ChemicalElement, Article*. Complex descriptions are built from simpler descriptions by means of the special symbols called *concept constructors*. For instance, the complex description of the form *Father* $\sqcup$ *Mother* intuitively represents the concept *Parent*, with *Father* and *Mother* treated as the appropriate atomic descriptions, and with the symbol $\sqcup$ denoting the union of concepts. We will use the letter A to denote an atomic concept description and letters C or D as symbols of any concept descriptions. The letter R will stand for the name of any role. The expression of the form $C(x)$ means that an individual x is an instance of a concept C. The term $R(x, y)$ represents a pair of individuals $\langle x, y \rangle$ belonging to a role R. Expressions of the first form are called *concept examples* and expressions of the second form are referred to as *role examples*.

The semantics of concept and role descriptions is formally defined by means of an interpretation I, which consists of the interpretation domain Δ^I and the interpretation function. The domain of the interpretation is the smallest, non-empty set of all the instances of all the considered concepts. The interpretation function is used to assign a subset of Δ^I to every concept description, and to assign a binary relation, contained in $\Delta^I \times \Delta^I$, to every role description. We say that an interpretation I *satisfies* a concept description or a role description if it assigns a non-empty set to it. The interpretation I of an expression E over a domain Δ^I is represented by the symbol E^I. Table 1 comprises all concept constructors of the $\mathcal{ALC}$-language.

In the first column a constructor name is given, the second column (*Syntax*) holds a scheme of the relevant concept description and the third column (*Semantics*) contains an interpretation of the description. All symbols appearing in the table are used with their usual meaning in the first order logic and the set theory. Moreover, there are two special concept descriptions, namely $\top$ (*top*) and $\bot$ (*bottom*). The first one denotes the most general concept, that is $\top^I = \Delta^I$ while the second represents the least general concept,

Table 1. Concept constructors of the $\mathcal{ALC}$-language

Constructor name	Syntax	Semantics
negation	$\neg\, C$	$\Delta^I \setminus C^I$
intersection	$C \sqcap D$	$D^I \cap C^I$
union	$C \sqcup D$	$D^I \cup C^I$
value restriction	$\forall R.C$	$\left\{ a \in \Delta^I \mid (\forall b)\, \langle a,b \rangle \in R^I \rightarrow b \in C^I \right\}$
existential quantification	$\exists R.C$	$\left\{ a \in \Delta^I \mid (\exists b)\, \langle a,b \rangle \in R^I \wedge b \in C^I \right\}$

i.e. $\perp^I = \emptyset$. Here we give some more examples of $\mathcal{ALC}$ concept descriptions with their informal interpretation: $\neg LivingObject$ (a non-living object), $(Employee \sqcap Student)$ (a working student), $\exists hasChild.(Mother \sqcup Father)$ (a grandparent), $\neg Woman \sqcap \exists hasChild.Person$ (a grandfather), $\forall hasChild.Boy$ (a person having no other children than boys).

It should be noted, that a well-known extension of the $\mathcal{ALC}$-language is the $\mathcal{ALCN}$-language. It enriches the former language with so-called *number restrictions*, i.e. two new concept constructors. They enable to create descriptions of the form $(\geq nR)$ and $(\leq nR)$ representing sets of all instances, which are in the relation R with at most (or, respectively, at least) n individuals.

The DL formalism enables to express theories with finite sets of axioms describing relationships among concepts and among roles. The very meaningful in practice is the theory called *terminology*, which axioms take the form of $A \equiv C$, where the symbol $\equiv$ represents the equality of sets. Every axiom of this type is called a *definition* of the concept A. We say that the interpretation I satisfies the definition $A \equiv C$ iff A^I and C^I are the same sets of individuals. Moreover, it is assumed that every concept can be defined only once in the given terminology. We say that the interpretation I *satisfies* the terminology T, or equivalently, that I is a *model* of T, if I satisfies every definition contained in T. A concept C is *satisfiable* with respect to a terminology T if there exists a model of T which satisfies C. In other case C is considered to be *unsatisfiable* with respect to T. In particular, a terminology may be an empty set. A concept which is (un)satisfiable with respect to an empty terminology is called an (un)satisfiable concept.

The inference process in a DL system consists in verification of hypotheses concerning various properties of concepts and roles. These hypotheses are used to denote the satisfiability of a concept, the *equivalence* or *disjointness* of concepts and the *subsumption* relationship between concepts. The formal definitions of subsumption, equivalence and disjointness are given below.

1. A concept C is *subsumed* by a concept D with respect to a terminology T if C^I is included in D^I for every model I of T.
2. Two concepts C and D are *equivalent* with respect to a terminology T if C^I and D^I are identical sets for every model I of T.

3. Two concepts C and D are *disjoint* with respect to a terminology T if C^I and D^I have no common elements for every model I of T.

It is important that each of these hypotheses can be reduced to the problem of checking unsatisfiability of a concept (with respect to the empty terminology). The reduction consists of the two steps. In the first step one of the following rules should be applied to a given hypothesis.

1. A concept C is satisfiable with respect to a terminology T iff it is not unsatisfiable with respect to T.
2. A concept C is subsumed by a concept D with respect to a terminology T iff a concept $C \sqcap \neg D$ is unsatisfiable with respect to T.
3. Concepts C and D are equivalent with respect to a terminology T iff a concept $(C \sqcap \neg D) \sqcup (\neg C \sqcap D)$ is unsatisfiable with respect to T.
4. Concepts C and D are disjoint with respect to the terminology T iff a concept $C \sqcap D$ is unsatisfiable with respect to T.

The second step is called *an elimination of a terminology* [2]. For every concept C a concept C' may be constructed, such that it is an unsatisfiable concept iff C is unsatisfiable with respect to the given terminology T. In order to obtain a concept C' from C, one has to replace all the atomic concepts appearing in C by their definientia taken from T. This process should be repeated subsequently until the concept C' is gained, which contains no atomic concept defined in T.

The inference system requires the transformation of the concept description into the *negation normal form* (NNF), before this concept is tested for unsatisfiability. In NNF form the negation symbols occur only in front of the atomic concept subdescriptions [2]. The process of unsatisfiability checking may be done via a tableau-based algorithm, which may be summarized as follows (see [2] for details).

During an inference process a tree Tb is constructed, which is called a *tableau* for historical reasons. The root of a tree Tb takes a form of the set $\{C(x)\}$, where C represents a tested concept and x is a symbol of some individual. The other nodes of the tree Tb may be obtained by applying the following inference rules to their parents.

1. If a node w contains an expression $(C \sqcap D)(x)$ but it does not contain expressions $C(x)$ and $D(x)$ then create a successor w_1 for w, such that $w_1 = w \cup \{C(x), D(x)\}$.
2. If a node w contains an expression $(C \sqcup D)(x)$ but it does not contain neither $C(x)$ nor $D(x)$ then create two successors w_1 and w_2 for w, such that $w_1 = w \cup \{C(x)\}$ and $w_2 = w \cup \{D(x)\}$.
3. If a node w contains an expression $(\exists R.C)(x_1)$ and there is no individual name x_2, such that a set $\{C(x_2), R(x_1, x_2)\}$ is included in w then create a successor w_1 for w, such that $w_1 = w \cup \{C(x_3), R(x_1, x_3)\}$, where x_3 is a new individual name not occurring in w.

4. If a node w contains an expression $(\forall R.C)(x_1)$ and an expression $R(x_1, x_2)$ and it does not contain an expression $C(x_2)$ then create a successor w_1 for w, such that $w_1 = w \cup \{C(x_2)\}$.

The leaves of a tree Tb are nodes that fulfill at least one of the following conditions.

1. A node contains a contradiction (called *clash*) if it contains a set $\{D(y), \neg D(y)\}$, where the letter D has a usual meaning and the letter y denotes some individual.
2. No inference rule can by applied to the node.

The path of the tree Tb ended by a leaf satisfying the first condition is called a *closed path*. If all the paths of the tree Tb are closed then the tree is called a *closed tableau*. A concept C is unsatisfiable iff one can construct a closed tableau for it.

Below we present an example, based on [2,8], in which we show a process of constructing the tableau for a false hypothesis. Therefore, the concept description corresponding to the hypothesis is satisfiable and thus, the resulting tableau is not closed.

Example 1. Let us assume that letters N, I and h denote, respectively, the description of the concept *NicePerson*, the description of the concept *IntelligentPerson* and the description of the role *hasChild*. Thus, the expression $(\exists h.N \sqcap I)$ can be interpreted as a description of the parent who has at least one nice and intelligent child. Whereas, the description $(\exists h.I) \sqcap (\exists h.N)$ symbolizes a concept of the parent having at least one intelligent child and at least one nice child, however it does not have to be the same person. In consequence, the hypothesis that $(\exists h.I) \sqcap (\exists h.N)$ is subsumed by $(\exists h.N \sqcap I)$ does not hold. This proposition can be verified by inference concerning unsatisfiability of the concept $\neg(\exists h.N \sqcap I) \sqcap (\exists h.I) \sqcap (\exists h.N)$ which has the following negation normal form: $(\forall h.\neg N \sqcup \neg I) \sqcap (\exists h.I) \sqcap (\exists h.N)$. The tableau, constructed during the inference process, is shown in Figure 1. The symbol

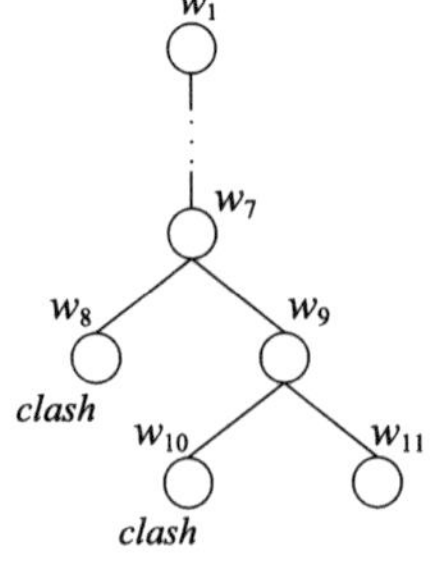

Fig. 1. An example of a tableau

'...', written vertically, denotes a non-branching path, which consists of nodes $w_2, w_3, \ldots, w_6$. Nodes of the tableau are labeled by the following sets.

$$w_1 = \{((\forall h.\neg N \sqcup \neg I) \sqcap (\exists h.I) \sqcap (\exists h.N))(x)\} \qquad \% \text{ r-1,}$$
$$w_2 = \{(\forall h.\neg N \sqcup \neg I)(x), ((\exists h.I) \sqcap (\exists h.N))(x)\} \qquad \% \text{ r-1,}$$
$$w_3 = \{(\forall h.\neg N \sqcup \neg I)(x), (\exists h.I)(x), (\exists h.N)(x)\} \qquad \% \text{ r-3,}$$
$$w_4 = \{h(x, x_1), I(x_1), (\forall h.\neg N \sqcup \neg I)(x), (\exists h.N)(x)\} \qquad \% \text{ r-3,}$$
$$w_5 = \{h(x, x_1), I(x_1), h(x, x_2), N(x_2), (\forall h.\neg N \sqcup \neg I)(x)\} \qquad \% \text{ r-4,}$$
$$w_6 = \{h(x, x_1), I(x_1), h(x, x_2), N(x_2), (\neg N \sqcup \neg I)(x_1),$$
$$(\forall h.\neg N \sqcup \neg I)(x)\} \qquad \% \text{ r-4,}$$
$$w_7 = \{h(x, x_1), I(x_1), h(x, x_2), N(x_2), (\neg N \sqcup \neg I)(x_1), (\neg N \sqcup \neg I)(x_2)\} \ \% \text{ r-2,}$$
$$w_8 = \{h(x, x_1), I(x_1), h(x, x_2), N(x_2), \neg I(x_1), (\neg N \sqcup \neg I)(x_2)\} \qquad \% \text{ clash,}$$
$$w_9 = \{h(x, x_1), I(x_1), h(x, x_2), N(x_2), \neg N(x_1), (\neg N \sqcup \neg I)(x_2)\} \qquad \% \text{ r-2,}$$
$$w_{10} = \{h(x, x_1), I(x_1), h(x, x_2), N(x_2), \neg N(x_1), \neg N(x_2)\} \qquad \% \text{ clash}$$
$$w_{11} = \{h(x, x_1), I(x_1), h(x, x_2), N(x_2), \neg N(x_1), \neg I(x_2)\}$$

A comment placed after the symbol '%' indicates an inference rule, which is applied to the given node. In particular case a comment has a form of the *clash* mark pointing out a contradiction in the node. It should be observed, that a path ended by the node w_{11} is not closed. Thus, the considered concept is not unsatisfiable and the label of the node provides the interpretation, which satisfies this concept.

3 Constraint Programming Paradigm

The constraint programming approach is the successor of the declarative computational model where knowledge is formulated by means of constraints imposed on entities (objects). In this approach a problem is described by a set of formulas (i.e. constraints) with variables ranging over some initial finite domains, which are usually represented by sets of non-negative integers. Solving a constraint satisfaction problem consists in finding a variable assignment which satisfies all considered formulas.

The problem of finding two small integers, such that a sum of them and the product are the same takes, in constraint programming, an extremely simple form. If X and Y represent these two positive integers lesser than 10 then the problem specification is as follows: X::1#9, Y::1#9, Z::1#81, X+Y=:Z, X*Y=:Z.

Every constraint imposes a specific connection among particular variables and it can be regarded as a partial information about the solution of the constraint satisfaction problem (values of variables). During the calculation this partial knowledge is used to perform local deductions on it. As new information (in the form of constraints) is added some values of variables are excluded from their domains. This process is named constraint propagation and corresponds to the narrowing of the finite domains. In the example with sum and product of two integers constraint propagation calculates the new domain for Z, namely Z::2#18. But constraint propagation is not a sufficient (complete) method and a controlled search is inevitable. When no more local deduction is possible a search step is performed that consists in splitting a problem into complementary subproblems, each with additional constraints. In our example computation is continued with additional constraints X=1 (X::1#1) and X ≠ 1 (X::2#9) respectively. The decomposition process (distribution) leads to the emergence of a search tree with branches representing distinct partial calculations of the given problem. Interleaving of the before mentioned two inference methods is the basic mechanism of efficient constraint programming systems and it is called propagate-and-search (propagate-and-distribute) approach. In the search tree for the example only one leaf contains the solution (value assignment), namely X=2, Y=2, Z=4.

In the problem description and calculation a key role is played by shared variables, which appear in different constraints and act as communication channels between them. Accordingly to the significance of possible values of variables constraints are divided into two categories. The first category, so-called *basic constraints*, represents knowledge about consistent, potential values of variables within their domains. These constraints are depicted as subsets of the finite domains, for instance X::3#7, Y::5#6 or Z::2#2, equivalent to Z=2. The conjunction of all basic constraints constitutes a *constraint store*. A solution of constraint satisfaction problem is found when values of all problem variables are determined. We say that the constraint store determines a variable X if this store entails X=d for some value d from the domain.

The second category comprises *non-basic constraints* representing relationships among variables, such as X+Y≥Z or X^2+Y^2=Z^2. These constraints are imposed by *propagators*. A propagator is a computational agent, which implements an operational semantics of the appropriate non-basic constraint. It observes actual domains of relevant variables and tries to narrow them as much as possible, until the fixpoint is reached. More formally, a propagator imposing a constraint C on a constraint store, which holds a constraint S, can *tell* (*propagate*) a new basic constraint B to this store if B is *entailed* by conjunction $S \wedge C$ and is consistent to the store. The updated store holds the conjunction $S \wedge B$. Results of constraint propagation do not depend on the order in which the propagators tell information to the store.

A propagator is *failed* if there is no variable assignment, which satisfies both the constraint imposed by this propagator and the constraint store. A propagator ceases to exist (i.e. it disappears) if it is *entailed* - every variable assignment that satisfies the constraint store satisfies also the propagator (i.e. the imposed constraint). In other cases a propagator is *stable*.

A computational architecture that consists of a constraint store and propagators connected to the store is called a *computation space* (or simply a *space*) [9]. A space may be *failed*, *stable*, or *solved* due to different states of its propagators. When a space is created, the constraint propagation starts and continues until the space becomes stable - all propagators in this space are stable. If one of the propagators is failed, so is the whole space. A space is solved if it contains no propagators (all propagators disappear).

It is possible that a space becomes stable, but is neither failed nor solved. As no further propagation can be done, the new spaces are created, which retain the same set of solutions. Distribution of a space S is related to a basic constraint, say B (or to a set of *alternatives*), which is told to a constraint store once in positive $(S \wedge B)$ and once in negative $(S \wedge \neg B)$ form. When a choice of B is appropriate, constraint propagation may be triggered in both new spaces. The most straightforward strategy of distribution (a *naive* strategy), in the context of finite domains, consists in defining B in the form of $X = d$, where X is any variable with a non-singleton domain and d is a value from this domain. Selection of a variable with the least number of values (a *first-fail* strategy) is often used as a powerful refinement of the naive strategy.

The interleaving of constraint propagation and space distribution yields a complete computation model for finite domains problems. In terms of computation spaces a tree of spaces (a *search tree*) represents every problem. A shape of this tree is entirely defined by the distribution strategy and its leaves correspond to solved or failed spaces. In order to find a problem solution one has to explore the search tree. Since this task is domain independent, the special *search engines* are provided with constraint programming systems.

Computation spaces are useful abstractions that encapsulate constraint-based computations. Every space is uniquely referred to by its name. A careful selection of operations on spaces is of a great importance for programming constraint satisfaction problems. In Mozart system one can find operations that create new spaces, clone stable spaces, inject computations into existing spaces, merge spaces with the root (top-level) space, block until space becomes stable or commit a distributable space to alternatives of its distributor.

It should be remarked, that problems from the considered domain may be solved also by some other approaches - one of them is a known *logic-algebraic method* (LAM) [7]. However, this method can be described in terms of constraint programming with Boolean variables (taking on only two values) standing for simple formulas (simple properties) from LAM. Constraints (relations) among variables take a form of logical formulas (facts in LAM).

4 Inference Process as Parallel Constraint Solving

As said before, every path of the tableau Tb represents an attempt of building an interpretation satisfying a concept description, which occurs in the root of the tableau. Every node, which ends a non-closed path, describes one possible interpretation. All paths of the tableau Tb may be constructed mutually independently, i.e. no information has to be exchanged between nodes which do not belong to the same path. Therefore, the process of constructing the Tb tree can be easily parallelized.

In this paper we propose to achieve this effect by the use of constraint programming methodology, particularly by parallel constraint solving technique, which is supported by the Mozart programming system. Hence, we presume that a tableau is modeled by a search tree and they are both created in a similar way. Furthermore, in our approach, concepts are represented by sets of non-negative integers, that is, individuals being concept examples are encoded as numbers. Consequently, relations between concepts and individuals (e.g. a membership of an individual to a concept) can be expressed by standard propagators defined in *FS* (*Finite Set*) library of the Mozart system. This leads to the following general realization assumptions.

1. Spaces of the search tree contain variables representing atomic concepts, which occur in the tableau. Every variable domain is a set of non-negative integers. More precisely, a variable is constrained by a pair of sets where the first element consists of individuals which are known to be examples of the considered concept while the second element is a set of individuals not being the concept examples. The root space initially contains no variables.
2. Every application of an inference rule to a node of the tableau corresponds to one step of the appropriate distribution strategy (described in the sequel). It may result in the insertion of new constraints (possibly with new variables) to a relevant space. Moreover, if a node v of the tableau has exactly one successor w then new constraints are inserted to the same space for both nodes v and w.
3. A node of the tableau which contains clash is represented by a failed space.

It should be noted, that in general, one space of the search tree may stand for more than one node of the tableau. More precisely, vertices of the search tree (i.e. spaces) can be bijectively mapped to appropriate sets of nodes of the tableau. Every set of this type consists of nodes forming one non-branching subpath; in particular case it may be a singleton set. This dependency follows from the fact, that an application of inference rules, which do not result in branching of the tableau, may be interpreted as an injection of subsequent constraints to the same space.

In every step of the search tree construction one space is distinguished as a *current space*, to which injection of constraints and propagation is applied by default. One has to observe, that if the tree is constructed in parallel by a few processes, then a concept of the current space is local for each of them.

The structure of the search tree is determined by the distribution strategy specified by the procedure `DistrNodes`, which is implemented in the Oz language [10]. The simplified declaration of this procedure is given below. For the sake of readability, we extend the Oz notation by meta-expressions, which are replaced in the implementation by appropriate Oz construction with the same meaning. Expressions of the object language (i.e. Oz) are set in typewriter font, while metalanguage symbols are in italic font. Additionally, Oz braces are boldfaced to distinguish them from metalanguage braces. The metalanguage contains elements described in section 2 (such as concepts and roles and examples of both of them) and symbols, which are commonly used in logic and in the set theory. Moreover, slightly abusing the notation of concept and role examples, we introduce meta-procedures and meta-functions (e.g., $NoIndiv$, $IsAtom$) with the semantics given below.

```
proc {DistrNodes Concs Roles VarsI VarsO}                        % 1
  if  A(x) ∈ Concs ∧ IsAtomic(A) ∧ NoConstr(A(x))               % 2
  then                                                           % 3
    {FS.include x var(A)}                                        % 4
    AddNewVar(VarsI, var(A), VarsIa)                             % 5
    {DistrNodes Concs \ {A(x)} Roles VarsIa VarsO}              % 6
  elseif ¬A(x) ∈ Concs ∧ NoConstr(¬A(x))                        % 7
  then                                                           % 8
    {FS.exclude x var(A)}                                        % 9
    AddNewVar(VarsI, var(A), VarsIa)                             % 10
    {DistrNodes Concs \ {A(x)} Roles VarsIa VarsO}             % 11
  elseif (C ⊓ D)(x) ∈ Concs ∧ (C(x) ∉ Concs ∨ D(x) ∉ Concs)    % 12
  then                                                           % 13
    {DistrNodes {C(x), D(x)} ∪ Concs Roles VarsI VarsO}         % 14
  elseif (C ⊔ D)(x) ∈ Concs ∧ C(x) ∉ Concs ∧ D(x) ∉ Concs       % 15
  then                                                           % 16
    choice                                                       % 17
       {DistrNodes {C(x)} ∪ Concs Roles VarsI VarsO}            % 18
    [] {DistrNodes {D(x)} ∪ Concs Roles VarsI VarsO}            % 19
    end                                                          % 20
  elseif (∃R.C)(x) ∈ Concs ∧ NoIndiv(y, R(x, y) ∈ Roles ∧      % 21
         C(y) ∈ Concs)                                          % 22
  then                                                           % 23
    NewIndiv(z)                                                  % 24
    {DistrNodes {C(z)} ∪ Concs {R(x, z)} ∪ Roles VarsI VarsO}  % 25
  elseif (∀R.C)(x) ∈ Concs ∧ R(x, y) ∈ Roles ∧ C(y) ∉ Concs)    % 26
  then {DistrNodes {C(y)} ∪ Concs Roles VarsI VarsO}            % 27
  else VarsO = VarsI                                             % 28
  end                                                            % 29
end                                                              % 30
```

The procedure `DistrNodes` has four arguments. The first two of them are sets of expressions denoting concept examples ($Concs$) and, respectively, role examples ($Roles$). The set of variables standing for atomic concepts is represented by last two arguments. The third argument ($VarsI$) is an input set, while the last argument ($VarsO$) contains variables introduced to the search tree by the considered procedure. The sets are implemented as lists.

The activities, undertaken by the procedure `DistrNodes` in lines 2–11, are intended to process two kinds of elements of the set $Concs$, i.e. examples of atomic concepts and examples of negated atomic concepts. The meta-expression $IsAtomic(A)$ in line 2 gets a boolean value $true$ iff A represents an atomic concept description. It should be noted, that an analogous test is not performed for concept descriptions preceded by the negation constructor, since all the descriptions are in negation normal form. The meta-expression of the form $NoConstr(C(x))$, appearing in lines 2 and 7, where $C = A$ or $C = \neg A$, obtains a boolean value true iff the current space does not contain constraints regarding the concept example $C(x)$. Moreover, the variable representing the concept A is available as a value of the meta-expression $var(A)$.

Lines 4 and 9 contain procedure calls introducing propagators to the current space, which are defined in the standard library *FS* of the Mozart system. These propagators impose constraints stating that a given individual belongs (line 4) or, respectively, does not belong (line 9) to a set, which represents an atomic concept description. The meta-procedure $AddNewVar(VarsI, Var, VarsIa)$ returns the set of variables $VarsIa$, which is an input set $VarsI$ extended by a new variable Var.

Lines 12–14 of the procedure `DistrNodes` correspond to the inference rule for the intersection constructor; a meta-expression contained in line 12 describes conditions of the rule while the line 14 describes the adequate action.

Processing the concept description, which is built with the union constructor (lines 15–20) results in branching of the search tree. In other words, a current space is duplicated in two successors, which are considered separately in further processing. In the procedure `DistrNodes`, the branching effect is achieved by the use of the `choice` instruction (line 17). It groups together a set of alternative statements, which are to be executed independently.

The subsequent lines (21–25) represent the inference rule for the existential quantification constructor. The meta-expression of the form $NoIndiv(y, P)$ in line 21 obtains a boolean value *true* if there is no individual satisfying the condition P - in other case, it is false. The meta-expression contained in the line 24 is to be interpreted as a procedure call, which creates a new individual that has not occurred in the current space. As said before, individuals are represented by subsequent, non-negative integers beginning from 0.

Lines 26–27 correspond to the inference rule regarding the value restriction constructor. If no inference rule is applicable (line 28) then the output argument $VarsO$ is unified with the input set of variables $VarsI$.

The distribution strategy plays a key role in the proof procedure intended for verification of the input hypothesis Hyp. This procedure is generated by the function `MakeHypProof`. A simplified definition of the regarded function is given below; we use the same notation as for the procedure `DistrNodes`. The symbol \$, which stands for the procedure name in line 2, causes that the expression denoting the procedure (lines 2–8) becomes its identifier.

```
fun {MakeHypProof Hyp}                              % 1
  proc {$ Sol}                                      % 2
    {TransToUnsatisfiability Hyp C}                 % 3
    {EliminateTerminology C C1}                     % 4
    {TransToNNF C1 D}                               % 5
    NewIndiv(z)                                     % 6
```

```
    {DistrNodes {D(z)} ∅ ∅ Sol}                                   % 7
  end                                                             % 8
end                                                               % 9
```

Firstly (line 3), the hypothesis Hyp is reduced to unsatisfiability checking of the appropriate concept C. In the next step, the terminology elimination procedure (line 4) is applied to C and the result is transformed into the negation normal form (line 5), which is denoted by the concept description D. Finally, after creation of a new individual z, the proof procedure builds a search tree (line 7) for the concept example $D(z)$. In the output parameter of the procedure (`Sol`) the list of constrained variables corresponding to atomic concepts is returned if at least one of leaves of the search tree is a non-failed space. A non-empty list `Sol` represents an interpretation which satisfies the concept D. In other case, that is, when the list is empty (`nil`), the concept D is unsatisfiable and thus the hypothesis Hyp holds. The execution of the proof procedure generally consists in building a search tree in connection with constraint propagation at particular vertices (i.e. in spaces). In the Mozart system, this task is carried out by special objects (i.e. search engines) being instances of various subclasses of the class `Search`. Every subclass implements a specific strategy of the search tree exploration. Particularly, objects from the subclass `Search.parallel` are used for the parallel processing of the search tree. A computer creating an object of this type takes a role of the *manager*, which controls the computational process. The manager points out so-called *workers*, that is, processes intended for exploring the tree. It also specifies the number of such processes, which run on each of the machines from the computational environment. For example, the following command

```
E = {New Search.parallel init(w1:2#rsh w2:1#rsh)}
```

declares that the search tree will be explored in parallel by three workers. Two of them run on the machine `w1` while the third runs on the computer `w2`. In all cases the manager communicates with workers using the remote command interpreter `rsh`. The proof procedure is triggered by sending an appropriate message to the object being a value of the variable `E`.

Example 2. We show how the procedure `DistrNodes` is applied to check the unsatisfiability of the concept representing the hypothesis from Example 1. Therefore, we present below a sequence of values assigned to input arguments of the procedure during subsequent calls. For the sake of readability we continue to use the same metalanguage as in the procedure definition (only the numbers denoting individuals are given in the object language, i.e. in Oz). For the same reason, the concepts which do not participate in further inferences are omitted from the list $Concs$. Every tuple of arguments is accompanied by a comment (preceded by the '%' sign) indicating the current space name. At first, the procedure is called with one-element list $Concs$.

```
Concs = {((∀h.¬N ⊔ ¬I) ⊓ (∃h.I) ⊓ (∃h.N))(0)}              % v₁
Roles = ∅
VarsI = ∅
```

Then, the inference rule for the intersection constructor is applied three times, which results in th following three-element list *Concs*.

```
Concs = {(∀h.¬N ⊔ ¬I)(0), (∃h.I)(0), (∃h.N)(0)}          % v₁
Roles = ∅
VarsI = ∅
```

In the next step, the second element of the list *Concs* (with the existential quantification constructor) is processed. It is not used in further inferences and thus it is removed from the resulting list of concepts. Instead, the conclusions $I(1)$ and $h(0,1)$ are respectively added to lists *Concs* and *Roles*.

```
Concs = {I(1), (∀h.¬N ⊔ ¬I)(0), (∃h.N)(0)}          % v₁
Roles = {h(0,1)}
VarsI = ∅
```

Later on, the expression $I(1)$ is taken into account. During this process, a new variable representing the atomic concept I is created and the propagator `FS.include` adds the individual 1 to the variable domain. We denote an atomic concept with constraints imposed on its domain by an expression of the form of $Var :< Excl, Incl >$, where Var is a variable representing the concept and expressions $Excl$ and $Incl$ represent sets of individuals, which are known to be instances ($Incl$) or, respectively, are not instances ($Excl$) of a considered concept.

```
Concs = {(∀h.¬N ⊔ ¬I)(0), (∃h.N)(0)}          % v₁
Roles = {h(0,1)}
VarsI = {I :< ∅, {1} >}
```

The same two steps are applied to the element $(\exists h.N)(0)$ of the list *Concs*. This results in the following tuple of input arguments.

```
Concs = {(∀h.¬N ⊔ ¬I)(0)}          % v₁
Roles = {h(0,1), h(0,2)}
VarsI = {I :< ∅, {1} >, N :< ∅, {2} >}
```

Afterward, the inference rule for the value restriction constructor is executed two times. The premise $(\forall h.\neg N \sqcup \neg I)(0)$ is neglected in further considerations.

```
Concs = {(¬N ⊔ ¬I)(1), (¬N ⊔ ¬I)(2)}          % v₁
Roles = {h(0,1), h(0,2)}
VarsI = {I :< ∅, {1} >, N :< ∅, {2} >}
```

Later on, the rule for the union constructor is applied to the first element of the list *Concs*. In consequence, two new search spaces are created, namely v_2 and v_3. Computations in these spaces can be performed in parallel by an appropriate search engine.

```
Concs = {(¬I)(1), (¬N ⊔ ¬I)(2)}          % v₂
Roles = {h(0,1), h(0,2)}
VarsI = {I :< ∅, {1} >, N :< ∅, {2} >}
```

```
Concs = {(¬N)(1), (¬N ⊔ ¬I)(2)}          % v₃
Roles = {h(0,1), h(0,2)}
VarsI = {I :< ∅, {1} >, N :< ∅, {2} >}
```

After the atomic concept examples (i.e. $(\neg I)(1)$ and $(\neg N)(1)$) are processed, it turns out that the space v_2 is failed since it contains the variable I, for which the individual 1 is both included and excluded from its domain. Hence, the further computations are realized only in the space v_3.

```
Concs = {(¬N ⊔ ¬I)(2)}                                    %  v2
Roles = {h(0,1), h(0,2)}
VarsI = {I :< {1}, {1} >, N :< ∅, {2} >}
```

```
Concs = {(¬N ⊔ ¬I)(2)}                                    %  v3
Roles = {h(0,1), h(0,2)}
VarsI = {I :< ∅, {1} >, N :< {1}, {2} >}
```

The computations consist in the analysis of the expression $(\neg N \sqcup \neg I)(2)$, which leads to the creation of subsequent new spaces, i.e. v_4 and v_5.

```
Concs = ∅                                                 %  v4
Roles = {h(0,1), h(0,2)}
VarsI = {I :< ∅, {1} >, N :< {1,2}, {2} >}
```

```
Concs = ∅                                                 %  v5
Roles = {h(0,1), h(0,2)}
VarsI = {I :< {2}, {1} >, N :< {1}, {2} >}
```

The space v_4 is failed because it encloses contradictory constraints imposed on the variable I, but the space v_5 does not contain any contradiction. This means that the input hypothesis is not unsatisfiable since it is satisfied by the interpretation represented by the list `VarsI` handled in the space v_5.

The search tree, representing the tableau from Figure 1, is depicted in Figure 2. The correspondence relation between vertices of the tree and nodes of the tableau, denoted by the symbol '$\leftrightarrow$', is as follows: $v_1 \leftrightarrow \{w_1, \ldots, w_7\}$, $v_2 \leftrightarrow \{w_8\}$, $v_3 \leftrightarrow \{w_9\}$, $v_4 \leftrightarrow \{w_{10}\}$, $v_5 \leftrightarrow \{w_{11}\}$. Expressions, printed inside vertex ovals, denote variables representing atomic concepts with constraints imposed on their domains.

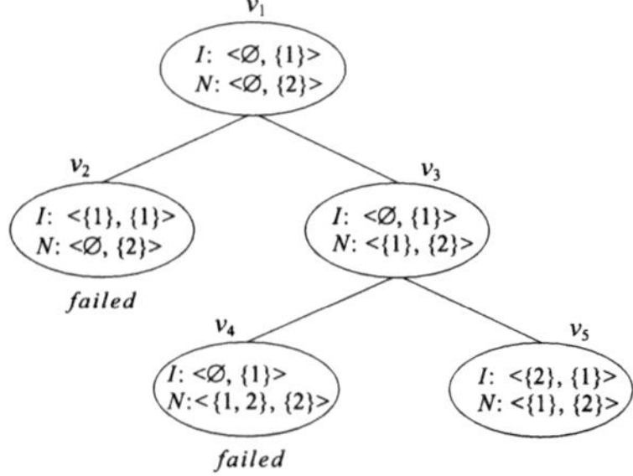

Fig. 2. A search tree for the tableau from Fig. 1

5 Final Remarks

Description Logics are useful formalism for representing knowledge and reasoning in information systems. Nowadays Semantic Web community and necessity of ontology-based processing [13] additionally confirm their utility. So, an efficient inferring in DLs is a task of particular importance. Studying a tableau-based inference algorithm for DLs, we find it similar to constraint programming approach with a search tree as the computational model. A method of transforming nodes of a tableau into vertices of the search tree is proposed. The implementation of the tableau algorithm in the constraint programming system Mozart with parallel search engines provides new capabilities of processing. With this satisfying results for $\mathcal{ALC}$ language we plan to extend the method for DLs with $\mathcal{ALCN}$ language. The research has been supported by Polish Ministry of Science and Higher Education under Grant N516 015 31/1553.

References

1. Apt KR (2003) Principles of Constraint Programming. Cambridge University Press, Cambridge
2. Baader F, McGuinness DL, Nardi D, Patel-Schneider PF (eds) (2003) The Description Logic Handbook: Theory, implementation, and applications. Cambridge University Press, Cambridge
3. Beth WE (1958) Completeness Results for Formal Systems. In: Proc. Int. Congress of Mathematics, Cambridge:281–288
4. Bonacina MP (2000) A taxonomy of parallel strategies for deduction. Annals of Mathematics and Artificial Intelligence 29(1–4):223–257
5. Brachman J, Schmolze JG (1985) An overview of the KL-ONE knowledge representation system. Cognitive Science 9(2):171–216
6. Brzykcy G, Meissner A (2005) A parallelisation of an inference process for Description Logics with $\mathcal{ALC}$ language. In: Proc. of CMS'05, Vol. 1, ONT, Kraków: 271–280
7. Bubnicki Z (2000), Learning processes in a class of knowledge-based systems. Kybernetes 29(7/8): 1016–1028
8. Meissner A (2004) System automatycznej dedukcji dla logiki deskrypcyjnej z językiem $\mathcal{ALCN}$. Studia z automatyki i informatyki 28/29:91–110. In Polish
9. Schulte C (2000) Programming Constraint Services. PhD Thesis, der Universität des Saarlandes, Saarbrücken
10. Van Roy P, Haridi S (2004) Concepts, Techniques, and Models of Computer Programming, The MIT Press, Cambridge, MA, USA
11. Wooldridge M (2002) An Introduction to MultiAgent Systems. John Wiley&Sons, New York
12. Mozart Consortium (2004) The Mozart Programming System, `http://www.mozart-oz.org`
13. Web Ontology Language (OWL) Guide Version 1.0, `http://www.w3.org/TR/owl-guide`
14. World Wide Web Consortium, Semantic Web, `http://www.w3.org/2001/sw/`

Applications of Genetic Algorithms in Realistic Wind Field Simulations

R. Montenegro, G. Montero, E. Rodríguez, J.M. Escobar, and
J.M. González-Yuste

Institute for Intelligent Systems and Numerical Applications in Engineering,
University of Las Palmas de Gran Canaria, Campus Universitario de Tafira,
35017 Las Palmas de Gran Canaria, Spain, `rafa@dma.ulpgc.es`

Mass consistent models have been widely use in 3-D wind modelling by finite element method. We have used a method for constructing tetrahedral meshes which are simultaneously adapted to the terrain orography and the roughness length by using a refinement/derefinement process in a 2-D mesh corresponding to the terrain surface, following the technique proposed in [14, 15, 18]. In this 2-D mesh we include a local refinement around several points which are previously defined by the user. Besides, we develop a technique for adapting the mesh to any contour that has an important role in the simulation, like shorelines or roughness length contours [3, 4], and we refine the mesh locally for improving the numerical solution with the procedure proposed in [6].

This wind model introduces new aspects on that proposed in [16, 19, 20]. The characterization of the atmospheric stability is carried out by means of the experimental measures of the intensities of turbulence. On the other hand, since several measures are often available at a same vertical line, we have constructed a least square optimization of such measures for developing a vertical profile of wind velocities from an optimum friction velocity. Besides, the main parameters governing the model are estimated using genetic algorithms with a parallel implementation [12, 20, 26]. In order to test the model, some numerical experiments are presented, comparing the results with realistic measures.

1 Introduction

The society has been becoming aware of environmental problems and nowadays it appreciates the use of renewable energies. Along last years the use of wind power for producing electric energy has augmented considerably. So companies of this sector are requesting more and more sophisticated tools which allow them to face the competitive and demanding market. Wind models are tools that allow the study of several problems related to the atmosphere, such

R. Montenegro et al.: *Applications of Genetic Algorithms in Realistic Wind Field Simulations*,
Studies in Computational Intelligence (SCI) **102**, 165–182 (2008)
`www.springerlink.com`

as, the effect of wind on structures, pollutant transport, fire spreading, wind farm location, etc.

Diagnostic models are not used to make forecasts through integrating conservative relations [23]. Therefore they are also called kinetic models [9]. This models generate wind fields that satisfy some physical conditions. If mass conservation law is the only imposed equation, we are defining a mass consistent model. The relative simplicity of diagnostic models makes them attractive from the practical point of view, since they do not require many input data and may be easily used. Pennel [22] checked that, in some cases, improved mass consistent models such as NOABL and COMPLEX obtained better results than other dynamic models which are more complex and expensive. However, we have to take into account that diagnostic models neither consider thermal effects nor those due to pressure gradients. As a consequence, problems like sea breezes can not be simulated with these models unless such effects are incorporated into the initial wind data using observations made in selected locations [8,21]. So diagnostic models have been designed for predicting the effects of the orography on the steady average wind, i.e., average wind in intervals from 10 minutes to 1 hour. There exists a wide range of diagnostic models which have been used by scientists in problems of meteorology and air pollution.

The primitive 2-D diagnostic models did not consider the terrain orography and the vertical profile of the wind. They built an interpolated wind field taking into account only the distance from the nodes to the measurement stations and then they solved the two-dimensional elliptic problem arising from the discretization in a plane; see [33] for a 2-D adaptive finite element model with mixed formulation. Nowadays, in problems defined over complex terrain, to have high quality meshes is essential for the discretization of the studied domains. Most of the existing models uses to work with regular meshes. This strategy is impracticable for problems with complex terrain since the size of the elements must be very small in order to capture the digital information of the map. In this case, there would be regions with no fine details where such small element size would not be necessary. This would finally lead us to larger linear systems of equations and higher computational cost for solving them. Our first 3-D models (see [16,17]) had some of these limitations since the meshes used for defining the terrain surface were uniform. In [20], we presented a new finite element model that uses adaptive nonstructured meshes of tetrahedra with elements of small size where it is necessary but maintaining greater elements where such level of discretization is not required. The resulting 3-D mesh also contains more nodes near the terrain surface, where we need more precision. In a postprocess, the mesh is smoothed and, if necessary, untangled by using the Escobar et al algorithm [3] in order to improve its quality. In addition, a local refinement procedure was proposed for improving the numerical solution [6]. Finally, although mass consistent models are widely used, they are often criticized because their results strongly depend on some governing parameters. This parameters are generally approximated using empirical criteria. Our model includes a tool for the parameter estimation based on genetic algorithms [20].

In order to check the model with realistic data, the company *Desarrollos Eólicos S.A.* (DESA) has provided us with technical support about digital terrain maps related to orography and roughness length, as well as measurements of wind and turbulence intensity in some anemometers located in Lugo (Spain). This work deals with the procedures required for inserting all of this information in the wind modelling. In section 2, we present the improvements to the adaptive discretization of the 3-D domain, including an adaptive procedure to capture the orography and roughness information simultaneously and additional local refinements in different regions of the terrain. The developed technique for inserting new information about wind measures at different heights and turbulence intensities is described in section 3. Several ideas about the parameter estimation in the wind field model is summarized in section 4. Numerical experiments with examples that illustrate all the new possibilities of our wind model are presented in section 5. Besides, a parameter estimation realistic problem is solved by using genetic algorithms for an episode along a day. Finally, we summarize the conclusions of this work and the topics that need further research.

2 Adaptive Discretization of the Domain

In this section we introduce several improvements which have been implemented in our adaptive mesh generation code: mesh adaption to terrain orography and roughness length and local refinement in the surrounding of the measurement stations or any other control point.

2.1 Mesh Adaption to Terrain Orography and Roughness Length

The mesh generation process starts with the determination of nodes allocated on the terrain surface. Their distribution must be adapted to the orographic and roughness characteristics in order to minimize the number of required nodes. The procedure first builds a sequence of nested meshes $T = \{\tau_1 < \tau_2 < \ldots < \tau_m\}$ from a regular triangulation τ_1 of the rectangular region which is studied, such that the level τ_j is obtained by a global refinement of the previous level τ_{j-1} with the 4-T Rivara's algorithm [25]. Every triangle of level τ_{j-1} is divided into four subtriangles inserting a new node in the middle point of the edges and connecting the node inserted in the longer edge with the opposite vertex and with the other two new nodes. Thus, in the mesh level τ_j there appear new nodes, edges and triangles that are defined as corresponding to level j. The number of levels m of the sequence is determined by the degree of discretization of the terrain, i.e., the diameter of the triangulation τ_m must be of the order of the spacial step of the digital map that we are using (the spacial step of the roughness length map is often greater or equal to that of the orographic map). In this way, we ensure that this regular mesh is able to capture all the orographic and roughness information by an interpolation of the heights and roughness length in the nodes of the mesh. Finally, we define a new sequence $T' = \{\tau_1 < \tau_2' < \ldots < \tau_{m'}'\}, m' \leq m$, applying the derefinement algorithm [5, 24]. In this step, two derefinement parameters ε_h and ε_r are introduced and they determine the accuracy of the approximation to the terrain surface and to its roughness length, respectively. The absolute difference between the height obtained in any point of the mesh $\tau_{m'}'$ and the corresponding exact height will be lower than ε_h. A similar condition is established for the roughness and ε_r. A node could be eliminated only if it verifies the two derefinement conditions simultaneously.

2.2 Local Refinement Around Control Points

The resulting mesh adapted to orography and roughness is not always enough to ensure a prescribed accuracy of the numerical model in some regions of the domain and they may require a finner discretization. We have solve this problem by refining the terrain surface mesh in those regions such that the nodes inserted inside them are not eliminated after the derefinement procedure. The vertical spacing function and the 3-D Delaunay triangulation algorithm, that complete our 3-D mesh generator (see [14, 15, 18]), will produce a tetrahedral mesh refined around the selected regions. So, the user can define the form and location of these regions and the number of additional triangles subdivisions to be carried out inside them in order to obtain the required element size.

3 Improvements to the Wind Model

Our wind model has been improved in order to consider the additional information that currently may be available at measurement stations. On the one hand, we have usually different stations located in the same tower for minimizing costs and fixing the wind profiles. Thus, the computation of the friction velocity, which was directly computed from a single wind velocity measured at a station, must be obtained from several measures. For this purpose, a least square approximation is carried out. On the other hand, these stations usually provides measures of the turbulence intensity which is related to the atmospheric stability of the region. So, the knowledge of the range of turbulence intensity will allow us to select the stability class. Following the Pasquill model for the atmospheric stability [27] and defining new ranges of turbulence intensity, a new table for Pasquill stability classification is built.

3.1 New Computation of the Friction Velocity

We consider a log-linear profile [10] in the planetary boundary layer, which takes into account the horizontal interpolation [16], the effect of roughness length on the wind speed and direction, and the atmospheric stability (neutral, stable or unstable) following the Pasquill classification. In the surface layer a logarithmic wind profile is constructed,

$$\mathbf{v}_0(z) = \frac{\mathbf{v}^*}{k}\left(\ln\frac{z}{z_0} - \Phi_m\right) \qquad z_0 < z \le z_{sl} \tag{1}$$

where $\mathbf{v}_0$ is the wind velocity, $k \simeq 0.4$ is the von Karman constant, z is the height of the considered point over the terrain level, z_0 is the roughness length, Φ_m is a function depending on the atmospheric stability and z_{sl} is the height of the surface layer. The friction velocity $\mathbf{v}^*$ is obtained at each point from the interpolation of measures at the height of the stations z_e (*horizontal interpolation*),

$$\mathbf{v}^* = \frac{k\,\mathbf{v}_0(z_e)}{\ln\frac{z_e}{z_0} - \Phi_m(z_e)} \tag{2}$$

Evidently, if n measures were available in a vertical line, the above equation would yield n different friction velocities,

$$\mathbf{v}_i^* = \frac{k\,\mathbf{v}_{0_i}(z_{e_i})}{\ln\frac{z_{e_i}}{z_0} - \Phi_m(z_{e_i})} \qquad i = 1,\ldots,n \tag{3}$$

In order to obtain the optimum value of $\mathbf{v}^*$, we solve a least square problem involving the wind velocities measured at different height and considering that the friction velocity is not a function of the height. Consider

$$A_i = \frac{1}{k}\left(\ln\frac{z_{e_i}}{z_0} - \Phi_m(z_{e_i})\right) \qquad i = 1,\ldots,n \tag{4}$$

such that

$$\mathbf{v}_{0_i}(z_{e_i}) = \mathbf{v}^* A_i \qquad i = 1,\ldots,n \tag{5}$$

If $\mathbf{v}_{s_i}$ is the measured velocity at the i-th station, then the function to be minimized is,

$$F_{obj} = \sum_{i=1}^{n}\left(\mathbf{v}_{0_i}(z_{e_i}) - \mathbf{v}_{s_i}(z_{e_i})\right)^2 = \sum_{i=1}^{n}\left(\mathbf{v}^* A_i - \mathbf{v}_{s_i}(z_{e_i})\right)^2 \tag{6}$$

whose minimum is obtained for the following friction velocity,

$$\mathbf{v}^* = \frac{\sum\limits_{i=1}^{n} A_i \mathbf{v}_{s_i}(z_{e_i})}{\sum\limits_{i=1}^{n} A_i^2} \tag{7}$$

Table 1. Pasquill Stability Classification depending on the surface wind speed and the isolation. Strong isolation corresponds to a sunny afternoon of the middle-summer in England; slight isolation is related to same conditions in middle-winter. Nighttime means the time from one hour before the sunset to one hour after the sun rises. Neutral class D should be used also, independently of the wind speed, for clouded sky along the day or the night, and for any condition of the sky during the hour before and after the nighttime.

Pasquill stability class					
	Isolation			Nighttime	
Surface wind				$\geq 4/8$	$\leq 3/8$
speed (m/s)	Strong	Moderate	Slight	Clouds	Clouds
< 2	A	A-B	B	-	-
$2 - 3$	A-B	B	C	E	F
$3 - 5$	B	B-C	C	D	E
$5 - 6$	C	C-D	D	D	D
> 6	C	D	D	D	D
For A-B, take the average of the values of A and B, etc.					

3.2 Atmospheric Stability Versus Turbulence Intensity

The atmospheric stability may be characterized by using the Pasquill stability classification of table 1. It considers the following classes for stability: A (extremely unstable), B (moderately unstable), C (slightly unstable), D (neutral), E (slightly stable) and F (moderately stable) [27].

The anemometers generally provides measures of the intensity of turbulence that may help to complete the information about the class of atmospheric stability in the studied region. The intensity of turbulence i is defined as the square root of the sum of variances σ_u^2, σ_v^2, σ_w^2, of the three components of the velocity u_0, v_0, w_0, respectively, divided by the average wind velocity that has been measured,

$$i = \frac{\sqrt{\sigma_u^2 + \sigma_v^2 + \sigma_w^2}}{|\mathbf{v}_0|} \tag{8}$$

However, only measures of speed variations are often available but not of the wind direction. In such cases, equation (8) is reduced to,

$$i = \frac{\sigma_{v_0}}{|\mathbf{v}_0|} \tag{9}$$

where σ_{v_0} represents the standard deviation of the measured wind speeds.

While an unstable atmosphere implies a high level of turbulence, with a range of turbulence intensities between 0.2 and 0.4 approximately, a stable atmosphere, with a small or almost null turbulence, is characterized by intensities from 0.05 to 0.1 [13]. In Table 2, the above relations of the turbulence

Table 2. Pasquill stability classification taking into account the surface wind speed and the turbulence stability.

	Pasquill stability class						
	Isolation				Nighttime		
Surface wind speed (m/s)	$i > 0.35$	$0.35 \geq i > 0.25$	$0.25 \geq i > 0.15$	$0.15 \geq i$	$i > 0.075$	$0.075 \geq i > 0.03$	$0.03 \geq i$
$|v_0| < 2$	A	B	B	B	F	F	F
$2 \leq |v_0| < 3$	A	B	C	C	E	E	F
$3 \leq |v_0| < 5$	B	B	C	C	D	E	E
$|v_0| \geq 5$	C	C	C	D	D	D	D

intensity and the atmospheric stability have been considered in order to define the Pasquill stability class.

4 Parameters Estimation with Genetic Algorithms

Genetic algorithms (GAs) are optimisation tools based on the natural evolution mechanism [1, 12, 30]. They produce successive trials that have an increasing probability to obtain a global optimum. This work is based on the model developed by Levine [11]. It is a standard genetic algorithm code (*pgapack* library), with string real coding. The most important aspects of GAs are the construction of an initial population, the evaluation of each individual in the fitness function, the selection of the parents of the next generation, the crossover of those parents to create the children, and the mutation to increase diversity.

Two population replacements are commonly used. The first, the generational replacement, replaces the entire population each generation [7]. The second, known as steady-state, only replaces a few individuals each generation [29, 31, 32]. In our experiments, initial population has been randomly generated and we use *iteration limit exceeded* as stopping criterion. Each population consists of 40 individuals, being replaced 4 of them each generation. Here we considered 250 generations. The selection phase allocates an intermediate population on the basis of the evaluation of the fitness function. We have chosen the stochastic universal selection scheme (SU) [11]. The crossover operator takes bits from each parent and combines them to create a child. Uniform crossover operator (U) is used here. It depends on the probability of exchange between two bits of the parents [28]. The mutation operator is better used after crossover [2]. It allows to reach individuals in the search space that could not be evaluated otherwise. When part of a chromosome has been randomly selected to be mutated, the corresponding genes belonging to that part are changed. This happens with probability p. This work deals with two mutation operators. The first is of the form $\nu \leftarrow \nu \pm p \times \nu$, where ν is the existing allele value, and p is selected from a Gaussian distribution (G). The second operator (R) simply replaces ν with a value selected uniformly randomly from the initialisation range of that gene.

The fitness function plays the role of the environment. It evaluates each string of a population. This is a measure, relative to the rest of the population, of how well that string satisfies a problem-specific metric. The values are mapped to a nonnegative and monotonically increasing fitness value. In the numerical experiments with this wind model [20], we look for optimal values of α, ε, γ and γ'. Specifically, the so called stability parameter α determines the rate between horizontal and vertical wind adjustment. For $\alpha >> 1$ flow adjustment in the vertical direction predominates, while for $\alpha << 1$ flow adjustment occurs primarily in the horizontal plane. Thus, the selection of α allows the air to go over a terrain barrier or around it. We search the optimum in $\left[10^{-2}, 10^{2}\right]$. The second parameter to be estimated is the weighting coefficient ε ($0 \leq \varepsilon \leq 1$) involved in the horizontal interpolation of wind measurements. For $\varepsilon \to 1$, the importance of the *horizontal distance* from each point to the measurement stations is greater, while $\varepsilon \to 0$ signifies more importance of the *height difference* between each point and the measurement stations. The parameter γ is related to the height of the planetary boundary layer. There exist different versions of where to search for this parameter. The interval $[0.15, 0.4]$ considered in our simulations includes all the proposed search spaces. Finally, the parameter γ' appears in the computation of the mixing height for stable atmosphere. Several authors have proposed that the value of γ' should be searched in the surroundings of 0.4. More details and references about the discussion of these parameters can be found in [20].

We propose to minimise the following fitness function which is defined as the average relative error of the wind velocities given by the model with respect to the measures at the reference stations,

$$F(\alpha, \varepsilon, \gamma, \gamma') = \frac{1}{N_r} \sum_{n=1}^{N_r} \frac{|\mathbf{v}_n - \mathbf{v}(x_n, y_n, z_n)|}{|\mathbf{v}_n|} \tag{10}$$

where $\mathbf{v}(x_n, y_n, z_n)$ is the wind velocity obtained by the model at the location of station n, $\mathbf{v}_n$ is the measured wind and N_r is the number of reference stations.

5 Numerical Experiments

We present several applications to show the improvements carried out in our wind model. All experiment were run on a XEON precision 530, except the parameter estimation problem which was solved using a cluster of PCs.

5.1 Surface Mesh Adaption to Orography and Roughness

The studied three-dimensional domain is located in a region of Lugo, Spain, at $43°N$ of latitude and it is defined by four points of UTM coordinates $A(609980, 4799020)$, $B(626000, 4799020)$, $C(626000, 4813040)$ and $D(609980, 4813040)$, respectively. The height of the top is $4000\ m$. A digital topographic map was provided by *DESA* on a mesh of element size $20 \times 20\ m$. The X axis corresponds to East direction and the Y one to North. Thus, we are working with a region of $16020 \times 14020\ m$. The minimum and maximum heights are $420\ m$ and $1020\ m$, respectively. Figure 1 represents the heights of the terrain. The measurement stations and the control points has been approximately plotted, such that from North to South we can find E243, E208, E212, E242, E206 and E283. Tables 3 and 4 contain their coordinates, respectively. Roughness is an essential factor on the atmospheric stratification, and therefore, on the characteristics of the resulting wind profile. Figure 2 shows the roughness length of the terrain which were supplied by *DESA*. We remark that some stations and control points are closed to contours of the roughness. In this case, the roughness length values are $0.03\ m$, $0.05\ m$, $0.08\ m$, $0.3\ m$ and $0.8\ m$.

Table 3. Coordinates in m of the measurement stations.

Station	UTM-E	UTM-N	Height
E206	615396	4805218	924.8
E208	616917	4807256	945.0
E212	617423	4806382	895.0

Table 4. Coordinates in m of the control points.

Control point	UTM-E	UTM-N	Height
E242	618290	4806136	873.2
E243	616629	4808235	947.0
E283	617473	4804111	849.0

Starting from a regular mesh of the rectangular region with element size of $1 \times 1\ km$ approximately, five global refinements are carried out using 4-T Rivara's algorithm [25]. With this number of refinement steps, we obtain a mesh with an element size about $31\ m$. In order to improve the discretization near the stations and control points, five additional local refinements are applied inside six circles with centre at the stations and control points, respectively,

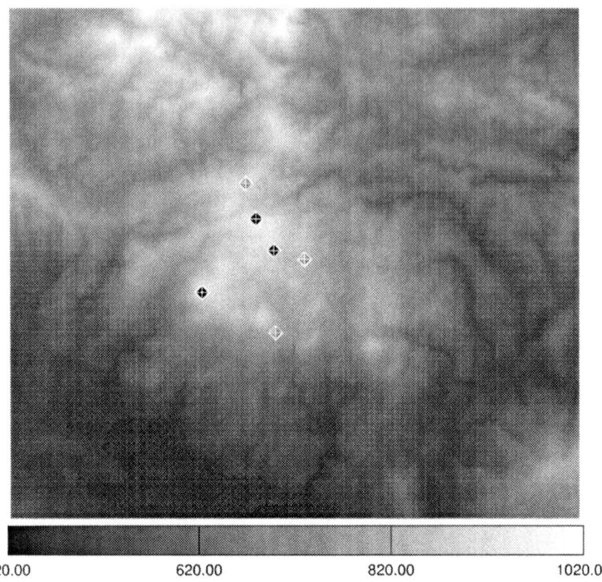

Fig. 1. Topographic map of the studied region in Lugo. From North to South, we can see the stations or control points E243, E208, E212, E242, E206 and E283.

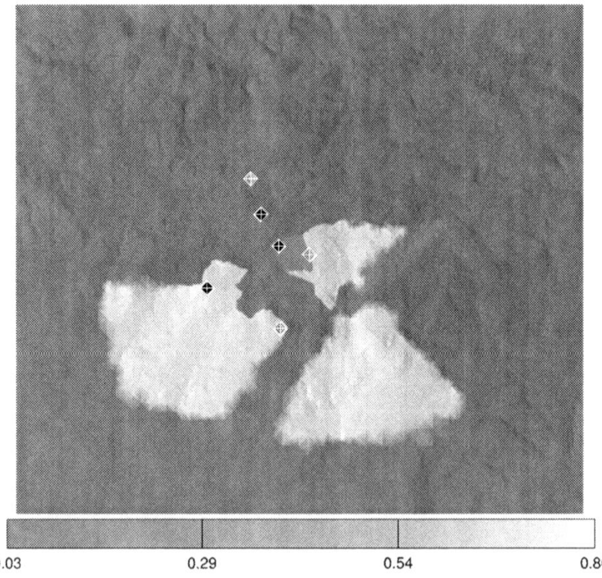

Fig. 2. Roughness length map of the studied region in Lugo with the station and control points.

and diameter 200 m. This produces a local element size about 1 m. Once we have interpolated the height and the roughness length in the nodes of these refined two-dimensional mesh, we use the derefinement algorithm [5, 24] described in section 2.1 with $\varepsilon_h = 10$ and $\varepsilon_r = 0.01$, keeping in any case the nodes located inside the circles. In figure 3 we can see the resulting triangulation of the terrain surface. The corresponding three-dimensional mesh, see figure 4, contains 102662 nodes and 515812 tetrahedra.

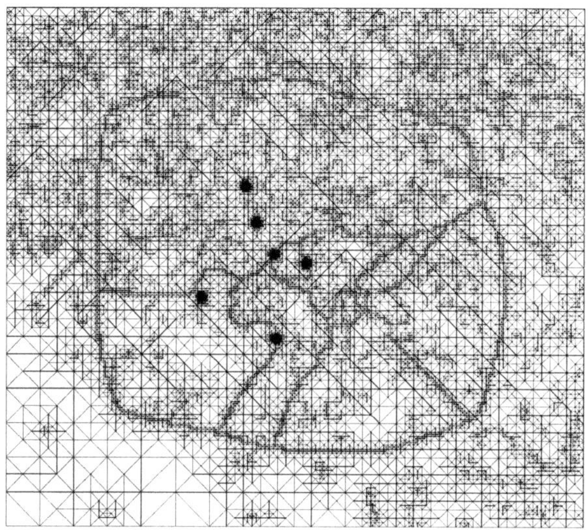

Fig. 3. Triangulation of the terrain simultaneously adapted to orography and roughness corresponding to the studied region in Lugo.

5.2 Surface Mesh Adaption to Contours

In many cases of environmental modelling, there are some contour lines which determinate certain characteristics of the studied region. For example, in wind simulation the well definition of shore lines or roughness contours may be very important for obtaining accurate results. Thus the mesh must be adapted following these contours such that they are represented by edges of the mesh. The procedure uses ideas from the smoothing of surface triangulation algorithm presented in [4]. As example, figure 5 shows an adaptive mesh, related to a region of the north west of Gran Canaria Island, to the shore line (plotted by points in black).

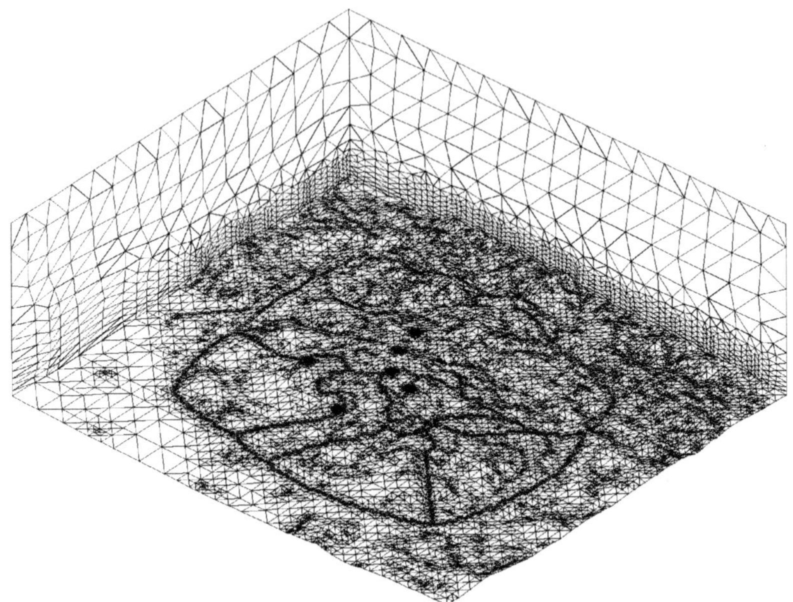

Fig. 4. Adaptive 3-D mesh corresponding to the studied region in Lugo.

5.3 Wind Simulation in a Realistic Episode

Using the mesh constructed for the Lugo application in the first numerical experiment, we have made a simulation for an episode along the March 21, 2003. The first step is to estimate the main parameters of the model and, then, apply the wind model using the estimated values. Next, the wind velocity is checked in the control points, where the wind data is supplied by *DESA*.

Parameter Estimation Along a Day

We have taken into account the table 2 for determining the stability class from the available turbulence intensity values. For the studied day, we have obtained neutral conditions. So we must estimate the stability parameter α, the weighting parameter ε related to the horizontal interpolation of wind velocities and the parameter γ involved in the computation of the planetary boundary layer; see, e.g., [20]. The estimation has been carried out each hour (24 computations). We have applied genetic algorithms to solve these parameter estimation problems, where the fitness functions are defined in terms of the relative velocity errors obtained by the model at the measurements stations. It is evident that, in order to avoid spurious solutions, more than 20 repetitions for parameter setting of each hour should be done. This fact would obviously imply an important increasing on the computational cost of the parameter

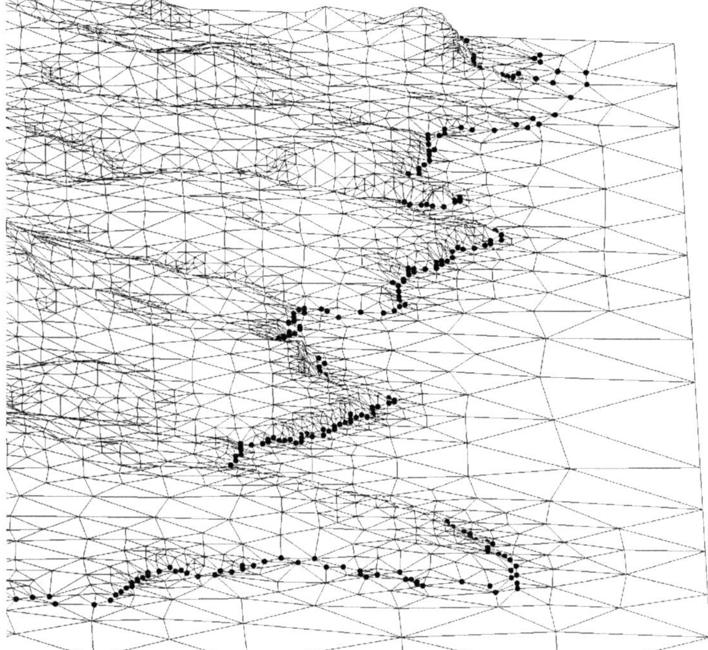

Fig. 5. Region defined in the north west of Gran Canaria Island. Contour plots and adapted mesh.

estimation process, even more if we take into account that each evaluation of the fitness function supposes the resolution of a finite element problem (in this case, about one hundred thousand unknowns). Nevertheless, from our previous experience in this kind of wind simulation problems, we have observed that just one computation is enough for reaching a good solution.

In figure 6 we can see the evolution of the values of the three parameters along the episode. The values of ε are practically constant and approximately equal to 1. This means that only the horizontal distance has effect on the horizontal interpolation. This result is agreed with the orographic characteristics of the studied domain. Likewise the values obtained for γ are closed to 0.15, that is, the lower limit for this parameter which is related to low planetary boundary layers. However, the stability parameter α varies in the interval 8-20. This range of values makes the wind predominantly flow more over the obstacles than around them.

Comparison of the Model Results with Empirical Data

Once the main parameters are estimated, we start the wind modelling along the selected episode using the obtained values. For March 21, 2003, only measures from E242 and E283 were available. Figures 7 and 8 show the wind

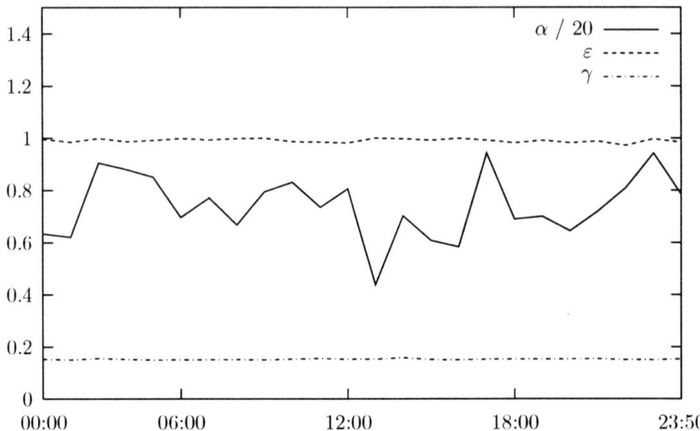

Fig. 6. Results of the estimation of α, ε and γ along the studied episode (March 21, 2003)

speeds obtained with the model and the reference values measured at the control points E242 and E283, respectively. More details of the errors of computed winds with respect to the measured wind may be seen in table 5. We remark that the average errors at the measurement stations are small as expected. The average error is 27.24% at control point E242 and 4.94% at E283.

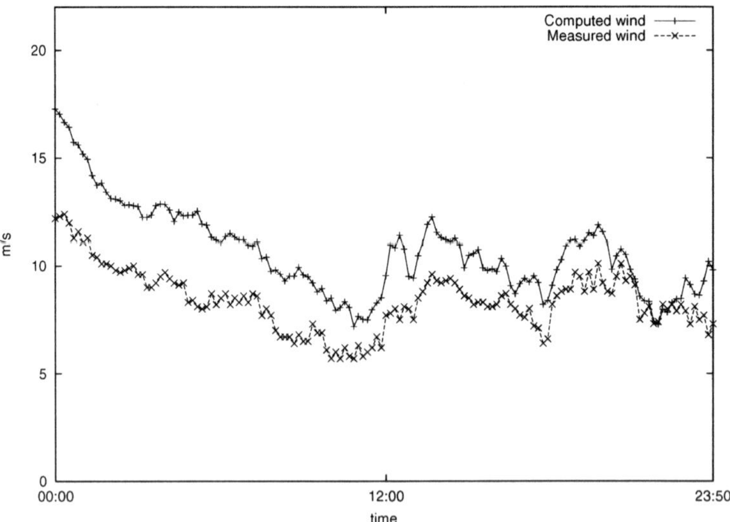

Fig. 7. Comparison of the wind velocities measured at the control station E242 (March 21, 2003).

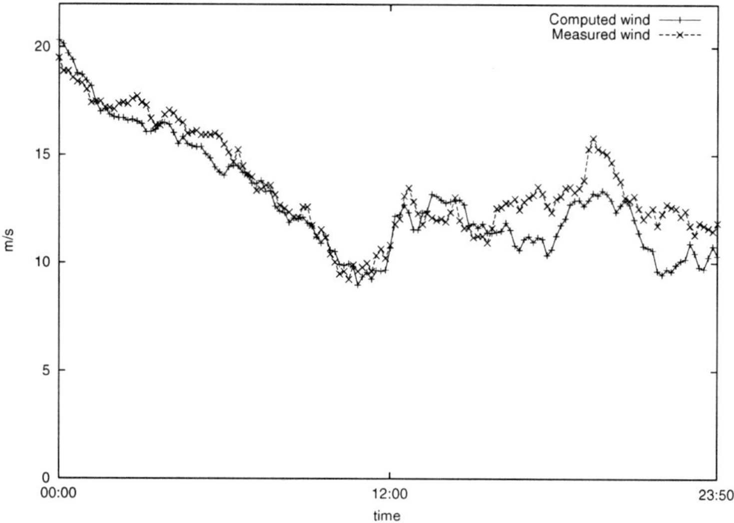

Fig. 8. Comparison of the wind velocities measured at the control station E283 (March 21, 2003).

Table 5. Error of the computed wind at stations and control points.

Stations and control points	Average measured wind	Average computed wind	% average error	Maximum absolute error	Minimum absolute error
E206 (49 m)	15.37	15.50	0.81 %	0.46	0.01
E208 (15 m)	8.57	8.98	4.74 %	1.25	0.00
E208 (30 m)	9.25	9.92	7.21 %	1.36	0.05
E212 (15 m)	8.46	8.44	0.20 %	0.63	0.00
E212 (30 m)	9.02	9.85	9.25 %	1.60	0.31
E242 (40 m)	8.40	10.69	27.24%	5.09	0.09
E283 (49 m)	13.62	12.95	4.94 %	3.04	0.02

6 Conclusions

We have proposed a technique for constructing tetrahedral meshes which are simultaneously adapted to the terrain orography and the roughness length. The use of our refinement/derefinement process in the 2-D mesh corresponding to the terrain surface allows us to obtain meshes that are accurately adapted to different functions as well as are locally refined around several points. These characteristics of the generated meshes are very important in the wind simulation since, on the one hand, the quality of the representation of both orography and roughness is critical for obtaining accurate results with the model, and on the other hand, the local refinement at the stations and control points is essential for inserting the wind data of the stations or recovering such data at any required point.

Some improvements have been carried out in the construction of the initial wind based on the horizontal interpolation of wind measures and vertical extrapolation in stratified atmosphere. The optimization of the friction velocity for several measures in the same tower allows to minimize the differences between the constructed vertical profile of wind and the measures. However, though such differences are small, further research is needed in order to construct new wind profiles that exactly satisfy all the available measures of wind velocities. In addition, the inclusion of observations of turbulence intensities has made the model to be able of automatically updating the suitable wind profile as function of the corresponding stability class.

The periodic updating of the main parameters of the model has proved to be fundamental for reducing the errors of the computed wind. However, further considerations should be taken into account in future works for a better performance of the model. For example, a finer map of roughness, a more sophisticated horizontal interpolation of wind velocities and a greater number of measurement stations well distributed over the studied region, will help to reduce errors at points like E242 where the roughness may not be well approximated. Finally, in order to obtain an accurate wind field in zones with very steep slopes, the mesh should be adapted to the contour lines, since a change in the direction of edges in the mesh may strongly affect the computed wind.

Acknowledgement. The work has been partially supported by the Spanish Government (Ministerio de Educación y Ciencia) and FEDER, grant number CGL2004-06171-C03-02/CLI. The authors are also grateful to Ignacio Láinez and Antonio Ruiz for their technical support provided under the scope of the collaboration agreement signed by the University of Las Palmas de Gran Canaria and Desarrollos Eólicos (DESA) in January 2005. The wind data correspond to a DESA's wind farm located in Lugo.

References

1. Bäck T, Fogel DB, Michalewicz Z (1997) Handbook of evolutionary computation. Oxford Univ. Press, New York-Oxford
2. Davis L (1991) Handbook of genetic algorithms. Van Nostrand Reinhold
3. Escobar JM, Rodríguez E, Montenegro R, Montero G, González-Yuste JM (2003) Simultaneous untangling and smoothing of tetrahedral meshes. Comp Meth Appl Mech Eng 192:2775–2787

4. Escobar JM, Montero G, Montenegro R, Rodríguez E (2006) An algebraic method for smoothing surface triangulations on a local parametric space. Int J Num Meth Eng 66:740–760
5. Ferragut L, Montenegro R, Plaza A (1994) Efficient refinement/derefinement algorithm of nested meshes to solve evolution problems. Comm Num Meth Eng 10:403–412
6. González-Yuste JM, Montenegro R, Escobar JM, Montero G, Rodríguez E (2004) Local refinement of 3-D triangulations using object-oriented methods. Adv Eng Soft 35:693–702
7. Holland J (1992) Adaption in natural and artificial systems. MIT Press
8. Kitada T, Kaki A, Ueda H, Peters LK (1983) Estimation of vertical air motion from limited horizontal wind data - A numerical experiment. Atmos Environ 17:2181–2192
9. Lalas DP, Tombrou M, Petrakis M (1988) Comparison of the performance of some numerical wind energy siting codes in rough terrain. In: European Community Wind Energy Conference, Herning, Denmark
10. Lalas DP, Ratto CF (1996) Modelling of atmospheric flow fields. World Scientific Publishing, Singapore
11. Levine D (1994) A Parallel Genetic Algorithm for the Set Partitioning Problem. PhD Thesis, Illinois Institute of Technology / Argonne National Laboratory
12. Michalewicz Z (1994) Genetic algorithms + data structures = evolution problems. Springer Verlag, Berlin-Heidelberg-New York
13. Mikkelsen T (2003) Modelling of pollutant transport in the atmosphere. MAN-HAZ position paper, Risø National Laboratory, Denmark
14. Montenegro R, Montero G, Escobar JM, Rodríguez E, González-Yuste JM (2002) Tetrahedral mesh generation for environmental problems over complex terrain. Lect N Comp Sci 2329:335–344
15. Montenegro R, Montero G, Escobar JM, Rodríguez E (2002) Efficient strategies for adaptive 3-D mesh generation over complex orography. Neural, Parallel & Scientific Computation 10:57–76
16. Montero G, Montenegro R, Escobar JM (1998) A 3-D diagnostic model for wind field adjustment. J Wind Engrg Ind Aer 74-76:249–261
17. Montero G, Sanin N (2001) Modelling of wind field adjustment using finite differences in a terrain conformal coordinate system. J Wind Engrg Ind Aer 89:471–488
18. Montero G, Montenegro R, Escobar JM, Rodríguez (2003) Generación automática de mallas de tetraedros adaptadas a orografías irregulares. Rev Int Mét Num Cálc Dis Ing 19(2):127–144
19. Montero G, Montenegro R, Escobar JM, Rodríguez E, González-Yuste JM (2004) Velocity field modelling for pollutant plume using 3-D adaptive finite element method. Lect N Comp Sci 3037:642–645
20. Montero G, Rodríguez E., Montenegro R, Escobar JM, González-Yuste JM (2005) Genetic algorithms for an improved parameter estimation with local refinenent of tetrahedral meshes in a wind model. Adv Engrg Soft 36:3–10
21. Moussiopoulos N, Flassak Th, Knittel G (1998) A refined diagnostic wind model. Environ Soft 3:85–94
22. Pennel WT (1983) An Evaluation of the Role of Numerical Wind Field Models in Wind Turbine Siting. Batelle Memorial Institute, Pacific Northwest Laboratory, Richland, Washington

23. Pielke R (1984) Mesoscale meteorological modeling. Academic Press, Inc., Orlando, Florida
24. Plaza A, Montenegro R, Ferragut L (1996) An improved derefinement algorithm of nested meshes. Adv Eng Soft 27:51–57
25. Rivara MC (1987) A grid generator based on 4-triangles conforming. Mesh-refinement algorithms. Int J Num Meth Eng 24:1343–1354
26. Rodríguez E, Montero G, Montenegro R, Escobar JM, González- Yuste JM (2002) Parameter estimation in a three-dimensional wind field model using genetic algorithms. Lect Notes in Comp Sci 2329:950–959
27. Seinfeld JH, Pandis SN (1998) Atmospheric chemistry and physics. From air pollution to climate change. John Wiley & Sons, Inc., New York
28. Spears W, DeJong K (1991) On the virtues of parametrized uniform crossover. In: Proceedings of the Fourth International Conference on Genetic Algorithms
29. Syswerda G (1989) Uniform crossover in genetic algorithms. In: Proceedings of the Third International Conference on Genetic Algorithms
30. Vose M (1999) The simple genetic algorithm. MIT Press, Cambridge, Massachusetts
31. Whitley D (1988) GENITOR: A different genetic algorithm. In: Rocky Mountain Conference on Artificial Intelligence
32. Whitley D (1989) The GENITOR algorithm and selection pressure: Why rank-based allocation of reproductive trials is best. In: Proceedings of the Third International Conference on Genetic Algorithms
33. Winter G, Montero G, Ferragut L, Montenegro R (1995) Adaptive strategies using standard and mixed finite elements for wind field adjustment. Solar Energy 54:49-56

Methodologies and Technologies for Rule-Based Systems Design and Implementation. Towards Hybrid Knowledge Engineering

Grzegorz J. Nalepa[1]

Institute of Automatics, AGH University of Science and Technology, Al. Mickiewicza 30, 30-059 Kraków, Poland, gjn@agh.edu.pl

Summary. A practical design of non-trivial rule-based systems requires a systematic, structured and consistent approach. The paper focuses on selected issues in RBS knowledge engineering. Some ideas on combining knowledge engineering with software engineering are discussed. Furthermore, results of RBS design tools survey are enclosed. In the paper an original design and implementation methodology for RBS is also presented. It has been developed in the MIRELLA project. It is a top-down hierarchical design methodology, based on new knowledge representation methods (XTT and ARD), on-line logical system analysis in Prolog, and XML-based knowledge encoding. Basing on the experience with XTT-based methodology, as well as tools supporting it, the paper discusses an extended hierarchical design methodology for RBS. A preview of the HEKATE project, which aims at developing a hybrid knowledge engineering methodology is also given.

1 Introduction

Knowledge-based systems (KBS) are an important class of intelligent systems originating from the field of Artificial Intelligence [17]. They can be especially useful for solving complex problems in cases where purely algorithmic or mathematical solutions are either unknown or demonstrably inefficient.

Building real-life KBS is a complex task. Since their architecture is fundamental different from classic software, classical software engineering approaches cannot be applied efficiently. Some specific development methodologies, commonly referred to as *knowledge engineering*, are required.

In AI *rules* are probably the most popular choice for building knowledge-based systems, that is the so-called rule-based expert systems [4, 5, 7]. Rule-based systems (RBS) are used extensively in practical applications, especially in domains such as automatic control, decision support, and system diagnosis. They constitute today one of the most important classes of KBS.

G.J. Nalepa: *Methodologies and Technologies for Rule-Based Systems Design and Implementation. Towards Hybrid Knowledge Engineering*, Studies in Computational Intelligence (SCI) **102**, 183–198 (2008)
www.springerlink.com

A rule-based expert system consists of a knowledge base and an inference engine. The knowledge engineering process aims at designing and evaluating the knowledge base, and implementing a proper inference engine. The process of building the knowledge base involves the selection of a knowledge representation method, knowledge acquisition, and possibly low-level knowledge encoding. In order to create an inference engine a reasoning technique must be selected, and the engine has to be programmed.

During the engineering process a number of problems occur. Particular problems concern the selection of a knowledge representation formalism as well as the actual design of an appropriate rule base. As the number of rules exceeds even relatively very low quantities, it is hard to keep the rule-base consistent, complete, and correct. These problems are related to knowledge-base verification, validation, and testing [21,22]. The selection of appropriate software tools and programming languages is non-trivial either.

This paper is devoted to discussing the most important differences between knowledge engineering (see Sect. 2) and classic software approaches [19] (see Sect. 3). When it comes to practical system implementation, current RBS development is heavily dependent on software engineering tools, which enforce certain design patterns not suitable for knowledge engineering. This is why the paper aims at identifying possible areas of cooperation between software and knowledge engineering approaches in Sect. 4. Section 5 identifies the most important issues in RBS knowledge engineering process and presents possible approaches. Then in Sect. 6 an overview of selected design and implementation tools is presented. Practical design of non-trivial rule-based systems requires a systematic, structured and consistent approach. In this paper an original design and implementation methodology for rule-based systems is discussed in Sect. 7. It is a top-down hierarchical design methodology, based on new knowledge representation methods (XTT and ARD), on-line logical system analysis in Prolog, and XML-based knowledge encoding. It is supported by a prototype CASE tool called Mirella. Basing on the experience with XTT-based methodology, as well as tools supporting it, in the Sect. 8 an extended hierarchical design methodology for RBS is discussed. At the end a preview of the HEKATE project, which aims at developing a hybrid knowledge engineering methodology, is also given.

2 Knowledge Engineering Approach

What makes KBS distinctive is the separation of knowledge storage (the knowledge base) from the knowledge processing facilities. In order to store knowledge, KBS use various knowledge representation methods, which are *declarative* in nature. In case of RBS these are *production rules*. Specific knowledge processing facilities, suitable for specific representation method used, are then selected. In case of RBS these are logic-based inference engines.

The knowledge engineering (KE) process in case of RBS involves two main tasks: knowledge base design, and inference engine implementation. Furthermore, some other specific tasks are also required, such as: knowledge base analysis and verification, and inference engine optimization. The performance of a complete RBS should be *evaluated* and *validated*. Classic expert systems books [5] represent this process as shown in Fig. 1. While this process is specific to expert systems in general, it is usually similar in case of other KBS.

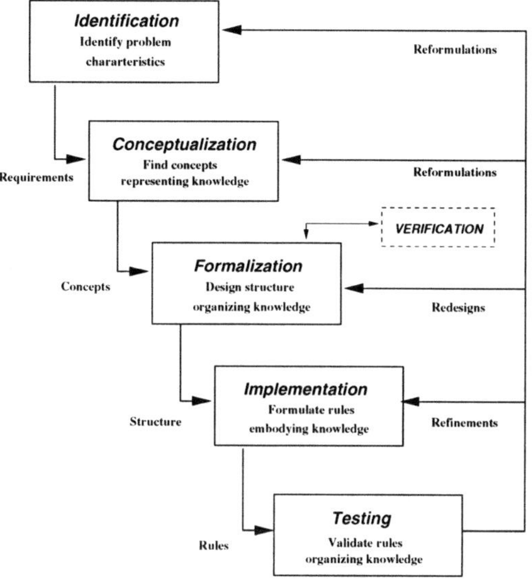

Fig. 1. Classic knowledge engineering process (Liebowitz 1998)

What is important about the process, is the fact that it should *capture* the expert knowledge and *represent* it in a way that is suitable for processing (this is the task for the knowledge engineer). The actual structure of a KBS does not need to be system specific – it should not "mimic" or model the structure of the real-world problem. However, the KBS should capture and contain the knowledge about the real-world system. The task of the programmers is to develop processing facilities for the knowledge representation.

It should be pointed out, that in case of KBS there is no single universal engineering approach, or universal modelling method (such as UML in software engineering). Different classes of KBS may require a specific approach, see [2, 5, 7, 20]. Having outlined the main aspects of KBS development, it can be discussed how they are related to classic software engineering methods.

3 Software Engineering Approach

Software engineering (SE) is the domain where a number of mature and well-proved design methods exist. They address needs of specific classes of business software. In software engineering the software development process and life cycle is represented by several models. One of the most common is called *the waterfall model* [19] and is shown in Fig. 2. In this process a number of development roles can be identified: users and/or domain experts, system analysts, programmers, testers, integrators, and end users. What makes this process different from knowledge engineering, is the fact, systems analysts in general try to *model* the *structure* of the real-world information system in the structure of computer software system. So the structure of the software corresponds to some respect to the structure of the real-world system. The task of the programmers is to encode and implement the model (which is the result of the system analysis) in some lower-level programming language.

The most important difference between software and knowledge engineering, is that the former tries to model how the system works, while the latter tries to capture and represent what is known about the system.

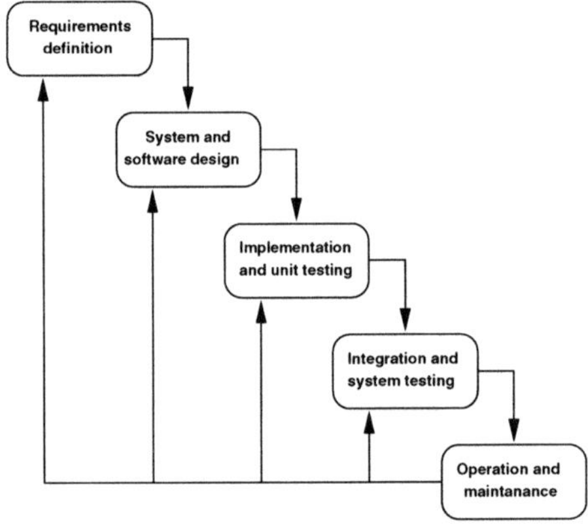

Fig. 2. Classic waterfall software life cycle (Sommerville 2004)

4 Heterogeneous Development Methodology

Historically, there has always been a strong feedback between software engineering and computer programming tools. At the same time these tools have been strongly determined by the actual architecture of computers themselves. For a number of years there has been a clear trend for the software engineering to become as implementation-independent as possible. Modern software engineering approaches tend to be abstract and conceptual.

On the other hand, knowledge engineering approaches have always been device and implementation-agnostic. The actual implementation of KBS has been based on some high level programming languages such as Lisp or Prolog. However, modern knowledge engineering tools heavily depend on some common development tools and programming languages, especially when it comes to user interfaces, network communication, etc.

It could be said, that these days software engineering becomes more knowledge-based, while knowledge engineering is more about software engineering. This opens multiple opportunities for both approaches to improve and benefit. Software engineering could adopt from knowledge engineering: advanced conceptual tools, such as declarative knowledge representation methods, knowledge transformation techniques based on existing inference strategies, as well as verification, validation and refinement methods. This trend is already visible in the *business rules* approach [16, 23]. *Model-Driven Architecture* (MDA) is a new software engineering paradigm that tries to provide a unified design and implementation method and appropriate tools for the declarative specification of business logic [9].

In order to improve and better integrate with existing software knowledge engineering could adopt: programming interfaces to existing software systems and tools, interfaces to advanced storage facilities such as databases and data warehouses, modern user interfaces, including graphical and web-based ones.

This paper is written from the knowledge engineering point of view. This is why the following sections focus on different ways of improving the KE process in case of RBS.

5 Rule-Based Systems Design Issues

In RBS development knowledge engineering is essentially a process of construction. As it was pointed out in Sect. 2, it involves two main tasks: knowledge base (rule base) design, and inference engine implementation.

5.1 Rule Base Design

The first decision that has to be made is one concerning knowledge representation method. It is widely recognized that there is no single formalism suitable to represent knowledge for all purposes. A variety of formalisms and structures is needed to represent knowledge. In the field of rule-based expert systems the *knowledge representation method* is a systematic way of "encoding" what an expert knows about some domain. However "encoding" means here rather "describing" then "encrypting" [4].

Some of the issues arising in knowledge representation are: syntax, semantics, expressive adequacy, reasoning, completeness and other consistency issues, real-world knowledge, control, flexibility. Different representations address these issues in different ways [2]. While there are numerous knowledge representation methods, the logic-based ones are essential to the theory and practice of rule-based systems and expert systems in general.

Although propositional calculus is a simple logical system, it can serve as a practically useful language for encoding rule-based systems. Further, both analysis and design of such systems is relatively simple. The most basic logical form of *propositional rules* is as follows (see [7]): $p_1 \wedge p_2 \wedge \ldots \wedge p_n \longrightarrow h$. This form of a rule is logically equivalent to a Horn clause, provided that all the literals are positive. A more complex rule may contain conclusion part composed of several propositions. In forward-chaining systems rules are applied by checking if their preconditions are satisfied. Whenever a rule is fired, its conclusion is added to the current knowledge base. Propositional rule-based systems can take various visual forms incorporating some structural representation; most important are: *decision tables* and *decision trees* [7].

Decision tables are an engineering way of representing production rules. Conditions are formed into a table which also holds appropriate actions. Classical decision tables use binary logic extended with "not important" mark to express states of conditions and actions to be performed.

The main advantage of decision tables is their simple, intuitive interpretation. One of the main disadvantages is that classical tables are limited to binary logic. In some cases the use of values of attributes is more convenient. A slightly extended tables are *OAV tables* (OAT). OAV stands for Object-Attribute-Value (OAT – Object-Attribute-Value-Table, see [7]).

Decision trees are an important representation, since the tree-like representation is readable, easy to use and understand. The root of the tree is an entry node, under any node there are some branching links. The selection of a link is carried out with respect to a conditional statement assigned to the node. The evaluation of this condition determines the selection of the link. The tree is traversed top-down, and at the leaves final decisions are defined.

Formal *ontologies* are an important knowledge representation method, used extensively in some new implementations of Web-oriented intelligent systems. Recently ontologies gained a precise semantic interpretation with the definition of OWL DL (description logics), which is currently extended by horn-clause rules (see [3] for a current proposal for SWRL/RuleML).

In expert system practice there are several other knowledge representation methods used. Their logical interpretation is not always as direct as decision rules, tables, or trees. However, they do have many applications as a valuable conceptualization tool. These includes: graphs, and conceptual graphs, semantic networks, and frames, see [5, 20] for more details.

5.2 Rule Base Encoding

On the low level rules have to be encoded in a format ready for processing. Inventing a new, specific rule format, may seem the most straightforward approach. It gives developers a lot of freedom when it comes to the implementation. However, it poses problems when interfacing with existing systems.

From the KE point of view, it is desirable to adopt some general standard. However, from the SE point of view, it might be desirable to adopt specific issues of the particular application.

A more common and reasonable approach consists in choosing an expert system shell, and using a predefined rule format. It simplifies the implementation, however it determines the system architecture. It can for example enforce certain inference strategy. More on this is elaborated in Sect. 6.3.

Encoding rules in some high level logic programming language such as Prolog is – to some degree – a good combination of the two above. Prolog allows inventing any rule format, while providing high-level inference strategies, see Sect. 6.2 for more details.

The development of the Web and recent W3C Semantic Web initiative make encoding rules for web applications an important issue. Encoding rules in an XML-based format, such as RuleML (`www.ruleml.org`) is often the best solution in such a case.

5.3 Rule Base Analysis

Rule-based expert systems technology is being applied to critical tasks and complex problem-solving. This is why there are concerns about its dependability. A proper system development cycle, as well as a rigorous verification and validation (*V&V*) can provide an appropriate level of quality and safety [22].

The verification and validation of expert systems are still a maturing field, so there is no apparent consensus among researches on a single definition. The following definitions may be found in [21]:

- *Verification* checks well-defined properties of an expert system against its specification; it can focus on the knowledge base or the inference engine.
- *Validation* checks whether an expert system corresponds to the system it is supposed to represent.
- *Testing* is the examination of the behavior of a program by executing the program on sample data sets.
- *Evaluation* focuses on the accuracy of the system knowledge.

In case of mission-critical RBS applied as control systems a formal verification is essential [7] in order to provide certain level of system safety.

5.4 Inference Engine Development

This stage involves choosing inference strategy for rule analysis. Two most general types of inference are: forward chaining and backward chaining. Furthermore, combinations of the two types can be applied. The most typical strategy is to use forward chaining as a general control strategy, while at some stages, if detailed goals are to be inferred, backward chaining is employed.

Depending on the rule encoding chosen an inference engine may be already provided. It is the case with expert systems shells. If Prolog rules are chosen, built-in Prolog backward-chaining approach can also be used directly.

Today a number of tools are freely available for an RBS developer. They support different phases of RBS design, implementation and analysis. Selected examples are described in the following section.

6 Selected Development Tools Overview

The modern tools available to assist in building expert systems can be divided into several categories discussed in the following subsections.

6.1 Conventional Programming Languages

Conventional programming languages, e.g. ANSI C, do not support programming paradigm suitable for expert systems. Their *procedural* approach does not match very well the *declarative* nature of an expert system. Using these languages, a development of expert systems, while possible, is very difficult.

Object-oriented languages could be considered higher level languages. There is a smaller *semantic gap* between expert systems and languages such as: Java, Smalltalk or Eiffel. This is why they are sometimes chosen as expert system implementation tools. Languages such as Python, or Ruby have been gaining a growing acceptance due to their fast prototyping capabilities.

It can be concluded that it is more common to choose conventional languages as low-level implementation tools, while using higher level tool such as *expert system shell* to build a knowledge base.

Java is a classic object-oriented programming language. However, it has become a language of choice for many Web-related AI projects. Currently there is a number of Java-based tools for expert systems, see Sect. 6.3. It is worth noting that a standardization effort (*JSR 94: Java Rule Engine API*) is currently undertaken to formulate a standard Java Rule Language.

6.2 AI Programming Languages

For many years *Lisp* has been a language of choice for *symbolic* computation. Features of Lisp [4] are: programs are represented by list structures, and primitive operations are operations on lists. Lisp is the foundation of many expert systems and shells, such as *CLIPS*. In last decades it was extended in many ways, including object-oriented framework *CLOS*.

There are, however, problems with Lisp. The main problem is that lists have limited knowledge representation capabilities. Another is that no strong programming methodology has emerged from Lisp-based tools. There is a number of different dialects of Lisp language too.

Prolog is both flexible and powerful, with strong logical foundations. It has facilities for both knowledge representation and processing. Opposed to Lisp which is a symbolic language, Prolog is a declarative one. However, it does have dual semantics, both declarative and procedural. It is well-suited to symbolic rather than numerical problems. Since there is only a small *semantic gap* between expert-systems and Prolog, the language is an ideal tool for practical development of these systems.

The Prolog language is based on predicate logic. Prolog clauses are Horn clauses from the logical point of view. In order to find solutions (satisfy goals) Prolog uses the resolution rule and unification. Prolog is studied in detail in [1]. It has some important features to support logic-based reasoning. The Prolog inference engine uses backward-chaining with backtracking and recursion.

Meta programs treat other programs as data. They are used to help in both understanding and building knowledge-based systems [20]. Prolog is almost unique in the extent to which it can serve as its own meta-language. A Prolog program can create new goals, examine itself, and modify the inference engine, blurring the distinction between program and data. Prolog-based meta-interpreters are ideal to build forward-chaining inference engines.

6.3 Selected Expert System Shells

CLIPS is one of the most common expert system development tools (`www.ghg.net/clips`). It supports multiple reasoning and conflict resolution strategies. CLIPS is an expert system shell, so it does not provide any tools supporting the design of the knowledge base.

Jess is a *Java Expert System Shell* (`jessrules.com`). It is inspired by CLIPS but implemented in Java. Compared to CLIPS it adds several features and offers superior performance. It is easy to integrate with Java-based web-enabled applications. It plays an important role in the JSR 94 effort.

jDREW (`www.jdrew.org`) is a deductive reasoning engine for clausal first order logic written in Java and well integrated with the Web. Knowledge-based systems can use jDREW as an embedded reasoning engine through its various APIs. jDREW can be easily deployed as part of a larger Java system.

The *Algernon* (`algernon-j.sf.net`) rule-based inference system is implemented in Java and interfaced with Protege ontology editor. It performs forward and backward rule-based processing of frame-based knowledge bases, and stores and retrieves information in ontologies and knowledge bases. It is aimed at integration with Semantic Web projects.

6.4 Selected Design Environments

Sphinx [8] is an integrated development environment for expert systems development. It uses backward-chaining inference engine, contains a shell (PC-Shell) and several design tools, such as CAKE, which supports the process of knowledge base design and simple verification.

KbBuilder [18] is an integrated environment for designing and verifying Sphinx knowledge bases. The approach is oriented towards backward-chaining systems based on simple attributive language. Furthermore, its verification capabilities are limited to local properties of the so-called decision units.

Mandarax (`mandarax.sf.net`) is an open source Java class library for deduction rules. It provides an infrastructure for defining, managing and querying rule bases. Mandarax includes open APIs to interface with relational databases and XML, in particular RuleML. *Oryx* is a graphical user interface application to design and maintain Mandarax knowledge bases.

XpertRule (`www.attar.com`) supports developing rule-based systems. It uses a simple visual knowledge builder which maps knowledge modules to decision trees, which are main knowledge representation units. It also provides additional features, such as fuzzy reasoning.

VisiRule (`www.lpa.co.uk`) is a visual design tool for developing expert systems. A principal idea is to support the designer by a graphical flowchart representing the decision logic. The chart can be automatically translated into a lower level logic-based representation, processed in Prolog. The most important feature is the support for the visual design of the knowledge base; it is however, limited to decision trees. VisiRule does not provides means to validate or evaluate the knowledge base.

Drools (`www.drools.org`) is a framework for building forward-chaining expert systems, with the use of the Rete algorithm. It is implemented and Java, and integrated with Java building tools. It generates source in a selected language, from a conceptual description encoded in XML. This description includes declarative parts (rules) and embedded procedural code in the target language. The tool does not offer any verification or evaluation facilities.

7 Mirella Project

In [10] results of a research and evaluation of multiple RBS design methods, supported by development tools have been presented. A conclusion has been drawn, that existing methods and tools have some serious limitations located in the following areas: knowledge representation, formal analysis and verification, and design support tools. Most important limitations concerning the knowledge representation methods consist in using system-specific knowledge representation formalisms. This results in restricted application area, and scalability problems. With respect to the practical analysis approaches, the following problems have been identified: late verification problem, inefficient development cycle, and lack of integrated software framework.

Available design approaches do not offer integrated computer development tools (*CASE*) supporting the RBS building process at *all* stages – from the design to implementation. Such methods support mainly subsequent stages of the *conceptual design*, while direct technical support for the logical design and during the implementation phase is mostly limited to providing a context-sensitive, syntax checking editors, or simple wizards that support the design.

Practical design of non-trivial RBS requires a systematic, structured and consistent approach. Such an approach is usually referred to as a *design methodology*. To overcome limitations outlined above, a new approach to RBS design process, supported by an integrated CASE tool, has been proposed [10].

It is a top-down hierarchical design methodology, based on the idea of meta-level approach to the design process. It includes three phases: conceptual, logical, and physical. It provides a clear separation of logical and physical (implementation) design phases. It offers equivalence of logical design specification and prototype implementation, and employs XTT, a new hybrid knowledge representation. The methodology is supported by a CASE tool.

The main goal of the methodology is to move the design procedure to a logical level, where knowledge specification is based on the use of abstract rule representation. The design specification can be automatically translated into a low-level code, including Prolog and XML, so that the designer can focus on logical specification of safety and reliability. On the other hand, selected system properties can be automatically analyzed on-line during the design, so that its characteristics are preserved. The generated Prolog code constitutes a prototype implementation of the system. Since it is equivalent to the visual design specification it can be considered an executable specification.

These ideas are the basis for the MIRELLA Project, `mirella.ia.agh.edu.pl`. The goals of the project are: to fully develop and refine the design process outlined above, extend its' application areas onto different real-life RBS, and provide computer tools and methods supporting this process. So far the following elements have been developed: the XTT knowledge representation method, the concept of an integrated design process, a prototype Mirella CASE tool. They have been all described in detail in [10]. Some of the applications of these ideas were presented in [11, 12]. Further developments include ARD conceptual design [7, 13]. All of these are shortly introduced below.

7.1 EXtended Tabular-Trees

The main idea behind the new visual knowledge representation language called *eXtended Tabular-Trees* [10] aims at combining some of the existing approaches, namely decision trees and decision tables building a special hierarchy of Object-Attribute-Tables [6, 7]. It allows for a hierarchical visual representation of the OAT tables linked into tree-like structure, according to the control specification provided. XTT as a design and knowledge representation method offers transparent, high density knowledge representation as well as a formally defined logical, Prolog-based interpretation, while preserving flexibility with respect to knowledge manipulation. On the *machine readable level* XTT can be represented in an XML-based *XTTML (XTT Markup*

Language) suitable for import and export operations; it can also be translated to XML-based rule markup formats such as *RuleML*.

7.2 Integrated Design Process

The *eXtended Tabular-Trees*-based design method introduces possibility of on-line system properties analysis and verification, during the system design phase. Using XTT as a core, in [10] an integrated design process, covering the following phases has been presented:

1. *Conceptual modeling*, in which system attributes and their functional relationships are identified; during this design phase the ARD modelling method is used. ARD stands for *Attribute-Relationship Diagrams* [7, 13]. It allows for specification of functional dependencies of system attributes using a visual representation. Using this model the logical XTT structure can be designed. ARD can be represented in an XML-based *ARDML* (*ARD Markup Language*) suitable for data exchange operations, as well as possibly transformations to other diagram formats.
2. *Logical design with on-line verification*, during which system structure is represented as XTT hierarchy, which can be instantly analyzed, verified (and corrected, if necessary) and even optimized on-line. The XTT hierarchy can also be represented in XML, using the XTTML format.
3. *Physical design*, in which a preliminary Prolog-based *implementation* is carried out. A RuleML translation of the XTT rule base is also available.

Using the predefined XTT translation it is possible to automatically build a prototype. It uses Prolog-based meta-language for representing XTT knowledge base and rule inference (also referred to as XTT-PROLOG).

7.3 Mirella CASE Tool

A prototype CASE tool for the XTT method called MIRELLA [10] has been developed. It supports XTT-based visual design methodology, with an integrated, incremental design and implementation process, providing the possibility of the on-line, incremental, verification of formal properties. Logical specification is directly translated into Prolog-based representation providing an executable prototype, so that system operational semantics is well-defined. In the MIRELLA Editor the specification looks as in Fig. 3.

7.4 Meta-Level Features

The approach is based on the idea of a knowledge representation method which offers the *design and implementation equivalence* by a direct XTT $\rightarrow$ Prolog mapping. Using a visual design method the designer can focus on building the system structure, since the prototype implementation can be *dynamically generated* and *automatically analyzed*. The approach discussed herein offers strict,

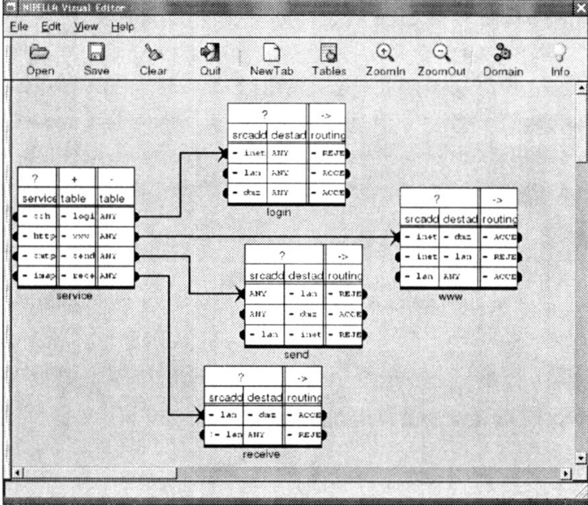

Fig. 3. RBS design session in Mirella

formal description of system attributes and structure, creates a framework for integrating the design and verification process, and supports the design and verification process by an integrated CASE tool.

In this way, it is possible to assure that some safety-critical system properties such as attribute domains, and basic system structural logical constraints are preserved during the design process.

7.5 Lessons Learned from Mirella

Expressive knowledge representation is needed in order to truly support the design. Mirella is focused around XTT, which proved to be a valuable tool, allowing for designing different classes of RBS. Addressing *all of the design phases* is an important issue addressed in Mirella. In order to successfully build real-life systems it is necessary to formulate a complete design methodology, covering design stages from conceptual analysis to the implementation, including verification. *On-line* formal verification allows for assuring system characteristics during the design, and keeping them up the the implementation. *Integration* of design phases is needed in order to truly support the designer, and preserve system characteristics during the design process. Easy to use *visual CASE tool* not only is important for the design support but also is necessary for the adoption of the new design methodology.

Mirella in its current state was successful as a proof of concept. However, after more than two years of development some possible areas of extension and improvement have been identified, such as: business rules support, automatic knowledge acquisition facilities, optional backward-chaining, possibly fuzzy rules support, and even more extended verification capabilities [14].

8 Towards Hybrid Knowledge Engineering

Basing on the experiences with the MIRELLA project a refined RBS design methodology is put forward. It addresses three design phases described in Sect. 7.2, that is: conceptual, logical, and physical design. It also addresses three important aspects of the design models used, that is:

- *visual representation*, which is valuable for both the design support and the human interaction,
- *knowledge encoding*, which is based on XML and is useful for automatic models transformations,
- *executable code*, which is based on Prolog representation of RBS.

An outline of this approach is shown in Fig. 4.

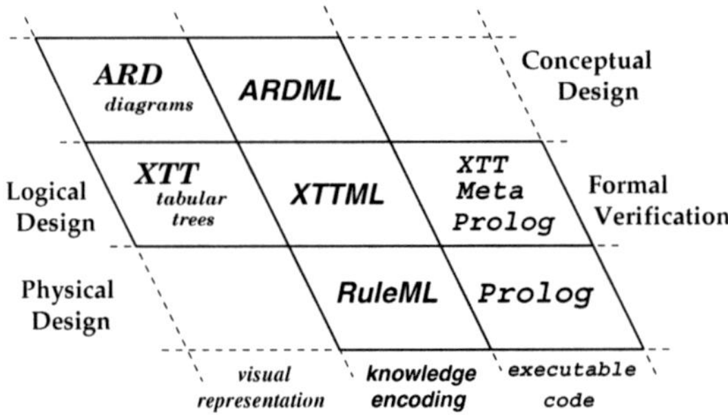

Fig. 4. Hierarchical design methodology

It is hoped, that if refined, this methodology could provide universal modelling methods for RBS design and implementation. The HEKATE projects aims at applying this methodology to practical design and analysis of intelligent systems. Main goals of the HEKATE project are to:

- develop an extended, hierarchical methodology for practical design, analysis and implementation of selected software classes,
- build computer CASE tools package supporting this methodology,
- test the approach on illustrative software examples.

The projects focuses on wide class of software, namely two very different classes, that is:

- general business software based on the so called *business logic*,
- low-level control software, possibly for the embedded control systems, based on a *control logic* [15].

A principal idea in this approach is to describe the logic behind the software using advanced knowledge representation methods. The logic would be expressed with use of a Prolog-based representation. The logical, Prolog-based core would be then embedded into a business application, or embedded control system. The business or control applications can be developed with some classic programming languages such as Java or C. The HEKATE project should eventually provide a coherent runtime environment for running the combined Prolog and Java/C code.

Hekate is currently (fall 2006) in a very early development stage. See the project webpage at `hekate.ia.agh.edu.pl` for more up to date information on the project progress, tools and technologies.

9 Concluding Remarks

In the paper RBS knowledge engineering issues, methodologies and selected tools have been discussed. They have been contrasted with some aspects of software engineering. The paper also discusses and advanced design methodology developed in MIRELLA project which aims at combining classic knowledge engineering methods with software engineering approach. While MIRELLA is work in progress it already created some valuable results such as XTT knowledge representation method, along with on-line Prolog verification approach. It is hoped that the HEKATE will develop these concepts into a complete hierarchical design and implementation methodology for both business software and rule-based systems.

References

1. Bratko I (2000) Prolog Programming for Artificial Intelligence. Addison Wesley, 3rd edition
2. Hopgood AA (2001) Intelligent Systems for Engineers and Scientists. CRC Press, Boca Raton London New York Washington, D.C., 2nd edition, ISBN 0849304563
3. Horrocks I, Patel-Schneider PF, Bechhofer S, Tsarkov D (2005) Owl rules: A proposal and prototype implementation. Journal of Web Semantics, 3(1):23–40
4. Jackson P (1999) Introduction to Expert Systems. Addison–Wesley, 3rd edition, ISBN 0-201-87686-8
5. Liebowitz J (ed) (1998) The Handbook of Applied Expert Systems. CRC Press, Boca Raton ISBN 0-8493-3106-4
6. Ligęza A, Wojnicki I, Nalepa GJ (2001) Tab-trees: a case tool for design of extended tabular systems. In: Mayr HC et al. (eds), Database and Expert Systems Applications, volume 2113 of Lecture Notes in Computer Sciences, pp. 422–431. Springer-Verlag, Berlin
7. Ligęza A (2006) Logical Foundations for Rule-Based Systems. Springer-Verlag, Berlin, Heidelberg
8. Michalik K (2003) Zintegrowany Pakiet Sztucznej Inteligencji Sphinx 4.0. AITech Artificial Intelligence Laboratory, Katowice, Poland
9. Miller J, Mukerji J (2003) MDA Guide Version 1.0.1. OMG

10. Nalepa GJ (2004) Meta-Level Approach to Integrated Process of Design and Implementation of Rule-Based Systems. PhD thesis, AGH University of Science and Technology, AGH Institute of Automatics, Cracow, Poland, September 2004
11. Nalepa GJ, Ligęza A (2003) Designing reliable web security systems using rule-based systems approach. In: Menasalvas E, Segovia J, Szczepaniak PS (eds), Advances in Web Intelligence. First International Atlantic Web Intelligence Conference AWIC 2003, Madrid, Spain, May 5–6, 2003, volume LNAI 2663 of Lecture Notes in Artificial Intelligence, pp. 124–133, Berlin, Heidelberg, New York, Springer-Verlag
12. Nalepa GJ, Ligęza A (2004) Markup-languages-based approach to knowledge management and representation. In: Nycz M, Owoc ML (eds), Pozyskiwanie Wiedzy i Zarządzanie Wiedzą, number 1011 in Prace Naukowe Akademii Ekonomicznej im. Oskara Langego we Wrocławiu, pp. 332–339, Wrocław, 2004. Akademia Ekonomiczna im. Oskara Langego we Wrocławiu.
13. Nalepa GJ, Ligęza A (2005) Conceptual modelling and automated implementation of rule-based systems. In Szmuc T, Zieliński K (eds), Software engineering : evolution and emerging technologies, volume 130 of Frontiers in Artificial Intelligence and Applications, pp. 330–340. IOS Press
14. Nalepa GJ, Ligęza A (2006) Prolog-based analysis of tabular rule-based systems with the xtt approach. In: Sutcliffe GCJ, Goebel RG (eds), FLAIRS 2006 : proceedings of the nineteenth international Florida Artificial Intelligence Research Society conference : [Melbourne Beach, Florida, May 11–13, 2006], pp. 426–431, FLAIRS. – Menlo Park, Florida Artificial Intelligence Research Society, AAAI Press
15. Nalepa GJ, Zięcik P (2006) Integrated embedded prolog platform for rule-based control systems. In: Napieralski A (ed), MIXDES 2006 : MIXed DESign of integrated circuits and systems : proceedings of the international conference : Gdynia, Poland 22–24 June 2006, pp. 716–721, Ł odź
16. Ross RG (2003) Principles of the Business Rule Approach. Addison-Wesley Professional
17. Russell S, Norvig P (2002) Artificial Intelligence: A Modern Approach. Prentice-Hall, 2nd edition
18. Simiński R (2002) Dynamiczna weryfikacja poprawności baz wiedzy w procesie ich weryfikacji. PhD thesis, Instytut Podstaw Informatyki PAN, Warszawa
19. Sommerville I (2004) Software Engineering. International Computer Science. Pearson Education Limited, 7th edition
20. Torsun IS (1995) Foundations of Intelligent Knowledge-Based Systems. Academic Press, London, San Diego, New York, Boston, Sydney, Tokyo, Toronto
21. Vermesan A (1998) The Handbook of Applied Expert Systems, chapter Foundation and Application of Expert System Verification and Validation. CRC Press
22. Vermesan A, Coenen F (eds) (1999) Validation and Verification of Knowledge Based Systems. Theory, Tools and Practice. Kluwer Academic Publisher, Boston
23. von Halle B (2001) Business Rules Applied: Building Better Systems Using the Business Rules Approach. Wiley

XML Schema Mappings Using Schema Constraints and Skolem Functions*

Tadeusz Pankowski[1,2]

[1] Institute of Control and Information Engineering,
Poznań University of Technology, Poland
[2] Faculty of Mathematics and Computer Science,
Adam Mickiewicz University, Poznań, Poland
email: tadeusz.pankowski@put.poznan.pl

Summary. A schema mapping is an executable specification describing transformation of data structured under different schemas. In this paper we discuss the problem of automatic generation of XML schema mappings using information provided by schemas and correspondences between schemas. Mappings are specified in a mapping language XDMap whose constructs are based on Skolem functions. We use Skolem functions with text-valued arguments from a source instance to create nodes in the target instance, and to specify functional dependencies between some values. First, using constraints defined within a schema, an algorithm produces the automapping representing this schema. Next, (auto)mappings can be combined by means of some operators delivering more general mappings between schemas provided that a correspondence between schemas is given. While a mapping is executed, some missing data can be inferred based on constraints encoded in the mapping specification.

1 Introduction

A schema mapping is a specification that describes how data structured under a source schema is to be transformed into data structured under a target schema [8]. Recently, this problem has received considerable attention in the context of data exchange [3,7], data integration [12,18], schema evolution [14], P2P databases [5,19], life science databases [22] or e-commerce [6,21], where data comes from many different sources with different schemas.

In this paper we are continuing our work on automatic generation of mappings and transformations of XML data [6,17,19,20,22]. The proposed method is based on *constraints* defined within XML schema: *keys, key references*, and *value dependencies*. We show how these constraints can be formalized and how they can be used to generate *automappings*. An automapping

---

*The work was supported in part by the Polish Ministry of Science and Higher Education under Grant N516 015 31/1553.

T. Pankowski: *XML Schema Mappings Using Schema Constraints and Skolem Functions*,
Studies in Computational Intelligence (SCI) **102**, 199–215 (2008)
www.springerlink.com

maps a schema onto itself. Automappings can be next combined giving arbitrary mappings between different schemas by means of *Match*, *Compose*, and *Merge* operators over (auto)mappings [19]. In order to express unique relationships between data we use Skolem functions in specification of mappings in XDMap. We also show how some missing or incomplete data, which are not given explicitly in sources, can be deduced based on value dependency constraints while a mapping is executed.

To define mappings we assume that key and some value constraints are specified within schema (using XML Schema [27] notation). We show how an automapping may be automatically generated from these constraints. It is significant in our approach that the constraints are specified outside the mapping by means of constraint-oriented notation. The generated automapping preserves these constraints. In contrast, in other mapping languages (e.g. in [28]) constraints must be explicitly encoded in the mapping language. This can make difficulties for future management when schemas evolve.

It is commonly accepted that the basic relationships between a source and a target relational schemas can be expressed as a source-to-target dependencies (STD) [2,9,14,15]. In [3] $STDs$ are adopted to XML data in such a way that if a certain pattern occurs in the source, another pattern has to occur in the target. In our approach, the main idea of using $STDs$ consists in specifying how nodes in a target instance depend on key paths, how these key paths correspond to paths in sources, and how target values depend on other values. So, our approach is more operational and uses DOM interpretation of XML documents. To generate the instance of a target schema from instances of source schemas, we use the idea of *chasing* [2,28].

The paper is organized as follows. In the following section we discuss and propose some definition formalizing notions which are used in the rest of the paper. Next, the language XDMap [19] is discussed. The language is used to specify XML schema mappings. We analyze its syntax and semantics. Then, we propose an algorithm for generating automappings. Finally, we show how missing data can be discovered while a mapping is executed. The last subsection concludes the paper.

2 Skolem Functions, Constraints, XML Trees and XML Schemas

A *Skolem function* returns a uniquely defined values for its arguments. Each of its invocation without arguments generates a new object. If it is invoked more than once for the same arguments it creates a new object only by the first invocation, by consecutive invocations it returns the identifier of the object created by the first evaluation. A concept of using Skolem functions for creation and manipulation object identifiers has been previously proposed in ILOG [11] and in [1, 10]. Recently, Skolem functions have been also used to schema mappings. For example, in Clio [23] they are used for generating missing target values if the target element cannot be null (e.g. components of keys), in [28] are used in a query rewriting based on data mapping.

In XDMap we consider Skolem functions with text-valued arguments, and they are used in two contexts:

1. To compute a string value for the given function name and its arguments. If f is a Skolem function name and $a_1, \ldots, a_m$ are string values, then the value of the Skolem term $f(a_1, \ldots, a_m)$ is the string "$f(a_1, \ldots, a_m)$", called a *term value*, obtained as the concatenation of the function name, its arguments, parenthesis, and separating commas. Further on quotation marks surrounding term values will be omitted.

2. To generate a node (a node identifier) in created resulting XML (data) trees. In this context a Skolem term $F_P(a_1, \ldots, a_m)$ will be used to express one-to-one relationship between a tuple $(a_1, \ldots, a_m)$ of key paths values from a source XML tree and the set of nodes labeled with P in a target tree.

We view an XML data as an ordered node-labeled unranked tree (XML tree). We assume that except for simple text values also Skolem term values may be assigned to text (leaf) nodes.

Let *Lab* be a countably infinite set of labels (names), $\mathcal{F}$ be a countably infinite set of Skolem function names, *Str* be a set of string values, *Term* be a set of term values of the form $f(a_1, \ldots, a_m)$, where $f \in \mathcal{F}$ and $a_i \in Str$, and *Did* be a set of document identifiers (URI addresses or file names). Attributes are modeled as elements.

Definition 1. *Let $L \subset Lab$ be a finite set of labels. An XML tree is a tuple*

$$I = (r, N^e, N^t, child, \leq, \lambda, \nu, \delta), \tag{1}$$

where:

1. *r is a distinguished root node, N^e is a finite set of element nodes, and N^t is a finite set of text nodes;*
2. *$child \subseteq \{r\} \cup N^e \times N^e \cup N^t$ – an acyclic binary relation introducing tree structure into (r, N^e, N^t) such that:*
   - *the root has only one child (this child is the outermost element), i.e. $(r, n) \in child \wedge (r, n') \in child \Rightarrow n = n' \wedge n \in N^e$;*
   - *each element node must have a child, i.e: $(n, n') \in child \wedge n' \in N^e \Rightarrow \exists n''(n'' \in (N^e \cup N^t) \wedge (n', n'') \in child)$;*
3. *$\leq$ – a total ordering relation on the set of nodes;*
4. *$\lambda : N^e \to L$ – a function labeling element nodes, the label $l = \lambda(n)$ is the type of n;*
5. *$\nu : N^t \to Str \cup Term$ – a function labeling text nodes with their text values (i.e. string or term values);*
6. *$\delta : \{r\} \to Did$ – a function assigning the document identifier to the root.*

In Fig. 1 there are three XML trees I_1, I_2 and I_3. The meaning of labels are: author (A), name (N) and university (U) of the author; paper (P) title (T), year (Y) of publication and the conference (C) where the paper has been presented. Elements labeled with R and K are used to join authors with their papers. Root nodes are not shown explicitly but we assume that they precede the outer most elements and are labeled with I_1, I_2, and I_3, respectively.

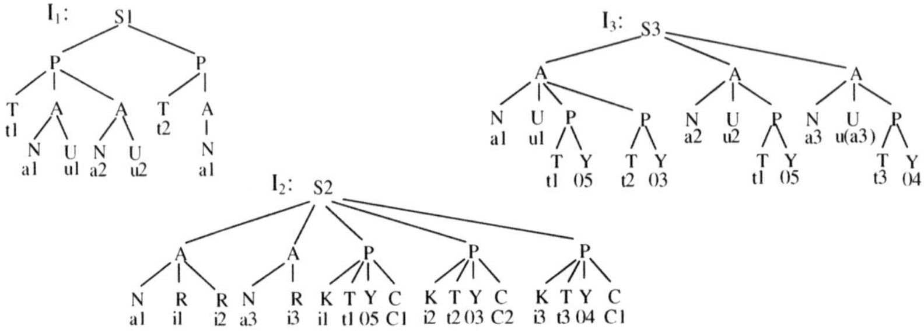

Fig. 1. Sample XML trees

XML schema defines both a structure and constraints for XML data. We will consider the following three kinds of constraints: *keys*, *key references*, and *value dependencies*.

1. *Key constraints.* We use a formal approach to keys for XML proposed by Buneman *et al.* [4]. A *key constraint* (or *key*) is an expression of the form

$$\kappa = (P, (P', (P_1, \ldots, P_k))),$$

   where $P/P'/P_i$ is a valid path for every $i = 1, \ldots, k$. The path P is called the *context path*, P' is called the *target path*, and $P_1, \ldots, P_k$ are called the *key paths* of κ. When $P = \epsilon$, we call κ an *absolute key* (then ϵ denotes the root), otherwise κ is called a *relative key*. In general, a path P over a set L of labels is an expression with the XPath syntax [26]: $P ::= \epsilon \mid l \mid P/l$, where ϵ is the empty path, $/l$ abbreviates `child::l` and selects the l element children of the context node. An XML tree I satisfies a key κ, denoted by $I \models \kappa$, if any subtree denoted by P' in the context determined by P is uniquely identified by the tuple $(s_1, \ldots, s_k)$ of text values of the tuple $(P_1, \ldots, P_k)$ of key paths. For example: $I_1 \models (/S1/P, T)$, $I_1 \models (/S1/P, (A, (N)))$ $I_1 \models (/S1/P/A, (U, (\epsilon)))$, where the last expression indicates that the node of type U is identified only by itself.

2. *A key reference* or *keyref* is an expression of the form

$$\rho = \kappa \ \textbf{ref} \ \kappa',$$

   where $\kappa = (P, (P', (P_1, \ldots, P_k)))$ and $\kappa' = (P, (P'', (P_1', \ldots, P_k')))$ are both key expressions. The key reference κ **ref** κ' defines a *foreign key* κ that refers to a *primary key* κ'. An XML tree I satisfies ρ, denoted $I \models \rho$, if:

- $I \models \kappa'$, i.e. κ' is a key satisfied in I, and
- for each node n in the set determined by P and for any value $(s_1, \ldots, s_k)$ of the key paths $(P_1, \ldots, P_k)$ in a node n' in the set reachable from n via P', there is exactly one node n'' in the set reachable from n via P'' in which $(P_1', \ldots, P_k')$ has the value equal to $(s_1, \ldots, s_k)$. For example: $I_2 \models (\epsilon, (/S2/A, (R)))$ **ref** $(\epsilon, (/S2/P, (K)))$.

3. *A value dependency* is an expression of the form

$$\tau = P/l/P' = f(P_1, \ldots, P_n),$$

that defines functional dependency between a text values denoted by $P/l/P'$ and a tuple of text values determined by the tuple $(P/l/P_1, \ldots, \ldots, P/l/P_n)$ of paths. An XML tree I satisfies a a value dependency τ, $I \models \tau$, if:

- $I \models (P, (l, (P_1, \ldots, P_n)))$,
- a text value of the path $P/l/P'$ functionally depends on the tuple of values determined by $(P/l/P_1, \ldots, P/l/P_n)$.

We use a Skolem function name $f \in \mathcal{F}$ to distinguish two different dependencies having the same set of determining paths. For example: $I_1 \models /S1/P/A/U = u(N)$, $I_2 \models /S2/P/Y = y(T)$, $I_2 \models /S2/P/C = c(T)$, $I_3 \models /S3/A/U = u(N)$, $I_3 \models /S3/A/P/Y = y(T)$.

Definition 2. *An XML schema over $(L, \mathcal{F})$ is a tuple*

$$\mathbf{S} = (top, Seq, Key, Keyref, Valdep), \tag{2}$$

where:

- *top is the distinguished label of the outermost element, $top \in L$,*
- *Seq is a function from L to regular expressions over $L - \{top\}$ defined by the grammar $e ::= \epsilon \mid l \mid e|e \mid ee \mid e? \mid e+ \mid e*$;*
- *Key assigns a key constraint $(P, (l, (P_1, \ldots, P_k)))$ to any label $l \in L - \{top\}$;*
- *Keyref assigns key references, $(P, (P'/l, (P_1, \ldots, P_k)))$ **ref** $Key(l')$, to some labels in $l \in L - \{top\}$.*
- *Valdep assigns a set of value dependencies, $P/l/P' = f(P_1, \ldots, P_n)$, to some labels in $l \in L - \{top\}$.* □

In Fig. 2 there are three XML data schemas (schema trees) $\mathbf{S}_1, \mathbf{S}_2$, and $\mathbf{S}_3$. Instances of these schemas are XML trees I_1, I_2 and I_3 in Fig. 1, respectively. These schema trees specify only structural part of an XML schema.

The full description of a schema corresponding to $\mathbf{S}_1$ from Fig. 2 written in XML Schema is presented in Fig. 3.

Definition 3. *An XML tree $I = (r, N^e, N^t, child, \leq, \lambda, \nu, \delta)$ conforms to the XML schema $\mathbf{S} = (top, Seq, Key, Keyref, Valdep)$, denoted by $I \models \mathbf{S}$, if:*

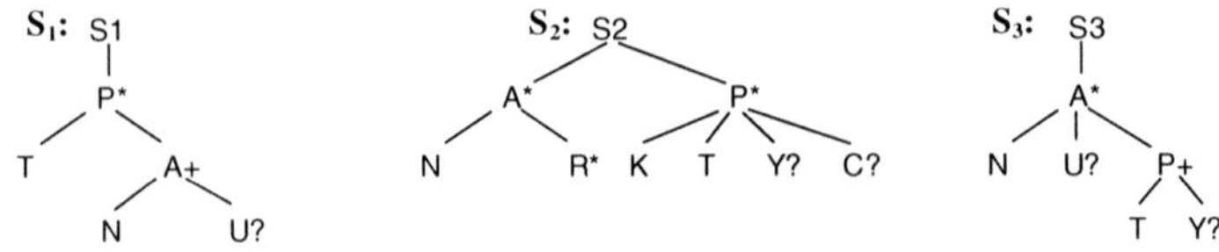

Fig. 2. Sample schemas describing structural part of instances from Fig. 1

```
<xs:schema xmlns:xs="http://www.w3.org/2001/XMLSchema">
 <xs:element name="S1">
    <xs:complexType>
    <xs:sequence>
        <xs:element ref="P"/>
    </xs:sequence>
    </xs:complexType>
 </xs:element>
 <xs:element name="P">
    <xs:complexType>
    <xs:sequence>
        <xs:element name="T" type="xs:string"/>
        <xs:element ref="A" minOccurs="1" maxOccurs="unbounded"/>
    </xs:sequence>
    </xs:complexType>
    <xs:key name="PKey">
    <xs:selector xpath="."/>
    <xs:field xpath="T"/>
    </xs:key>
 </xs:element>
 <xs:element name="A">
    <xs:complexType>
    <xs:sequence>
        <xs:element name="N" type="xs:string"/>
        <xs:element name="U" type="xs:string" minOccurs="0"/>
    </xs:sequence>
    </xs:complexType>
    <xs:key name="AKey">
    <xs:selector xpath="."/>
    <xs:field xpath="N"/>
    </xs:key>
    <xs:valdep>
    <xs:dependent xpath="U"/>
    <xs:function name="u"/>
    <xs:argument xpath="N"/>
    </xs:valdep>
 </xs:element>
 </xs:schema>
```

Fig. 3. Schema of S_1 in XML Schema extended with `<xs:valdep>` declaration

1. *For the root node r, $\delta(r)$ is defined, and for any text node $n \in N^t$, $\nu(n)$ is defined.*
2. *If $(r, n) \in child$, then $\lambda(n) = top$.*
3. *For any node n in I with children $(n_1, \ldots, n_m)$ such that $n_1 < \ldots < n_m$, if $\lambda(n) = l$, then the sequence $\lambda(n_1) \ldots \lambda(n_m)$ is a word of the language defined by the regular expression $Seq(l)$.*
4. *For any label $l \in L$*
   - *if a key constraint $Key(l)$ is defined for l, then: $I \models Key(l)$,*

- *if a key reference $Keyref(l)$ is defined for l, then: $I \models Keyref(l)$,*
- *if a value dependency $\tau \in Valdep(l)$, then: $I \models \tau$.*

3 XML Schema Mapping – Basic Ideas

In a general setting of relational data exchange [8,9,15], a schema mapping is a triple $\mathcal{M} = (\mathbf{S}, \mathbf{T}, \Sigma)$, where $\mathbf{S}$ and $\mathbf{T}$ are, respectively, source and target schemas, and Σ is a set of formulas of some logical formalism over $(\mathbf{S}, \mathbf{T})$.

The formulas in Σ are often specified using *source-to-target tuple-generating dependencies* [2], or *STDs*, that express the relationship between $\mathbf{S}$ and $\mathbf{T}$. They have been used to formalize relational data exchange by Fagin et al [8,9]. They have also been investigated as GLAV assertions in data integration scenario [12]. A *STD* is a first-order formula of the form [2]

$$\forall \mathbf{x}(\phi_S(\mathbf{x}) \Rightarrow \exists \mathbf{y} \psi_T(\mathbf{x}, \mathbf{y})), \tag{3}$$

where $\phi_S(\mathbf{x})$ is a conjunction of atomic formulas over $\mathbf{S}$, and $\psi_T(\mathbf{x}, \mathbf{y}))$ is a conjunction of atomic formulas over $\mathbf{T}$.

In data exchange the following problem is considered [3,8]: given an instance I over the source schema $\mathbf{S}$, find an instance J over the target schema $\mathbf{T}$ such that the pair $\langle I, J \rangle$ satisfy the *STDs* in Σ. Such an instance J is called a *solution* for I under $\mathcal{M}$.

In general, a mapping specification in XDMap conforms to the general form of *STDs* and has the form [19]:

$$\forall \mathbf{x}(G(\mathbf{x}) \wedge \Phi(\mathbf{x}) \Rightarrow \exists \mathbf{y} C(\mathbf{x}; \mathbf{y}) \wedge \Delta(\mathbf{x}; \mathbf{y})),$$

where $G(\mathbf{x})$ and $\Phi(\mathbf{x})$ are conjunctions of atomic formulas over a source, and $C(\mathbf{x}; \mathbf{y})$ and $\Delta(\mathbf{x}; \mathbf{y})$ are conjunctions of atomic formulas over a target. $G(\mathbf{x})$ defines source variables over source tree, $\Phi(\mathbf{x})$ restricts values of variables, $C(\mathbf{x}; \mathbf{y})$ specifies constraints (value dependencies) over target tree, and $\Delta(\mathbf{x}; \mathbf{y})$ defines child-parent relationships between nodes and, if necessary, also text values for text nodes in the target tree.

Definition 4. *Let a source schema $\mathbf{S} = (\mathbf{S}_1, \ldots, \mathbf{S}_n)$ be a sequence of source XML schemas over $(L_{\mathbf{S}_i}, \mathcal{F}_{\mathbf{S}_i})$, respectively, and $\mathbf{T}$ be a target XML schema over $(L_{\mathbf{T}}, \mathcal{F}_{\mathbf{T}})$. A mapping $\mathcal{M}_{\mathbf{S},\mathbf{T}}$ in XDMap from $\mathbf{S}$ into $\mathbf{T}$ is defined as follows:*

$$\mathcal{M}_{\mathbf{S},\mathbf{T}} ::= (G, \Phi, C, \Delta)(\mathbf{x}; \mathbf{y}) := \textbf{foreach } G(\mathbf{x})$$
$$\textbf{where } \Phi(\mathbf{x})$$
$$\textbf{when } C(\mathbf{x}; \mathbf{y})$$
$$\textbf{exists } \Delta(\mathbf{x}; \mathbf{y})$$

where

1. *$G(\mathbf{x})$ is a list of variable definitions over source schemas in $\mathbf{S}$, a definition of a variable x is an expression of the form: x <u>in</u> P or x <u>in</u> x'/P;*

2. $\Phi(\mathbf{x})$ *is a conjunction of restrictions of the form:* $x = x'$ *over* $\mathbf{x}$;
3. $C(\mathbf{x}; \mathbf{y})$ *is a list of target value dependencies of the form:* $x = f(\mathbf{x})$ *or* $y = f(\mathbf{x})$, *where* $x \in \mathbf{x}$, $y \in \mathbf{y}$;
4. $\Delta(\mathbf{x}; \mathbf{y})$ *is a conjunction* $\delta_1 \wedge \cdots \wedge \delta_m$ *of formulas of the form* $F_{P/l}(\mathbf{x}'; \mathbf{y}')$ **in** $F_P(\mathbf{x}''; \mathbf{y}'')/l[\textbf{with } z]$, *where*

   - $F_{P/l}(\mathbf{x}'; \mathbf{y}')$ *is a Skolem term, where* P/l *is a path in* $\mathbf{T}$;
   - $F_P(\mathbf{x}''; \mathbf{y}'')/l$ *is a target path expression, where* l *is a label in* $\mathbf{T}$;
   - $(\mathbf{x}'; \mathbf{y}') \subseteq (\mathbf{x}; \mathbf{y})$, $(\mathbf{x}''; \mathbf{y}'') \subseteq (\mathbf{x}'; \mathbf{y}')$, $z \in (\mathbf{x}'; \mathbf{y}')$.

A mapping from $\mathbf{S}_1$ into $\mathbf{S}_3$ is given in Fig. 3.

$$
\begin{aligned}
\mathcal{M}_{13}(x_T, x_N, x_U; y_Y) =& \\
&\textbf{foreach } (x_T, x_N, x_U) \underline{\textbf{ in }} \mathbf{S}_1 \\
&\textbf{where true} \\
&\textbf{when } x_U = u(x_N), y_Y = y(x_T) \\
&\textbf{exists} \\
&\quad F_{/S3}() \textbf{ in } F_{()}()/S3 \\
&\quad F_{/S3/A}(x_N) \textbf{ in } F_{/S3}()/A \\
&\quad F_{/S3/A/N}(x_N) \textbf{ in } F_{/S3/A}(x_N)/N \textbf{ with } x_N \\
&\quad F_{/S3/A/U}(x_N, x_U) \textbf{ in } F_{/S3/A}(x_N)/U \textbf{ with } x_U \\
&\quad F_{/S3/A/P}(x_N, x_T) \textbf{ in } F_{/S3/A}(x_N)/P \\
&\quad F_{/S3/A/P/T}(x_N, x_T) \textbf{ in } F_{/S3/A/P}(x_N, x_T)/T \textbf{ with } x_T \\
&\quad F_{/S3/A/P/Y}(x_N, x_T, y_Y) \textbf{ in } F_{/S3/A/P}(x_N, x_T)/Y \textbf{ with } y_Y
\end{aligned}
$$

Fig. 4. Mapping $\mathcal{M}_{13}$ from $\mathbf{S}_1$ into $\mathbf{S}_3$

Note that the definition of text variables, **foreach** (x_T, x_N, x_U) $\underline{\textbf{in}}$ $\mathbf{S}_1$, abbreviates in fact the following definition:

foreach x_{S1} $\underline{\textbf{in}}$ $/S1, x_P$ $\underline{\textbf{in}}$ $x_{S1}/P, x_T$ $\underline{\textbf{in}}$ x_P/T,
$\qquad x_A$ $\underline{\textbf{in}}$ $x_P/A, x_N$ $\underline{\textbf{in}}$ $x_A/N, x_U$ $\underline{\textbf{in}}$ x_A/U.

Some variables, for example x_{S1}, x_P, x_A, are auxiliary variables local in the **foreach** clause, while another, for example x_T, x_N, x_U, are source text variables global in the mapping. The variable y_Y is a target variable defined in the **when** clause as a function of x_T. The variable y_Y is defined only in the **when** clause because instances of $\mathbf{S}_1$ do not provide data about year of publication. However, we know that the year of publication of a paper depends on the title of the paper. This is denoted by the value dependency constraint: $y_Y = y(x_T)$.

Assuming that a schema denotes a set of all its instances, a mapping is an n-ary function that maps a tuple of source instances to a target instance (we will omit subscripts of $\mathcal{M}$ if they are clear from the context):

$$
\mathcal{M} : \mathbf{S}_1 \times \ldots \times \mathbf{S}_m \rightarrow \mathbf{T}, \tag{4}
$$

where for any tuple $(I_1, \ldots, I_n)$ such that $I_i \models \mathbf{S}_i$, $\mathcal{M}(I_1, \ldots, I_n) = J \models \mathbf{T}$.

We will pay a special attention to *automappings* which are identity mappings from a schema onto itself, i.e. a mapping $\mathcal{M}$ is the automapping over $\mathbf{S}$ iff $\mathcal{M} : \mathbf{S} \to \mathbf{S}$, where for each $I \in \mathbf{S}$, $\mathcal{M}(I) = I$.

4 Semantics of Mapping Rules

We propose a semantics for the XDMap in the operational way. We state how, for a given set $(\Omega, \leq)$ of bindings of variables in $(\mathbf{x}; \mathbf{y})$ into text values of an XML tree, determined by the **foreach/where/when** clauses, a set of mapping rules is executed and how the resulting target instance tree is created. The result tree must be ordered, so the encoding of the tree's nodes is a challenging issue. To encode ordering of the XML tree's node a variaty of order encoding methods is possible [13, 16]. We will use the Dewey order encoding, where each node n is assigned a vector $Pos(n)$ that represents the path from the document's root to n. For example, if $Pos(n) = 0.1.3.2$ then n is the second child of a node n', where n' is the third child of the outermost element that is the first element of the document (0 represents the root). The Dewey vector provides information about both the relative position of a node within children of the node's parent, and the absolute position of the node within the document (document's XML tree).

We will construct the Dewey vector using the fact that bindings $\omega \in \Omega$ are totally ordered. Then we assume the following position function:

- $Pos(r) = 0$, if r is the root,
- $Pos(n) = Pos(n').\omega$, where n' is the parent of n and ω is a binding for variables $\mathbf{x}; \mathbf{y}$ in an expression $F_P(\mathbf{x}; \mathbf{y})$ used to generate the node n. If the set of variables is empty (that is the case for the outermost element) we assume that ω equals 1.

Semantics for XDMap is defined as follows:

Definition 5. *Let* $\mathcal{M} = (G, \Phi, C, \Delta)(\mathbf{x}; \mathbf{y})$ *be a mapping over a schema* $\mathbf{S}$ *and* I *be an instance of* $\mathbf{S}$. *A target instance* $J = \mathcal{M}(I)$,

$$J = (r, N^e, N^t, child, \leq, \lambda, \nu, \delta), \tag{5}$$

is obtained by means of the semantic functions $\mathbb{E}, \mathbb{M}, \mathbb{R},$ *and* $\mathbb{N}$ *in the following way:*

1. $r = \mathbb{N}(F_{@doc}())$, *and* $\delta(r) = @doc$, $Pos(r) = 0$.
2. $n = \mathbb{N}(F_{P/l}(\mathbf{x}; \mathbf{y}))(\omega) = F_{P/l}(\omega(\mathbf{x}); \omega(\mathbf{y}))$, $n \in N^e$, $\lambda(n) = l$.
3. $\mathbb{R}(F_{P/l}(\mathbf{x}; \mathbf{y})$ **in** $F_P(\mathbf{x}'; \mathbf{y}')/l$ [**with** x''])$(\omega) =$
  $\mathbb{R}(\mathbb{N}(F_{P/l}(\omega(\mathbf{x}); \omega(\mathbf{y}))$ **in** $F_P(\omega(\mathbf{x}'); \omega(\mathbf{y}'))/l$ [**with** $\omega((x'')$])
  $if \quad n = \mathbb{N}(F_{P/l}(\mathbf{x}; \mathbf{y})(\omega), n' = \mathbb{N}(F_P(\mathbf{x}'; \mathbf{y}')(\omega), v = \omega(x''),$
  $then(n', n) \in child, Pos(n) = Pos(n').\omega,$
    $n'' = newId() \in N^t, (n, n'') \in child, \nu(n'') = v, Pos(n'') = Pos(n).1$

4. $n \leq n' \Leftrightarrow Pos(n) \leq Pos(n')$.
5. $\mathbb{M}(\Delta(\mathbf{x};\mathbf{y}))(\Omega) = \{\mathbb{R}(\delta(\mathbf{x};\mathbf{y}))(\omega) \mid \delta \in \Delta, \ \omega \in \Omega\}$.
6. $\mathbb{E}(\mathcal{M}(\mathbf{x};\mathbf{y}))(I) = \mathbb{M}(\Delta(\mathbf{x};\mathbf{y}))(((G,\Phi,C)(\mathbf{x};\mathbf{y}))(I)) = \mathbb{M}(\Delta(\mathbf{x};\mathbf{y}))(\Omega)$, *where Ω is a totally ordered set of bindings of variables $(\mathbf{x};\mathbf{y})$ determined by (G,Φ,C).* $\square$

According to the definition of semantics, a new XML tree is created as follows:

1. The execution of $F_{@doc}()$ produces a new root node r. The node is associated with a provided name @*doc* that is a unique name (URI or file name) of the newly created document. The root node r gets the ordering number 0 and precedes the first node of the created tree.
2. Execution of $F_{P/l}(\omega(\mathbf{x});\omega(\mathbf{y}))$ produces a new element node of type l.
3. The expression $\mathbb{R}(n, n'/l, [\, v \,])$ acts as follows: n is assumed to be a child of type l of n', the order position of n is set to $Pos(n').\omega$; if the third optional argument v is given, then a new text node n'' is created, n'' becomes a child of n, and v is assigned to n'' as its text value, the order position of n'' is set to $Pos(n).1$;
4. The total ordering on nodes coincides with the total ordering on Dewey vectors.
5. The execution from p. (3) is carried out for any mapping rule $\delta \in \Delta$ and for any binding $\omega \in \Omega$.
6. Execution of $\mathcal{M}(I)$ proceeds in two phases. First, a set Ω of bindings is determined, and then Ω is used in execution of the set Δ of mapping rules by the semantic function $\mathbb{M}$.

5 Using Constraints to Generate Automappings

Information provided by an XML schema $\mathbf{S}$ my be used to generate the automapping that transform instances of $\mathbf{S}$ onto themselves, Algorithm 1.

Algorithm 1 *Automapping generation from a schema*
Input : An XML schema $\mathbf{S} = (top, Seq, Key, Keyref, Dep)$ over $(L, \mathcal{F})$.
Output : Automapping
$$\mathcal{M} = (\mathbf{foreach}\ G, \mathbf{where}\ \Phi, \mathbf{when}\ C, \mathbf{exists}\ \Delta)\ over\mathbf{S}.$$

1. $G := \{x_{top}\ \underline{\mathbf{in}}\ /top\}; \Phi := \{\mathbf{true}\};$
   $C := \emptyset; \Delta := \{F_{top}()\ \mathbf{in}\ F_{(@doc)}()/top\};$
   $\mathbf{x}_{top} := (x_{/top})$
2. **foreach** $l \in L$ **begin**
   case $Key(l)$ **of**
   $(P, (l, (P_1, \ldots, P_m))),\ m > 0:$
   $G := G \cup \{x_{P/l}\ \underline{\mathbf{in}}\ x_P/l, \ldots, x_{P/l/P_m}\ \underline{\mathbf{in}}\ x_{P/l}/P_m\}$

$$\mathbf{x}_{P/l} := \mathbf{x}_P \circ (x_{P/l/P_1}, \ldots, x_{P/l/P_m})$$
$$\Delta := \Delta \cup \{F_{P/l}(\mathbf{x}_{P/l}) \text{ in } F_P(\mathbf{x}_P)/l\}$$
$(P, (l, \epsilon)) :$
$$\qquad G := G \cup \{x_{P/l} \underline{\text{ in }} x_P/l)$$
$$\qquad \mathbf{x}_{P/l} := \mathbf{x}_P \circ (x_{P/l})$$
$$\qquad \Delta := \Delta \cup \{F_{P/l}(\mathbf{x}_{P/l}) \text{ in } F_P(\mathbf{x}_P)/l \text{ with } x_{P/l}\}$$
endcase
if $Keyref(l) = (P, (P'/l, (P_1, \ldots, P_k)))$ **ref** $(P, (P''/l', (P'_1, \ldots, P'_k)))$
   then
$$\Phi := \Phi \cup \{x_{P/P'/l/P_1} = x_{P/P''/l'/P'_1}, \ldots, x_{P/P'/l/P_m} = x_{P/P''/l'/P'_m}\}$$
foreach $P/l/P' = f(P_1, \ldots, P_m) \in Valdep(l)$
$$\qquad C := C \cup \{x_{P/l/P'} = f(x_{P/l/P_1}, \ldots, x_{P/l/P_m})\}$$
end

For example, Algorithm 1 generates the automapping $\mathcal{M}_{33}$ for the schema $\mathbf{S}_3$ (Fig. 5).

$$
\begin{aligned}
\mathcal{M}_{33}(\mathbf{y}) = \ &\textbf{foreach } (y_N, y_U, y_T, y_Y) \underline{\text{ in }} \mathbf{S}_3 \\
&\textbf{where true} \\
&\textbf{when } y_U = u(y_N), y_Y = y(y_T) \\
&\textbf{exists} \\
&\quad F_{/S3}() \textbf{ in } F_{()}()/S3 \\
&\quad F_{/S3/A}(y_N) \textbf{ in } F_{/S3}()/A \\
&\quad F_{/S3/A/N}(y_N) \textbf{ in } F_{/S3/A}(y_N)/N \textbf{ with } y_N \\
&\quad F_{/S3/A/U}(y_N, y_U) \textbf{ in } F_{/S3/A}(y_N)/U \textbf{ with } y_U \\
&\quad F_{/S3/A/P}(y_N, y_T) \textbf{ in } F_{/S3/A}(y_N)/P \\
&\quad F_{/S3/A/P/T}(y_N, y_T) \textbf{ in } F_{/S3/A/P}(y_N, y_T)/T \textbf{ with } y_T \\
&\quad F_{/S3/A/P/Y}(y_N, y_T, y_Y) \textbf{ in } F_{/S3/A/P}(y_N, y_T)/Y \textbf{ with } y_Y
\end{aligned}
$$

Fig. 5. Automapping $\mathcal{M}_{33}$ over $\mathbf{S}_3$

In (a fragment of) the definition of $\mathbf{S}_2$ (Fig. 6), the schema specifies the *key* and *keyref* relationships between the K child element of the P element (the *primary key*) and the R child element of the A element (the *foreign key*). For this schema, Algorithm 1 generates the automapping $\mathcal{M}_{22}$ given in Fig. 7.

Mappings can be combined by means of some operators giving a result that in turn is a mapping. We have defined three operations: *Match, Compose*, and *Merge* in [19]. Some of these operations require specification of a *correspondence* between paths of schemas under consideration. Establishing the correspondence is a crucial task in definition of data mappings [24].

```
<xs:schema xmlns:xs="http://www.w3.org/2001/XMLSchema">
<xs:element name="S2">
\ldots
</xs:element>
<xs:element name="A">
    \ldots
    <xs:keyref name="RKeyref" refer="KKey">
    <xs:selector xpath="."/>
    <xs:field xpath="R"/>
    </xs:keyref>
</xs:element>
<xs:element name="P">
    \ldots
    <xs:key name="KKey">
    <xs:selector xpath="."/>
    <xs:field xpath="K"/>
    </xs:key>
</xs:element>
</xs:schema>
```

Fig. 6. Schema of $\mathbf{S}_2$ expressed in XML Schema language

$$\mathcal{M}_{22} = \textbf{foreach } (z_N, z_R, z_K, z_T, z_Y, z_C \underline{\textbf{ in }} \mathbf{S}_2$$
$$\textbf{where } z_R = z_K$$
$$\textbf{when } z_K = k(z_N, z_T), z_Y = y(z_T), z_C = c(z_T)$$
$$\textbf{exists}$$
$$F_{/S2}() \textbf{ in } F_{()}()/S2$$
$$F_{/S2/A}(z_N) \textbf{ in } F_{/S2}()/A$$
$$F_{/S2/A/N}(z_N) \textbf{ in } F_{/S2/A}(z_N)/N \textbf{ with } z_N$$
$$F_{/S2/A/R}(z_N, z_K) \textbf{ in } F_{/S2/A}(z_N)/R \textbf{ with } z_K$$
$$F_{/S2/P}(z_K) \textbf{ in } F_{/S2}()/P$$
$$F_{/S2/P/K}(z_K) \textbf{ in } F_{/S2/P}(z_K)/K \textbf{ with } z_K$$
$$F_{/S2/P/T}(z_K, z_T) \textbf{ in } F_{/S2/P}(z_K)/T \textbf{ with } z_T$$
$$F_{/S2/P/Y}(z_K, z_Y) \textbf{ in } F_{/S2/P}(z_K)/Y \textbf{ with } z_Y$$
$$F_{/S2/P/C}(z_K, z_C) \textbf{ in } F_{/S2/P}(z_K)/C \textbf{ with } z_C$$

Fig. 7. Automapping $\mathcal{M}_{22}$ over $\mathbf{S}_2$

6 Using Value Constraints to Infer Missing Data by Executing Mappings

In Fig. 8 there is an executable mapping, that integrates, by means of the *Merge* operator, instances of schemas $\mathbf{S}_1$ and $\mathbf{S}_2$ under the schema $\mathbf{S}_3$. Now, we focus on the problem of discovering missing values in the process of mapping execution. The discovery is achieved using some inference rules over (partial) bindings of variables.

Execution of $Merge_{\mathbf{S}_3}(\mathbf{S}_1, \mathbf{S}_2)(I_1, I_2)$ consists of the following four steps:

$Merge_{S_3}(\mathbf{S}_1, \mathbf{S}_2) =$
    foreach (x_T, x_N, x_U) <u>in</u> $\mathbf{S}_1, (z_N, z_R, z_K, z_T, z_Y)$ <u>in</u> $\mathbf{S}_2$
    where $z_R = z_K$
    when $x_U = u(x_N), y_Y = y(x_T), v_U = u(z_N), z_Y = y(z_T)$
    exists
        (1) $F_{/S3}()$ **in** $F_{()}()/S3$
        (2) $F_{/S3/A}(x_N)$ **in** $F_{/S3}()/A$
           $F_{/S3/A}(z_N)$ **in** $F_{/S3}()/A$
        (3) $F_{/S3/A/N}(x_N)$ **in** $F_{/S3/A}(x_N)/N$ **with** x_N
           $F_{/S3/A/N}(z_N)$ **in** $F_{/S3/A}(z_N)/N$ **with** z_N
        (4) $F_{/S3/A/U}(x_N, x_U)$ **in** $F_{/S3/A}(x_N)/U$ **with** x_U
           $F_{/S3/A/U}(z_N, v_U)$ **in** $F_{/S3/A}(z_N)/U$ **with** v_U
        (5) $F_{/S3/A/P}(x_N, x_T)$ **in** $F_{/S3/A}(x_N)/P$
           $F_{/S3/A/P}(z_N, z_T)$ **in** $F_{/S3/A}(z_N)/P$
        (6) $F_{/S3/A/P/T}(x_N, x_T)$ **in** $F_{/S3/A/P}(x_N, x_T)/T$ **with** x_T
           $F_{/S3/A/P/T}(z_N, z_T)$ **in** $F_{/S3/A/P}(z_N, z_T)/T$ **with** z_T
        (7) $F_{/S3/A/P/Y}(x_N, x_T, y_Y)$ **in** $F_{/S3/A/P}(x_N, x_T)/Y$ **with** y_Y
           $F_{/S3/A/P/Y}(z_N, z_T, z_Y)$ **in** $F_{/S3/A/P}(z_N, z_T)/Y$ **with** z_Y

Fig. 8. A mapping specifying merging of $\mathbf{S}_1$ and $\mathbf{S}_2$ under $\mathbf{S}_3$

1. *Determining a set Ω of bindings and a set Ω'_Ω of dependent bindings.*
   Variable specifications in the **foreach** clause over schemas $\mathbf{S}_1$ and $\mathbf{S}_2$ are
   computed against instances I_1 and I_2, respectively, and produce two sets
   Ω_1 and Ω_2 of partially defined bindings. By Ω we denote the union of
   Ω_1 and Ω_2. By Ω'_Ω we denote a set of dependent bindings for dependent
   variables (specified in the **when** clause). A binding $\omega'_\omega \in \Omega'_\Omega$ binds a term
   value to a dependent variable, e.g. $\omega'_\omega(x_U) = u(\omega(x_N))$. The set Ω of all
   bindings for all variables, and the set Ω'_Ω of dependent bindings for all
   dependent variables, are shown in Fig. 9(1). Bindings in Ω are partial
   functions because some bindings for some variables may be undefined –
   we denote this by $\bot$. For example, $\omega_3(x_U) = \bot$.
2. *Expanding bindings from Ω.*
   If $\omega \in \Omega$ and $\omega(x) = \bot$, i.e. ω is not defined for x, then we assume

$$\omega(x) := \omega'_\omega(x). \tag{6}$$

   In this way we assign a term value to a variable for which there is no
   explicit binding. For example, $\omega_3(x_U) := \omega'_{\omega_3}(x_U) = u(\omega_3(x_N)) = u(a1)$.
   In Fig. 9(2) there is the result of expanding bindings from Ω.
3. *Resolving term values in bindings.*
   In this step we try to discover text values for these variables to which
   term values have been assigned. We say that such variables have *missing
   values*. To achieve this the following inference rule is applied:

$$\omega'_{\omega_1}(x_1) = \omega'_{\omega_2}(x_2) \Rightarrow \omega_1(x_1) := \omega_2(x_2) \tag{7}$$

For example, in this way we can obtain that $\omega_1(y_Y) = 05$. This is obtained as follows: $\omega'_{\omega_1}(y_Y) = \omega'_{\omega_4}(z_Y)$, so using the rule (7) we have $\omega_1(y_Y) := \omega_4(z_Y) = 05$. Note that $\omega_6(v_U)$ can not be resolved. The resolved set of bindings is shown in Fig. 9(3).

(1) Bindings $\Omega = \Omega_1 \cup \Omega_2$:

Ω	x_T	x_N	x_U	y_Y	z_N	z_R	z_K	z_T	z_Y	v_U
ω_1	$t1$	$a1$	$u1$	$\perp$						
ω_2	$t1$	$a2$	$u2$	$\perp$						
ω_3	$t2$	$a1$	$\perp$	$\perp$						
ω_4					$a1$	$i1$	$i1$	$t1$	05	$\perp$
ω_5					$a1$	$i2$	$i2$	$t2$	03	$\perp$
ω_6					$a3$	$i3$	$i3$	$t3$	04	$\perp$

Dependent bindings Ω'_Ω:

Ω'_Ω	x_U	y_Y	v_U	z_Y
ω'_{ω_1}	$u(a1)$	$y(t1)$		
ω'_{ω_2}	$u(a2)$	$y(t1)$		
ω'_{ω_3}	$u(a1)$	$y(t2)$		
ω'_{ω_4}			$u(a1)$	$y(t1)$
ω'_{ω_5}			$u(a1)$	$y(t2)$
ω'_{ω_6}			$u(a3)$	$y(t3)$

(2) Set Ω of bindings after expanding:

Ω	x_T	x_N	x_U	y_Y	z_N	z_R	z_K	z_T	z_Y	v_U
ω_1	$t1$	$a1$	$u1$	$y(t1)$						
ω_2	$t1$	$a2$	$u2$	$y(t1)$						
ω_3	$t2$	$a1$	$u(a1)$	$y(t2)$						
ω_4					$a1$	$i1$	$i1$	$t1$	05	$u(a1)$
ω_5					$a1$	$i2$	$i2$	$t2$	03	$u(a1)$
ω_6					$a3$	$i3$	$i3$	$t3$	04	$u(a3)$

(3) Set Ω of bindings after expanding and resolving:

Ω	x_T	x_N	x_U	y_Y	z_N	z_K	z_T	z_Y	v_U
ω_1	$t1$	$a1$	$u1$	05					
ω_2	$t1$	$a2$	$u2$	05					
ω_3	$t2$	$a1$	$u1$	03					
ω_4					$a1$	$i1$	$t1$	05	$u1$
ω_5					$a1$	$i2$	$t2$	03	$u1$
ω_6					$a3$	$i3$	$t3$	04	$u(a3)$

Fig. 9. Determining, expanding and resolving a set Ω of bindings during execution of the mapping $Merge_{\mathbf{S}_3}(\mathbf{S}_1, \mathbf{S}_2)$ on the pair of instances (I_1, I_2)

After preparing an expanded and resolved set Ω of bindings, the **exist** clause of the mapping can be executed. For the mapping in Fig. 8 the execution proceeds as follows (we discuss only some representative mapping expressions):

(1) Two new nodes are created, the root r and the node n of the outermost element of type $/S3$, as results of Skolem functions $F_{()}()$ and $F_{/S3}()$, respectively. The node n is a child of type $S3$ of r.
(2) A new node n' for any distinct value of x_N is created. Each such node has the type $/S3/A$ and is a child of type A of the node n created by $F_{/S3}()$.

(3) For any distinct value of x_N a new node n'' of type $/S3/A/N$ is created. Each such node is a child of type N of the node created by invocation of $F_{/S3/A}(x_N)$ in (2) for the same value of x_N. Because n'' is a leaf, it obtains the text value equal to the current value of x_N.

(4) Analogously for the rest of the specification.

As the result, we obtain the instance I_3 depicted in Fig. 1.

7 Conclusion

In the paper the problem of schema mapping is considered, which occurs in many data management systems such as XML data exchange, XML data integration or e-commerce applications. Our solution to this problem relies on the automatic generation of semantics-preserving schema mappings. We discussed how automappings could be generated using schema constraints, such as keys, key references and value dependencies, defined in XML Schema. The mapping specification language XDMap is discussed. Mapping rules in XDMap are defined in conformity with source-to-target data generating dependencies. Skolem functions are used in these rules to express both functional dependencies between some text values in the target instance, and between tuples of key path values in the sources and subtrees in the target. Mappings between two schemas can be generated automatically from their automappings and correspondences between schemas. Automappings represent schemas, so operations over schemas and mappings can be defined and performed in a uniform way. We show how constraints on values can be used to infer some missing data. Our techniques can be applied in various XML data exchange scenarios, and are especially useful when the set of data sources change dynamically (e.g. in P2P environment) [5, 25] or when merging data from heterogeneous sources is needed [22]. The method proposed in the paper is under implementation in a system for semantic integration of XML data in P2P environment using schemas and ontologies [6, 19].

References

1. Abiteboul S, Buneman P, Suciu D (2000) Data on the Web. From Relational to Semistructured Data and XML. Morgan Kaufmann, San Francisco
2. Abiteboul S, Hull R, Vianu V (1995) Foundations of Databases. Addison-Wesley, Reading, Massachusetts
3. Arenas M, Libkin L (2005) XML Data Exchange: Consistency and Query Answering. PODS Conference 2005, 13–24
4. Buneman P, Davidson SB, Fan W, Hara CS, Tan WC (2003) Reasoning about keys for XML. Information Systems, 28(8), 1037–1063
5. Calvanese D, Giacomo GD, Lenzerini M, Rosati R (2004) Logical Foundations of Peer-To-Peer Data Integration. In: Proc. of the 23rd ACM SIGMOD Symposium on Principles of Database Systems (PODS 2004), 241–251

6. Cybulka J, Meissner A, Pankowski T (2006) Schema- and Ontology-Based XML Data Exchange in Semantic E-Business Applications. Business Information Systems 2006, Lecture Notes in Informatics, Vol.85, 429–441
7. Fagin R, Kolaitis PG, Miller RJ, Popa L (2002) Data Exchange: Semantics and Query Answering. ICDT 2003, Lecture Notes in Computer Science 2572, Springer, 2002, 207–224
8. Fagin R, Kolaitis PG, Popa L (2005) Data exchange: getting to the core. ACM Trans. Database Syst., 30(1), 2005, 174–210
9. Fagin R, Kolaitis PG, Popa L, Tan WC (2004) Composing Schema Mappings: Second-Order Dependencies to the Rescue. PODS 2004, 83–94
10. Fernandez MF, Florescu D, Kang J, Levy AY, Suciu D (1998) Catching the Boat with Strudel: Experiences with a Web-Site Management System. SIGMOD Conference 1998, 414–425
11. Hull R, Yoshikawa M (1990) ILOG: Declarative Creation and Manipulation of Object Identifiers. VLDB 1990, 455–468
12. Lenzerini M (2002) Data Integration: A Theoretical Perspective. PODS 2002, 233–246
13. Li Q, Moon B (2001) Indexing and Querying XML Data for Regular Path Expressions. In: Proc. of the 27th International Conference on Very Large Data Bases. VLDB 2001, Rome, Italy, 361–370
14. Melnik S, Bernstein PA, Halevy AY, Rahm E (2005) Supporting Executable Mappings in Model Management. SIGMOD Conference 2005, 167–178
15. Nash A, Bernstein PA, Melnik S (2005) Composition of Mappings Given by Embedded Dependencies. PODS 2005, 172–183
16. O'Neil P, O'Neil E, Pal S, Cseri I, Schaller G, Westbury N (2004) ORDPATHs: Insert-Friendly XML Node Label. In: Proc. of the 2004 ACM SIGMOD International Conference on Management of Data, 2004, 903–908
17. Pankowski T (2004) A High-Level Language for Specifying XML Data Transformations. Advances in Databases and Information Systems ADBIS 2004, Lecture Notes in Computer Science 3255, Springer, 2004, 159–172
18. Pankowski T (2005) Specifying Schema Mappings for Query Reformulation in Data Integration Systems. Atlantic Web Intelligence Conference – AWIC'2005, Lecture Notes in Computer Science 3528, Springer, 2005, 361–365
19. Pankowski T (2006) Management of executable schema mappings for XML data exchange. Database Technologies for Handling XML Information on the Web, EDBT 2006 Workshops, Lecture Notes in Computer Science 4254, Springer, 2006, 264–277
20. Pankowski T (2006) Reasoning About Data in XML Data Integration. Information Processing and Management of Uncertainty in Knowledge-based Systems, IPMU 2006, Vol. 3, Editions EDK, Paris, 2506–2513
21. Pankowski T (2005) Integration of XML Data in Peer-To-Peer E-commerce Applications. 5th IFIP Conference I3E'2005, Springer, New York, 481–496
22. Pankowski T, Hunt E (2005) Data Merging in Life Science Data Integration Systems. Intelligent Information Systems, New Trends in Intelligent Information Processing and Web Mining, Advances in Soft Computing, Springer Verlag, 2005, 279–288
23. Popa L, Velegrakis Y, Miller RJ, Hernández MA, Fagin R (2002) Translating Web Data. VLDB 2002, 598–609
24. Rahm E, Bernstein PA (2001) A survey of approaches to automatic schema matching. The VLDB Journal, 10(4), 2001, 334–350

25. Tatarinov I, Halevy AY (2004) Efficient Query Reformulation in Peer-Data Management Systems. SIGMOD Conference 2004, 539–550
26. XML Path Language (XPath) 2.0, W3C Working Draft: 2002. `http://www.w3.org/TR/xpath20`
27. XML Schema Part 1: Structures: 2004. `http://www.w3.org/TR/xmlschema-1`
28. Yu C, Popa L (2004) Constraint-Based XML Query Rewriting For Data Integration. SIGMOD Conference 2004, 371–382

Outline of Modification Systems

Josep Lluís de la Rosa, Albert Figueras, Christian Quintero,
Josep Antoni Ramon, Salvador Ibarra, and Santiago Esteva

ARLab — Agents Research Lab, EASY XIT center of CIDEM at the University of
Girona, Campus Montilivi, Building P4, 17071 Girona, Spain. peplluis@eia.udg.cat

Summary. This paper tries to understand the keys necessary for a new approach
for automatic control. It starts by analyzing its history and identifying the symptoms
that occur once and again when new paradigms, theories or breakthrough inventions
came up. Then, it analyses the symptoms of today and discusses whether they match
any of the previous symptoms in the past. Then, yet another theory is proposed
here, the *modification systems*, which joins the benefits of Automatic Control and
Agents Metaphor: The *modification systems*, which are designed as a generalization
of control systems and situated agents, where anybody does not control a system but
modifies a system by some multidimensional change of its original behavior toward a
desired target behavior. We show some examples and case studies of their behavior,
which as a potential generalization of automatic control give the background to
conceive further tools to design more simple but powerful controllers.

1 Introduction

This paper is the result of yet another analysis of the past and the future
of automatic control and artificial intelligence, in the general trend set by
many important names in the automatic control community [1, 3, 4, 9]. One
can see, by looking backwards in its history, how research in the two areas
cyclically entered into a cul-de-sac until an invention or a new theory leapt
into great advances. We feel, and particularly agree with [4], that research
in automatic control and artificial intelligence are again entering into a cul-
de-sac because the state of the art shows little evidence of real progress and
a great deal of minor contributions to industrial applications, the thing that
"really matters". In the last 40 years, a lot of theoretical material is being
developed, with thousands of architectures and models having been proposed
without representing breakthrough advances, perhaps excepting fuzzy logic,
neural networks, predictive control and the digital computing control. The fact
is that many "temptative" theories are not applied, and are kept applied for
nice toy examples [1,3] that only work in laboratories, or even worse, are only
sustainable on paper or, more modernly, on a PowerPoint presentation. And in

J.L. de la Rosa et al.: *Outline of Modification Systems*, Studies in Computational Intelligence
(SCI) **102**, 217–233 (2008)
www.springerlink.com

history, there are other examples of cul-de-sac. For example after the birth of self regulated machines, when engineers and scientists tried to improving the Watt's regulator without understanding the "separated" essence of control, or in the birth of calculator machines when they tried to improve their calculus capacity without understanding the "separated" essence of programming.

This paper proposes a reconsideration of automatic control, as defined by Murray in [4], "Control refers to the use of algorithms and feedback in engineered systems. At its simplest, a control system is a device in which a sensed quantity is used to **modify** the behavior of a system through computation and actuation". Then this paper starts in section 2 its history to identify the symptoms that again and again are occurring before the introduction of a new paradigm, theory or breakthrough invention. Thus, in section 3 there will be analysis if these symptoms happen today. Then section 4 comes with yet another approach that joins the benefits of Automatic Control and Agents Metaphor that is proposed: the *modification systems*, a generalization of control systems and situated agents, where there is no *control* of a system but its *modification* by some multidimensional change of its original behavior toward a desired target behavior. An illustrative case study is presented in section 5 and some considerations of what new facilities this approach would represent for the design of control systems is introduced in section 6. Finally, some conclusions and future work are shown in section 7.

2 Retrospective Look Back at the Control History

The most significant control development during the 18th century was the steam engine governor, which was first used early in 1789. The original Watt governor had several disadvantages: it provided only proportional control and hence exact control of speed at only one operating condition (this led to comments that it was "a moderator, not a controller"); it could operate only over a small speed range; and it required careful maintenance. The first 70 years of the 19th century saw extensive efforts to improve on the Watt governor, and thousands of governor patents were granted throughout the world. Many were for mechanisms designed to avoid the offset inherent in the Watt governor. This was the birth of integral control.

At that time, the governor was yet another *modification* of the steam machine to make it stable and controllable. It was simply that. The important fact is that, in the dawning of the industrial revolution, there was the need for improved machines, the steam machines principally to power the textile industry, and engineers modified the machines in the proper way. There was no control, simply a mechanical modification of the machines to make them profitable for industry.

The little but powerful advances in the 19th century were introducing a separate essence of the governor with respect to the machines (see Fig. 1): it had a *measure* device (the balls), it had an *actuator* device (the transmission

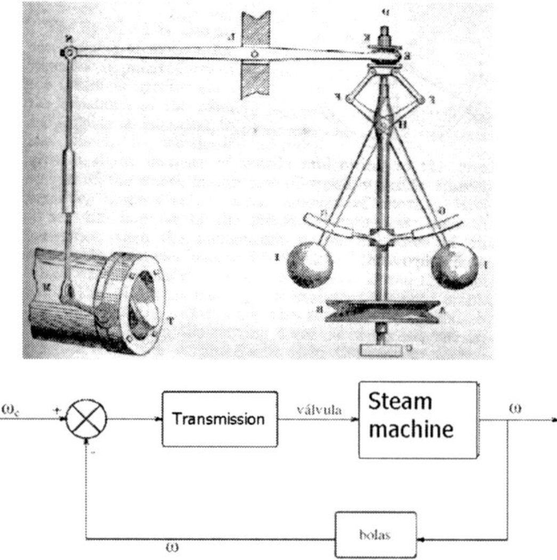

Fig. 1. The separate essence of control of the Governor

to the valve), and finally it had a logic unit of control, in terms of proportional or integral control. So, the *symptoms* that at the end of the 18th century led to the necessity of control were:

1. A challenge: need of *regulated* machines
2. An Invention: the governor
3. The lack of theory: to change working conditions of the governor is difficult

Therefore, there was a need for control theory but when the "control loop" theory came up in the late XIX century, it was **complicated**, not practical, and reduced to "nothing relevant" or "**yet the same, it is already done**" But then when it was finally understood and used in the first mid XX century, that theory flourished into great valuable results. However, in industrial applications there was a strong need for very simple controllers, easy to understand by engineers and especially easy to maintain. This is what we call the KISS method in control (keep it simple, stupid). This is a very typical human reaction that for example put extreme opposition to the new Foxboro Stabilog PI Temperature regulator of the mid 1950's because this device was not understood, and therefore was not used!

The growth in performance of control systems was progressively decreasing during the second half of the 20th century although this period produced much of the existing control theory corpus. This is what we call *the success but progressive saturation of control theory* as stated in section 2. The new theory in the recent years, again, is not used by engineers [3, 4].

3 Are the Symptoms the Same Today?

From the several states of the art: [2,3,8,11] we can see the following symptoms in the early XXI century:

1. New challenge: machines are not isolated, but interconnected
2. Saturation of control theory: more complex control technique does not mean more performance
3. Invention: intelligent agents and other emergent soft computing enter in the arena
4. Lack of theory: no real examples for the application of the new invention

   Additionally

1. There is strong industrial need for simplicity (KISS)
2. Reductionism and conservatism still prevents the advance of new control approaches (Fuzzy, soft computing, etc).

Thus, we claim that the 19th century symptoms happen again in the 21st century: new requirements from the industry that need of new inventions from the control and agent theory [11] to meet the requirements, and new paradigms that are poorly understood. However, were does control theory stand today? Accordingly to [4] an examination of a typical issue of the *IEEE Transactions on Automatic Control* reveals a wide gap between the theory and real-world problems. Increasingly, control is becoming task-oriented, especially in the realm of robotics. By contrast, classical control -as reflected in the *Transactions*- is set-point-oriented. It is a sobering thought that much of control theory as it is taught today is of little, if any relevance to task-oriented control.

Here follows the need for research effort reorientation. As again mentioned in [4], "The tradition of a rigorous use of mathematics combined with a strong interaction with applications has produced a set of tools that are used in a wide variety of technologies, but the opportunities for future impact are even richer than those of the past, and the field is well positioned to expand its tools for use in new areas and applications; the pervasiveness of communications, computing, and sensing will enable many new applications of control but will also require substantial expansion of the current theory and tools; and the control community must embrace new, information-rich applications and generalize existing concepts to apply to systems at higher levels of decision making".

Then, as stated by [5], for future industrial competitiveness, new types of competence and system solutions are needed. The use of control-based approaches in analysis and design of embedded (computing) systems is one promising approach. Furthermore, low-level real-time technology will be combined with high-level aspects, concerning programming, networking, safety, security, simulation and control. The use of these technologies in the implementation of complex controllers will necessarily be based on development tools and methodologies that provide support for design, implementation, verification and deployment in a holistic manner. Extensions of the Unified Modeling Language (UML) will potentially provide us with methods to model embedded systems in such a way that we shall be able to determine properties such as responsiveness, schedulability or resource requirements of real-time systems already in early design phases. When complexity increases, engineers rely on practice-proven, effective designs. Capturing design knowledge is a critical issue familiar to control engineers, who repeatedly re-use control designs that are well known and well documented.

However, when software is involved in the final implementation many other factors, not well documented in the control design textbooks, must be taken into account. It is not easy to concisely document the control and the software parts of a controller in a coherent, integrated way. An interesting methodology that can help us with this knowledge capture task is the use of design patterns for the systems in our field. These patterns will contribute to the effective sharing of the best design knowledge, and will serve as a basis for effective systems. These are issues to be dealt with by the modification systems.

4 Definition of Modification Systems

Let us have a look at the definition of control by Murray in [4], "Control as defined in the report refers to the use of algorithms and feedback in engineered systems. At its simplest, a control system is a device in which a sensed quantity is used to **modify** the behavior of a system through computation and actuation".

Let us keep this definition in mind, since to **modify** the behavior of the system is meant to improve its behavior according a set of performance parameters such as stability, regulation, control effort, precision, overshoot, etc. However, the modification of a system can be developed in a much broader sense than so far proposed by the automatic control community. The C2 paradigm of "Computers for Control" is shifting towards the C4 paradigm of "Computers, Communication and Cognition for Control" providing an integrated perspective on the role computers play in control systems and control plays in computer systems [8]. This change is mainly due to new developments in computers and knowledge management, and the rapidly emerging field of telecommunications providing a number of possible applications in control. Control engineers will have to master computer and software technologies to be able to build the systems of the future, and software engineers need to use control concepts to master the ever-increasing complexity of computing systems. This is the time to talk about agents.

Let us have a look at the definition of intelligent agents. Luck in [2] talks about *agents metaphor* as an agent based inspired approach to engineering. This metaphor includes all definitions of agency as for example the strong agency definition by [7] such as "An agent is a piece of software which works autonomously by taking advantage of other agents within the same ill-structured environment. It is situated, that is it senses, thinks, interacts, and has goals, intentions, beliefs, desires..." A weak and more practical definition of agency is given by the same authors [7] as "An agent is a **situated piece of software** which have social skills", and there are radically weak definitions like [8] as "An agent is a **pointer** to itself and to the rest of the world", added to which is in [9] that "From the control point of view, an agent is the next generation of controllers". The latter statement needs our following completion: "although it is still failing expectations". However, we are going to take advantage of the agent metaphor [2,11] to redefine the control by going back to the origin: the modification of machines.

Given a machine or system

$$S1 \text{ with some undesired behavior}$$

We obtain

$$S2 = M(S1), \text{ with some new desired behavior}$$

Definition: A Modification System is an orthonormal base of Cartesian Modifiers

(1, 0, 0) the interaction Cartesian Modifier unit
(0, 1, 0) the awareness Cartesian Modifier unit
(0, 0, 1) the control Cartesian Modifier unit

Then, a modifier M is a linear combination of the orthonormal base of Cartesian Modifiers

$$M = I \cdot (1,0,0) + A \cdot (0,1,0) + C \cdot (0,0,1)$$

Being I, A, $C \varepsilon \Re$, the degree of modification on each dimension: interaction, awareness and control. The higher value I, A, or C means the higher modification that the modifier is introducing into the machine at every **dimension**.

In the literature, so far modifiers were only **automatic controllers (0, 0, C)**, that is, one dimensional modifiers. Now, we propose two more dimensions: **awareness (0, A, 0) and interaction (I, 0, 0)**.

From the architecture point of view, how can these modification dimensions act? Can they interact? The fact is that every modification dimension acts directly to the machine without the need of the existence of the other dimensions. Let us argue this new feature. In the current state of the art, the majority of intelligent systems architectures, as asserted by [8] are of three levels, where every level reports to the superior level and takes the information from the lower one. The actuation goes from the superior level that takes

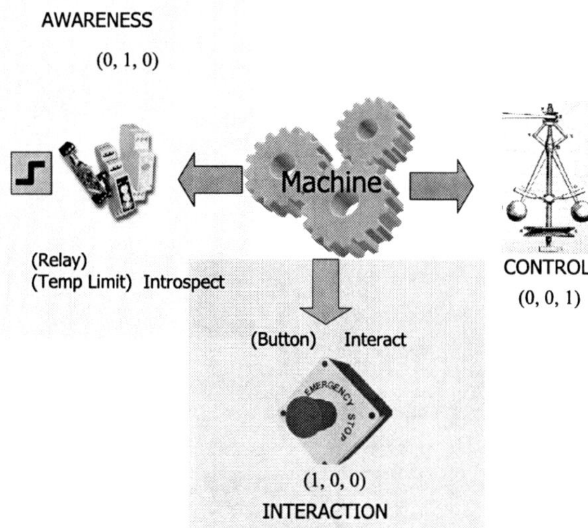

Fig. 2. The implementation of Cartesian Modifiers

decisions and interacts with other agents, and lets the lower levels take decisions such as convert decisions into tasks and then execute those tasks in the control level.

In 1986 Brooks [10] published a new proposal that slightly broke this strict 3-levels architecture: this was the subsumption architecture, where all levels try to gain control of the control and the machine.

Thus, this architecture can evolve towards the modification systems by considering that a modifier M can be represented by a vector (I, A, C), being the units of I, A, and C the multiplication factor that M modifies the behavior of the machine with respect to the basic Cartesian one-dimensional Modifiers, that we call the Cartesian Modifiers. Remember that they are the following: the unit of I is (1, 0, 0), the unit of A is (0, 1, 0), and the unit of C is (0, 0, 1).

In the following example one can see how a machine is operated directly by a simple gain feedback loop (control C), by a speed limit (awareness A) that introduces extra meaning, and by a safe button (interaction I) that lets an engineer or another system stop the machine in case of danger. Fig. 2 shows how they directly modify the behavior of the machine.

Now, let us take a very simple example. Imagine a machine (a traditional steam engine) that given an input in open loop (OPEN LOOP) burns at a temperature $T > T_0$, which is the set point. A solution to keep the machine stuck to that temperature set point T_0 could be to tell a man to keep an eye to the machine and press a button (BUTTON) to stop the machine when it burns much too high, and starts it again to reach that temperature. This may be not the "best" solution, but it is a solution that is applied to many machines in meet industries. One may think of another solution that could be to install a temperature limiter (TEMP LIMIT) so that the machine smoothes

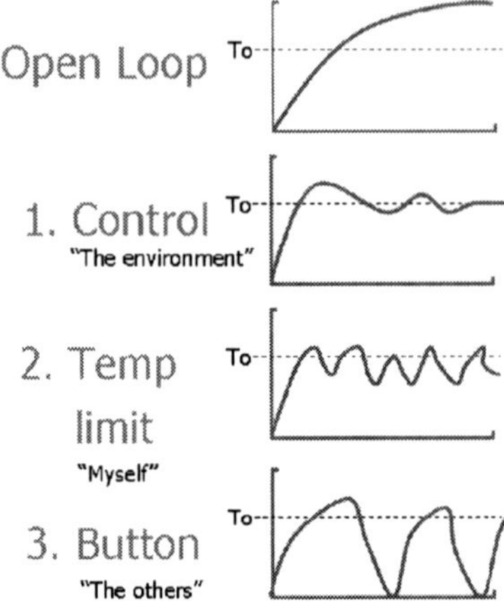

Fig. 3. The effects of the three approaches (dimensions) to a machine

the temperature when it crosses or approaches the temperature limit. This is very common in the manufacturing industry. Finally, others may propose to apply a regulator (CONTROL) that will finely tune the machine temperature around the set point T0. This is common in the process industry. Fig. 3 shows how the machine behaves differently according to the action of every approach, trying to modify the behavior of a machine to reach a set point T_0.

From this example one can claim that control is better, but it is only a partial view, since not only precision is used as performance measure, but many others, as for example, security. Any of these previous approaches are applicable, and even can be combined. Here follows that the dimensional representation requires that every dimension of the modifier could actuate directly to the machines similarly to the several layers of Brooks subsumption architecture. Here follows the representation of the subsumption architecture, PID, and switch control into the 3 dimensional spaces of modification systems (see Fig. 4).

5 Case Study

Let us design modification systems to control machines (robots) which have to cooperate (interact) to play robotic soccer games. The RoboCup soccer test bed [6] which is devoted to foster compared research in advanced technologies and new concepts, is our laboratory used to get the proof of concept of our theory. Particularly this paper is result of analyzing the results obtained in [13]. The design of modifiers is crucial not only for the performance of every single robot, but for the performance of a whole robotic team ruled by a given modifier M_i. The performance of every modifier is measured as a ration of won games versus a fixed amount of games that play the robotic team versus

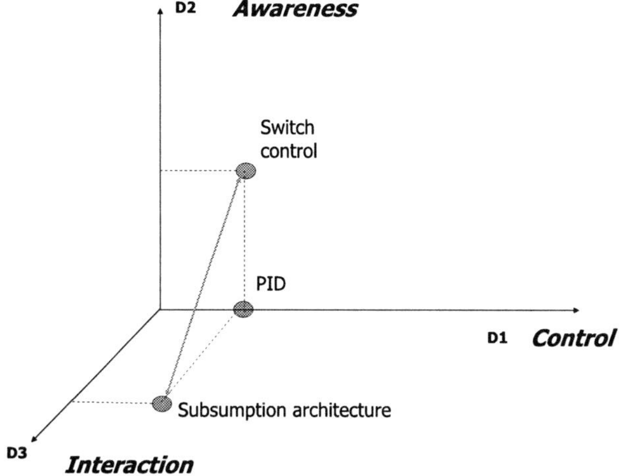

Fig. 4. Distance between the switch control and subsumption architecture

a blind opponent robotic team that behaves with the sufficient difficulty to make up clear differences among the 8 modifiers [13], or among the same 8 modified teams each other. The initial values of the coordination parameters for each modifier in the 30 games experiment series were randomly set at every game.

The Cartesian Modifiers of the robots' behavior that we design for this example are:

- **Control.** This modifier measures the **position** the machine within the football match environment, in a particular implementation of position and speed control. The control signal is a value related to the distance between the current location of each agent and the location of the ball. Such knowledge is regarding to the environment, and represents the physical situation of each agent in the environment. In this sense, the calculation of the control signal (u) is of the distance between a determinate agent j with respect a given goal g as Equation 1 shows.

$$u_{(j,g)} = (1 - d_{j,g}/dmax_g)u\varepsilon[0, 1] \tag{1}$$

- **Awareness.** This modifier is implemented by **introspection**, and is able

  to analyze its physical body and determine what tasks are executable according to its physical capabilities. It is a particular implementation of Awareness. The knowledge about the physical agents' bodies (introspection) is obtained through the representation of them on a capabilities base. Introspection Coefficient $IC\varepsilon[0, 1]$ represents the knowledge of the modifiers about the physical capabilities of the machines bodies (soccer robots) to perform any proposed task. In particular, the introspection process is performed by using neural networks taking to account the environment conditions (e.g. agents' locations, target' locations) and tasks requirements (e.g. achieve the target, avoid obstacles). The greater IC represents a good agent' performance. Particularly, a neural network's structure with two neural networks with an intersection between them has been proposed. Thus, the training of the first neural network takes into account the agent's initial position and the position of a proposed point to obtain the agent's time to achieve the proposed target. The second neural network uses the time rate calculated by the first neural network and the desired final angle to know the capability of each agent to perform any proposed task.

- **Interaction.** Modifiers ground their decision on the result of the interactions with other agents, to get a web of trust on the other cooperative agents to drive the future decisions of the robot (machine). Trust represents the social relationship among modifiers that rule the interaction and behavior of the machines (soccer robots). A trust coefficient T takes into account the result of the past interactions of a modifier with other modifiers that rule other robots (other machines). The modifier evaluates the performance of the proposed task based on the $T\varepsilon[0,1]$. Equation 2 shows the T reinforcement calculus if goals are reached. Otherwise, using Equation 3 T is penalized if goals are not reached.

$$T_{(j,s)} = T_{(j,s)} + \Delta A_{(s,\zeta)} \tag{2}$$

$$T_{(j,s)} = T_{(j,s)} - \Delta P_{(s,\omega)} \tag{3}$$

High T values represent the more trusted modifiers that rule the other robots regarding oneself; $\Delta A_{(s,\zeta)}$ and $\Delta P_{(s,\omega)}$ are the awards and punishments given by the scene s respectively and ζ is the number of awards in the scene s and ω is the number of penalties in the scene s.

Considering the classification above, 8 **Modifiers** are synthesized by means of all the combinations of the Cartesian Modifiers) that result from the 3 dimensions of the Modification System, as depicted in the following Table 1.

Let us explain the different type of implemented modifiers.

- *Random:* The robots move with the basic unmodified behavior.
- *Autistic:* The modifier takes its decisions only according to its physical capabilities, that is, only awareness, and more concretely, introspection or self-awareness.
- *Situated:* In this case, the decision-making structure of the modifier is based on its position in the environment and not using other modifiers, and let the onboard control to develop the task, normally a free trajectory task.
- *Sociable:* The modifier only bases its decision on the result of the interactions with other MS of other machines (robots).

Table 1. Eight Modifiers for the case study

Modifier M_i	(alias)	Interaction (Trust)	Awareness (Introspection)	Control (Position)
$(0, 0, 0)$	0 — Random (no modification)	0	0	0
$(0, 0, 1)$	1 — situated	0	0	1
$(0, 1, 0)$	2 — autistic	0	1	0
$(0, 1, 1)$	3 — individualist	0	1	1
$(1, 0, 0)$	4 — sociable	1	0	0
$(1, 0, 1)$	5 — negotiator	1	0	1
$(1, 1, 0)$	6 — samaritan	1	1	0
$(1, 1, 1)$	7 — totally situated	1	1	1

- *Individualistic:* This modifier is the linear combination of autistic and situated modifiers, with which the modifier executes its goals without the need for interaction with other modifiers. The modifier only uses the knowledge of the physical capabilities of its body and its position in the environment to make decisions.
- *Negotiator:* This modifier is the linear combination of situated and sociable modifiers. With this combination the modifier intention is to exploit its social status and its position in the environment. Somehow it will only work with those modifiers who benefit more from its cooperation.
- *Samaritan:* The modifier knows the physical capabilities of the machine (robot) and makes them public with the intention of finding other agents with related goals.
- *Totally situated:* This modifier takes its decisions using the linear combination of 3 one-dimensional modifiers (4-social, 2-autistic and 4-situated), that is, it is the combination of the three dimensions (control, awareness and interaction).

The implemented architecture is PAULA [13], where all modifiers actuate directly to the robots (the machines), without layers.

Here follows the results of 30 games per modifier M_i, as Table 2, 3, and 4 show. Since the initial conditions are random, the estimated amount of games that make measure of the resulting won games reaches the 99.5% confidence by 30 games.

From tables 2 and 3 a quite uniform behavior pattern appears. There are four clear zones, being the first and last positions for the totally situated (7) and the random (0) modifiers respectively. Then, there are two middle zones, the upper-zone is of ranks 2, 3 and 4, and the lower-zone is of ranks 5, 6 and 7. In the lower zone there are the one dimensional modifiers 2-autistic, 1-situated and 4-sociable, and in the upper zone there are the two-dimensional modifiers 6-samaritan, 3-individualist, 5-negotiator. Almost no change in the individual rank of the modifiers happens in both experiments series, except

Table 2. Classification compared to a blind opponent

Rank	Modifier M_i	(alias)	Won Games W1	W1 (%)
1	(1, 1, 1)	7 — totally situated	21	70.0%
2	(1, 1, 0)	6 — Samaritan	16	53.3%
3	(0, 1, 1)	3 — individualist	14	46.7%
4	(1. 0, 1)	5 — negotiator	13	43.3%
5	(0, 0, 1)	1 — situated	12	40.0%
6	(0, 1, 0)	2 — autistic	10	33.3%
7	(1, 0, 0)	4 — sociable	9	30.0%
8	(0, 0, 0)	0 — random	4	13.3%

Table 3. Classification of a league among all the teams ruled by the 8 modifiers

Rank	Modifier M_i	(alias)	Won Games W2	W2 (%)
1	(1, 1, 1)	7 — totally situated	25	89.3%
2	(1, 1, 0)	6 — samaritan	21	75.0%
3	(0, 1, 1)	3 — individualist	19	67.9%
4	(1. 0, 1)	5 — negotiator	16	57.1%
5	(0, 1, 0)	2 — autistic	13	46.4%
6	(0, 0, 1)	1 — situated	9	32.1%
7	(1, 0, 0)	4 — sociable	7	25.0%
8	(0, 0, 0)	0 — random	2	7.1%

for the swap of 1-situated and 2-autistic as the only exception. Here follows a
preliminary conclusion: the composition of modifiers increase the performance
as the result of higher modification, not only in the control dimension but
on the other dimensions. Let us analyse the contributions of every single
dimension through all the modifiers in table 3 and then a representation of
average results in table 4.

Table 4. Analysis of every C,A,I contribution

	Modifier M		Won Games W1	Won Games W2
	$Control(x, x, 0) = \neg C$	39/120		43/120
without	$Awareness(x, 0, x) = \neg A$	37/120		34/120
	$Interaction(0, x, x) = \neg I$	40/120		43/120
	$Control(x, x, 1) = C$	59/120		69/120
with	$Awareness(x, 1, x) = A$	61/120		79/120
	$Interaction(1, x, x) = I$	58/120		69/120

The summary of the results are depicted in a graph in Fig. 5, where W(X)
represents the *average* number of won games in tables 2, 3 and 4 from the two
series of experiments:

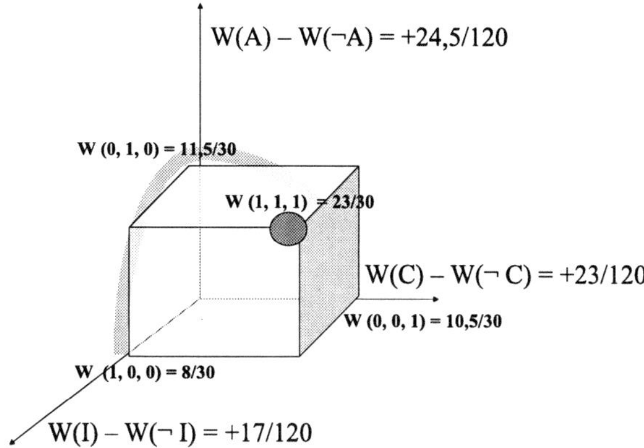

Fig. 5. A summary of the 8 modifiers from Tables 2, 3 and 4

The modification of the machine behavior on every dimension contributes with better performance, as along with $+23/120$ average won games average in control dimension, $+24,5/120$ average won games in awareness dimension, and $+17/120$ average won games in interaction dimension. The global contribution of the three dimensions of modification is much higher compared to the absence of modification ($24/30$ vs. $4/30 = +20/30$ average won games) or higher than the simple control modifier ($23/30$ vs. $10.5/30 = +12.5/30$ average won games).

As Table 2 shows, there is somewhat an equivalence or even an improvement of the $(1, 1, 0)$ modifier vs. the $(0, 1, 1)$ or the $(1, 0, 1)$. Equally, $(0, 0, 1)$ is analogous to $(0, 1, 0)$ but both are better than $(1, 0, 0)$. This is a first hint to make a conjecture about the possible orthogonal nature of the Cartesian modifiers definition. This is in the following section outlined because it gives interesting *equivalence properties* that we foresee they will be useful for the future design of controllers or holons in automation, according to the KISS philosophy of control engineers.

6 Equivalence Hyperplane Hypothesis

To show this idea, and for the sake of simplicity, let us design modifiers only using the following two dimensions, *Control* and *Awareness*, as follows:

- Modifiers *M1* and *M2*, are characterized by the couples $(0, \alpha1, \beta1)$ and $(0, \alpha2, \beta2)$ and the performances *P1* and *P2* respectively.

Is it possible the following equivalence?

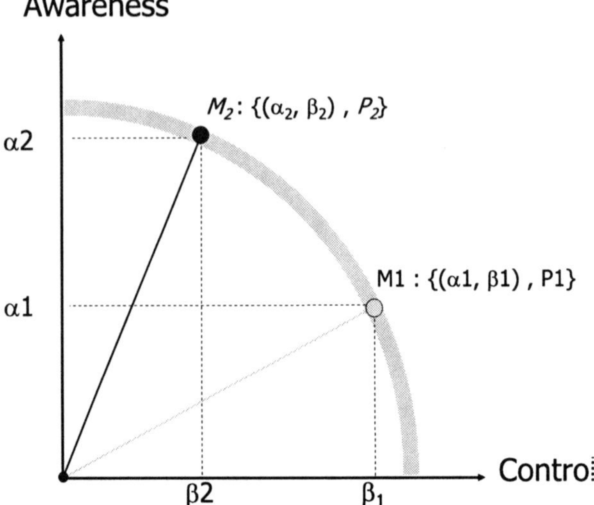

Fig. 6. Equivalence of Modification Systems Hypothesis

$$If\,(P1 \approx P2) \Rightarrow \exists(\alpha1, \beta1) \wedge \exists(\alpha2, \beta2)/(\alpha2 \geq \alpha1) \wedge (\beta2 \leq \beta1)$$

as depicted in Fig. 6?

This is an equivalence hyperplane (a curve in this bidimensional space), where there are equivalent modifiers all along the line. The equivalence means that those infinite M_i modifiers modify the machines exactly the same and, potentially, they all give analogous performance.

If this equivalence hypothesis was true then, based on the equivalence hypothesis one could infer that:

$$High\,(\uparrow)\,Awareness\,modifier \Rightarrow Low(\downarrow)\,Control\,modifier$$

We claim that this equivalence will impact on the design of control systems, powering the existing KISS philosophy in control design while achieving the highest performance ever by designing proper modifiers bye using the other two dimensions (awareness and interaction) rather than keeping with control dimension.

Experiment: Let us design the modifier for passing a ball between two soccer robots. This task is depicted in Fig. 7

The success of this task is measured in its execution as when the second robot touches the ball with regards to the number of trials. A simple feedback controller (one dimensional modifier M_1) gives a performance of 76.7% of success (as Fig. 8 shows).

Following the KISS philosophy, let us keep the same controller but increase the awareness of this modifier by introducing introspection. This gives a new modifier M_3 with a higher performance of 81.4%. This would mean that, to gain the same performance, we had to increase the modification of the control

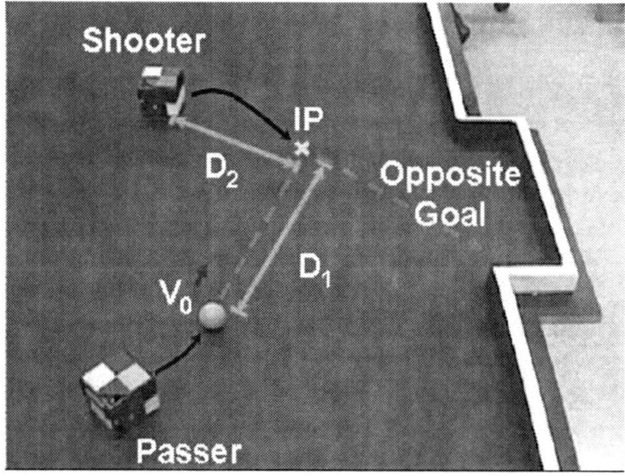

Fig. 7. The passing task

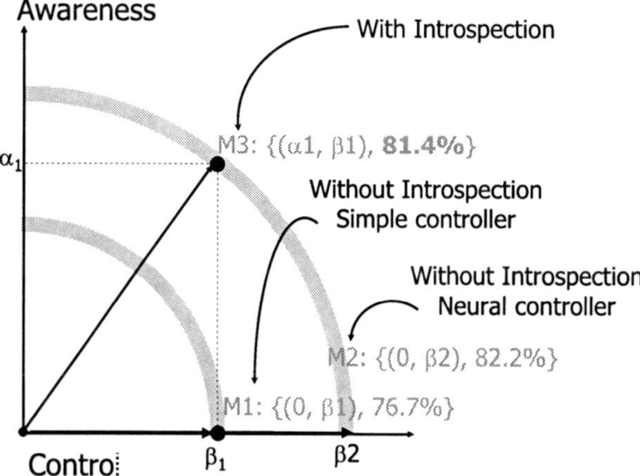

Fig. 8. Experimental test of the equivalence hyperplane to design new modifiers without sophisticating the control dimension

dimension without awareness. And this is done with modifier M_2, that is a more sophisticated modifier focused on the control dimension (introducing a soft computing algorithm) that after its tuning gives a performance of 82.2%. Here follows that now we may have more simple alternatives to work instead of developing more complicated control dimension modifiers, and they consist in working with other dimensions different than control that let improve the performance of the machine while keeping the simplicity of the M_1 simple control modifier. Thus, we have empirically shown how equivalent controllers can be designed by developing the other dimensions than control.

7 Conclusions and Future Work

The history of control from the very initial ages to nowadays and several surveys authored by some big names in automatic control theory and practice, has motivated the need for reorientation of research in control. In this way we have presented the modification systems theory. It consists of a multidimensional representation of the different levels that directly modify (actuate) the behavior of a system, namely, *control, awareness* and *interaction.* This will give new hints to engineers to design control systems. In the future, control engineers may keep KISS design in the control dimension, by explicitly introducing awareness and interaction. This approach presents the design of modifiers as a holistic approach compared to the design of control systems. Some examples of the impact of the new dimensions that already existed in history are shown together with some suggestions about not only the orthogonality of the three dimensions, but the possible existence of equivalence hyper planes.

The future work of this paper is to rewrite the history of automatic control from the point of view of modification systems, create more formal definitions of the modification systems, namely the orthonormality of the Modification Systems, prove the equivalence hyperplane hypothesis, define the basis for the orthogonal units of modification systems and eventually create a new design methodology of modification systems that will preserve the KISS principle to be applied with success in the future industry. Yet another theory!

Acknowledgement. This manuscript was submitted on September 5, 2006, reviewed on December 5, 2006, accepted on January 8, 2007. This work was supported in part by the Grant TIN2006-15111/Estudio sobre Arquitecturas de Cooperación from the Spanish government and by EU project No. 34744 ONE: Open Negotiation Environment, FP6-2005-IST-5, ICT-for Networked Businesses.

References

1. Zahed L (1996) The Evolution of Systems Analysis and Control: A Personal Perspective. IEEE Control Systems Magazine, June 1996, pp. 95–98
2. Luck M, McBurne P, Shehory O, Willmott S (2005) Agent Technology: Computing as Interaction. A Roadmap for Agent Based Computing, Compiled, written and edited by M. Luck, P. McBurney, O. Shehory, S. Willmott and the AgentLink Community
3. Bennet S (1996) A Brief History of Automatic Control. IEEE Control Systems Magazine, June 1996, pp. 17–25
4. Murray R, Astrom KJ, Boyd S, Brockett RW, Stein G (2003) Future Directions in Control in an Information-Rich World. IEEE Control Systems Magazine, Apr. 2003, vol. 23, no. 2, pp. 20–33. Previous version Available: `http://www.cds.caltech.edu/~murray/cdspanel`

5. Halang WA, Sanz R, Babuska R, Roth H (2005) Information and Communication Technology Embraces Control. Status Report prepared by the IFAC Coordinating Committee on Computers, Cognition and Communication, World IFAC Congress
6. Asada M, Kuniyoshi Y, et al. (1997) The RoboCup Physical Agent Challenge. First RoboCup Workshop in the XV IJCAI-97 International Joint Conference on Artificial Intelligence, pp. 51–56, http://www.robocup.org
7. Wooldridge M, Jennings NR (1995) Intelligent Agents: Theory and Practice. The Knowledge Engineering Review, vol. 10:2, pp. 115–152
8. Sanz R, Escasany J, López I (2001) Systems and Consciousness. In: Proc. Conf. Towards a Science of Consciousness (TSC)
9. Sanz R, Holland O, Sloman A, Kirilyuk A, Edmondson W, Torrance S. (2005) Self-aware Control Systems. Research Whitepaper for the Bioinspired Intelligent Information Systems Call, IFAC 2005
10. Brooks RA (1991) A robust layered control system for a mobile robot. IEEE Journal of Robotics and Automation, 2 (1), pp. 14–23, 1987. And a new version in: Brooks RA "New Approaches to Robotics". Science, vol. 253, September 1991, pp. 1227–1232
11. Jennings NR, Bussmann S (2003) Agent-Based Control Systems. Why Are They Suited to Engineering Complex Systems? IEEE Control Systems Magazine, Jun. 2003, vol. 23, no. 3, pp. 61–73
12. Hall KH, Staron RJ, Vrba P. (2005) Experience with Holonic and Agent-Based Control Systems and Their Adoption by Industry. Holonic and Multi-Agent Systems for Manufacturing, vol. 3593/2005, pp. 1–10
13. Ibarra S, Quintero MC, Busquets D, Ramón J, de la Rosa J, Castán J. (2006) Improving the Team-work in Heterogeneous Multiagent Systems. Situation Matching Approach. Frontiers in Artificial Intelligence and Applications - AI Research and Development, ISSN 0922-6389, vol. 146, pp. 275–282, October 2006, IOS Press

Software Metrics Mining to Predict the Performance of Estimation of Distribution Algorithms in Test Data Generation

Ramón Sagarna[1] and Jose A. Lozano[2]

[1] Intelligent Systems Group, University of the Basque Country, 20018 San
   Sebastián, Spain `ccbsaalr@si.ehu.es`
[2] Intelligent Systems Group, University of the Basque Country, 20018 San
   Sebastián, Spain `ja.lozano@ehu.es`

Test data generation for a software system is a difficult and costly task. Thus,
given a program, it would be highly desirable to choose the most adequate
approach. A first step towards this is the prediction of the performance of
a test data generator. We conduct a preliminary study on the suitability of
software metrics for this problem in the context of a generator based on Estimation of Distribution Algorithms (EDAs). EDAs are a family of evolutionary
algorithms that have been previously applied to the test data generation problem in software testing with promising results. More precisely, we analyze the
adequacy of Data Mining techniques for predicting whether the EDAs based
generator is able to fulfill branch coverage or not. Results offer interesting
conclusions on the predictive capability of software metrics and show Data
Mining as a powerful field to be further investigated in this area.

1 Introduction

Considering the crucial role software plays nowadays, quality assurance becomes a main issue for the industry in the field. A typical way of improving quality during the software development process is to focus on the most
complex modules and to assign them larger resources. In order to determine
complexity, software metrics that measure some property of the module are
employed, e.g. the number of code lines in a method.

Another important tool to attain quality is testing. This element from the
software's life cycle is the primary way used in practice to verify the correctness
of software. In fact, 50% of the project resources are usually committed to this
phase [2]. Among the problems related to testing, the automatic generation of
the input cases to be applied to the program under test is especially difficult.
A common strategy for tackling this task consists of creating test inputs that

R. Sagarna and J.A. Lozano: *Software Metrics Mining to Predict the Performance of Estima-
tion of Distribution Algorithms in Test Data Generation*, Studies in Computational Intelligence
(SCI) **102**, 235–254 (2008)
`www.springerlink.com`   © Springer-Verlag Berlin Heidelberg 2008

fulfill an adequacy criterion based on the program structure. That is, adequacy criteria come defined by the entities revealed by the program source code. For example, entities such as the branches the flow of control can take from a conditional statement define the branch coverage criterion, i.e. every program branch must be exercised. In order to know the level of completion attained by a set of test cases, a coverage measure provides the percentage of exercised structural entities.

In the last few years, a number of approaches under the name of Search Based Software Test Data Generation (SBSTDG) have been developed, offering interesting results [18]. SBSTDG tackles the test case generation as a search for the appropriate inputs by formulating an optimization problem. This problem is then solved using metaheuristic search techniques.

Given a program, it would be highly desirable to know in advance how difficult it is for a generator to obtain the test cases. In other words, we would like to predict the performance of test case generation for a given program. This way, efforts could be addressed towards the most promising strategy or the most adequate parameter values for a generator.

In this work, the suitability of software metrics to represent the difficulty of the test data generation process is studied in the context of SBSTDG. This issue was already dealt with in a previous work where conclusions on the adequacy of the metrics were not very encouraging [13]. Here, we propose the application of Data Mining techniques, which allow for a more sophisticated analysis and in-depth exploration of data.

More precisely, as a preliminary case study, we concentrate on test data generation for branch coverage of numerical calculus functions written in the C programming language. These functions were extracted from the book "Numerical Recipes in C. The Art of Scientific Computing" [20]. The generator follows a SBSTDG strategy that employs Estimation of Distribution Algorithms (EDAs) as the search technique [15]. A database was built where the C functions were the instances and the set of variables consisted of several software metrics and the coverage level reached by the EDAs based generator. Then, supervised classification techniques were used in order to build a model that, given a set of metric values for a function, discerns whether complete branch coverage is attained by the generator or not. Results for several classifiers were compared and interesting conclusions arise on the suitability of software metrics and Data Mining methods to deal with this problem.

The rest of the chapter is arranged as follows. The next two sections outline a number of related works and our underlying motivation. We continue with the description of the EDAs based test data generator. Then, the database used in the study is explained and the supervised classification techniques applied are briefly commented on. The following section shows the analysis performed and its results. We finish with conclusions and some ideas for future work.

2 Related Work

The application of Data Mining or software complexity metrics in testing is not new.

The work of some researchers has concentrated on the identification of the fault-prone modules in a system. This way, testing or development efforts could focus on the most risky components. For instance, Wong et al. [27] investigate the adequacy of three metrics to predict whether a C function is likely to have faults. The study considered the metrics in isolation, their union and their intersection. Basically, the methodology was to obtain metric values for a faulty version of a software system. A function was considered to be fault-prone if its metric value was below some cut-off point. Four cut-off points were treated. For each one, the number of files with a function below the point was counted. Then, the correct version of the system was used to calculate the predictive precision of a metric on the basis of the number of fault-prone files with "true" faults. They concluded metrics were suitable since, as the cut point value increased (lower metric complexity), precision became lower. Another interesting result regarding the metrics comparison was that intersection of the three offered the highest precision values.

In other articles, the goal has been to identify the points of a system where faults "hide". Brun and Ernst [3] propose a method for obtaining the properties of a program that are likely to indicate errors. Properties are obtained from program executions and refer to the observed values on relationships among variables, e.g. `out[1]` $\leq$ `in[1]` or `x = a`. These properties are then characterized by some attributes. After building a database of properties from faulty and corrected versions of a program, Machine Learning was used to elicit a classification model. More precisely, experiments were conducted on two learning algorithms: support vector machines [23] and decision trees [21]. The resulting model aimed at discerning whether a given property was fault leading or not. Thus, this technique could be used to help users in testing their programs.

Clearly, the most related work to ours we have found in the literature is the paper by Lammermann et al. [13]. There, the capability of software metrics to represent the complexity of test data generation was also studied in the context of SBSTDG. Taking complexity as the relation between coverage and number of generated inputs during the process, 13 programs were ranked. Then, the dissimilarity between this ranking and the one associated to a metric was calculated. In a first experiment, up to seven metrics based on the source code structure were considered in isolation, concluding that one metric was not able to express complexity. A second trial was to take several metrics into account. For this, the authors define a new metric to be the weighted sum of the other metrics, and they use Genetic Algorithms [10] to search for the best weight combination. Once again, results were not successful, since no improvement was achieved with regard to the best result with one metric. It is argued that, as the reason for this poor performance, rather than structural, dataflow characteristics have a major influence on SBSTDG complexity. To support this argument, some dataflow characteristics were removed from programs and experiments were repeated, obtaining acceptable results. Although these authors tackle the same problem as in this chapter, important differences exist between both works. We will explain them in detail in Section 3.

3 Motivation

Automatic test data generation is a hard task, as input cases must conform to the test type and its requirements. Approaches based on SBSTDG have been offering promising results and, hence, they constitute nowadays a serious alternative to accomplish this task [18]. Since it is a costly process, it would be worthwhile to have in advance an idea of the performance of a given SBSTDG generator. This could help, for instance, to choose the most appropriate generator for a program or to find the parameter values offering best results.

In some phases of the development process, software metrics are used to measure the complexity of a module. Thus, it could be thought that software metrics might also be used to measure the difficulty of a SBSTDG approach. The work by Lammermann et al. [13] studied this idea. Applying relatively straightforward methods (a rank for the metrics in isolation and a linear model for metrics combination) they concluded that structural metrics were not suitable. However, this conclusion is doubtful since the study was based on simple models which might have been unable to capture significant information from data. Moreover, each metrics model was validated against the same data from which the model was obtained, and only 13 programs and 7 measures were considered.

Considering the conclusion of the work by Lammermann et al. correct for simple models, software metrics might still be a suitable alternative if a more intricate relation with SBSTDG is found. Thus, we propose to go a step further by using Data Mining in the study. Data Mining allows for a more in-depth exploration of data and it may manage useful non-trivial information. More precisely, we concentrate on EDAs for the generation of test cases, as they have offered excellent results in this problem [22]. As a case study, numerical calculus functions [20] and the branch coverage testing criterion are chosen. Branch coverage is a classical criterion included in many testing standards. Therefore, our objective is to study the capability of software metrics to predict whether the EDAs based generator fulfills complete branch coverage or not. This involves building a database with the metrics values and the coverage level obtained for each function. Then, a supervised classification problem is tackled. We first learn a one metric variable classifier using the 1R method [11]. Next, all the metrics are considered by using Bayesian network classifiers [9]. The Bayesian network classification paradigm involves two interesting properties regarding the model: the dependency level between variables can be controlled and interpretability is intuitive. These properties allow us to analyze the relations between metrics in the model and shed some light on their influence for discrimination.

4 Test Data Generation through EDAs

The approach used to generate test data for branch coverage via EDAs conforms to a SBSTDG strategy which is also followed by other works [25]. Next, we outline this generator; for a detailed description the reader is referred to the works by Sagarna and Lozano [22] and Wegener et al. [25].

The basic idea is to tackle the test case generation as the resolution of a number of function optimization problems. That is, the scheme in Figure 1 is followed.

Repeat while untreated branches exist
$b \leftarrow$ Select objective branch to exercise
Obtain input optimizing function for b

Fig. 1. Test case generation scheme.

In the first step, a yet untreated branch is selected, usually with the help of a graph reflecting the program branches, e.g. a control flow graph [8]. This branch is then marked as an objective.

The second step consists of solving the following optimization problem: given the search space Ω formed by the program inputs and a function $h :$ $\Omega \longrightarrow \mathbb{R}$, find $x^* \in \Omega$ such that $h(x^*) \leq h(x) \ \forall x \in \Omega$. $h(x)$ is formulated so that if an executed input exercises the objective, the value is minimum, otherwise the value is proportional to the proximity of x to the objective coverage. Thus, in order to obtain the function value of an input, it must be previously executed on an instrumented version of the program which will provide the necessary information.

In our case, $h(x)$ is calculated upon the conditional statement **COND** associated to the objective branch in the code. If x doesn't reach **COND**, then a condition distance d_c is calculated [25] by considering the execution path followed by x in the control flow graph. Denoting by v_c the vertex in the control flow graph representing **COND**, and by v_n the nearest vertex to **COND** in the path followed by x, $d_c(v_c, v_n)$ is the minimum number of branching vertices straying from the subpath between v_c and v_n. By contrast, if **COND** is reached but the objective branch is not covered, then a distance d_e to fulfilling the expression $A\,\mathbf{OP}\,B$ in **COND**, where **OP** denotes a comparison operator, is considered [24]. If A_x and B_x are numerical representations of the values taken by A and B in the execution, $d_e(A_x, B_x) = |A_x - B_x| + K$, with $K > 0$. Thus, according to these two distances and maintaining the notation, the following objective function is formulated:

$$
h(\boldsymbol{x}) = \begin{cases} d_c(v_c, v_n) & \text{if } \mathbf{COND} \text{ not reached} \\[2mm] \frac{d_c(A_{\boldsymbol{x}}, B_{\boldsymbol{x}})}{L + d_c(A_{\boldsymbol{x}}, B_{\boldsymbol{x}})} & \text{if } \mathbf{COND} \text{ reached but branch not attained} \\[2mm] 0 & \text{otherwise} \end{cases} \qquad (1)
$$

where $L > 0$ is a previously defined constant. Notice that L is employed to ensure that the function value when $\mathbf{COND}$ is not reached surpasses the value when $\mathbf{COND}$ is reached but the branch is not attained.

The resolution of each of these function minimization problems is then sought by means of an EDA.

4.1 Estimation of Distribution Algorithms

Estimation of Distribution Algorithms (EDAs) [15] [16] constitute a family of population based evolutionary algorithms. The main difference of these metaheuristics with regard to others from evolutionary computation is that individuals are not created through the classical recombination operators. Instead, EDAs obtain new individuals by sampling a probability distribution, previously estimated from the set of selected individuals.

In the context of evolutionary algorithms with discrete domain, an individual with n genes can be considered as an instantiation $\boldsymbol{x} = (x_1, x_2, \ldots, x_n)$ of a n-dimensional random variable $\boldsymbol{X} = (X_1, X_2, \ldots, X_n)$. Given the population of the l-th generation, D_l, the S selected individuals, D_l^{Sel}, constitute a data set of S cases of $\boldsymbol{X} = (X_1, X_2, \ldots, X_n)$. EDAs estimate the joint probability distribution of $\boldsymbol{X}$, $p(\boldsymbol{x}) = p(\boldsymbol{X} = \boldsymbol{x})$, from D_l^{Sel}.

A pseudocode for the abstract EDA is presented in Figure 2. In this figure, the distribution of the l-th generation is represented by $p_l(\boldsymbol{x}) = p(\boldsymbol{x}|D_{l-1}^{Sel})$.

$D_0 \leftarrow$ Generate M individuals randomly
Repeat for $l = 1, 2, \ldots$, until stopping criterion met
   $D_{l-1}^{Sel} \leftarrow$ Select $S \leq M$ individuals from D_{l-1}
   $p_l(\boldsymbol{x}) = p(\boldsymbol{x}|D_{l-1}^{Sel}) \leftarrow$ Estimate probability of an individual being in D_{l-1}^{Sel}
   $D_{l-1}^{New} \leftarrow$ Sample N individuals from $p_l(\boldsymbol{x})$
   $D_l \leftarrow$ Build next population with M individuals from D_{l-1} and D_{l-1}^{New}

Fig. 2. Pseudocode for the abstract EDA.

The key point of EDAs is how to estimate $p_l(\boldsymbol{x})$. The computation of all the parameters is unviable, therefore it is factorized by following a probability model. According to this, two major branches can be distinguished in the field to date. On the one side, a number of approaches depart from a fixed model and concentrate on parameters learning. In contrast, algorithms in the other branch do not specify the model in advance and, in consequence, both the structure and the parameters need to be learnt at each generation. Nevertheless, these EDAs assume a complexity level on their models that, in some cases, limits the possible dependencies among the variables $X_1, X_2, \ldots, X_n$. Thus, regarding this complexity, these EDA instances may be classified as those where model variables are independent, those where bivariate dependencies are considered and those where no restriction is made on the variable interrelations. Given a problem, a strongly limited model will not reflect any existing dependency between two variables, however, the cost of the estimation of $p_l(\boldsymbol{x})$ is relatively low. On the other hand, a less restrictive model is able to show dependencies between variables although its computational cost may be too expensive.

EDAs where variables in the model are independent factorize the n-dimensional joint probability distribution as a product of n univariate probability distributions, that is:

$$p_l(\boldsymbol{x}) = \prod_{i=1}^{n} p_l(x_i). \tag{2}$$

For example, the Univariate Marginal Distribution Algorithm [19] estimates $p_l(x_i)$ as the relative frequencies of x_i in dataset D_{l-1}^{Sel}.

Bivariate EDAs make use of second order statistics to estimate the probability distribution. Hence, apart from the probability values, a structure that reflects the dependencies among the variables must be given. The factorization carried out by the models in this category can be expressed as follows:

$$p_l(\boldsymbol{x}) = \prod_{i=1}^{n} p_l(x_i|x_{j(i)}) \tag{3}$$

where $X_{j(i)}$ is the variable, if any, on which X_i depends.

EDAs with no restriction on the variables in the model estimate the joint probability distribution by means of probabilistic graphical models [4]. The factorization associated with this type of EDAs is as follows:

$$p_l(\boldsymbol{x}) = \prod_{i=1}^{n} p_l(x_i|\boldsymbol{pa_i}) \tag{4}$$

where $\boldsymbol{pa_i}$ are the instantiations of $\boldsymbol{Pa_i}$, the set of variables on which X_i depends.

In the Estimation of Bayesian Network Algorithm (EBNA) [14], the factorization of the joint probability distribution is given by a Bayesian network learned from D_{l-1}^{Sel} [4].

In this work, test data generation was performed using a bivariate EDA named TREE [15]. This is an adaptation of the Combining Optimizers with Mutual Information Trees (COMIT) algorithm [1]. In COMIT, $p_l(\boldsymbol{x})$ is estimated through the Maximum Weight Spanning Tree algorithm [5]. This algorithm constructs a tree structured model that minimizes the Kullback-Leibler cross-entropy between $p_l(\boldsymbol{x})$ and its factorization. Once an estimation of $p_l(\boldsymbol{x})$ is obtained, COMIT samples a number of individuals from it and selects the best to be the initial solutions of a local search. The resulting individuals are then used to create the new population. In TREE, this local search step is eliminated and, thus, the next population is obtained directly from $p_l(\boldsymbol{x})$.

4.2 An Execution Example of the Test Data Generator

To sum up, the preprocessing required to automate the generation of test data following the scheme in Figure 1 should be noticed. Figure 3 illustrates this by showing an example program and the elements to be induced from it: the control flow graph and the instrumented program version. The reduced box on the right represents the information supplied by a hypothetical execution of the instrumented program.

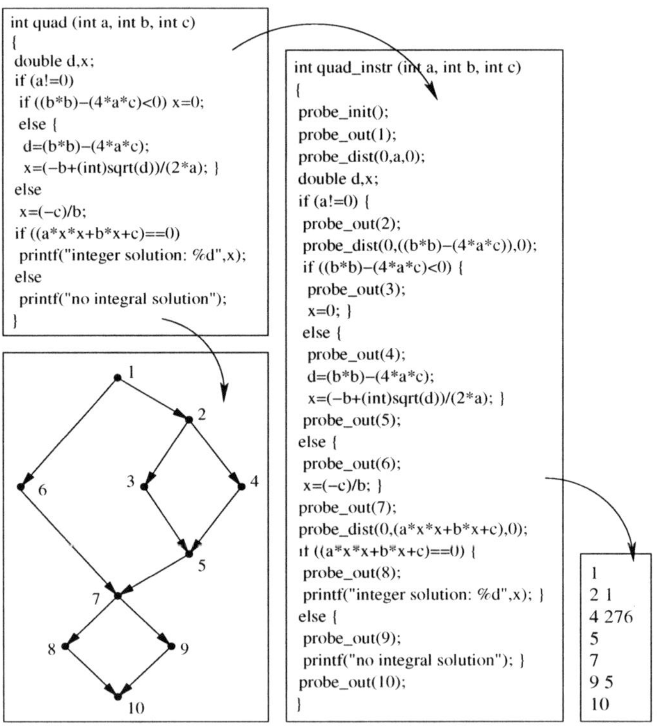

Fig. 3. Example of source code, control flow graph, instrumented version, and output information.

The graph is used to select the next objective branch whose coverage will be pursued, for example, branch $(2, 3)$. Then, an optimization problem is to be solved using an EDA. An individual in the EDA is a bit string representation of the program input, i.e. three integers in the example.

Each individual (input) generated during the search is executed on the instrumented program version in order to elicite its objective function value. The instrumentation results shown in the reduced box of Figure 3 correspond to input $(1, 20, 31)$. Each line of the box contains the traversed basic block (numbered as in the graph) and, if the previous block had a condition with an expression $A\,\mathbf{OP}\,B$, the value of $|A - B|$ in the execution. Using this information, the value of the condition distance (d_c) shown in equation 1 can be obtained. However, this is not necessary, as input $(1, 20, 31)$ reaches the condition of branch $(2, 3)$. Hence, according to equation 1 and taking $K = 1$ and $L = 1000$, $h(1, 20, 31) = \frac{276+1}{1000+276+1} = 0.2169$.

Once the EDA finishes, a new round of the scheme in Figure 1 is performed until complete coverage has been reached or every branch has been treated.

5 Metrics to Predict Performance of EDAs Based Test Data Generation

Given a set of software metrics, we aim at predicting whether an EDAs based test data generator fulfills branch coverage or not. For this, we formulate a supervised classification problem. That is, given a database of instances characterized by a set of attribute values and labeled, induce a classifier to predict the label of a new instance. According to this, two elements are required: a database and an induction method. Here, we describe the database.

In our case, instances are a set of C functions obtained from the book "Numerical Recipes in C. The Art of Scientific Computing" [20]. In a number of functions, some of the parameters took complex types that made the representation of an input difficult. Hence, we restricted ourselves to functions with parameters of numerical type, character type or arrays of these. All in all, 125 functions were collected; they can be consulted in the Appendix.

Attributes in our database are software metrics. More precisely, in this work, we concentrate on measuring structural properties of the source code. Therefore, for each function, up to 43 structural metrics values were obtained from a number of software metrics tools. The code properties covered involve measurements on the number of lines, the kind of statements, keywords and control flow graph. A detailed description of the metrics, together with the software they come from, is included in the Appendix.

The label variable of an instance is the branch coverage attained by the EDAs based generator. This implies that, for each function, test cases were obtained using this generator. The EDA chosen was TREE, as it offered the best performance in a previous work [22]. Coverage results may vary significantly depending on the parameter values of the EDA. Thus, in order to avoid misleading conclusions, we fixed parameter values for all the functions. Population size was set to 100 and half of the population was selected at each generation. The new population was created in an elitist way. The stopping criterion was reaching a maximum number of generations (100) or a maximum number of inputs in the whole process (100000). Once the generator finishes, we label the function with the achieved coverage value.

Once the database is built a classifier needs to be learnt from it. The next section describes the model induction techniques applied in this study.

6 Classification Model Induction Methods

Following the study of Lammermann et al. and in order to investigate the predictive capability of one metric, a model induction technique that builds a one variable classifier was used. Such a technique is 1R [11]. 1R is a relatively simple method that incorporates a specific discretization procedure for numerical attributes. In this discretization, the range of values of the attribute is divided into several disjoint intervals. The number of intervals comes defined by a threshold parameter that sets the minimum number of instances in the same class for each interval (except the rightmost); a common value for this parameter is 6. Once all attributes are nominal, a one variable decision tree is built with each of them. Then, they are ranked according to the accuracy of the decision tree on the training set and the best attribute is selected for the classification. It should be remarked that, although 1R leads to a simple classifier, high accuracy levels are achieved in many databases, making it able to compete with more complex models [11].

We are also willing to study the adequacy of combining different metrics for the classification task. Bayesian network classification models were chosen for this [9]. In this paradigm, class label C and attributes $X_1, X_2, \ldots, X_n$ are seen respectively as a random variable and a n-dimensional random variable $\boldsymbol{X} = (X_1, X_2, \ldots, X_n)$. Accordingly, c and $\boldsymbol{x} = (x_1, x_2, \ldots, x_n)$ denote their instantiations. The classifier consists of a Bayesian network [4] that factorizes the joint probability distribution $p(\boldsymbol{X}, C)$. Since the following holds,

$$p(C|\boldsymbol{X}) = \frac{p(C, \boldsymbol{X})}{p(\boldsymbol{X})} \propto p(\boldsymbol{X}, C), \tag{5}$$

we can use the factorization to assign an instance $\boldsymbol{x}$ the label c maximizing $p(c|\boldsymbol{x})$.

The model induction implies a Bayesian network has to be learnt from the database. Such a network comes defined by a pair $(S, \boldsymbol{\theta})$ where S is a directed acyclic graph representing the (in)dependencies between the variables and $\boldsymbol{\theta}$ is the set of conditional probability values needed to define the joint probability distribution. As regards the level of the restrictions imposed on the network structure, several alternatives can be distinguished. In the case of naive Bayes, the whole structure is fixed in advance by assuming that all the attributes are conditionally independent given the class; therefore, only the parameter values need to be elicited. An example structure of naive Bayes is depicted on the left side of Figure 4. Another possibility is to induce an unrestricted Bayesian network from data, that is, to learn both the structure S and the parameters θ. A usual way to determine θ is through their maximum likelihood estimates. For structure learning, instead, several options exist. A common strategy consist of evaluating each structure with a score function value calculated from data and using a search method to obtain the best structure. For example, a popular approach is to employ a hill climbing algorithm where, at each step, one dependency is added or deleted from the structure; to select this dependency the $K2$ scoring function can be used [6]. An important aspect of this approach is that the maximum number of parents a variable can have is given as a parameter. The right side of Figure 4 shows an example structure of a *general* Bayesian network classifier.

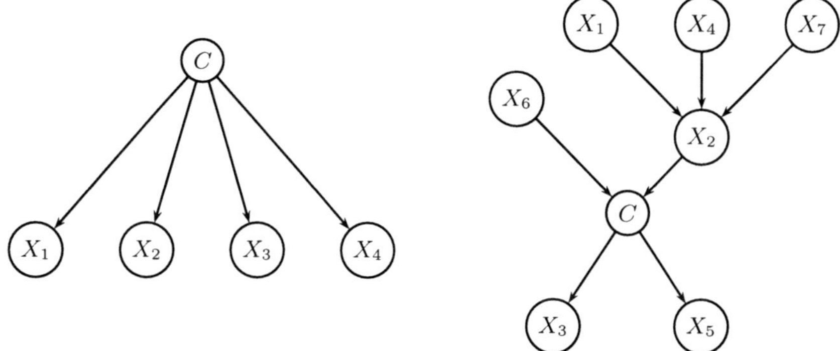

Fig. 4. Structure examples for a Naive Bayes (left) and a general Bayesian network (right) classifier.

The reason for choosing Bayesian network classifiers can now be clearly understood. On the one hand, during model induction, the dependency level between attributes can be explicitly controlled, either by fixing the structure (naive Bayes) or by limiting the number of parents (the general case). On the other hand, once the model is built, its graphical structure reflecting relationships between attributes allows for an intuitive interpretation.

7 Study and Results

Our study on the adequacy of software metrics to predict branch coverage with the EDAs based test data generator involves two topics: classification with one metric and classification with all the metrics.

Classifiers were built from the database by means of the methods previously introduced. These are implemented in the Weka software environment

[26], which was the tool employed for this task. However, before any other step, the database needed to be preprocessed.

7.1 Database Preprocessing

Implementation of numerical calculus functions often follows a particular style that affects the characteristics of the source code. For example, it is usual to have a small number of comments. In consequence, a number of metrics in our set took useless values and had to be removed. Specifically, no multiway conditional statement was in our functions, so the *CyclomaticModified* metric became a copy of *Cyclomatic*. For the same reason, the value of *NumSwitch* and *NumCase* was 0. Constant values were also obtained for *LineInactive*, *NumReturn*, *NumConst*, *NumEnum* and *NumDefault*. All of them were then removed from the database, resulting in a final set of 35 metrics.

In order to avoid disturbing influences in the induction which could lead to a biased model, we also eliminated outlier instances. After this step, 112 instance functions were left in the database.

Both metrics and functions deleted are marked in bold in the Appendix.

Preprocessing was finished with the discretization of the coverage label. Originally, coverage is a numerical variable (representing a percentage), but our objective is to classify an instance as fulfilling branch coverage or not. Hence, '1' was assigned to value 100 and '0' to the rest. This resulted in 74 instances labeled with '1' and 38 with '0'. Notice that this proportion is not surprising, since, in general, SBSTDG generators often show a high quality performance [18] [25]. Therefore, we decided not to balance the database as we believe the current proportion is quite representative of the real world situation.

7.2 Classification with One Metric

We depart from the conclusion by Lammermann et al. [13], that is, structural metrics are not able to represent complexity of a SBSTDG approach in a relatively straightforward manner. In order to check if this conclusion applied to our set of metrics, coverage values found in the database were faced with metrics values. To be precise, for each metric, a plot was generated with the coverage values attained by the functions in the x-axis and the metric values in the y-axis. If a coverage value was obtained by several functions, then the average of the metric values for those functions was depicted. Figure 5 shows these plots. Labels were removed since the purpose of the graphics is to show the "shape" of the curves for each metric.

Assume, with no loss of generality, that a metric takes growing values to represent increasing complexity. If a "simple" relation between coverage and a metric existed, e.g. a linear combination or an inverse proportion, then curves in Figure 5 should be nondecreasing or nonincreasing. Clearly, this is not the case.

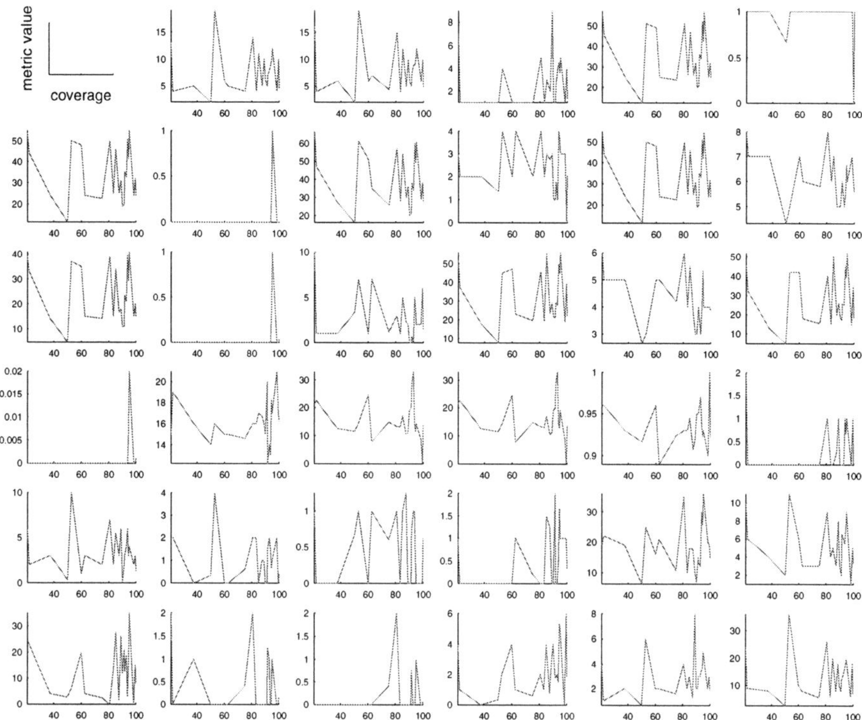

Fig. 5. Curves for the coverage and the software metrics in the database.

Supposing a straightforward method is not enough to solve our classification problem, we applied the 1R algorithm to build a one metric classifier. Notice 1R takes all the metrics into account and selects the best for classification; in other words, the best one metric classifier (according to the algorithm) is induced.

The accuracy of the classifier is calculated on the basis of the number of instances whose labels were predicted correctly. To estimate this accuracy, a ten-fold cross-validation was applied, considering Kohavi's advice [12].

A number of values from 0 to 10 were experimented for the threshold parameter of the 1R. The best accuracy was obtained with a value of 6. The algorithm returned a decision tree classifier built on metric *Cyclomatic*, which led to a 75.89% accuracy. This value is not high enough to make reliable predictions, however, it by no means involves a poor performance, as almost 76 out of 100 functions would be correctly classified. Taking a closer look at the results, the rate of true positives was 0.53 for class '0' (non-complete coverage) and 0.88 for class '1' (complete coverage). Thus, a little more than half of the functions not fulfilling branch coverage are correctly identified, lowering the overall performance of the classifier. This implies some of the leaves in the *Cyclomatic* decision tree misclassify by assigning a '1' label to almost a half of the functions. Therefore, it seems that, for *Cyclomatic*, instances in our database are not distributed over the intervals so that those corresponding to class '0' can be clearly distinguished. A possibility to improve this behavior might be to use several metrics for classification. This is the next topic of our study.

7.3 Classification with the Whole Metrics Set

Taking all the metrics to the model would allow us to consider the underlying structure of the database, which could be beneficial for the classification. However, a broad variety of alternatives are possible for combining the metrics in the model. We adopted the Bayesian network classification paradigm, as the different approaches in it allow us to manage a range of complexity levels on the attributes interrelations. Hence, both, Naive Bayes and general Bayesian network classifiers, were employed in our study. The simplest case is Naive Bayes, since it assumes no dependency exists between metrics. By contrast, the general Bayesian network's model is able to capture different orders dependencies. We explored several complexity levels for this last model by employing a number of values for the maximum number of parents for a metric. Therefore, a hill climbing strategy, together with the K2 scoring function, was chosen for the learning of the Bayesian network.

Apparently, Naive Bayes classifiers improve their performance if variables are discrete [7]. In consequence, prior to model induction, it is common to discretize continuous attributes in the database. In fact, the Weka environment assumes all attributes are nominal for the Bayesian network models. We discretized metrics in the database following a simple strategy, the equal-width intervals method [7]. Values for the number of intervals of each metric attribute were fixed by examining their associated histograms in Weka.

Once again, the accuracy of a classifier was estimated following a ten-fold cross-validation.

Table 1 shows results of the classifiers obtained. We restricted complexity of the general Bayesian network model by setting the maximum number of parents for a metric to 2, 3, 4, 5, 6, 8 and 10. The first row offers the accuracy of the classifier, while the second (TP_0) and third (TP_1) rows present the true positive rate for class label '0' and '1' respectively.

Table 1. Accuracy and true positives rates for the Bayesian network classifiers.

	Naive Bayes	General Bayesian network Classifier						
		2	3	4	5	6	8	10
Accuracy	70.54	82.14	78.57	78.57	76.79	76.79	76.79	76.79
TP_0	0.58	0.74	0.74	0.74	0.71	0.71	0.71	0.71
TP_1	0.77	0.87	0.81	0.81	0.80	0.80	0.80	0.80

It can be noticed in Table 1 that the best overall accuracy is reached by the general Bayesian network classifier with two parents at most. As the maximum number of parents increases, accuracy gets lower. The common interpretation for this phenomenon is that the model accurately captures the underlying structure in the particular data and loses generalization capability when the number of parents grows, i.e. the model overfits data. On the other hand, the naive Bayes classifier leads to an even lower accuracy. The reason for this could be the opposite to the general case, that is, the strong assumption of independence between metrics biases the prediction ability of the classifier.

Comparing these results to those of the 1R, a difference can be appreciated. The 75.89% of 1R is clearly outperformed by the best value of 82.14% in Table 1, which is a more acceptable accuracy for prediction. Indeed, all the general Bayesian network classifier accuracy results beat 1R's. These differences can be understood by looking at the true positives rates. The true positives for class '0' (TP_0) are all higher than in the 1R, even for the naive Bayes. By contrast, true positives for class '1' (TP_1) are slightly lower than in the 1R classifier, although the differences are not so high as to neutralize the improvement in TP_0. This means that, using Bayesian network classifiers, difficulties of 1R can be partially overcome and, hence, predicting on the basis of several metrics improves the performance when employing just one metric.

An interesting property of the Bayesian network paradigm is that its structure allows for an easy interpretation of the model. We explored the structure resulting from the general Bayesian network classifiers in order to discern the most significant metrics for the classification. This can be made through the concept of Markov blanket. The Markov blanket of a variable X in a structure consist of X's parents, X's sons and the parents of these sons. Moreover, for classification, it holds that only the Markov blanket of the class variable is taken into account. In our case, the Markov blanket of the class variable was composed of metrics *NumIf* and *NumBranch* for two parents at maximum, and *LineBlank*, *NumIf* and *NumBranch* for the remaining classifiers. Thus, we are using just two or three of the metrics in the database for prediction and we could forget about the others. Accordingly, we built a naive Bayes classifier with only *NumIf* and *NumBranch*, and the obtained accuracy was 83.93%. The true positives rates were 0.71 for class '0' and 0.91 for class '1', which constitute a significant improvement with regard to the case using all the metrics. This is not so surprising since, regarding at Figure 5, it can be appreciated that many of the curves are similarly shaped. Therefore, it could be guessed many of the metrics are highly correlated and, indeed, we believe attribute selection could be highly beneficial for the classification. Notice that this is an interesting issue, as it allows us to concentrate on just a small set of the metrics offered by a software metrics tool, which often is a large number.

8 Conclusions

Predicting the performance of the test data generation for programs is highly desirable. We conducted a preliminary study on the suitability of software metrics for this problem in the context of an EDAs based generator for numerical calculus functions. Since simple methods seem not to be successful, we explored the possibility of using Data Mining techniques by transforming the problem into a classification task. The analysis involved two topics: classification with just one metric and classification by combining several metrics.

Apropos the first topic, the 1R induction method determined that metric *Cyclomatic* offered the best results. However, these suggest that one metric is not able to make reliable predictions of performance. By contrast, the use of Bayesian network classifiers on several metrics offers a noticeable improvement on the accuracy and appears to overcome some of the deficiencies of the one metric case. We also observed that the best Bayesian network classifier predicted on the basis of just two metrics: *NumIf* and *NumBranch*. Therefore, differences with the one metric case are not just quantitative, but also qualitative. According to the underlying structure of the database, *NumIf* and *NumBranch* together predict more accurately than *Cyclomatic*.

The sieve of metrics implied by Bayesian network classifiers gives a clue to the possible directions of future work. Attribute selection could be very beneficial, both to improve accuracy and to help understand the relations between metrics and performance of the test data generation task. In fact, we believe one of the interesting contributions of this work is the use of a database of software metrics for this task. The encouraging preliminary results open the door to the employment of more advanced Data Mining techniques in this field.

Acknowledgement. This work was supported by the ETORTEK and SAIOTEK projects of the Basque Government, as well as the TIN2005-03824 project of the Spanish Ministry of Education and Science. R. Sagarna received additional support from the Department of Education, Universities and Research of the Basque Government under the program of Researcher Education.

The authors also wish to thank Rubén Armañanzas, Jose Luis Flores and Guzmán Santafé for their valuable comments during the preparation of this chapter.

Appendix

The list below gives the names of the functions taken as instances in our database. Functions deleted during the database preprocessing step (Section 7.1) are marked in bold.

The functions are: **adi**, *avevar*, *badluk*, **bcucof**, *bcuint*, *bessi*, *bessi0*, *bessi1*, *bessj*, *bessj0*, *bessj1*, *bessk*, *bessk0*, *bessk1*, *bessy*, *bessy0*, *bessy1*, *betacf*, *betai*, *bnldev*, *caldat*, *cel*, *chder*, *chebev*, *chebpc*, *chint*, *chsone*, *chstwo*, *cntab1*, *cntab2, correl, cosft, crank, ddpoly,* **des**, *eigsrt, el2, elmhes, erf, erfc, erfcc, eulsum, evlmem, factln, factrl, fit, fixrts, fleg, flmoon, four1, fpoly, ftest, gamdev, gammln, gammp, gammq, gasdev, gauleg,* **gaussj**, *gcf, gser,* **hqr**, **hunt**, *indexx, irbit2,* **jacobi**, *julday, kendl1, kendl2, kstwo, laguer, locate, ludcmp, mdian1, mdian2, memcof, moment, mprove, pcshft, pearsn, piksr2, piksrt, plgndr, poidev, polcoe, polcof, poldiv, polin2, polint, probks, qcksrt, qroot, ran0,* **ran1**, *ran2, ran3,* **ran4**, *rank, ratint, realft, rofunc, shell, simp1, simp2, simp3, sinft, sncndn,* **solvde**, *sort, sort2, sort3,* **sparse**, *spear, splie2, splin2, svbksb,* **svdcmp**, *toeplz, tptest, tqli,* **tred2**, *tridag, twofft, vander* and *zroots*.

Next, software complexity metrics employed as attributes in the database are described. They are sorted by the software metrics tool associated. Metrics deleted in the database preprocessing step (Section 7.1) are marked in bold.

Tool: UnderstandC++

Metrics:

Cyclomatic cyclomatic complexity
CyclomaticModified modified cyclomatic complexity
CyclomaticStrict strict cyclomatic complexity
CyclomaticEssential essential cyclomatic complexity
Lines number of lines in the function
LineBlank number of blank lines in the function
LineCode number of source code lines in the function
LineComment number of lines containing a comment in the function
FileLines number of lines in the file containing the function
FileLineBlank number of blank lines in the file containing the function
FileLineCode number of source code lines in the file containing the function
FileLineCodeDecl number of lines with declarative code in the file containing
   the function
FileLineCodeExe number of lines with executable code in the file containing
   the function
FileLineComment number of lines with a comment in the file containing the
   function
LineInactive number of lines inactive from the preprocessor view
LinePrep number of preprocessor lines
CountStmt number of declarative plus executable statements in the code
CountStmtDecl number of declarative statements in the code
CountStmtExe number of executable statements in the code
CommentToCode number of lines with a comment to number of lines with
   code ratio

Cyclomatic complexity (CC) related metrics are based on the definition by McCabe [17]. The CC of a control flow graph G with m arcs and n vertices is $m - n + 2$; it refers to the number of independent paths through G. In the Modified CC, the l arcs of a multiway decision vertex (that with *outdegree* > 2) are counted as one. In the Strict CC, one is counted for each || or && found in the source code condition associated with a decision vertex (that with *outdegree* > 1) in G. The Essential CC measures the amount of unstructured code in a function. For this, G is reduced by contracting the vertices and arcs representing structured programming primitives, and then calculating CC for the reduced graph; the reduction proceeds from the deepest level of nesting outward, until the graph cannot be reduced any further.

Tool: CodeAnalyzer

Metrics:

AvgLineLength average line length
CodeToCommWhite code lines number to comment plus whitespace lines number ratio
CodeToWhite code lines number to whitespace lines number ratio
CodeToTotal code lines number to total number of lines ratio

Tool: RSM 6.52

Metrics:

NumIf number of `if` keywords in the code
NumElse number of `else` keywords
NumSwitch number of `switch` keywords
NumCase number of `case` keywords
NumWhile number of `while` keywords
NumDo number of `do` keywords
NumFor number of `for` keywords
NumBreak number of `break` keywords
NumReturn number of `return` keywords
NumGoto number of `goto` keywords
NumConst number of `const` keywords
NumEnum number of `enum` keywords
NumDefault number of `default` keywords
NumString number of literal strings
NumPar number of parenthesis
NumBrace number of braces
NumBracket number of brackets

Tool: Metre 2.3

Metrics:

MaxDepth maximum depth of a control structure

Tool: EDAs based test data generator

Metrics:

NumBranch number of branches in the code

Values for this metric were obtained using our test data generator.

References

1. Baluja S, Davies S (1997) Combining multiple optimization with optimal dependency trees. Technical Report CMU-CS-97-157, Carnegie Mellon University
2. Beizer B (1990) Software Testing Techniques. Van Nostrand Rheinhold, New York
3. Brun Y, Ernst MD (2004) Finding latent code errors via machine learning over program executions. In: Proceedings of the 26th International Conference on Software Engineering. IEEE Computer Society, Los Alamitos CA
4. Castillo E, Gutiérrez J, Hadi A (1997) Expert Systems and Probabilistic Network Models. Springer, Berlin Heidelberg New York
5. Chow C, Liu C (1968) Approximating discrete probability distributions with dependence trees. IEEE Transactions on information Theory 14:462–467
6. Cooper G, Herskovits E (1992) A Bayesian method for the induction of probabilistic networks from data. Machine Learning 9:309–347
7. Dougherty J, Kohavi R, Sahami M (1995) Supervised and Unsupervised Discretization of Continuous Features. In: Proceedings of the 12th International Conference on Machine Learning. Morgan Kaufmann, San Francisco CA
8. Fenton NE (1985) The structural complexity of flowgraphs. In: Alavi Y, Chartrand G, Lesniak L, Lick DR, Wall CE (eds) Graph Theory with Applications to Algorithms and Computer Science. John Wiley & Sons, New York
9. Friedman N, Geiger D, Goldszmidt M (1997) Bayesian Network Classifiers. Machine Learning 29(2):131–164
10. Goldberg DE (1989) Genetic Algorithms in Search, Optimization and Machine Learning. Addison Wesley, Reading MA
11. Holte RC (1993) Very Simple Classification Rules Perform Well on Most Commonly Used Datasets. Machine Learning 11:63–91
12. Kohavi R (1995) A Study of Cross-Validation and Bootstrap for Accuracy Estimation and Model Selection. In: Proceedings of the 14th International Conference on Artificial Intelligence. Morgan Kaufmann, San Francisco CA
13. Lammermann F, Baresel A, Wegener J (2004) Evaluating Evolutionary Testability with Software Measurements. In: Proceedings of the Genetic and Evolutionary Computation Conference. Springer, Berlin Heidelberg New York
14. Larrañaga P, Etxeberria R, Lozano JA, Peña JM (2000) Combinatorial optimization by learning and simulation of Bayesian networks In: Proceedings of the Sixteenth Conference on Uncertainty in Artificial Intelligence. Morgan Kaufmann, San Francisco CA
15. Larrañaga P, Lozano JA (2002) Estimation of Distribution Algorithms. A New Tool for Evolutionary Computation. Kluwer Academic Publishers, Boston Dordrecht London

16. Lozano JA, Larrañaga P, Inza I, Bengoetxea E (2006) Evolutionary Computation: Advances in Estimation of Distribution Algorithms. Springer, Berlin Heidelberg New York
17. McCabe TJ (1976) A Complexity Measure. IEEE Transactions on Software Engineering 12(2):208–220
18. McMinn P (2004) Search-based software test data generation: a survey. Software Testing Verification and Reliability 14(2):105–156
19. Mühlenbein H (1998) The equation for response to selection and its use for prediction. Evolutionary Computation 5(3):303–346
20. Press WH, Flannery BP, Teukolsky SA, Vetterling WT (1988) Numerical Recipes in C. The Art of Scientific Computing. Cambridge University Press, Cambridge New York Melbourne
21. Quinlan JR (1993) C4.5: Programs for Machine Learning. Morgan Kaufmann, Los Altos CA
22. Sagarna R, Lozano JA (2005) On the performance of Estimation of Distribution Algorithms applied to Software Testing. Applied Artificial Intelligence 19(5):457–489
23. Schölkopf B, Burges CJC, Smola A (1998) Advances in Kernel Methods - Support Vector Learning. MIT Press, Cambridge MA
24. Sthamer H (1996) The automatic generation of software test data using genetic algorithms. PhD Thesis, University of Glamorgan, Pontyprid, Wales, Great Britain
25. Wegener J, Baresel A, Sthamer H (2001) Evolutionary test environment for automatic structural testing. Information and Software Technology 43:841–854
26. Witten IH, Frank E (2005) Data Mining: Practical machine learning tools and techniques. Morgan Kaufmann, San Francisco CA
27. Wong WE, Horgan JR, Syring M, Zage W, Zage D (2000) Applying design metrics to predict fault-proneness: a case study on a large-scale software system. Software - Practice and Experience 30:1587–1608

Design and Analysis of Rule-based Systems with Adder Designer

Marcin Szpyrka[1]

AGH University of Science and Technology, Al. Mickiewicza 30, 30-059 Kraków
`mszpyrka@agh.edu.pl`

Monitoring and control systems are an important class of embedded systems. They check sensors providing information about the system's environment and take actions depending on the sensor reading. An important part of such a system is a control process that makes decisions based on collected data. The control process may be implemented to use a rule-based system to make decisions. This paper focuses on the design and analysis of such rule-based systems for embedded control systems.

The presented approach is being developed so as to facilitate designing of Petri nets' models of embedded real-time systems. RTCP-nets (Real-Time Coloured Petri nets, see [10]) are used as the modelling language. They are a result of adaptation of timed coloured Petri nets (see [3]) to modelling and analysis of embedded systems. RTCP-nets enable modelling of embedded systems incorporating a rule-based system. The paper focuses on designing of rule-based systems that can be included into an RTCP-net model.

A rule-based system can be represented in various forms, e.g. decision tables, decision trees, extended tabular trees (XTT, [7]), Petri nets ([2]) etc. An interesting comparison of different forms of rule-based systems can be found in [5]. Rule-based systems can be also developed in various forms. First of all, a set of decision rules can be explicitly given by a designer. On the other hand, a set of decision rules can be generated from the acquired data automatically ([8,9]). A more detailed presentation of the current state-of-art can be found in [4].

In most basic versions, a rule-based system for control or decision support consists of a single-layer set of rules and a simple inference engine; it works by selecting and executing a single rule at a time, provided that the preconditions of the rule are satisfied in the current state. A rule-based system can be represented as a single decision table with rows labelled with rules' numbers and columns labelled with attributes' names. Each cell in such a decision table contains a single atomic value of the corresponding attribute. Such decision tables are often called *attributive decision tables with atomic values of attributes* (see [1,5,6]).

M. Szpyrka: *Design and Analysis of Rule-based Systems with Adder Designer*, Studies in Computational Intelligence (SCI) **102**, 255–271 (2008)
`www.springerlink.com`   © Springer-Verlag Berlin Heidelberg 2008

Encoding decision tables with use of atomic values of attributes only is not sufficient for many realistic applications. If the domains of attributes contain more than several values it may be really hard to cope with the number of decision rules. To handle the problem one can use formulae instead of atomic values of attributes. In such a case, a cell in a decision table will contain a formula that evaluates to a boolean value for conditional attributes, and to a single value (that belongs to the corresponding domain) for decision attributes. The result of this approach is a decision table with generalised decision rules (or rules' patterns). Each generalised decision rule covers a set of decision rules with atomic values of attributes (simple decision rules). Therefore, the number of generalised decision rules is significantly less than the number of the simple ones.

The goal of the approach presented in this paper is to prepare such a decision table to be included into an RTCP-net model. To achieve a reasonable level of a rule-based system quality the set of rules must be designed in an appropriate way. Moreover, it should satisfy some properties such as: completeness, consistency and optimality. The approach is supported by a computer tool called *Adder Designer*. The tool allows designing tables with both simple and generalised decision rules. Moreover, it is equipped with transformation algorithms that allow users to convert a decision table with generalised decision rules into a table with simple ones and to glue two or more simple rules into a generalised one. Finally, Adder Designer enables users to verify selected qualitative properties of decision tables.

1 Decision Tables with Atomic Values of Attributes

The basic form of a decision rule is as follows:

$$IF < preconditions > \\ THEN < conclusions >, \tag{1}$$

where $< preconditions >$ is a formula defining when the rule can be applied, and $< conclusions >$ is the definition of the effect of applying the rule; it can be a logical formula, a decision or an action.

Let $\mathcal{A}$ denote a set of attributes selected to describe important features of the system under consideration, i.e., conditions and actions, $\mathcal{A} = \{A_1, A_2, \ldots, A_n\}$. For any attribute $A_i \in \mathcal{A}$, let D_i denote the domain (finite set of possible values) of A_i. It can be assumed that D_i contains at least two different elements. The set $\mathcal{A}$ is divided into two parts. $\mathcal{A}_c = \{A_{c_1}, A_{c_2}, \ldots, A_{c_k}\}$ will denote the set of *conditional attributes*, and $\mathcal{A}_d = \{A_{d_1}, A_{d_2}, \ldots, A_{d_m}\}$ will denote the set of *decision attributes*. For the sake of simplicity it will be assumed that $\mathcal{A}_c$ and $\mathcal{A}_d$ are non-empty, finite, and ordered sets. Therefore, a decision rule with atomic values of attributes (a simple decision rule) takes the following form:

$$
(A_{c_1} = a_{c_1}) \wedge \ldots \wedge (A_{c_k} = a_{c_k}) \implies \\
(A_{d_1} = a_{d_1}) \wedge \ldots \wedge (A_{d_m} = a_{d_m}),
\tag{2}
$$

where $a_{c_i} \in D_{c_i}$, for $i = 1, 2, \ldots, k$ and $a_{d_i} \in D_{d_i}$, for $i = 1, 2, \ldots, m$.

A set of simple decision rules can be represented as a *simple decision table*. To construct such a decision table, we draw a column for each conditional and decision attribute. Then, for every possible combination of values of conditional attributes a row should be drawn. We fill cells so as to reflect which actions should be performed for each combination of conditions. Let $\mathcal{R} = \{R_1, R_2, \ldots, R_l\}$ denote the set of all decision rules. A general scheme of such a decision table is as follows:

$$
\begin{array}{c|ccc|ccc}
 & A_{c_1} & \ldots & A_{c_k} & A_{d_1} & \ldots & A_{d_m} \\
\hline
R_1 & a_{1c_1} & \ldots & a_{1c_k} & a_{1d_1} & \ldots & a_{1d_m} \\
R_2 & a_{2c_1} & \ldots & a_{2c_k} & a_{2d_1} & \ldots & a_{2d_m} \\
\vdots & \vdots & \ldots & \vdots & \vdots & \ldots & \vdots \\
R_l & a_{lc_1} & \ldots & a_{lc_k} & a_{ld_1} & \ldots & a_{ld_m}
\end{array}
\tag{3}
$$

An example of a simple decision table with two conditional and one decision attribute is presented in Table 1.

Table 1. Example of a simple decision table

	A	B	C
R_1	1	a	*off*
R_2	1	b	*on*
R_3	2	a	*off*
R_4	2	b	*on*
R_5	3	a	*on*
R_6	3	b	*on*
R_7	4	a	*on*
R_8	4	b	*on*

Domains for these attributes are defined as follows:
$D_A = \text{int with } 1..4,$
$D_B = \text{with } a|b,$
$D_C = \text{bool with } (\textit{off}, \textit{on}).$

The domains are defined using statements typical for coloured Petri nets ([3]). D_A is a subset of integers, D_B is an enumerated type with two values a and b, and D_C is a Boolean type, where *off* and *on* stand for *false* and *true* respectively.

2 Decision Tables with Generalised Decision Rules

Let's consider the set of decision rules presented in Table 1. If the value of attribute B is equal to b the decision is allways equal to on. Therefore, instead of the rules R_2, R_4, R_6 and R_8, we can take only one rule: $(B = b) \Longrightarrow (C = on)$. The new rule is said to *cover* the rules R_2, R_4, R_6 and R_8. On the other hand, if the value of attribute A is equal to or greater than 3, the decision is also allways equal to on. Thus, instead of the rules $R_5, \ldots, R_8$, we can take the rule: $(A \geq 3) \Longrightarrow (C = on)$.

A formula for an attribute $A_i \in \mathcal{A}$ in a rule $R_j \in \mathcal{R}$ will be denoted by $R_j(A_i)$. To every attribute $A_i \in \mathcal{A}_c$ there will be attached a *variable A_i* that may take any value belonging to the domain D_i. A formula $R_j(A_i) \equiv A_i$ (for conditional attributes) is a shorthand for $R_j(A_i) \equiv A_i \in D_i$ and it always evaluates to *true*. Table 1 can be represented in the following *condensed* form:

Table 2. Generalised decision table – version 1

	A	B	C
R_1	A	$B = b$	on
R_2	$A \leq 2$	$B = a$	off
R_3	$A \geq 3$	$B = a$	on

The decision table with generalised decision rules presented in Table 2 is not the only possible transformation of the Table 1. Another interesting possibility of transformation is presented in Table 3. The only difference between the tables is the modification of the rule R_3. After this modification, both rules R_1 and R_3 can be applied in some states. Such a situation is not treated as a mistake.

Table 3. Generalised decision table – version 2

	A	B	C
R_1	A	$B = b$	on
R_2	$A \leq 2$	$B = a$	off
R_3	$A \geq 3$	B	on

It is evident that transformation of a simple decision table into a corresponding generalised decision table is ambiguous. The final version of such a table is dependent on subjective decisions of a designer. Regardless of this, a generalised decision table has to fulfill the following requirements: Each cell of a decision table should contain a formula, which evaluates to a boolean value for conditional attributes, and to a single value (that belongs to the corresponding domain) for decision attributes.

Decision rules providing a decision or conclusion will be called *positive rules*. Sometimes it is necessary to state in an explicit way that the particular combination of input values (values of conditional attributes) is impossible or not allowed. Such combinations of input values are represented as *negative rules*. For negative rules values of decision attributes are omitted. When necessary, $\mathcal{R}^+$ and $\mathcal{R}^-$ will be used to denote the subset of positive and negative rules respectively.

Table 4. Example of a decision table

	A_{c_1}	A_{c_2}	A_{c_3}	A_{d_1}	A_{d_2}
R_1	$A_{c_1} < 4$	$A_{c_2} = on$	$A_{c_3} = a \vee A_{c_3} = c$	$A_{c_1} + 2$	*on*
R_2	A_{c_1}	$A_{c_2} = on$	$A_{c_3} = b \vee A_{c_3} = d$	3	*off*
R_3	$A_{c_1} = 5$	A_{c_2}	$A_{c_3} \neq b$	2	$\neg A_{c_2}$
R_4	$A_{c_1} > 2$	A_{c_2}	$A_{c_3} = b \vee A_{c_3} = c$	$A_{c_1} - 1$	*on*
R_5	$A_{c_1} = 2$	A_{c_2}	$A_{c_3} = a$	4	*on*
R_6	A_{c_1}	$A_{c_2} = off$	$A_{c_3} = d$		
R_7	$A_{c_1} \leq 3$	$A_{c_2} = off$	A_{c_3}		

An example of a decision table with 3 conditional and 2 decision attributes, and 5 positive and 2 negative rules is presented in Table 4. Domains for these attributes are defined as follows:
$D_{c_1} = D_{d_1} =$ int with 1..5,
$D_{c_2} = D_{d_2} =$ bool with (off, on),
$D_{c_3} =$ with $a|b|c|d$.

To be usefull a generalised decision table should satisfy some qualitative properties such as completeness, consistency (determinism) and optimality. A decision table is considered to be *complete* if for any possible input situation at least one rule can produce a decision. A decision table is *deterministic* if no two different rules can produce different results for the same input situation. The last property means that any dependent rules were removed. Formal definitions of these properties are presented below.

Definition 1. *A* transition function *is a function φ that assigns to each conditional and decision attribute a single value from the attribute domain.*

Definition 2. *A transition function φ is said to* satisfy *the conditional part of a rule $R_j \in \mathcal{R}$ ($\varphi \cong R_j | A_c$) iff each formula $R_j(A_i)$, where $A_i \in \mathcal{A}_c$, evaluates to* true, *for values the function φ assigns to attributes.*

A transition function φ is said to satisfy a rule $R_j \in \mathcal{R}^+$ ($\varphi \cong R_j$) iff φ satisfies the conditional part of the rule R_j and each formula $R_j(A_i)$, where $A_i \in \mathcal{A}_d$, evaluates to $\varphi(A_i)$, for values the function φ assigns to attributes.

Any transition function cannot be said to satisfy a negative rule.

A transition function φ can be represented as a sequence of its values, e.g., $\varphi = (1, on, a, 2, on)$. The set of all transition functions compatible with the decision table scheme will be denoted by Φ. Let the following transition functions be given:

$\varphi_1 = (2, on, b, 3, off)$,

$\varphi_2 = (5, off, c, 2, on)$,

$\varphi_3 = (4, off, a, 1, on)$.

The following relationships hold:

$\varphi_1 \cong R_2$,

$\varphi_2 \cong R_3$,

$\varphi_2 \cong R_4|\mathcal{A}_c$, but $\neg(\varphi_2 \cong R_4)$,

$\forall R_j \in \mathcal{R}: \neg(\varphi_3 \cong R_j|\mathcal{A}_c)$.

Definition 3. *The set $\mathcal{R}$ is complete iff for any transition function φ there exists a rule $R_i \in \mathcal{R}$ such that φ satisfies the conditional part of the rule R_i, i.e.:* $\forall \varphi \in \Phi \, \exists R_i \in \mathcal{R}: \varphi \cong R_i|\mathcal{A}_c$.

The considered set of decision rules is not complete. There is not any decision rule such that the transition function φ_3 satisfies the conditional part of it.

Let Φ^+ denote the set of all transition functions such that for any $\varphi \in \Phi^+$ and for any decision rule $R_j \in \mathcal{R}^-$, the transition function φ does not satisfy the conditional part of R_j. Φ^+ is a set of transition functions that determine the set of all allowed input states.

Definition 4. *The set $\mathcal{R}$ is consistent iff for any transition function $\varphi \in \Phi^+$, and any two rules $R_i, R_j \in \mathcal{R}^+$ if φ satisfies the rule R_i, and φ satisfies the conditional part of the rule R_j, then φ satisfies the rule R_j, i.e.:* $\forall \varphi \in \Phi^+ \, \forall R_i, R_j \in \mathcal{R}^+: (\varphi \cong R_i \wedge \varphi \cong R_j|\mathcal{A}_c) \Rightarrow \varphi \cong R_j$.

The considered set of decision rules is not consistent. The transition function φ_2 satisfies the rule R_3 and the conditional part of the rule R_4 but it does not satisfy the rule R_4.

Definition 5. *Let $\mathcal{R}$ be a complete and consistent set of decision rules. A rule $R_i \in \mathcal{R}^+$ is independent iff the set $\mathcal{R} - \{R_i\}$ is not complete. A rule $R_i \in \mathcal{R}^+$ is dependent iff the rule is not independent. The set $\mathcal{R}$ is semi-optimal iff any rule belonging to the set $\mathcal{R}$ is independent.*

The semi-optimality should be verified after a set of rules is complete and consistent. The verification algorithm is presented in Fig. 1.

3 Adder Designer

Manual analysis of a decision table can be time-consuming even for very small sets of decision rules. *Adder Designer* supports design and analysis of both

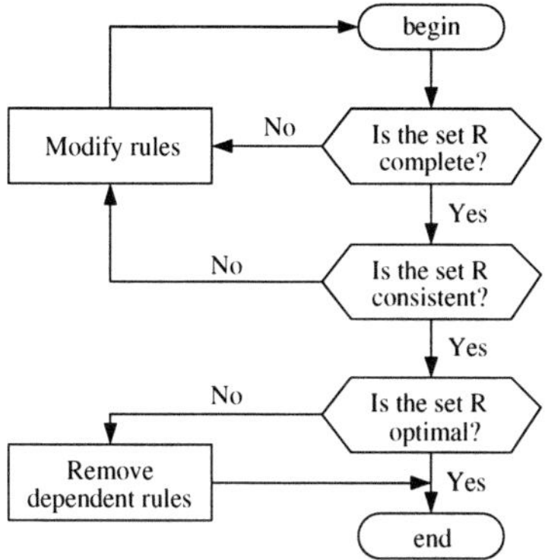

Fig. 1. Scheme block of the verification procedure

simple and generalised decision tables. Adder Designer is a free software covered by the GNU Library General Public License. It is being implemented in the GNU/Linux environment by the use of the Qt Open Source Edition. Qt is a comprehensive C++ application development framework. It includes a class library and tools for cross-platform development and internationalisation. The Qt Open Source Edition is freely available for the development of Open Source software for Linux, Unix, Mac OS X and Windows under the GPL license. Code written for either environment compiles and runs with the other ones. *Adder Tools home page*, hosting information about the current status of the project, is located at `http://adder.ia.agh.edu.pl`. An example of Adder Designer session is shown in the Fig. 2.

The proposed approach to the design of decision tables consists of a few steps. It is first necessary to define attributes selected to describe important features of the system under consideration. There are possible three types of domains: integer, boolean and enumerated data type. Moreover, a new domain may be defined as an alias for already defined one. Secondly, it is necessary to choose conditional and decision attributes. Each attribute can be used twice. Finally, the set of decision rules should be defined. For positive rules, each cell in the corresponding row must be filled. On the other hand, for negative rules the cells that correspond to decision attributes must stay empty.

All commands used for the design of decision tables are gathered in *Table* and pop-up menu. The menus are shown in Fig. 3 and Fig. 4 respectively.

The verification stage is included into the design process. At any time, during the design stage, users can check whether a decision table is complete, consistent or it contains some dependent rules.

Let $\mathcal{R} = \{R_1, R_2, \ldots, R_l\}$ be a set of decision rules and Φ denote the set of all transition functions. Let $\sim_c$ be an equivalence relation on Φ, such that:

$$\varphi_1 \sim_c \varphi_2 \Leftrightarrow \forall A_i \in \mathcal{A}_c : \varphi_1(A_i) = \varphi_2(A_i). \tag{4}$$

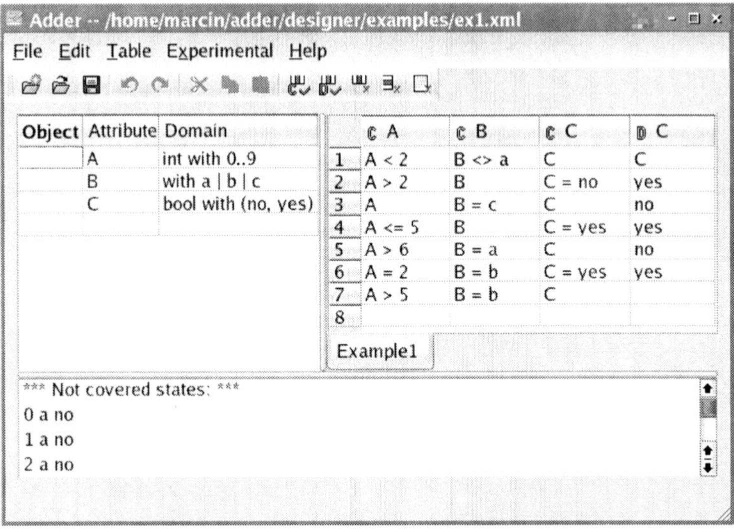

Fig. 2. Example of Adder Designer session

Fig. 3. Table menu

Restriction of the set Φ to the set of conditional attributes is defined as follows:

$$\Phi|\mathcal{A}_c = \{\varphi|\mathcal{A}_c : \varphi \in \Phi\}. \tag{5}$$

Let $\psi \in \Phi|\mathcal{A}_c$, $\varphi \in \Phi$ and $\psi = \varphi|\mathcal{A}_c$.

$$\psi \cong R_i|\mathcal{A}_c \Leftrightarrow \varphi \cong R_i|\mathcal{A}_c. \tag{6}$$

Adder Designer uses methods based on colour Petri nets theory to check completeness, consistency and semi-optimality (see [11]). A simplified representations of these algorithms are as follows:

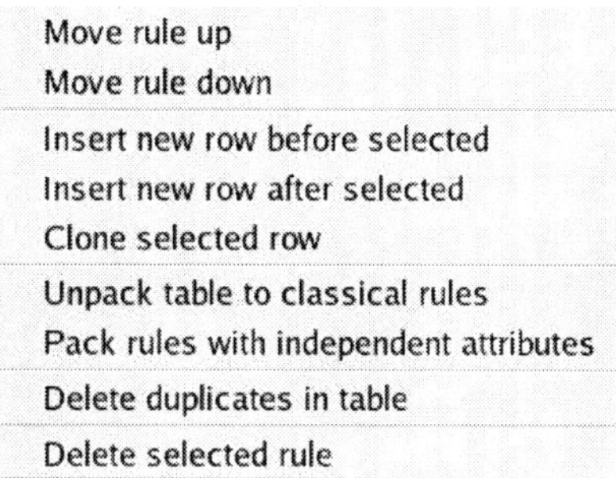

Fig. 4. Pop-up menu

Completeness

$\Psi := \emptyset$;
for all $\psi \in \Phi|\mathcal{A}_c$ **do**
   covered := false;
   for all $R_i \in \mathcal{R}$ **do**
     if $\psi \cong R_i|\mathcal{A}_c$ **then**
       covered := true;
     end if
   end for
   if covered = false **then**
     $\Psi = \Psi \cup \{\psi\}$;
   end if
end for
if $|\Psi| > 0$ **then**
   Not covered states: Ψ;
end if

The result of completeness analysis is a list of input states (combinations of values of conditional attributes) that are not covered by decision rules. For the decision table presented in Fig. 2 the report of completeness analysis is as follows:

```
*** Not covered states: ***
0 a no
1 a no
2 a no
2 b no
6 a yes
Table is not complete.
```

Let $\psi \in \Phi|\mathcal{A}_c$ and let $R_i \in \mathcal{R}^+$ be a positive decision rule such that $\psi \cong R_i|\mathcal{A}_c$. $\psi_i^* \in \Phi$ will be used to denote a transition function such that:

$$\forall A \in \mathcal{A}_c: \psi(A) = \psi_i^*(A) \wedge \psi_i^* \cong R_i. \tag{7}$$

Consistency

```
Θ := ∅;
for all ψ ∈ Φ|A_c do
   R' := ∅;
   Ψ := ∅;
   for all R_i ∈ R do
      if ψ ≅ R_i|A_c then
         R' := R' ∪ {R_i};
         Ψ := Ψ ∪ {ψ_i^*};
      end if
   end for
   if |R'| > 1 ∧ R' ∩ R⁻ = ∅ ∧ |Ψ| > 1  then
      Θ = Θ ∪ {R'};
   end if
end for
if |Θ| > 0 then
   Not consistent sets of rules: Θ;
end if
```

After consistency analysis users receive a list of sets of inconsistent rules. Each such a set of rules is labelled with an input state that is covered by the rules and results of applying these rules for the state are also presented. A part of consistency analysis report for the table presented in Fig. 2 is as follows:

```
*** Not consistent  sets  of  rules: ***
State:   0  c  yes
R1:    yes
R3:    no
R4:    yes
State:   1  c  yes
R1:    yes
R3:    no
R4:    yes
 ...
```

Semi-optimality

```
R' := R;
for all ψ ∈ Φ|A_c do
   S := ∅;
   for all R_i ∈ R do
      if ψ ≅ R_i|A_c then
         S := S ∪ {R_i};
      end if
```

```
  end for
  if |S| = 1  then
     R' := R' - S;
  end if
 end for
 if |R'| > 0 then
    Dependent rules: R';
 end if
```

The result of semi-optimality analysis is a set of dependent rules. For the considered decision table such a set contains the rule R_6 only.

In addition to this the commands *Unpack table to classical rules* and *Pack rules with independent attributes* are used to convert a generalised decision table into a simple one and vice versa.

Adder Designer uses XML format to store projects. A piece of XML code describing a decision table is presented below:

```
<attribute name="A" domain="int with 0..9"/>
...
<table name="Example1">
  <conditionalAttributes>
    <attribute name="A"/>
    <attribute name="B"/>
    <attribute name="C"/>
  </conditionalAttributes>
  <decisionAttributes>
    <attribute name="C"/>
  </decisionAttributes>
  <rules>
    <rule>
      <formula expression="A &lt; 2"/>
      <formula expression="B &lt;&gt; a"/>
      <formula expression="C"/>
      <formula expression="C"/>
    </rule>
    ...
  </rules>
</table>
```

The XML format may be used by another tools to generate input files for Adder Designer.

4 Examples

Two examples of decision tables are presented in this section. The first one was designed for a traffic lights control system and is used to decide, which

state should be displayed next. The second example concerns a firewall system
and is based on the one presented in [7].

4.1 Traffic Lights Control System

Let's consider a traffic lights control system for crossroads presented in Fig. 5.

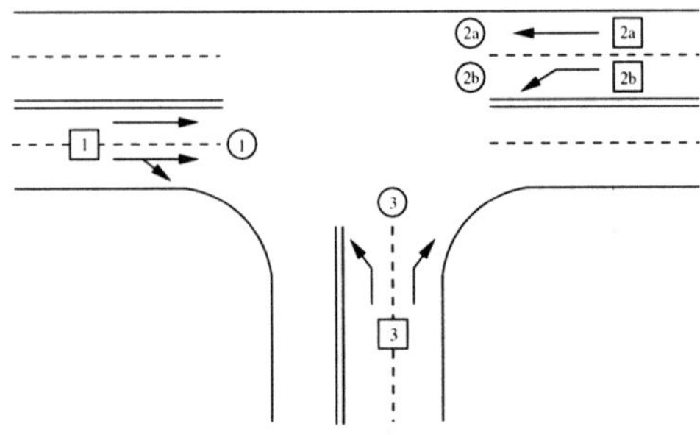

Fig. 5. Crossroads model

Table 5. Acceptable traffic lights' states

State number	Lights 1	Lights 2a	Lights 2b	Lights 3
1	green	green	red	red
2	red	green	green	red
3	red	red	red	green

The system should take into consideration the traffic rate in the input
roadways. All roads are monitored, monitors are drawn as rectangles, and
four traffic lights are used (drawn as circles). Three different traffic lights'
states are possible and are presented in the Tab. 5. The system works in the
following way. If there are some vehicles at all input roadways, the states 1,
2 and 3 are displayed sequentially. If there is no need to display a state, the
state is omitted and the next state is displayed. The four monitors are used
to determine the state to be displayed next.

Five conditional and five decision attributes are used to describe the system. The attribute S stands for the current (conditional one) and the new
state (decision one). Attributes $T1$, $T2a$, $T2b$ and $T3$ stand for the states of
monitors, while attributes $L1$, $L2a$, $L2b$ and $L3$ stand for the states of traffic
lights. A state of an monitor denotes the number of vehicles waiting in the
corresponding input road. Domains for these attributes are defined as follows:
$D_S = $ int with 1..3,

$D_{T1} = D_{T2a} = D_{T2b} = D_{T3} = $ int with 1..5,
$D_{L1} = D_{L2a} = D_{L2b} = D_{L3} = $ with $green|red$.

The decision table designed for the traffic light's driver is presented in Tab. 6.

Table 6. Decision table for the traffic lights control system

Conditional attributes					Decision attributes				
S	T1	T2a	T2b	T3	S	L1	L2a	L2b	L3
$S = 1$	$T1$	$T2a$	$T2b > 0$	$T3$	2	*red*	*green*	*green*	*red*
$S = 1$	$T1$	$T2a$	$T2b = 0$	$T3 > 0$	3	*red*	*red*	*red*	*green*
$S = 1$	$T1$	$T2a$	$T2b = 0$	$T3 = 0$	1	*green*	*green*	*red*	*red*
$S = 2$	$T1$	$T2a$	$T2b$	$T3 > 0$	3	*red*	*red*	*red*	*green*
$S = 2$	$T1 > 0$	$T2a$	$T2b$	$T3 = 0$	1	*green*	*green*	*red*	*red*
$S = 2$	$T1 = 0$	$T2a$	$T2b$	$T3 = 0$	2	*red*	*green*	*green*	*red*
$S = 3$	$T1 > 0$	$T2a$	$T2b$	$T3$	1	*green*	*green*	*red*	*red*
$S = 3$	$T1 = 0$	$T2a > 0$	$T2b$	$T3$	2	*red*	*green*	*green*	*red*
$S = 3$	$T1 = 0$	$T2a$	$T2b > 0$	$T3$	2	*red*	*green*	*green*	*red*
$S = 3$	$T1 = 0$	$T2a = 0$	$T2b = 0$	$T3$	3	*red*	*red*	*red*	*green*

4.2 Firewall Control System

Let us consider an example of computer network design, presented in Fig. 6. It is a typical configuration for many security-aware small office, or company networks. The network is composed of three subnetworks: LAN (local area network), DMZ (the so-called *demilitarized zone*), and INET (Internet connection). The subnetworks are separated by a *firewall* having three network interfaces.

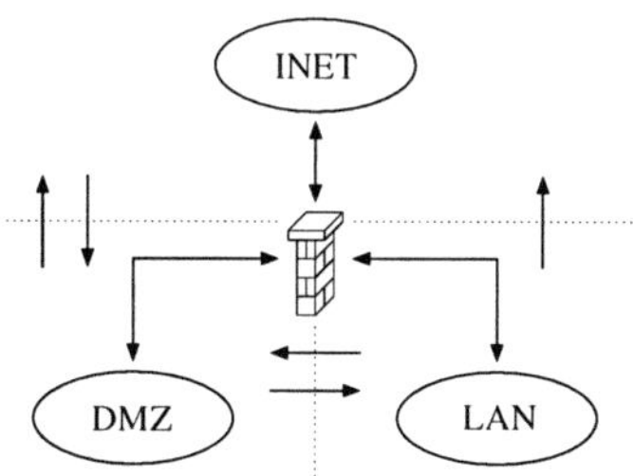

Fig. 6. Network firewall configuration

The firewall controls the input and output and decides whether the request should be accepted or rejected. Decision table for such a firewall system contains three conditional (service, source address, destination address) and one decision attribute (routing). The attribute *Service* stands for a type of the net service, attributes *Srcaddr* and *Destaddr* are connected with source and destination IP addresses respectively, and the attribute *Routing* stands for the final routing decision. Domains for these attributes are enumerated data types and they are defined as follows:

$D_{Service} = $ with *ssh* | *smtp* | *http* | *imap*;

$D_{Srcaddr} = $ with *inet* | *dmz* | *lan*;

$D_{Destaddr} = D_{Srcaddr}$;

$D_{Routing} = $ with *accept* | *reject*.

A complete decision table for the firewall system (presented in Table 7) contains eleven positive and four negatives rules.

Table 7. Decision table for the firewall system

Conditional attributes			Decision attributes
Service	**Srcaddr**	**Destaddr**	**Routing**
Service = http	*Srcaddr = inet*	*Destaddr = dmz*	*accept*
Service = http	*Srcaddr = inet*	*Destaddr = lan*	*reject*
Service = http	*Srcaddr = lan*	*Destaddr*	*accept*
Service = smtp	*Srcaddr*	*Destaddr = lan*	*reject*
Service = smtp	*Srcaddr*	*Destaddr = dmz*	*accept*
Service = smtp	*Srcaddr = lan*	*Destaddr = inet*	*reject*
Service = imap	*Srcaddr = lan*	*Destaddr = dmz*	*accept*
Service = imap	*Srcaddr ≠ lan*	*Destaddr*	*reject*
Service = ssh	*Srcaddr = inet*	*Destaddr*	*reject*
Service = ssh	*Srcaddr = lan*	*Destaddr*	*accept*
Service = ssh	*Srcaddr = dmz*	*Destaddr*	*accept*
Service = http	*Srcaddr = dmz*	*Destaddr*	
Service = http	*Srcaddr = inet*	*Destaddr = inet*	
Service = imap	*Srcaddr = lan*	*Destaddr ≠ dmz*	
Service = smtp	*Srcaddr ≠ lan*	*Destaddr = inet*	

5 RTCP-nets

RTCP-nets [10] are an adaptation of CP-nets to modelling and analysis of embedded systems. They are suitable for modelling systems incorporating a rule-based system. A special form of hierarchical RTCP-nets called *canonical form* has been defined to speed up and facilitate drawing of the models. RTCP-nets in canonical form are composed of four types of subnets with precisely defined structures: primary place pages, primary transition pages, linking pages, and D-nets. The general structure of an RTCP-net in canonical form is shown in Fig. 7.

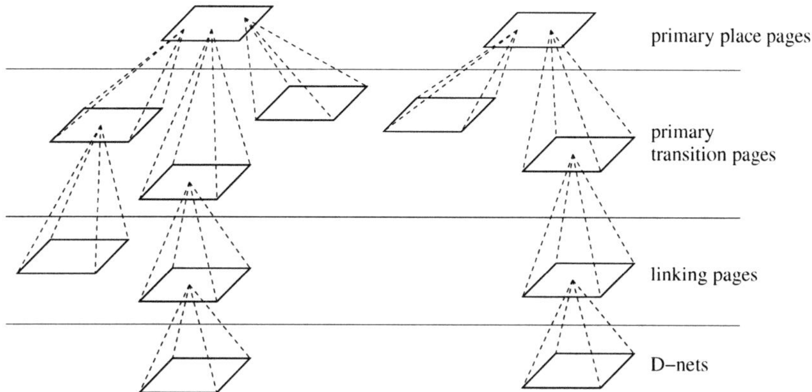

Fig. 7. General structure of an RTCP-net in the canonical form

D-nets belong to the most bottom level of a model and are used to represent rule-based systems in a Petri net form. A D-net is a non-hierarchical coloured Petri net that represents a set of decision rules. It contains two places: a *conditional place* (input place) for values of conditional attributes and a *decision place* (output place) for values of decision attributes. Each positive decision rule is represented by a transition and its input and output arcs. A token placed on the conditional place denotes a sequence of values of conditional attributes. Similarly, a token placed on the decision place denotes a sequence of values of decision attributes. It should be underlined that decision tables and D-nets are equivalent forms of decision rules representation. An algorithm of transformation of a decision table into a D-net can be found in [11].

D-net for the decision table presented in Table 6 is shown in Fig. 8. It should be underlined that users are not assumed to know anything about Petri nets to use the Adder Designer for the design and verification of decision tables. However, the tool allows users to generate a part of an RTCP-net model automatically.

6 Summary

Adder Designer, a tool for design and analysis of rule-based systems in the form of generalised decision tables, has been presented in the paper. The tool is equipped with a decision table editor and verification procedures. A survey of decision tables properties and the corresponding verification algorithms has been also presented.

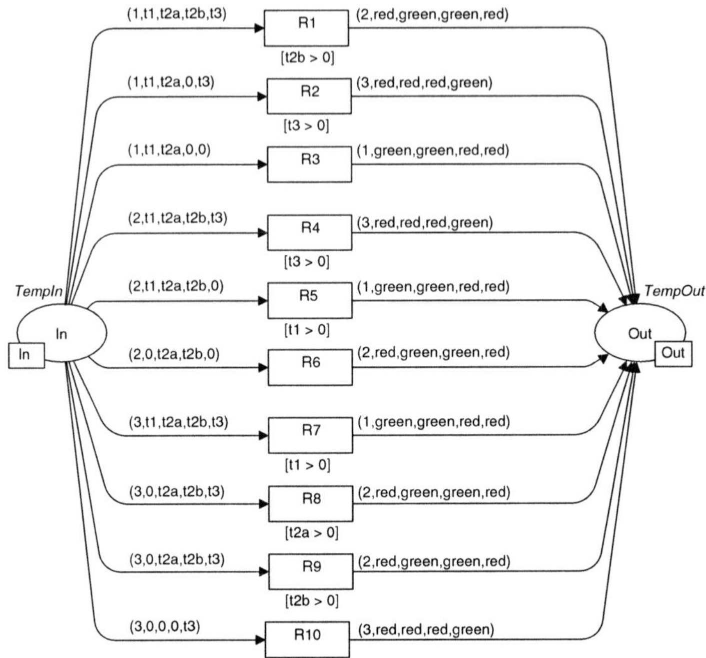

Fig. 8. D-net for the traffic lights control system

Some of the algorithms are based on typical Petri nets analysis methods. The considered decision tables may be automatically transformed into an equivalent Petri net form called D-net. A D-net is a non-hierarchical coloured Petri net. D-nets may be used both to specify external system behaviour and to model a rule-based system. In the second case, D-nets constitute the bottom layer of an RTCP-net model (a Petri net model). For more details see [10].

Development of the tool is still in progress. The source code of is organized carefully and new modules can be included without any problems. Our future plans will focus on the development of new and faster algorithms for verification and optimization of decision tables. Moreover, Adder Designer is released under the GPL license and everyone may develop his own verification procedures and include them into the tool.

References

1. Davis AM (1988) A comparison of techniques for the specification of external system bahavior. Communication of the ACM 31(9):1098–1115
2. Fryc B, Pancerz K, Suraj Z (2004) Approximate Petri nets for rule-based decision making. In: Komorowski J., Tsumoto S. (eds) Proceedings of the 4th International Conference on Rough Sets and Current Trends in Computing, RSCTC 2004. LNAI 3066:733–742

3. Jensen K (1992-1997) Coloured Petri nets. Basic concepts, analysis methods and practical use. Vol. 1–3 Springer, Berlin Heidelberg New York
4. Liebowitz J (1998) The handbook of applied expert systems. CRC Press
5. Ligęza A (2006) Logical foundations of rule-based systems. Springer, Berlin Heidelberg New York
6. Macaulay LA (1996) Requirements Engineering. Springer, Berlin Heidelberg New York
7. Nalepa GJ, Ligęza A (2004) Designing reliable web security systems using rule-based systems approach. In: Menasalvas E, Segovia J, Szczepaniak PS (eds) Advances in Web Intelligence: first international Atlantic Web Intelligence Conference AWIC 2003. LNCS 2663:124–133
8. Pawlak Z (1991) Rough sets. Kluwer Academic Publishers
9. Peters JF, Skowron A, Suraj Z (2000) An application of rough set methods in control design. Fundamenta Informaticae, 43(1-4):269–290
10. Szpyrka M (2004) Fast and flexible modelling of real-time systems with RTCP-nets. Computer Science, 6:81–94
11. Szpyrka M, Szmuc T (2006) D-nets – Petri net form of rule-based systems. Foundations of Computing and Decision Sciences, 31(2):157–167

A Query-Driven Exploration of Discovered Association Rules

Krzysztof Świder[1], Bartosz Jędrzejec[1], and Marian Wysocki[1]

Rzeszów University of Technology,
W. Pola 2, 35-959 Rzeszów, Poland
`kswider@prz-rzeszow.pl`, `bartoszj@prz-rzeszow.pl`,
`mwysocki@prz.rzeszow.pl`

Summary. The paper concerns the presentation phase of a knowledge discovery process with use of association rules. The rules, once obtained, have normally to be explained and interpreted in order to make use of them. The authors propose an approach based on the employment of Predictive Model Markup Language (PMML) to facilitate an environment for the systematic examination of complex mining models. The PMML is an XML application developed by the Data Mining Group dedicated to data analysis models. We start with a short description of PMML, and show an example of an automatically encoded mining model. Then XQuery language is involved to demonstrate how to explore a model by querying its PMML structure. Preliminary results for a real association rule model are presented in the final part of the paper. Three approaches are considered: (1) simple direct querying of the PMML structure of the discovered model, (2) interactive browsing the rule base using set-theoretic operations, (3) automatic query formulation with genetic programming.

1 Introduction

The growing interest in XML technology rapidly led to a large number of applications. The technology is now recommended by the World Wide Web Consortium [5] and intensively exploited and propagated by many commercial companies. The significant feature of XML is its application potential, not only for WWW-specific purposes, but also in such areas as: electronic data interchange, databases and languages for special purposes. XML is a meta-language with numerous applications like languages covering several areas including: chemistry, multimedia, navigation, music etc. Each of these languages is provided by document type definition, necessary to build the correct structures. One of the special XML applications is Predictive Models Markup Language (PMML) [20], developed by Data Mining Group[1]. The language

---

[1] http://www.dmg.org/

K. Świder et al.: *A Query-Driven Exploration of Discovered Association Rules*, Studies in Computational Intelligence (SCI) **102**, 273–288 (2008)
`www.springerlink.com`

is typically aimed to specify statistical and data mining models enabling the exchange of models between different applications.

Data mining is a rapidly growing research area relating to the use of artificial intelligence and machine learning methods in advanced data analysis. The data is usually stored in large repositories including relational databases, data warehouses, text files, web resources etc. The result of the mining procedure is a kind of knowledge base, called a mining model, containing useful patterns and characteristics derived from the data. Such models are frequently complex and difficult to interpret – therefore considerable efforts are generally required to evaluate mining results. In a typical case, clever visualization techniques are involved to present the previously generated patterns with the aim to enable better insight into mining results. However, when a model becomes complex, it is not always easy to make exhaustive use of a picture and give it a clear description.

Association rule mining is an essential data mining problem, widely used in many realworld applications. Often a rule mining system generates a large number of rules while only a small subset of them are really useful in particular employment. Therefore the derived large set of association rules should be reduced to be more concise and easier to analyze. The new rules are expected to be more compact (with small sets of attributes), non-redundant and easy to apply. This problem was detected early on and actively investigated in numerous studies [8, 11–13, 22]. There are two main approaches encountered in the literature concerning the reduction of data mining models. The first method is based on a reduction of the source data through the imposition of restrictions on the data attributes. In this case, the mining process is executed only on data which satisfy these constraints. The examples of such an approach are the DMQL language [8] and the MINE Rule Operator [17, 18]. Setting appropriate threshold parameters plays a crucial role in using these tools. Too strict restrictions may result in useless mining models and in the need to start the mining process from the beginning.

In the second approach model reduction is obtained through the selection of appropriate rules from the model built. The query language MSQL [12, 13] and the SMARTSKIP System [10] are the representative tools. MSQL Language enables generation and processing of the generated association rules. The SMARTSKIP System gives an opportunity to store the association rules in specially prepared relational tables, and to select rules from the tables using SQL queries. This method is more promising because of the possibility of multiple querying the association rule sets and analyzing the resulting subsets of rules without building new models.

The aim of our work is to develop methods and tools with the aim of exploring mining models (currently limited to association rules) stored as PMML-files. We use a query-driven method empowered by regular XML querying technology. A considerable advantage of such an approach is the natural ability to analyze the models while preserving their portability between different applications. The PMML standard is widely supported by commonly used commercial data mining tools [3,19]. Section 2.1 in this paper is a short description of PMML. Section 2 is aimed at discovering an association rule model and obtaining a PMML form of the rules derived from the real data. In section 3 the model is explored using XQuery technique, recently recommended by the World Wide Web Consortium (W3C) as a standard for query language for XML structures. The compact syntax as well as close similarity of XQuery to the SQL language, make it easy to interpret and analyze queries used in the rule exploration process. Three approaches are considered: (1) simple direct querying of the PMML structure of the discovered model, (2) interactive browsing the rule base using set-theoretic operations, (3) automatic query formulation with genetic programming. In the first method the model discovered is reduced by specifying new thresholds for support and confidence, the two popular interestingness measures for association rules. The second method uses a querying process controlled by the analyst, who iteratively puts domain specific constraints into the next query to proceed, taking into account current results. In the last approach the synthesis of the query is performed by a computer on the basis of the assessment of the resulting rule sets with respect to a criterion defined by the user. The query, considered as an XQuery program is modified through the mechanisms of genetic programming in order to improve the quality of the selected association rules. Application of genetic programming to query reformulation was studied before in a different context (see e.g. [6,15]). To the best of the authors' knowledge, using it to select valuable sets of association rules from large rule models seems to be a new proposal.

2 Discovering Association Rule Model

This section is aimed at discovering an association rule model and obtaining the PMML form of the rules. The example model derived from real data will be explored using XQuery language in the next section.

2.1 Data Mining Models in PMML

The Predictive Model Markup Language is an XML application providing a standard text-based format for statistical and data mining models. The essential benefit of such an approach is the possibility to share models between PMML-compliant applications, which denotes the portability of data mining models between mining tools from different vendors. The language is a result of the successful activity of the Data Mining Group, an organization, which defines new standards for data mining technology and groups a number of commercial institutions including IBM, Microsoft, Oracle, SAS, SPSS and others. In PMML version 2.1 [20] the XML Schema standard was applied to define document structure. The structure is shown in Fig. 1.

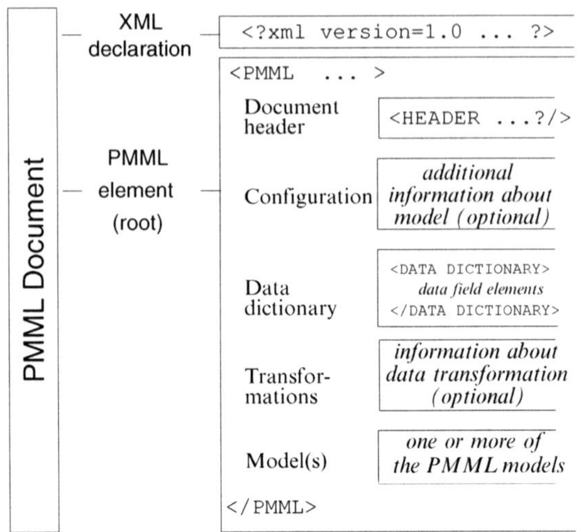

Fig. 1. The general structure of PMML document

Each PMML document starts with an XML declaration. The main element in Fig. 1 named PMML contains attributes defining the location of the document schema. The essential part of the PMML document defines one or more of the models allowed in version 2.1 of the language. The current edition allows most of popular statistical and data mining models commonly used for data analysis.

2.2 Association Rules

Association rules are an important type of data mining models [2,9]. In general, an association rule is an expression in the form $A \Rightarrow B$, where A and B are sets of items (usually called itemsets) taken from some universal set of items I, and $A \cap B = \varnothing$. The rules are normally obtained by analyzing a set D of task relevant data consisting of a large number of records T often referred to as transactions. Each transaction is composed of items from I. For any transaction T and an itemset A, we say that T contains A, if $A \subseteq T$. The percentage of transactions in D that contain an itemset A is usually referred to as support of the itemset. An itemset that satisfies minimum support is called frequent itemset.

There are two popular interestingness measures for an association rule $A \Rightarrow B$ i.e.: support and confidence. They are defined as follows:

1. The rule $A \Rightarrow B$ holds in the transaction set D with support *sup*, if *sup* is the percentage of transactions in D that contain both A and B.
2. The rule $A \Rightarrow B$ has confidence *conf* in the transaction set D, if *conf* is the percentage of all transactions containing A, that also contain B.

Minimum support and minimum confidence thresholds are commonly used to control the process of mining association rules. The classic example of association rule mining is market basket analysis, but many other interesting application areas exist, e.g.: business management, click-stream analysis etc.

2.3 Mining the Data

Although a vast amount of data is presently stored in various data repositories, it is not always easy to get the right data to experiment with. In many cases the only way out is to use one of the numerous Web resources publishing various kinds of data. To obtain required association models we used real data from the Fatal Accident Reporting System published by U.S. National Highway Traffic Safety Administration[2]. The data is a collection of records reporting fatal crashes, i.e. those that resulted in the death of a person, within the 50 States including the District of Columbia and Puerto Rico. We considered only the selected data referred to years 1998 and 1999, i.e. over 74.000 records.

Assume we want to use the association model to discover characteristic association relationships between a predefined set of external conditions connected with an accident. To obtain models as simple as possible the number of attributes we took into consideration is limited to the following:

- Weather ("No Adverse Atmospheric Conditions", "Rain", "Sleet", "Snow", "Fog", "Rain and Fog", "Smog", "Smoke", "Blowing Sand or Dust", "Unknown");
- Harmful Event ("Overturn", "Fire/Explosion", "Immersion", "Gas Inhalation", "Fell from Vehicle", "Injured in Vehicle", "Other Non-Collision", "Pedestrian", etc. – totally 50 values);
- Light Condition ("Daylight", "Dark", "Dark but Lighted", "Dawn", "Dusk", "Unknown");
- Speed Limit ("No Speed Limit", 01-99 – Speed Limit in Miles Per Hour, "Unknown").

Establishing the minimum support threshold to 0.3 we obtained a rather slight model where the total number of association rules was limited to 4. The central part of the generated PMML-code of the model is an AssociationModel element in the form:

```
<AssociationModel minimumConfidence="0.0"
      numberOfRules="4" avgNumberOfItemsPerTA="5.0"
      minimumSupport="0.3"
      numberOfTransactions="74150"
      numberOfItemsets="6"
      modelName="ACCIDENT_ASSOCIATED_COND"
      functionName="associationRules"
      numberOfItems="4"
```

---

[2] ftp://ftp.nhtsa.dot.gov/FARS/

```
        maxNumberOfItemsPerTA="5">
   <MiningSchema>
    <MiningField name="item" outliers="asIs"
      usageType="active" />
   </MiningSchema>
   <Item value="55 mph" id="1" />
   <Item value="Daylight" id="2" />
   <Item value="Vehicle in Transport" id="3" />
   <Item value="No Adverse Atmospheric Conditions" id="4" />
   <Itemset numberOfItems="1" support="0.3327" id="1">
      <ItemRef itemRef="1" />
   </Itemset>
   <Itemset numberOfItems="1" support="0.5063" id="2">
      <ItemRef itemRef="2" />
   </Itemset>
   <Itemset numberOfItems="1" support="0.3977" id="3">
      <ItemRef itemRef="3" />
   </Itemset>
   <Itemset numberOfItems="1" support="0.8811" id="4">
      <ItemRef itemRef="4" />
   </Itemset>
   <Itemset numberOfItems="2" support="0.4516" id="5">
      <ItemRef itemRef="2" /> <ItemRef itemRef="4" />
   </Itemset>
   <Itemset numberOfItems="2" support="0.3470" id="6">
      <ItemRef itemRef="3" /> <ItemRef itemRef="4" />
   </Itemset>
   <AssociationRule confidence="0.8920" support="0.4516"
       consequent="4" antecedent="2" />
   <AssociationRule confidence="0.5126" support="0.4516"
       consequent="2" antecedent="4" />
   <AssociationRule confidence="0.8724" support="0.3470"
       consequent="4" antecedent="3" />
   <AssociationRule confidence="0.3938" support="0.3470"
       consequent="3" antecedent="4" />
 </AssociationModel>
```

The model consists of four major parts: attributes, items, itemsets and rules.
The attributes describe the essential model parameters as: minimum support,
minimum confidence, number of items, number of itemsets, number of rules
etc. The rules included in that simple model are:

```
"Daylight" ==> "No Adverse Atmospheric Conditions"
       sup=0.4516 conf=0.8920
"No Adverse Atmospheric Conditions" ==> "Daylight"
       sup=0.4516 conf=0.5126
"Vehicle in Transport" ==> "No Adverse Atmospheric Conditions"
       sup=0.3470 conf=0.8724
"No Adverse Atmospheric Conditions" ==> "Vehicle in Transport"
       sup=0.3470 conf=0.3938
```

Then we obtained two further models for reduced values of support threshold using the same data. These models are more complex – therefore – instead of listing their PMML code, we summarized all the three models in Table 1.

Table 1. The example association models

Parameter	Model 1	Model 2	Model 3
Number of transactions	74,150	74,150	74,150
Minimum support	0.3	0.06	0.007
Minimum confidence	0	0	0
Number of items	4	14	40
Number of itemsets	6	38	269
Number of rules	4	54	560

The minimum support thresholds of models 2 and 3 are fixed to be 'less restrictive', therefore, they contain more rules to interpret. The problem is: how to work effectively with such complex models and particularly how to discover the useful knowledge from the large number of generated association rules? An attempt to find a possible way out with the use of PMML code will be presented in the next section.

3 Querying the Rule Base

As confirmed in Table 1, setting fewer minimum thresholds for support and confidence values leads to more complex association models. In this section such a large model will be managed by querying its PMML structure according to the general schema shown in Fig. 2.

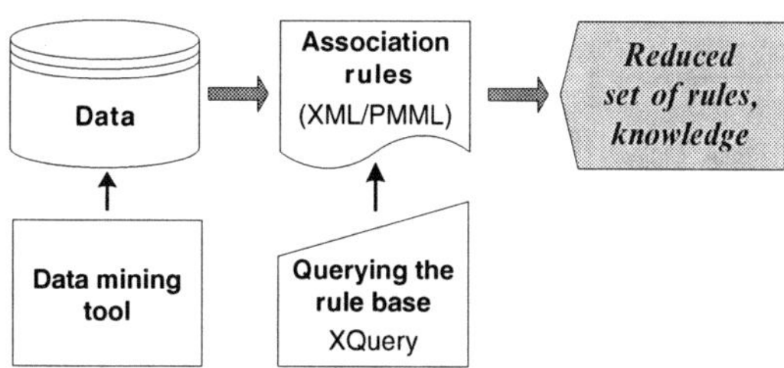

Fig. 2. Querying association rules

A number of techniques and tools have been introduced in the last few years in the area of querying semi-structured data and XML [1]. The recent

W3C recommendation for XQuery 1.0 [4] is certainly the most promising proposition of query language for XML structures. In our opinion the close similarity to SQL, its compact and easy to understand syntax are the main advantages of XQuery, making it suitable for our investigations. An additional feature is the availability of embedded functions, enabling set-theoretic operations (union, difference, intersection) on the query results. In section 3.1 we use the XQuery demonstrator from Altova[3] in order to accomplish simple queries of a discovered set of rules. The preliminary tests are quite simple and yield rather evident results but they adequately show the potential capabilities of our approach. In the following two subsections we introduce our two applications developed for querying the rule base: the interactive browser (in 3.2) and the tool for automatic query formulation and management using genetic programming (in 3.3).

3.1 Simple Direct Querying

As the introductory examination we will show in particular, that the rules of Model 1 appearing in Table 1 can be easily received by querying the PMML structure of Model 3. Assuming that the structure was formerly stored in a file named assoc_model_3.xml, this can be accomplished by the following XQuery statement:

```
for $i in doc("assoc_model_3.xml")//AssociationRule
    [@support>=0.3]
return $i
```

This query looks up the rule base of the input complex model and extracts the rules with support greater than or equal to 0.3. The result is a set of four AssociationRule elements representing the four rules identical to those contained in Model 1. Thus we are able to 'cut down' the discovered association rule model just specifying new minimum values for support and confidence thresholds in the query condition, *instead of repeating the whole mining process*. Although appropriate thresholds for support and confidence in mining associations normally allow finding the reasonable amount of 'interesting' rules, the problem with choosing the right values of the thresholds is not straightforward. Specifically, fixing them too high, we tend to loose some interesting patterns in resulting model, while – underestimated, they lead to large number of rules, which are difficult to manage for the analyst. Querying the association rule model allows to change the threshold values for already existing set of rules.

The main (and only) result of the querying action in the previous example was the reduction of the complex model. In the following test we will show, that the proposed querying method is much more powerful and the proper query operations are able to discover interesting properties using the entire structure of an initial complex model. In order to demonstrate this, the Model

---

[3] Actually we used the Altova XMLSpy Home Edition – available free via: http://www.altova.com/download_components.html

3 in Table 1 will be queried with the aim of finding information about the most frequent itemsets in the model. The required query is:

```
for $itemset in doc("assoc_model_3.xml")//AssociationModel/Itemset
order by $itemset/@support descending
return $itemset
```

and the resulting itemsets can be formatted as follows:

```
"No Adverse Atmospheric Conditions"; sup=0.8811
"Daylight"; sup=0.5063
"Daylight"|"No Adverse Atmospheric Conditions"; sup=0.4516
"Vehicle in Transport"; sup=0.5063
"Vehicle in Transport"|"No Adverse Atmospheric Condition";
                                                   sup=0.3470
"55 mph"; sup=0.3327
"Dark"; sup=0.2986
"55 mph"|"No Adverse Atmospheric Conditions"; sup=0.2903
"Dark"|"No Adverse Atmospheric Conditions"; sup=0.2606
"Daylight"|"Vehicle in Transport"; sup=0.2591
...
```

All frequent itemsets of Model 3 were extracted from the model structure and the resulting list was ordered by support values in descending order. In order to preserve space only the itemsets with support value over 0.25 are listed. It is remarkable that even the results of such a simple query give several interesting indications about the data under consideration. In particular we notice that the item "No Adverse Atmospheric Conditions" is contained in 5 of the first 10 most frequent itemsets. Moreover the itemset containing only this item is the absolute leader, appearing in more than 88% cases. This simple discovery gives an evident, but appealing, suggestion that most of the fatal accidents were not caused by any abnormal weather conditions. In addition one can be encouraged to further investigate the model. For example, the existing rule base could be filtered with special attention to the rules containing the item "No Adverse Atmospheric Conditions". This becomes feasible with the tool introduced in the next subsection.

3.2 Browsing with Constraints

In the following we make an assumption, that in many real situations the association rules obtained by mining algorithm are used by practitioners, who combine the mining results with domain knowledge and their current needs. For example they could try to limit the space of interest to the rules referring to an existing, previously defined, problem. In order to face the problem we propose a kind of interactive browser, which explores the rule base to seek the most valuable rules. In this case the querying process consists of a number of steps – each controlled by an analyst, who puts domain specific constraints into the next query to proceed. Consequently, at each step, some specified operations are applied to the current set of rules in order to extract the reduced number of most interesting rules. The collection of available operations is currently limited to set-theoretic operations: union, intersection and difference. If any consecutive operation leads to a subset of rules, which does not satisfy user expectations, it is possible to abandon any number of the most recent steps and reestablish the queries specifying new constraints. Thus the search for valuable rules potentially traverses a graph structure as shown in Fig. 3.

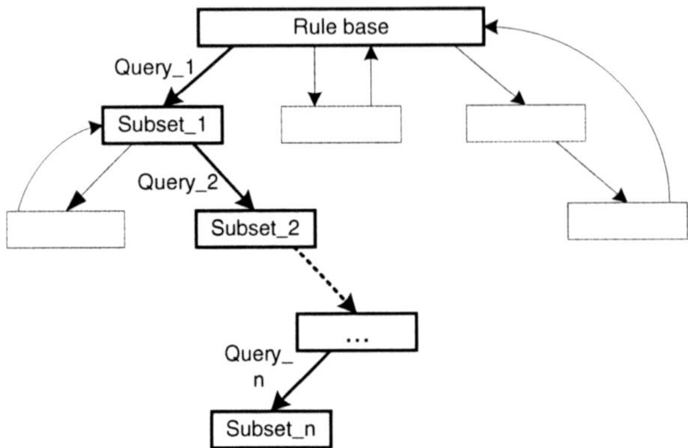

Fig. 3. Browsing the rule base

The node "Rule base" represents the original set of all mined rules, while any other node stands for a subset of rules obtained as a result of the preceding query. The edges denote queries, aimed at reducing a set of rules by involving some problem-specific constraints. Another feature (not shown in Fig. 3) is that for any node representing a reduced number of rules there generally exist more then one path connecting the node with the "Rule base".

In favor of a practical verification of the idea of interactive rule browsing a prototype application was implemented and tested for real data. The program was prepared in Java and used a predefined set of XQuery functions. As an example let us consider another rule base obtained for data describing accidents introduced in section 2.3. Now the number of attributes under consideration is extended to 10 (Road Function Class, First Harmful Event, Manner of Collision, Relation to Junction, Relation to Roadway, Trafficway Flow, Number of Travel Lanes, Speed Limit, Light Condition and Atmospheric Condition). With the minimum support threshold set to 0.14 and the minimum confidence to 0.3 we obtained a model with a total number of 144 attribute-value pairs (items), 228 itemsets and 1476 rules (Model 4). Further, we used our interactive Association Rule Browser (Fig. 4) to find an essential subset of rules with two specific items ("Daylight", "No Adverse Atmospheric Conditions") in antecedent.

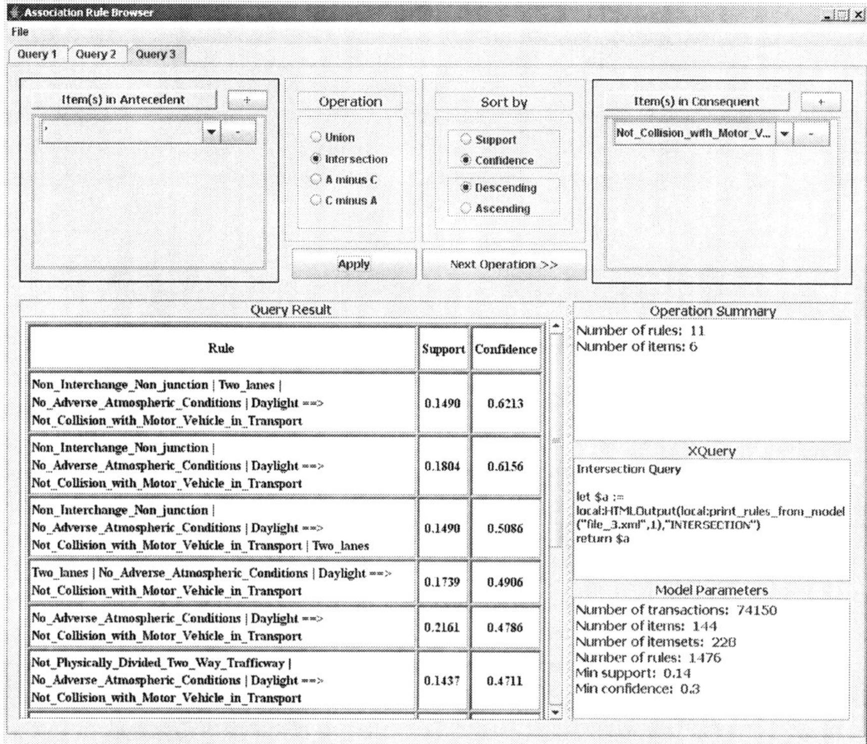

Fig. 4. The interactive Association Rule Browser

The selection of interesting rules was completed in 3 steps detailed in Table. 2.

Table 2. The operations arranged for the example model

Step	Operation	Number of rules
1	Starting with the whole rule base extract rules with "Daylight" in antecedent	411
2	Consider the subset of rules from step 1 and extract rules with "No Adverse Atmospheric Conditions" in antecedent.	78
3	Consider the subset of rules from step 2 and extract rules with "Not Collision with Motor Vehicle in Transport" in consequent.	11

All operations are feasible by the straightforward application of the set-theoretical intersection to the specified sets of rules. After Step 2 was completed, an additional constraint was specified with the aim to focus on rules containing "Not Collision With Motor Vehicle in Transport" in consequent.

Six of the eleven resulting rules (sorted by confidence value) are visible in the Query Result window in Fig. 4. An interesting feature expressed by the rules is the fact, that up to 60 percent of fatal accidents, which occur in daylight and "good weather", happen without colliding with an other motor vehicle.

3.3 Automatic Query Formulation with Genetic Programming

As it is obvious from the preceding sections, the result of mining the association rules depends on the query (sequence of queries) applied to the basic association model. It is an interesting problem, how queries that return valuable rule sets can be automatically developed. We propose a solution where the synthesis of the query is performed by a computer program on the basis of the assessment of the resulting rule sets with respect to a criterion defined by the user. The approach we suggest considers the query as an XQuery program obtained with genetic programming (GP) [14]. Through the mechanisms of GP, the query is modified in order to improve the quality of the selected association rules. GP is closely related to genetic algorithms (GA) [16]. The principal difference is that the representations of the solution to the problem that are mutated and combined are programs rather then encoded strings (chromosomes). The programs are represented in the form of expression trees whose nodes are functions, variables or constants (see Fig. 5). The nodes that have subtrees are nonterminals and they represent functions with the subtrees as their arguments. Variables and constants are always leaves of the tree.

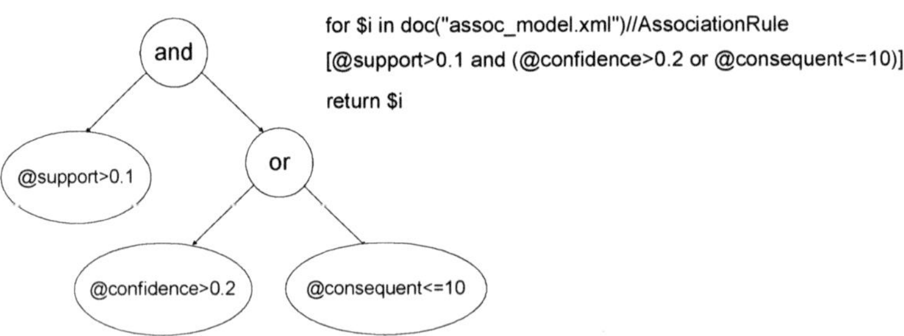

Fig. 5. Sample query expressed by the XQuery statement and its representation in the form of tree

The most relevant functions are the arithmetical and logical operators and the if – then function. Like GA, GP begins with a set of randomly generated idividuals (programs), called the population. Each program is rated by the objective function or (in GA terminology) the fitness function that should return higher values for better programs. In our case better programs determine queries that select better rule sets. Thus the programs are evaluated through the assessment of the resulting rule sets. GP iteratively transforms a population to a new generation. The simulated evolution is based on the Darwinian principle of the reproduction and survival of the fittest. Elements of the next generation are created by recombining (crossover) randomly chosen parts (subtrees) from two selected trees and/or by randomly mutating a randomly chosen part of one selected tree. The probability of being chosen for reproducing is proportional to the fitness score.

The application of GP to query reformulation was studied before in [15] in a different context of information retrieval. The authors considered Boolean query design via relevance feedback incorporated, in part, via user defined measures over a trial set of documents. As such a measure they used precision (the percentage of documents retrieved that were relevant), recall (the percent of relevant documents retrieved), and a linear combination of both. Further works related to the application of GP to information retrieval, especially focused on query definition are reviewed in [6]. To the best of the authors' knowledge applying GP to select valuable sets of association rules from large rule models seems to be a new proposal.

In our introductory experiments we used the following objective (fitness) function to evaluate the rule sets returned by a program (query)

$$ F = \frac{\sum_Q SW_{ai} + \sum_Q SW_{ci}}{\sum_M SW_{ai} + \sum_M SW_{ci}} \cdot \frac{\text{number of significant rules in } Q}{\text{cardinality of } Q} \tag{1} $$

where: M – set of all rules in the model, Q – rule set returned by the query, $SW_{ai} = \sum_{Z_i} A_{ai}$ – weight of the i-th frequent set Z_i in the rule antecedents, $SW_{ci} = \sum_{Z_i} A_{ci}$ – weight of the i-th frequent set Z_i in the rule consequents, A_{ai} (A_{ci}) – weight of the i-th item in the antecedent (consequent) of the rule, with $A_{ci}, A_{ai} \in N \cap <0; 9>$.

The user sets a non-zero weight to the item he or she prefers in the rule. The first factor on the right-hand side of the criterion (1) expresses this preference. The weights equal to zero denote unimportant items for the user. The other nine values have been chosen based on the result in [21] suggesting that in order to reasonably maintain the consistency when deriving priorities from paired comparisons, the number of rates under consideration must be less or equal to nine. The second factor takes into account the percentage of *significant rules* in the whole set of rules returned by the query. Significant rule is the rule containing at least n items with non-zero weight in antecedent and/or consequent. The number n is arbitrarily set for any particular experiment. Experiments have been performed using our GP application prepared in Java. For example, we used the rule Model 4 from section 3.2 and we assigned weights equal to 9 to the items: "No Adverse Atmospheric Conditions", "Daylight" (in antecedents), and "Not Collision with Motor Vehicle in Transport" (in consequents), whereas other items received weights equal to 0. Fig. 6 shows an examplary run of the fitness function characterizing the best individuals in consecutive GP operations performed taking populations of 500 individuals represented by trees of depth lower than 6, and the limit of generations equal to 20. The best query

```
for $i in doc("assoc_model_4.xml")//AssociationRule
[((((((@consequent<=161) and (@consequent<=45))
and ((@confidence<=0.6919) or (@confidence>=0.8087)))
and ((@antecedent<=221) or ((@consequent>251)
or (@consequent>=223)))) and (((@confidence<=0.6941)
or ((@confidence<0.8326) and (@consequent<11)))
or ((@confidence<=0.6435) or (@consequent<=45))))
and (((@antecedent>211) or ((@antecedent<47)
or (@confidence<0.5482))) and ((@confidence<0.5255)
 and (@antecedent>=216)))]
return $i
```

returned 5 rules. All the rules appeared in the resulting set in section 3.2.

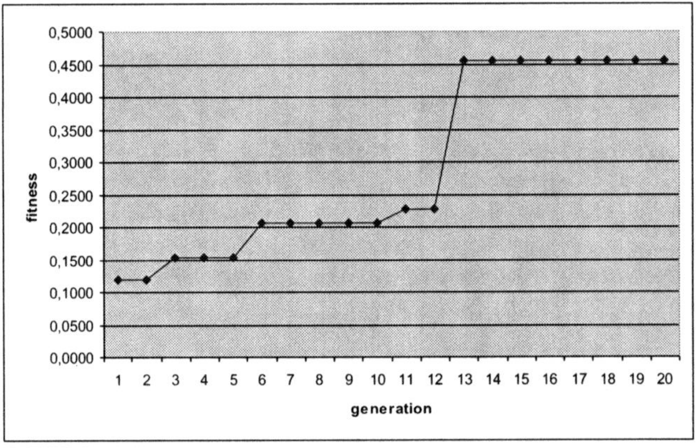

Fig. 6. The best objective (fitness) value vs generations

Assigning maximal weights to some particular items, and zeroes to others, we should expect that the genetic programming method yields a query similar to what we would independently formulate using the distinguished items similarly as in section 3.2. The proposed objective function is more powerful because it makes possible to assign degrees of importance to items we expect to have in resulting association rules. Other criteria e.g. based on distance measures [7] or changing patterns [13] are also possible. In the first case the automatically obtained queries would return clustered (dispersed) rule sets. In changing patterns a kind of support (confidence) sensitivity respectively to particular item(s) of a rule could be examined.

4 Conclusions

In our research we used PMML standard – a language defined within XML. The essential feature of the PMML format is its ability to accumulate the symbolic and portable structure for statistical and data mining models. The informal introduction to PMML in this paper was completed by clarifying an example of a simple association rule model. Further, we proposed and demonstrated a new approach for data mining models investigation – by querying their PMML form. In order to accomplish this, we involved the XQuery language for the association rule model. Three approaches were considered: (1) simple direct querying the PMML structure of the discovered model, (2) interactive browsing the rule base using set-theoretic operations, (3) automatic query formulation with genetic programming. In the first method the discovered model was reduced by specifying new thresholds for support and confidence, the two popular interestingness measures for association rules. The second method used a querying process controlled by the analyst, who iteratively put domain specific constraints into the next query to proceed, taking into account current results. In the last approach we made a new proposal to perform the synthesis of the query with genetic programming on the basis of the assessment of the resulting rule sets with respect to a criterion defined by the user. We suggested such a criterion and presented results of an experiment. The method seems promising. However, formulation of the criterion (criteria) to evaluate the association rule sets returned by automatically generated queries is an interesting problem that remains open for further research.

Acknowledgments

The authors thank the anonymous reviewers whose critical remarks and suggestions significantly affected the final version of the text.

This work was partly supported by the grants N516 026 31/2545 and N516 016 32/1938 from the Polish Ministry of Science and Higher Education.

References

1. Abiteboul S, Buneman P, Suciu D (1999) Data on the Web. From Relations to Semistructured Data and XML. Morgan Kaufmann
2. Agrawal R, Imieliński T, Swami A (1993) Mining Association Rules between Sets of Items in Large Databases. Proc. ACM SIGMOD International Conference on Management of Data, Washington, USA:207–216
3. Baragoin C, Chan R, Gottschalk H, Meyer G, Pereira P, Verhees J (2002) Enhance Your Business Applications. Simple Integration of Advanced Data Mining Functions. IBM Corporation. Available via `http://www.redbooks.ibm.com/redbooks/pdfs/sg246879.pdf`

4. Boag S, Chamberlin D, Fernandez MF, Florescu D, Robie J, Simeon J (2007) XQuery 1.0: An XML Query Language. Available via `http://www.w3.org/TR/xquery/`

5. Bray T, Paoli J, Sperberg-McQueen CE, Maler E, Yergeau F (2004) Extensible Markup Language (XML) 1.0 Third Edition. W3C Recommendation. Available via `http://www.w3.org/TR/2004/REC-xml-20040204/`

6. Cordon O, Herrera-Viedma E, Lopez-Pujalte C, Luque M, Zarco C (2003) A Review on the Application of Evolutionary Computation to Information Retrieval. Int. J. of Approximate Reasoning 34(3):241–264

7. Gupta GK, Strehl A, Ghosh J (1999) Distance Based Clustering of Association Rules. Proc. of ANNIE 1999. ASME Press 9:759–764

8. Han J, Fu Y, Wang W, Koperski K, Zaiane O (1996) DMQL: a data mining query language for relational databases. In SIGMOD'96 Workshop on Research Issues in DMKD, Montreal, Canada

9. Han J, Kamber M (2001) Data Mining. Concepts and Techniques. Morgan Kaufmann

10. Hipp J et al. (2002) Efficient Rule Retrieval and Postponed Restrict Operations for Association Rule Mining. Pacific-Asia Conference on Knowledge Discovery and Data Mining:52–65

11. Imieliński T, Mannila H (1996) A database perspective on knowledge discovery. Communications of the ACM 39(11):58–64

12. Imieliński T, Virmani A (1999) MSQL: A Query Language for Database Mining. Data Mining and Knowledge Discovery 3(2):373–408

13. Imieliński T, Virmani A, Abdulghani A (1999) DMajor – Application Programming Interface for Database Mining. Data Mining and Knowledge Discovery 3(2):347–372

14. Koza JR et al. (2003) Genetic Programming IV Routine Human-Competitive Machine Intelligence. Kluwer Academic Publishers

15. Kraft DH, Petry FE, Bucles BP, Sadasivan T (1994) The Use of Genetic Programming to Build Queries for Information Retrieval. Proc. of the 1994 IEEE World Congress on Computational Intelligence:468–473

16. Michalewicz Z (1996) Genetic Algorithms + Data Structures = Evolution Programs. Springer–Verlag, Berlin Heidelberg

17. Meo R et al. (1996) A New SQL–like Operator for Mining Association Rules. The VLDB Journal:122–133

18. Meo R et al. (1998) A tightlycoupled architecture for data mining. In ICDE'98. IEEE Computer Society Press:316–322

19. Oracle (2003) Oracle9i Data Mining. Oracle Corporation. Available via `http://otn.oracle.com/products/bi/pdf/o9i2dm_ds.pdf`

20. PMML (2003) Predictive Model Markup Language (PMML) Project Page. Available via `http://sourceforge.net/projects/pmml`

21. Saaty TL (1996) The Analytic Hierarchy Process. McGraw Hill, New York 1980, reprinted by RWS Publications, Pittsburgh

22. Tuzhilin A, Liu B (2002) Querying Multiple Sets of Discovered Rules. In Proc. of the 8th ACM SIGKDD Int. Conf. On Knowledge Discovery and Data Mining. SIGKDD'02:52–60

A Universal Tool for Multirobot System Simulation*

Wojciech Turek[1], Robert Marcjan[1], and Krzysztof Cetnarowicz[1]

Institute of Computer Science, AGH University of Science and Technology
wojciech.turek@agh.edu.pl, marcjan@agh.edu.pl, cetnar@agh.edu.pl

Summary. The software tool presented in this paper is an advanced, distributed simulation environment for robot modelling, programming, and control program testing. It offers accurate, three-dimensional simulation of kinematics and dynamics of complex mechanical constructions, and ability of creating control programs in a convenient programming language. The article presents general architecture of the system, its abilities, and the process of distributed simulation, together with some results of experiments and examples of possible applications. The system was developed at the Institute of Computer Science, AGH University of Science and Technology and is used for research on multirobot systems.

1 Introduction

Research into robotics is one of the most interesting challenges of modern science and technology. This wide discipline interconnects number of different issues like material technology, analogue and digital electronics, wireless communication, control software creation, artificial intelligence and many others. Therefore design, building and programming of a new robot are very complex and sophisticated tasks, which usually take years to complete. In addition inevitability of number of prototypes and unforeseen problems during building process makes this kind of research extremely expensive. Some of these issues can be solved by use of simulation tools.

This article describes the RoBOSS Simulation System, which is an advanced software tool for robot modelling, simulation and control programs testing.

2 Applications of Simulation in Robotics

An advanced simulation software could significantly support both building and programming of robots. An accurate model of designed robot could be very

---

*This work was partially supported by the grant MEiN Nr 3 T11C 038 29

W. Turek et al.: *A Universal Tool for Multirobot System Simulation*, Studies in Computational Intelligence (SCI) **102**, 289–303 (2008)
www.springerlink.com

useful in testing construction stability, estimating required actuators abilities and analysing ranges of robot's movement. This sort of test can quickly remove many mistakes from a design.

One of the most important advantages of virtual world is that it can exceed all restrictions of reality. Simulated model of a robot can be built of materials that are very light and resistant, its engines can be extremely powerful, it can also be equipped with fault-proof and perfectly accurate sensors.

Another very important application of robot simulator is control software development. First of all it is far safer to perform tests of control programs in virtual reality because even a small mistake can cause a very serious damages to the robot and its environment. Moreover, created simulation model can be easily simplified or modified to expose every interesting feature of the real robot. The robot can also be placed in any required environment in order to test its abilities.

What is more, many sophisticated control algorithms are very difficult to test on real robots. For instance, genetic algorithms or neuronal networks require long process of learning, that can be performed automatically by a simulation system. In addition time speed in a simulated world can be multiplied in order to reduce total learning time.

One of the most interesting issue concerning robots control algorithms is management of groups of cooperating mobile machines and multiagent systems development [19]. Due to large number of devices required, it is also one of the most expensive disciplines. Therefore usage of proper, high-performance simulator seems to be the only possible solution.

Described applications require the following features of an universal simulation system:

- three-dimensional virtual world
- ability of representing complex mechanical constructions, like mobile and stationary robots
- realistic and reliable simulation of rigid body kinematics and dynamics
- collision detection, friction and bounciness simulation
- support for modelling of typical robots subassemblies (actuators, sensors)
- reasonable units of required parameters and simulation results that allow simple usa of physical and mathematical equations
- convenient programming interface, various languages
- ability of simultaneous simulation of multiply robots
- adjustable accuracy of calculations and high performance

3 Available Tools

There are several software tools, that claim to be robot simulators, but very few of them meet any of the requirements listed above. Despite of careful research, authors were unable of finding a simulator, that would fulfil all of those.

Commercial applications of assembly arms require CAD tools, that are used in assembly lines design process. One of the most advanced tools of this kind is "EASY-ROB" [3], which offers libraries of robots models manufactured by ABB, KUKA, Bosh and Staubli. There are a few simpler tools for simulation of serial-link manipulators, like a "Robotics Toolbox for MATLAB" developed by Peter Corke [4], or "Robotect" designed by Ophirtech [5]. These tools offer kinematics, dynamics and trajectory generation algorithms, but do not support advanced modelling capabilities needed for mobile robots simulation.

There are many software tools designed to develop control programs for a particular model of robot. For instance, at least two simulators ("Khepera Simulator" [7], "Easybot" [8]) were implemented for the Khepera [6] robot, which is one of the most popular research mobile robot. Obviously, none of these programs can be called universal.

Another group of tools, that are designed for particular kinds of robots, are robot soccer simulators. The idea of robots playing soccer was introduced by professor Jong-Hwan Kim, who founded the Federation of Robot-soccer Association (FIRA [17]) on 1995, and became very popular over the last decade. The most advanced robot-soccer simulators are: MiS20 [9], UberSim [10], SimuroSot Simulator [17] and The RobotCup Soccer Simulator [12].

Very few programs offer functionalities of three-dimensional simulation of kinematics and dynamics for sophisticated mechanical constructions. The most advanced of those seems to be the "Webots 5" [15] simulator. It is a commercial tool based on a open-source dynamics library, supporting modelling of stationary, mobile and even flying robots. It simulates various sensors and effectors and other typical robot subassemblies. It is a suitable tool for robot design and simple control program testing, but it cannot be used for simulation of numerous groups of mobile robots due to performance issues.

Two other tools, that can simulate 3D kinematics and dynamics, are an open-source project Player/Stage/Gazebo [14] and Yobotics Simulation Construction Set [13]. The Yabotics is a commercial program which can simulate only one robot at a time. The Gazebo (3D version of the Player/Stage project) does not support robot modelling.

The lack of an open-source tool, that would be suitable for research into multirobot systems encouraged authors to develop a new, universal robot simulator.

4 RoBOSS Simulation System

RoBOSS is an universal, distributed robot simulation system, that supports modelling and testing of any mechanical structure consisting of rigid bodies. It offers accurate, three-dimensional simulation of kinematics and dynamics of robots, and ability of creating control programs in convenient programming language. Distribution of necessary calculations guaranties high performance, that is needed during simulation of complex models.

4.1 A Subject of Simulation

A typical robot consist of the following kind of elements:

- rigid parts
- actuators, that connect parts
- sensors

Therefore a main subject of simulation for RoBOSS system is a mathematical model of a group of three-dimensional rigid objects.

Every moveable rigid objects is considered to be a part of a robot. A robot can consist of one or more connected parts, that can change their position and orientation in a virtual world according to rules defined by connections. Each connection can have a number of actuators attached that can force parts to move against each other. A robot can be equipped with a number of various sensors.

Part of a robot has two sets of features. Those which are constant during simulation process can be called part's properties:

- shape and size
- mass
- friction
- bounciness

A group of features which are variable during simulation are called part's state:

- shape and size
- rotation
- linear velocity
- angular velocity

Each actuator must define two parts that are connected. There can be various type of parts connections, differing in a number of blocked degrees of freedom.

Two parts that are not joined, have six degrees of freedom (movement and rotation around each of three axis). For example, a window attached to a wall has only one degree of freedom, which allows it to rotate around its hinges.

Every non-blocked degree of freedom can be equipped with a proper type of engine, that can force parts to move against each other. Engine can be described by maximum force and velocity that is possible to achieve.

Due to wide variety of sensors, that can be used by robots, it is very difficult to define an universal description format of a sensor. The only common feature is a part that the sensor is connected to. Other properties depend on individual sensor features.

A simulated world is composed of a set of robots and a group of unmoveable rigid shapes, that create a static environment for robots. Environmental shapes are similar to robots parts, that do not have mass. Their state remains constant during simulation process.

A state of a simulated world is a set of robots states. A robots state is a set of its parts states. An initial configuration and one state of simulated world determine a comprehensive information about every simulated object.

Described set of elements and their parameters, that define simulated world (figure 1), is probably insufficient in many cases, therefore it is very important to allow its further development. RoBOSS system stores simulation definition data using the XML language which is a very convenient solution for storing hierarchical data.

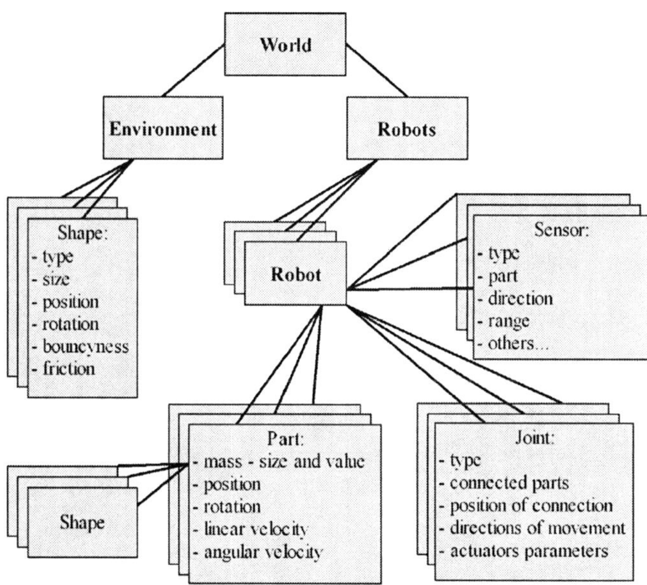

Fig. 1. Simulation subject definition structure

Whole simulated world can be defined by a single tree and stored in a single XML file. There are two basic groups of nodes: robots and environmental objects. Robots definition consists of three types of nodes representing Parts, Joints (connections and actuators) and Sensors. In order to reduce size of the file, every robot definition can be stored in a separated file. This allows user to include many copies of robot without repeating its definition.

Described file format can be easily developed by adding appropriate nodes and attributes to the tree.

4.2 System Architecture

The system is composed of several separated, cooperating programs that communicate using local area network. There are three types of programs: a Controller, Simulators and Clients. It is possible to run all these programs on one computer, however high performance can only be achieved by using a cluster of connected machines. The elements of the system and the connections between them are shown in figure 2.

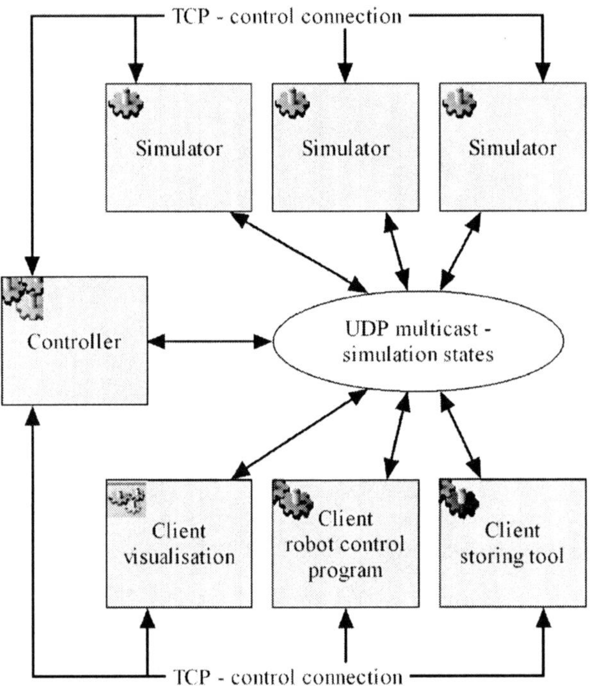

Fig. 2. Architecture of the RoBOSS system

The Controller is the most important element of the system. It is responsible for storing a simulation definition and state, supervising and synchronizing other programs and controlling simulation settings. It allows user to modify many simulation parameters, save or load simulation state or start additional Simulators on remote computers.

Main calculations of simulated robots behaviour are performed by Simulators. Number of Simulators connected to one Controller is not restricted.

The Controller is dynamically distributing computations between available Simulators, so that each of them is equally loaded.

Any other program that is participating in the simulation is considered to be a Client. The most important type of a Client is robot control program, but there can be others, like visualisation or storing tool.

The simulation system is designed to work under the MS Windows operating system. Its components are written in C++ and C#, and reqire .NET Framework 1.1 or newer. The 3D visualisation uses Direct3D technology, it requires DirectX 9.0 libraries.

Communication between programs is based on two network protocols:

- Connection-oriented control protocols
- Connectionless simulation protocol

The first is used for initializing new programs and synchronizing the simulation process. The second is based on multicast UDP protocol and is used to publish a simulation state changes. This solution allows Clients to receive complete and most recent information without additional communication.

4.3 Clients

Programs, that cooperate with the simulation system can be written in any programming language — network interface allows them to communicate with the controller and the simulators. The easiest way of writing a client is to use a communication library, that implements both necessary protocols. The communication library provides high level application programming interface, and can be used by programs written in C++, Managed C++, Visual Basic and C#.

Each client, has access to complete information about the simulation state, therefore tools like visualisation do not require any special protocol or treatment. Clients can alter robots state and behaviour by using a set of orders, which are divided into four levels:

- direct modification of robots position and orientation
- direct modification of robots linear or angular velocity
- adding forces or torques to robots parts
- setting orders for robots actuators

The communication library is also responsible for calculating values returned by sensors. Provided that it has all necessary information, it can successfully perform the task, relieving the Simulators and the network. There are four types of sensors supported:

- laser rangefinder
- odometer
- GPS
- gyrocompass

Implementation of another types of sensors is in progress.

4.4 Simulation Process

The simulation of rigid body system is a continuous process. Each Simulator periodically calculates and publishes state changes of assigned robots that occurred during given, very short time step. Calculation of state changes consists of following steps:

- collision detection, creation of temporary connections between colliding objects
- solving a problem of moving every rigid body

Collision detection algorithm is based on hierarchical detection of bounding objects intersections. Two types of bounding objects, shown in figure 3, are used:

- sphere, that surrounds maximum possible size of robot, regardless of its current state,
- axis aligned box, that is calculated as necessary for a robot and each of its parts.

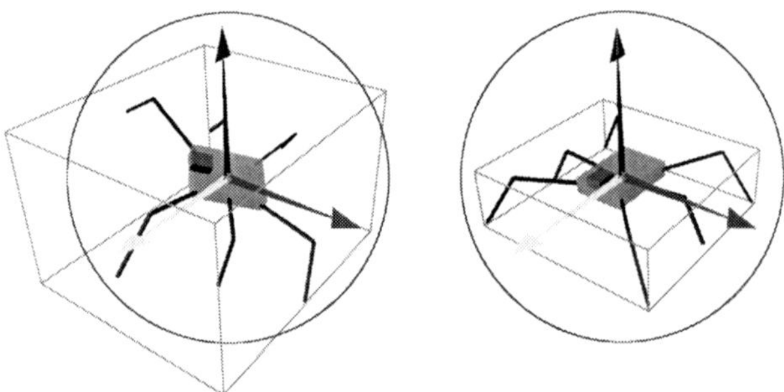

Fig. 3. Axis aligned bounding box and bounding sphere

A radius of the bounding sphere is calculated once, at the beginning of the simulation. It requires analysing of the robot architecture - maximum range of movement of its parts must be estimated. If there is a part in a robot definition, that is not joined to any other part, the radius of the bounding sphere is infinite.

Usage of bounding objects can significantly reduce number of necessary collision tests. Only one comparison is required to determine, if two bounding spheres collide, six comparisons are needed for bounding boxes test. Usually very few pairs of parts remain after this procedure. Shapes associated with those are tested against each other. If a collision is detected, a temporary connection between colliding parts is created, and appropriate reaction forces are added.

Each connection between robots parts (temporary or permanent) is a restriction in a number of degrees of freedom. Every part must be moved according to its velocity and attached forces, without violating any of these restriction, which is a complex computational problem. Main calculations are based on Open Dynamics Engine [16], an open-source library for simulating rigid body dynamics, written by Russell Smith. The library is usually used by games developers, but due to its high accuracy, performance and open-source license it can be successfully used in robot simulation.

All modified states of robots are published using the simulation protocol. Afterwords all orders received from the Controller and Clients are carried out, and another simulation cycle is being started.

4.5 Distribution of Computations

Fast simulation of a numerous group of complex robots exceeds abilities of a single computer. To overcome this problem RoBOSS simulator can perform parallel computations using a cluster of computers connected with a local area network.

Used simulation algorithm requires simultaneous calculations of all temporarily or permanently connected parts. It is safe to assume, that all parts of a single robot are always colliding (they are connected directly or indirectly), therefore a single robot is considered to be an unbreakable unit of computations. The Controller is dynamically distributing simulated robots among Simulators using an algorithm that has two aims:

- balancing of Simulators load
- reduction of duplicated computations

Duplication of computations occurs when colliding robots are assigned to different Simulators. This situation forces both Simulators to calculate state changes of both robots. The algorithm of robots distribution is based on analysis of their position and velocity and prediction of possible collisions. The Controller dynamically divides robots into potentially colliding groups (called "islands"), estimates a computational cost of each island and distributes these island among available Simulators. Performance of computers in the cluster is also taken under consideration by measuring time of computations and confronting it with the estimated cost.

This approach gives the best results for groups of robots that are distributed in a large environment and do not create numerous islands. Maximum number of Simulators that can be efficiently used is restricted by the number of islands - when all robots are colliding (directly or indirectly) the are all assigned to one Simulator. This is the most pessimistic situation, which should not occur, provided that robots are designed to avoid collisions. Even if the pessimistic situation takes place, total performance of the cluster is not worse than the performance of a single computer. In an optimistic situation, time needed by a cluster of n computers for a single simulation step equals $c + \frac{t_1}{n}$, where c is a constant time needed for collision detection and state publication, and t_1 is a time needed by a single Simulator to calculate a state changes in a single simulation step.

4.6 Performance Tests

Number of tests have been carried out in order to verify efficiency of described method of computations distribution. A robot used was a simple model of a car built of four wheels and a body. A controlling algorithm was based on a range-finder sensors that provided information about the closest objects. It allowed the robot to manoeuvre between static obstacles without any collisions.

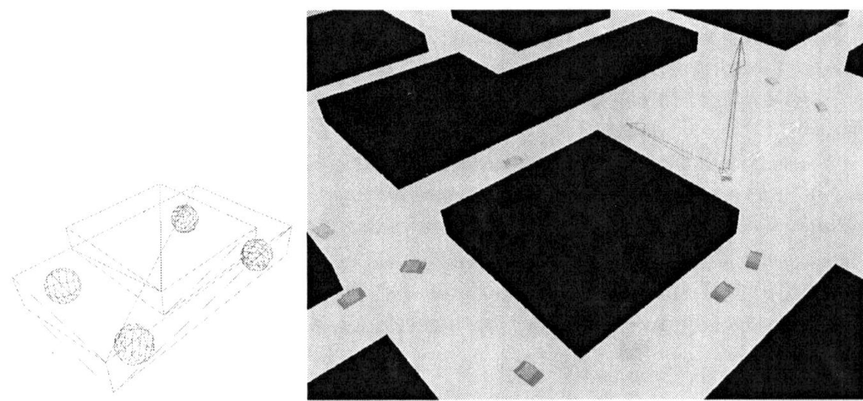

Fig. 4. Model of a car and its environment

A group of fifty robots was situated in a model of a small city, shown in figure 4. Robots were driving around, causing a random number of collisions. The simulation system was running on a cluster of fourteen computers, twelve of them were used for calculations, one for visualization and one for Controller program. The number of islands in the simulated model was very close to the number of robots and much bigger than the number of simulators, therefore each simulator was approximately equally loaded. During the experiment one simulator was detached from the system every 20 seconds.

Table 1. Efficiency of calculations distribution.

Number of simulators	1	2	3	4	5	6	7	8	9	10	11	12
Efficiency of distribution	1	0.95	0.94	0.94	0.73	0.83	0.82	0.67	0.77	0.73	0.71	0.65

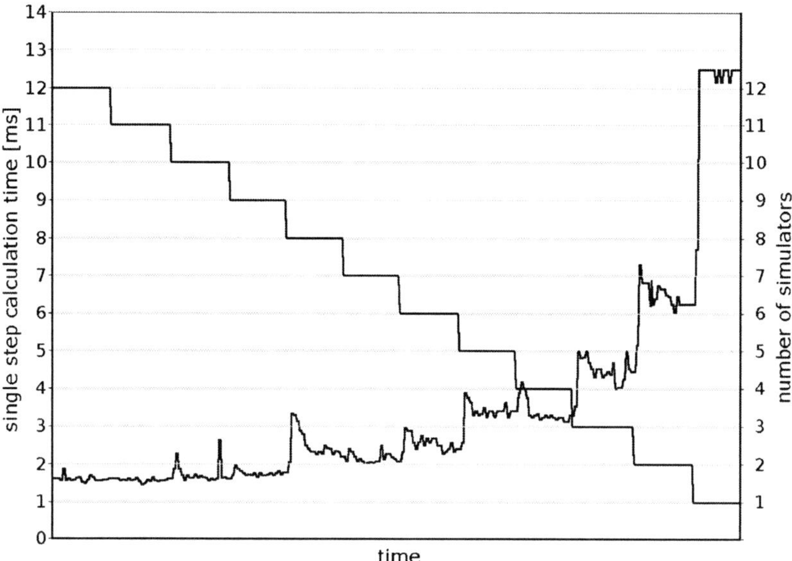

Fig. 5. Relation between number of simulators and the single step calculation time.

The diagram presented in figure 5 shows the relation between the number of simulators connected to the Controller and the time required to calculate a single step of simulation. The peaks of step calculation time, that occur after a simulator is disconnected, show that the time needed by the system to recalculate distribution parameters is very short.

Efficiency of calculations distribution is quite satisfactory - use of a cluster of computers can significantly increase general performance of the simulation system.

5 Examples of Applications

RoBOSS simulation system is capable of simulating behaviour of almost every mechanical construction. Following examples show applications of the system in design and testing of some typical types of robots.

5.1 Industrial Arms

Industrial arms are most common commercial application of robots [2] [18]. They are used in most modern assembly lines, due to reliability, precision, speed and repeatability. These are also the most important parameters that describe abilities of an arm.

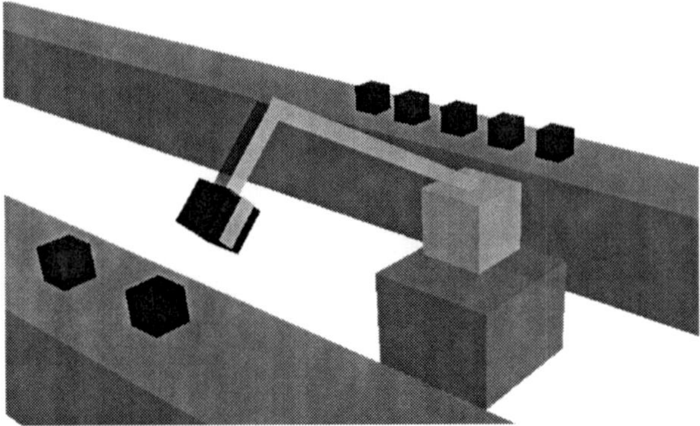

Fig. 6. Model of four-degrees-of-freedom industrial arm

RoBOSS simulator is a suitable tool for design and programming of complex industrial arms. It can be used to estimate necessary number of degrees of freedom, required sizes of parts and powers of actuators. Moreover it allows user to design whole assembly line, predict required space and avoid collisions between arms.

Figure 6 shows a four-degrees-of-freedom arm equipped with a simple gripper. Its task is to move boxes from one conveyor belt to another. The diagram presents changes in position of a gripper controlled by a simple, sensor-based algorithm.

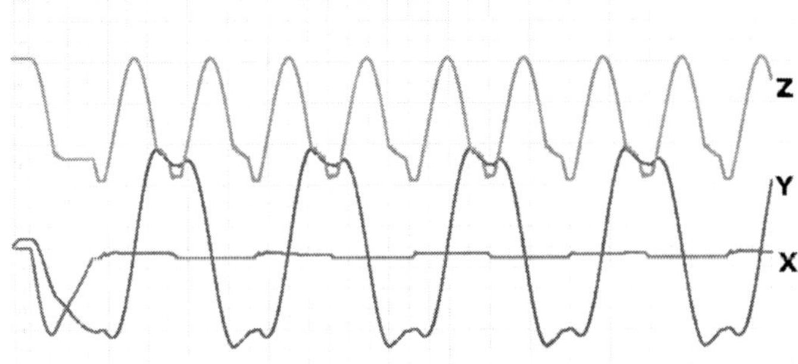

Fig. 7. Tests of accuracy and repeatability

The diagram in figure 7 was created by RoBOSS Analyser program. It can be used to estimate both precision and repeatability of the arm. After a few experiments with control program it was possible to determine all properties and abilities of the robot.

5.2 FIRA MiroSot Robots

Another example of RoBOSS system application is mobile robots modelling. Devices, that were used during experiments are FIRA MiroSot soccer-playing, differential drive robots [1] [17]. These semi-autonomous vehicles are equipped with two engines, a battery, a programmable microcontroller, and a radio receiver. They are only 7.5 centimetres long, and their maximum speed can reach 5 m/s.

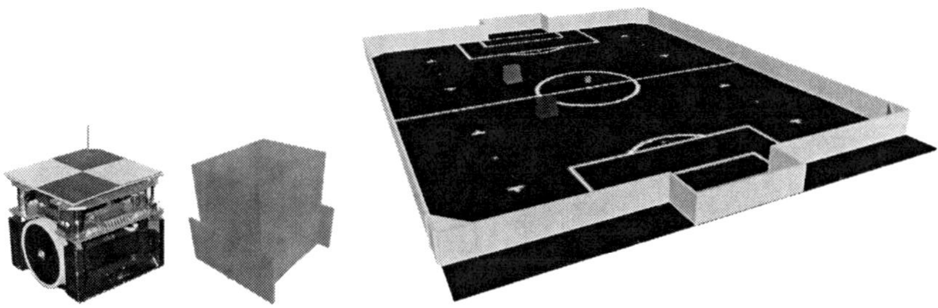

Fig. 8. FIRA MiroSot robot [17], its model, and a simulated pitch

Two values had to be calibrated in order to achieve realistic behaviour of simulated robots:

- power of robots engines
- friction between wheels and a pitch

A number of tests have been carried out and the results show that the difference between simulation and real robots behaviour are similar to those between two real robots. Final version of the model is very useful for control programs development - several groups can work independently because constant access to hardware is no required. Moreover, there is no risk of damaging the robots by using an untested program.

5.3 Spider Robot

One of the most interesting types of mobile robots are legged machines. This type of propulsion is very complicated in design and assembly, but also very flexible - legged creatures can reach places that are inaccessible for wheeled machines. Research into this matter are very popular, but also very expensive. The simulator can be very useful during the design process and may help to avoid some costly mistakes.

The following example presents a model of a Spider robot - six-leg walking machine. The aim of this model is to estimate required powers of actuators and

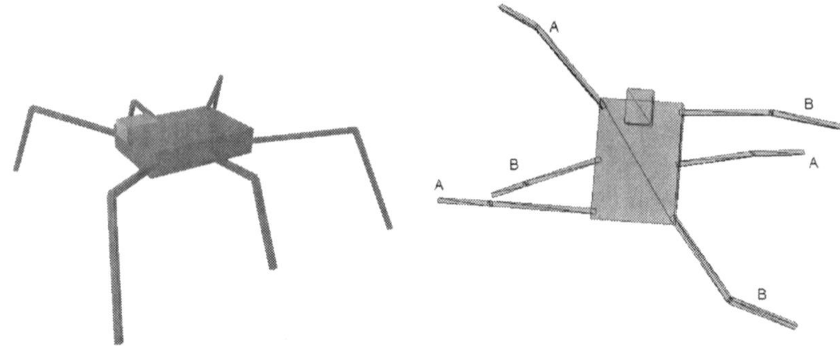

Fig. 9. Spider robot

create a control program, that will allow the spider to move fast and stable. Defined model is 40cm long, each of its legs has three degrees of freedom.

During the movement, the spider stands on three of his legs. When legs marked as A are raised and moved forward, legs B are held low and moved backward. After reaching desired position, functions of legs are switched - spider stands on legs A and moves legs B forward. This simple method required a number of tests to reach an optimal coordination between legs, that would result in a stable movement of the spider. The final effect is quite impressive - the robot is able to move with maximum speed of $2\frac{m}{s}$, when required angular velocity of his legs is smaller than $\frac{0.8\pi}{s}$.

Unfortunately, due to high powers of effectors used in this model, it is impossible to build a real robot - sources of energy and engines would exceed the mass and the size assumed.

6 Conclusions and Further Work

Described simulation system can be a very useful tool in robot design and prototyping. It boost a process of new robot creation and significantly reduces required resources. The RoBOSS system simulates 3D kinematics and dynamics of complex, rigid-body based constructions, supports many types of sensors and actuators and provides a convenient programming interface.

Unlike the other advanced 3D robot simulators, RoBOSS can offer high performance due to distribution of computations and adjustable accuracy of computations. Therefore it can be used in research into complex multirobot systems.

The system is being continuously improved. A camera sensor, simulated radio communication and a new, fully 3D visualization will be added soon. Support tools for storing and analysis of performed simulations will be developed. Moreover new types of real robots controllers will be implemented in the communication library to allow transparent switching between simulation and hardware.

References

1. Kim JH, Vadakkepat P (2000) Multi-agent systems: A survey from the robot soccer perspective. Journal of Intelligent Automation and Soft Computing, vol. 6, r 1
2. Arkin R (1998) Behavior-Based Robotics, MIT Press, ISBN 0-262-01165-4
3. Anton S (2006) EASY-ROB simulator documentation, `http://www.easy-rob.de/`
4. Corke P (1996) A Robotics Toolbox for MATLAB. IEEE Robotics and Automation Magazine, vol. 3, nr 1, pp.24–32
5. Ophirtech (2005) EASY-ROB simulator documentation, `http://www.ophirtech.com/`
6. K-Team Corporation (2006) Khepera II robot specification, `http://k-team.com/robots/khepera/`
7. Michel O (2005) Khepera Simulator documentation, `http://diwww.epfl.ch/lami/team/michel/khep-sim/`
8. Iwe H (2005) Easybot simulator documentation, `http://easybot.htw-dresden.de/`
9. Fikkert W, Dollen H (2005) MiS20 simulator documentation, `http://hmi.ewi.utwente.nl/MiS20/`
10. Weloso M, Browning B, Go J (2005) UberSim simulator documentation, `http://www.cs.cmu.edu/~robosoccer/ubersim/`
11. Iwe H (2005) Easybot simulator documentation, `http://easybot.htw-dresden.de/`
12. The Robocup Soccer Simulator Maintenance Group (2005) RobotCup Soccer Simulator documentation, `http://sserver.sourceforge.net/`
13. Pratt J, Krupp B, Paluska D, Morse C (2005) Yabotics simulator documentation, `http://yobotics.com/simulation/simulation.htm`
14. Gerkey B, Vaughan R, Howard A (2003) The Player/Stage Project: Tools for Multi-Robot and Distributed Sensor Systems. Proceedings of the International Conference on Advanced Robotics (ICAR 2003) Coimbra, Portugal, June 30 – July 3, pp. 317–323
15. Michel O (2005) Webots simulator documentation, `http://cyberboticspc1.epfl.ch/cdrom/common/doc/webots/reference/reference.pdf`
16. Smith R (2005) Open Dynamics Engine, `http://ode.org/` , `http://ode.org/ode-latest-userguide.html`
17. Federation of International Robot-soccer Association (2005) Specification of soccer-playing mobile robots, `http://www.fira.net/`
18. Murphy R (2000) An Introduction to AI Robotics, MIT Press, ISBN 0-262-13383-0
19. Browning B, Bruce J, Bowling M, Veloso M (2005) STP — Skills, tactics and plays for multi-robot control in adversarial environments. IEEE Journal of Control and Systems Engineering, nr I1, pp. 33–52

Advancing Dense Stereo Correspondence with the Infection Algorithm

Francisco Fernández de Vega[1], Gustavo Olague[2], Cynthia B. Pérez[2], and Evelyne Lutton[3]

[1] University of Extremadura `fcofdez@unex.es`
[2] CICESE Research Center `olague@cicese.mx`
[3] INRIA Rocquencourt, Complex Team

Dense stereo matching is a well-know problem of great interest for attaining 3D reconstruction in stereo vision. Although different techniques have been traditionally employed for solving the problem, given its difficulty -the problem is considered as belonging to NP-hard problem class- any improvement is always welcome, both in computing resources required for solving the problem, and also on the time required for attaining the reconstruction. In this chapter, we describe the Infection Algorithm, a new technique that embodies ideas borrowed from epidemic algorithms and also cellular automata. The idea is to present a new strategy inspired by natural processes that demonstrates how cellular automata could be applied for dense stereo matching. The new algorithm is capable of deciding for each of the image pixels to be processed, which of the cellular automata available has to be employed, so that the quality of reconstruction is maximized. Texture, geometry as well as other attributes from images are considered. Some experiments on real stereo pairs are performed, and the results are compared with previous approaches which demonstrate it as reliable for image matching.

1 Introduction

During the last few years, nature-inspired algorithms have proved to be powerful techniques for tackling a wide set of optimization problems: Cellular Automata, Evolutionary Algorithms, Particle Swarm Optimization, Ant Colony Optimization, Inmune Systems, to name just a few, are being applied for efficiently solving difficult real-life problems [1]. Among the problems that are addressed, image processing field provides a number of optimization problems that are still awaiting for new efficient approaches. On the other hand, this kind of problems provides great challenges to artificial life and bioinspired approaches.

F.F. de Vega et al.: *Advancing Dense Stereo Correspondence with the Infection Algorithm*,
Studies in Computational Intelligence (SCI) **102**, 305–324 (2008)
`www.springerlink.com`   © Springer-Verlag Berlin Heidelberg 2008

In this chapter we describe a new algorithm aimed at solving one of these problems, namely 3D reconstruction in stereo vision. Basically, dense stereo matching consists in determining pairs of pixels that projected on at least two images and which belongs to the same physical 3D point [10]. This problem is considered to be one of the most interesting and difficult problems in stereo vision [4,5]. The inherent difficulty of the problem comes from the ambiguities produced when the acquisition process for the stereo pair takes place: geometry, noise, lack of texture, occlusions and saturation increase the difficulty of the problem.

This problem has been one of the main subjects in computer vision research. Although no general solution for the problem exists, researchers are confident with the existence of an algorithm capable of solving the problem. We present a new nature-inspired technique that borrows ideas from cellular automata field, and virus propagation over populations. The *infection algorithm*, that was first described in [14], employes an epidemic automaton that propagates the matched pixels as an infection over the whole image, so that at the end of the process the contents of two images are matched. The algorithm also provides the rendering of 3D information allowing the generation of synthetic images in order to show the same scene from novel viewpoints.

We will show that different epidemic automata can be applied for analysing images, each of them solving the problem with different degrees of accuracy. We also take into account the computing effort required for solving the problem, and consequently the time that can be saved during the process.

This chapter is organised as follows: we first present in Section 2 a description of the correspondence problem. Section 3 describes the main ideas employed for designing the new algorithm. Section 4 presents the basic infection algorithm and the main results obtained. Section 5 shows an improvement on the main algorithm that makes use of evolution and the new results obtained. Finally section 6 presents our conclusions.

2 Problem Statement

Computational Stereo consists of recovering the three dimensional features of a scene from multiple images taken from different viewpoints. One of the main problems in this process -that is known as the correspondence problem or stereo matching- is to find the corresponding points between a pair of images. We consider here that both images have been taken by a moving camera, with the hypothesis that a unique three-dimensional physical point is projected into a unique pair of image points (see Figure 1). The movement between the images is a translation with a small rotation along the x, y and z axes respectively: $T_x = 4.91\ mm$, $T_y = 114.17\ mm$, $T_z = 69.95\ mm$, $R_x = 0.84°$, $R_y = 0.16°$, $R_z = 0.55°$.

Stereo mathching is so complex that usually turns intractable. Several constraints and assumptions are usually exploited for efficiently solving the

Fig. 1. A couple of stereo images taken at the EvoVision Laboratory -left and right stereo images. Occlusion, textureless areas, sensor saturation and optical constraints are problems appearing in the images.

problem. The main assumption used by stereo algorithms is that any single three-dimensional physical point projects to a unique pair of image points in the two observing cameras. If both points are located by the algorithm, then it is possible to reconstruct the three-dimensional scene. Nevertheless, the problem becomes intractable because of several reasons, and there is no close solution to this problem: specularities, lack of textures or occlusions produce ambiguities. For instance, in Figure 1, the wooden house together with the calibration grid appears clearly while the puppet appears only on the right image, and the file organizer behind the box with the children face, appears only on the left image. This kind of problems are examples of the visibility constraint: the visibility of an object from a particular viewpoint could fail if it is occluded by either some part of another object or self-occlusion. If the object lies outside the field-of-view limits of the sensor, the visibility could also fail. Furthermore, matching textureless areas is also an ill-posed problem, since many points of these regions would match almost all points of the corresponding region in the other image of the stereoscopic pair (see for instance the EvoVision house in Figure 1).

Dense stereo disparity is a simplification of the problem in which the pair is considered to be rectified and the images are taken on a linear path with the optical axis perpendicular to the camera displacement. In this way, the problem of matching is evaluated indirectly through a univalued function in disparity space that best describes the shape of the surfaces in the scene. The performance of the infection algorithm has been evaluated according to the proposed methodology by Scharstein and Szeliski, in which the quality is measured by percentages of bad matching. The error is computed as the percentage of pixels far from the true disparity by more than 1 pixel. For a comprehensive discussion on the problem, we refer the reader to the survey by Scharstein and Szeliski [15].

We perform our experiments on the benchmark Middlebury database. This database includes 4 stereo pairs, named *Tsukuba, Sawtooth, Venus* and *Map*. It is important to mention that the Middlebury test is limited to specific camera geometries and closely spaced viewpoints.

2.1 Correlation Function and Main Constraints

Researchers have traditionally employed the *correlation function* for measuring similarities between image windows of fixed size. The input for this function is a stereo pair of images, I_l (left) and I_r (right). Correlation consists of a search process that provides a measure of correlation to identify corresponding pixels on both images. A similarity criterion is maximized within a search region.

It works in the following way: Let p_l with image coordinates (x, y), and p_r with image coordinates (x', y'), be pixels in the left and right image, $2W + 1$ the width (in pixels) of the correlation window, $\overline{(I_l(x, y))}$ and $\overline{(I_r(x', y'))}$ are the mean values of the images in the windows centered on p_l and p_r, $R(p_l)$ the search region in the right image associated with p_l, and $\phi(I_l, I_r)$ a function of both image windows. The ϕ function is defined as the Zero-mean Normalized Cross-Correlation (ZNCC) in order to match the contents of both images.

$$\phi(I_l, I_r) = \frac{\sum_{i,j\in[-W,W]}[AB]}{\sqrt{\sum_{i,j\in[-W,W]}A^2 \sum_{i,j\in[-W,W]}B^2]}}$$
$$A = (I_l(x + i, y + j) - \overline{I_l(x, y)})$$
$$B = (I_r(x' + i, y' + j) - \overline{I_r(x', y')})$$

$$(1)$$

2.2 Main Constraints

Stereo matching has many complex aspects that turns intractable the problem. In order to solve the problem a number of constraints and assumptions are exploited which take into account occlusions, lack of texture, saturation, as well as field-of-view. In this way, we have considered the following constraints for minimizing false matches:

- Epipolar Geometry: Given a feature point in the left image, the corresponding point in the right image must lie on the corresponding epipolar line. This constraints transforms the two-dimensional search space to one-dimesional space. This constraint has demonstrated to be reliable once epipolar geometry is known.
- Ordering: if $p_l \leftrightarrow p_r$ and $p_l' \leftrightarrow p_r'$ and if p_l is to the left of p_r then p_l' should also be to the left of p_r' and vice versa. This means that the ordering of features is preserved between both images.
- Orientation: The orientation of the epipolar line that characterizes pixels are considered together with gray values of pixels.

3 The Matching Problem as an Infection Process

The algorithm presented here employs the concept of natural viruses in the search of correspondences on real stereo images. In order to do so, we employ cellular automata and epidemic algorithms for expanding information through the image. This section reviews concepts related with both cellular automata and epidemic algorithms.

3.1 Artificial Epidemics

The infection algorithm has considered two main aspects. When a scene is observed, not everything in front of the observer is processed simultaneously. Instead of, the observer focuses first his attention to some parts which keep the interest on the scene. We thus believe that many of the attention process is applied to only some parts, while guessing the rest of the scene by employing the information extracted from the important parts. This process of guessing some information based on local regions could be implemented by spreading the image matching and model building.

Spreading information over pixels from an image can be considered similar to the invasion of a disease into a population. This phenomenon that frequently appears in nature, has been a concern for human beings for centuries, and has been studied during the last fifty years. The goal has focused on developing ideas about how a given disease spreads into a population, and if possible, to decide how to vaccinate people for stopping the propagation of the disease.

Probably, the first attemp to formulate a mathematical model for disease propagation is due to Lowell Reed and Wade Hapton in the 1920's [6], and the model was called SIR (Susceptible/Infective/Recovered) model. Within this model, the population is divided into three classes, according to their status in relation to the disease:

- Susceptible: The individual is free of the disease, but can catch it.
- Infective: The individual features the disease and can transmit it to others.
- Recovered: The individual has recovered from the disease and cannot longer pass it on.

The disease spreading is considered a probabilistic process: some probability values per unit of time are employed to calculate transitions from a given state to another one. On the other hand, some contact patterns among individuals from the population are required for modeling the phenomenon.

The original model was revised later, and several researchers have developed exact solutions for this timely called *epidemic models* [12].

Epidemic models have been applied to a series of problems, including the study of behaviors in computer networks [11], [9]. More recently, these models have been also employed for describing interactions through social, technological and biological networks. The new approach, called *small-world networks*, studied the diffusion of information on a given network featuring a given set of properties [7].

Epidemic algorithms can be also studied from a different point of view: given that the models are employing transition rules that each individual may feature, and that some authors have studied and modelled disease spread by means of cellular automata [8]. We describe next the *infection algorithm* that makes use of both concepts, *epidemic models* and *cellular automata.*

3.2 Cellular Automata

Cellular automata (CA) are a class of spatially and temporally discrete mathematical system, that exhibit local interaction and usually synchronous dynamical evolution [2]. They have been succesfully employed for modelling a wide variety of physical systems featuring cooperative behavior [3]. CA consititute a very interesting approach to carry out realistic high performance simulations, because of its intrinsic parallel nature.

When CA are employed, space is represented by a discrete lattice of cells, so that each of the cells concurrently interacts with its neighborhood according to a defined vicinity. Some transition rules define this interaction, and the set of rules must be appropriately selected for the problem that is modelled by means of the CA. CA are said to exhibit emergent behavior, because they feature global responses which emerge from the local behavior defined by local transition rules. This emergent phenomena is one of the best-known features of CA.

As we will describe later, we employ CA and transition rules to determine state of individual pixels from image. The cellular automaton will perform image matching, so that at the end of the process, corresponding pixels from two images will be found. The cellular automaton will make use of some external information, such as the constraints previously described, and correlation information, which will be computed from images.

Therefore, by joining the basics of epidemic algorithms, as well as local interaction of cells within CA, we will describe next the infection algorithm which can be regarded as a new approach to the problem of image matching.

4 The Infection Algorithm

The infection algorithm attempts to find correspondences on real stereo images. Although other algorithms exists aiming at the same goal, the new algorithm try to perform the same process of guessing the maximum number of correspondences, while it improves the quality of matching.

We model the matching process as an epidemic process based on transition rules that are applied to pixels distributed along a 2D lattice. The automaton makes use of emergent behavior, by means of simple rules defining local interaction within pixels in a given vicinity. These rules produce a global behavior that emerges from the image corresponding process. Thus, let us define some notations:

Definition 1. Epidemic Cellular Automata: Our epidemic cellular automaton can be formally introduced as a quadruple $E = (S, d, N, f)$ where $S = 5$ is a finite set composed of 4 states and the wild card $(*)$, $d = 2$ a positive integer, $N \subset Z^d$ a finite set, and $f_i : S^N \rightarrow S$ an arbitrary set of (local) functions, where $i = \{1, \ldots, 14\}$. The global function $G_f : S^L \rightarrow S^L$ is defined by $G_f(c)_v = f(c_v + N)$. Also, it is useful to mention that S is defined by the following sets:

- $S = \{\alpha_1, \varphi_2, \beta_3, \varepsilon_0, *\}$ a finite alphabet,
- $S_f = \{\alpha_1, \beta_3\}$ is the set of final output states,
- $S_0 = \{\varepsilon_0\}$ is called the initial input state.

Our epidemic cellular automaton has 4 states that are defined as follows:

- Let α_1 be the Explored (Sick) state, that represents the cells that have been infected by the virus (it refers to the pixels that have been computed in order to find their matches).
- ε_0 be the Not-Explored (Healthy) state, that represents the cells which have not been infected by the virus (it refers to the pixels which remain in the initial state).
- β_3 be the Automatically Allocated (Immune) state, that represents the cells which cannot be infected by the virus. This state represents the cells which are immune to the disease (it refers to the pixels which have been confirmed by the algorithm in order to automatically allocate a pixel match).
- φ_2 be the Proposed (Infected) state, that represents the cells which have acquired the virus with a probability of recovering from the disease (it refers to the pixels which have been guessed by the algorithm in order to decide later the better match based on local information).

Our algorithm employes several lattices around the whole image. Each lattice contains the state of each studied pixels during the whole run. Thus, each of the lattices contains a pixel in the center of a small window placed around an image neighborhood. Each lattice is of size 7x7 with 9 closed neighbors and 16 external ones. At the beginning of the process some cells from the automaton, pixels from the image, are proposed as seeds. The seeds act like infection nucleous, which try to expand the virus by surronding neighboring cells. These seeds can be considered as the *points of interest* that are usually considered in images when the 3D reconstruction problem is tackled [15]. Actually, seeds are selected within the infection algorithm by also analysing the points of interest.

In this way, the infection algorithm evaluates every cell from the lattice, and updates their values by applying the CA rules, by taking into account the states of the cell and the considered neighborhood.

In the infection algorithm, we observe that healthy cells have not been affected by the virus -nothing about the state of the pixel was decided yet. Infected individuals are cells which have received the virus but haven't developed the disease -we have controversial information about the pixel. Sick cells are the ones for which it was impossible to recover from the disease -and therefore the value of the pixel had to be effectively computed. Finally immune cells, are the ones that although received the virus -we have information from neighbors- were capable of avoiding the disease -all the information from neighbors were in agreement and we could decide the value without computing it. In all the cases, the transition from a given state to another one is performed according to the set of rules that will be described later.

The algorithm works first by calculating the projection matrix and the fundamental matrix. We consider calibrated systems using a calibration grid from which we calculate a set of correspondences with the high-accurate corner detector described in [13]. The information obtained is used to calibrate rigorously the stereo system. We have to mention that because the correspondence problem is focused in a calibrated system using real stereo images, we work now for simplicity with a scene that is rigid: the camera is moving but not the scence.

Each cell within the cellular automaton contains: the cell state, their corresponding coordinates, and the angle of the epipolar line.

The set of rules are included in a file, which allows to easily exchange the behavior of the algorithm by simply changing the rules within the file. When a rule is to be applied to a pixel, it calculates the epipolar line that corresponds to the studied pixel using the fundamental matrix information. The correlation window is defined and centered along the epipolar line when the search process is started. The correlation is the main measurement that is used to identify which pixel on the right image correspond to the studied pixel of the left image. In the precise time when the central cell is automatically allocated the algorithm verifies if the angle of the epipolar line and the orientation constraints on the explored neighborhood have the same values with respect to the central cell. As a result, the algorithm saves the coordinates of the corresponding pixel, otherwise the algorithm calculates the pixel. On the other hand, when the central cell has been proposed we only mark the cell as infected. Finally, once the search is finished, the algorithm saves a synthetic image, which is the projection of the scene between the two original images.

The main steps of the infection algorithm could be summarized as follows:

1. First, all pixes in the photographs are initiated to the state Healthy.
2. Then, we put the seeds in the cellular automata by extracting the pixels of maximum interest from the photographs. Given that the value of these pixels have been really computed, we consider them as sick cells.
3. We apply the transition rules to any pixel in the photograph -cells in the CA. Only the pixels whose state is Healthy or Infected can change their state when rules are applied.

4. Finally, if any pixel in the photograph is in a state different from Immune, or sick, go to step 3; otherwise, all the pixels have been computed or succesfully guessed, and the process finishes.

Figure 2 shows the application of the infection algorithm to a given pixel from a couple of stereo images. It shows also five lattices that we have used in the evolutionary infection algorithm. The first two lattices correspond to the images acquired by the stereo cameras. The third lattice is used by the epidemic cellular automaton in order to process the information. The fourth lattice corresponds to the reprojected image, while the fifth lattice (Canny image) is used as a database in which we save information related to contours and texture. In this work, we are interested in providing a quantitative result to measure the benefit of using the infection algorithm. We decide to apply our method to the problem of dense two-frame stereo matching.

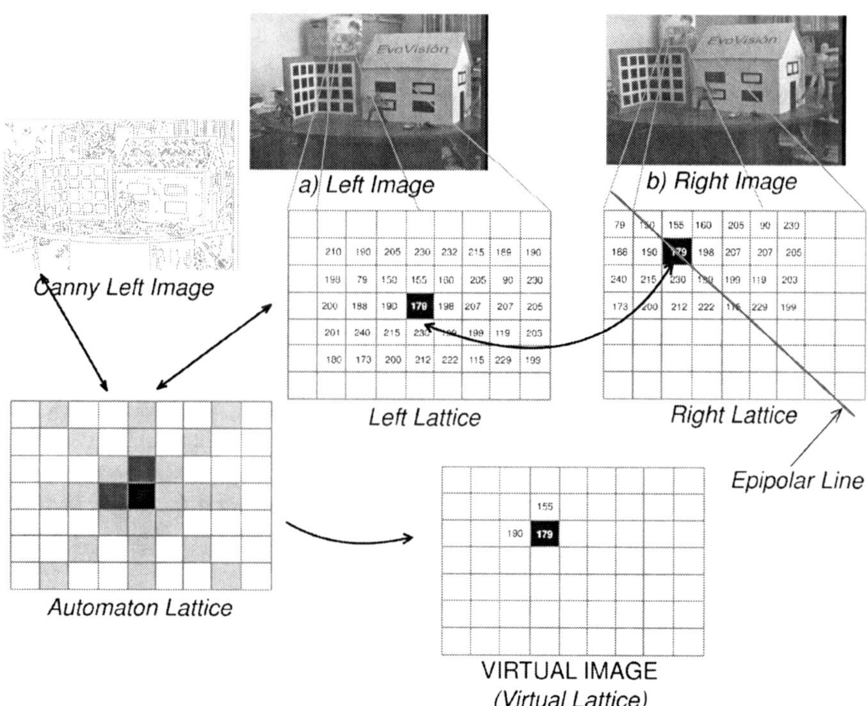

Fig. 2. The infection algorithm. A cellular automaton relates each pixel in the left image to the corresponding one on the right image. Canny and virtual images during the correspondence process are shown. The blue color represents the Sick (Explored) state and the black color represents the Healthy (Not-Explored) state.

Figure 3 shows two transition graphs that were encoded by an expert, and included within the main infection algorithm. As we will see later, each of the graphs combined with the main algorithm provide different accuracy results and fast labellings. Figures 4 and 5, describe the basic transition rules applied by the cellular automata within the considered neighborhood.

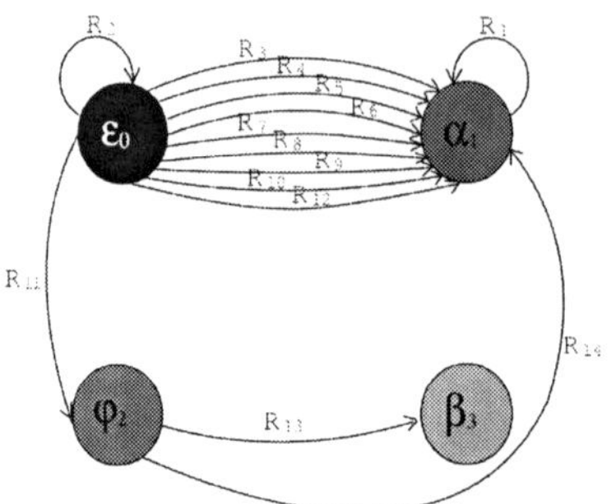
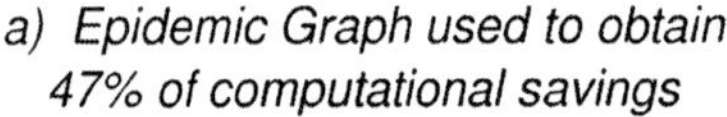

a) *Epidemic Graph used to obtain 47% of computational savings*

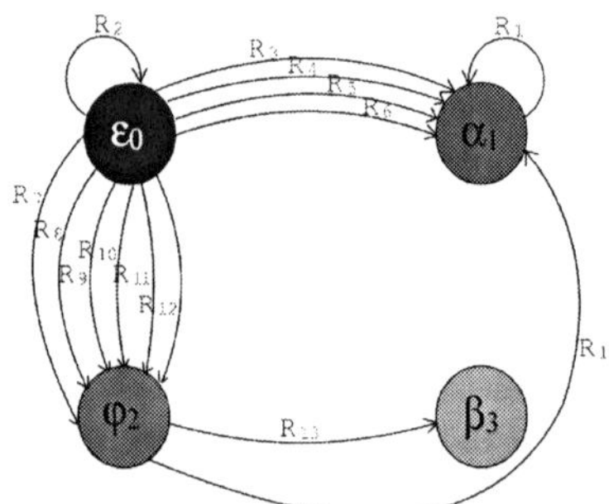

b) *Epidemic Graph used to obtain 99% of computational savings*

Fig. 3. Epidemic graphs designed by an expert, and employed within the infection algorithm.

4.1 Results Obtained by the Infection Algorithm

The infection algorithm was implemented under the Linux operating system on an Intel Pentium 4 at 2.0Ghz with 256Mb of RAM. We have used libraries programmed in C++, designed specially for computer vision, called VxL (Vision x Libraries). Figure 6 shows the results that were obtained when the algorithm was applied to the couple of images shown in figure 1. The final two images compare the result obtained by our algorithm with those obtained with an exhaustive search. We saved 47 % of calculation while also obtained slightly better results than a exhaustive search.

Similarly, figure 7 shows the results obtained with different transition graphs. Quality of results depends on the computing savings performed by each of the graphs. It would be of interest to find a balance between the quality of results and computing savings. Given that different areas of the image are easier while other more difficult, in the next section we will describe how different graphs can be considered simultaneously for improving the quality of results.

5 Evolving Infection

As described above, different results were obtained when applying each of the different set of rules, varying from higher quality/more computing time to

R_1: CC(α_1) ---> ACCION(α_1)

R_2: CC(ε_0) ^LU(ε_0) ^CU(ε_0) ^RU(ε_0) ^LR(ε_0) ^RR(ε_0) ^LD(ε_0) ^CD(ε_0) ^RD(ε_0) ---> ACCION(ε_0)

R_3: CC(ε_0) ^LR(ε_0) ^RR(ε_0) ^LD(ε_0) ^CD(ε_0) ^RD(α_1) ---> ACCION(α_1)

R_4: CC(ε_0) ^CU(ε_0) ^RU(α_1) ^RR(α_1) ^CD(ε_0) ---> ACCION(α_1)

R_5: CC(ε_0) ^LU(α_1) ^CU(ε_0) ^LR(α_1) ^CD(ε_0) ---> ACCION(α_1)

R_6: CC(ε_0) ^LU(α_1) ^CU(α_1) ^RU(α_1) ^LR(ε_0) ^RR(ε_0) ---> ACCION(α_1)

R_7: CC(ε_0) ^EXP(3) ^EXT(3) ---> ACCION(α_1)

R_8: CC(ε_0) ^EXP(3) ^AUT(3) ---> ACCION(α_1)

R_9: CC(ε_0) ^CU(ε_0) ^LR(α_1) ^RR(ε_0) ^LD(α_1) ^CD(φ_2) ---> ACCION(α_1)

R_{10}: CC(ε_0) ^CU(ε_0) ^LR(ε_0) ^RR(α_1) ^CD(φ_2) ^RD(α_1) ---> ACCION(α_1)

R_{11}: CC(ε_0) ^EXP(3) ---> ACCION(φ_2)

R_{12}: CC(ε_0) ^PROP(3) ---> ACCION(α_1)

R_{13}: CC(φ_2) ^EXP(3) ---> ACCION(β_3)

R_{14}: CC(φ_2) ^PROP(3) ---> ACCION(α_1)

LU( ), CU( ), RU( ), LR( ), RR( ), LD( ), CD( ) and RD( ): These functions evaluate the corresponding location in the neighborhood.

ACCION( ): This function provides the output state of the rule.

EXP( ): This function represents the number of explored pixels in the close neighborhood.

PROP( ): This function represents the number of proposed pixels in the close neighborhood.

AUT( ): This function represents the number of automatically allocated pixels in the close and external neighborhoods.

EXT( ): This function represents the number of automatically allocated pixels in the external neighborhood.

Fig. 4. This table summarizes the fourteen rules that we have used in the infection algorithm.

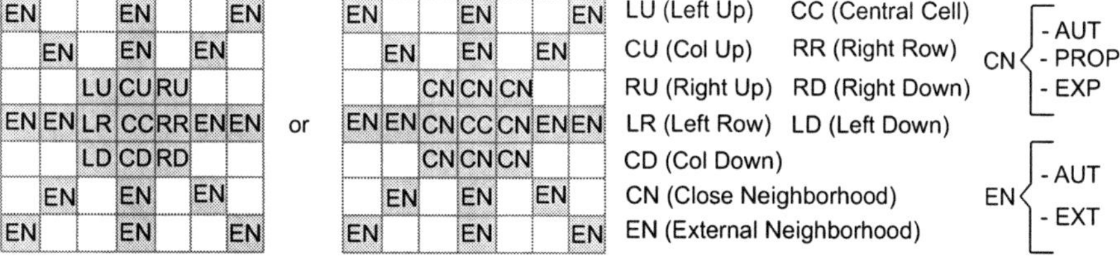

Fig. 5. This figure shows the layout of the neighborhood used by the cellular automata.

lower quality/less computing time. Four different epidemic cellular automata where employed with results varying from 47 % of effort saving to 99%. But it would be of interest to have the possibility of combining several set of rules, different CAs, so that we could combine the benefit of each of the automaton. Hence, the algorithm not only finds a match within the stereo pair, but it provides an efficient and general process using geometric and texture information. The new algorithm would be more general, combining

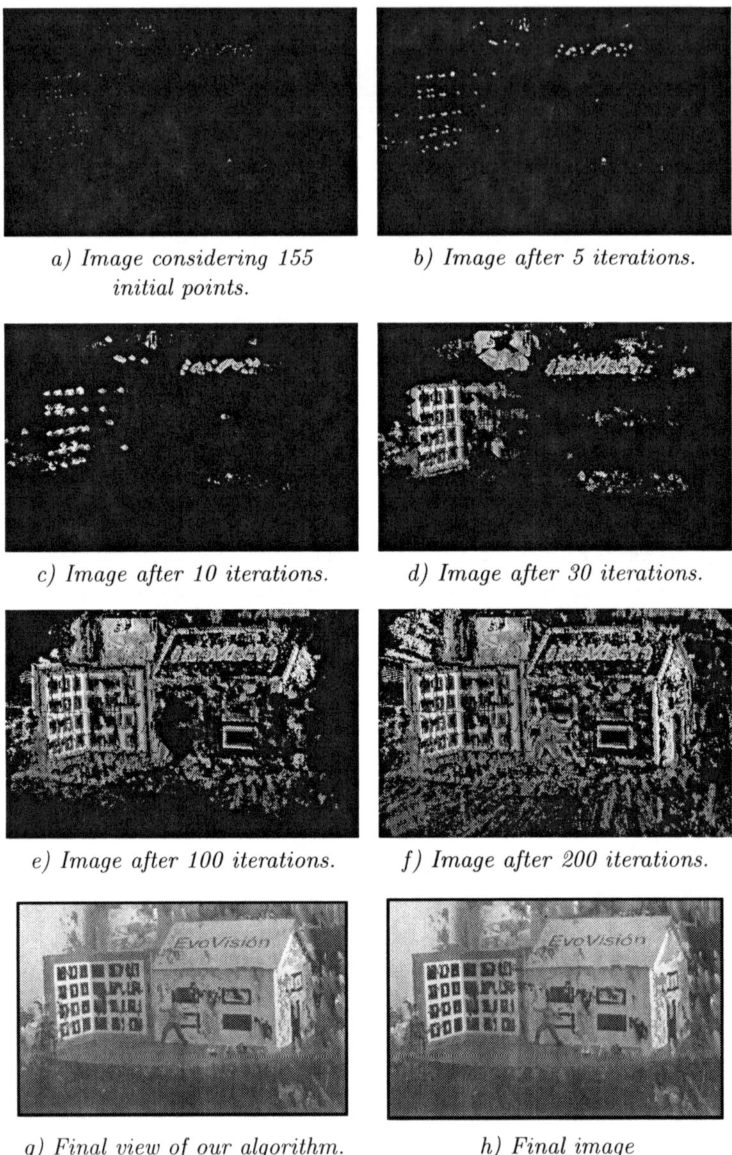

a) Image considering 155 initial points.

b) Image after 5 iterations.

c) Image after 10 iterations.

d) Image after 30 iterations.

e) Image after 100 iterations.

f) Image after 200 iterations.

g) Final view of our algorithm.

h) Final image with an exhaustive search.

Fig. 6. These 8 images show the evolution of our algorithm within a synthetic view. The new image represents a new viewpoint between the two original images. The last two images show the final step of our algorithm. We apply a median filter to achieve the new synthetic image g), as well as to obtain image h), which is product of an exhaustive search. We compared images g) and h): the quality is slightly better on g), while saving 47% of calculations.

a) Final view saving 47 %
of calculations.

b) Final view saving 78 %
of calculations.

c) Final view saving 99 %
of calculations.

d) Final image
with an exhaustive search.

Fig. 7. Results obtained in different experiments, each of them employing a different transition graph and rule set.

the best of each partial image, and could be used with any pair of images with little additional effort to adapt it.

The new version of the algorithm attempts to provide a balance between the exploration and exploitation of the matching process. It employes two set of rules, because each one provides a particular characteristic from the exploration and exploitation standpoint. The 47% epidemic cellular automaton, we called A, provides a particular strategy for working on areas where matching is easier to find. On the other hand, the 99% epidemic cellular automaton, called B, provides a strategy for working on areas where the matching process is hard to achieve. Figure 8 shows a summary of the Evolutionary Infection algorithm.

Figure 9 shows a set of experiments where the evolutionary Infection Algorithm (Figure9d) was compared with the previous version of the Infection Algorithm (Figure9a,b,c). Figure9a is the result of obtaining 47% of computational effort savings, while Figure 9b is the result of obtaining 70% of computational effort savings and Figure 9c shows the result to obtain 99% of computational effort savings. Figure 9d presents the result that we obtain with the new algorithm. Clearly, the final image shows how the algorithm

```
1. BEGIN
2. Calculate Projection Matrices MM1 and MM2 of the Left and Right image.
3. Calculate Fundamental Matrix (F) Left to Right.
4. Coding of rules matrices 47% and 99%.
5. Process the left  image with the canny edge detector.
6. Iniatiate epidemic automata to Healthy(Not-Explored)  state.
7. Corner detection of the left image with O&H or K&R detectors.
8. Obtain the left image with canny edge detector.
9. While the number of Immune (Automatic) and Sick (Explored) states be
     different at time t and t+1 step.
10.DO
11.     FOR col=6 to N
12.         FOR row=6 to M
13.             neighbors=countKindNeighbor(row,col)
14.             pos=statesPosition(row,col)
15.             stateCell=automata[row][col].cell
16.             contEdge=findEdgeProbability(row,col)
17.             IF(contEdge>umbral)
18.                 action=findRule(stateCell,neighbors,pos,Mrules47)
19.             ELSE
20.                 action=findRule(stateCell,neighbors,pos,Mrules99)
21.             ENDIF
22.             SWITCH(action)
23.                 1: EXPLORED
24.                 2: PROPOSED
25.                 3: Analyse the inclination of the epipolar line
26.                     IF the inclination = to the inclination of at least three
                            neighbors
27.                         AUTOMATICALLY ALLOCATED
28.                     ELSE
29.                         EXPLORED
30.                     ENDIF
31.             ENDSWITCH
32.         ENDFOR
33.     ENDFOR
34. ENDDO
35. Virtual image obtained.
36. END
```

Fig. 8. Pseudo-code for the evolutionary infection algorithm.

combines both epidemic cellular automata. We observe that the geometry is preserved with a nice texture reconstruction. We also observe that the new algorithm spends about the same time employed by the 70% epidemic cellular automaton with a slightly better texture result.

a) Final view with 47% savings.

b) Final view with 70% savings.

c) Final view with 99% savings.

d) Final view with evolution of 47% and 99% epidemic automata.

Fig. 9. Results of different experiments in which the rules were changed to contrast the epidemic cellular automata.

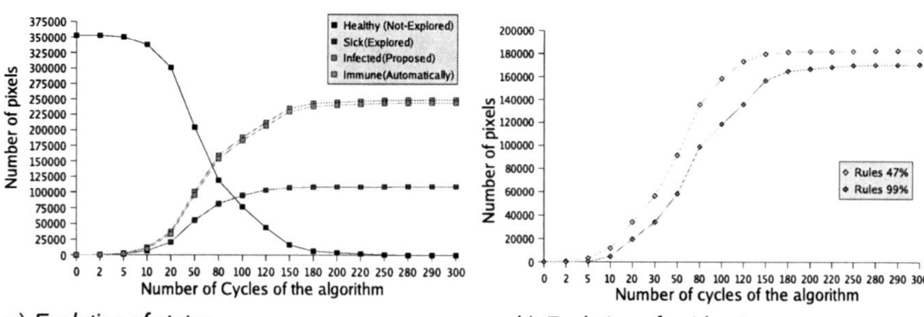

a) Evolution of states.

b) Evolution of epidemic automata.

Fig. 10. Evolution of the epidemic cellular automata to solve the dense correspondence matching.

Figure 10a shows the behavior of the evolutionary infection algorithm that corresponds to the final results of Figure9d. On the other hand, Figure 10b describes the evolution of the epidemic automata. Finally, we decided to test the Evolutionary Infection Algorithm with the standard test used in the computer vision community [15]. Test data along with ground truth is available at authors website that can be used as test bed for quantitative evaluation and comparison of different stereo algorithms. We employ grayscale version of the images, even if this represent a drawback with respect to the final result.

Figure 11 shows on the first colum the four left images belonging to the benchmark set, as well as the computed disparity maps that were obtained with our algorithm (we apply 0% computing savings for obtaining the best map). The results are comparable to other algorithms that make similar assumptions: grayscales images, windows based approach, and pixel resolution [16–18]. In fact, the infection algorithm has been recently included in the Middlebury database. In order to further improve the results, we enhanced the quality of images with an interpolation approach [19].

According to Table 1 these statistics are collected for all unoccluded image pixels (shown in column *all*), for all unoccluded pixels in the untextured regions (shown in column *untex.*), and finally for all unoccluded image pixels close to a disparity discontinuity (shown in column *disc.*). In this way, we obtain the new disparity images shown in Figure 12 together with the ground truth. These results show that the same algorithm could be ameliorated if the resolution of the original images is improved. We provide the results to illustrate the quality of the results that can be achieved with the infection algorithm. However, the infection algorithm was realized to explore the field of artificial life using the correspondence problem. Therefore, the final judgment should be made also from the standpoint of the ALife community. In the near future, we expect to use the Evolutionary Infection Algorithm in the search of novel vantage viewpoints.

Table 1. Our result on the Middlebury database. Note that the *untex.* column of the *Map* image is missing because this image does not have untextured regions.

Infection Algorithm	Tsukuba			Sawtooth		
	all	untex.	disc.	all	untex.	disc.
Original	8.90(37)	7.64(37)	42.48(40)	5.79(39)	4.87(39)	34.20(40)
1st Interp.	7.95(36)	8.55(37)	30.24(38)	3.59(34)	1.31(30)	24.24(37)

	Venus			Map		
	all	untex.	disc.	all		disc.
Original	5.33(34)	6.60(32)	41.30(39)	3.33(37)		32.43(39)
1st Interp.	4.41(34)	5.48(32)	32.94(38)	1.42(25)		17.57(36)

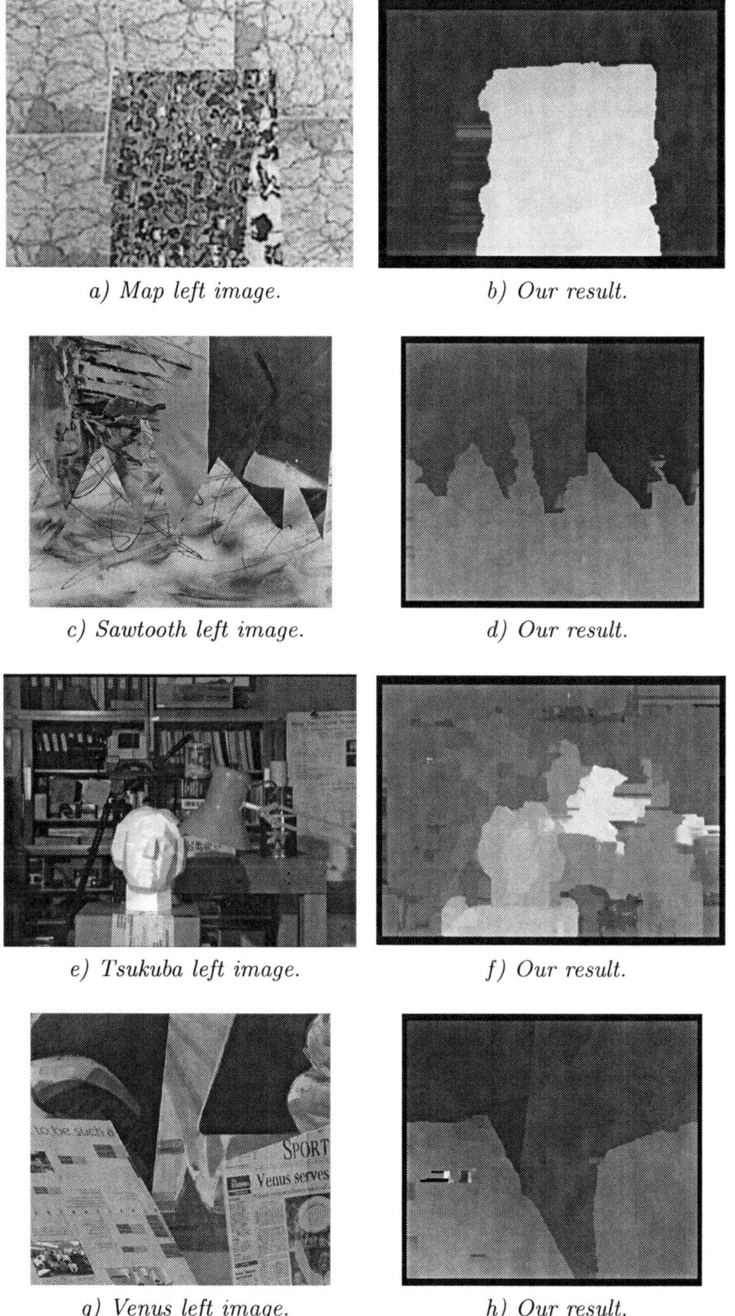

a) Map left image.

b) Our result.

c) Sawtooth left image.

d) Our result.

e) Tsukuba left image.

f) Our result.

g) Venus left image.

h) Our result.

Fig. 11. These four pair of images were obtained from the Middlebury stereo matching web page.

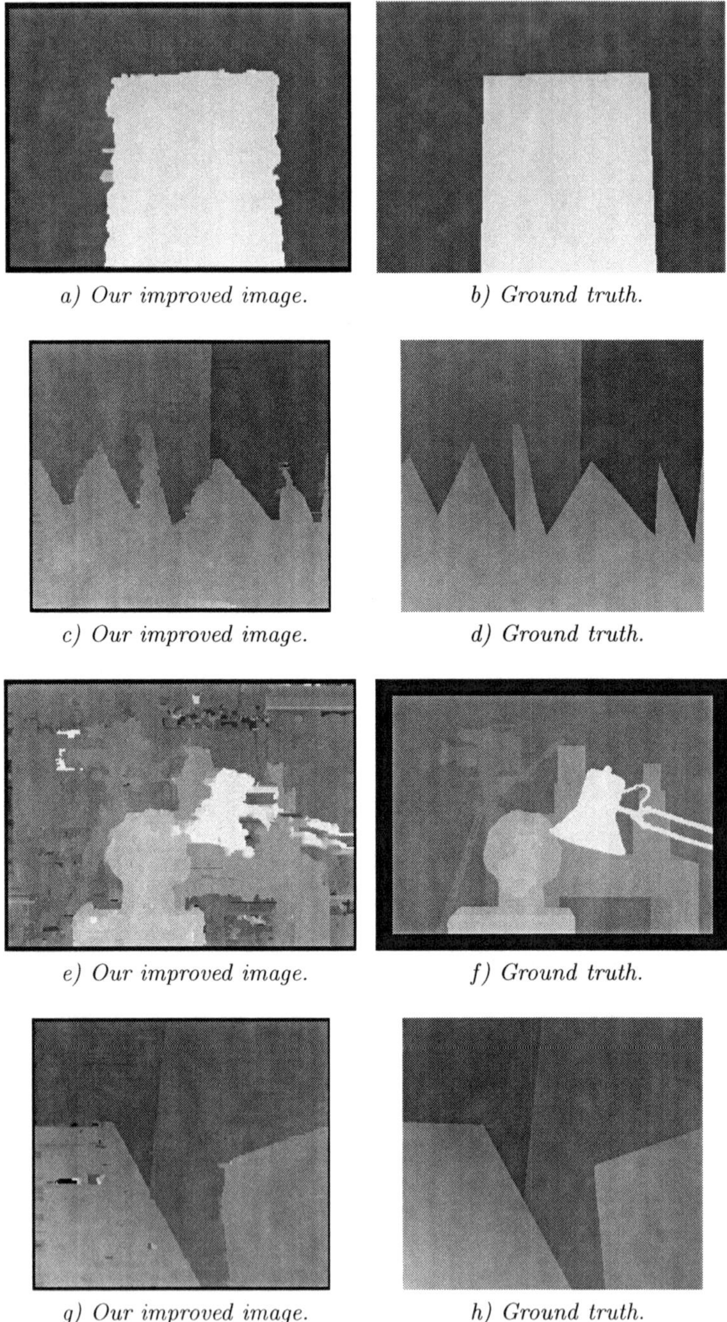

a) Our improved image.

b) Ground truth.

c) Our improved image.

d) Ground truth.

e) Our improved image.

f) Ground truth.

g) Our improved image.

h) Ground truth.

Fig. 12. The final computed disparity maps are shown to illustrate the quality of the infection algorithm.

6 Conclusions

This chapter has presented an approach to the problem of dense stereo matching and dense stereo disparity by means of an evolutionary infection algorithm.

The complexity of the problem reported in this chapter and its solution can provide a number of new ideas for addresing the problems presented from an artificial life point of view. The evolutionary infection algorithm avoids the employment of knowledge currently included in the standard algorithms.

The comparison with a standard test bed provides enough confidence that this kind of approaches can be considered as part of the state-of-the-art.

Acknowledgement. This research was funded by Spanish Ministry of Science and Technology, Project Oplink, and CONACyT through the LAFMI project 634-212.

References

1. Olariu S (2006) Handbook of Bioinspired Algorithms and Applications. Crc Press Llc
2. von Neumann J (1966) Theory of Self-Reproducing Automata. University of Illinois Press, Urbana
3. Chopard B, Droz M (1998) Cellular automata modeling of phisical systems. Cambridge University Press
4. Dhond U, Aggarwal K (1989) Structure from Stereo. IEEE Transactions on Systems and Man and Cybernetics. Vol. 19(6):1489–1509
5. Luo Q, Zhou J, Yu S, Xiao D (2003) Stereo Matching and oclusion detection with integrity and illusion sensitivity. Pattern Recognition Letters. 24(2003):1143–1149
6. Abbey H (1952) An Examination of the Reed Frost Theory of Epidemics. Human Biology, 24:201–233
7. Watts DJ (1999) Small Worlds. Princeton University Press
8. Maniatty W, Szymanski B, Caraco T (2001) Parallel Computing with Generalized Cellular Automata. Nova Science Publishers, Inc
9. Ganesh AJ, Kermarrec AM, Massoulie L (2001) Peer-to-Peer Lightweight Membership Service for Large-Scale Group Communication. Networked Group Communication. pp. 44–55
10. Brown MZ, Burschka D, Hager GD (2003) Advances in computational stereo. IEEE Trans. on Pattern Analysis and Machine Intelligence. 25(8), 993–1008
11. Maniatty W, Szymanski BK, Caraco T (1993) Epidemics modeling and simulation on a parallel machine. In: Proceedings of the International Conference on Applied Modelling and Simulation, pp. 69–70, Vancouver, Canada, July 1993. IASTED, Calgary, Canada
12. Kermack WO, McKendrick AG (1927) A Contribution to the Mathematical Theory of Epidemics. In: Proceedings of the Royal Society of London. Series A, Containing Papers of a Mathematical and Physical Character, Vol. 115, No. 772 (Aug. 1, 1927), pp. 700–721

13. Olague G, Hernández B, Dunn E (2003) High Accurate Corner Measurement using USEF Functions and Evolutionary Algorithms. Applications of Evolutionary Computing. EvoWorkshops 2003, LNCS 2611, pp. 410–421
14. Olague G, Fernández F, Pérez CB, Lutton E (2004) The infection algorithm: An artificial epidemic approach to dense stereo matching. In: Yao X et al. (eds), Parallel Problem Solving from Nature VIII. Lecture Notes in Computer Science 3242, pp. 622–632. Birmingham, UK: Springer-Verlag
15. Scharstein D, Szeliski R (2002) A Taxonomy and Evaluation of Dense Two-Frame Stereo Correspondence Algorithms. International Journal of Computer Vision, Springer
16. Birchfield S, Tomasi C (1998) Depth discontinuities by pixel-to-pixel stereo. International Conference on Computer Vision 1998
17. Shao J (2001) Combination of stereo, motion and rendering for 3D footage display. IEEE Computer Vision and Pattern Recognition 2001 Stereo Workshop. International Journal of Computer Vision 2002
18. Sun C (2001) Fast stereo matching using rectangular subregioning and 3D maximum-surface techniques. IEEE Computer Vision and Pattern Recognition 2001 Stereo Workshop. International Journal of Computer Vision 2002
19. Legrand P, Levy-Vehel J (2003) Local regularity-based image denoising. IEEE International Conference on Image Processing, pp. 377–380

Johnny Washington

Evolution, History and Destiny

Letters to Alain Locke (1886–1954) and Others

PETER LANG
New York • Washington, D.C./Baltimore • Bern
Frankfurt am Main • Berlin • Brussels • Vienna • Oxford

Library of Congress Cataloging-in-Publication Data

Washington, Johnny.
Evolution, history and destiny: letters to Alain Locke
(1886–1954) and others / Johnny Washington.
p. cm. — (Studies in literary criticism and theory; vol. 13)
Includes bibliographical references and index.
1. Afro-Americans—Ethnic identity. 2. Afro-Americans—History. 3. Afro-
Americans—Social conditions. 4. Afro-American intellectuals. 5. Social evolution.
6. Human evolution. 7. Fate and fatalism. 8. Afro-American philosophy.
I. Title. II. Studies in literary criticism and theory; v. 13.
E185.625 .W37   973.04’96073—dc21   00-035636
ISBN 978-0-8204-4970-8
ISSN 1073-2004

Die Deutsche Bibliothek-CIP-Einheitsaufnahme

Washington, Johnny:
Evolution, history and destiny: letters to Alain Locke
(1886–1954) and others / Johnny Washington.
–New York; Washington, D.C./Baltimore; Bern;
Frankfurt am Main; Berlin; Brussels; Vienna; Oxford: Lang.
(Studies in literary criticism and theory; Vol. 13)
ISBN 978-0-8204-4970-8

The paper in this book meets the guidelines for permanence and durability
of the Committee on Production Guidelines for Book Longevity
of the Council of Library Resources.

© 2002, 2008 Peter Lang Publishing
29 Broadway, 18th floor, New York, NY
www.peterlang.com

Printed in the United States of America

Contents

Acknowledgments

At this time I wish to thank the following persons, without whose assistance this volume would not have been possible: Dale Miles, and Tracey K. Parker, graduate students, respectively, at Southwest Missouri State University, who proofread the entire manuscript; Kwee-Hock Cha, Yeoh Chiang-Tatt and Jay Jenkins of Southwest Missouri State University, who assisted me in formatting the manuscript; John Bryant who also read the manuscript and offered valuable suggestions; and Lisa Dillon of Peter Lang Publishing, who in kind ways provided useful suggestions. Moreover, I owe thanks to Drs. Abhay Tilak, Manohar A. Tilak, and Richard J. Douthart, who offered valuable criticism and suggestions. Finally, special thanks go to my lovely wife, Loutisha, and our three children, Latanya, Laurie, and Johnny, Jr. I alone, however, am responsible for any defects this book may possess.

Further, I am grateful to the following for permission to reprint:

Dr. Richard J. Douthart, for permissions to reprint some pieces of correspondence resulting from several years of correspondence between the two of us.

Dr. Manohar A. Tilak, for permission to reprint some of his notes, pieces of correspondence, and some other sources from a decade-lone collaboration between the two of us.

The editor of the journal, *Science and Culture* for permission to reprint the article entitled, "The Evoluon Theory: A General Unified Theory of Evolution," Vol. 45, April 1979.

The editor of the journal, *The Accelerator, Official Publication of the American Chemical Society,* for permission to reprint the essay, "New Postulation on the General Theory of Evolution, Evolvability, Direct Phase Evolution, Evoluon, and Rate of Evolutionary Change," September-October, 1973.

Dr. Abhay Tilak, the editor or publisher of the book, *Infinities to Eternities* (1998), by Manohar A. Tilak, that includes the last two-mentioned articles.

Professor Kuhn, Dr. Sargent, II., Dr. Siekevitz, and Dr. Kelkar for permission to reprint pieces of correspondence.

Prologue

The 21st century harbors many challenges and possibilities that have become more evident following the September 11, 2001 terrorist attacks on the U.S. Some of the fundamental principles of democracy are now being threatened, and the basic values of Christians, Hindus, Jews, and Muslims as well as some other religions of the world are being revisited. Political and military alliances among the nations of the world are shifting. A war is occurring in Afghanistan. The political-military tension between Pakistan and India is heightened; both countries have nuclear weapons positioned to engage in reciprocal assaults. Anti-American sentiment is growing in Indonesia, the Philippines and other parts of the world where many Muslims seek to wage a "Holy War" against the U.S. Consequently the U.S. is enjoying greater unity among its citizens, in the concerted effort to route terrorists. However, America is not an isolated island among the nations of the world. Opportunities have now arisen so that the nations of the world can approximate the unity in the world community that is enjoyed in the U.S., a unity that has the possibility of flourishing in the democracies of the world. A companion of democratic forms of government, the Destiny model which I shall explore in this work, and which I spent over twenty-five years developing, provides a framework in allowing for such unity to be approximated. In our current or post-terrorism world, following the September 11, 2001 terrorist attacks on the U.S., we will need a new political or diplomatic forum that will make a greater effort to approximate the ideals embedded in the UN's Charter or in the teachings of Dr. Martin Luther King, Jr. The Destiny model calls attention to what may be called social democracy, the concerns of which focuses on the respective cultural, religious and ethnic ensembles of the people of the world, especially the people in economic deprived countries in Asia, Africa, Latin America and the Middle-East, where women are beginning to exercise greater freedom. As it shall become evident shortly, the Destiny model offers hope.

I can trace my initial interest in the notion of Destiny to my early high school years, during the early stages of the Civil Rights era of the 1950s.

After my seeing photographs in *Ebony* magazine, in which sheriff's deputies were using dogs to attack Dr. Martin Luther King, Jr., it occurred to me that African Americans would be more effective in their struggle if they had a greater sense of purpose, a sense of Destiny. As I followed the Civil Rights Movement it became evident to me that although Dr. Martin Luther King, Jr. and Malcolm X, as well as members of the Black Panther Party, each used "Destiny," none of these individuals clearly defined Destiny. It was at this early stage that I made a commitment to myself that I would offer a definition of the term "Destiny." It took years for me to arrive at a working definition of Destiny because I had to deal with two concerns: first, I had to get around the problem of racism posed by the Manifest Destiny doctrine initially advocated by Missouri Senator Thomas H. Benton in the mid-19th century, which put forth the doctrine that because the white race was superior to others, it was the God-given duty of the white race to spread "civilization" to the Pacific Ocean and beyond. Furthermore, any group such as Native Americans or African Americans that tries to block such efforts would be destroyed. Second, I had to be prepared to deal with any objections people may raise who may accuse me of advocating a form of pre-determinism or fate, of which the views of the religious group, the Calvinists, were representative. The Calvinists held that a select few individuals were predestined to go to heaven. Their final Destiny was predictable in a manner similar to the predictions concerning the movements of the planets as subsequently described by the Newtonian paradigm. When I returned from my travels in Nairobi, Kenya, and Dakar, Senegal during August 1991, the African American newspaper, the *Palm Beach Gazette*,[1] published some of my articles in which I recounted my travels in Africa. Between June 1997 and April 1998, it published "The African Destiny Series" which included materials that are now incorporated into this manuscript entitled, *Evolution, History and Destiny: Letters to Alain Locke (1886–1954) and Others*. As in the "The African Destiny Series," the materials in this manuscript, presented in the format of letters or epistolary form, are addressed to individuals such as W.E.B. DuBois (1868–1963), Alain Locke (1886–1954), William Leo Hansberry (1894–1965), and Malcolm X (1925–1965), respectively. Among the individuals mentioned here, the public is perhaps most familiar with the views of Malcolm X. In light of this, I believe he needs no introduction.

Thus W.E.B. DuBois' paternal grandfather, Alexander DuBois, was of

French origins; he had immigrated to New York in the eighteenth century. Born in Haiti in 1825, DuBois' father Alfred moved to the U.S. in the late 1850s. When William Edward Burghardt DuBois was born in Great Barrington, Massachusetts in 1868, the Emancipation Proclamation had been issued only five years prior to 1868, and the U.S. Civil War had ended only three years earlier. Undoubtedly the consequences of these events influenced DuBois' life and outlook on the world. He dedicated most of his life researching the problems and aspirations that affected the African continent or the African American community. To this end, he seemed to have become aware during his teen-age years of his mission to become an intellectual giant, so that he could be effective in his calling. As Francis L. Broderick in his book,[2] *W.E.B. DuBois*, wrote:

> In the Negro community DuBois came to hold a special place. As a member of one of the oldest families, as the only Negro in his high-school class of twelve, as one of the two of three students who would go on to college, and as the local correspondent for a Negro newspaper, he took on seriousness and self-importance all out of proportion to his sixteen years. Already DuBois was fascinated by the record of his own intellectual development. At age fifteen, he was gathering and annotating his collected papers. (*DuBois*, p. 4)

It was during his elementary school years at Great Barrington when DuBois first became aware that he was Black and different from his white school mates. In his book, *The Souls of Black Folk,* he described an incident in which one of his white school mates refused to give him a visiting card because he was Black. Her refusal was accented by a glance, a certain look that he described thus:

> In a wee wooden schoolhouse, something put it into the boys' and girls' heads to buy gorgeous visiting cards—ten cents a package—and exchange. The exchange was merry, till one girl, a tall newcomer, refused my card,—refused it peremptorily, with a glance. Then it dawned on me with a certain suddenness that I was different from the others; or like, mayhap, in heart and life and longing, but shut out from their world by a vast veil.[3]

Between 1869 and 1873 DuBois' mother worked as a domestic servant. During 1884 he graduated from high school, and a year later in 1885, the year that his mother died, he entered Fisk University in Nashville, Tennessee, where he studied German, Greek, Latin, classical literature, philosophy, ethics, chemistry, and physics. It was during his Fisk years that he heard, for the first time, Black folk songs and the spirituals. During 1889, DuBois received a B.A. degree from Fisk University; the same year he entered Harvard University where he studied philosophy with William James and George Santayana, economics with Frank Taussig, and history with Albert Bushnell Hart. Between 1892 and 1893 he studied in Europe at the University of Berlin, Germany. During this time he traveled extensively throughout Europe. *The Suppression of the African Slave Trade to the United States of America, 1638–1870*[4] was the title of his Ph.D. degree dissertation; he received his Ph.D. degree from Harvard University in 1895. *The Souls* was one of his first major publications. During 1910 he helped organize the National Association for the Advancement of Colored People (NAACP); during 1919 he helped organize, in Paris, the first Pan-African Congress. Intense ideological conflict arose between DuBois and Booker T. Washington, a conflict that persisted between the turn of the last century and up to the time of Washington's death in 1915. The two differed on principles around which to organize the education of African American youths. Washington held that the focus should be on educating the masses in skills suitable for making a living; DuBois, by contrast, held that the focus should be on educating what he called the "Talented Tenth," a select group of African American youths to be educated in the higher culture. In addition to the Pan-African movement, or the NAACP, Dr. DuBois was an inspiration to such movements as the Harlem Renaissance, and *Negritude*. He published over ten books and hundreds of articles. He taught for a number of years at Atlanta University. In 1963 at the age of ninety-five Dr. DuBois became a citizen of Ghana, where he died during August 27, 1963, on the eve of the Civil Rights march on Washington.

Alain Locke's own autobiographic sketch is instructive here.

> Philadelphia, with her birthright of provincialism flavored the urbanity and her petty bourgeois psyche with the Tory slant, at the start set the key of the paradox; circumstance compounded it by decreeing me as a Negro, a dubious and doubting sort of American and by reason of the racial inheritance making me more of a pagan than a Puritan, more of a humanist than a pragmatist.
>
> Verily paradox has followed me the rest of my days: At Harvard, clinging to the genteel tradition of Palmer, Royce and Munsterberg, yet attracted by the disillusion of Santayana and the radical protest of James; again in 1916 I returned to work under Royce but was destined to take my

doctorate in Value Theory under Perry. At Oxford, the Austrian philosophy of value; socially Anglophile, but because of race loyalty, strenuously anti-imperialist; universalist in religion, internationalist and pacifist in world-view, but forced by a sense of simple justice to approve of the militant counter-nationalisms of Zionism, Young Turkey, Young Egypt, Young India, and with reservations even Garveyism and current-day "Nippon over Asia." Finally a cultural cosmopolitan, but perforce an advocate of cultural racialism as a defense counter-move for the American Negro, and accordingly more of a philosophical mid-wife to a generation of younger Negro poets, writers, artists than a professional philosopher.

Small wonder, then, with this psychography, that I project my personal history into its inevitable rationalization as cultural pluralism and value relativism, with not too orthodox reaction to the American way of life.[5]

Alain Locke was born September 13, 1886; he was the only child of Pliny Ishmael Locke (1853–1922) and Mary Hawkins Locke (1853–1922). Alain Locke's family tree from which he descended included free, educated Blacks, on both his mother's and father's sides. Alain's grandfather attended Cambridge University in England, and he taught school in New Jersey as well as in Liberia. Two years following the Civil War, Alain's father graduated from the Institute for Colored Youth, in Philadelphia, where he later became the principal. Alain Locke graduated from Central High School in Philadelphia, at the turn of the previous century, in 1902, a year before the publication of *The Souls*. Two years later he graduated from Philadelphia School of Pedagogy, where he earned a B.A. degree. Between 1904 and 1907, Locke studied at Harvard University where he received a B.A. degree in philosophy. As his autobiographical statement indicates, Locke's Harvard teachers included William James, Josiah Royce, George Santayana, Ralph Barton Perry, as well as Hugo Munsterberg, some of the same teachers with whom DuBois studied. Between 1907 and 1909 Locke attended Oxford University as a Rhodes Scholar, (the first African American Rhodes Scholar). Between 1910 and 1911, he attended the University of Berlin where he focused on the works of Franz Brentano, Alexius Meinong, and Christian Freiherr von Ehrenfels. During this time he had the opportunity to study with Henri Bergson in Paris. Also between 1910 and 1911, Locke met for the first time Booker T. Washington who at the time was on a tour of Europe. During 1912 Locke accompanied Booker T. Washington on a tour of the U.S. south. Locke received his Ph.D. degree from Harvard University in 1918. In 1912 he began teaching at Howard University, where he remained until his retirement in 1953. He died June 9, 1954, on the eve of U.S. Civil Rights Movement. Locke edited or authored over ten books; one of his most celebrated books includes *The New Negro*.[6] His publications extend to a variety of fields.

In addition to philosophy, he published on topics dealing with music, drama, poetry, literary criticism, African art, aesthetics, education and African American culture generally. His main contributions to philosophy lay in the area of axiology as formulated in his essay, "Values and Imperatives." (*American Philosophy*, pp. 313–333) Like DuBois, Locke played a major role in inspiring the Harlem Renaissance movement as well as the *Negritude* movement. He was an admirer of Booker T. Washington. But Locke seemed to have had an ambivalent attitude towards the views of Marcus Garvey. Rarely did Locke or DuBois mention the views of William Leo Hansberry, whose biographical sketch I shall briefly outline below.

Born in 1894, it was during 1916 when William Leo Hansberry himself received his initial inspiration to study African culture and civilization through the acquaintance with Dr. DuBois' book, *The Negro*, focusing especially on the chapter dealing with the political and cultural achievements of ancient or Medieval Africa. This was an area to which Hansberry devoted his career researching. Moreover, just as Alain Locke's parents had been educators the same is true on Hansberry's father's side. Hansberry's father may be regarded as having been a member of the educated elite class. His father was a professor at Alcorn A&M. College in Mississippi, and as a role model he no doubt exerted the earliest influence on the young Hansberry. When his father died, he left his son a large library with many books on ancient history. Hansberry himself wrote:

> I acquired, while still quite young, a deep interest in the stirring epic of human strivings in the distant and romantic age…and by the end of my freshman year in college at old Atlanta University I had become, largely through independent reading…something of an authority on the 'glory that was Greece' and the 'grandeur that was Rome.'[7]

Hansberry's early interest in the study of African history led him to the discovery of the degree in which such history had been distorted or neglected by previous European historians or scholars. His biographic sketch indicates William Leo Hansberry's earliest quest for a college education began at Atlanta University. This, however, led to frustration. He was unable to acquire adequate knowledge of African history at Atlanta University, and thus he subsequently transferred to Harvard University which allowed him to pursue courses in history and anthropology. It will be recalled that Harvard University was where both Alain Locke and DuBois received their respective Ph.D. degrees, and both Hansberry and Locke have ties with Howard University. In 1921 Hansberry received his B.A. degree from Harvard University, and the following year he joined the Howard University faculty, where Hansberry began offering African Studies courses that subsequently led to his establishing an African Studies Program, one of the first in the U.S. "Negro Civilizations of Ancient Africa"; "The Ancient Civilizations of Ethiopia"; and "The Civilization of West Africa in Medieval and Early Modern Times"—these were the titles of the first three courses included in the program. It is reported that Professor Hansberry faced many difficulties and setbacks by the Howard University administration, as well as discouragement from some of his colleagues, who urged the university to abolish the African Studies program that he was attempting to establish. It is ironic that at the time, Howard University, a predominantly Black institution, had little interest in African American specific courses. Some report that in spite of detractors, and the limited financial support he received, Professor Hansberry persisted in his efforts. It is reported that he provided a great deal of his personal finances in keeping the program viable. In addition to his heavy teaching duties, and many other assignments, Professor Hansberry found time to publish some of his research results. These included: i) "Sources for the Study of Ethiopian History," *Howard University Studies in History,* 1930, vol. II; ii) "A Negro in Anthropology," *Opportunity,* 1933, vol. XI; iii) "African Studies," *Phylon,* 1944, vol. V; iv) "Imperial Ethiopia in Ancient Times," *The Ethiopian Review* (Addis Ababa, Ethiopia: August 1944), vol. I; v) "Ethiopia in the Middle Ages," *The Ethiopian Review* (September and November, 1944), vol. I; vi) "The Historical Background of African Art," *Howard University Gallery of Art,* 1953; vii) "Africa and the Western World," *The Midwest Journal,* 1955, vol. VII; viii) "Indigenous African Religions," *Africa Seen by American Negro Scholars,* (Présence Africaine, 1958); ix) "African Kush, Old Ethiopia and the Balad es Sudan," *Journal of Human Relations,* 1960, vol. VIII; x) "Africa: The World's Richest Continent," *Freedomways,* 1963, vol. III; xi) *Africana at Nsukka* (Viking Press, 1964); xii) "Ethiopian Ambassadors to Latin Courts and Latin Emissaries to Prestor John," *Ethiopia Observer* (Ethiopia and Britain, 1965), vol., IX, no. 2; and, xiii) "W.E.B. DuBois' Influence on African History," *Freedomways,*

8

1965, vol. V, no. 1. (*Pillars*, pp., 22–23). Hansberry received his M.A. degree from Harvard in 1932. He played a major role in the Pan-African movement as early as 1927, and in 1934 he organized the Ethiopian Research Council. Indeed he may be regarded as a principal forerunner of what today is known as Afrocentricism, and, as in the works of DuBois, and Alain Locke, his research provided the groundwork for the *Negritude* movement, the Harlem Renaissance movement and the African decolonization movements. Like Locke, Hansberry remained at Howard University throughout his entire academic career. This is precisely where his experiences and Locke's converged. To my knowledge, Alain Locke, who wrote extensively about African culture and history, never mentioned Hansberry's contributions to this area of research. Although there were, by contrast, degrees of tension between Locke and DuBois, the two co-authored an article together and worked together on other projects. I know not what DuBois' attitude was towards Hansberry. I believe that DuBois' and Locke's attitudes toward Marcus Garvey were ambivalent. I also draw on the philosophic views of Johann Gottlieb Fichte (1762–1814), John Dewey (1859–1952), and Henri Bergson (1859–1941), to mention only a few. Most of these Destiny "letters" are addressed to Alain Locke. Although these letters are addressed to deceased individuals, such letters are not works of fiction. Rather their contents are factual. I was inspired by the works of Lancelot Law Whyte who in one of his books, *Focus and Diversions*,[8] addressed several letters to the astronomer Johannes Kepler (1571–1630). Included below is a sample of Whyte's letter writing format:

> My dear Kepler
>
> I address these thoughts to you, not in the belief that they will reach you, though that would delight me, but because it helps me to do so. Of all those that I have admired, you may best understand what I have to say, for you did not separate aesthetic emotion from mathematical science. Your sensuous delight in geometrical forms yielded insights into the algebra of natural processes. More than that, of all the great thinkers who, following Aristotle, recognized a pervasive formative process, a *facultas formatrix*, in nature, you alone were also a creator of mathematical physics. Thus the fancy and presumption of writing this letter may be excused, not merely for my devotion to you, but because I thereby find a focus, almost a justification, for my habit of trying to view deep matters in a long perspective. My thoughts acquire shape in the illusion that I am telling you what your work has meant to me.
>
> But first let me explain how I discovered you…

I was also inspired by the teachings of Saint Paul who wrote letters to the churches of Asia Minor. As Plato effectively communicated his views in dialogue form, I am attempting to do so in the letter-writing form. These letters are intended to constitute narratives of aspects of the human condition, throwing light on the dimensions of the process of evolution as well as Destiny. At the outset, such variety of letters may appear as fragments or disjointed pieces of history. Shortly it will become evident that in spite of the various letters, there is an underlying unity, a unifying fountain from which their contents flow, echoing a common theme (Bergson) that reflects the larger macronarratives. These macronarratives are the basis on which the Destiny model rests. Further, because of the generality of the Destiny model, it is undecidable as to whether any existing community in fact approximates it. However, any political model that contradicts the Destiny model is inadequate or wrong. Most scholars familiar with Kant's works regard him as a religious person who merely denied knowledge of the existence of God as such in order to make room for faith in the Christian God. Bear in mind that the Destiny model extends beyond the Christian God. There are people in the world besides Christians or Jews. Already in India there are a billion people, most of whom are Hindus; in China, whose population includes some one billion people, there are millions of Confucians. I also intend for my works to be read by Muslims in Africa and other parts of the world that include over eight hundred million Muslims. Many Africans are neither Muslims nor Christians; yet they may find appealing the Destiny model. It is anticipated that the Destiny model will appeal to people in diverse regions of the world.

Moreover, I intended for the Destiny model to be neutral with regards to any given religious affiliation. The connection between Destiny and the realm of transcendence leads many to attribute to transcendent Destiny exclusive religious significance. This is a mistake, as it will become evident as the model unfolds.

I defined Destiny as an ideal of unity towards which a people strives in successive generations. Destiny pertains to groups, that is, human groupings; "fate" is applicable to the isolated individual. I distinguished between transcendent Destiny and immanent Destiny. If the former pertains to ideal unity, the latter-mentioned pertains to the marketplace and the political realms that are characterized by disunity, conflict, diversity, and strife. Precisely what constitutes one's immanent Destiny? It has to do with problems, issues, or projects with which one is seriously or sincerely concerned,

10

the pursuit of which usually involves struggles. In his axiological schema Dr. Alain Locke construed values, which he examined in terms of attitudes, preferences or feelings, as hinging on a polarity: the positive pole in effect includes good, beauty, right and creativity; its negative pole includes bad, evil, ugly, wrong and inhibitions. Similarly, our immanent Destiny hinges on what may be called a Destiny polarity: its negative pole includes war, hunger, starvation, disease, poverty, racial strife, inhibitions, stagnation; its positive pole includes peace, well-being, health, prosperity, unity, freedom, harmony, and creativity. Moreover Kant held that the a-historic ideals of God, Freedom, and Immortality are regulative ideals located outside of this world, ideals we are compelled to postulate in the noumena, transcendent world, in order to make life meaningful. However, our categories of the understanding are limited to the phenomena world as it appears. We cannot know the transcendent, noumena world. Similarly I hold that there is a transcendent Destiny located in the noumena world about which we can have no objective knowledge, but our immanent Destiny is characterized by conflict and strife. I want to offer an explanation by way of illustration as to why we cannot know directly our transcendent.

The transcendent realm's main roads or interstate highways are eternally blocked off, under perpetual construction or repairs. Thick fog lingers over bridges. Its road signs are barely legible, due to the intense luminescence of its street lights. We can, however, seek ways to enter the outskirts of our transcendent Destiny, approach its rest-stop area. This can be done by means of side roads, by-passes, trails and detours. Such routes require turning attention to such aesthetic forms as blues/jazz music, art, poetry, literature and cinema.

These are juncture points in the landscape of our immanent Destiny that points us in the directions of our transcendent Destiny, wherein lies transcendent unity of which our democratic ideals are paradigmatic. Although there are four Destiny modes, the ethnic mode, the national mode, the world mode, and the cosmic mode, for the sake of this discussion, the ethnic Destiny mode will be emphasized. There are several thousand ethnic groups in the world, and each group may be regarded as passengers on a given boat. However, again, for the sake of this discussion, we will focus on four boats representing, respectively, four ethnic or national groups, say, Americans, Kenyans, east Asian Indians, and Afghans. The setting is New York City's harbor,—New York City is a paradigm of the American melting-pot ideal. Moreover, the four boats leave simultaneously New York

City's harbor loaded with the above-mentioned passengers, bound for San Francisco. The passengers in a given boat can communicate among themselves, and communicate by cellular phones with the passengers in the other boats. This scenario demonstrates that the various ethnic groups have a common origin and a common destination, (Destiny). The ocean on which they sail represents the commonality of their immanent Destiny mode and the sun or the "North Star" represents the commonality of their transcendent Destiny. If a fanatic from, say, boat A, crosses over and hijacks boat B resulting in boat B or all four boats being run aground, this would disrupt the journey for each of the boats. The Navy or other peacekeeping forms would have to be brought in to restore order in the shipping lanes, in the respective societies.

An alternative and preferable scenario would be one in which each of the four boats reaches its desired destination without delays, disruptions, or accidents. This scenario will most likely prevail if the passengers on the respective boats came to the recognition that there is a transcendent ideal of unity that is over and above the rail lines, or shipping lanes, over and above their immanent Destiny modes. However this transcendent Destiny cannot be directly known. One has to go on faith, and make use of the indirect routes described above, routes whose streets or lanes enjoy intense luminescence. They are flooded with light, and the feeling of brotherhood is pervasive. Socrates, Jesus, Mohammed, Buddha, Confucius, and Moses, collectively and individually, stand as beacons of hope, as conductors on this transcendent journey.

The world of phenomena or the immanent world is subjected to the laws of cause and effect; the noumena world is subjected to the laws of freedom. Human beings are members of both worlds, and the noumena world is the cause or ground of the phenomena world. Granted that Kant provided a complicated argument to explain the causal connections between the noumena and the phenomena worlds, I will not dwell on Kant's argument at this stage.

Kant raised the following questions regarding human Destiny: What can we know? What can we do? And what can we hope for? To which he replied that we can know the world as it appears, we can perform our moral duties, and we can hope to be rewarded in the afterlife. It will be remembered that Newton in effect held that he had no explanation as to the nature of gravity in and of itself. Rather he could only offer an explanation as to the *effect* of gravity. In a similar vein I hold that although we cannot know the nature of our transcendent Destiny, we can discern its *effect* that is exhibited or felt in the immanent realm in the course of our striving and struggling. We can struggle against elements of resistance, and we can hope to approximate our transcendent Destiny through which, alone, genuine unity is enjoyed. The universe struggles to overcome struggles, an aim that is ultimately futile or without the possibility of complete success. Certainly, biological evolution is characterized by such struggles. And each Destiny mode is grounded in evolution in terms of which the cosmic Destiny mode of the given continuum, the natural, physical world resulted from evolution, beginning with the big bang. When human beings came on the scene they began living in small clans, or tribes, or ethnic groups; these in turn during the past few centuries in the Western world evolved into nation-states, nationalities, and nation-states are beginning to be transformed into a world community, (world Destiny). As will be seen, the Destiny model I am advocating has nothing to do with religious, racial, ethnic or political ideologies or doctrines such as the values included in Manifest Destiny. Nor is it associated with any particular economic doctrine such as communism or socialism or "dictatorship of the proletariat."

The ethnic mode pertains to the experience of an ethnic group such as African Americans, Mexican Americans, Seminoles, Cherokees, Luo people, or Jews; the national mode pertains to a nation-state such as the U.S., Brazil, Panama, Guyana, Ghana, Japan, France, New Zealand, or Kenya; the world mode pertains to the world community, as echoed in the United Nation's ideal or the world community about which Dr. Martin Luther King, Jr. spoke. The Internet and interlocking world economies and other processes are moving the world closer to oneness. The world mode is a synthesis of the ethnic and the national Destiny modes. The cosmic mode pertains to the physical cosmos that the physicist, biologist, or chemist investigates. Although there are four Destiny modes, strictly speaking there is only one transcendent Destiny, by which alone we can approximate the ideal of unity. On this point I believe Marcus Garvey would agree with me, in the spirit of Pan-Africanism.

Each Destiny mode co-asserts its own Destiny schemata; and each Destiny mode may be regarded in part as a complex adaptive system described by Murray Gell-Mann, a Nobel Prize winner in physics, in his book, *The Quark and the Jaguar.*[9] By Destiny schemata I mean the underlying, basic principles of a society expressed as effectively compressed complex information, by which a given society organizes itself and functions. Alain Locke, rarely, if ever, used the term "schemata," or "schema." However, his axiology model, as formulated in his essay, "Values and Imperatives" may be construed as a schema. As will be seen, Gell-Mann holds that the schemata that guide his complex system play a prescriptive as well as a descriptive role. Later I shall say more about Gell-Mann's notion of schemata or complex adaptive systems.

In this work, I also introduce a novel notion, "Destinicity." By Destinicity I have in mind a primordial, creative principle manifest within the cultural expressions of a people, in their music, art, religion, science, and philosophy as they approximate their immanent or transcendent Destiny. Destinicity is a synthesis of the ethnicity elements as well as the Destiny elements of a people; hence, the word, "Destinicity." Further, Destinicity includes the economic activities of a people, a people's local economy as well as the world economies. In fact, a group approximates its Destiny by working, struggling, fighting, gaming, and loving. Here, I am paralleling Hegel's or Fichte's notion of Spirit or Bergson's notion of *élan vital.* "Destinicity" is also informed by the African notion of *Nommo,* the word, or cultural forms as examined in Janheinz Jahn's book, *Muntu: The New African Culture,*[10] where he shows that the creative, metaphysical force in African culture is connected to the blues as a musical form.

The academic field whose parameters I have developed is called Destinicity Studies; it includes not only the African/African American experience but the human condition generally. It focuses on the past, present and future possibilities of a people and regards experience as a coherent unity, rather than a set of fragmented historic happenings. It will become evident that I am seeking to ground the Destiny of a people in their ethnic or tribal or local, cultural experiences, to capture the creative spirit that informs their respective group identities. For example, when I focus on the African American experience, I dwell on their contributions to blues, jazz, spirituals and other cultural forms. My focusing on the local or ethnic experiences of a people is what distinguishes the Destiny model from the model offered by the Medieval Islamic thinkers, Judaic-Christian thinkers, Eastern philosophers, African philosophers/sages or the teachings of European idealist philosophers such as Kant, Fichte, and Hegel. In the spirit of the Science of Complexity that is paradigmatic of the Postmodern era, I seek to draw attention to the ontological status of cultural, religious, and ethnic minorities, to the diverse voices that they represent. I also incorporate the voices and perspectives of feminist thinkers, including ethicist Virginia Held as well as Bernice Johnson Reagon, who stresses the role of caring, loving, and mothering. The Destiny Studies curriculum includes not only Afrocentric courses and issues but also the multicultural, ethnic-specific courses. A core course in such a curriculum could be organized around a course entitled Introduction to Destiny Studies that describes the model, contrasting it with existing models such as multiculturalism and Afrocentrism. Such a course could also consider the merits or demerits of each alternative model. Clearly the Destiny model itself could be useful in exploring many social issues. As recent as August 30, 1999, the controversy resurfaced regarding whether or not evolution or creationism is to be allowed to be taught in the Kansas public schools. It will be seen that the Destiny model offers a viable forum in terms of which such controversial matters could be debated or examined. The Destinicity (or Destiny) Studies model is also applicable to such areas as Nursing, Political Science, Social Work, Sociology/Anthropology, Environmental Studies, Women Studies, Gender Studies, Education, Ethnic Studies, Philosophy and Artificial Intelligence Studies. Moreover Destinicity Studies gives consideration to unity, and it introduces into the curriculum or enquiring process (research) the normative dimensions. It was Hegel who held that knowledge of history informs us of our Destiny. The October 10, 1997 issue of *The Chronicle of Higher Education* included an article entitled, "An Era of Painful Self-Examination for Many Intellectuals in Africa." As the title suggests, the point was made that some African intellectuals, including those who were interviewed for the above-mentioned article, are undergoing an identity crisis as well as a Destiny crisis. The source of the crisis is attributed in part to the fact that since independence, African intellectuals have subscribed to ideas derived from the Western world that have induced in some African intellectuals a sense of inferiority and low self-esteem. The article goes on to point out that such a crisis is pervading the entire continent. The fact that the interviewees are not demanding the importation of more wheat, rice, computers or fax machines into their country is significant, although I believe these would be welcomed. Rather, they are demanding a thorough consideration on their part of the notion of the African Destiny and identity grounded in their history.

Chapter 1 shall focus on ethnic appellations ascribed to African Americans: "Negro," "Colored," "Afro-American," "Black," and "African American" are among these. I suggested a novel appellation "Africantude" as a spin-off from *Negritude*, a once popular appellation. I also examine various dehumanizing images such as the "black sambos," as well as images of the "Aunt Jemima" and "Uncle Moses" types that were pervasive up until the 1960s. As one peruses this chapter, a hypothetical scenario involving the field AI (artificial intelligence) is readily available that invites the question as to what would be the outcome of placing on a planet such as Mars, say, a million robots in which the history of African Americans is "uploaded," such that the these robots assume an ethnic identity. What would constitute the identity of these robots? What criterion would be used in determining their identity? Would they be considered African Americans? How would their Destiny be related to the Destiny of earthly African Americans or any other group whose history may have been uploaded? The issues regarding AI, dissociated from the issue of ethnicity, is considered in Appendix A and B, respectively, where a "Being" exercising artificial intelligence in the Cosmic Destiny mode is described. Letters in chapter 1 as well as chapter 2 shall be devoted exclusively to Dr. DuBois. Between 1990 and 1994, I traveled to such places as: Merida, Mexico; Nairobi, Kenya; Dakar, Senegal; Tegucigalpa, Honduras; Managua, Nicaragua; San Jose, Costa Rica; Ahmadabad, India; and Lagos, Nigeria, where I gave lectures focusing on Destiny. In most of the countries the problems of poverty, political oppression, and economic-social crises were noticeable. Many of those who heard my presentations at these conferences appeared to be eager to grasp the issues. They felt that the Destiny model could shed light on the problems in their respective countries.

In my presentations I stressed that effective economic-cultural development plans cannot be exclusively imposed from the top, from such an organization as the World Bank, or International Monetary Funds, but have to arise from the various ethnic or local cultural groups seeking to affect their own respective immanent Destiny modes. During Franklin D. Roosevelt's administration, he established various social-economic programs to bring the country out of an economic depression. Among the programs were the Works Progress Association (WPA) and the Civilian Conservation Corps (CCC) that were quite effective. A variation of what may be called the Roosevelt Plan was developed by the late Sam McGhee, who was president of the local Minority Contractors Association in Delray Beach, Florida. Inspired not only by the Roosevelt Plan but also by what may be called the

Booker T. Washington model, Mr. McGhee's goal was to rebuild a blighted area in the heart of the African American community section of Delray Beach. In his opinion, this rebuilding process would take place block by block, step by step. To this end, he sought to build or purchase a center to train contractors, business people, and students, to recreate a model block. Once created, the model block would be showcased to highlight their talents, and in the process encourage renovating of rest of the city, block by block. Mr. McGhee was inspired by the Destiny model which he considered in implementing his plan. I had the opportunity to have had discussions with Mr. McGhee when he was developing his plan. Although the McGhee model was intended for the urban environment, it is easily extendable to rural communities or to the international context.

On September 21, 1998, world leaders, including President Clinton, as reported by the Springfield, Missouri Public Radio, met at New York University to discuss a new political model, called the Third Way, which seeks to offer an alternative to traditional democratic liberalism, on the one hand, and communism, on the other. The Third Way focuses not so much on world economies but on local economies. Such a model is aimed at stimulating economies in the inner cities in such areas as Miami, Chicago, and Detroit, in rural areas as well as grassroots economies in the so-called Third World countries where the instruments of advanced forms of capitalism that would facilitate taxation, banking, and investing are not adequately in place. Many attributed Russia's current financial woes to the absence or immaturity of such instruments. Similar conditions exist in many countries in Africa, Asia, the Caribbean, and Latin America. If the advanced industrialized nations are now entertaining the Third Way (community economic) model mentioned above, the Destiny model offers what I call the Fourfold Way, in recognition of the four Destiny modes; the first three Destiny modes have their own economies. The Fourfold Way links a people's economic activities, manifest in the forms of struggles and counter-struggles, to their immanent or transcendent Destiny, whereas the Third Way does not make that Destinicity linkage.

Self-Efficacy[11] is the title of Albert Bandura's book in which he holds that certain beliefs can be self-efficacious in a manner analogous to some religious or patriotic views that energize groups or individuals without resulting in fanaticism or terrorism. Thus, the Destiny model is equally efficacious. It empowers the individual or group that embraces it. Space does not permit me to develop the notion here. However, I mentioned Bandura in this context to indicate that his model may be useful in assisting people in developing the economic bases of their communities. Here I will briefly telescope, by way of anticipating, some of the main points that shall be brought to the fore in chapters 1 through 6, that is, Part I of this work under discussion.

On the whole, inasmuch as chapter 1 is concerned with the problem of ethnic identity, it constitutes the sub-theme for the entire work, a sub-theme that is concomitant to the controlling theme; namely, Destiny, that manifests itself in the world as a journey whose paths involve many setbacks and impediments. It will be recalled that Hegel's book, *Phenomenology of Spirit*, is concerned with the problem of knowledge, including self-knowledge.

This quest for knowledge involves a journey of the Spirit that becomes divided, an unhappy, frustrating condition that it seeks to overcome so as to return to the primordial unity it once enjoyed. In the course of the unfolding of the Spirit or consciousness, as it journeys through history, seeking to understand itself, it faces many perplexities, disillusions, and obstacles that result in inadequate knowledge. Certainty becomes elusive.

A perpetual struggle for the Spirit is to overcome such inadequate knowledge, and to achieve complete, rational knowledge found only in the Absolute. Similarly, letters two and three here in chapter 1 set the tone for the letters that follow. As was indicated, in these letters the various appellations that have been ascribed to Blacks in the past are examined; an alternative, novel appellation, or "Africantude" is proposed. Yet one wonders whether such a proposal is adequate, in light of letter four included in chapter 1. Here we come upon the false images, (false self-knowledge) associated with the "Aunt Jemima" and "Uncle Moses" types. Most African Americans found such images objectionable; some whites, by contrast, found them amusing, a source of entertainment.

Chapter 2 highlights my traveling adventures in Africa, Asia, and Central America. Inasmuch as I engaged in these traveling activities, it is as though I was playing the role of an actor in a grand novel. In parodying Hegel, my traveling adventures may be construed as an "embodiment" of the "World Spirit" (Hegel) in search of its true identity; in this context, World Spirit is tantamount to World Destinicity. As it unfolds, it achieves a greater degree of complexity, of freedom and spirit. The World Destinicity actualizes itself in part in chapter 3 by striving to determine its bearings in the schema of historic time, seeking to determine the markers or bounda-

18

ries of modernity/postmodernity. Such a quest forces the World Destinicity to plunge deep into African history; this occurs in chapter 4. Here the focus is on exploring hypotheses regarding the origin(s) of the human species; the origins of iron technology; the origins of ancient religions and philosophies in Egypt and Ethiopia, as well as the origins of the great kingdoms in Medieval West Africa. Chapter 4 ends by examining the views of the Muslim philosophers, Avicenna and Averroës, regarding the existence of God, and the related augment offered by Avicenna concerning human Destiny. I shall show the inadequacy of his argument concerning God and Destiny. In chapter 5 the World Destinicity returns to the U.S., where the Destiny model is strengthened by elaborating on the transcendent/immanent distinction. In doing so, certain views of Whitehead who offers a description of God are incorporated into the Destiny model. On the basis of Whitehead's view, God's presence is manifest as processes in the field of evolution. Also in this chapter the racist doctrine of Manifest Destiny shall be examined and rejected as an inadequate Destiny model. In seeking what may be regarded as an antitoxin to Manifest Destiny, World Destinicity embraces Josiah Royce's principle of loyalty and community, within which self-esteem can be nurtured. The views of Royce are combined with the views of Fichte, who emphasized the idea that working and struggling is a way of approximating one's Destiny and enjoying self-esteem. Chapter 6 begins by considering certain key activists, Martin L. King, Jr. and Malcolm X, of the Civil Rights Movement. Martin Luther King, Jr., and to a lesser degree Malcolm X, may be regarded as an embodiment of the World Destinicity.

In addressing letters to Malcolm X, attention is brought to the Black Panther Party, a revolutionary group of African Americans who advocated armed rebellion in pursuit of their authentic identity, and political freedom. The letters to Malcolm X are followed by letters to Alain Locke, in which, among other issues, the problem of racial insanity is considered. When considering the issue of racial insanity, it becomes evident that World Destinicity suffers what may be called an "identity crisis," one of the worst crises it suffers during its journey in approximating transcendent unity. Such an identity crisis may result in either total madness or creative genius. But there is hope. In the next to the last letter in chapter 6, two letters are addressed to the church. It is in this context that I suggest that the religious experience, including Pentecostalism, combined with the Destiny model may prohibit or reduce social or group insanity. Also important is the caring ethics mentioned in chapter 2.

The Nation's "Founding Fathers," influenced by the values of European Enlightenment, the roots of what is called "modernity" urged white Americans to reject their ethnic identity, and to embrace the "melting-pot" doctrine. Whereas African Americans embraced their ethnic identity, from which flow great thinkers, artists, entertainers, and athletes. I suspect that the recent school shootings and other fanatical behavior among white, male youths throughout this country is indicative in part of their searching for their ethnic or cultural identity.

Most African Americans are not confused about their ethnic or cultural identity in the manner suffered by whites, and thus the former do not go on shooting sprees as just described. It was Hegel who in effect said that, paradoxically, the slave had a greater sense of self-knowledge than the slave-master. African Americans rejected the false identity that whites imposed on them but African Americans embraced their cultural heritages generally. Many of the authors on whose views I shall focus in chapter 3 are interested in determining the marks of modernity. However, when I considered this problem, I shifted the focus to determine the mark(s) of postmodernity.

I hold that Einstein is representative of Western postmodernity; that Booker T. Washington and Alain Locke are representative of the first phase of African American postmodernity; that Martin L. King, Jr. (and perhaps to a lesser degree Malcolm X) are representative of the latter phase of African American postmodernity; and Newton is representative of Western modernity. During the past few decades some scholars in such areas as Literary Criticism, Religious Studies, and Theology have given attention to exploring the issue of modernity and postmodernity. Paul Heelas is among these scholars, as evident by the title of his book, *Religion, Modernity and Postmodernity*.[12] In the passage below he comments on what he considers to be a defining feature of modernity:

> One of the great marks of modernity, it has been claimed, is that it is characterized by a number of differentiations. Accordingly, it comes as no surprise to find theories such as Scott Lash (1990) claiming that the great mark of the postmodern is dedifferentiation. The situation, however, is considerably more complicated than this in that both differentiating and dedifferentiating processes are taking place within both modernity and postmodernity. (*Religion, Modernity*, pp. 3–4)

The author added:

> Modernity, we have seen, can be thought of as an amalgam of various differentiations and dedifferentiations. Furthermore, differentiation can never be total: boundaries, if they were to become too strong, would make social life impossible. Equally, dedifferentiation can never be comprehensive: for the same or the whole to exist there must be something different. And each process elicits the other. Stephen Toulmin, in *Cosmopolis: The Hidden Agenda of Modernity* (1990), claims that the Enlightenment search for universal was elicited by the Thirty Year War—an event which serves to highlight the dangers of difference. Conversely, it might be argued, the weight of uniformity can encourage the proliferation of particular individual and cultural identities. (*Religion, Modernity*, pp. 3–4)

The Destiny model seeks to provide a means by which a balance can be achieved between the differentiation and dedifferentiation tendencies. Here the focus is on determining what may be called the boundary conditions of modernity, that is, when the modern world began and what are the distinguishing marks of the ancient world, the Medieval Age, modernity, and postmodernity, although my focus is on the feature of postmodernity. What features determine the African American sense of modernity or postmodernity? What features determine the European sense of modernity or postmodernity? What role did Booker T. Washington or Alain Locke play in informing African Americans of their Destiny within the context of modernity/postmodernity?

Here it will be seen that the World Destinicity is torn between modernity on the one hand and postmodernity on the other. The phenomenon of change or time is what both modern and postmodern eras have in common, but each differs in its perspective of change. In the book, *Modern and Postmodern Strategies*,[13] Monika Kilian called attention to features that distinguish modern tradition from the postmodern era with regard to the process of change. Kilian's schema is represented below:

Modern	*Postmodern*
change within stasis	change without referential marker
(Hegelian) sublation	proliferation
unification & totality	multiplicity & non-commensuration
implosion	explosion
continuity	rapture
The World	many worlds
immutable	creative

Clearly, scholars such as Drs. Alain Locke and W.E.B. DuBois fit neatly into the Postmodern camp in Philosophy that was founded by Nietzsche. Nietzsche, Kierkegaard, Marx, and Freud, respectively, were forerunners of postmodernity. The Destiny model draws on both the modern and post-modern perspectives. Moreover, the Destiny model seeks to strike a balance between transcendent unity and immanent diversity, contrast or conflict, and between freedom and necessity. The effort to achieve such a balance is one of the main concerns of the Destiny model. It is just as universal as the Kantian categorical imperative or the ideals embedded in the U.S. Declaration of Independence. An advantage of the model is that it provides equal attention to particularities on the one hand and universal dimensions on the other. It merits repeating that chapter 4 shall be devoted to exploring in historic perspectives the ancient or Medieval kingdoms of Ethiopia (Kush, Meroe, and Axum), as well as ancient or Medieval kingdoms of West Africa. This portion of the series of letters relies much on the works of the African American historian William Leo Hansberry. A question I posed to Hansberry was: How did the ancient or Medieval Christian Ethiopians' sense of Destiny differ from their Muslim counterparts in West Africa? Another question I raised was: Who were the ancient Ethiopians (Kushites) and what was the nature of the kinship between themselves and the ancient Egyptians? In another context, I criticized the Muslim philosophers Avicenna (980–1037), and Averröes (1126–1198), for paying insufficient attention to the African traditions and values. As was mentioned, Dr. Alain Locke published a number of works focusing on African art and culture. In this regard I discussed Locke's contributions in my book, *Alain Locke and Philosophy*.[14] As I researched Locke's materials during the past two decades, Ms. Esme E. Bhan, of the Moorland-Spingarn Research Center at Howard University, was quite helpful. Some of the sources relied upon in this portion of the series are indicated in the notes.[15]

"Johnny Washington Explores The African Destiny! Philosophy Professor writes to Alain Locke—Commending him on his endeavors and tackles the possibility that the Africans have one destiny!" read the weekly headline of the *Palm Beach Gazette* as it published the series. In the spirit of Pan-Africanism, Africans and the people of African origins have one transcendent Destiny whose "event horizon" can be discerned, if not fully known, with great lucidity, by means of entering into dialogues with the African ancestors, the African gods, as well as the Mother Spirit that soar above the African continent under the night sky, above the Baba tree, above the Nile River and above the morning dew. The Mother Spirit reveals itself in visions, poems, dreams, dance, the drum, states of trance, as well as scientific or philosophic discourse. In the Vedic tradition of India, the word "OM" is regarded as the creative force by means of which the cosmos came into being. In the biblical tradition a variation of the creative, spiritual force is expressed in the Scripture: "In the beginning was the Word, and the Word was God, and the Word was with God" (Gospel of John). It is in the vein of the "the Word" that I attempted to communicate in the form of "letters" with the ancestors such as W.E.B. DuBois, Alain Locke, William Leo Hansberry and Malcolm X, the sons of Africa. In certain traditions in Africa it is believed that the "word" or human mode of communicating through the natural language is a means by which material reality can become energized or activated. While Kant may be correct in holding that we can have no objective knowledge of the transcendent world, he admitted that we can *think* about such a world. It follows that one is in a position to think about the messages revealed by the Mother Spirit, the concerning of which is a manifestation of the universal *élan vital*. More research is required to determine more clearly the relation between the levels of self-knowledge an individual or a people may acquire and the degree to which the individual or a people approximates the Destiny ideal.

In the context of discussing the African Destiny, the question often arises as to what degree did the ancient Africans have a sense of transcendence, comparable to, say, the principle of transcendence attributed to the Christian, Jewish, or Muslim God? Many who are unfamiliar with African history or culture deny that the Africans had a sense of transcendence, since, the conjecture is often made, in most of traditional African societies, before the arrival of Christianity or Islam, the emphasis was on tribal religions or gods and the concomitant ancestral spirits or deities. As to the question of transcendence, we can shed light on it by considering the ancient Africans' conception of god. *An African Classical Age* is the title of historian Christopher Ehret's book,[16] in which he places in perspective the transcendence question, to which he gives an affirmative reply. Ehret relies on linguistic evidence, or etymology, in compiling his research on African history. He wrote:

> The derivation of the word *°Nyàmbé* is an issue that has not been much studied so far. But one proposal, especially deserving of further consideration, is that it comes from an old Niger-Congo verb root the early Bantu

reflex of which was *-àmb- "to begin." That etymology identifies *Nyàmbé as "Beginning," or in other terms, as First Principle, a conception in keeping with the view of *Nyàmbé as Creator that one must attribute to the early Bantu and presumably to their earlier Niger-Congo forebears. (*An African*, p. 159)

In Ehret's analysis, the god *Nyàmbé was believed to have resided in the sky or sun. This reference to sky or sun seems to attribute to *Nyàmbé a transcendent status. Ehret claims that later in the last millennium B.C. a remarkable shift occurred in the East Africans' conception of God. They began to think of God as having not so much a transcendent status but as assuming an immanent status. *Nyàmbé became intimately related to various tribes and ancestral spirits. *African Religions and Philosophy*[17] is the title of John S. Mbiti's book in which he is explicit about the attribution of transcendent as well as immanent status of traditional God in Africa:

> For most of their life, African peoples place God in the transcendental plane, making it seem as if He is remote from their daily affairs. But they know that He is immanent, being manifest in natural objects and phenomena, and they can turn to Him in acts of worship, at any place and any time. The distinction between these related attributes could be stated that, in theory God is transcendental but in practice He is immanent. (*African Religions*, p. 33)

The sky God *Nyàmbé or the immanent-transcendent God about which Mbiti speaks is another name for the Mother Spirit that informs the African Destiny of which I spoke above. A larger issue that Ehret was dealing with in his book has to do with the role that Eastern Africans played between 1000 B.C. and 400 A.D. in contributing to what may be called the immanent Destiny of the world, its cultural and economic dimensions. In indicating the purpose of his book Ehret wrote:

> This book, *An African Classical Age: Eastern and Southern Africa in World History, 1000 B.C. to A.D. 400*, brings to light 1,400 years of social and economic history across Greater Eastern Africa. Its geographical scope, the Eastern side of the continent of Africa from Uganda and Kenya on the north to the Cape of Good Hope at the far south, forms a complex of regions larger than the whole of mainland Europe, excluding only the Russian Republic. Its telling reveals a history driven by the clash of strikingly different economies and cultures and invigorated by the opportunities and the challenges of technological change. The major currents in that history

link its events, both indirectly and directly, to the wider developments of world history in those times. (*An African*, p. xv)

Ehret also emphasized that the Eastern Africans made a number of contributions to the immanent Destiny of the world: crop productions or farming, and iron-workings, which influenced world events, stand out.

Kant as well as Fichte, Hegel, or Marx grounded social struggles in history. It should be evident by now that the notion of struggle is a keystone in the Destiny model, a notion that is elucidated in chapter 5. Such an act, or process of striving reflected in the definition of Destiny (or Destinicity) implies freedom. One can safely attribute the notion of striving to all forms of complex adaptive systems, especially living organisms. Animals strive to escape predators, acquire food, or sexual mates. They do not do so on the basis of a plan that they themselves have freely formulated. Animals' behavior is instinct driven. By contrast, the striving of human beings hinges on plans by which they organize their lives; such plans constitute the narrative of their lives, individually and collectively. Animals cannot generate narratives at whose center is the notion of the self, time, the past, present, and future.

Striving here also is analogous to Husserl's notion of intentionality, a distinctive feature of human awareness. Intending and striving are one and the same. They allow one to become aware of the realm of transcendence, although object knowledge of such a realm is beyond reach. Additionally, authentic freedom is grounded in the noumena world where its transcendent modality of Destiny is posited as an erotic ideal that lures us onward to unite with it.

Due to our biological makeup, we instinctively seek unity. Our striving for unity is grounded in our biological evolution. It may have a basis in the biological machinery of our brain. My claim here is in harmony with the views of Sherwin B. Nuland. In his book, *How We Live*, he wrote:

> In my view, the nearness of destructive chaos against which we can never cease struggling creates in us the instinctive need for order, and every fiber in our being is somehow aware of this. The rhythm and tempo of our bodies tells it to us. The cells of our innermost structure speak in not-to-be-denied voices to the depths of our consciousness, bringing inexpressible knowledge that we can survive only by making *unity* of the multitrillion processes of metabolic activities. To do otherwise is lethal [Emphasis added].[18]

A goal that includes paying off one's house mortgage, or paying one's child's way through medical school, inasmuch as this may be regarded as a serious concern, pertains to one's immanent Destiny in which one's sense of survival, self-esteem, or security may be perpetually threatened or heightened. Transcendent Destiny within which is found the principle of ideal unity may be regarded as lying between, on the one hand, one's serious concerns, and, on the other, one's ultimate concern, in seeking a relationship with Destiny. This position parallels Paul Tillich's definition of the religious experience in seeking a relationship with God who is defined as our *ultimate* concern, beyond which we can have no higher concerns. The notion of transcendent Destiny is one notch below the notion of transcendent God. Through such disciplines as Political Science, Sociology, Anthropology, History, Philosophy, Psychology, and the physical sciences generally, we can investigate our immanent Destiny, to which the Kantian categories of the understanding are applicable. Although we cannot obtain objective, scientific knowledge of our transcendent Destiny, we can approximate it through struggles and counter-struggles. When we struggle to overcome resistance in whatever form it occurs, we enhance our self-esteem, one of the highest principles we can embrace. The act of struggling allows us to anticipate our transcendent Destiny. Admittedly a thinker such as Frantz Fanon (who was a Hegelian) in his book, *The Wretched of the Earth*, devoted attention to the African struggle for independence from colonial rule. However, his model was inadequate in that he placed too much emphasis on violence, and he regarded the transcendent realm, where world unity may be approximated, as objectionable. Dr. Martin Luther King, Jr., by contrast, accepted the principles of non-violence as well as the notion of transcendence, as a means of dealing with the problem of self-esteem and freedom. I define freedom as the capacity to act in concert, to act with others, in the course of which, to struggle against all elements of resistance or opposition. Freedom is exercised in the degree that we struggle against opposition, in approximating the ideal of unity. The Afrocentrism model is often considered in the African Destiny consideration. The trouble with this model is that it focuses on the African/African American experiences almost to the exclusion of the experiences of other ethnic or national groups. The Internet, interlocking economies, international travels and commerce—each plays a major role in moving human groupings closer to the world community envisioned by the Destiny model. World peace involves urging the world to adopt certain principles along the lines that Alain Locke advocated: tolerance, reciprocity, parity, pluralism, and relativism, each of which may be construed as an ingredient in certain social schemata to a given complex adaptive system (Gell-Mann). Such principles, Locke believes, shall enable us to strike a balance between the elements of ethnic or cultural diversity and ethnic or cultural unity. (Unfortunately the Internet is also providing a forum of a growing number of hate groups instigating racial or religious intolerance). Many scholars are of the opinion that in the not too distant future, nation-states shall disappear, or rather that they shall be gathered up into the world community (Marx).

As to the question of how the Destiny model relates to other traditions in which Destiny ideals are prominent, let us at the outset consider the Vedic tradition and the law of karma. It is widely held that one's Destiny is determined by the law of karma. Certain forms of yoga or meditation were designed to enable the individual to meet the challenges associated with his or her Destiny. One must struggle in this life to attempt to overcome or escape the chain of birth. The Vedic notion of Destiny focuses on re-birth, or reincarnation. In many of the African traditions the stress is on predestination of the isolated individual who "receives" his or her Destiny before birth (see E. Bolaji Idowu's book, *Olódùmarè: God in Yoruba Belief*).[19] Hegel held that the individual's or society's Destiny is "driven" as it were by the World Spirit that unfolds dialectically. Although a notion such as struggle plays a prominent role in the Destiny model, I do not intend for it to be construed as adversarial or mean spirited. It will be remembered that Dr. W.E.B. DuBois organized the first chapter of his book *The Souls* around the notion of striving, and the entire book deals with the manner in which African Americans had to work and struggle throughout the centuries in the U.S. Such a striving results in a futile effort to overcome the split in the consciousness of African Americans at the individual and the collective levels. My work reads like a Sunday school newspaper, in contrast to the works of someone such as Frantz Fanon as expressed in his book, *The Wretched of the Earth*. In it he advocated open violence against the colonial oppressors. Fanon, like Dr. DuBois, was also a follower of Hegel's teaching. Both DuBois and Fanon inspired not only the African independence movement but also the U.S. Civil Rights movement. Instead of emphasizing activities of the sort mentioned here, Jean-François Lyotard in effect stresses the importance of what he calls gaming in dealing with one's Destiny. What is important in gaming, as in playing basketball or baseball or chess, are the rules of a given game that establish the basis for justice among those participating in the game. If a given ethnic or cultural group desires to stay in the game played out in the larger community, it must generate a set of rules in coordinating its members' behavior. The rules associated with gaming prohibit the practices of violence and protect the principles of justice and respect. The gaming rules are embedded in the group's narrative or history.

Chapter 6 shall deal with a number of diverse issues, one of which is striking. This has to do with two letters devoted to the church community. In the "Church Letters" I urged the respective congregations to adopt the Destiny model, inasmuch as it was in harmony with the Christian ideals of unity and universal peace. It so happens to be in harmony with the basic values of many other religions. I also attempted to show that just as Christianity enjoins that its principles are applicable in solving social problems, also I maintained that the Destiny model that incorporates features of Pragmatism and other sources is equally applicable to approaching many problems we face today. I go on to suggest that we establish an African/African American Destinicity Studies Center (AAADSC) or A^3DSC in which we examine issues and problems affecting the immanent Destiny of the people of African origins and people generally. Moreover, I recommend that during each African American History month (February), we celebrate what I call African Destinicity Day, to be effective the year 2005.

Chapters 7, 8 and 9 shall focus on the works of Dr. Manohar A. Tilak who has developed a novel model, called the "Evoluon Theory," that describes the evolution of the universe. The model describes the various stages of evolution beginning with the first and simplest stage, the big bang, characterized by the feature of naked singularity. Tilak called each such stage an "evoluon." Evoluon 1, (Evl1) represents the first stage of evolution; the next stage of evolution, Evoluon 2, (Evl2), is characterized by duality, manifest as electron-positron, or wave-particle, or matter-antimatter occurrences; the third stage, Evl3, is characterized by the three-dimensional configuration of the atom, where ordinary matter arose; Evl4 is represented by the four nucleotides, essential for life, in the DNA molecule; Evl5 is exhibited by the arrival of human beings, whose behavior is guided by the brain, or mind, in the field of evolution. According to Tilak there are three additional evoluons, Evl6, Evl7, and Evl8. At these levels, beginning at Evl6 already on the horizon, a new Being(s) will dominate the field of evolution, and in effect the evolution of human beings will be left behind. This new Being is representative of the Cosmic Destiny mode. What I find appealing about the Evoluon Theory is its feature of universality and its eightfold schema for categorizing experience. It includes the entire scope of existence, beginning with the big bang scenario that is believed to have occurred between 15 and 20 billion years ago, and extending into the future some 80 billion years or beyond. The bridge between the Destiny model and the Evoluon Theory lies with the cosmic Destiny mode. As was mentioned, the Evoluon Theory was developed by Tilak in the 1970s. Some twenty years later Tilak and I began our collaborative works, in which I assisted in bringing the Evoluon Theory more in line with certain developments in Western philosophic traditions. Tilak invented Evoluon Theory, and I as well as Dr. Richard J. Douthart, who was one of Tilak's colleagues at Ely Lilly Corporation in the early 1970s, and others, assisted Tilak in developing Evoluon Theory. Douthart returned to examining the Evoluon Theory after many years of absence, and he is now reinterpreting or formalizing it with respect to developments in contemporary science. In Part II of this work, I shall provide a more thorough introduction of Tilak and Douthart, respectively. When Tilak and I first began our collaborative works in the early 1990s, I recorded on videotape many one hour-long interviews with Tilak at Florida Atlantic University in Boca Raton, Florida.[20] Below is the copy of a letter reflecting the reaction of an individual who has perused the taped recordings.

Dear Professor Washington:

I write this letter as an ardent student of life. I recently had a chance to review the tapes you have prepared in association with Dr. Manohar Anant Tilak. In all moderation, they constitute the Fifth Veda. The material contained in these tapes transcends all ancient and contemporaneous attempts of conveying the most fundamental knowledge. Upanishads have upheld the view point that all abstract knowledge can become available to a human being only through "experience" and can not be communicated by way of written or spoken word. Your tapes have made it possible. An ordinary person, blessed only with enough curiosity, will find it an enormously indulging experience to go through the tapes and at the end will find himself a totally selfless, timeless entity. The derivation of joy is the one and only natural consequence. I humbly submit to you that the work you gentlemen have done makes me disinterested in whatever I have known or heard and whatever I will hear. This contentment is invaluable. The truth and only truth, in its entirety is available in your creation. It is interesting that the work has happened in a rudimentary language such as English. From now on the value of English has increased. While associ-

ated with the body, the Atman has no special powers. Yet the future scenario of reincarnation is available from the Law of Karma and Atman automatically is reincarnated in a suitable world as a transcendental entity. Professor Washington, I salute your efforts.
Very Truly Yours,
Ravindra V. Tilak
July 25, 1993

In Gell-Mann's view of a complex adaptive system, schemata are the means by which the process of evolution accumulates information, analogous to the information included in the complex DNA molecule, information that complex living systems, organisms, in turn use in guiding their present and future behavior. (*The Quark*, p. 69)

In introducing the notion schemata and related notions, Gell-Mann has clearly demonstrated some similarities between complex adaptive systems such as an ecological system, an economy, a computer and consciousness. But he has not adequately explained how human self-consciousness is a unique feature of the world. Here I shall briefly consider the notion of consciousness, and later go on to suggest that in the future it may be possible for robots to achieve a "moral" conscience as well as an ethnic or cultural identity.

Here I shall briefly pause to shed light on the nature of consciousness by considering it within the context of the activity of thinking. In considering this problem, I find informative Edward O. Wilson's views.[21] I like the metaphor in which he regards the self as an actor in a drama where various scenarios are being decided; in this context, "scenarios" are comparable to Gell-Mann's notion of schemata. Wilson wrote:

The self is not an ineffable being living apart within the brain. Rather, it is the key dramatic character of the scenarios. It must exist, and play on center stage, because the senses are located in the body and the body creates the mind to represent the governance of all conscious actions. The self and the body are therefore inseparably fused: The self, despite the illusion of its independence created in the scenarios, cannot exist apart from the body, and the body cannot survive for long without the self...

The self, an actor in a perpetually changing drama, lacks full command of its own actions. It does not make decisions solely by conscious, purely rational choice. Much of the computation in decision making is unconscious—strings dancing the puppet ego. Circuits and determining molecular processes exist outside conscious thought. They consolidate certain memories and delete others, bias connections and analogies, and reinforce the neurohormonal loops that regulate subsequent emotional re-

> sponse. Before the curtain is drawn and the play unfolds, the stage has already been partly set and much of the script written. (*Consilience*, p. 130)

My take on this matter regarding the self is as follows. If Wilson suggested that there is only one self involved in a given drama during a particular time, I am of the opinion that there are at least two conscious selves involved when thinking is exercised. The drama is largely constituted by a narrative, and the narrative is woven out of the dialogues that result from the conversations that take place between the two selves that emerge out of the original unity of an individual's consciousness. At the outset there is primordial unity, a unity that in turn gives rise to a duality, that is, two selves that are played out in the above-mentioned drama. When thinking is not exercised, one particular self comes to the fore; when thinking is exercised, at least two selves come to the fore. I am trying to show the relation between the self, thinking, and self-consciousness. Further, I believe that Hegel is right in saying that self-consciousness involves the quest for recognition by the other, a quest that often involves struggles or conflicts. This position is similar to points raised by Gregory S. Paul and Earl Cox in their book, *Beyond Humanity*.[22] I found the book's first paragraph, focusing on suffering (struggling), quite telling: "First we suffer, then we die. This is the great human dilemma. In the ancient dawning of human self-reflective awareness, a new and terrible consciousness of painful suffering and death stunned slowly evolving human minds. It was in fact raw horror—the fear of fear itself." (*Beyond Humanity*, p. xiii) Some of the views explored in the last few chapters of this present manuscript are foreshadowed in Tilak's book, *Infinities to Eternities*.[23] Here I need to express a brief disclaimer. In *Infinities to Eternities* many of the issues, especially those pertaining to mysticism, the Vedic religion and related issues, are outside of the scope of my area of competence, and thus they are Tilak's own effort arrived at independently of my own. In Appendix A of this work, *Evolutioin, History and Destiny...*,I have included Tilak's previously published article, "New Postulate on the General Theory of Evolution, Evolvability, Direct Phase Evolution, Evoluon, and the Rate of Evolutionary Change" (1973). In Appendix B, I have included Tilak's previously published article, "The Evoluon Theory: A General Unified Theory of Evolution" (1979). In these works a focus is on the ways in which human evolution is now threatened as a result of the macro-evolutionary-phase-change (MEPC) that the Earth is undergoing exponentially.

Tilak predicts that the MEPC that is now occurring will give rise to a new being or beings with a life and Destiny of its own, that this new being will be dissociated from the biological basis of life. Rather it will emerge out of its interacting, extrovertly, with nature, after having acquired an entirely artificial "life" process; further, it will evolve, and take on a career and Destiny of its own, independently of human beings. Murray Gell-Mann, in his above-mentioned book, seemed to have envisioned a new being on the horizon similar to the one about which Tilak speaks. Gell-Mann suggested that future computers may enable people to become interconnected neurologically, such that a new "being" will emerge. He wrote:

> Some day, for better or for worse, such interconnections might be possible. A human being could be wired directly to an advanced computer, (not through spoken language or an interface like a console), and by means of that computer to one or more other human beings. Thoughts and feelings would be completely shared, with none of the selectivity or deception that language permits…My friend Shirley Hufstedler says that being wired up together is not something she would recommend to a couple about to be married. I am not sure that I would recommend such a procedure at all (although if everything went well it might alleviate some of our most intractable human problems). But it would certainly create a new form of complex adaptive system, a true composite of many human beings. (*The Quark*, p. 21)

In a similar vein Paul and Cox speculated about the future thus:

> The hypothesis [held by Paul and Cox] of the Extraordinary Future presumes that conscious thought, similar to but not necessarily identical to the sort humans possess, can be achieved by a machine artifact of human ingenuity. This is the Primary Assumption. It also presumes that identities can be transferred from natural mind machines into artificial ones. This is the key assumption. If these assumptions are incorrect, then the Extraordinary Future is in big trouble. Before we can decide whether these assumptions are sound, we must first consider some basic questions concerning brain structure and function, the mind-brain connection, consciousness, and the nature of identity. (*Beyond Humanity*, p. 130)

Tilak was prophetic in his announcement some 30 years ago of the coming of the new being(s). In this connection a serious question naturally arises. What sort of moral principle shall guide the conduct of this new being(s)? At the outset I defined Destiny in a way that it included only human beings. However, Douthart suggested that the Destiny model is equally applicable to AI agents, that is, robots of the future. I agree. In this light, the Destiny definition needs to be restated thus: Destiny is an ideal of unity towards which any intelligent or rational agents strive in successive generations. Earlier I mentioned the notion of caring, along with similar notions offered by certain feminist writers. If we include caring as an ingredient of the Destiny model, it may be more effective as a moral criterion for human beings as well as AI agents. I already indicated that I agreed with Lyotard that gaming is also a way of actualizing one's potential or approximating one's Destiny; Josiah Royce emphasized the importance of one committing one's self to a cause, that is, we exercise freedom, autonomy, and efficaciousness when we commit ourselves to a cause by subscribing to the principle of loyalty; Dewey in effect says we approximate our immanent Destiny when, in following the scientific method, we seek to resolve problematic situations; in the process, our identity undergoes transformation, and achieves greater richness. Clearly it may be the case that the features of the models offered by either of the thinkers, together with the caring principle may be a basis for developing what may be called an AI ethics, as new beings emerge in the field of biological and "artificial" evolution. As we approach the twentieth-first century, the Destiny model will offer a ray of hope for this century and beyond. As we face the new millennium it is paramount that we seriously consider the issues raised here. Many challenges await us in the twenty-first century and beyond. Features of the Destiny model such as the one offered here can serve as signposts. A comment made by Henri Bergson in his book, *Creative Evolution*,[24] captures the threefold theme, Evolution, History and Destiny of my manuscript:

All organized beings, from the humblest to the highest, from the first origins of life to the time in which we are, and in all places and in all times, do but evidence a single impulsion, the inverse of the movement of matter, and in itself indivisible. All the living hold together, and all yield to the same tremendous push. The animal takes its stand on the plant, man bestrides animality, and the whole of humanity, in space and in time, is one immense army galloping beside and before and behind each of us in an overwhelming charge able to beat down every resistance and to clear many obstacles, perhaps even death. (*Creative Evolution*, p. 30)

Part I
History, Ethnicity and Destiny

Chapter 1

Ethnic Appellations and Conflicts

African American Opera, the Blues, and *Ebonics*

Dr. W.E.B. DuBois (1868–1963)   Destinicity Letter #1

Re: Opera, the Blues, and *Ebonics*

Dear Dr. DuBois:

It is a pleasure for me to inform you of the concert that I attended, performed by African American artist George Shirley, along with pianist Howard Watkins, here at Southwest Missouri State University during February 4, 1997. His appearing was part of our African American Heritage Month celebration that extended the entire month of February. I believe that during your time only one week was set aside to celebrate the contributions of African Americans. I enjoyed the Shirley performance. Listening to his performance was a big contrast to the blues, no less than the spirituals, forms of music that I grew up on, in west Alabama, near the state line of Mississippi. During my youth the major blues artists included Muddy Waters, John Lee Hooker, Lightnin' Hopkins, Sunny Boy Williamson and others. Yet as I listened to Shirley's performance, I could, in many of his pieces, detect a thread of commonality between certain songs lifted to the operatic level and the blues, and the spirituals, a description of which you provided so well in your book, *The Souls*. George Shirley's audience was especially delightfully moved when his performance focused on work or convict songs. He sang these songs in a manner that added special significance to the entire occasion. During those moments one knew that one was in contact with an art form that was squarely grounded in the African American tradition, associated with the cotton fields, prayer meetings, and juke-joints, where gin and corn liquors were bought and sold, and where the African American immanent Destiny has its roots. The audience, mostly white, loved the spirituals integrated into this formal mode of expressiveness.

Ironically, Shirley's eloquent performance occurred the evening of February 4, around the time the O.J. Simpson (a former African American football star who allegedly murdered his white wife) verdict was being announced in the media, and President Clinton was delivering his State of the Union address. In trying to decide whether to watch the TV announcement of the O. J. Simpson verdict, watch President Clinton deliver his State of the Union Address or attend the George Shirley concert, I decided to do the latter. My mind settled on music that evening. I think I made a good choice. The following morning, a reporter from Channel 10 TV, here in Springfield, Missouri, phoned me, seeking to schedule an appointment to interview me regarding the O.J. Simpson verdict. She interviewed me at 1:00 p.m. on February 5, 1997, and it was aired the same day during the 6:00 p.m., and again at 10:00 p.m. news hours. While we are on this topic of interviews, I wish to inform you about a radio talk show I heard during December 20, 1996. Here the listeners, mostly white, were calling in about the issue of teaching *Ebonics*, a dialect that exhibits the idioms of African Americans in public schools. How I got involved in this is that about two weeks earlier, the producer of the show indicated to me in a phone conversation that she wanted to interview me, that is, have me visit the radio station or contact the radio station by way of phone, where listeners would be calling in, live, on December 20, 1996, at 8:00 a.m., to discuss Kwanzaa. I agreed to participate. When I arrived at my office the morning of December 20, to participate by way of phone in the radio discussion, (it was quite cold that morning, around 7 degrees F.,) for some strange reason, the radio producer never called me, as she previously indicated that she would do.

Nevertheless, I decided to tune in to the radio station in question, while I graded my final exams. When I tuned in, I discovered that they had switched the topic, without my previous knowledge, and the discussion was not on Kwanzaa, but on *Ebonics*. In the minds of many of the individuals who called in to this radio station on this cold December morning, five days before Christmas, certain Blacks, including Jesse Jackson and Louis Farrakhan, are the ones who are racists, not the white people in the Ozarks or in other parts of this country. Many regarded *Ebonics* as a form of separatism, and separatism in their unenlightened views is bad or objectionable. I found the radio discussion somewhat amusing, to say the least. During February 4, 1997, I informed my students in my African American Studies course about the contents of the *Ebonics* radio program, and my comments stimulated much debate among students.

That many of the listeners who called in to that radio station to discuss *Ebonics* appeared to me to have a quite narrow perspective on matters pertaining to race is not surprising to me. The racist doctrine of Manifest Destiny originated here in Missouri in the 1840s, it will be recalled. In view of the *Ebonics* debate reflected in the media here in the Ozarks, the people in this region of the country can benefit from a broader perspective that the Destiny model offers or the wholesome perspective your many works, including your book, *The Souls,* offers. Whether *Ebonics* is a form of racial separatism or not is a matter of debate, but we need not revert to a dogmatic attitude in examining the issues. Besides, I believe that it was you, yourself, who held that voluntary segregation or separation is not in itself inherently evil. Rather, forced segregation or separation is evil. This attitude about *Ebonics* described above is evident of racial antagonisms pervasive in this country.

I should add that on February 24, 1997, here at Southwest Missouri State University I convened a panel entitled, "Strivings and Struggles in the Thought of W.E.B. DuBois" to examine the first chapter of your above-mentioned book. In addition to myself, the other panelists were student Jelani A. Logan, Drs. Alaine S. Hutson, and Cecil A. Poe. Dr. Jack Knight assisted me in convening this panel. I led the discussion by informing the audience of the degree to which your notion of double-consciousness was influenced by the philosophy of Hegel. I criticized you for offering no effective solutions in overcoming this duality of consciousness that renders the African American identity problematic.

True, during your extended career you intermittently embraced communism as a remedy to many of the problems that Blacks suffered. I went on to inform the audience that you should be commended for playing a major role in the founding of Pan-Africanism and the NAACP. Both the NAACP and Pan-Africanism are viable today, although each has suffered setbacks and losses. Neither has solved this double-consciousness problem that you described so eloquently in your book, *The Souls.* You will probably be surprised to gather that the Soviet Union no longer exists, and communism throughout the world is virtually dead, feeble or limp.

Your Talented Tenth model still inspires hope to a dedicated few. But I am not sure whether it speaks to many in the hip-hop generation surrounded by drugs, unemployment, and violence, the environment out of which springs rap music, *Ebonics,* and other modes of cultural expressions. The Destiny model attempts to reconcile elements from your views as well as from the views of Booker T. Washington. Not only that, but the Destiny model is also informed by the respective views of Zora N. Hurston, Ida B. Wells, Marcus Garvey and Alain Locke's ideas as well as many others. By copy of this letter I am informing Alain Locke and others of the issues I raised in this letter. I will write more at another time.

Sincerely,

Dr. Johnny Washington, Ph.D.   February 26, 1997

CC. Dr. Alain Locke

Dr. Martin Luther King, Jr.

Marcus Garvey

"African American," "Black," and "Africantude"...

Dr. W.E.B. DuBois (1868–1963)   Destinicity Letter #2

Re: Destinicity and Identity: Part I

Dear Dr. W.E.B. DuBois:

In perusing your works I came across a letter by student Roland A. Barton presumably addressed to you. In this letter Barton indicated that he found objectionable the use of the appellation "Negro." Your reply, along with Barton's letter, was published in a 1928, issue of *The Crisis* magazine, the official publication of the National Association for the Advancement of Colored People.

To place the debate in perspective, I shall cite Barton's letter in its entirety and cite part of your letter of reply. The respective letters are included in Eric J. Sundquist's work, *The Oxford W.E.B. DuBois Reader*.[1]

The Name "Negro"

Dear Sir:

I am only a high school student in my Sophomore year, and have not the understanding of you college educated men. It seems to me that since *The Crisis* is the Official Organ of the National Association for the Advancement of Colored People which stands for equality for all Americans, why would it designate, and segregate us as "Negroes," and not as "Americans."

The most piercing thing that hurts me in this February *Crisis*, which forced me to write, was the notice that called the natives of Africa, "Negroes," instead of calling them "Africans," or "natives."

The word, "Negro," or "nigger," is a white man's word to make us feel inferior. I hope to be a worker for my race, that is why I wrote this letter. I hope by the time I become a man, that this word, "Negro," will be abolished.
Roland A. Barton. (*The Oxford*, p. 70)

Barton seemed to have found acceptable "American" or "African," and he came close to introducing "African American" or "American African"; what he found problematic was "Negro" or "nigger." Dr. DuBois, when you died in 1963, the debate regarding the ethnic identity of Blacks was beginning to resurface. As early as 1966, Stokely Carmichael [who died during the week of November 20, 1998] urged African Americans to reject "Negro" or "Afro-American" or "Colored" and to adopt "Black." In this context, he urged many Blacks to register to vote in order to experience political power. Thus, the birth of the slogan "Black Power." Many of us who witnessed the 1960s thought that this identity crisis was a new phenomenon. But as the two above-mentioned letters show, this debate extends to the 1920s and beyond. Since you gave such a lengthy reply to Barton's letter, I shall not cite your letter in its entirety. Rather, I will comment on some of its main points.

First, you admonished Barton not to mistake names for things. As you wrote:

> Do not at the outset of your career make the all too common error of mistaking names for things. Names are only conventional signs for identifying things. Things are the reality that counts. If a thing is despised, either because of ignorance or because it is despicable, you will not alter matters by changing its name. If men despise Negroes, they will not despise them less if Negroes are called "colored," or "Afro-Americans."
> (*The Oxford*, p. 70)

On this score, apparently some African Americans during the mid-1960s through the present would have taken issue with you. As I mentioned, a few years following your death, they adopted "Black," and some twenty years later, a debate arose as to which was preferable, "Black" or "African American." I myself, and I think I can speak for many Blacks, use these two appellations interchangeably. However, I will indicate shortly why I believe that either of these notions "Black" or "African American" is inadequate. We now need a novel term to reflect the meaning and identity of the people of African origins who are now living in Africa or other parts of the world. While I am on the subject, let me mention that a new form of musical expression called rap has arisen in this country in which "nigger" (and even "hoa" in referring to Black women) is widely employed. This was a point I was making in my first letter to you. Rap music is quite popular among youths, both Black and white. You, yourself, and Dr. Alain Locke held in high regard the spirituals and blues, which, you two argued, arose from the Black folk tradition. I often wonder what would have been your evaluation of rap music. In your Barton letter you went on to raise this question: "But why seek to change the name? 'Negro' is a fine word. Etymologically and phonetically it is much better and more logical than 'African' or 'colored' or any of the various hyphenated circumstances." (*The Oxford*, p. 70) As progressive as you were on matters pertaining to Blacks, I am surprised that you spoke so approvingly of "Negro." To modern readers, this statement by you seems almost reactionary. But you were not alone in your opinion. Booker T. Washington, Alain Locke, Madam C.J. Walker, Ida B. Wells, Marcus Garvey and others apparently found "Negro" acceptable. It is ironic that you all found it acceptable but Barton did not. It seems as though he was over half of a century ahead of his times. I suspect that he was caught up in the spirit of the Harlem Renaissance, which, during the times his letter was written, was at its apex. And you, along with Dr. Alain Locke, and others, being of the older generation, were unable to think of the matter in the manner similar to Barton, because the consciousness that pervaded you and your contemporaries was informed or shaped by the social history through which "Negro" or "Colored" had undergone many twists and turns. In the 1920s, it will be recalled, Dr. Locke introduced the term "New Negro" but it didn't catch on. Clearly the media has played a major role, either positively or negatively, in shaping this element of identity. The radio and movie media were in their early stages in the 1920s. I suspect that the "Old Negro" was representative of the Booker T. Washington era. Dr. DuBois, in your Barton reply you made reference to the career or history of "Negro," together with its related "Colored" thus:

> Historically, of course, your dislike of the word Negro is easily explained: "Negroes" among your grandfathers meant black folk; "Colored" people were mulattoes. The mulattoes hated and despised the blacks and were insulted if called "Negroes." But we are not insulted—not you and I. We are quite as proud of our black ancestors as of our white. And perhaps a little prouder. What hurts us is the mere memory that any man of Negro

descent was ever so cowardly as to despise any part of his own blood. (*The Oxford*, p. 71)

In ending your letter, Dr. DuBois, you reverted to the point that it is not the name that counts, for names are merely signs pointing the way. What really counts is the thing: "Come on, 'Kid,' let's go get the Thing!" (*The Oxford*, p. 71) It is clear that you were attempting to move the Kid (who apparently was representative of some Blacks) away from the linguistic quibbles to the world of work, struggle and action through which we can seek self-knowledge. In following Fichte and Hegel who without doubt, influenced you, I urged Blacks and people in general to approximate their Destiny through working and struggling to overcome oppositions. I realize that my views here paralleled your own as evident by the first chapter of your celebrated book, *The Souls*. When I published my book, *A Journey*, I devoted a chapter to the issue of ethnic appellations. Here I left undecided which was preferable, "Black" or "African American." Rather I focused on the social or cultural dynamics that make possible the appellation shifting Blacks have undergone: "Negro," "Colored," "New Negro," "Afro-American," "Black," and "African American." Yet it is clear that you did not see any import to this appellation debate, Barton's views to the contrary notwithstanding. You recalled the *Negritude* movement of the 1930s that was inspired by your own works and those of Professors Alain Locke, William Leo Hansberry and others. If the *Negritude* movement placed emphasis on the aesthetic side of the Black experience, I am now seeking an appellation that pays equal attention to the economic side of the Black experience without harboring the prefix "Negri" as in *Negritude* that evokes the connotation "nigger." By Destiny I mean an Ideal of unity towards which a people strives in successive generations. Destinicity combines the ethnicity and the immanent Destiny ideals of a people. I shall elaborate on this point in the sequel. Destiny allows for generality or universality, a normative feature that permits respect of the cultural difference or identity of each ethnic group. "Black" is merely a descriptive term that does not connect a people to the lands or cultural heritage that may nurture their sense of identity. As was suggested in the Prologue, a more adequate appellation that I wish to introduce is "Africantude" (it can be expressed in the form, "Africantude people," the "Africantude community," the "Africantude movement," "Africantude Destiny," or "Africantude Destinicity." These suggest the solidarity of the people of African origins, in contrast to a notion such as "Diaspora" that denotes the disunity of a people. Clearly the identity or appellation of a people is closely connected to their Destiny, a point that you suggested in much of your writings. That is why I find it so hard to believe that you indicated to Barton that this identity or ethnic name issue was no big deal. Thus, I am in total disagreement with you. I am on the side of Barton who sagaciously posed to you a good question. I believe Barton would find appealing "Africantude." This novel appellation along with Destinicity stresses unity of a people, either at the individual or community level. These two novel notions I believe also address the double-consciousness problem you called attention to in your book, *The Souls*. I am now bringing my letter to a close. I will write more later.

Sincerely,

Dr. Johnny Washington, Ph.D.   February 28, 1997

Pragmatism and Ethnic Meaning

Dr. W.E.B. DuBois (1868–1963)   Destinicity Letter #3

Re: Destinicity and Identity: Part II

Dear Dr. DuBois:

This is a follow-up on my February 28, 1997, letter to you in which I revisited pieces of correspondence between yourself and student Roland A. Barton that were published in a 1928 issue of *The Crisis* magazine. In his letter in question to you, Barton desired to know why the people of African origins (here I will use interchangeably the "people of African origins," "Blacks," and "African Americans") accepted the appellation "Negro," a notion that he seemed to have despised. It seemed as though Barton was bordering on prophecy when he suggested that he thought "African" was preferable to "Negro." However, in your reply to him, you indicated that he was plowing in the wrong field in seeking to reject the "Negro" appellation. In my February 28, 1997 letter to you I told you how I felt about the matter, so I will not repeat my views here, except to say now as I said previously, that it seemed to me that you were wrong and Barton was right,— right in questioning the significance or appropriateness of "Negro." This is especially the case, in view of the fact that a few decades following Barton's letter to you, African Americans rejected altogether "Negro" and adopted "Black," followed a few decades later by "African American." It is a bit strange that neither you nor any other progressive race-leader adopted "African American." Today "Black" and "African American" are largely synonymous. What is the source of the shifting from one ethnic appellation to another? I venture to claim that this shifting is due in part to what may be called the cultural saturation factor, the saturation of cultural complexity. The notion of saturation of complexity will become clearer later in this work. I should add in this context that a few years following your death in 1963, two significant African American leaders came to the fore, Dr. Martin Luther King, Jr., and Malcolm X. Apparently Dr. Martin Luther King, Jr. accepted the prevailing appellations that denoted the identity of the people of African origins, but Malcolm X seemed to have found it problematic. He often employed the reference "the so-called Negro," in referring to the people of African origins in the U.S. "Little" was his surname that he rejected, because, in his view, it was ascribed to his fore-parents by the slave master. Malcolm replaced "Little" with "X," and he assumed the name "Malcolm X." Can you imagine our celebrating the holiday of Dr. Martin Luther X, Jr? One wonders why Malcolm X did not make the complete shift and suggest that "African X," or "Afro-X," to replace "African American." If one probes deeply into the matter, the word "African" or "Africa" might also have a suspect origin. However, it is the case that, inspired by Malcolm X's teachings, many African Americans began borrowing names from the African languages in naming their offspring, and in renaming themselves. Further, some U.S. colleges and universities have adopted the Afrocentric model of education; many universities around the world have African Studies Programs.

It may seem to you that I am belaboring this Barton letter, and in fact I am so doing, because it enables us to put in historic perspective this identity debate. My aim in this letter is to identify what I believe to have been the philosophic basis for Barton's argument that is embedded in my previous letter to you. Yet I recognize that Barton was not a college philosophy student, for he was only a high school sophomore. As I reflected on his views, I can discern a philosophic ring to his point of view with regard to the issue of ethnic identity, an issue that you brushed aside as without significance. When you replied to Barton's query, you admitted that names and words have meanings, with no serious consequences. You went on to suggest that the real meaning was to be found in the world of practices and things. On this point I agree, in part, with you. Certainly meaning is constituted by things in the world that involve work and practices, as you suggested. Meaning is also derived from the ways in which the mind employs certain schemata that include appellations in dealing with the world. To the degree that your views placed one squarely within the world of things and practices you were a Pragmatist, echoing the views of such individuals as Charles Sanders Peirce, John Dewey, William James, and Josiah Royce. To the degree that Barton placed emphasis on the conceptual aspect of experience on appellations, he was, for the sake of this discussion, within the Kantian tradition. This description is also applicable to my own views. It was the German philosopher Immanuel Kant who influenced the entire modern philosophic tradition, of which, in addition to yourself, Charles S. Peirce, John Dewey, Josiah Royce, C.I. Lewis, and Alain Locke were parts. Kant urged that the mind imposes meaning on the world. He went so far to say that by virtue of its *a priori* categories including space, time, substance, causality, totality, plurality, and unity, among others, the mind constitutes the world. That is, the mind "creates" the world of things, which you, in your Barton letter, regarded as primary reality.

In my book, *A Journey*, I traded on the distinction that Dr. Alain Locke's classmate C.I. Lewis made between sense meaning and linguistic meaning. I did so to demonstrate how the appellation "Black" or "African American" enables African Americans to enjoy meaning in life, and I believe Barton would appreciate this distinction. In Lewis' view, linguistic meaning pertains to the semantics of words found in, say, a dictionary, where the meaning of a given word is exhibited through the meaning of another word. For example, "vixen" means a "female fox"; or the meaning of "Venus" is connected by what we mean by "the morning star." However, sense meaning is deeper as well as broader than linguistic meaning. Sense meaning pertains to the Kantian schema that is an *a priori* mental structure in terms of which experience acquires unity and significance in accordance with the principle of generality. It also involves our sense of time, that is, the past, present, and future that make possible our acquiring a sense of Destiny, in both the immanent and transcendent sense.

Pragmatism also subscribes to a principle of intersubjectivity that is concomitant with the scientific method. The principle of intersubjectivity is constituted by the enquirers of the scientific community. A problematic situation creates doubt in the enquirers' mind. The enquirers conduct investigations and experiments in accordance with the scientific method, with the aim of removing the doubt or the problematic situation. If the doubt in turn is replaced by a belief that is shared by the community of enquirers, who exchange ideas and perspectives in arriving at their belief, the belief achieves objectivity and meaning. This version of Pragmatism was represented by people such as Peirce, James, and Dewey. Royce, who was an Idealist, was also interested in Pragmatism, and its concomitant principle of intersubjectivity. Like Dewey and others, he felt that the practical effect resulting from our acting on a belief is the ultimate test in determining the meaningfulness of a belief, provided that the belief is shared by what he called a community of interpreters, whose role is similar to the community of enquirers. I venture to maintain that the principle of intersubjectivity associated with Pragmatism is also paramount in enabling us to acquire sense meaning of ethnic appellations under discussion. The community of enquirers or interpreters also plays major roles in our constituting the sense meaning of a given ethnic appellation or, the sense meaning of immanent or transcendent. The scientific method may be useful in assisting us in interpreting information embedded in the various Destiny schemata, or the schemata pertaining to our immanent world, mentioned in the Prologue. The communicative or dialogue dimension of the Pragmatic method is useful in encountering our transcendent Destiny. It is also the case that a certain Postmodernist thinker such as Richard Rorty stressed the role of dialogue in constituting the identity of a people. In a similar vein, Jean-François Lyotard in his book, *The Postmodern Condition,* emphasized the role of what he calls gaming in constituting the identity of a people. Already, in the Prologue, I briefly considered Lyotard's view.

Moreover, with Peirce, no less than Royce, I hold that symbols are important in rendering experience meaningful. I am speaking not so much about symbols associated with linguistic meaning but with what may be called sense-meaning symbols; the same is true in regard to Destiny and Destinicity. "Black," "African American," "Africantude" are symbols on which sense meaning hinges, symbols that require interpretation by each generation of people. Clearly Barton was calling for a reinterpretation by his generation of "Negro," "native," and "African." Inasmuch as you and Barton were in different generations, the meaning you two, respectively, assigned to these appellations were on a collision course. In one context the sense meaning of a given word can be negative, in another context, positive. The sense meaning assigned to "nigga" (it should be noted that some rap artists changed "nigger" to "nigga") by some whites was intended to de-

46

mean Blacks, to reduce them to mere things without purpose or dignity. Apparently some whites allowed themselves to be elevated to a false level of superiority by calling Blacks, "nigger." One ethnic group is harmed by "nigga" and the other group gets the illusion of benefit by "nigger." As I mentioned in my previous letter to you, many rap artists today also employ "nigga" in such a way as to heighten the African American's sense of self-esteem. I know this seems paradoxical, but it is so, just as they use "bitch" and "hoa" (or "whore") with the intention to make a Black woman feel good about herself. The hip-hop generation is seeking hope. Destiny posits hope. Thus, Dr. DuBois, it seems to me when Barton was querying you regarding the objectionable features of "Negro" he was searching for the sense meaning of his ethnic identity. In replying to him, however, you never moved beyond the linguistic meaning. In this regard your reply was weak and uninformative.

Sincerely,

Dr. Johnny Washington, Ph.D.   March 17, 1997

Perceptions and Images of African Americans

Dr. W.E.B. DuBois (1868–1963)   Destinicity Letter #4

Re: Scholar's Black Collectibles and Memorabilia

Dear Dr. DuBois:

Dr. Kenneth W. Goings[2] who in 1991 joined the Florida Atlantic University History faculty has a unique scholarly interest. He examines, records, peruses and collects so-called "Black Collectibles and Memorabilia," a subject in which you had an interest, inasmuch as it relates to the issue of African American identity. As you might recall, in my first few letters to you, I examined your notion of African American identity. This theme has recurred repeatedly in my pieces of correspondence to you. Dr. Goings' collection enables us to consider this identity issue from a perspective apparently held by whites in which a given object or scene is construed or designed to exaggerate the features of African Americans, or conveys the impression that African Americans are of an inferior order, an impression that is often intended to amuse whites and insult Blacks. In this letter to you I will not engage you with philosophic discussions. Rather, as the old adage has it, "A picture is worth a thousand words." The *Palm Beach Gazette* has already published a few items from Dr. Goings' collection depicting the image about which I have been speaking. I am sitting here wondering what would have been your reactions to these pictures. Would you have laughed? Or would you have cried? How did you feel about white people collecting these types of items? Moreover, in your opinion, how did these items shape whites' (and Blacks') attitudes about Blacks? Please describe your feelings on this matter in your next letter to me. Although Dr. Alain Locke did not focus on the so-called Black collectibles as such, he threw light on the matter when he discussed the minstrel shows in which white performers in Black makeup entertained whites. As you were well aware, certain commercially produced collectible items were made in the distorted image of Black people from the 1880s to the 1950s. Such collectible items included advertising cards, postcards, household items such as Aunt Jemima and Uncle Moses salt- or pepper-shakers, mammy cookie jars, and even toys and games. Evidently there is a great interest among whites in such items. According to Dr. Goings, the items that are now regarded as Black collectibles were, as recently as the 1960s, used by many white people in a twofold manner: to promote certain products in the market-place, and as a propaganda tool to reinforce stereotypical views of Blacks as inferior beings. "The usage of these items" says Dr. Goings, "was quite widespread during the 1880s through the 1950s." In this connection, Dr. Goings noted a neat comparison: that the time frame of what may be called the *Black Collectible Era* coincided with the time frame when European powers were "carving out" the African states, so that the Europeans could more effectively maintain colonial rule over the Africans. It has already been observed that in the middle 1950s, certain African nations began to strive for independence. Similarly, in the U.S., by the time the Civil Right Movement shifted into a higher gear, the practice of manufacturing and displaying these collectibles was virtually ended. Many businesses stopped using these as promotional items. Whereas, prior to the 1960s, they were sold in so-called "dime stores," drug stores, department stores, and groceries.

Sincerely,

Dr. Johnny Washington, Ph.D.   April 12, 1997

Chapter 2

Africa, India, and Central America

Strife in Somalia and Ethiopia

Dr. W.E.B. DuBois (1868-1963)   Destinicity Letter #5

Re: Social Strife in Somalia and Ethiopia

Dear Dr. W.E.B. DuBois:

I believe you have by now received several letters from me. I do not expect a reply. Yet I feel the need to write you, nevertheless. I have always held you in the highest esteem as a person, and I respect you as a scholar. In fact you (and Dr. Alain Locke and Mr. Booker T. Washington) have been my role models over the years. Writing you is a way of informing people about some important issues. In one of my letters to you I forwarded a copy to Dr. Alain Locke, alerting him to the fact that in the sequel, I will address letters to him as I am now writing to you. In my previous letters to you I focused mainly on issues pertaining to ethnic appellations and the concomitant element of ethnic identity. Now, in this letter I will shift my focus to the issue of colonialism. Yet the "colonialism" letter that follows is related to the previous ones inasmuch as it is evident that colonial rule adversely affected the identity of the colonized, whether in Africa, Asia or Latin America. First, I wish to bring you up to date on what happened during the past few decades that gave rise to the political-social strife resulting largely from the consequences of colonialism in East Africa, especially Somalia as well as Ethiopia. I am bringing this issue of the effect of colonialism in Africa to your attention in part because you spent much of your long career examining this problem. Your own words and deeds inspired many Africans who strove for political independence. Admittedly most African countries, with the exception of South Africa and a few others, received independence by the end of the 1960s. The act of decolonization alone did not solve Africa's problems, Frantz Fanon's views to the contrary notwithstanding. I mentioned Fanon, because I think you were knowledgeable about his works. I trust also that you remember Aimé Césaire, the author of the book, *Discourse of Colonialism*.[1] He was one of the founders of the *Negritude* movement that you and Dr. Alain Locke inspired. In harmony with your own views, both Fanon[2] and Césaire were among those who dedicated their lives to the task of routing colonialism off the continent of Africa. Although the statement below was written by Césaire, it just as well could have been written by you. Here I am praising you as well as Césaire:

> First we must study how colonization works to *decivilize* the colonizer [the European], to *brutalize* him in the true sense of the word, to degrade him, to awaken him to buried instincts, to covetousness, violence, race hatred, and moral relativism; and we must show that each time a head is cut off or an eye put out in Vietnam and in France they accept the fact, each time a little girl is raped and in France they accept the fact, each time a Madagascan is tortured and in France they accept the fact, civilization acquires another dead weight, a universal regression takes place, a gangrene sets in, a center of infection begins to spread…(*Discourse*, p. 13)

Similarly, Fanon attempted to understand the causes and consequences of colonialism; he explored the role of violence in establishing and maintaining colonialism. But for him the instruments of violence within the colonial context can be doubled edged: although the colonizer used violence to exert rule, the colonized or oppressed must use violence against the oppressor to overcome oppression. No other options were available.

I understand you visited Africa in 1923. Whether this was your first trip to Africa I do not know. I also understand that Dr. Alain Locke visited Africa (Sudan) in the 1920s. I traveled to Africa for the first time in 1991, when I visited Kenya, a country that borders on Somalia. It was around this time also that Ethiopia was involved in internal strife. In Kenya I saw many Ethiopian refugees. The conflict under discussion in Ethiopia as well as in Somalia had ramifications not only in Kenya but also throughout East Africa. The situation was so bad in Somalia that in 1992, the U.S. Marines were sent to restore order. They were complaining that they were having difficulty in carrying out their mission because most of the roads were in disrepair. Of course, much of the information we obtained from the media sources about the plight in Somalia during the time was superficial and, in some instances, outright misleading. Some of this information resulted in reinforcing prejudices and misconceptions about Somalia. One common misconception was that many of the people of Somalia, of which the "war lords" were held up as representative, were merely tribes fighting among themselves. We need to remind ourselves that the Africans in Somalia were not unique in suffering civil wars. Dr. DuBois, you will recall that in the U.S. we had our own Civil War in the mid-nineteenth century, precipitated by the issue of slavery. You, yourself, were born in 1868, a few years following the Emancipation Proclamation. During the U.S. Civil War over 600,000 lives were lost. In the history of Africa, it is worth noting, rarely has an army associated with one of its countries or ethnic groups, invaded another country beyond the borders of Africa. Nor did the Africans invent firearms or other means of mass killing. Seen in this light, most of the wars on the continent of Africa were limited in aim and scope, a fact that is praiseworthy. With the exception of conquests of the Almoravids (Berbers) in the eleventh century A.D., Africans, for the most part, were not colonizers in foreign lands. Yet, as in the views of Césaire and Fanon, this "colonization factor," as exercised by Europeans, has to be kept in mind, in understanding the political-economic situation out of which arose the crises in Somalia, Ethiopia and other parts of East Africa. You were more knowledgeable about African history than I am. Nevertheless, allow me to recount in outline the relevant events connected with the history of the crises. During the early history of Islam, the Arabs colonized much of western Africa, and parts of Eastern Africa, including what is now Somalia. When Europeans were experiencing the so-called Renaissance or Reformation, between 1300 and 1500, the people of Somali origin and the Arabs fought the Christians, who had established a stronghold in Ethiopia during the early phases of the Christian era.

During 1884, the so-called "Berlin Conference" convened in Berlin. Dr. DuBois, I know that you were aware of this notorious conference, as you frequently made references to it in your works informing the world of how detrimental colonialism was to Africa. At that conference, Africa was formally divided among the European powers. It should be noted that their "informal" colonial ventures in Africa extend several centuries prior to the 1800s. The major European powers, including England and France, quarreled over the regions of Africa. The country that is now Djibouti was ruled by the French during the 1800s. Around this time, the British dominated the area that is now the northern part of Somalia. During the Second World War, the Italians conquered the neighboring country of Ethiopia; and during the Second World War, the former used poisonous gas on the latter.

52

Before the conquest of Ethiopia by the Italians, Ethiopia had existed throughout the ages without being subjected to the rule of another country, not even Egypt. During the Second World War, Ethiopian farmers, along with the British troops, drove the Italians out of Eastern Africa. In 1960, during high points of the U.S. Civil Rights Movement, Somalia achieved independence from European colonial rule. In 1969, the Somali government shifted to military rule, headed by Siad Barre. One of his main goals was to transform Somalia into a socialist or communist state. This prompted him to establish ties with the then-Soviet Union. With such an alliance established, this initially placed Barre at odds with the U.S. However, in 1974, certain political events developed in the region that prompted the Soviet Union to vacillate in its commitments to Somalia: In 1974, members of a then-Marxist movement in Ethiopia achieved control of Ethiopia's government, and thereby routed Haile Sellassie. Such an act effectively put an end to Ethiopia's monarchical form of government.

Prior to the "Ethiopian Revolution" of 1974, Ethiopia was on friendly terms with the U.S. But the political events that developed in 1974, the newly established Marxist rule in Ethiopia, forced Ethiopia to be abandoned by the U.S. Thus, Ethiopians turned instead to the then-Soviet Union for support. The Soviet Union in turn abandoned Somalia, as it were, and supported Ethiopia.

Confronted with the fact of having been abandoned by the Soviet Union, people of Somalia turned to the U.S. for military and economic support, which the U.S. generously provided, as recently as 1989. It is reported that Barre was a brutal dictator, comparable to the Somozas of Nicaragua. For example, it is reported that in July 1989, Barre massacred some 450 of his political opponents in Mogadishu. This, then, is a brief outline of the political situation of Somalia during the 1990s, which may provide an explanation as to why there was so much violence, hunger, starvation and social strife in Somalia and East Africa generally. Clearly, in recent history, Somalia was a chessboard, where the U.S. and the Soviet Union, like France and England before them, played out their power struggles against each other. The Super Powers "won" or "broke even" and Somalia suffered the consequences: hunger, starvation, disease, social strife, and the U.S. Marines.

During a March 19, 1997, Public Radio news report, it was announced that Mrs. Hillary Clinton was, during the time of the announcement, visiting Cape Town, South Africa, and that she was scheduled to spend an additional two weeks visiting other countries on the continent. (President Clinton is in his second term in Office, and he, himself, has not yet visited Africa). In this same newscast, it was reported that many African leaders were, at the time, meeting in Nairobi to discuss the current crisis in Zaire, where, within the past few months, over a million people have been massacred. While I was traveling from Kenya to Senegal in 1991, my plane briefly landed in Zaire, where as early as 1991, I could discern the volatile situation in central Africa. No, things have not gotten much better in East Africa since 1991.

Sincerely,

Dr. Johnny Washington, Ph.D.   March 20, 1997

Dr. W.E.B. DuBois (1868-1963)   Destinicity Letter #6

Re: A Journey to Africa:

Dear Dr. DuBois:

This is the lead letter to a series of letters I will write you in which I will recount my travel experiences in Africa. In this letter I will place the issues in the perspective aspects of geography, in view of fact that many people have a limited knowledge of Africa's geography or history. As early as the 1890s you expressed a deep interest in Africa, as evident by the title of your Ph.D. dissertation, *The Suppression of the African Slave-Trade to the United States of America, 1638-1870*, (1896). You were a pioneer in African Studies. The contents of this letter are mainly for the benefit of those who are not so knowledgeable about the African experience. The population of Africa is over 600,000,000. In Nigeria alone there are more than 120,000,000 people. Indeed Africa is an exceedingly large continent, second only to Asia in landmass. Among Africa's many countries, Sudan is the largest in area, covering 967,500 square miles. Slightly smaller than the state of Texas, Kenya covers 224,961 square miles. Despite its size and magnitude, Africa is one of the poorest continents on the planet. It was reported that at least 30,205,000 Africans were at risk of starvation in 1991, and at that time 240,000,000 suffered lingering hunger. Sudan, Ethiopia, Somalia, Malawi, Mozambique, Angola, Niger, Liberia and Burkina Faso and Mauritania— these are the worst off African countries. My report is not so much about starvation in Africa. Rather, in what follows I will recount my travel experience in Kenya and Senegal. And I will place in perspective historical and sociological facts about Africa in general, Kenya and Senegal in particular. Kenya is bounded on the East by Somalia and the Indian Ocean; on the west and northwest by Uganda and Sudan, respectively; on the north by Ethiopia; and the south by Tanzania. Like south Florida's climate, Kenya's coastal area is hot and humid, and includes a mixture of beaches, mangrove swamps, and coconut palms. But the terrain in Kenya is more diversified than in south Florida. Kenya has mountains, whereas Florida has none. Also in Kenya there are dry plains and rich, fertile highland. In fact, one of the highest peaks of land in Africa is Mount Kenya (in Kenya) whose apex is 17,058 feet above sea level. During my trip to Kenya, I also had the pleasure of visiting beautiful snow capped Mount Kilimanjaro (19,340 feet). As I traveled by car through Kenya, from Nairobi to Nakura, I observed the variations in the climate, the terrain and vegetation. During this journey, I crossed the Equator, which divides Kenya into nearly equal halves. Moreover, I came through the Great Rift Valley, in Kenya, where many authorities maintain, human evolution began approximately three million years ago. [It is sad to report that during August 7, 1998, a huge bombing occurred in Kenya and Tanzania, where hundreds of people were wounded or killed, an act that is believed to have been done by terrorists]. My entire tour of the countryside covered more than 200 miles one way. I had the opportunity to observe close-up the beauty and diversity of Kenya's environment and warm, friendly people. Within the highland regions where most of my traveling took place, the temperatures average about 67 degrees, with a yearly rainfall that ranges from 40 to 50 inches. The soil and water are quite suitable for farming. Corn (or maize) is a basic food for Kenyans. Their chief cash crops include coffee and tea. Tourism is also an integral part of the economy. The official language in Kenya is Swahili, by which some forty (40) tribal or ethnic groups communicate among themselves. Dr. DuBois, I am uncertain whether you spoke an African language, or, which African language you would have urged African Americans or Africans to adopt or promote. I find it interesting that the Bantu language group reflects many of the ontological or metaphysical principles found in Western philosophy. Furthermore, some music scholars go so far to say that the structure of blues or jazz music is imbedded in the Bantu language. To my knowledge in your book, *The Souls*, needless to say, you traced the blues and jazz forms of expression to African cultures but you never linked this thesis in a systematic way to any particular African language. I understand that before your death you were compiling an African encyclopedia, a project that has yet to be completed. In the major cities such as Nairobi, Mombassa, and Nakura, many of the people speak English through which national or international commerce is conducted. Each of the 40 tribes or ethnic groups possesses its own language. Thus, Swahili, and to a lesser degree, English is the major language for inter-ethnic group communication. The practice of Christianity is wide spread in Kenya. In many places, traditional religion was routed by the Europeans. The driver of the tourist car in which I traveled was from the Kikuyu (or Gikuyu) ethnic group, the largest in Kenya. About two million Kenyans constitute the Kikuyu group. (Professor H. Odera Oruka, the coordinator of the 1991 World Conference of Philosophy in Kenya was of the Luo group). Kalenjin, Kamba, and Luhya are other sizable ethnic groups in Kenya. A number of Kenyans (about 5 to 10 per cent) live nomadic lives, attending to crops and livestock—goats, sheep and cattle. The Massai, tall, slender individuals, is one of Kenya's most well known nomadic groups. Many rural Kenyans with whom I came in contact speculated that I was of Massai descent. Others were of the opinion that I came from a tribe that lived in the Lotikipi Plain, where the borders of Sudan and Kenya connect. In the early history of Africa, the Massai distinguished themselves as great warriors. During their early contacts with the Europeans, it is reported that they deliberately infected the Massai people with diseases and the Massai population was nearly destroyed. Kenya has a population of some 26,000,000, about 98 percent of whom are Black Africans. The city of Nairobi, which was founded by the British at the turn of the last century, includes approximately one million people.

I almost forgot to mention historic West Africa. By 300 A.D. a number of ethnic groups in West Africa began undergoing political transformations, resulting in greater centralized control of the region. Such transformations subsequently gave rise to the kingdoms that included Ghana, Mali and Songhay. The former reached its apex during the 1000s A.D. It was around this time also that China achieved its cultural apex. With the downfall of Ghana, there arose in the 1200s the Kingdom of Mali whose boundary lines included what are now called Gambia, Guinea, Senegal, Mali and Mauritania. It is ironic that when the African people achieved glory during the 1000s through the 1300s, Europeans were in the "Dark Ages." Today, history has again reversed itself. During the past few centuries, Europeans have enjoyed excessive economic and cultural developments, while much of

Africa is facing starvation. An explanation for this disparity between the condition of contemporary Africa and Europe lies in the fact that for centuries, Europeans exploited Africa's natural and human resources, it goes without saying. Some five hundred years ago (1492) Columbus "discovered" the "New World" and nearly a century latter (in 1588) the practice of African slave trading was well entrenched. The beginning of the slave trade marks Europe's colonial presence in Africa; most of the New World slaves were taken from West Africa. From the latter part of the 1500s onward, their colonial rule became more pervasive throughout the continent. Dr. DuBois, you have already written extensively about the slave trading effect, and thus I will not belabor this point. During the latter parts of the nineteenth century European governments began to claim for themselves parts of Africa. That is, they began to carve out parts of Africa. At the turn of the century, European governments reached an agreement as to the manner in which the continent of Africa would be divided among themselves (into some fifty non-sovereign states). Moreover, it was also decided at this time which European governments—Belgium, France, Germany, England, Italy, or Portugal—would benefit from and exert colonial rule over certain African nations. This was a time when the consequences of New World slavery were in their heydays. It seemed as though Europeans in Europe and in the Americas attempted to double-team or gang-up on, as it were, Africans: Colonialism hemmed-in Africans in Africa, and slavery hemmed-in the people of African origins in the New World. This resulted in what I have called the "African bad destiny," like a bad battery that saps a car's electrical energy. Slavery rendered African Americans a "rotten Destiny;" its core was filled with worms. Following the Second World War, many African countries began to struggle for independence. In 1954, the year that Dr. Martin Luther King, Jr. initiated the Civil Rights Movement in the U.S., the Algerians desired independence from French rule, and rebelled against the French. Needless to say, it was a long drawn out struggle that resulted in independence in 1962. Ghana received independence from Great Britain in 1957. Today, the Black Africans in South Africa have recently achieved independence, with Nelson Mandela as its first Black president. Some say that he is performing well. Inspired by the political agitation of Jomo Kenyatta and other members of the so-called Mau Mau movement, Kenya gained independence from Great Britain in 1964, and Kenyatta was installed as the first president of the Republic of Kenya. The current president of Kenya is Mr. Moi, whose portrait hangs on almost every public facility in Kenya.

Although Kenya purports to subscribe to a democratic form of government, many observers maintain that with Kenya's one-party system, it is more of an autocratic form of government. Many of the participants of the 1991 World Conference of Philosophy attested to the observation that Kenya appeared to be quite repressive. While we were in Nairobi during the above-mentioned conference, it became quite evident to most observers that Kenya's politically repressive atmosphere effectively discouraged free expression. This conviction was shared by most of the members of the International Development Ethics Association (IDEA), who attended the conference. The president of IDEA echoed the sentiments of IDEA's members by addressing an open letter to the media. While it is the case that the military is not dominant in Kenya's government as it is in many other African countries such as Zaire, clearly the Kenyans do not seem to enjoy political freedom to the same degree as the people in a few other African nations. Following my visit to Kenya, I traveled to Senegal, where I presented a paper at the Thirteenth Annual Caribbean Studies Association. This twofold trip allowed me to get a broader view of contemporary African society. In the sequel, more attention will be devoted to recounting my visit to Senegal.

Sincerely,

Dr. Johnny Washington, Ph.D.   April 5, 1997

Dr. W.E.B. DuBois (1868-1963)   Destinicity Letter #7

Re: A Journey to Africa: Part I

Dear Dr. DuBois:

I visited Africa for the first time during July 1991, during which I attended a World Conference of Philosophy (WCP) conference, held at the Kenyatta International Conference Center in Nairobi, Kenya, July 21-25, 1991. When I returned from that trip, I began publishing a series of articles recounting my visit to Africa, in the *Palm Beach Gazette*, at the invitation of Mr. Lee Ivory, the publisher. I am pleased that Mr. Ivory was at that time interested in my articles.

Prior to my publishing such articles in the *Palm Beach Gazette*, little attention by the white or African American media was devoted to Africa. Indeed, I am pleased to say that since 1991, the *Palm Beach Gazette* has kept Africa in the forefront of the news. It focused not so much on negative aspects of Africa, with all of the emphasis of starvation and civil wars. Rather, Mr. Ivory seems to be equally interested in providing a more balanced picture of Africa. In this regard, the *Palm Beach Gazette* needs to be commended. I believe this viable and courageous newspaper has your blessings. Many of this newspaper's editorial articles are bold and provocative, in the spirit of many of your own editorials. In one of my previous letters to you, Dr. DuBois, I focused on Somalia. In this and a few other ones, I shall present to you in these epistolaries my traveling experience in Kenya as well as in Senegal. I first became aware of the plans for the above-mentioned Nairobi conference during May 1989. Professor H. Odera Oruka, now deceased, who, from January 1989 through July 1989, was a Visiting Scholar at Earlham College in Richmond, Indiana invited me to Earlham College in May 1989 to give a public presentation.

I first met Professor Oruka at the international African/African American Philosophy Conference that was held at Haverford College during 1982. During my Earlham visit, the late Professor Oruka called to my attention his plans to convene this 1991 conference. As we discussed these plans in May 1989 he not only extended to me a special invitation to attend the Conference, but he included me in a more formal manner in its planning. He assigned me the duties as the North American Coordinator of this conference, and urged me to invite other individuals throughout the world to attend. Moreover, he made clear that although the conference would appeal to scholars in diverse fields, nevertheless its basic orientation would be philosophic, having as its general theme: *Philosophy, Man, and the Environment.* Certain specific themes included: i) *Traditional and Modernity, and Their Impact on Social Change and Social Structure*; ii) *State Sovereignty, Regional Identities and Cross-National Co-operation*; iii) *Human Rights and the Concept of the Rights of the Peoples;* iv) *The Role of Government vs. That of Businesses*; v) *Law and Justice;* vi) *Dependence Between North and South*; vii) *Phenomenology Today*; viii) *Women and Philosophy*; ix) *Are There Women Rights?;* and x) *Philosophy and World Hunger.* Dr. DuBois, I believe any of these themes would have aroused your interest. The same is true of Dr. Alain Locke's. Let me tell you about a related conference in which I participated.

During July 1989, the International Development Ethics Association (IDEA) convened its Second International Conference in Merida, Mexico.

There I presented a paper entitled, "An Outline of An Economic Ethics for Developing Countries." During this conference I was selected to be a member of the IDEA's Executive Board, and during several of its business meetings at this conference plans were made for IDEA's future conferences. In planning for the Third International IDEA conference, Nairobi was considered as a possible site. Since Professor Oruka was still in the U.S. (in connection with his appointment at Earlham College), he was scheduled to attend this Merida conference, but unfortunately he had conflicting obligations and was unable to attend. Had he attended, I believe we could have persuaded the IDEA officials to allow its next major conference to be held in Kenya. In any event, it was subsequently decided that the IDEA's Third International conference would be held at the *Universidad Nacional Autonoma*, June 21-27, 1992 in Tegucigalpa, Honduras, the results of which I shall describe to you in subsequent letters. The conviction was that since many of IDEA's members reside in Latin America, a country such as Honduras would be desirable. And in light of this, that it would be much easier for Latin Americans to obtain financial support to attend a conference within that hemisphere rather than to travel to Africa, or some other parts of the world. Further, it was agreed that because of the distance between the U.S. and Africa, it would be difficult to convene an IDEA conference on the continent of Africa. It would, however, be plausible to hold in Africa a "mini-conference." That was the thinking of the group. Thus, when the plans unfolded for the WCP, this seemed to have provided ample opportunity for the IDEA's mini-conference to be held in Africa. Such a conference would involve allowing IDEA to convene several sessions within the context of the WCP, a plan which IDEA adopted. Further discussions of the possibility of holding such a conference in Africa was conducted at the December 1989, Eastern Division meeting of the American Philosophical Association (APA), which met in Atlanta. The Executive Council of IDEA, of which I am a member, met during that APA convention, and the WCP was an agenda item. Subsequently, two other individuals of IDEA, together with myself, constituted a subcommittee to establish a theme(s) for the mini-conference in Africa. We decided to offer several sessions, each of which would be 90 minutes in length, involving no more than three participants. These sessions would be under the rubrics: *Ethics, Economics and Development,* and *Ethics, Technology and Development.* As one can easily see, such themes were narrowly focused and as a result, the individuals who responded to the invitation to attend the WCP in connection with IDEA were limited, as was anticipated. In selecting these two narrowly focused themes, the intention was to provide issues that would be compatible with those of the WCP, without being repetitive or conflicting. On October 1, 1990, I communicated with Professor Oruka by fax to inform him of the tentative plans of IDEA, followed by a more detailed proposal that he found acceptable; it was sent to him on December 3, 1990. Most of my pieces of correspondence to Professor Oruka were carried out by fax transmittal; partly because of the difference in time zone, (Nairobi is seven hours ahead of Florida), it was extremely difficult to communicate by phone. It takes a letter at least fourteen days to get to Kenya from the U.S. On a few occasions, the fax line to Nairobi was busy, and I would have to wait for two to three days before I could get through. Professor Oruka never faxed me, but he phoned me several times. Rarely did I phone him. Because of the shift in time zones, I found it much easier to send him faxes. It is relatively inexpensive in the U.S. to place a phone call or fax to Kenya, as you can imagine. But it is extremely expensive to place a phone call or fax from Kenya to the U.S. When I was in Kenya, I called home and spoke about three minutes. The bill was $50. in U.S. currency. As the IDEA conference director, one of my tasks was to encourage people to submit paper proposals. Since I already had the mailing list of IDEA's participants of the 1989 Merida conference, I relied on it, along with others, in inviting individuals to attend this WCP.

On October 1, 1990, I faxed Professor Oruka (who had by now returned to Kenya) a progress report, in preparing for a more thoroughly developed IDEA proposal: That as of October 1, 1990 I had received commitments from approximately eighteen individuals who wished to participate in the WCP, under the aegis of IDEA. That, moreover, I anticipated that the number of participants would increase to approximately 30 to 40 individuals. Also in this report I mentioned the IDEA's themes, as indicated above. By the time the March 15, 1991 paper deadline (for paper proposal to be submitted to me) arrived, I had in fact received commitments from some thirty individuals.

Between the time that IDEA began to make initial plans for the WCP and the March 15, 1991 deadline, a number of negative events occurred that, I believe, discouraged individuals from attending the WCP: i) political unrest in Kenya; ii) the Gulf War; and, iii) and economic recession in the U.S. Some economists might see a causal relation between ii) and iii).) The latter mentioned prompted many colleges and universities, including Florida

Atlantic University, to reduce, if not eliminate temporarily, their travel funds. Clearly such negative factors discouraged some people from attending the WCP. On several occasions during the height of the political unrest in Kenya, Professor Oruka phoned me from Nairobi to assure me that it was "safe" to come to Kenya. In the initial plans, approximately 2,000 people were expected to attend the WCP, but largely because of the above-mentioned negative factors, that number was reduced to approximately 500 individuals, fifteen of whom were associated with IDEA. A few Kenyan women, it is worth noting, were actively involved in the WCP: they presented papers and participated in discussions. But to my knowledge, no Kenyan women were included as keynote speakers and most of the conference planning was done by men. During one of the general sessions, a few of the women who presumably were associated with the University of Nairobi, expressed opposition to the theme: "Philosophy, Man, and the Environment." They had problems with the notion *Man* which they felt should be changed to *People*. The problem with *Man* they rightly held, is that it belittles or ignores the role of women in African society. Reactions of that type suggest that African women are achieving a new level of political awareness and self-assertion. Ali A. Mazrui was one of the WCP keynote speakers; he was born in Mombasa, Kenya but during the WCP he taught at State University of New York at Binghamton. He "stirred" the political atmosphere at the conference, and indeed throughout Kenya, on a numerous occasions when he openly criticized the Kenyan government for restricting political freedom of the people. Some people claim that the Kenyan women are oppressed in a twofold manner: on the one hand, by virtue of the male dominated tradition, and on the other, the current repressive political system. Indeed the women in most African countries have a difficult time, a point that was hotly debated during certain sessions of the conference. Since you devoted much attention to the notion of double-consciousness in describing the African American situation in North America, I have a question for you: does this description regarding double-consciousness apply also to Africans generally, and if so, do African women suffer double-consciousness to the same degree as African men? Similarly in what ways, if any, do the people of African origins in the Caribbean or Latin America also suffer double-consciousness? A report has it that regardless of the culture, women generally experience pain and pleasure different from the way men experience these. I am assuming here that double-consciousness involves pain, frustration or disappointment. There are degrees of pain and pleasure. Permit me to comment on the above question. When I visited Africa I observed a problem analogous to the one you described in your book, *The Souls*. But in the African context I think it is appropriate to construe this identity issue not in term of double-consciousness. Rather, I believe here we have what may be called a "triple-consciousness" problem, although I know my peculiar term is unHegelian. (It would take us to far out of the park to open up the debate regarding the identity problem that Blacks suffer in a country such as Brazil). It seems to me that the African is struggling with a threefold identity problem: first, his/her tribal (or ethnic) consciousness; secondly, his/her, national consciousness; and thirdly, his/her African consciousness. He/she is torn between expressing loyalty to his/her ethnic group, and the nation-state, say, Kenya or Ghana, which is an artificial construction imposed on the African situation. Then, there is the "outside world," that is, Europe or the U.S., with which he/she becomes familiar through CNN TV, the British Broadcasting System Radio, the Internet, other media sources, as well as through tourists from abroad. These force the African to regard himself/herself as someone who is utterly different from (or "inferior" to) those in other parts of the world. Dr. DuBois, I believe you were knowledgeable about this triple-consciousness problem, brought on largely by colonialism, just as, on your analysis, in the U.S. slavery and segregation induced in African Americans a double-consciousness that often resulted in "invisibility" (Ralph Ellison). Another question: Do certain African or African American women, or women of African origin, experience invisibility similar to the way that apparently African American or African men experience invisibility? Still another question: Do white men or women experience double-consciousness or invisibility?

Finally, what is the relation between double-consciousness, invisibility, and self-esteem? You tried to approach the double or triple-consciousness problem by introducing Pan-Africanism, an organization that you assisted in founding, aimed at establishing political and cultural unity of the people of African origin. In any event, I will close now. I learned much about Africa during my 1991 visit and by reading your works.

Sincerely,

Dr. Johnny Washington, Ph.D.   April 7, 1997

CC: Dr. Alain Locke

Booker T. Washington

Marcus Garvey

Dr. W.E.B. DuBois (1868-1963)   Destinicity Letter #8

Re: A Journey to Africa: Part II

Dear Dr. DuBois:

In this letter I will return to a theme I considered in my first three and last letters to you. The theme to which I keep returning pertains to ethnic or racial identity: Who are Black people? What relation do they have to Africans? I realize I am repeating an issue I raised in Destinicity letters two and three when I considered various ethnic appellations. When I attended the World Conference of Philosophy in Nairobi, Kenya during July 1991, I had the opportunity to observe the Africans' reactions to this question of ethnic identity. The issue arose for me when I was asked to fill out the immigration forms in preparation to go through customs at the Kenyatta Airport in Nairobi. (The immigration agent himself was presumably a Kenyan). The form to be completed required that one identify one's nationality. I did so by indicating that I was an "African American." In responding to the query about nationality, I deliberately chose not to use the "appropriate" term "American." Instead, I used "African American." When the immigration agent examined my form, he initially seemed puzzled. He demanded clarification. But in this quest for clarification, he was not rude; in fact he was quite polite. He wanted to know precisely what I meant by "African American." He wanted to know whether I meant that I was an indigenous African who resided in America but was at the time returning on a visit to Africa. In an attempt to clarify the issue, I told the agent by "African American" I meant that I was a Black American whose ancestors were brought to America as slaves a long time ago. But such a reply apparently was not satisfactory. He seemed equally puzzled by the appellation "Black American." Undoubtedly he was acting on the assumption that if he could connect my identity to a particular geographic or tribal location, he could become clearer about who I was. Thus, he asked me what tribe was I from? To really throw him off guard and to inject a little humor in to the situation, I told him I was from "tribe Alabama." By "tribe Alabama" I was referring to the State of Alabama in the U.S., where I was born and reared. Once I gave him such a reply, he permitted me to go on through the customs line. I think that the customs agent finally figured out that I was not from one of the local tribes. Among the people with whom I traveled to Nairobi was an indigenous African originally from Ghana but who now lives in the U.S. I

64

informed him of the brief interrogation I had undergone with the immigration agent. I queried my fellow African traveler, Professor Kwasi Wiredu, as to whether the immigration agent actually knew what "African American" or "Black" meant. Or, whether the agent was just giving me a "hard time." My fellow traveler's reply was that the agent probably did not know what those two above-mentioned appellations meant. Rather, he perhaps was more familiar with "Negro," or "American" or "U.S. Citizen." Clearly, the sense of race among Africans on the continent of Africa differs from the sense of race felt among African descendants in the U.S., whom I shall refer to in this context as "Blacks." Here in the interest of brevity, I will not elaborate on the difference between racial awareness and tribal awareness. While it is the case that Blacks in the U.S., since the 1960s, have acquired this heightened sense of affinity for African culture, a good question is whether such a sense is reciprocal, that is: to what extent have the Africans acquired a psychological or cultural appreciation for the African American experience in the U.S.?

During my traveling in both Kenya and Senegal it became evident to me that the Africans regarded me as an American, and only incidentally as an African descendant. In my book *A Journey*, and elsewhere I left undecided as to which was the most appropriate appellation, "Black" or "African American." In fact, I went on to offer a novel appellation "Africantude," which I believe would be suitable for your Pan-African model. I believe that the Destiny model is more powerful, simpler, and richer than your Pan-African model.

Sincerely,
Johnny Washington, Ph.D.   April 12, 1997

Dr. W.E.B. DuBois (1868-1963)   Destinicity Letter #9

Re: A Journey to Africa: Part III

Dear Dr. DuBois:
I have already recounted to you in a prior letter the results of the 1991 WCP that was held in Kenya. Now, I will develop this issue more fully. The WCP provided the opportunity to examine a number of diverse themes, and topics. An important issue, which I felt should have been discussed but

which was not adequately considered, pertained to the notion of national Destiny in African societies. I know you would be interested in this topic, since much of your own works revolved around this theme. It seems to me that this inattention to the Destiny theme created what may be called an intellectual void in the WCP. I am here using "national Destiny" interchangeably with "collective vision" or "national goal" of a people. Already you are familiar with my definition of Destiny; thus, I will not elaborate on it here. However, it is worth noting in this context that Dr. Martin L. King, Jr. implicitly accepted Hegel's notion of Destiny, when the former incorporated into his own thinking the dialectical law of history, whose goal is self-knowledge and freedom. As a Third World Country, Africa is—and indeed ought to be—concerned with the development of its material and cultural resources. But, as I will demonstrate shortly, before Africans can get clear about their development priorities, they first need to get clearer about their continental Destiny, that is, the African Destiny. Some observers have noted that the WCP devoted too much attention to environmental issues. This attention attracted scholars such as Drs. Moravcsik and Føllesdal, two of my former Stanford University teachers. (By the way, Dr. Moravcsik was a student of C.I. Lewis who was Dr. Alain Locke's classmate at Harvard University, where Lewis's teachers included Josiah Royce, who taught both you, and Dr. Alain Locke). It did seem somewhat odd that the WCP, convening in Kenya, would devote so much attention to environmental issues. Was Africa the appropriate place to devote so much discussion to environmental problems? That question arises because the convening of the African conference within which environmental issues were given much attention makes it seem that Africans themselves are to blame for the air and water pollution, soil erosion, and the destruction of the rainforest. Without doubt, many of these problems in Africa were brought on by the colonial ventures of the Europeans, who possessed the technology to degrade and exploit the natural and social environments on the continent of Africa. I attended a session in which scholar presented a paper that focused on the views of the American philosopher John Rawls's book, *A Theory of Justice,* in explaining the injustice involved in the destruction of the Brazilian rainforest. The question that arose in my mind was: How did all this relates to the African situation? I am not saying that environmental issues are not important. Certainly they merit consideration. My point is that one wonders whether the WCP in Kenya was the appropriate place to devote so much attention to such issues. Besides the environmental problems, the notion of economic development was provided adequate consideration at the WCP. [This was made possible by the mini-conference of the International Development Ethics Association (IDEA) conference, of which I was the convener]. It would have been desirable had equal attention been devoted to examining the notion of national Destiny, a consideration of which would have placed in a brighter light what is required to develop efficiently Africa's economic and cultural resources. My presentation lasted about 30 minutes. When I finished presenting my paper, a few minutes were made available for questions and answers. I entertained a number of queries and comments, two of which stand out. First, an African student made the comment that he felt uncomfortable with my definition of "Destiny." That in most cases when people speak of Destiny it is connected with a religious doctrine or with some political ideology that can result in the establishment of a repressive political state or a state of affairs that heightens tribal or racial tension.

My reply was that, apart from the Hegelian notion to which I alluded, the Destiny model did not include any religious or political ideology. Rather, I had in mind merely an abstract ideal of freedom that is to serve as a galvanizing force. A similar critical comment was directed to me by a European scholar. He said that he was uneasy with the Destiny model, an uneasiness introduced by my connecting Destiny with the notion of race or ethnicity. He added that such a conception of Destiny could easily degenerate into a situation similar to the one that arose in Nazi Germany in the 1930s and 1940s, where the false sense of racial superiority was advocated by the German government. My response to the student's comment was equally applicable here: that no political or religious ideology, comparable to that to which the Germans once subscribed, was associated the Destiny model. On the contrary my sense of Destiny was intended to inspire hope, so that Africans and the people of African origins could constitute their own Destiny. My views here are in the spirit of your own. The foregoing observations I have made about the WCP were not intended to convey the impression that the WCP was without any merits. Rather, what I intended to do was to examine the strengths and weaknesses of the July 1991 WCP, and to suggest ways in which it could have been more effective. Thus, I would like to see a similar conference convened in Africa that focused exclusively on your works dealing with, say, the Talented Tenth model, and the related ideal of Pan-Africanism, within the schema of African Destiny. I wish at this time to thank the late Professor H. Odera Oruka and his staff for making that Conference possible. As I mentioned in the October 17, 1991 issue of the *Palm Beach Gazette,* many observers have noted that the positive effects of that conference are still being felt throughout the continent of Africa. I recently received a public announcement that during August, 1998, the WCP will convene in Boston. If I get the opportunity to attend that conference, I will present my notion of the African Destiny. I have developed it more fully since the 1991 WCP.

Sincerely,

Dr. Johnny Washington, Ph.D.   April 8, 1997

Dr. W.E.B. DuBois (1868-1963)   Destinicity Letter #10

Re: A Journey to Africa: Part IV

Dear Dr. DuBois:

This is a continuation of my story concerning my July, 1991 traveling experience in Kenya. As I might have mentioned in my previous pieces of correspondence to you, I had the opportunity to see much of Kenya's countryside when I attended the World Conference of Philosophy (WCP), July 21-25, 1991. I wished you could have accompanied me. I did, however, enjoy the company of Dr. Rudy Mattai and his wife Barbara whom I had met when I was teaching at the University of Tennessee, Martin, where the three of us were on the faculty. The Mattais, who had traveled to Africa several times before, and I toured Kenyan countryside by way of a private tour guide, arranged by Rudy who by the way is originally from British Guyana. During my traveling through Kenya, I was struck by the widespread and intense deprivation that the people suffered. Most of our traveling was done by automobile. We began one of our journeys on what appeared to have been a paved road that lasted for approximately 25 miles, after which its conditions deteriorated, intermittently, to the point that its "pot holes," some wide enough to contain the front-end of a Volkswagen, were dispersed every 100 yards or so. In certain places the roads would be without pavement altogether. Consequently, we had in many instances to reduce our speed to 5 to 10 miles per hour. In other areas, however, the roads were good. As in Nairobi and in other Kenyan cities, there are many signs of starvation in Kenya's countryside, where in most instances there was no running water or electricity. In Kenya as in many other African countries, most of the people live not in the cities but in rural areas, where Western medical services and supplies are quite limited. Such problems are compounded by the factors of starvation and hunger. In Somalia, for instances, such factors are made worst by social strife, as I mentioned in one of my previous letters to you. During the year 1991, in Kenya alone some 40,000 people died of starvation. On the west coast of Africa in Gambia, a tiny country largely surrounded by Senegal, in 1991 there were many thousands deaths brought on by starvation.

It has been observed that in many cases when the U.S. provided food to the African countries, it did so with "strings attached." Apart from the starvation aspects, many parts of Kenya possess a transcending, awe-evoking serenity, pervasive throughout the region. I was inspired by the beauty of Kenya's lands, with the diversity of plants and animals. What was most impressive was the beauty and kindness one feels among the people. In traveling throughout the countryside, some of the people with whom I came in contact reminded me of the qualities of kindness exhibited by the people I knew when I was a child growing up in rural Alabama. Some reminded me of Aunt Mattie, and Uncle Lee. One lady resembled my first grade teacher. A farmer reminded me of the local blacksmith who was a friend of my father during my youth. Besides being delighted by the presence of the wonderful people in Kenya, I was as I indicated, inspired by Kenya's natural beauty that is embedded in and dispersed throughout nature. Let me briefly share with you a particular inspiring moment, still vivid in my mind, that I experienced while visiting Kenya's Nakura National Park, about 100 miles north of Nairobi. The experience I will here briefly describe bordered on the realm of the transcendent. We arrived in the park late, around 7:00 p.m., slightly before closing time, before the sunset. Thus, we arrived at a moment between the setting sun and the rising moon at twilight. The tour guide took us to one of the main attractions, the "Pink Lake." "Pink Lake" was an appropriate name for the lake, since it was the habitat for pink flamingos and other water birds. At the time we were there some 10,000 flamingos gathered on the lake, and their presence turned it virtually pink, the scenery of which was accentuated by the setting sun and rising moon. I remember very clearly, as we stood on this lake's edge, a hippopotamus, partly submerged in the lake, stood motionless for several minutes (at least a half hour) with its mouth opened wide. That particular hippopotamus was surrounded by at least ten others, as though they were protecting her, or engaged in some hippopotamus rituals. Encircling us nearby were the antelopes, zebras, monkeys, colorful ostriches, storks and other wild creatures. Yet in the middle of this wildlife scenery, there seemed to have been a certain harmony, the likes of which one is unable to experience in south Florida or in the U.S. generally. At that time I saw no signs of struggle or aggression among the animals. Rather, a high degree of harmony seemed to radiate from the concentration of different species of animals, the grassy lands, the bluish sky, and the pinkish water. Such a harmony seemed to have been reinforced by the calmness, stillness, and connectedness of the people, the plants, the animals, the serenity of the lake, the seemingly red, saucer-shaped rising moon, and the penny-colored, setting sun. The swift stilt-legged, as though spring-loaded, antelopes, moved in contrast to all this calmness. As we stood in the park, the gray dust on the lonely trail on which we drove was beginning to settle in the background. About 300 feet away a few European tourists gazed at the Pink Lake. It seemed that the temperature, humidity and wind and all other natural elements struck a perfect balance, under the ideal temperature conditions: 72 degrees Fahrenheit. No mosquitoes, no snakes, no frogs, no spiders. (I have a deep, irrational fear of snakes, even the harmless ones. While in this seemingly ideal setting, I still had the instinct to watch out for snakes). It was at this precise moment that I felt a certain tranquility, a harmony of harmonies, a oneness with nature, the likes of which I never felt before. As I was standing there absorbing all of this, I said to myself: would not it be nice if I could transport this environment back to the south Florida area with its traffic jams, thieves, fax machines, video games, lottery games, dog racing tracks, car phones, voice mail, e-mail, liquor stores, churches and prostitutes. It was precisely at that moment in the park that I seemed to have a glimpse of what the philosopher Kant called the noumena realm, within which are included the eternal ideals of God, Freedom and Immortality. This is the realm that informs the Destiny of the individual and the human race.

A scientific hypothesis has it that human evolution began in Kenya or in and around the Great Rift Valley in East Africa some 3.5 million years ago. I would be surprised and disappointed to find out that such a hypothesis was wrong. I am convinced that with all that beauty and harmony, something significant got started in this region of Africa millions of years ago. Dr. DuBois, did you experience a similar feeling when you visited Africa? In *The Souls* you connected the spirituals and the blues to the African heri-

tage. I believe you are right in this observation. When I listen to the blues on occasion it seems to me as though I am transported to the realm of harmony similar to the experience I enjoyed in the Kenyan park. It is unfortunate that today much of the land of Kenya and indeed throughout Africa is undergoing rapid deterioration—air or water pollution and the destruction of the rainforests—a deterioration whose intensification was largely brought on by the European colonialists.
Sincerely,
Dr. Johnny Washington, Ph.D   April 9, 1997

Dr. W.E.B. DuBois (1868-1963)   Destinicity Letter #11

Re: A Journey to Africa: Part V

Dear Dr. DuBois:

If you are not tired of hearing about my 1991 trip to Kenya, in what follows I will provide you more details. Bear with me. The day was July 26, 1991, following the World Conference of Philosophy (WCP), that convened in Nairobi, Kenya. It was around 10:00 a.m., and the weather was cool, with misty overcast. In various instances, we had to turn on our wipers, to clear the moisture that accumulated on the windshields, as we drove from Nakura National Park in Kenya, to what I now call the "fog scene," an area on the edge of a portion of the Great Rift Valley in Kenya. I will provide in the discussion below a more precise description of the site of the Great Rift Valley. At the point where I stood, its walls were some 8,000 feet deep. The Great Rift Valley, like the Grand Canyon in the state of Arizona in the U.S., is an unusual geological site. Both originated in prehistoric times. Like the Grand Canyon, certain points of the Great Rift Valley are a mile deep, and some twenty miles wide, the distance between Boca Raton, Florida and Fort Lauderdale, Florida. But the analogy between these two geological sites ends here. Whereas the Grand Canyon, at whose bottom flows the Colorado River, is approximately 277 miles long, the Great Rift Valley, which is constituted by a system of valleys, covers an area of more than 4,000 miles. It extends from Syria, through the Dead Sea, (which is 1,310 feet below sea level), the Sea of Galilee in the Middle East, through Kenya in East Africa, and through Mozambique in Southeastern Africa.

The Great Rift Valley is the result of a combination of geologic events, including volcanic eruptions and the formation of mountains during the earliest career of the Earth. In many areas along the Valley, one finds the most fertile land on the planet.

I traveled to the "fog scene" by way of the same car and driver by which I had traveled during the previous day, when Drs. Rudy and Barbara Mattai and I visited the Nakura National Park. During the morning of the July 26, the Mattais and I went on separate journeys. They stayed in East Africa for another week; later that day I departed for Senegal. Thus, in the absence of Barbara and Rudy, when I traveled to the Great Rift Valley, I was accompanied by the driver only, who claimed that he was a member of the Massai tribe, of which he was proud. I missed the company of the Mattais; Rudy, who knew quite well the "African psychology," often assisted me in negotiating transactions in the marketplace. I felt abandoned when Rudy and Barbara's path departed from my own.

As we drove to the Great Rift Valley's site in question, we were greeted by vendors, about twenty-five men, all of whom were from the Massai tribe, one of the largest tribes in Kenya. They wanted to sell me various items, including an ashtray in the amount of $60.00. One boasted to me that they often sold to Europeans ashtrays and similar items for $60.00 or more. I said to myself: "You will not sell me an ashtray for $60.00." First, I was reluctant to buy such items, because it was clear that they were trying to overcharge me. In an effort to convince me that I should buy a spear, about five vendors surrounded me and began withdrawing the swords and spears from the scabbards, indicating to me the degree of perfection with which these items were designed. Made of steel, the swords were between two and three feet in length. Can you imagine how apprehensive I may have felt, with this shiny, steel sword pointed directly at me, as I stood on the edge of 8,000 feet of foggy cliff? While all of this negotiating was going on, my driver whom I had used as a "broker," as a go-between myself and the vendors "disappeared," leaving me momentarily to defend myself. His disappearance, for about fifteen minutes, I think, was part of the sales gimmick employed to induce me to make purchases without the benefit of a "broker." I did not buy the sword, but I did at a reasonable price buy a spear and two small elephant figurines.

When we first approached the site of the Great Rift Valley, one of the vendors who appeared to play the role of a curator, greeted me. Then he uttered the phrases, "This is where it all began." And he repeated: "This is where it all began." Since the engulfing fog was so thick, I could barely see twenty feet ahead. The curator continued: "You are at the site of the Great Rift Valley. This is the general location where human evolution got started millions of years ago." My driver who by this time had returned to the scene gladly informed me that this general vicinity of the Great Rift Valley was also the homeland of his tribal ancestors, the Massai people, who were the direct descendants of the "original man," from whom human beings have derived. The fog scene was noteworthy in itself in that as I stood on the wall of the cliff, I was 8,000 feet above the floor of the Valley. But due to the heavy fog, I could not see adequately the beauty of the scene below. When I was told that the Valley below marks the general location from which the human species evolved, I really felt that I had gone back into time, a time when the human species was in the earliest phases of its career approximately 2.5 millions years ago. Intermittently a break in the fog occurred, and I could momentarily get a glimpse of the magnitude of the cliff over which I attempted to gaze. Frequently the thought occurred to me that, if this is the place where it all began, and if I could only "see" through the fog, I could satisfy one of my life-long curiosities, expressed in the question: Where on Earth is the "Garden of Eden" located? As I gazed across the "void," I had the feeling that perhaps I was in the general vicinity of the "original position," what cosmologists called the point of "naked singularity." Certain cosmologists are of the opinion that the universe arose out of a single point of energy-matter, concentrated in a form about the size of a thumbtack, called "the point of singularity." The point of singularity is the original point of energy preceding the big bang that resulted in the formation of the cosmos between fifteen and twenty billion years ago. When I approached the Great Rift Valley in Kenya, the thought occurred to me that I was now at a privileged position to deepen my understanding of the place and role of human beings in the cosmic schema of things.

Through the study of biological evolution, I have tried to identify the cultural or biological "singularity" of the human species. By visiting the Great Rift Valley, it seemed as though I was able to "witness," if not "see," what appeared to have been the point of singularity of human evolution. It is a well-established scientific fact that human evolution began in East Africa. Another analogy may be drawn between the point of "singularity" which I have associated with the Great Rift Valley and the point of singularity of the cosmos. Scientists can approximate the age of the universe by employing radio telescopes and other instruments. Many scientists hold the view that there exists an element of intense, thick radiation, called "black body radiation" near the point of singularity, about 19 billion years back in time, if we assume that the big bang occurred some 20 billion years ago. That this black body radiation is so thick that no known instrument can penetrate its wall to actually "see" the point of singularity. Beyond the point of singularity, all logical or causal relations are reversed or they exhibit results that are totally different than those of which we are familiar. I am attempting to hold that beyond the point of singularity lies the realm of transcendent Destiny and of Nature's Destiny.

The analogy between black body radiation and the above-mentioned fog is clear. It seems that this black body radiation in the cosmos was playing a concealing role comparable to the "fog" that was trapped in the Great Rift Valley at the time of my visit. (Obviously such fog is not always trapped in the Great Rift Valley; it just happened to be present during the time I made my visit). Dr. DuBois, in your book, *The Souls*, in the spirit of Kantianism, you spoke of a Veil beyond which our sense perceptions cannot reach. Commentators hold that your notion of a Veil is inspired by Kant's notion of a noumena world about which we can have no objective knowledge. However, for Kant this noumena world includes the ideals of God, Freedom and Immortality. By contrast, in your view this noumena world, seen in the political context, includes violence and terror usually directed against Blacks by whites. In any event it seems as though this fog served as a Veil analogous to the one of which you speak. You went on to speculate that behind this Veil lay violence and terror that are directed against Blacks. Similarly it may well be the case that behind the fog that I confronted lay violence or the effects of the violence and terror on which colonialism once rested. I do not believe I am overstating this Veil/fog analogy. You, yourself, were good at drawing neat analogies, and I am here merely adopting your technique. Yes, I am parodying your works in this regard. Allow me to continue relating my story. During the previous day when I was in the Nakura National Park, it seemed that nature was *cooperating* with me; I could appreciate the natural beauty manifest in my presence. When I visited the site of the Great Rift Valley, it seemed that nature was *working against* me, preventing me from gaining insight into the site(s) from which human beings may have originated. This is a great contrast to my Nakura Park experience that previous day. On that occasion, it appeared that I had gained insight into the Destiny of the human race and of nature generally. I was in a position similar to the illuminating dream experience of

74

Dr. Martin Luther King, Jr. who had gone to the mountaintop. But at the "fog scene," it appeared that my experience was just the opposite of Dr. King's dream. A few hours later I boarded a plane for Dakar, Senegal, where I was confronted with certain historical facts about the African slave trade, which extends as far back as 1588, nearly a century following Columbus' "New World" discovery.

Sincerely,

Dr. Johnny Washington, Ph.D.   April 9, 1997

Traveling in Kenya, Senegal, and Nigeria

Dr. W.E.B. DuBois (1868-1963)   Destinicity Letter #12

Re: A Journey to Africa: Part VI

Dear Dr. DuBois

In the December 5, 1991 issue of the *Palm Beach Gazette,* I recounted what I called my "fog scene" experience at a site on the Great Rift Valley, in Kenya, East Africa. The story continues thus. It was on July 26, 1991 when I spent about thirty-five minutes at the site negotiating with local vendors. One of my main motives for visiting the site was to appreciate the scenery of the valley below; the point where I stood included an 8,000 feet high cliff. The thickness of the fog at the time prevented my seeing much beyond twenty-five feet. At the time, I presumed that within a few hours, the fog would evaporate and I could enjoy the scenery. But I had no time to tarry, since my flight to Dakar, Senegal, West Africa was scheduled to depart from Jomo Kenyatta International Airport in Nairobi within a few hours, at 12:45 p.m. When I left the fog scene, I was driven directly to the airport, where I arrived around 12:00 noon. I traveled to Nairobi from New York a week earlier by Pan Am airlines. This firm is now defunct. Pan Am had no flights from Nairobi to Dakar. Nor could one fly non-stop from Nairobi to Dakar, or fly directly from East to West Africa. On the way to Dakar, I flew from Nairobi to Ivory Coast's capitol city Abidjan, on Ethiopian Airlines flight 951. It arrived at Abidjan at 6:15 p.m., local time. In Abidjan at 11:00 p.m., I boarded Air Afrique flight RK 49 and arrived at Dakar at 1:30 a.m., July 27, 1991. I would be curious to know what mode of transportation you used in traveling to Africa during 1923. On a more serious note, to what degree your attitude towards Africa was altered following your visit to Africa. In this regard, one of your biographers Eric J. Sundquist makes the following observation:

> It was his [DuBois'] 1923 visit to Africa as an official envoy to Liberia that afforded DuBois the realization, as he now recalls it, that the "income-bearing value of race prejudice was the cause and not the result of theories of race inferiority" both in the colonized world and in America during the decades since the collapse of the cotton kingdom. (*The Oxford*, 94)

After traveling to Africa, you began to regard racial prejudice towards Africans as having grounds in economic oppression of Africans and the people of African origins. During my travel from East to West Africa I stayed in Abidjan for five hours; the entire journey from Nairobi to Dakar lasted more than twelve hours. (The distance between Nairobi and Dakar covers some 4,000 miles). There were certain intermediate stops between Nairobi and Abidjan, each of which lasted about 30 minutes. The first of these was Kinshasa, the capitol of Zaire. Currently, that country is experiencing political unrest and instability. In Zaire one could at the time almost smell a pending *coup d'état*, (an observation confirmed by a December 5, 1991 *Associated Press* report). And as I have reported repeatedly in my letters to you, currently Zaire and the entire region is undergoing serious crises. When I was traveling in Zaire in 1991, this element of political unrest was evident to me by the presence of the many soldiers who mingled at the Kinshasa airport. Some soldiers boarded the plane to go to other points in Africa. When we stopped in Kinshasa, it seemed as if our plane had landed at a military camp. There were armed or uniformed soldiers all over the place. It is worth noting that Zaire's President Mobutu Sese Seko has exerted autocratic rule over this former French territory for over thirty years. Many observers regard Mobutu, (who has recently undergone surgery for cancer in Paris), as the source of the problems in Zaire, whose population includes some 35 million people. Now, there is a concerted effort to overthrow his rule. There is strife, including mass killings and starvation, in Rwanda and throughout the region. I believe Mobutu's days are approaching a countdown. By the time this letter reaches the *Palm Beach Gazette*, he will be overthrown, I venture to claim. Following the stop in Zaire, the next stop was Togo's capitol city Lomé, where we again stayed for about thirty minutes. As in Kenya, and Zaire, the political situation in Togo is also unstable.

On December 3, 1991, a *coup d'état* was attempted in this small country to overthrow President Eyadema. I do not know what the military or political connections are between Zaire and Togo. It seemed to me as I traveled through these countries during July 26, 1991, that many of soldiers who got on the plane in Zaire departed the plane in Togo. This small country is bordered on the east by Benin, on the west by Ghana, north by Burkina Faso, and south by the South Atlantic Ocean. At the risk of boring you, I will continue recounting my story. It was ironic that as the Ethiopian plane on which I traveled paused in Zaire and Togo, the "soft," relaxing music that was played on the plane was of the Country and Western type. It seemed to me that among the songs I heard was one by Conway Twitty and another by Willie Nelson. This reminded me of the times when I lived in Nashville, Tennessee. When I heard such music, I recall thinking: "How strange it is, to be in the heart of Black Africa, and on an African airline, to listen to Conway Twitty's music." In Kenya, by contrast, I was accustomed to hearing, *via* taped recordings, R&B artists such as Michael Jackson and Whitney Houston. Indeed there is much "soul" in Nairobi, or so it seems. During 1991, I heard no rap music in Africa. I presume it is common in Africa today. I know you as well as Dr. Alain Locke liked the spirituals, jazz, and blues.

The Ivory Coast, an area about the size of New Mexico (U.S.), is a former French colony. The dominant language in Ivory Coast is French. Thus, most of the vendors at the concession stands in the airport spoke mainly French. Similar to Senegal's currency, Ivory Coast's currency is tied to the French franc. Again, this was a neat contrast with the activities of the Kenyan marketplace, where most transactions are conducted in English, and the currency is tied to the British shilling. Since I had a five-hour delay in Abidjan, I entertained the possibilities of going into the city of Abidjan but decided not to make the effort. In order to go into town, I suspect I would have had to go through the "red-tape" required by customs and security to pass through the airport. To avoid the hassle, I sat in the Abidjan Airport for five hours, where I had the chance to talk to others, and catch up on my reading and correspondence. It would have been desirable if I had a laptop computer with the capacity to receive or send faxes and e-mail messages. Most of the concession stands at the airport were operated by African men who spoke French. The French-speaking vendors at the airport did not seem to have much patience for someone such as myself who spoke no French. Either I was impatient with them or they were impatient with me;

such tension was in part due to, I believe, the cultural differences between the African American and "African French" cultures. In fact, some of the vendors at the Abidjan airport acted rude towards me as I sought clarification as to the manner in which the currency was to be converted from the U.S. dollar to local currency. It seemed to me every time I completed a currency conversion, I lost a little money in the transaction. As I waited in the Abidjan airport, I got into a heated discussion with a person (whom I will refer to as a fellow traveler) who was on the flight with me from Nairobi to Dakar. I had met him briefly at the World Conference of Philosophy (WCP). He came to a session in which I presented a paper that examined the notion of African Destiny. While in Abidjan, I began sharing with the fellow traveler my Kenyan experiences, including my observation about the ways in which African women seemed to be exploited as laborers. The upshot of the conversation between this fellow traveler and myself is reflected in the following remarks. I told him that I was struck by the seemingly oppressive conditions under which the African women, in comparison with the men, labored, especially those who lived in the rural areas. Women perform much of the farm work. Additionally, it seemed unbearable the ways that these women have to carry heavy loads on their backs or heads, sometimes utilizing both, simultaneously. Of course I was for the first time observing such activities in a live setting. But this was nothing new for these women who have been laboring under such conditions for centuries. Often one would see a woman balancing as much as a sixty-pound load on her head, while carrying a baby on her back. I gathered that it was customary for a woman to carry such a load as far as ten miles one-way. The heavy load under which she struggles and suffers includes such items geared to fulfilling life's necessities—food, water, fuel (usually wood) and animal feed (vegetation). Without doubt, the African women do more than their share of the laboring activities. This fact is attested by an observation made in the September/October 1991 issue of the magazine, *Mother Jones:* "Women do three-fourths of Africa's agricultural work. In 1982, just .05% of all UN spending on African agriculture went to programs for rural women" (p. 40).

I continued the conversation with my fellow traveler with the remarks that during my entire stay in Kenya, I saw only one donkey and only one farm tractor, and the latter sounded as though it needed a tune-up. When I toured the rural areas of Kenya and witnessed the ways that African women labored and toiled, the evil of colonialism or its consequences was at once driven home to me. I saw clearly the negative effects of colonialism, how it had distorted and degraded African women, attempting to rob them of beauty, love and hope. The negative effects of colonialism have lingered over Africa like a fog or a persistent cough. Colonialism, together with other negative social forces in Africa, has conditioned the African woman to fulfill the functions assigned to donkeys and tractors. She has to perform back-distorting tasks, as well as perform her traditional "housework" roles, preparing food, rearing children, etc. This was an observation that I shared with my fellow traveler originally from Senegal, who did not seem to be interested in hearing what I had to say. I went on to say to my fellow traveler that inasmuch as African women as laborers represented an aspect of African tradition that is being threatened by postmodernity, it would be desirable if one could devise a way to reduce the burden of labor imposed on them, without at the same time destroying altogether the African traditional values associated with child rearing and family life. My fellow traveler's initial reactions to my comments were in the form of anger and resentment. He declared: "You, as most African Americans, are romanticizing the African past. Such a past is oppressive not only to the African woman but to all Africans!" He added that I was "seeking to preserve the oppressive African past. But that history now needs to be destroyed." He reminded me that the European colonial presence had "done good" to Africa by seeking to reduce poverty and disease, and to improve the educational opportunities. To which I retorted: "Perhaps there is some misunderstanding between the two of us." This misunderstanding, I insisted, is brought on in part by our cultural differences. I know that this was a vague reply, but this was the way I dealt with this sensitive issue. I deliberately gave a vague reply. I also informed him that I, as well as many other African Americans had a genuine interest in Africa's development, in her Destiny.

As he talked, he became very emotional towards me. At that point it seemed as though an emotional fog drifted between us, effectively blocking communication. After this intense moment of emotional expression, I replied calmly by saying, "Wait a minute, wait a minute. Let's get some understanding here. African history belongs just as much to me as it does to you. But here we will not argue about ownership." Then, I posed to him the question, as to what he thought should be the role of African Americans in Africa. His reply was quite abrasive and negative. He said that as far as economic development was concerned, African Americans could best help Africans by remaining in the U.S. and staying out of Africa. I did not realize how controversial such a discussion about the role of the African woman was. My fellow traveler seemed to have had some questions about my idea of introducing technology in Africa to ease the pain and suffering under which the African women labored. Now that I have thought about it, his reasons in part made sense. His reasons were that because of the traditions that have defined the differentiating roles of African men and women, whatever new technology is introduced to ease the burden of the women, the men in effect would "bump" the women for the easier, more attractive jobs. Consequently, the African men, to a greater degree than the African women, would benefit from any labor saving technology. That is, any jobs utilizing modern technology to help the African women would be taken over by the African men, and women would continue to be stuck with labor-intensive jobs. That was his opinion of the matter. Indeed throughout Africa one sees from the standpoint of Westerners women laboring under dehumanizing conditions. In the hotels in which I stayed in Nairobi as well as in Dakar, most of the house keeping tasks in the hotels and restaurants were performed by men. The men worked at registration desks and cash registers. The women, by contrast, labored in the fields, in the marketplaces, and in the homes, night and day. As I have since reflected on the plight of African women, my mind has settled on the conviction that inasmuch as Africa depends heavily on the labor of women, they indeed play a major, decisive role in Africa's economic-cultural development, in Africa's Destiny. Consequently any and all development initiatives supported by the United Nations or any other sources, must give due considerations to the role of African women, with the aim, among others, towards eliminating the dehumanizing conditions under which they labor.

At 11:00 p.m. we boarded Air Afrique flight RK 49 for Dakar. One thing that stood out in my mind concerning this flight was the fact that they served us much food. In traveling on Pan Am Airlines, at mealtime, they would provide the passengers only one tray of food at a particular time. Whereas, on Air Afrique they would serve the passengers double of everything, and most of the meals were representative of the African dish. One other thing impressed me about this flight was the fact that as we were approaching Dakar, the airline stewardesses reminded the travelers that the seat-belt, and smoking regulations and other requirements governing the safety of passengers were enjoined by the U.S. Federal Aviation Administration, to which we as passengers were compelled to comply. I arrived in Dakar around 1:30 a.m. July 27, 1991. The following day my wife Loutisha, who flew in from Miami *via* New York, joined me in Dakar. We stayed at the Teranga Hotel, in which we could see the Island of Gorée on which sat the nefarious slave house.

Sincerely,

Dr. Johnny Washington, Ph.D.   April 10, 1997

Dr. W.E.B. DuBois (1868-1963)   Destinicity Letter #13

Re: Journey to West Africa: Part VII

Dear Dr. DuBois:

During my Summer 1991 visit to Senegal, West Africa, I stayed at the Teranga Hotel in the capitol city Dakar. That hotel was comfortable but quite expensive. A breakfast meal could cost as much as $20.00; the average dinner was $25.00. Due to the expensive meals at the hotel, my wife Loutisha and I found a local Vietnamese restaurant that we visited frequently. When I left the July 1991 World Conference of Philosophy in Nairobi, Kenya I flew directly to Dakar. My main purpose for going to Dakar was to attend the 13th Annual Conference of the Caribbean Studies Association that convened at the Teranga Hotel, from July 28 through July 31, 1991. Among the participants of the Caribbean Conference were Professors Ja A. Jahannes, Anna Maria-Evans, Alfred Young, and Angela Young. Loutisha and I often joined Angela and Alfred to go out to eat and tour the city. Since I had lived as a soldier in Vietnam during the 1960s, I had become accustomed to the Vietnamese food. I was surprised in seeing how easily Loutisha, Angela and Alfred adapted so well to the food, which I found quite delightful. The menu often included pigeon, but I was afraid to try that dish. It is reported that in some Asian countries, snake, cat or dog meats are available on certain menus. The Vietnamese food is in many ways comparable to that of the Chinese. It is quite unsafe to travel alone in a foreign country such as Senegal, it goes without saying. The local vendors can easily detect an outsider who can become a victim of street swindling or pocket picking. In Senegal as well as in Kenya, the moment one walks out of one's hotel room, a drove of vendors might follow one for blocks (or miles) in the attempt to sell one a trinket or a doll. In certain quarters of the city one can also find prostitutes; some hold that the practices of prostitution as well as homosexuality were introduced into Africa by the Europeans during the colonial era. Near the Teranga Hotel, there were at least 30 security guards whose job in part was to "protect" the guests from the vendors. Many of the people of Senegal feel that the American tourists are rich. The vendors in Senegal were not allowed to penetrate the Taranga Hotel's grounds. Once outside the hotel's parking lot, one was in what may be called the vendors' territory. If one were interested in the items that the vendor sought to sell, one's first and foremost tasks would be to exercise skills in negotiating. In most cases, during the bargaining process one could get the vendors to lower the price considerably. Many of them appear to be out to overcharge the Americans, regardless of one's ethnic origins. Dr. DuBois, if you were to travel to certain parts of Africa today, I believe you would find appalling the manner in which many African communities are catering so much to the tourist industry. Many African economies are relying almost exclusively on the revenues generated by the activities of tourists, mostly from Europe and to a lesser degree the U.S. I doubt whether you would approve of the manner in which many of the African economies operate. For example, in some parts of Kenya, one finds tea plantations, some of which consist of many square miles. Near some plantations established by the Europeans, one can find starvation. Additionally, many African governments are ruled by military dictators.

During July 28, 1991 (which was a Sunday), Loutisha, Angela, Alfred and myself walked a few blocks from our hotel to visit an area called "The Market," which reminded me of a huge yard sale, within which everything one could imagine to be bought or sold could be found. In fact The Market also reminded me of the flea markets that were held on Maxwell Street on the west side of Chicago. About three-fourths of a mile from the hotel in which we stayed, The Market was concentrated in approximately a 10-block area, which included about 300,000 people—shoppers, vendors, spectators, and tourists, among others.

It required a great deal of skill to be able to navigate as a pedestrian through the river of people who flowed endlessly down the seemingly ancient streets of Dakar. The Market included a section in which many meat items were bought and sold. Over an extensive area one could see tons and tons of raw beef, chicken, goat, fish and what seemed to have been pork, although 80% of the people of Senegal are Muslims.

Much of the raw meats were lying there in the open air without the benefit of ice or refrigerators. The entire scene was incredible. Paralleling the slabs of raw meat, one found tons and tons of dry produce, including rice, beans, corn and peanuts. (It should be noted that Senegal has peanuts as one of its major cash crops). In the next row of goods one finds enough fish to feed a semi-vegetarian eating army. It is ironic that here in The Market in Dakar one sees so much food available to be bought and sold in a capitalist economy and yet there is so much hunger and starvation throughout the continent. Along with the fish, corn, beans and slabs of goat meat, there were the artisans, including shoemakers, seamstresses, leather and wood workers, medicine men, among others, who produced items for the marketplace. Any Americans who visit a country such as Senegal for the first time must be prepared to face a cultural shock. I think such a trip is indeed worth the effort, nevertheless. Slave trading in the "New World" had one of its points of origins on Gorée Island, two miles off the coast of Dakar. I could nearly see the "Mansion of the Slaves," from my hotel window. This "Mansion" was a building in which slaves were quartered, and stripped of their manhood or womanhood.

The small island that is some 300 meters wide and 900 meters long, was established as a slave depot by the Portuguese settlers in 1588. Gorée was one of the most infamous slave depots on the coast of Africa. Between 1588 and the end of the slave trading era, millions of African slaves were shipped through this sinister place.

Senegal has a rich history that should be of interest to any American, especially African Americans. During August 3-8, 1992, when I attended the Pan-African Think Tank in Badagri, Nigeria, (about 50 miles from Lagos, the capitol of Nigeria), I also had the opportunity to visit another former slave depot in Badagri, whose history is similar to Gorée's.

Moreover, when I was in Badagri, I along with other Pan-African Congress Think Tank participants, had the opportunity to visit the then-sitting king and queen of Badagri. Our purpose was to present to them the results of our deliberations concerning the establishing of an Agenda for Africans and the people of African origins worldwide.

Sincerely,

Dr. Johnny Washington, Ph.D.   April 11, 1997

PS. I hope I did not leave you with a bad impression that Africa is beyond hope. It is the case that Africa is suffering many problems; yet many possibilities lie ahead for Africa. New technological developments are occurring that will address many of its problems in medicine, agriculture and education. The future is bright for Africa.

Traveling in Ahmadabad (India)

Dr. W.E.B. DuBois (1868-1963)   Destinicity Letter #14

Re: Traveling in India

Dear Dr. DuBois:

As early as Spring 1989, I received a letter of invitation from Dr. Yeager Hudson of Colby College, Maine, to participate in the VIII International Social Philosophy Conference that he would be convening in Ahmadabad, India, December 28-31, 1991. In Ahmadabad I visited the Headquarters from which Mahatma Gandhi launched his campaign to liberate India from the colonial rule of England. In the Gandhi Headquarters that now includes a museum and library, some extracted passages from the works of Dr. Martin Luther King, Jr. are displayed among Gandhi's memorabilia.

Indeed, Dr. Martin Luther King, Jr. is highly regarded among the people of India as he was inspired by the teachings of Gandhi. His principle of non-violence was basic to the social reformative measures that Dr. King introduced in the U.S. On December 25, 1991, after I shared with my family the moments during which we opened Christmas (Kwanzaa) presents and rejoiced in the sacred occasion, at around 10:30 a.m. I went to Fort Lauderdale-Hollywood Airport, to board American Airlines' 12:45 p.m. flight for New York, John F. Kennedy International Airport. The city of Delhi, with a population of more than 30 million people, was one of the areas for which I was destined. In New York, I joined a number of other travelers who were to attend the same conference to which I was headed. I arrived in JFK Airport at 5:01 p.m. Then, I transferred to British Airways' flight 172, which departed at 9:15 p.m. for London. (Keep in mind that London is five hours ahead of U.S. Eastern time; consequently, when it was 9:15 p.m. in New York, it was 2:15 a.m. in London). The plane arrived in London at 9:30 a.m. on December 26, and it was not scheduled to leave for Delhi until 1:45 p.m. the same day. We stayed in London for five hours, in preparation for the journey to Delhi. As we took off from Gatwick London Terminal, the pilot described the flight path he would take to Delhi. From London we would fly over Budapest, Hungary; across a tip of the Black Sea; across Istanbul, Turkey; Tehran, Iran and into Delhi, a total of some 4,200 miles. The journey was half way around the world. One has to keep in mind that when one flies from the Western hemisphere to the Eastern hemisphere, to such a place as India, one crosses what is called the International Date Line. In so doing, one loses a day, but gains twelve hours. For example, in traveling to India, I left New York at 9:15 p.m., December 25, 1991. I arrived in Bombay at 4:30 a.m. December 27, 1991; this was the case, in spite of the fact that I only had a scheduled five-hour delay in London. Similarly, I left Bombay at 2:15 a.m. January 2, 1992, and arrived in Fort Lauderdale at 10:57 a.m. January 2, 1992.

Yes, you would not believe it. During my journey to India I again had a serious encounter with fog, similar to the problem that I faced during my July 1991 visit to Kenya, Africa. During this fog encounter in India, it was so dense that it interfered with our flight plans to Ahmadabad, India. Although we were scheduled to land in Delhi, December 27, 1991 at 4:30 a.m., due to the intensity of the fog, the flight from London was diverted to Bombay. (The distance between Delhi and Bombay is about 750 miles; and between Delhi and Ahmadabad is 485 miles). At around 2:30 a.m. when we were some 1,000 miles past Tehran, the pilot announced on the loudspeaker: "I am sorry to inform you, but due to the heavy fog at the Delhi airport, I have been instructed to divert the plane away from Delhi to Bombay. Moreover, it is unlikely that the fog at the Delhi airport will be lifted before 12 noon today." Thus, about two hours later, we landed at Bombay International Airport. We were instructed not to leave the plane, since the Bombay airport was not at the time prepared to accommodate us. We stayed on the plane on the runway in Bombay until 8:30 a.m. In the meantime, for the first two hours that we sat still at Bombay airport, the pilot said intermittently that he was optimistic that "soon we will be given the permission to continue our journey to Delhi." Finally, at around 8:30 a.m. the pilot informed us the orders from London were in effect: that we were to unload the plane, and go to breakfast at the Bombay airport's restaurant, where we were to wait for further instructions. While at the airport I got for the first time a good taste of an Indian breakfast. It was also revealing to find that India, being a country that frowns upon the eating of pork and beef, and the eating of meat in general, would allow the serving of bacon. Some Indians wasted no time in getting into the line that included bacon. The Indians have a strange way of preparing coffee; the coffee is brewed together with the milk. It took me a while to get used to this type of coffee. Tea is made in a similar manner. At around 11:45 a.m. we again loaded airport busses and were taken back to the British Airways, bound for Delhi, where we were to go through customs. The flight to Delhi lasted about an hour and half. We arrived there about 1:30 p.m. at which time the fog still lingered. After we went through customs, we had to shift from India's international to its national airport, and fly directly to Gujarat, not by Air India that flies its international routes, but Indian Air that flies its national (domestic) routes. We arrived at Hotel Karnavati on Ashram Road in Ahmadabad, Gujarat, December 27, 1991 at around 9:00 p.m.

While we sat at on the runway in Bombay, I witnessed several scenes that left impressions on me; two of the scenes merit recounting here, because I believe you would be interested in my story: the first part I will mention I regarded as a positive experience, the other as a negative one. To overcome the boredom generated by our waiting on the Bombay runway, I placed the earphones on my head and turned through the channels, stopping at one within which Whitney Houston was being interviewed by British Airways *International Pop Chart* commentator Phillip Schofield. (This of course was a prerecorded interview). I gathered from the interview to my surprise that Dionne Warrick was Whitney Houston's aunt. Moreover, the interviewer queried Houston as to what was unique about her music. Her reply was that she liked to sing songs in which the individual's struggle results in victory—"All the Man I Need" was one of the songs she had in mind. I like the notion of "struggle," as this is compatible with your own views reflected in your book, *The Souls,* and in the Destiny model.

The Whitney Houston interview, it seemed to me, incidentally, was followed by a movie that was a parody of the celebrated movie *Gone with the Wind.* In this movie, one of the main characters was a Black woman who fits the description of the Aunt Jemima type—fat, jovial, docile, and dumbacting. What was peculiar about the character was the fact that whenever she was around white people, she would "inflate" twice her size, and assume the Aunt Jemima type of behavior. Whenever she was not in the presence of white people, she would deflate to her normal size (about the size of Diana Ross) and assume the behavior of a very intelligent person. Some people found this *Gone with the Wind* parody quite amusing. This 747 model aircraft in which we were traveling was filled with passengers, mostly Indians, a few Europeans, and Americans; I was the only African American. As the movie played, I could hear a few of them chuckle. But it was not funny to me, not one bit. In fact, I found the movie quite offensive, as it projected a demeaning image of African Americans. When I was in Bombay, however, I saw a few Africans. I have no idea how many Africans live in India. It is also the case that many people of Indian origin live in Africa, it goes without saying. Here I was, the day after Christmas, stranded on a runway in Bombay with mostly Indians as my fellow travelers, and watching a movie in which African Americans were portrayed in a negative, demeaning manner. Do not jump to any conclusions at this stage. I am not blaming the Indians. In fact the ones with whom I came in contact treated me quite nicely. And I am looking forward to returning to India some day.

I am blaming not the Indians but British Airways for showing this *Gone with the Wind* parody. Couple this with the fact that this incident took place on an airline associated with the country that once colonized most of Africa, India, Latin America and other parts of the world. British Airways routes extend worldwide. Assuming that such a movie was played in most of its airlines, then this means that Aunt Jemima image is disseminated worldwide. The following day, I mentioned to one of the conference participants who is a European American, Dr. Donald A. Crosby, that I was disturbed by the movie under discussion. As we sat on the runway, I asked him whether he had seen the movie in question. At first, he said that he had no idea what I was speaking of, that he had seen no such movie. Then, I went on to describe the scene in the movie calling attention to the Black woman, a main character in the movie, who was "inflatable" or "deflatable." Suddenly my fellow traveler Dr. Crosby, from Colorado State University, a colleague of Dr. David Crocker, (the president of International Development Ethics Association), recognized the scene that I was describing. Dr. Crosby's initial reaction to my query was that he "found the movie amusing." I told him that I was embarrassed or angered by the movie. My reaction prompted him to reflect more closely on the matter. Thus, he went on to say, "Now that I have thought about it from your perspective as a Black person, I can see how you might have been offended by the movie." We both converged in the conviction that Europeans, (including whites in the U.S.) have used movies and other media sources to provide degrading images of African Americans. Dr. DuBois, I wonder whether I was overreacting to this movie incident? Am I am being too sensitive? In my fourth Destinicity letter I mentioned that Dr. Goings who at the time was a faculty member at Florida Atlantic University, has spent much of his career collecting memorabilia of the Aunt Jemima type items. He has researched ways in which racial attitudes are conveyed through what he calls "Black Collectibles" and other memorabilia, comparable to the above-mentioned movie. Thus, I am convinced that British Airways needs to reassess its entertainment policies and practices to determine the merits or demerits of materials presumably intended for amusements. The title of the paper I presented at the India Conference under discussion was, "Obligation and Normativity: An Alain Lockean Perspective." Below is a summary of that paper whose theme I developed more fully in my book, *A Journey*.

> An analysis of notions such as "freedom," "rights," and "obligation" (Dharma) usually leads into a discussion of social pressures, exerted by customs and traditions. John Dewey regarded the normative notions associated with social pressures as customary morality.
>
> Kant, it will be recalled, focused on the view of rationality in order to explicate such normative notions. Central to his ethics is the notion of duty. The moral law that is purely rational presents itself to the human will as a *rational necessity*. One is obligated to do one's duty, to bring one's will in conformity to the moral law, simply because it is one's duty to do so. No other justification is required. Henri Bergson (one of Alain Locke's Paris teachers) reversed, as it were, Kant's normative schema; for Bergson sought to analyze normative notions such as freedom and rights by considering what he called the "closed society," within which moral duty is comparable to the instinctive behavior of ants in the ant-hill or the behavior of bees in the bee-hive. That is, the behavior of human beings in the closed society and that of the insect is driven by the necessitating laws of biology, of life itself. Such a morality involves social conformative in contrast to the freedom and openness of the morality found in the open society. An analysis of normativity and obligation that differs from any of the above is offered by the African American philosopher Alain Locke (1886-1954). In his essay, "Values and Imperatives" and in other works, Locke explicated the obligatory nature of norms by examining the psychology and dynamics of the valuing process. If for Kant the moral law presents itself as a rational necessity, for Locke such a law presents itself as a *psychological necessity*. Locke offered a psychological explanation of the obligatory nature of norms. Thus the views of Dewey, Kant, and Bergson will be contrasted with the views of Alain Locke. One objective will be to show how Alain Locke's views of obligation shed light on the nature of ethnic, racial and national conflicts and what is required to approximate the principle of harmony.

Indeed my presentation stimulated much reaction from the audience. I took some good pictures of my sessions and other scenes in India. The *Palm Beach Gazette* has copies of some of the photos that it might share with its readers. Yes, Dr. DuBois, your name came up repeatedly at this conference, especially during the session in which I presented my paper.

This should not surprise you, as people often associate many of your views connected with your Talented Tenth model, with those of Dr. Alain

Locke who also believed in the Talented Tenth. Here at this Indian conference, the focus was on the philosophic issue pertaining to freedom and obligation. As I write this letter to you, I am presently recollecting vividly a concern that the conference participants, especially the Indians, both Hindus and Muslims, raised. They were disturbed by the fact that Dr. Alain Locke did not devote sufficient attention to the notion of transcendence. The question of transcendence came up in the context of my comparing the views of Dr. Locke to those of Buddha, (who was of Indian origin). I indicated to the conference participants that neither Buddha, Confucius, nor Dr. Alain Locke, emphasized transcendence, although the latter subscribed to the Baha'i faith that includes degrees of transcendence. Much of the Gita is dealing with the importance of engaging in battles, regardless of the prospect of victory or defeat. The conference also focused much attention on the teachings of Dr. Martin Luther King, Jr. As I mentioned to you previously, I had the pleasure of visiting Gandhi's museum, within which is included Dr. King's memorabilia.

Sincerely,

Dr. Johnny Washington, Ph.D.   April 11, 1997

Traveling in Honduras and Nicaragua

Dr. W.E.B. DuBois (1868-1963)   Destinicity Letter #15

Re: Social Strife in Central America: Part 1

Dear Dr. W.E.B. DuBois:

In a previous letter to you I recounted my travels to Kenya, where I met people from Somalia and Ethiopia. Similarly, in this letter and the one that follows, I will reflect on my 1992 trip to Central America. I will revert to addressing a series of letters to you regarding my African traveling. In the 1980s, during the height of the recent civil wars in Central America, the image of the country of Nicaragua was firmly implanted in the American mind. Oliver North assisted toward that end. I feel compelled to inform you of my traveling in Central America, in view of the initiative you took in founding Pan-Africanism.

Thus it is clear beyond doubt that such an initiative is evident of your commitment to the people of African origins worldwide, including those in

Central America. From June 23, through June 28, 1992, I had the opportunity to visit a neighboring country of Nicaragua; namely, Honduras. I spent most of my time in its capitol city, Tegucigalpa. Here, I attended the III International Development Ethics Association (IDEA) conference, of which Drs. David Crocker and Ramon Romero were the conveners. As the name suggests, IDEA's purpose was to examine alternative ethical models for developing nations. Held at *Antiguo Paraninfo De La Universidad* Tegucigalpa, Honduras, the theme of the IDEA conference was: *The Ethics of Ecodevelopment: Culture, The Environment, and Dependency.* This conference included some 110 participants worldwide, together with some 70 local, grassroots participants. Included among the various sub-themes of the conference were: i) *Socio-Economic-Political Systems and Ecodevelopment;* ii) *Cultural Identity and Development;* iii) *Economical Institutions and Ecodevelopment;* iv) *Ecofeminism and Ecodevelopment;* v) *Culture and Ecodevelopment in India;* vi) *Culture, the Economy, and Ecodevelopment;* vii) *Ethical Norms and Political Power;* viii) *Education and Ecodevelopment: Problems and Perspectives;* ix) *Sub-themes: Environment: Biodiversity and Cultural Diversity;* x) *Biodiversity and Ecodevelopment;* and xi) *Model of Sustainable Agriculture…*

During a June 26 session entitled, *Substantive Ethics: Principles for Ecodevelopment* I presented my paper "Destiny, Development and African American Philosophy." Dr. DuBois, I believe my views here echo many of yours. The contents of my paper may be summarized as follows: The object of this paper was to consider ways in which African American philosophy could be applied in addressing problems of the African American community, of which the recent Los Angeles riot was a symptom. I maintained that the causes of such problems lay partly in the fact that African Americans lacked an adequate sense of Destiny. I proposed that in the U.S., an African American Center for Applied Philosophy to be established in which the issues of economic-cultural development, informed by the Destiny model, becomes the main focus. I offered three models for consideration.

The Enterprise Zone Program model, the neoFranklin D. Roosevelt model, and the Booker T. Washington model. Although I indicated the inadequacy of each model, I expressed preference for the latter two mentioned models. Each of these models stresses self-help in developing an economic base for the community. In fact, I suggested that it would be desirable, drawing on the Hegelian dialectics, to achieve a synthesis of the neo-Franklin D. Roosevelt model and the Booker T. Washington model. My introducing Booker T. Washington's model here is not to stir hard feelings between the two of you. You two differed on issues. I am aware of the conflict between you and Mr. Washington, a conflict some scholars hold was in the long run beneficial to the Black community and to the nation generally. Because of such conflict, alternative political models, informed by the respective views of the two of you, came to the fore. Therein lay in part the merits of the conflict between you. I know I sound Hegelian on this score. Let me return to the issue at hand. Between June 28, 1992 and July 5, 1992, I visited war-torn Nicaragua. Here certain people found appealing the enterprise zone model. In the full-fledged paper in question, I examined in detail the merits or demerits of each of the development models. On June 27, the IDEA participants took a field trip to a rainforest, located in the mountain regions of Honduras. It was a beautiful site. During the laborious two-hour journey, we saw what appeared to have been the wreckage of a U.S. warplane that apparently had been engaged on a clandestine operation when the crash occurred. The local tour guide said that the plane crashed in 1989, while the U.S. military was on a mission to provide supplies to the *Contras*. No one survived the crash, it was reported. Although Honduras was not directly involved in the war between the *Contras* and the Sandinistas, the report is that invariably certain areas of southern Honduras were used as supply depots for the *Contras*. And occasionally battles between the *Contras* and the Sandinistas erupted in Honduras near Nicaragua's border.

Ironically, in Honduras more so than in Nicaragua, the appearance of a police state was at the time quite clear. In Tegucigalpa one found armed soldiers on nearly every other street corner. The prevailing opinion is that human rights violations are widespread in Honduras. Hence the import of the military to suppress the possibility of any mass revolts. As I was walking the streets one day, during the times under discussion, I saw an American who said that the soldiers stopped him the previous night and demanded to see his passport. He said that he had left his passport in his hotel room. Once the soldiers discovered this, they "fined" him $50.00 for being without his passport. I am confident you would, correctly, have construed this incident as outright "extortion." I understand that such corrupt practices are widespread not only in Central America, but throughout Africa, Asia, and other parts of the world as well. I know you can relate to my point here as you, yourself, spent most of your life exposing such social injustices brought on largely by economic oppression. One found no such military presence in Managua; at least this was the case during my visit. Things might have changed much in Nicaragua by now. Yet a feeling of political uncertainty seemed to have pervaded Managua. In Managua one got the impression that such a people have undergone serious and prolonged political upheavals with a social price. Such an impression was well founded. In 1972, for example, an earthquake virtually destroyed downtown Managua, and its downtown area remains to be rebuilt. The 1972 earthquake resulted in some 20,000 casualties. Between 1972 and 1990, Nicaragua lost some 90,000 people, 70,000 of whom were casualties of insurrections or wars. On June 26, 1992, our group tour included a visit to a shrimp farm in Honduras that was the center of controversy. You, yourself, were often regarded by many as a controversial man. I am wondering to myself what would have been your reactions to the following controversy. It has to do with a large corporation that purchased an undeveloped shrimp area that had for years supported the livelihood of some 5,000 families in this region. The bio-shrimp farm corporation used modern technology and chemicals to grow shrimps on a rapid basis for commercial purpose. In the process, it restricted the rights of the local people to have access to the shrimps in their natural habitat. The corporation in question wanted to expand; whereas, the local people wanted to discourage such expansion so as to protect their livelihood, natural environment and traditions. Clearly, this conflict or controversy represented a clash of competing economic development models. What a moral dilemma! Whether or not they have solved this problem, I do not know. I do believe that the cooperative model, in which poor people buy and sell goods through a collective effort, that is, your version of Marxism, a model that you offered as a solution to the economic problem that Black Americans faced, would have been a step in approaching the above-mentioned problem. You formulated this model so well in your book, *Dusk of Dawn* (1940). Certain political groups, including the Sandinistas in Nicaragua, experimented with communism, of the version you proposed. Perhaps you will be surprised to know that with computers that can perform a trillion calculations per second, together with other forms of technology, this very technology is causing greater moral dilemmas than the one indicated above. A few weeks ago, scientists in Scotland announced that they have successfully cloned a sheep. Some are of the opinion that this capacity to clone not only animals but also human beings could threaten our immanent Destiny. On June 28, 1992, a few other IDEA participants and I left Tegucigalpa for Managua, where we subsequently gave a series of presentations at the National University of Nicaragua as well as at the University of Central America, both of which are located in Managua. Pedro Mercado Sanders, who participated in the Tegucigalpa IDEA confer-

92

ence, also was a fellow traveler to Nicaragua, his home country. At the Tegucigalpa airport I had the opportunity to speak with Pedro. Professor Ofelia Schutte, a native of Cuba, served as the translator for us, since Pedro spoke Spanish and no English. I gathered from our discussions that his ancestors included a mixture of Africans and indigenous peoples.

Thus, he maintained that he was interested in tracing his African roots. This much he knew: that at the turn of the century, his grandfather, who was an African American, migrated from Florida to Nicaragua to be employed by a U.S. fruit company. His grandfather (whose sir name was Sanders) settled in Nicaragua and married an indigenous woman. Pedro is a descendant of that union. I do not know to which Indian ethnic group Pedro is related. I am sure you are already knowledgeable about the fact that some 10% of the population of Nicaragua is of African origin, and many emigrated from Jamaica. Most of them live on the Nicaraguan Atlantic Coast region. I cautiously queried Pedro as to his relation to the Sandinista Revolution. However, before I posed the question to him, I made it clear that I was not employed by the U.S. government as a law enforcement agent and thus he could speak freely with me. His reply was that he was now the director of a Regional Council, a political position comparable to a governor or mayor in the U.S. However, I wish to end this letter by reverting to this question of ethics and technology, a theme that I shall explore more thoroughly when I focus on biological evolution. One wonders how such technology, as in the sheep-cloning discovery, might effect Black people generally. Will this new biogenetic technology be controlled by the powerful and wealthy?
Sincerely,
Dr. Johnny Washington, Ph.D.   March 21, 1997

Dr. W.E.B. DuBois (1868-1963)   Destinicity Letter #16

Re: Strife in Central America

Dear Dr. DuBois:
    This is a follow-up on my March 21, 1997 letter in which I focused on my 1992 travels in Honduras. In this letter I will focus more on my travels in Nicaragua, one of the largest countries in Central America. [During the

last week of October, 1998, Hurricane Brit devastated much of the continent of Central America where over 12,000 people were killed, hundreds of thousands were injured and several million homes were destroyed]. In addition to this, the entire region is currently undergoing profound social and political strife, brought on by many factors, including: the 1972 Earthquake; the Insurrection of 1978, (when Pedro J. Chamorro was killed); and, the recent *Contras*-Sandinistas War. In the main, between 1972 and 1990, Nicaragua lost some 90,000 people. I understand now that the U.S. government is considering selling fighter planes to certain Latin American governments, including the government of Chile. Already you are knowledgeable about the history of Nicaragua. Yet allow me to make this historic observation. I am sure you are familiar with my writing style by now. I have a knack for placing an issue in historic perspective. Some Native Americans, the Nicarao tribe, lived in Nicaragua since time immemorial. It goes without saying, however, that Columbus' "discovery" of the Americas in 1492 let loose a chain of negative events that subsequently resulted in nearly the annihilation of the Native Americans in that region. It is reported that Columbus himself landed in Nicaragua in 1505 and claimed the country for Spain. It just so happened that I had the opportunity to visit this region some 500 years following the landing of Columbus. While I am on the subject of "discovery" it is instructive to note that Professor Ivan Van Sertima in his book, *The African Presence in Ancient America: They Came Before Columbus* (1976), is persuasive in his argument that the Africans arrived in the Americas before Columbus. But when the Africans arrived, unlike the arrival of the white man, they did not kill or route the Native Americans. Dr. DuBois, I am interested in your opinion on this matter, and I will research your works to ascertain your opinion, in view of the fact that the debate as to who came first is ongoing. In the early 1500s the Spaniards made contacts with the Nicarao people, many of whom the Spaniards hated, humiliated, raped, killed, loved, or intermarried.

Now, Nicaragua's population includes mainly the *mestizos,* a people whose ancestors include Indians and Europeans. There are well over three million of them. In my previous letter I mentioned the African presence in this region. As I mentioned elsewhere, I visited Managua with a contingent of the International Development Ethics Association (IDEA) members. These included, Ofelia Schutte, *Filosofia, Universidad de Florida*, USA; Rodney Peffer, *Filosofia, Universidad of San Diego*, USA; Ken Aman, *Filosofia y Religion, Montclair State*, USA; Edna Crocker, USA; and David Crocker, *Filosofia, Colo-*

rado State Universidad, USA. Dr. Crocker is quite knowledgeable about your works as well as the works of Booker T. Washington. I am also urging IDEA members, including Dr. Crocker, to assume a greater interest in the new field of enquiry I am opening, Destinicity Studies.

While in Managua, the IDEA contingent, of which I was a part, gave a series of presentations at the National University of Nicaragua at Managua (founded in 1812), and the Central American University. In my estimation, the latter-mentioned institution was better off financially, as evident by its well-kept grounds, physical facilities and faculty morale. Within the past decade or so much funding was diverted from Nicaragua's educational institutions and other sources, to provide financial support for its protracted civil war. The two universities we visited were concerned with similar problems, one of which merits mentioning: curricula re-orientation, as they referred to their endeavors. The faculty members of the respective universities wanted to know how to make Philosophy more practical so as to facilitate the cultural and economic development of the country. They said the war in question had increased the financial burden of the country. It had prompted debate about education and political reformative measures, thereby forcing the people to grapple with the question as to how they planned to utilize the limited resources. Certainly such social strife prompted them to become more aware of their immanent Destiny. The people in turn looked to the universities for answers. In our visiting these universities we were mainly in contact with the Philosophy faculty members, as most of us of the IDEA team were philosophers. Thus, the scholars of Nicaragua were at that time interested in a central question: how to make the Philosophy curricula practical? They wanted to know whether it would it be preferable to stress Logical Positivism, Anglo-American Philosophy, or an alternative model? It follows that my presentation at the National University of Nicaragua was mainly an extended version of the paper I presented at the IDEA Conference in Honduras. In this paper, as I mentioned, I attempted to combine the views of Franklin Roosevelt and Booker T. Washington. Dr. DuBois, as you know, it is difficult to discuss the views of Mr. Washington without considering your own views. In this context your philosophic views naturally were brought to the fore. It merits repeating the title of my paper presented in Central America: "Destiny, Development and African American Philosophy." When we finished our presentations at the above-mentioned institution, the audience wanted to know what sort of development program our respective institutions offered?

Such a question somewhat placed me in an awkward situation. It was indeed ironic that here I was advising the Nicaraguan people how to restructure their philosophy curricula to meet the needs of economic development, but to my knowledge my own institution, then in 1992, Florida Atlantic University, offered no program that dealt with Third World Development as such. The only thing that I could think about that came close to such a curricula was FAU's Certificate Program in Ethnic Studies that I established. I explained that while this program did not deal with economic development as such, it was concerned with understanding the cultural and ethnic experiences of the descendants of Africans, Asians, Hispanics, and Native Americans in the U.S. As you already are aware, the U.S. ties in Nicaragua extends to the turn of the century, when the former proposed to build a canal through the latter. The proposal was subsequently rejected by the people of Nicaragua. Later a canal was built by the U.S. in Panama.

During 1912, the U.S. sent its marines to Nicaragua to protect the former's interest. The U.S. Marines remained present in Nicaragua intermittently until 1933, the year that General Augusto Cesar Sandino was assassinated by General Anastasio Somoza. Although Sandino was assassinated in 1933, prior to then he projected an ideal that inspired the creation of the *Frente Sandinistas de Liberacion National* (FSLN) party, which is named in Sandino's honor. Carlos Fonseca, Tomas Borge and Silvio Borge founded the party in 1961. I had the opportunity to visit the gravesites of Carlos Fonseca and other revolutionaries. The FSLN, now known as the Sandinistas, achieved victory over dictator Luis Somoza in 1979, who was assassinated in 1980 in Paraguay. It is reported that Somoza castrated or beheaded many of his political opponents.

The U.S. Marines' presence in Nicaragua produced rebel General Augusto Cesar Sandino, who, between 1927 through 1933, attempted to drive the marines out of Nicaragua. In opposition to Sandino, the U.S. in turn produced the Nicaraguan army called the "National Guard," of which the U.S. trained General Anastasio Somoza was placed in command. Between 1937 and 1979, the Somoza "dynasty" exerted dictatorial powers over the country. Between 1978 and 1979 a political struggle erupted between the Sandinistas and the Somoza forces, backed by the U.S. In the early 1980s, the Somoza forces came to be known as the *Contras,* in whose support Oliver North is reported to have raised money.

A great inspiration to the Sandinista Movement, in 1984 Daniel Ortega was elected as President of Nicaragua, an individual whom the then-

President Reagan found objectionable. In Reagan's view, he was too much of a communist. Dr. DuBois, considering the fact that you intermittently adopted communism, I would be interested in your assessment of the Sandinistas movement that regarded itself as following the teachings of Karl Marx. I trust that some of your followers will respond to my query. It should be noted that the Sandinistas lost the election in 1989; after which, Nicaragua experimented with a coalition government, into which both the *Contras* and the Sandinistas were included. While in Managua, we had the opportunity to meet Miguel d' Escoto, the former Foreign Minister of Nicaragua. We were introduced to Escoto through Professor Ken Aman [who died during Spring 1998, it is sad to report]; the two were students together, studying to be priests, at Meryknoll Seminary, in the early 1960s. Escoto is also the founder of Orbis Books. Miguel d' Escoto was an impressive individual, quite dignified. On July 2, 1992, we visited Escoto at his "Palace" on whose walls hung beautiful paintings, and the rooms were filled with expensive, luxurious furniture. I had the feeling that I was at the U.S. White House. I understand that Escoto, as a priest and as a government official, had been criticized for living in such luxurious place while the masses went without basic necessities. Our meeting with Escoto lasted an hour and fifteen minutes. We discussed many issues, including the role of democracy in Central America; the Reagan Administration; the Peace-Making role of Costa Rica; and the tactics of non-violence. While talking with Escoto, it became clear that some of his political views seemed to be paradoxical. As Foreign Minister during the Sandinistia administration, he regarded violence as a means by which national policies were to be implemented; now that the Civil War has ended, and he is no longer in position of authority, he held, with Martin L. King, Jr. that the use of violence is wrong and should never be employed. On several occasions during the meeting Escoto cited with approval the views of Dr. Martin Luther King, Jr., whom he regarded as a role model, to which the entire region of Central America should subscribe.

In this context I interjected into the conversation the fact that Dr. King was highly influenced by your works, especially your book, *The Souls*. Just a side point: coincidentally, the year that Dr. King gave his celebrated "I Have a Dream" speech in 1963, is the same year that you died in Ghana, where you are buried. I am happy you lived to see some of the fruits of your efforts in contributing to the cause of Blacks.

It is reported that prior to 1972, prior to the earthquake that virtually destroyed downtown area of Managua, Escoto was an anti-Marxist; when he became involved in the Sandinista movement in 1972, he became a full-fledged Marxist, a position that was incompatible with the Catholic Church. When we visited Escoto in 1992, already he was suspended from the priest-hood, and the process was underway by the Pope to expel Escoto from the religious society. As I traveled through Nicaragua, I observed that the cost of living there is quite expensive. The supermarkets are well stocked with goods; but, due to the economic embargo imposed by the U.S., most peo-ple were unable to purchase such goods. It would be interesting to see to what extent Escoto can persuade the Nicaraguan people to embrace the ideals of Dr. Martin Luther King, Jr., who advocated non-violence, in approximating the ideals of social justice. I was unable to determine the man-ner in which Escoto and his comrades regarded Malcolm X.

Sincerely,

Dr. Johnny Washington, Ph.D.   March 25, 1997

Strife in Yugoslavia

Dr. W.E.B. DuBois (1868-1963)   Destinicity Letter #17

Re: Yugoslavia

Dear Dr. DuBois:

In a subsequent letter to you, I shall speak about the issues of terror-ism. I shall underscore the manner in which whites had employed terrorism against African Americans. In calling this to your attention, I knew that I was not exploring a novel or striking issue. When we think of colonialism within the modern era, we usually limit the notion to places such as Africa, Latin America and Asia. But as Aimé Césaire pointed out in his book, *Dis-course on Colonialism,* colonialism, together with the terrorism it harbors, had a boomerang effect, such that it recoiled back onto the European society through the actions of such individuals as Hitler, Stalin, and Tito. For ex-ample, Hitler invaded Yugoslavia in 1941, where he established concentra-tion camps to kill Jews in Serbia, during which time an ethnic cleansing program was initiated. The recent war in Bosnia involved ethnic cleansing, similar to a practice that is currently taking place in Zaire among the warring ethnic groups. I suspect that some of the Africans are adopting the Bosnian ethnic cleansing model. The various ethnic/religious groups in Yugoslavia and other parts of Eastern Europe are also struggling with what may be called the ethnic identity factor, it goes without saying.

Dr. DuBois I know it is the case that in some of your works you praised Stalin and other communist leaders. Not only that, but during the 1950s, you visited Russia as well as China. It is not clear what your motives were for visiting these countries. Some scholars hold that because you traveled to these countries, and because of your Marxist views, the U.S. government "black balled" you, as it did Paul Robeson, who also became quite "friendly" with the then-Soviet Union. I do not believe Dr. Alain Locke ever became "friendly" with China or the Soviet Union. I am not sure whether you were knowledgeable about the many terrorist crimes that Stalin allegedly committed against the Soviet people. He forced them to adopt his version of communism that many regarded as tantamount to totalitarianism, which hinges on terrorism. Certainly the Europeans, in the views of Césaire, had difficulties in believing that they themselves had become victims of colonialism imported to Europe by Hitler, Stalin or Tito. That is what I meant by saying that colonialism had a boomerang effect. Africa was the land in which modern colonialism was implemented, and it subsequently swung its way back to Europe through the warped words and deeds of Hitler, Stalin and Tito; the latter ruled Yugoslavia for 35 years. Hitler committed suicide during the Second World War. Following the Second World War, as you might recall, Germany was divided between East and West, and then in the early 1990s, Germany was re-unified. The Berlin Wall was dismantled. Currently the economy of Germany is thriving. And Europe is now proposing a single currency for Europeans. As for Stalin, he later died in office, and during the 1990s, the Soviet Union was dissolved, in part due to ethnic strife. Today the situation is still quite incendiary in the former Yugoslavia and throughout the former Soviet Union. This is the case apart from the fact that the United Nations peacekeeping forces are currently stationed in the former Yugoslavia. At this very moment, I understand that the U.S. is threatening to send troops to Albania to reduce civil strife. It seems that "contained" or "limited" wars are occurring on most of the continents, including Europe, Asia, and Africa. In one of my previous letters to you, I mentioned the civil unrest in Ireland, the Middle East and in Central America. Here at home the U.S. government is taking people off welfare, an action that will harm many Blacks, Hispanics, Asian Americans,

Native Americans and poor whites. I venture to claim that if some social net is not soon installed, this welfare issue is going to result in much social strife in this country. Believe me.

As you know from your knowledge of high school European history classes, colonial rule was attempted in the region of Yugoslavia in many instances throughout history, three of which I will briefly consider: the ancient Roman era; the 1400s; and what may be called "the Tito era." People lived in the region that we now call Yugoslavia over 100,000 years ago. Some authorities believe that the early people in that region hundreds of thousands of years ago originated in Eastern Africa. I believe that you, yourself, subscribed to the view that all races have their roots on the continent of Africa. During the times of the birth of Jesus, the Romans colonized the region of the former Yugoslavia. Over 100 miles wide at certain points, the Adriatic Sea is what divided Italy and the former Yugoslavia. You as well as Dr. Alain Locke are quite knowledgeable about the history of this region, as the two of you often traveled to Europe; the two of you studied in European universities; and the two of you were deeply grounded in the teachings of Fichte, Kant, Hegel and other European philosophers. (I had the opportunity of visiting Germany, England, and France during 1991). Already I mentioned the interest you expressed in the teachings of Karl Marx. Dr. Alain Locke was not a Marxist, I am here to remind you. This in part is what distinguished your views from Dr. Alain Locke's, whose views I have embraced.

I anticipate that I will bore you by briefly recounting this history, but for the sake of those who may read this letter allow me to do so, nevertheless. I promise you I will be brief. Much of the ethnic unrest that is felt in the region of the former Yugoslavia today has its origins in the ancient Roman era, when the region was divided into East and West, in connection with the splitting up of the Roman Empire in 395 A.D. The western part of the Empire included the areas that we now call Croatia, Sloven, and parts of Bosnia. The eastern part of the Empire included the areas that we now call Macedonia, Montenegro and Serbia. The latter-two mentioned provinces are what constituted the Yugoslavia that emerged following the Second World War. Serbia dominates much of the region since the Bosnian war began and is widely regarded as the source of most of the problems. Further, the Serbs are regarded by many as the new colonizers of the region. During the times of the Romans, those in the western parts of the Empire subscribed to the Roman Catholic faith; those in the eastern parts subscribed to the Eastern Orthodox faith. Such religious-ethnic fragmentation is the source of much violence prevalent throughout the region today.

The Bosnian war began around 1991 and the region is unstable today, although President Clinton recently sent troops to put an end to the fighting. As you know, the people in that region of the world rarely enjoyed "autonomous" rule. During most of their history they have been dominated by foreign rule. By the 1400s, Austria as well as Hungary dominated the region. I indicated elsewhere that between the 1300s and 1500s the Arabs (Muslims) and the people of Somalia fought to drive the Christians out of Ethiopia. Around this time, the Turks, who were of the Islamic faith also set out to colonize the region that now includes Bosnia and Herzegovina. Today the presence of a proposed independent Muslim country in the "heart of Europe" frightens many people. The so-called ethnic cleansing is largely exerted by the Eastern Orthodox Serbs against the Muslims. Tito was of the Serb community. Then there are the Catholic elements to be reckoned with. With the exception of the short-lived Kingdom of the Serbs, Croats, and Slovenes that was established in 1918, following the end of the First World War, rarely has the region been unified. Tito tried to unify the region when he imposed communist rule. He was a dictator whose goal was to control the Destiny of the people under his rule. Hitler and Stalin had similar motives with regards to their respective countries.

Of course you knew about the former dictator Josip Broz Tito's attempt at unification that resulted in colonizing the region. The focus on Tito is instructive; his practices will enable us to understand the manner in which colonialism was imposed on whites, on Europeans, who were affected by the boomerang of colonialism, whose ramifications included the processes of so-called "ethnic cleansing." Moreover, would you believe that one year after the Bosnian war began, there were over 134,208 casualties? Just as they experienced ethnic cleansing and related acts of terrorism in Hitler's Germany, Stalin's Russia, and more recently in Bosnia, so also there is ethnic cleansing occurring in central Africa this very moment. This ethnic cleansing seems to be a worldwide phenomenon. And at the core of each ethnic cleansing program is the issue of ethnic identity of the victimized group. Thus, it goes without saying that African Americans are not the only group struggling with the ethnic identity issue. Many leaders are searching for answers to this problem. If you have any hints as to how we might go about approaching this problem, please "reveal" them to us. As I mentioned, there was an attempt at unification of the former Yugoslavia region

around the time of the Second World War, when Tito played a major role in a resistance movement to route Hitler and his army, as well as the then-King Peter of Yugoslavia. Tito and his forces were successful; they subsequently established communism in Yugoslavia, a form of communism that was independent of that on which the former Soviet Union was based. However, Tito attempted to remain on friendly terms with the Soviet Union as well as with the West. As in the case of Somalia, the U.S. became uncertain as to what sort of role it should play in putting an end to the senseless killing in the former Yugoslavia. During the early phases of his Administration, President Clinton seemed to have had difficulties in deciding whether to send to Bosnia the U.S. troops, or, whether to allow the United Nation's Cyrus Vance's peace proposal an opportunity to work. Last year a meeting was held in Dayton, Ohio to resolve the matter, to route terrorism in Yugoslavia. Contingents of U.S. troops, included in the United Nations forces, are in Bosnia today, and peace seems to be on the horizon.
Sincerely,
Dr. Johnny Washington, Ph.D.   April 4, 1997

What Makes a Fanatic?

Dr. W.E.B. DuBois (1868-1963)   Destinicity Letter #18

Re: What Causes Religious Fanatics?

Dear Dr. DuBois:

Needless to say a fanatic is someone who may be regarded as having lost his/her authentic identity, and fanatic behavior is a struggle on the part of the fanatic individual to recapture such identity. First, it merits mentioning the following incidents in the context of exploring the issue of fanaticism: i) the February 26, 1993 bombing of the World Trade Center; ii) the Davidian religious cult in Waco, Texas; iii) the March 10, 1993 anti-abortion protest killing of Dr. David Gunn; iv) the April 19, 1995 bombing of the Alfred P. Murray Federal Building in Oklahoma City, in which 168 people were killed; v) the 1996 prowling of the Unibomber; vi) the March 28, 1997 mass suicide of 39 individuals in San Diego; [and during August 7, 1998 synchronized bombing attacks occurred in Nairobi, Kenya and Dar Es Salaam, Tanzania that resulted in the killing of some 258 people, twelve of whom were Americans. There were some 5,500 injured. The East Africa as well as the September 11, 2001 bombing attacks was believed to have been sponsored by Osama bin Laden who is reported to be a known terrorist]. These are instances of fanatic behavior of individuals or groups. Following the February 26, 1993 bombing of the World Trade Center, I published an article entitled, "What Causes Religious Fanaticism?" in which I attempted to shed light on the etiology of fanatic behavior. It seems to me that we are now living during a time when fanatic behavior is on the rise. One is also reminded of the recent killings among middle school and high school youths who exhibit fanatic behavior by killing their parents, teachers or classmates. It is proper for me to revisit this article, and present it to you in the form of a letter. I believe you would be interested in this topic inasmuch as fanatic behavior is often employed in an effort to induce social change. You, yourself, were a spokesman for social change, but to my knowledge, you never advocated fanatic behavior or terrorism. You were a rational man who not only believed in the power of the ballot, but also the importance of education in bringing about social change. This in part might explain, I believe, why Dr. Martin L. King, Jr. was interested in your views. As in your own views and those of Dr. King, the Destiny model has no room for fanatics or terrorists. Destiny is intended to promote universal unity and harmony.

It is worth mentioning that the first three terrorist incidents mentioned above possess a common thread that connects these seemingly unrelated acts: each involved the killing of human beings; each involved what may be called "premeditated violence;" and each of the apparent perpetrators harbored certain convictions associated with religious or political fanaticism. Mahmoud Abouhalima, an alleged supporter of Sheik Omar Abdel-Rahman, was held as a prime suspect in connection with the 1993 terrorist bombing of the World Trade Center.

It was reported Mahmoud Abouhalima sought to overthrow the Egyptian government by force. After spending one year at the Teacher's College in Alexandria University, Abouhalima came to New Jersey. On February 28, 1993, high school dropout David Koresh, a leader of the Waco Branch Davidian cult, engaged federal law enforcement agents in a 45-minute gun battle that resulted in the killing of four officers, and the burning of the compound in which many of his followers were killed. The week prior to the killing of Dr. David Gunn, Michael F. Griffin attended a church service in which he asked the congregation to pray that Dr. Gunn "gives himself to God." Griffin allegedly fatally shot Dr. Gunn in the back "to save the lives of unborn babies." Yes, abortion is a hot topic in our society today. I am aware that you wrote articles focusing on the role of women in society, but I do not have any evidence that you focused exclusively in any of your writings on the issue of abortion. My guess is that you would have been against it. Certainly the African American Episcopal minister Alexander Crummell (1819–1898), your mentor, would have been against abortion. But he would not have been fanatical about the matter. In the April 1993 issue of *Take 5* magazine in an article entitled, "What's Wrong with Terrorism?" I threw light on the practice of terrorism by juxtaposing it with the element of civil disobedience, of which Socrates, Jesus and Dr. Martin L. King, Jr. were representative spokespersons. Similarly there are affinities between religious fanaticism and saintliness, a point that was made clear by one of your Harvard teachers William James (1842–1910). In his book, *The Varieties*,[3] James provided a psychological profile of the religious fanatic. A consideration of James' view might shed light on the above-mentioned cases and enable one to appreciate the gravity of the issue at hand today. According to James, the virtues that saints possess include purity, charity, sympathy and devoutness. Genuine saints are those who can balance such varying virtues without allowing any particular one of them excessively to prevail over the others. Such is the case provided that saints satisfy other criteria, an examination of which is beyond the scope of this letter. In recent times Dr. Martin Luther King, Jr. as well as Gandhi bordered on saintliness exemplifying well-poised characters. Unlike Jim Jones (in 1978 he persuaded some 900 of his followers to commit suicide by drinking poison) or David Koresh, neither King nor Gandhi induced his followers to drink poison or to be destroyed by fire, or attempt to ride an UFO to heaven on the tail of a comet. These cult members were paradigmatic cases of fanatics. I do not recall the exact number, but some Blacks were led to their death by Jim Jones, David Koresh, or Marshall Applewhite. Heretofore, African American cult "leaders" are rare. Since the Abolitionist era rarely has Blacks reverted to terrorism in attempting to induce social changes. Admittedly the Black Panther Party in the 1960s and in the early 1970s advocated a violent revolution, but they did so in the open. Terrorism takes place in hiding. Moreover, today's parents, educators and legislators are seeking answers as to why cult leaders, skinheads, anti-government activists and other fanatics tend to be appealing to many people.

The virtue of devoutness is a causative factor in fanatic behavior. The fanatics are those in whom the virtues of devoutness, combined with aggressive tendencies, absolute loyalty to a person, god, ideal, or a cause, become excessive and extreme, thereby dominating or excluding other interests. Such factors result in destroying or seriously distorting the individual's personal identity. When CNN interviewed one of the surviving Koresh cult members, the interviewer asked a member how she felt when she saw the fire destroying the compound? Her reply was that she regrets that she herself was not in the flames. The fire destroyed some 85 cult members, including 24 children, a tragedy that ended a 51-day stand-off. In contrast with the fanatics, the saints are the ones who as a result of psychological transformations have shifted from the natural mode of us worldlings characterized by all sorts of inhibitions, to a higher mode of spiritual excitement, in which most inhibitions are flung aside. The fanatics are in a similar mode, but due to excessiveness of mental faculties, their psyche conditions have become pathological. Fanatics are crazy. Really. "Spiritual excitement," in *The Varieties,* James wrote, "takes pathological forms whenever other interests are too few and the intellect too narrow…Fanaticism…is only loyalty carried to a convulsive extreme." (*The Varieties,* p. 265) I suspect that there are more fanatics than there are saints in the world. The fanatic is an embodiment of what may be called moral evil, evil caused by human beings. Later, I shall discuss what may be called physical or natural evil, manifest as natural disasters such as floods, earthquakes, hurricanes, droughts, heat waves, blizzards, etc. Dr. DuBois, you were a sociologist, and thus I believe you would agree with me in saying that there are more fanatics than saints in the world.

David Koresh was a high school dropout interested exclusively in "interpreting" the Bible to fit his own convictions. He set out to "re-write" the biblical book of "Revelation." He regarded himself as the "Messiah" to whom his followers were compelled to express absolute devotion. His aggressive behavior, and the aggression he evoked in law enforcement officials, resulted in Attorney General Janet Reno attempting to explain the "irrational behavior" of both cult members as well as law enforcement officials, who many claim were just as fanatic or irrational as David Koresh. If the Waco stand-off had lasted a few days longer, one wonders whether Koresh would had succeed in "converting" some of the FBI agents into his fold? Many had begun to participate in his "Bible study" classes, it was reported. Dr. DuBois, I am sure that you would agree with me on this point:

that many of the lynching practices and other terrorist acts that Blacks suffered in the U.S., were carried out by racial (or racist) fanatics who employed terrorism.

We are accustomed to regard terrorism as isolated phenomena that occur in places such as the Middle East. Yet, as your own research brought to light, some of the most vicious acts of terrorism occurred here on the U.S. soil, against African Americans. And some African American leaders are now seeking racial reparations for such acts of terror. In many instances governmental practices or public officials assisted the Ku Klux Klan and other hate mongering groups in committing terrorist acts against Blacks. Evidence for this is found in many incidents during the Civil Rights Movement of the 1960s, where often, the government, if it did not play an active role in exerting terrorism against Blacks, did nothing to protect Blacks from such terrorism.

Take the Rosewood, Florida incident of the 1920s, in which an entire town populated mostly by African Americans, was burned, or rather, completely destroyed, by mobs. A similar incident occurred in Tulsa in 1921. What factors, then, may be regarded as remedies to religious, political or racial, or cult fanaticism? Broad intellectual and aesthetic interests; worldliness; and consideration of others—these, your former teacher William James would say, are necessary virtues that might discourage the budding of fanatic behavior.

This remedy may be applicable to addressing the problems of school shootings by disturbed youths. No; all broad-minded people will not become saints.

Sincerely,

Dr. Johnny Washington, Ph.D.   April 3, 1997

What Makes a Terrorist?

Dr. W.E.B. DuBois (1868-1963)   Destinicity Letter #19

Re: What Makes a Terrorist?

Dear Dr. DuBois:

This is a follow-up on my April 3, 1997 letter to you, in which I attempted to relate a number of issues, including those associated with the activities of saintliness, cult leaders, fanatics and terrorists. Some fanatics turn out to be terrorists. Here in this letter, I will focus mainly on terrorism. When I met with my students in my Introduction to African American Studies class today, I told them that it was commendable that since the Reconstruction Era no Black leaders such as Frederick Douglass, Booker T. Washington, Marcus Garvey, Alain Locke, Ida B. Wells, Martin Luther King, Jr. or Malcolm X advocated the use of terrorism. True, it is reported that Ida B. Wells strapped two pistols on her sides as she traveled around the country investigating the practices of the lynching of Black men. She was merely protecting herself. And I do not blame her for that. Would you? Besides, she carried her guns in the light of day, in the public realm. The U.S. Constitution permits one to bear arms, does it not? I wish to take this opportunity to remind those who might read this letter to you of the following point, of which you, yourself, are already aware: that there are at least five ways to induce social-political changes within a society: *via*, education, political persuasion, bribery, civil disobedience, and force or violence. The latter-mentioned is often associated with revolution and war; these involve what may be called "high-intensity violence" to achieve a political objective. The acceptable rules of warfare prohibit directing violence towards the civilian population. You and Dr. Alain Locke wrote extensively about the First World War and the Second World War, and the soldiers that returned home from the wars. I can remember my parents talking about the Korean War, and I remember when the African American soldiers returned home. I was a member of the First Infantry Division during the Vietnam War, and I was also part of the Civil Rights Movement of the 1960s, a movement that your own works in part inspired. Like war and revolution, terrorism also employs force or violence to achieve certain political ends. The first two usually aim at overthrowing the existing power structure or authority within a given society. Terrorism, however, plows in a different field. In my April 3, 1997 letter to you I indicated how the larger society had often employed terrorism against Blacks. In that context, I focused on lynching. I forgot to mention the more recent incidents of the burning of Black churches throughout this nation. Rarely have the perpetrators been brought to justice and punished. Clearly, terrorism involves an act of protest against certain policies or practices perceived to be unjust. Insofar as they are acting outside of the acceptable norms of a given society, the terrorists are non-conformists. To the degree this is the case, they engage in acts that are comparable to those of individuals engaged in civil disobedience or nonviolent resistance. The difference between the nonviolent resisters and terrorists will be indicated shortly. In the Western world, terrorism is less acceptable than war and revolution. Regardless of its motives, terrorism in the Western world is commonly regarded as morally reprehensible. There are also just and unjust wars, a position that the "drum-major" of nonviolence, Dr. Martin L King, Jr. once held. Prior to his becoming a full-fledged pacifist, he held that certain wars were justified, provided that their aim was to route or eliminate a larger evil such as that which Hitler represented. I think Dr. Alain Locke was a pacifist. But you were not. You were difficult to categorize, as you were so fascinated by the Russian Revolution of 1917 and its resulting communism. And in some of your writings you seemed to have held our "Founding Fathers" in high regards. A similar remark can be made regarding your attitude towards Nat Turner, John Brown and other militant abolitionists. Many find it difficult to approve of a war of aggression; however, a war fought in self-defense is generally acceptable. But "self-defense" is ambiguous. The nation of the U.S. arose out of the Revolution of 1776. The two major wars of this century, the First World War and the Second World War, involved the U.S. to a great extent. Although it became quite unpopular, the Vietnam War lasted over a decade, and some 58,169 lives were lost, and some 304,000 were wounded. The recent Gulf War witnessed a revival of patriotism among many U.S. citizens.

What features does terrorism possess that render it unacceptable, morally objectionable to many Westerners? One argument against terrorism is that its acts of violence are usually directed indiscriminately against innocent civilians and property. The February 26, 1993 explosion in New York City is a case in point. In that instant some 1,000 civilians were injured, five people killed, and the World Trade Center rendered temporarily useless.

Civil disobedience may be regarded as a positive moral act of nonconformity that is acceptable by many members of the Western world. Socrates, Jesus, Gandhi, and Martin L. King, Jr. were representative of nonconformistism aimed at approximating the ideal of social justice. In so doing they violated the laws of their respective societies. But they did so without the use of violence or force. Their acts of non-conformity were a sort of protest carried out in the light of day, and they accepted the consequence of their noncompliance with laws. By contrast, the terrorists engage in what may be called "negative non-conformity." Unlike the non-conformists who employ no violence, the terrorists rely almost exclusively on violence, indiscriminate violence, to draw attention to a cause, a perceived injustice. In many non-Western countries, terrorism is supported by a given government. Whereas the non-conformists, who are nonviolent, carry out their acts in the open, to be effective the terrorists do so in a clandestine manner. They live in hiding. Whereas the nonviolent, non-conformists accept the consequences of their action, the terrorists, while concealing their personal identity, tend to "phone in" to the media to "get credit" for their actions. Moreover, the principle of reciprocity, analogous to the Golden Rule, is a key element in assessing the morality of terrorism, so as to bring to the fore its distinguishing features.

As John Dewey suggested in his book[4], *Theory of the Moral Life*, the principle of reciprocity would enjoin such a self-disclosure—by which responsibility can be assigned—and dialog, to occur. Rather than displaying their identity, as in the case of the individual such as Gandhi or Martin Luther King, Jr., the terrorists usually conceal their identity, as in the terrorist bombing of the World Trade Center in New York City. The revealing of one's identity involves suffering the consequences of one's action. I shall end this letter in the spirit of my April 3, 1997 letter to you: education, political persuasion, and civil disobedience seem to be sensible alternatives to terrorism. These require no moral justification, a point on which you and I are in agreement.

Sincerely,

Dr. Johnny Washington, Ph.D.   April 4, 1997

The Ethics of Caring: A Feminist Perspective

Dr. W.E.B. DuBois (1868-1963)   Destinicity Letter #20

Re: Destinicity and the Ethics of Caring

Dear Dr. DuBois:

In this letter I will consider the principle of caring within the philosophic context. It is appropriate to talk about caring in light of having dwelt extensively in some of my previous letters discussing issues pertaining to the horrors of colonialism, fanaticism and terrorism. As this discussion unfolds, I trust that it will become clear that the principle of caring may be regarded as a counterpoint to the practices of fanaticism and the related practices of terrorism that I explored in some of my previous letters to you. Some people hold that one way to reduce the excessive violent behavior among our youths who often go on killing sprees is to let the youths know that we care for them. In some of my public lectures I have suggested that a possible means of reducing violence among youth lies in urging educators to make a greater effort to integrate the caring principle into the school curriculum. You will recall that one of your former teachers, William James, illuminated the issue of fanaticism and offered ways to guard against it. Although the principle of caring was implicit in his solution to fanaticism, he never brought it to the fore.

I now feel compelled to introduce the notion of caring in my consideration of Destiny. My reason for doing so is that in the hand of a fanatic such as a Hitler, a Stalin or a Tito, the notion of Destiny can lead to the destruction of a people provided, that it is not guided by certain ethical principles that forbid harming people in the interest of mere domination or power. True, Fichte, Kant, Hegel and Nietzsche had a place for Destiny in their respective philosophic views. People such as Hitler, Stalin and Tito fanatically interpreted many of these philosophers' views of Destiny to the detriment of others.

Similarly the doctrine of Manifest Destiny as employed by the U.S. government was a misinterpretation of the German philosophers' views. We know how devastating the Manifest Destiny doctrine was for Native Americans, Hispanic Americans, and African Americans. Ironically the duties of the Buffalo soldiers, a contingent of African American soldiers, constituting a racially segregated unit of the U.S. Army, included implementing the Manifest Destiny policy. Inasmuch as a great deal of my works have focused on the notion of Destiny along with its economic and political factors, it might seem a bit odd that I shall now turn to the notion of caring, inspired by feminist thought. I always try to keep an open mind.

Already I have been accused of being too Euro-centric in my works, as I tend to draw on the views of certain European philosophers. To the degree that I do so, I am not out of harmony with people such as yourself, Drs. Alain Locke, and Dr. Martin Luther King, Jr., who were highly influenced by European philosophers.

As we know, Martin Luther King, Jr. also was highly influenced by the teachings of Gandhi. Additionally, I goes without saying, I am confident that certain strains of feminist thought as developed by white women as well as by African American women have much to offer philosophic dis-

110

course. You, yourself, wrote on women's issues. However, some feminists
believe that your attitude towards women were ambivalent or ambiguous.
Similar observations are made about Dr. Alain Locke who never married.
Allow me to digress here for a moment to inform you what a feminist, Joy
James, in her work, *Transcending the Talented Tenth*[5] has to say about your
views on this matter. I will quote at length her view:

> Du Bois's profeminist politics clearly marks his opposition to patriarchy
> and misogyny. Still, a masculinist worldview influences his writing to di-
> minish his gender progressivism. Du Bois rejected patriarchal myths about
> female inferiority and male superiority. Yet he holds on to a masculinist
> framework that presents the male as normative. Since masculinism does
> not explicitly advocate male superiority or rigid gender social roles, it is
> not identical to patriarchal ideology. Masculinism can share patriarchy's
> presupposition of the male as normative without its antifemale politics
> and rhetoric. Men who support feminist politics, as profeminists, may ad-
> vocate the equality or even superiority of women. For instance, Du Bois
> argued against sexism and occasionally for the superiority of women.
> However, even without patriarchal intent, certain works may replicate
> conventional gender roles. (*Transcending*, p. 35)

Joy James adds:

> Du Bois' fictional portraits of African-American women emphasize and
> romanticize the strength of Black women. They thus differ from his non-
> fiction writing regarding individual African-American women. Although
> Du Bois makes no chauvinistic pronouncements like the aristocratic ones
> characterizing his early writings on the Talented Tenth, his nonfiction
> minimizes Black female agency. Without misogynist dogma, his writings
> naturalize the dominance of Black males in African American political
> discourse. (*Transcending*, p. 36)

What I find appealing in threads of the feminist thought are certain principles such as feeling, love, loyalty and caring, which are emphasized and
which are usually exercised in the household or private realm. Destiny is
two-dimensional, involving the private and the public realms. We usually
associate Destiny with the public or objective dimension of the world. This
element of Destiny that is played out in the public dimension represents the
"hard" side of Destiny. The "soft" side of Destiny as manifest in its private
dimension, pertains to the household, and the controlling notion here is
caring. Moreover, the household is important because this is the realm
through which human birth occurs, and within which future generations,

newcomers, are prepared to strive and struggle, in approximating Destiny. While it is the case that struggles do occur within the private household, unlike the public struggles, those in the private realm involve what Karl Jaspers calls "loving struggles," involve dialogs and interpersonal interactions, where individuals respect one another as unique individuals. Often loving struggles and strife enrich the romantic, sexual interactions between the woman and man, and render caring more meaningful for the two of them, in the private domain. However, the development of this line of thought will take me too far off course. What I wish here to explore is the possibility only in outline of an ethical basis for the practice of caring exercised in the private dimension of Destiny. In so doing, we will rely heavily on Virginia Held's essay, "Feminism and Moral Theory."

As the title of her essay suggests, Held introduces into ethical discourse a feminist perspective. Here her point is not so much focused on the principle of caring that governs the relationship between a man and women in the household; rather the centrality of her thought includes the caring relationship between mother and child in the private realm. The implication of her view is that society needs to allow women to devote more time to the household, not the world of work, so that women can fulfill their proper role, in the realm of the household in rearing and caring for children. (I anticipate this position that is half developed here will upset many feminists who take issue with the view that women should be limited to the household).

Held, who is a white woman, points out the limitations or weaknesses of traditional ethical theories formulated by such thinkers as Aristotle, Kant, and Mill. She suggests that in the past, most ethical theorists were men who by virtue of the male perspective, offered views that neglected or ignored an important ethical realm; namely, the family, and its significant activities and relations. The moral attitudes of caring and trusting are the means by which such activities and relations are governed or nurtured. Held juxtaposed what she calls the marketplace model with the mothering model that governs the household or private realm. On the one hand we have the model to which men subscribe; and, on the other hand, there is the model to which women subscribe. According to Held, the marketplace model relies on the contractual principle, that is, the buyer-seller contract, to regulate the activities and relations among competing individuals. The marketplace is an arena comparable to Hobbes' state of nature where competition among individuals is intense and aggressive, where struggles and reciprocal assaults

112

often occur. Each is against every other, and life is nasty and short. In Hobbes' view, in the civil society, the sword that represents the authority of the state is always available to assure the enforcement of a contract. Thus, force, competition, violence, death and destruction, are characteristic of the marketplace that is closely allied to the political realm. Dr. DuBois, I believe you would agree with Hobbes on this point. Held rejects the marketplace model because, in her view, it fails to do justice to the nurturing, caring experience of women and the family generally. When Held[6] turned to the mothering model, she made this observation:

> It is certainly instructive to consider it, [the mothering model] at least tentatively, as paradigmatic. If this were done, the competition and desire for domination thought of as acceptable for rational economic man might appear as very particular and limited human connection, suitable perhaps, if at all, only for a restricted marketplace. (*Contemporary*, p. 73)

She added, "Such a relation of conflict and competition can be seen to be unacceptable for establishing the social trust on which public institutions must rest, or for upholding the bonds on which *caring*, regard, friendship, or love must be based" [emphasis added]. (*Contemporary*, p. 73) In additional to the contractual principle on which the marketplace model is based, Held elucidated the distinction between the self and the universal. In the traditional ethical continuum, on the one hand, there is the self, seen in isolation; and, on the other end of the continuum, there is the element of the universal, or humanity in general. Invariably, certain ethicists seek to establish a view in which the universal, in contrast to the self, is the cornerstone of ethics. Witness Kant's categorical imperative or Mill's principle of utility, or Rousseau's general will. Held claims that most traditional ethicists who for the most part have been men, not women, formulated theories according to which the individual is secondary, the universal is primary. In each of the theories, the individual is urged to ignore his or her own will, and instead, subscribe to a universal principle. Thus, Held's point seems to be on the mark. Between these two extremes, the individual and the universal, Held sought a middle position within which to ground ethical conduct. What she had in mind is the family, an arena that lies out side of the political order, and its auxiliary, the marketplace. The realm of the traditional family is constituted in such a way to include the mother, father and at least one offspring.

The activities of birthing, and rearing children and caring for them are basic activities that occur within the family. According to Held, such activities are governed not so much by a universal ethical standard or by egotistical ethics. Rather, these activities are inspired and informed by the mothering model. Mothers, in the act of mothering, are concerned with balancing the needs and desires not of an abstract universal representative of humanity in general. Since mothers, in the act of mothering, are concerned a great deal with the welfare of children, the children (or child) constitute what Held calls the particular others. Who are these "particular others?" Particular others, in Held's view, are the children in the realm of the family, with whom mothers, or those playing the mothering role, have entwined relations. Held described this domain of the particular others in this manner:

> In the domain of particular others, the self is already closely entwined in relations with others, and the relation may be much more real, salient, and important than the interests of any individual self in isolation. But the "others" in the picture are not "all others," or "everyone," or what a universal point of view could provide. (*Contemporary*, p. 74)

Held made it clear that the particular others need not be one's own offspring. The attitudes of caring and nurturing can be extended to children in Africa, or India, as well as to the children next door. Nor did Held want to say that the mothering model is applicable only to actual mothers. Rather, her point seems to be that a new model of ethics, offered by the practices of mothering, is needed to supplement (or replace) the traditional ethical theories governed by the marketplace model. My views regarding the ethics of caring associated with mothering is also in harmony with certain African American feminists' views as descried by Joy James mentioned above. Among the women whom she regarded as playing this mothering role are such individuals as, Bernice Johnson Reagon, Annie Devine, Septima Clark, Rosa Parks, Fannie Lou Hamer, and Malika Mae Adams. Dr. DuBois, I suspect that you were familiar with some of these names. I am sure you knew Ms. Rosa Parks. In describing Reagon's attitude of the mothering role in the struggle to overcome oppression, Joy James wrote, "Bernice Johnson Reagon places Black liberation and nationalism within the context of mothering." (*Transcending*, p. 120)

Dr. DuBois, I believe you and Reagon converged in the conviction regarding the role of Black women in mothering and nurturing the Black community. However, I believe you as well as Dr. Alain Locke would assign an equally important role to educators and educational institutions in "mothering" or nurturing the Black community, the Black nation. Perhaps you might recall, it was Dr. Alain Locke who in following Socrates, regarded himself as a philosophic midwife in enabling African Americans to bring forth thoughts and ideals. Thus, it does not take much to extend this particular others principle to problems facing the African American community and the nation generally. It follows that legislators or civic leaders need to be more sensitive, and adopt a more caring attitude, as they seek to reform Welfare, and Medicare and Medicaid programs. I should add that one of my former Stanford teachers, Professor Julius Moravcsik, has offered a viable model of ethics, in which the principle of sensitivity plays a prominent role. Further, I doubt whether the principle of caring is compatible with the effort to dissolve or abolish the practices of Affirmative Action. As we set out to overhaul these institutions, society tends to subscribe almost exclusively to the marketplace model, with all of its defects and merits. To this end, many of the "experts" on whom society rely include lawyers, or corporate executives, with M.B.A. degrees, many of whom have never been in touch with the toiling masses, including those in the African American community, where unemployment is high. People need good jobs so that they can support their family members, including the children as well as the elders. (Held focused mainly on children). Moreover, schools and other social institutions need to adopt a more caring attitude. Only in this way can we merge the ideals of democracy with our Destiny ideals, in accordance to which we can actualize our human potential. This caring notion can be extended beyond the private dimension to the public dimension of Destiny that includes the social as well as the natural environment. Moreover you passed away during the mid-1960s, during the early stages of the computer revolution. We now have the Internet System, a network of many computers that allow quick communications by means of the e-mail format. During your career you wrote hundreds of letters. I believe that you would really appreciate the e-mail system and the Internet generally. Some authorities predict that by the year 2030, computers will have developed the capacity to think. Moreover, during the past decade research in the area of brain studies, as well as artificial intelligence (AI), has exponentially increased. The physicist Murray Gell-Mann, who in the 1960s received the Nobel Prize researching the activities of elementary particles, including quarks, is now working in the area of information or complexity theory, and he predicts that in the not too distant future, robots, whose nature is representative of a complex adaptive system, will exhibit behavior similar to the behavior of human beings. Thus in this light, the claim may be made that robots, or any other form of AI, will have a Destiny. The Destiny model pertains not only to the social or cultural environment but to the physical environment—indeed the entire cosmos—as well. Can you imagine a region in space, say, Mars, colonized by some 10 million robots in which the African American identity or history has been "uploaded" and in which such robots would have the capacity to replicate or evolve on their own? In the distant future robots will develop an ethnic identity. If this is so, then it might be safe to say that they will have a sense of Destiny; if this is so, the Destiny model that includes the caring ingredient will serve as a normative guide for robotic behavior. This would be an advantage over Asimov's robotic ethics. I believe Professor Moravcsik would agree with me that the robots of the future need to exercise caring and sensitivity towards natural environment as well towards us human beings.

Sincerely,

Dr. Johnny Washington, Ph.D.   April 5, 1997

Chapter 3

Modernity and Postmodernity

Modernity/Postmodernity: Hume and Einstein

Dr. Alain Locke (1886–1954)   Destinicity Letter #21

Re: Modernity and Destinicity: Part I

> There are even cases in which a kind of pleasure is conditioned by a certain *rhythmic sequence* of little unpleasurable stimuli: in this way a very rapid increase of the feeling of power, the feeling of pleasure, is achieved. This is the case, e.g., in tickling, also the sexual tickling in the act of coitus; here we see displeasure at work as an ingredient of pleasure. It seems, a little hindrance that is overcome and immediately followed by another little hindrance that is again overcome—this game of resistance and victory arouses most strongly that general feeling of superabundant, excessive power that constitutes the essence of pleasure.
>
> —Friedrich Nietzsche, *The Will to Power*

Dear Dr. Locke:

Over a period of time now I have written Dr. W.E.B. DuBois a series of letters focusing on a variety of issues including ethnic appellations, traveling in Africa, India, and Central America. Other issues included Aunt Jemima images, fanaticism, terrorism as well as the ethics of caring. I especially felt comfortable writing to Dr. DuBois, since he, as I myself, was trained in academic Philosophy, although his forte was Sociology and political activism. Similarly the field of Philosophy was your own forte to which you remained committed during your entire career. In view of this I feel like exploring some issues at a higher philosophic level, during the course of which I will draw on your views. This is one of the reasons why I have addressed my letter writing activities to you in this portion of the series. In addition to the many articles I have published, I have written two books

118

focusing on your ideas, *Alain Locke and Philosophy,* and *A Journey.* An aim of this series of "letters" to you, included here in chapter four, is threefold: i) to explore the meaning of modernity or postmodernity; ii) to determine what modern or postmodern ideas influenced your views, as well as the views of African Americans generally; and, iii) to consider the possible directions African American postmodernity might take. A consideration of ii) might enable others to gain a deeper appreciation of your thoughts. In my April 11, 1997 letter to Dr. DuBois, where I described my 1991 trip to India, I indicated that during an Indian conference, I compared some of your views to the teachings of Buddha and Confucius. Here at the outset I shall draw parallels between your views and those of Albert Einstein (1879–1955). Already in the Prologue of this work I considered the similarities and differences between modernity and postmodernity; thus, there is no need to revert to that discussion.

We are on the threshold of a new millennium, the contemplation of which is mixed with anxiety as well as hope. Currently a generally held belief is that the global Internet will crash on January 1, 2000 if certain defensive measures are not taken. My thinking about the next thousand years makes me cognitive of the Destiny of the human race, of our place in the scheme of time, of history. Some two thousand years have elapsed since the birth of Jesus, whose birth constituted the centrality of Westerners' sense of history. We have a neat way of characterizing the time lines of history as ancient, Medieval, modern, and post-modern (contemporary) eras. Many hold that such historic divisions are arbitrary, that it is difficult to determine when a given historic era begins, or when it ends. I have my own theory as to why such a difficulty prevails. Let me briefly share with you my theory, and I will appreciate any reactions you might offer. I say "my theory" but actually I am drawing on the works of David Hume (1711–1776), with whose views you are quite familiar. As you know, it was David Hume who held that any meaningful idea has a corresponding antecedent impression, acquired through sense perceptions. For example, the idea of "house" is meaningful to me. During my previous experience, I perceived houses of various sizes, forms, shapes and colors. Such perceptions allowed me to obtain a valid idea of a house and similar empirical objects. Thus, according to Hume, in order to determine the meaning or validity of any idea, we must be able to establish the antecedent impressions from which the idea was derived. We are reminded that on the basis of Hume's analysis of causality, an idea such as "necessity," that is, the necessary connection between a cause and its effect, is without meaning or validity, because of our inability to trace its antecedent impressions back to the empirical world. Similarly, many hold that it is difficult to have a meaningful or valid idea of modernity or postmodernity, because we are unable to acquire an impression of modernity, postmodernity or for that matter, of any other historic era. I like Hume's notion of time that he defines in turns of empirical objects in space. Any object in space undergoes changes such as changes in color, shape or size, which are evident of successive moments, and these in turn are evident of moments of the phenomenon of time, a theory by which Einstein was influenced. Hume's novel notion of time hinges on his attempt to offer a theory that unifies the notions of matter, space and time.

Much of what Hume had to say about space and time paralleled Lee Smolin's views of these notions, formulated in his book, *The Life of the Cosmos* (1997). Hume rejected Newton's notion of space and time. First, Hume defined space in terms of a physical object, something that exhibits the features of extension. Space has to be defined in terms of that which is extended and the relations between items that enjoy extension. That is, space can be understood only by means of that which can be perceived and we perceive matter; we never perceive a "pure" vacuum. The same is true of time. An object that is extended "in space" is perceptible, by means of its colors, shape and size, etc., each of which undergoes changes, and these changes are the marks of time. It follows that space and time, for Hume, hinge on complex objects in the world, and these time-bounded objects are constantly undergoing changes. It follows that Smolin would find Hume's notion of space-time in part plausible. Hume, like Smolin, is rejecting Newton's notion of absolute space and time.

Moreover, Smolin is seeking to develop a theory of quantum cosmology that attempts to combine quantum theory with general relativity. In so doing, he offers a novel theory of space that draws on string theory, an aborning scientific theory. He goes on to say that it may well be the case that the cosmic strings have knots in them, and that perhaps such knotted strings constitute space, that is, the fabric of space. He is rejecting Newton as well as Einstein's paradigm. I wish to suggest that as a possible way to offer a model that "unites" space and time, it may be useful to construe the knotted strings as vibrating with a certain frequency and regularity to the extent that such an occurrence is the basis of our sense of time. The phenomenon of time is tied in with the frequency of the vibrating knotted strings, just as Hume defined time in terms of space, and both space and time hinge on extension, that is, a material, perceptible object, extended in "space" and which is subject to change, or succession. Of course we cannot perceive the cosmic strings in the sense that Hume perceived his extended object, but hopefully, in the future we can measure or detect the effect of the vibrating strings, just as we can feel or measure the effects of gravity. It is anticipated that one day we will actually be in a position to detect gravity particles or waves. Smolin does not explicitly define time in terms of space. Rather he defines time in terms of dynamic complexity, of change. Hume defines time along these lines as well. Drawing on both Smolin and Hume's ideas, I am trying to define time in relation to space, the existence of which rests on vibrating, knotted strings. These knotted strings have effects that are manifested in the realm of nature, in evolution, and in the Cosmic Destiny mode, just as Leibniz's monads have effects in the realm of nature. Hence, a connection between what I am seeking to propose and the Evoluon Theory.

Here I am going deeper into the philosophic area of epistemology or metaphysics, outside of your area of specialization. You seemed to have had a greater interest in such areas as Aesthetics, Ethics, and Social and Political Philosophy. In spite of our difficulty in determining whether or not modernity or postmodernity is a meaningful category, we Westerners or any other people do have a sense of history. In Hegel's view the Western sense of history began in ancient Greece. In the history of Western philosophy, the ancient period lies between the times of Thales of the 6th century B.C. and around the times of Marcus Aurelius (121–180 A.D.). Between the times of Augustine (354–430) and Nicholas of Cusa (1401–1464) lay the period called Medieval Philosophy. Rene Descartes (1596–1650) is regarded as the "founder" of modern philosophy, and Newton is regarded as the founder of modern science, on which hinges the ideals of European Enlightenment. The late phase of African American postmodernity is associated with the 1960s Civil Rights Movement, of which Dr. Martin Luther King, Jr. and Malcolm X were representative. The first phase of the African American postmodernity is represented by Booker T. Washington, and to a lesser degree, Dr. DuBois.

Just as the notions of modernity or postmodernity are difficult to define, so also is the related notion "time," a difficulty we briefly explored above. Again the question arises: what is time and how is it to be measured? Is it absolute or relative? Does it involve duration or successive moments (Bergson)? Most people would have difficulties in offering a scientific or metaphysical view of time. To what extent does the feature of time constitute an essential ingredient in our grasping what C.I. Lewis called the sense meaning of experience? As in the case of "necessity," we in the Western world at the practical level, organize our lives on the basis of time; we have a sense of the past, present, and future. It is one thing to have a sense of time, it is another to offer a theory as to the nature of time. First, allow me to proceed on the assumption that modernity or postmodernity is a meaningful concept. Secondly, that the sense of history and time varies from culture to culture. In India, the ancient belief is pervasive that the law of karma, together with the notion of re-birth, determines our previous and future lives; thus, time and history are cyclical. In the Western world, history is regarded as involving a linear process of directionality, with a past, present and future. In certain African societies, the past is primary reality. Moreover, in his book, *African Religions,* John S. Mbiti makes the point that because of the Africans' peculiar sense of time, they have a very limited sense of the future, that Africans place much value on the remote past. Some scholars maintain that because of the Africans' lack of an appreciation of the future, along with its concomitant sense of progress, this factor sheds light on the difficulties in introducing modern or postmodern technology into Africa.

Insofar as African Americans are included in the U.S., which is part of the Western world, their sense of time and history is comparable to that of other Westerners. Whatever analysis is employed in examining the modalities of social change in the Western world, such an analysis is applicable to African Americans. Yet, because of slavery, segregation and their consequences, African Americans are not altogether an integral part of the Western world. In fact when the U.S. Constitution was drafted, African Americans were excluded from the human community; each African American slave was regarded as three-fifths of an individual. Thus, because of the practices of exclusion, rejection, denial, and oppression of the African Americans, their sense of modernity, postmodernity, and the objective conditions that inform such a sense seem to be more complex than the sense exercised by their white counterparts, the Europeans, and the Euro-Americans. Stated another way, the manner in which modernity or postmodernity affected African Americans may be different from the way it affected Europeans and Euro-Americans. In this letter I shall venture to claim that modernity for Westerners occurred largely as a result of the scientific findings by the then-Cambridge University scientist J.J. Thompson on April 30, 1897 who discovered the electron, along with the discoveries by Albert Einstein and other scientists during the first three decades of the twentieth-century. Consequently, Einstein became the symbol of what I am calling postmodernity. The discovery of the electron enabled us to deepen our understanding of electricity, upon which modern civilization relies a great deal.

Without my dwelling on Einstein's analysis of the warping of space-time produced by objects approaching the speed of light, we can see that Einstein forced us to construe time itself in an unusual and unfamiliar perspective. He demonstrated that perceived time slows down as we approach the speed of light or when a massive stellar object distorts the space-time fabric. His twin's paradox is a case in point. Couple this with the fact that his research gave us the insight to achieve the capacity to split the atom; as was previously indicated, Lee Smolin thinks that insofar as Einstein rejected quantum theory, Einstein's model of space and time is inadequate.

Nevertheless, it was Einstein who taught us that the splitting of the atom was a practical possibility. The presence of nuclear weapons altered the elements of political power associated with nation-states. During the Cold War era, most of the so-called "Third World" nations were forced to align themselves with the U.S. or with the then-Soviet Union. Rarely do scientific discoveries have a profound and immediate effect on the masses, but Einstein's discoveries pertaining to special and general relativity theories did so. His theories forced the scientific community to reject or call into question the Newtonian model of the universe, when we begin to deal with objects approaching the speed of light or when we seek to convert matter to energy. Newton held that the universe operated in accordance with fixed, determined universal laws. He introduced a fixed, absolute frame of reference associated with the so-called ether to explain the behavior of matter, events, space and time. Einstein accepted Newton's universal determinism but denied the existence of an absolute frame of reference. In Einstein's view, space, time, motion, mass, velocity—each is relative to the observer. His results were formulated in his special relativity theory (1905) and general relativity theory (1915).

As a spin-off of these results, in the 1920s, Werner Heisenberg introduced quantum mechanics that rested on the uncertainty principle. However, Einstein himself rejected quantum mechanics. Without doubt the achievements of Einstein, together with the achievements of Heisenberg and others, represent what may be called the "primary phase" of postmodernity. Admittedly, the masses of people did not comprehend the theoretical aspects of the works of Einstein and others. Yet the practical applications of their works benefited the masses. Einstein himself was a symbol of postmodernity. Although the masses could not comprehend Einstein's scientific achievements, his presence, and his legend stimulated their imagination. Abraham Pais in his essay, "Knowledge and Belief: The Impact of Einstein's Relativity Theory" offers a vivid description of the Einstein legend thus:

> The Einstein legend began on…7 November 1919, when the *London Times* carried an article with the headlines: "Revolution in Science / New Theory of the Universe / Newtonian Ideas Overthrown / Space 'Warped.'" On 10 November the *New York Times* reported: "Lights All Askew in the Heavens / Stars Not Where They Seemed To Be, But Nobody Need Worry."[1]

Pais added:

> Einstein's science and the salesmanship of the press were necessary but not sufficient conditions for the creation of the legend. The essence of Einstein's unique position goes deeper and has everything to do, it seems to me, with the stars and with language. A new man appears abruptly. He carries the message of a new order in the universe. He is a new Moses come down from the mountain to bring the law and a new Joshua controlling the motion of heavenly bodies. He speaks in strange tongues but wise men aver that the stars testify to his veracity. (p. 158)

When the results of Einstein's special relativity theory were published in 1905, Dr. Alain Locke, you were nineteen years of age. It is worth noting that both you and Einstein were influenced by the philosopher-physicist Ernst Mach's principle of equivalence. Heinz R. Pagels in book, *The Cosmic Code*[2] describes the principle of equivalence: "The fact that we cannot physically distinguish a nonuniform motion like an acceleration from gravity is called the principle of equivalence—the equivalence of nonuniform motion and gravity." (*The Cosmic*, pp. 25–27). Using this principle, Einstein was able to demonstrate that gravity "created" by an accelerating object is equivalent to "real" gravity associated with an object such as the Earth or the sun. Similarly, you relied on the principle of equivalence to explain what may be called "symmetrical patterns" that cut across cultures, that is, to explain common elements within various cultures. You made specific references to Mach's views in the contexts of explaining stability and changes that cut across cultures. Within the realm of physics, the principle of equivalence allows scientists to deduce universal laws. Thus, it was such a principle that enabled Einstein to discover general relativity and enabled you to provide insight into the nature of culture and of axiology.

Located at the Moorland-Spingarn Research Center at Howard University is a piece of correspondence between you and Einstein. Similarly there are pieces of correspondence between you and Booker T. Washington. My claim here is that you bridged the gap between Western postmodernity, of which Einstein was representative, and African American postmodernity, of which Booker T. Washington was representative. Einstein seemed to have had little, if any, interest in value relativism. Just as Einstein rejected the absolutism inherent in Newtonian physics, you rejected absolutism inherent in ethics, religion, aesthetics, body politic and other fields. I believe your attitude towards the Evoluon Theory would be mixed, provided that it is construed as making metaphysical commitments.

Sincerely,

Dr. Johnny Washington, Ph.D.   April 14, 1997

Postmodernity: Dr. Alain Locke

Dr. Alain Locke (1886–1954)   Destinicity Letter #22

Re: Modernity and Destinicity: Part II

> It is not the satisfaction of the will that causes pleasure,…but rather the will's forward thrust and again and again becoming master over that which stands in its way. The feeling of pleasure lies precisely in the dissatisfaction of the will, in the fact that the will is never satisfied unless it has opponents and resistance—The happy man: a herd ideal.
>
> —Friedrich Nietzsche, *The Will to Power*

Dear Dr. Locke:

I wish to continue where I left off in my April 14, 1997 letter to you, in which I was, among other things, making comparisons between your own views and those of Einstein's relativity theory. I mentioned Einstein's works in this context because his scientific achievements, as your own philosophic achievements, represent a historic turning point in the schema of time, that is, history. I will demonstrate that the two of you may be regarded as marks of postmodernity. You were interested not only in the spirit of relativity theory and related results. You also held that the scientific method itself could be extended to the humanities, and it could thereby serve as a basis for what you called "scientific humanism." It is clear that you appreciated the results of modern science, in suggesting what directions value theory, that is, axiology, should take in guiding enquiry. The enquiry that you envisaged would enable us to have a greater appreciation of racial and cultural differences. A connection between your axiology and African Americans' sense of postmodernity is that you employed your axiology to illuminate race problems in the Americas as well as in Africa and other parts of the world. Aimé Césaire, Léopold Sédar Senghor, and other *Negritude* participants were highly influenced by your views as well as DuBois'.

Thus, on the basis of your claim, value absolutism, that is, values that are determined by a creed analogous to, say, the Kantian categorical imperative that denies the multiplicity or relativity of norms, and of ethnicity, provided the basis for a totalitarian form of government, of which Nazism was representative. In some of my previous letters to Dr. DuBois, I described the negative effects of colonialism as practiced by Hitler, Stalin, and Tito, whose respective governments exemplified dogmatism, cultural intolerance and violence in the highest degree. I certainly agree with you that absolutism is objectionable, a view I kept in mind as I formulated the Destiny model.

In your essay, "Values and Imperatives," you spoke of the possible advantages of value relativism that you contrasted with value absolutism. In your view the day would soon come when relativism will be accepted and absolutism will be rejected. At the turn of the century, scientists introduced the notion "field" (influence that implies a detectable limit) to explain the nature and relations between forces such as electricity and magnetism. In the late 1920s, the "field" concept allowed scientists to attempt to combine relativity theory with quantum mechanics. It is interesting to note that the notion "field" formed the cornerstone of your axiology; you relied on such a notion to provide an integrating principle to your value theory. It was Max Weber (1864–1920) who held that the disconnecting of truth, beauty and goodness that occurred in the modern world is a characteristic feature of modernity. The modern world became fragmented and disjointed. Throughout the Middle Ages Europeans sought unity in the Christian God. The Protestant Reformation played a major role in disrupting that unity. If

Kant sought transcendent unity in the practical postulates of God, Freedom and Immortality, Dr. Locke, you sought immanent unity in human consciousness or feeling that is the cornerstone of your axiology. The Destiny model seeks to restore the principle of transcendent unity that may enable us to become more fully integrated with the Spirit. I accept Hegel's view of the Spirit that he defines in terms of freedom. Moreover, he construes Spirit as a manifestation of universal consciousness, whose existence predates the existence of nature, the physical world. Thus, for Hegel, there is a close connection between time (history), freedom and Spirit. Hegel allows us to make an effective connection between history (evolution) and Spirit. Clearly Hegel's view provides a framework for postmodernity insofar as it incorporates the complexity of existence (in contrast to "reality") whose essence unfolds or evolves within the context of history. I am reminded of Douthart's claim that existence is that which evolves. Inasmuch as the Spirit evolves, it enjoys existence. In describing the processes of evolution, Douthart also considered the notion "saturated complexity." He did so in the context of explaining the manner in which the complexity of evolution shifted from one major phase to the next. When a given stage of evolution becomes saturated with complexity, such saturation induces a shift to the next stage, thus driving the system on to higher stages of complexity. At this point in evolution the human brain or consciousness represents the highest form of saturated complexity, which may be contrasted with the diffused complexity in the brain of a donkey or a goose. In contrast to the human brain, I venture to claim that heretofore all computers and other forms of artificial intelligence constitute diffused complexity, (although I am aware of the fact that Alan Turing with his Turing machine apparently blurred the distinction between a computer and a human being). Diffused complexity is characteristic of all forms of physical matter or energy.

As was suggested in "Values and Imperatives," you appealed to the "field" metaphor in an attempt to offer a description of the four value fields: the religious, the ethical-moral, the logical-scientific and the aesthetic-artistic value types. The unifying factor, in your view, was the notion of attitudes or feelings. Our values are constituted by feelings or attitudes, the unifying source of the value fields. When I regard a scene as having a religious value, the scene in question induces in me the feeling of exaltation. When I regard a scene as having moral or ethical value, the scene in question induces in me the feeling of tension or conflict. It follows that when my feeling or attitude shifts, the value content shifts. The value fields are dynamic and interrelated. Such a relation is analogous to the interconnection between the electrical field and the magnetic field. This dynamism allowed you to explain the conflicting, merging, transferring and diverging nature of the four value fields. Dr. Locke, you spoke of the four value fields; similarly, in physics there is the desire to provide a theory that describes the unification of the four fundamental forces of nature: the strong nuclear force, the weak nuclear force, electromagnetism, and gravity. Einstein spent the last twenty-five years of his life trying to offer a mathematical description of the unity of the four fundamental forces or fields. The difficulty is to connect gravity with the other three forces. Some predict that once the four forces are unified, our science will enter a remarkable new phase of development. Such a phase would perhaps signify a significant mark of postmodernity. Many are optimistic that a theory that effectively describes the unification of the fundamental forces will at some point in the future be developed. A reason for such optimism is that during the origin of the universe, some fifteen to twenty billion years ago, during the point of singularity, the four forces were at the outset unified, but subsequently became disunited. What appears to be a discontinuity of forces, following the big bang, many believe is merely an illusion. The unity, although imperceptible by ordinary scientific instruments, has always been present. Such a model that describes the unification of the forces focuses on the mere directionality of the Destiny of the universe, an observation Einstein and his successors have overlooked. This directionality unfolds along the asymmetrical arrow of time, within which is included the element of anticipation. Certainly Alfred N. Whitehead would undoubtedly agree with my hypothesis. Dr. Locke, this is just a side point: if you recalled, Bergson defined matter as a relaxation of energy, and he defined Spirit as a concentration of energy. Now, Dr. Locke, if we extend Douthart's or Bergson's distinction to your area of axiology, what would be the result? Are values diffused or saturated, relaxed or concentrated? In your axiology, you described values as enjoying a certain fluidity. Consider the value field, say, the religious value field that can shift to the logical or aesthetic value field. Is this shifting due to saturated complexity? A consideration of the novel notion Destinicity subsequently gave rise to another novel notion "World Destinicity," whose analogous notions include "Ethnic Destinicity," "National Destinicity," as well as the "Cosmic Destinicity," paralleling the ethnic Destiny mode, the national Destiny mode and the cosmic Destiny mode, respectively. In the Prologue I focused on the "World Destinicity," which, due to its generality, allows us to see things on a large scale. When a given ethnic system or schemata undergoes saturated complexity of information, it gives rise to "National Destinicity," which, in turn, after undergoing saturated complexity, gives rise to the "World Destinicity." The final outcome is the "Cosmic Destinicity" that will fulfill its role in Evl6, Evl7 and Evl8 where robots will prevail. Dr. Locke, I have often toyed with the idea of grounding your axiology within a historic or sociological context. I believe your axiology would have a greater explanatory power if one could appreciate how the various value fields underwent the process of evolution or historic development. Sincerely,

Dr. Johnny Washington, Ph.D.   April 15, 1997

Postmodernity: Booker T. Washington

Dr. Alain Locke (1886–1954)   Destinicity Letter #23

Re: Modernity and Destinicity: Part III

> The warlike and the peaceful.—Are you a man with the instincts of a warrior in your system? If so, a second question arises: are you by instinct a warrior of attack or a warrior of defense?

> —Friedrich Nietzsche, *The Will to Power*

Dear Dr. Locke:

In my last letter to you dated April 15, 1997 I mentioned the views of Alfred N. Whitehead, whose ideas I shall consider more thoroughly later in this work in which I elaborate on the novel notion Destinicity. You failed to make a distinction between transcendence and immanence. Like Dewey, you seemed to have thought that the realm of immanence is the only reality. You held that things go wrong when people allow absolutes to become the controlling norms, with detrimental effects. I am in a position to work around what may be called the absolutist problem, inasmuch as I place the absolutes, which I construe not as fixed but as dynamic, luring principles, grounded in freedom, in the realm of transcendence. This realm is open-ended and dynamic. It is instructive to consider a contrast between your own endeavors and those of the Confucians of ancient China. The Confucians viewed the norms of society as absolute, grounded in the cosmos.[3]

What is interesting is how the Confucians derived such absolutes. They began with the social order out of which the principles in question arose, and then projected such principles onto the cosmos, making it seem that the cosmos itself subscribed to order, ritual, custom, integrity and the like. A similarity between the Confucians' axiology and yours is this: in both views, the principle of reciprocity is at the forefront. According to the Confucians, the heavenly bodies obey the cosmic principle of reciprocity. The Confucians' principle here is analogous to the views of the astronomer Kepler who focused not so much on the notion of reciprocity but on the notion of harmony and proportionality. In his view the universe obeys the principle of harmony. However, he maintained that such a harmony is not altogether out there in the world. Rather, the harmony is in the human mind that subsequently imposes harmony onto the world. It could be the case that this principle of reciprocity, about which the Confucians speak, is a psychological schema, like Kant's *a priori* categories, that the mind in part projects onto the world. Already I have argued that the ethnic appellation such as "Black" or "African American" is an element within the Kantian schema that enables us to derive what C.I. Lewis called sense meaning, (in contrast to linguistic meaning), so that we can bestow meaningfulness on our lives. It is difficult to determine whether Mr. Booker T. Washington (who died in 1915, around the time Einstein was completing his works on general relativity) was an absolutist or a relativist. Mr. Washington seemed to have accepted the basic creeds of Christianity (absolutism), as well as the basic tenets of American pragmatism, the emphasis on practical results.

In your essay, "Booker T. Washington: Strategist of Southern Reconstruction,"[4] you described Mr. Washington as a transitional figure, who introduced a New Order in place of the Old Order. Clearly Washington played a decisive, pivotal role in determining the Destiny of African Americans. This point was well stated by Louis Lomax in his introduction to Mr. Booker T. Washington's book, *Up From Slavery*.[5] It was during the middle of the Civil Rights Movement (1965), Lomax indicated that if Abraham Lincoln affected the Destiny of Blacks, so did Mr. Booker T. Washington. African American postmodernity was distinct from so-called "Western postmodernity." If Einstein played a major role in lifting the curtain so as to allow Western postmodernity to unfold, would you not agree that Mr. Booker T. Washington played a major role in lifting the curtain to allow African American postmodernity to unfold? In making this claim I am suggesting that Mr. Washington, more so than Dr. DuBois, had a greater impact on the African American experience. Lomax suggested, but did not state, that Mr. Washington was the father of African American modernity, what I called "postmodernity."

Houston Baker is more to the point; he went so far as to attempt to establish the exact date and place that African American postmodernity was launched. The launching occurred September 18, 1895, when Mr. Booker T. Washington delivered his ten-minute address at the Atlanta Exposition in Atlanta, Georgia.[6] The phrase "master of form" is used by Baker in describing Mr. Washington. First, let us see what Baker means by "form." He defines form as a space in terms of which boundaries are established between the self and the other. (*Modernism*, p. 16) The point is that Mr. Booker T. Washington was effective as a leader in that he mastered the minstrelsy traditions that relied on artistic form.

Post-Medieval minstrelsy originated on the slave plantations in the U.S. South. On most plantations there was a group of talented slaves whose job it was to entertain the guests of their respective masters. During certain stages of the history of minstrelsy, the actors were white people who, with blackened faces, imitated and exaggerated the behavior of Blacks and portrayed them in a clownish manner. In your view, "Negro minstrelsy came to the American stage," before the 1830s by white actors, with "blackface" imitating Negroes or Blacks.[7] It was during this time that whites attempted to create a demeaning, dehumanizing image of African Americans, against which Mr. Booker T. Washington had to struggle in informing the Destiny of African Americans. I believe he could have been more effective had he made a greater effort to subscribe to a view of transcendence I mentioned in the opening of this letter.

Sincerely,

Dr. Johnny Washington, Ph.D.   April 16, 1997

Dr. Alain Locke (1886–1954)   Destinicity Letter #24

Re: Modernity and Destinicity: Part IV

> How does one become stronger?—By coming to decisions slowly; and by clinging tenaciously to what one has decided. Everything else follows.

The sudden and changeable: the two species of weakness. Not to mistake oneself for one of them; to feel the distance—before it is too late!

Beware of the good-natured! Association with them makes one languid. All associations are good that make one practice the weapons of defense and offense that reside in one's instincts. All one's inventiveness toward testing one's strength of will—To see the distinguishing feature in this, and not in knowledge, astuteness, wit.

—Friedrich Nietzsche, *The Will to Power*

Dear Dr. Locke:

In much of my correspondence to Dr. W.E.B. DuBois, I considered the issue of the identity of African Americans. In fact my first four letters to him dealt with the identity problem. We have returned to it by way of the minstrelsy tradition, of which I spoke toward the end of my last letter. In much of your writings on African American music, you threw much light on this topic. Some of my previous pieces of correspondence focused on Dr. Kenneth Goings' research on the Aunt Jemima images. A similarity exists between such images and the images associated with the minstrelsy tradition. Perhaps the Aunt Jemima image grew out of the minstrelsy tradition. On the basis of your position, it was during the latter part of the minstrelsy tradition that the pseudo-Negro image was created by white actors imitating Blacks. The insights here enable us to understand the history of the demeaning image or identity that whites forced upon Blacks. The stages of minstrelsy that you focused on were between 1850 and 1895 (*His Music*, p. 37) It will be remembered that the year 1895 was the year that Frederick Douglass died, and this was also when Mr. Washington became famous; it was also around this time that the Supreme Court in 1896 ruled that segregation was permissible.

On Houston Baker's observation, Mr. Washington became an effective leader mainly because he was able to adopt the minstrelsy form of expression to persuade the white as well as Black Americans to accept his policy of race relations and industrial education. By relying on a modified version of the minstrelsy form, Mr. Washington's message, in contrast to Dr. W.E.B. DuBois' message, was presented to the larger society in a non-threatening manner. Washington himself represented conciliation and compromise; however, people such as Mr. Frederick Douglass and Dr. DuBois appeared to pose a threat to whites. Although Baker regarded Mr. Washington as setting the stage for African American modernity, Baker attributes the real credit to you and other Harlem Renaissance activists. Already we have considered the pseudo-Negro that appeared on the scene during the minstrelsy tradition that whites regarded as representative of African Americans generally. Mr. Booker T. Washington quietly and cleverly attempted to employ the minstrelsy form to dispel the image of the pseudo-Negro created by whites to entertain themselves and to dehumanize Blacks. To that end, his book, *Up From Slavery* was widely received by both Northern and Southern whites. However, following the death of Mr. Washington, the New Negro, representative of the Black artists of the 1920s, came on the scene. The book *The New Negro* that you edited was intended to document the achievements and aspirations of these artists. Unlike Mr. Booker T. Washington, they rejected altogether the minstrelsy forms. The sounds, forms and images of these artists allowed them for the first time to express their true identity, reflective of the African American postmodernity that was appearing on the Destiny horizon. Such an identity became quite expressive during the Civil Rights Movement of the 1960s, at whose centers were individuals such as Dr. Martin Luther King, Jr., Malcolm X and others.

While you were studying in Europe in 1911, Mr. Washington happened to be in Europe at that time. It was during this time that you met Washington, apparently for the first time. Indeed, you seemed favorably impressed with him. From 1911, onward until the death of Mr. Washington, the two of you exchanged pieces of correspondence, most of which are now included in your (the Alain Locke) Collection at the Moorland-Spingarn Research Center at Howard University. Following your return to the United States, you accompanied Mr. Washington on a "fact finding" tour of the southern parts of the United States. In 1912, racial tension erupted in Jacksonville, Florida, one of the areas that you and Mr. Washington together visited. It is worth adding that a few years later a serious massacre occurred in Rosewood, Florida, where an entire community of Blacks was murdered by a white mob, and the property of Blacks was destroyed by fire set by that white mob.

"Booker T. Washington: Strategist of Southern Reconstruction," the essay that you wrote, but did not publish, provided insights about the associations between Mr. Washington and yourself, and the admiration you extended to him. I am curious as to why you did not bother to publish this wonderful essay on Mr. Washington? In this work, we find you commending Mr. Washington for the statesman-like qualities he exhibited in Europe, while advising the Europeans on how to deal with the problem of uplifting the status of women. In so doing, on the basis of your insight, Mr. Washington encouraged the practices of what may be called a "feminist elite," the European women on top must "pull-up" the women on the bottom. [In September of 1999, a conference was held in your honor at Howard University; its theme focused on gender and race].

Today certain feminists are quite active in advancing the cause of women. It is ironic that Mr. Washington opposed the elitist principle in considering the problems of African Americans, DuBois' views to the contrary notwithstanding. You credited Mr. Washington for initiating reformative measures in education that effected the entire nation, not just the Black community. As was suggested, following the death of Mr. Washington, the New Negro, in connection with the Harlem Renaissance, emerged. Many people regard you, along with Dr. W.E.B. DuBois, as the forerunner of the Harlem Renaissance, a movement that was energized by Marcus Garvey's words and deeds. It will be remembered that Garvey, like you, held Mr. Washington in the highest esteem.

As in the views of Louis Lomax, Houston Baker gives much credit to Mr. Washington in shifting the African American community into the new era. However, Baker thinks that the highest credit belongs to you and the Harlem Renaissance movement. A point that Baker makes is that your major role concerning the African American modernity was that you shifted the focus beyond the region of the South (Booker T. Washington), to the nation and to the world.

Sincerely,

Dr. Johnny Washington, Ph.D.   April 17, 1997

Transcendence, Unity and Self-Esteem

Dr. Alain Locke (1886–1954)   Destinicity Letter #25

Re: Modernity and Destinicity: Part V

> Men who are *destinies* [emphasis added], who by bearing themselves bear *destinies*, the whole species of *heroic* bearers of burdens: oh how they would like to rest from themselves for once! how they thirst for strong hearts and necks, so as to be free from what oppresses them, at least for a few

hours! And how vainly they thirst!—They wait; they look at everything that passes: no one approaches them with as much as a thousandth part of their suffering and passion, no one divines *in what way* they are waiting—At length, at length they learn their first piece of worldly prudence—not to wait any more; and soon another one: to be genial, to be modest, from now on to endure everyone, to endure everything—in short, to endure even a little more than they have endured so far.

—Friedrich Nietzsche, The Will to Power

Dear Dr. Locke:

Let me now describe in a bit more detail the Destiny model. In doing so in this context I will draw on the views of Kant as well as Karl Jaspers. In some of my letters to Dr. W.E.B. DuBois I provided my definition of Destiny, according to which it is an ideal toward which a people strives in successive generations. Certainly the transcendent ideal of Destiny is over and above the pragmatic concerns of any given ethnic group or nation, found in the immanent Destiny modality. Rather this transcendent ideal is that which urges all ethnic groups or nation-states to transcendent unity, a goal that may never be reached but one must struggle to approximate. Its aim is to lift our hearts and minds beyond the immanent Destiny modes, where disunity, manifest in the political arena and in the marketplace, occurs. This transcendent unity includes the principle of equality, by which each individual as well as ethnic, religious or cultural group is equal to every other; it is a principle that lifts every individual or group beyond the Hobbesian state of nature; it is a principle of equality that allows Muslims, Jews, Christians, Hindus, Buddhists, as well as Confucians, among others, to stand on the same footing with mutual recognition and respect; and it is a principle that urges people to respect not only cultural diversity and unity but also diversity and unity that are found within our natural environment. There is no need here to introduce the God hypothesis, the concerning of which usually engenders religious conflicts among the various religious groups. I believe certain feminist thinkers could also benefit from the Destiny model. It would place them in a better position to examine the so-called "women's issues" within a broad context so that cross-cultural comparison could effectively facilitate social science research. For example, in your essay, "The Need for a New Organon in Education" (1950), a basic theme is the importance of cross-cultural comparison. What were your feelings on feminist issues? Were your views similar to Dr. DuBois' or Mr.

Booker T. Washington's? In spite of the differences among the various ethnic or religious groups, and the differences among nations of the world, there are certain ideals that transcend the various cultures or groups. As was stated, one such ideal is that of transcendent Destiny, and there is a connection between Destiny, ethnic identity, dignity, equality and freedom. Destiny is the principle by which these are completely unified at the transcendent level. Already I mentioned the role that the principle of Destiny plays in unifying the elements of identity, equality and freedom that are exercised or nurtured in the community as pertaining to our immanent Destiny. Similarly the four Destiny modes: the ethnic, the national, the world and the cosmic are unified by our transcendent Destiny that we have the capacity to posit or postulate.

In the past, an examination of certain problems of the African American community has focused on a number of issues, including that of low self-esteem among its youths. A claim is that, due to self-doubt and low self-esteem, many African American youths lose interest in school or other social matters that enjoin commitments. What I wish to maintain is that a major problem in the African American community (and the country generally), is the lack of a sense of Destiny, the embracing of which inspires hope in this life. I wish to connect Destiny with certain psychological, or moral qualities such as, "resoluteness," "sincerity," "openness," "directionality," "purposefulness" and "caring." A goal is to urge people to regard the Destiny model in a way in which, when resolutely embraced, its ideals will become efficacious. Not only that but the Destiny model is posited to allow the principle of self- or group-efficacy to be effective in the lives of people so they can obtain some degree of control of their respective immanent Destiny modes. In one of his books[8], Albert Bandura is persuasive in making a connection between self-efficacy and destiny. He wrote: "Without commitment to common purposes that transcend narrow self-interests, the exercise of control can degenerate into personal and factional power conflicts. People must work together if they are to realize the shared *destiny* [emphasis added] they desire and preserve a habitable environment for generations to come." (*Self-Efficacy*, p. 2)

In light of the recent shooting sprees that have occurred in many middle schools and high schools across the country, I wish to emphasize that the principle of caring needs to be connected to the self-efficacy model. Granted you yourself, Dr. Locke, urged society to accept the principle of tolerance in approaching racial and social problems, such a principle alone is inadequate. Tolerance needs to be combined with the efficacious notion of caring, a notion that needs to be nurtured in the respective schemata of the Destiny modalities.

I know you were an admirer of Friedrich Nietzsche (1844–1900) who like Søren Kierkegaard (1813–1855) was a forerunner of existentialism. And I believe you as well as Dr. DuBois accepted in part Nietzsche's views of Destiny that are associated with his view of the will to power. Nietzsche was also an elitist, just as one finds the principle of elitism reflected in the works of Dr. DuBois and your own notion of the Talented Tenth. I do not know how you felt about existentialism in general. Nor do I know what your attitude towards Jaspers' views was. I myself can appreciate Jaspers' views of Destiny that are similar to Nietzsche's. I especially like Jaspers' notion of "loving struggle." I believe that you would find some of his views acceptable.

Jaspers held that one could reach transcendence or ultimate reality, and exercise the existential freedom that it harbors, to the degree that one embraced the principle of resoluteness, or sincerity, a position that is similar to Nietzsche's. I know you have problems with the notion of "transcendence" or of "ultimate reality" but Jaspers, like Nietzsche, has his own peculiar meaning of such a term.

The ancient Confucians also believed that one could reach ultimate reality by adopting the principle of sincerity or resoluteness through which one enters The Way, the Tao, which pertains to Ultimate Reality. Today such spiritual qualities, which may be strengthened if combined with the principle of caring, are not sufficiently stressed in the education of African American youth and society generally. I believe that a consideration of Destiny is important, insofar as it seeks to introduce a transcendent, unifying principle, an element of directionality and a stronger sense of identity into the lives of people. I am surprised that you did not devote more attention to the Destiny notion. I suspect that you felt that it was too much engaged, historically, with the notion of transcendence and its concomitant element of absolutism that you found objectionable. Then, there was the Manifest Destiny doctrine. My notion of Destiny is different from the one associated with traditional absolutism or with Manifest Destiny. The Destiny model I advocate seeks universal harmony and unity of all races and ethnic groups.
Sincerely,
Dr. Johnny Washington, Ph.D.   April 19, 1997

Chapter 4

African History and Destiny

African History: Some Preliminary Remarks

Professor William Leo Hansberry (1894–1965)   Destinicity Letter #26

Re: African Historic Perspective: Prologue: Part I

Dear Professor Hansberry:

Presently I am addressing this portion of letters to you inasmuch as you contributed to our understanding the African civilization. In this series of letters to you I wish to provide greater contents to my definition of Destiny. There is a close connection between the Destiny of a people and their historic development. For generations Ethiopia was regarded as a symbol of Destiny not only for many Africans in Ethiopia and other parts of Africa, but also for African Americans. Thus I shall allow the historic perspective to be the basis of this series of letters to you that shall unfold along the following themes: i) *African History: Some Preliminary Remarks,* ii) *Hypotheses of Biological Evolution;* iii) *Evolution: Iron Technology;* iv) *Ethnic Identity: Ancient Ethiopians;* v) *Ancient Philosophies: African and Greek;* vi) *Sage Philosophy;* vii) *Ancient and Medieval Kingdoms of Africa;* viii) *Queen of Sheba;* ix) *Christianity in Ethiopia;* x) *The Nine Saints of Ethiopia;* xi) *Prester John;* xii) *The World of Mansa Musa: Ghana, Mali, and Songhay;* xiii) *the World of Sunni Ali: Ghana, Mail and Songhay;* xiv) *the World of Askia the Great: Ghana, Mali and Songhay;* xv) *and, the Muslim Philosophers Avicenna and Averroës.*

If immanent Destiny is regarded as a landscape that includes the natural as well as the social environment, the above-mentioned themes are prominent features of the Destiny landscape. Some of these features represent crossroads, or junctures indicative of historic turning points and transformations. Since in the first few letters in this series we are here concerned with a preliminary discussion of the notion of origin, of the beginning of

138

human evolution, I shall commence this series of letters by briefly considering the origin of the universe associated with what scientists called the point of singularity of the big bang. This big bang hypothesis was debated by scientists during the early parts of the 1960s. When the opportunity arises in this series of letters to you, I shall consider your intellectual affinity to evolutionary scientists such Charles Darwin and the Leakeys, among others.
Sincerely,
Dr. Johnny Washington, Ph.D.   April 21, 1997

Hypotheses of Biological Evolution

Professor William Leo Hansberry (1894–1965)   Destinicity Letter #27

Re: Hypotheses of Evolution: Part II

Dear Professor Hansberry:

You are already knowledgeable about the facts of many things I shall say about Africa here, but I shall repeat them nevertheless for the benefits of others who may read this "letter." In your book, *Africana At Nsukka*[1] you in effect described a portion of East Africa as the point of singularity of human evolution; of course, I am in this context using "singularity" metaphorically. You also reminded us that in addition to your own research, many other sources held that Africa is the point of origin of human evolution and that the African civilization was not imported from other regions of the world. (*Africana*, p. 19) You went on to make the claim that the view that Africa was not the source of human biological evolution or civilization was vigorously perpetuated by whites during the New World slavery era. Their aim was to provide a rationale for enslaving Africans in the Americas or colonizing Africans on the continent of Africa. In the ancient world, the Europeans' attitude towards Africans, generally, was not condescending or negative, a point that another one of your Howard University colleagues, Dr. Frank Snowden, also made in his works. You, as well as Snowden, held that many ancient Greeks regarded Ethiopia as the cradle from which other cultures originated. For the most part, from pre-history on through the Middle Ages, Europeans as a whole possessed limited knowledge about Black Africa. Martin Birnal in his book, *Black Athena*, also converged on the path opened by you and Snowden.

Darwin's view that Africa constituted the singularity of biological and cultural evolution was confirmed a century later by Mary and Louis Leakey. During 1959, the Leakeys uncovered evidence that supported the "Africa hypothesis" with data from Olduvia Gorge located in Northern Tanzania, East Africa. They discovered a human skull which was assigned the name "Zinjanthropus" or the "Nut-Cracker Man," who scientists regarded as evident of the "Adam" of the human race. What is interesting about the Nut-Cracker Man is that he had the capacity to make and use tools. According to most scientists, the capacity to make and use tools is an important feature that distinguishes human beings from the lower primates. In 1960, however, a much more interesting discovery was made in Africa; namely, the discovery of "Homo-habilis" or the "Handy Man," whose features indicated that he was much more advanced than the Nut-Cracker Man who presumably ate plants only. The Nut-Cracker Man seemed to have been much smaller in size than the Handy Man, and his teeth indicated that he was a meat eater who relied on tools and weapons for obtaining food. The tools that the Nut-Cracker Man left behind were found throughout the world, although Africa is believed to have been the point of origin. The evidence discovered by paleontologist Donald C. Johanson also merits mentioning. During 1974 on the plains of the Ethiopia Afari Triangle, Johanson unearthed the 3.5 million-year old remains of a human-like creature whose body resembled that of a modern woman. He called her "Lucy." She was described as having had a small head and was believed to have walked upright. The significant question such a finding raised was: where do Lucy and her relatives fit into the scientific model that represents the path of human evolution? Now, the view is held by many that some 3.5 million years ago two paths of evolution became quite distinct: one path led to a small brained ape-like individual or human-like ape who never learned to use tools and thereby became extinct about a million years ago. Johanson claimed that modern human beings are the descendants of Lucy and her relatives. That Lucy's remains are significant because she and her relatives represented the point in the evolutionary field where the two paths diverged in Africa about 3.5 million years ago. Richard Leakey denied the significance of the Lucy discovery. He maintained that Lucy was merely another member whose ancestors and descendants evolved along the *Australopithecus afarensis* path which led ultimately to a dead end. In his view the splitting of the paths occurred over eight million years ago. Here we will not decide who has a more accurate view, Leakey or Johanson. Finally, Professor

Hansberry, let me make reference to Afrocentrism, a model which you assisted in establishing, although during your days it was not called Afrocentrism. Today Afrocentrist scholars go so far as to hold that in order to obtain a more balanced picture regarding Africa's role in the schema of biological and cultural evolution, we need to reshape the education curricula in such a way as to place Africa in the center of research and teaching. Heretofore we have placed Europe at the center of research and teaching, an approach that resulted in diminishing the significance of Africa's contributions to world civilization. These Afrocentrist scholars maintain that a more adequate approach to research and teaching hinges on the Afrocentrist model. As I previously suggested, you were a forerunner of Afrocentrism. The same is true of Drs. W.E.B. DuBois, Alain Locke, Frank Snowden and others. To this end, each year during African American History Month (February), we celebrate your contributions. You have made us proud of ourselves. We express great reverence towards you. You are symbolic of our Destinicity.

Sincerely,

Dr. Johnny Washington, Ph.D.   April 24, 1997

Evolution: Iron Technology

Professor William Leo Hansberry (1894–1965)   Destinicity Letter #28

Re: Agriculture and Iron in Ancient Africa: Part III

Dear Professor Hansberry:

Let me open this discussion by drawing a neat parallel. When I wrote to Dr. Alain Locke in a series of letters entitled "Postmodernity and Destinicity," I was interested in determining the boundaries of postmodernity, that is, the features(s) indicative of the beginning of the postmodern world. In writing this series of letters to you I am here focusing mainly on the antiquity, the ancient world as well as the Medieval Age. It is equally difficult for scholars to agree on where the ancient world began and ended, or where the Medieval Age began and ended. I will not attempt to decide these issues here. What will be clear is that such categories as "ancient," "Medieval," "modern" and "postmodern" are not universally applicable. It seemed that Africa reached one of its highest peaks of development during the so-called

Middle Ages, or Medieval Period, and during the this time in Europe, ignorance and "darkness" prevailed. Moreover, when Europeans shifted out of the Dark Ages, into the "Age of Enlightenment," they began the institution of slavery in the New World, and this, together with the subsequent effect of African colonialism, had a devastating effect on Africa. The notion of Enlightenment seems to be incompatible with the practice of slavery in the New World. Having made these preliminary observations, let me focus on pre-historic Africa. Africans relied on the hunting-gathering economy for thousands of years, until they entered into the Early Iron Age. Their usage of metal tools represented a major development in the history of African civilization and world civilization generally. As will be seen the usage of iron represented one of the most striking revolutionary achievements among human beings. Their discovering the method of smelting iron ore was perhaps comparable to the splitting of the atom, which was a mark of Western postmodernity, as I have indicated in my previous pieces of correspondence to Dr. Alain Locke. As early as the Sixth Century A.D., Africans began to transform iron-ore into tools that enabled them to build dwelling places, to produce food and to make weapons. Consequently communities were settled with well-defined political and economic structures. As was suggested, the development of iron technology marks a major juncture in the schema of evolution for Africans and subsequently the world. Most historians agree that the iron hoe and plow had a profound effect on African livelihood. As the historian Joseph E. Harris suggested, the iron technology shifted African societies from mere subsistence economy to a trading economy. Such an economy provided the foundation for the rise of such kingdoms as Medieval Ghana, Mali, and Songhay, and later stimulated the economies of Europe, Asia, the Middle East and other parts of the world. Joseph E. Harris provided an illuminating discussion of the trading routes that were associated with the above-mentioned economic activities. During the so-called Middle Ages, Ghana, Mali and Songhay were the major commercial centers, with trade routes connecting other African countries in the Sudan, Sahara and North Africa regions and beyond. Along the westernmost route, for example, leading from Ghana, through Walata, Audaghost and on to other routes to Tunis and beyond, the Berbers, Black Africans, and Arabs exported gold and other goods and imported from North Africa salt, cloth, jewels, and trinkets. Beliefs and ideas were also among the exchanges. Scholars are uncertain as to the manner in which knowledge about iron smelting originated and spread throughout African societies. Some sources

142

indicate that knowledge appeared first in North Africa and subsequently spread throughout the rest of Africa and the rest of the world. Iron tools and weapons dating back to 2,000 B.C. have been found in Egypt. Other scholars think that knowledge of iron smelting may have been introduced into West Africa by way of Carthage, where bronze jewelry has been found dating back to 600 B.C. Still others believe that the Berbers may have been one of the sources through whom iron technology was transmitted from North Africa to South of the Sahara; on the basis of recent findings subjected to radiocarbon testing, another hypothesis resulting from the Nok culture of central Nigeria calls attention to the contributions of the Nok people. Some authorities hold that the Nok people were making iron as early as 2,500 B.C. When England, having discovered the power of steam, initiated the Industrial Revolution in the mid-1800s, ores such as iron, steel and coal achieved great value, without which such a material transformation would not have been possible. If Africans had recognized the power of steam, in the context of their iron-smelting skills, one can only speculate as to what technological changes Africa and perhaps the world would have undergone. Professor Hansberry, I am aware that perhaps none of the views I have expressed here are surprising to you. However, I believe you will be pleased to gather that your views are in accord with the findings of contemporary scientific debates regarding cultural and biological evolution. I certainly have achieved a greater appreciation of African history by reading your works, as introduced to me by Joseph E. Harris.
Sincerely,
Dr. Johnny Washington, Ph.D.   April 25, 1997

Ethnic Identity: Ancient Ethiopians

Professor William Leo Hansberry (1894–1965)   Destinicity Letter #29

Re: Identity of Ancient Ethiopians: Part IV

Dear Professor Hansberry:

Frank Snowden, yourself and others observed that the term "Ethiopia" as employed by the ancient Greeks meant "burnt face" which, in the ancient world, was applicable to all of Black Africa, India and Asia. Later "Ethiopia" was used in a more restricted sense to denote the area south of

Egypt that is now Ethiopia. You, as well as Cheikh Anta Diop and some other scholars, contended that the early Egyptians were Black, contrary to the views of many other scholars. You and Diop stated that many of the scholars who wished to have us believe that the Egyptians were white were motivated to do so in order to give credit to whites for having created the ancient Egyptian civilization, in order to diminish the achievements of Black Africans. In an attempt to dispel the view that the Egyptians were white, Diop cited Herodotus who regarded the Egyptians as Black. The Egyptians often represented their gods as having Black features. I have already mentioned your works in which you claim that the Egyptian gods were of Ethiopian origins.

In your book,[2] *Africa &Africans,* you wrote, "…[it] was a custom whereby the Egyptians once each year carried the tabernacles of certain of their great gods to Ethiopia and, after celebrations there, brought the shrine back to Egypt—'as if the gods had returned out of Ethiopia.'" (*Africa & Africans,* P. 80) Henry Olela in his book, *From Ancient Africa to Ancient Greece,*[3] offered additional evidence in support of the claim that there were much interactions between the ancient Ethiopians and ancient Greeks. Already Basil Davidson has told us how the practice of iron-making among the ancient Africans influenced the development of European societies as well. Another question: How reliable is the biblical Old Testament, say, the book of First Kings in determining the Ethiopian historic identity, origin or Destiny?

Sincerely,

Dr. Johnny Washington, Ph.D.   April 25, 1997

Ancient Philosophies: African and Greek

Professor William Leo Hansberry (1894–1965)   Destinicity Letter #30

Re: Ancient Philosophies: African and Greek: Part V

Dear Professor Hansberry:

Scholars researching the history of ideas frequently write as though ancient Greek philosophers were original in formulating their world-views. Some scholars, however, are a bit more modest and admit that the Greeks were influenced by other people of the world, those with whom they had

144

social interactions in their Ionic colonies in Asia Minor, and the Archipelago. Clearly the Greeks had established relations with the Egyptians and Ethiopians among whom ideas flowed as far back as 3,200 B.C, a fact that seemed to be overlooked by many scholars. Many scholars with certain cultural biases, when discussing the origins of Western philosophy, fail to mention any contributions the Egyptians or Ethiopians may have made and instead give most credits to the Greeks. Yet some scholars have demonstrated with compelling evidence that the Greeks scholars "borrowed" certain ideas from the Egyptians, who had already developed sophisticated religious and philosophic ideas and other cultural resources before the Greeks knew about Egypt and Ethiopia. Along this line, Henry Olela appealed to Wendt, in supporting the claim that many of the Greek philosophers studied in Egypt, including the philosophers, "Thales, and Anaximander, the mathematician Pythagoras, the statesman, Solon…" (*Ancient Africa*, p. 61) This observation was confirmed by Cheikh Anta Diop; thus the title of his book, *The African Origin of Civilization.*[4]

These views of Olela and Diop are in harmony with your views and those of others. Now let us consider in a bit more detail the manner in which Homer, Thales and Anaximander were influenced by the views of the ancient Ethiopians. Although Homer was not regarded as a philosopher in the strict sense of the term, some of his ideas formulated in a mythological framework induced subsequent Greek philosophers to offer a much more "rigorous" rationalistic explanation for concerns similar to his own. Most of Homer's views were expressed in *The Iliad* and *The Odyssey.* These two works border on mythology. It is uncertain as to the exact date of Homer's birth, and little is known about his personal history. Some denied that a personal Homer existed. Most scholars believe that there was a personal Homer. You held that he spent some time in Egypt. Moreover that the Egyptians learned from the Ethiopians the use of alcohol. It was also from Egypt that Homer obtained presumably for the first time knowledge about the gods, knowledge that he apparently took back to Greece. (*Africa &Africans*, p. 71) Hence, in part therein lies the origin of the Greeks' idea of religion, philosophy and cosmology. Homer raised issues concerning the nature of reality. He arrived at the view that "water" was basic reality. As Olela indicated, Homer would have us believe that the unifying source of the universe is water. (*Ancient Africa*, p. 112) There is strong evidence that Homer was an Egyptian, and while living in Egypt, he got the idea that water was the basis of reality. The Egyptian world-view was organized around the Nile River on which they depended for crop irrigation and other sources. Water, so essential for the Egyptians, took on religious significance and undoubtedly Homer included this notion into his cosmology. Scientists who advocate evolution theory hold that water was essential for the evolution of life. Moreover, Professor Hansberry, you suggested that the Greek gods, for example, Zeus among others, were introduced into Greece from Ethiopia by Homer.

Thales lived in Miletus, a Greek colony in Ionia on the coast of Asia Minor. He is thought to have been born around 645 B.C. and died 70 years later. Some sources have it that he lived to be 90; that he was a Phoenician but was later forced into exile in Miletus. He is considered the first Greek philosopher who offered a rational or scientific reply to the question: What is the basis of reality? To this question, Thales argued that the answer was water. His explanation was that the world of nature was a living animal that possessed a soul. The Earth floats on a raft. Further, the world was not self-created but was created by gods who made the world out of water. In the biblical creation myth, God is said to have created Adam out of clay. One wonders whether Homer or any of the other pre-Socratics had a notion of human Destiny? I am aware that Aristotle impressed upon Alexander the Great who studied under Aristotle the notion that the Greeks were racially superior to any other people on Earth, and it was the duty and Destiny of the Greeks through Alexander to conquer the world. Apparently Alexander the Great believed his teacher, Aristotle, as Alexander did in fact conquer the known world, on whose shoulders "rode" the Destiny of the world (Hegel). His conquests included the Middle East, Europe and parts of India. I am not praising Alexander the Great; I am merely making an observation. Aristotle was talking about immanent Destiny primarily. Did the Greeks distinguish between immanent and transcendent Destiny, a distinction I have elucidated elsewhere in this series of letters? What was the ancient Ethiopian or Egyptian's notion of Destiny? Clearly the Egyptians linked the notion of Destiny to the afterlife. Similarly the Asiatic Indians tied their sense of Destiny to the cycles of reincarnation. The Ethiopians subsequently adopted the Christian view of the soul and the afterlife. I am seeking a notion of Destiny that focuses on our Earthly struggles and counter-struggles by which we approximate our transcendent Destiny.

Why did Thales come to regard water as basic reality? Some scholars suggested that in addition to the Homeric influence, Thales also visited Egypt and integrated the Egyptians' world-view into his philosophy. Strong

146

evidence also indicates that Thales also learned geometry while in Egypt and later took this concept back to Europe where the Greeks made it into an elaborate science. Certain Egyptian priests constituted a secret order called the "Mystery System" in which knowledge was recorded in hieroglyphics and in which views about society, human beings and the world and gods were examined, shared mainly among themselves, and not readily disseminated to the masses. Olela held that Thales was a member of the Mystery System, and as a member, he was not permitted to express his views in writing. This may explain why he, like Socrates, did little, if any, writing. Anaximander, the son of Praxiades, was also a Milesian and is thought to have been about 28 years younger than Thales. Anaximander was impressed with Thales' astronomic and mathematical works, but he opposed his metaphysics, the cornerstone of which was the notion of water. In Anaximander's view, the Earth was not an animal floating like a raft on the water. Rather, he saw the Earth as a solid cylindrical body floating freely in a surrounding medium which he regarded as basic reality, "the Boundless," out of which everything was made and which he identified with God who is deathless and imperishable. As in the view of Homer and Thales, Anaximander thought that there must be a basic reality, but in his view the notion of water was too specific. The problem with the water hypothesis is that if one began with the assumption that water was basic reality, one will have difficulties in explaining how the opposite of water, dryness, came into being. The next major Greek philosopher, Anaximenes, rejected Anaximander's notion of the Boundless and urged us to accept the notion that air is the basis of reality. Like Homer, Thales and Anaximander, Anaximenes' notion of air as basic reality is quite likely derived from the Africans. Olela observed that the Egyptian god Shu or Amon or Air-god was adopted by Greek thinkers such as Anaximander who in adopting the Egyptian approach to understanding the nature of ultimate reality accepted the Egyptian view that water, whose essence was elevated to the level of a god, was regarded as basic reality. (*Ancient Africa* p. 129) Since the Homeric era, which was shaped by the Egyptian world view, Western philosophers have been exploring the question of the nature of reality. It was Heidegger who said that one of the most difficult questions one could raise is: why is there something and not nothing? In his book, *The Two Sources of Morality and Religion*,[5] Bergson threw clear light on this question: "But the question presupposes that reality fills a void, that underneath Being lies nothing, that *de jure* there should be nothing, that we must therefore explain why there is *de facto* something. And this presupposition is pure illusion, for the idea of absolute nothing has not one jot more meaning than a square circle." (*The Two Sources*, p. 251) It seems as though this question, as it has been handed down to us through the Western tradition, was first raised by such thinkers as Homer, Thales, Anaximander, and Anaximenes; in seeking an answer, they presumably were influenced by African sources. On the basis of the foregoing discussion, it appears that the pre-Socratics as well as Socrates, Plato and Aristotle were influenced directly or indirectly by the African traditions. Many of the European philosophers were introduced to African philosophic ideas by way of sailors, travelers, scholars and merchants. Those who were directly influenced by the Africans visited Egypt and Ethiopia where they studied with the men and women of learning. In Olela's view, Homer, Thales, Anaximander, Pythagoras, Euclid as well as Plato, visited Egypt or Ethiopia. There is no compelling evidence that Aristotle, who is believed to have been born around 384 B.C., visited Egypt or Ethiopia. However, it is evident that one of his students, Alexander, traveled to Egypt or Ethiopia, from where he shipped back to Aristotle many volumes of books on a variety of subjects, including politics, education, metaphysics, logic, mathematics, astronomy and religion, on the basis of Olela's views. Thus there was much cultural exchange between Africa, including Egypt and Ethiopia, and Greece and Athens from around 3,200 B.C. to 430 A.D and beyond. During 430 A.D. the Vandals who were Arian Christians invaded North Africa and Spain and destroyed the learning centers of Egypt. This invasion was motivated in part by theological disputes that I shall consider shortly, based largely on your works.

Sincerely,

Dr. Johnny Washington, Ph.D.   April 25, 1997

Professor William Leo Hansberry (1894–1965)   Destinicity Letter #31

Re: Sage Philosophy: Part VI

Dear Professor Hansberry:

Since we have been discussing ancient philosophic traditions, it would be appropriate here to consider an African philosophic tradition. African civilizations included many diverse cultural or ethnic (tribal) strands, each of

148

which reflected its own philosophic tradition. Assuming, as I am here, that the contents of philosophic tradition was in part culturally dependent, I shall not attempt to provide a complete survey of African philosophies. Let me begin by reverting in passing to a pioneering work that is instructive regarding the nature of African philosophic traditions; namely, John S. Mbiti's book, *African Religions*. It is difficult here to render a precise definition of traditional African philosophy. Each culture harbors its own philosophic values and meanings. The diverse African philosophic traditions, generally, involve systems of symbols—myths, rituals, rites, beliefs, and ideals—rationally and emotionally based that offer explanations regarding the meaning of physical and spiritual realities and seek to construe these in ways that the individual is brought into harmony with himself, others and the natural environment. In this description I wish to include sages whose life task it is to articulate and clarify values associated with the above-mentioned systemic realities. Many European scholars, including Levy Bruhl, in the past examined what they called the "African mind." In comparing the so-called African mind with the European mind, Bruhl drew the conclusion that because Africans were "primitive people" they were incapable of thinking rationally and developing philosophic views, comparable to those developed by Plato, Descartes, Hume, Kant or Hegel. Bruhl held that the African mind did not develop beyond the level of pre-logic and magic. Bruhl and others like him supported the myth that Africans are inferior. Ironically Drs. DuBois, Alain Locke, Martin L. King, Jr., as well as myself and many besides, were influenced to a large degree by Hegel's teachings. The same is true of Marx, Nkrumah and Fanon. I do not know about Malcolm X. I do not know whether he also read Hegel's works. Hegel seemed to have had a very low opinion of Africans' intellectual ability. Although I am not trying to apologize for Hegel's seemingly racist views, it is not fair to single out Hegel here and call him a racist. He was a prominent European thinker who was merely reflecting the prevailing views of the times. Certainly he was ill-informed about Africa's history. Professor Hansberry, when in the sequel of these letters I recount your views of the achievements of the Africans during the ancient or Medieval era, including individuals who were engineers, statesmen (statespersons), scholars, philosophers, the views of Hegel and Bruhl become suspect. The North African born St. Augustine (354–430) provided an intellectual framework for Christendom. I have recently reread his works that examine the concept of time, a notion that commands my attention. This is evident by the portion of a series of letters to Dr. Alain

Locke in which I explored the boundary conditions of modernity or post-modernity. Much of what St. Augustine has to say about time provides a basis for outlining a theory of history. History as a discipline was your area of specialization. Additionally, some scholars go so far to say that the great Islamic scholar Averroës (1126–1198) was an African. Certain contemporary African philosophers, including Professors John O. Sodipo, Barry Hallen and the late H. Odera Oruka, have researched the contributions of what they called philosophic sages in Africa. Professor Oruka's research resulted in several publications. I have before me his work entitled, *Oginga Oding*.[6] I first met Professor Oruka in 1982, when he presented the early stages of his sagacity philosophy research at the First Africana Philosophy Conference, organized by Professor Lucius Outlaw, and held at Haverford College in Pennsylvania between July 12 and 23, 1982. I shall here cite a passage from the paper entitled, "Sagacity in African Philosophy" which he presented at that conference:

> [Sages] may not be wise (rational) in understanding or solving the inconsistencies of their culture nor the foreign innovations that encroach on it. In other words, they are the spokesmen of the people, but they speak what after all is known to almost every average person within the culture (pp. 6–7).

Professor Oruka made it clear that certain sages go beyond the prevailing norms of the group and offer their own philosophic world-view, which may or may not be compatible with tradition. When the sages engaged in philosophic activities in the ways here described, they were engaged in philosophic activities proper. Among the sages, mostly Kenyans, Professor Oruka interviewed in conducting his sage research were Chege Kamau, who was born in 1911 in Kihumbu-ini Location; Mzee Josiah Osuru, who was born around 1906 in Kapesur Village; Ali Mwitani Masero, who was born in 1939 in at Emalaha, Utonga; and, Stephen M'Mukindia Kithanje, who was born in 1922 in Meru. Below is a sample of a piece of research in which Ali Mwitani Masero was interviewed focusing on the issue of the equality between men and women:

Q. What of man and woman?
A. Well, as I said, they are *abandu*, they *think*. That is all. For a woman, since old—since the days of our ancestors—is below a man. Even if she is a district officer or president, she is below a man.
Q. But maybe it has been so out of ignorance, out of misconception!

150

A. Maybe, but not quite. A woman leaves her place of birth, like a bird of the bush, to go to a man's home. She is a migrant. How then can she be equal to the *host*, the man?

Q. But can we not interpret that more positively? For a person to leave her mother, people and land to begin a new life with a new people, does that in itself not imply great sacrifice, broad-mindedness and fortitude—qualities which should be rated as superior?

A. Maybe. That is how you think. But me, I wouldn't call it superiority. Our people do not call it that either. I think it is weakness; to be swayed and swept from your roots into the wilderness. Why can't a woman be principled and say: "If a man wants us to be partners, let him come to my father's home and I will build him a house there!"?

Q. But custom, society's laws, do not support this.
    Exactly. Those laws are mere affirmation to the truth that a man is above a woman.

Q. Give another reason why you think a woman is inferior to a man.

A. Well, women reveal their inner feelings too easily. (*Sage*, p. 95)

I make the claim that each society, whether in Africa, Asia, the Americas, or Europe, has sages who assist the respective societies in approximating their modes of Destiny.

Sincerely,

Dr. Johnny Washington, Ph.D.   April 25, 1997

Ancient and Medieval Kingdoms of Africa

Professor William Leo Hansberry (1894–1965)   Destinicity Letter #32

Re: Piankhy the Great, Pharaoh Taharqa, the Kingdoms of Kush, Axum, and Meroe: Part VII

Dear Professor Hansberry:

In this letter, together with the four others that follow, I shall devote attention to your views included in your book, *Pillars*. My aim here is to render your views to a larger lay audience. I hope that this series of letters will play a role in heightening our awareness about Africa's past achievements and give us inspiration to deal with the present plight that Africa is suffering. Thus, it is in this spirit that I draw on your works in writing these letters to you.

Before the times of Homer, there were many interactions between Egypt and the rest of Africa. Centuries earlier, Egypt and Ethiopia had already built splendid civilizations in which learning, technology, commerce and art flourished. For many centuries the Egyptians commanded, intermittently, international respect and influence. The Greeks had much catching-up to do. By the end of the second millennium B.C., Egypt had extended much influence in the region called Kush, an ancient Ethiopian kingdom between the third and fourth cataracts of the Nile. At that time the capital of Egypt was situated at Thebes. It is not clear when Kush became an independent kingdom, but many scholars believe that Kush arose as a kingdom with Napta as its capitol at the beginning of the first millennium B.C. Around that time also the political influence of Egypt as a world power was on the downswing, on the basis of historian Harris's view. As a consequence, in the eighth century B.C. Kush conquered Egypt, where five of its rulers were immortalized as the twenty-fifth Dynasty of the Pharaohs. The eighth century B.C. was one of the crowning points of the golden age of Kush. Piankhy the Great, an Ethiopian king, was one of the individuals who contributed to Kush's glory during that era.

In 676 B.C. during which time the Pharaoh Taharqa was in power, the Assyrians occupied Lower Egypt and around 666 B.C., they nearly destroyed the city of Thebes. Taharqa, whose power was seriously weakened by the Assyrian occupation, was pushed southward into Kush. Later, sometime during the sixth century B.C. (the exact date is uncertain) another Ethiopian kingdom, Meroe became the new capitol of Kush.

Professor Hansberry, you pointed out that Meroe played a key role in the development of iron technology, and it is believed that it was responsible for spreading the idea of iron making throughout the many parts of Africa, perhaps the world. Joseph E. Harris, as well as other scholars, held that it is quite likely that Meroe arrived at the idea of iron technology independently of outside influence. Its thriving years paralleled the thriving years of Alexandria, during the Ptolemaic era. In addition to the iron industry on which Meroe's economy flourished, it is believed that Meroe grew rich by exporting ivory, slaves, ebony and other goods. Eventually, the kingdom began to experience poverty that is believed to have been brought on by competition in trade from still another of Ethiopia's ancient kingdoms, Axum. You informed us that by the middle of the fourth century A.D. Axum's rulers, who were Christians, conquered Kush and burned the city of Meroe. Axum became the capital of the ancient kingdom of Ethiopia.

Competition in trade motivated the conquest. In today's society, competition is still a powerful motive. You indicated that during the early parts of the 1950s it was discovered that the civilization of Kush arose in the first millennium B.C. out of the intermingling of the Northeastern Ethiopians with the Arabians, the Sabean people of South Arabia, across the Red Sea. The Sabean people spoke a Semitic language. At the peak of its career, Axum whose capital was Adulis, controlled the major trade routes of Meroe and Egypt and exerted much political influence over the Red Sea regions, including Yemen which around 330 A.D. became part of Axum's empire. With the intention of making our discussion more concrete we will now focus on certain individuals—some familiar, other not so familiar, to Westerners—whose lives you and Harris suggested shaped Ethiopians' immanent Destiny. The Queen of Sheba, also known as Queen Makeda, King Ezana, Frumentius, the Nine Saints and Prester John shall be considered in turn.

Sincerely,

Dr. Johnny Washington, Ph.D.   April 28, 1997

Queen of Sheba

Professor William Leo Hansberry (1894–1965)   Destinicity Letter #33

Re: Queen of Sheba: Part VIII

Dear Professor Hansberry:

The Queen of Sheba is mentioned quite often in the Bible, a key source by which the Queen's legend has been handed down to the Christian west. In the biblical literature her name is often associated with King Solomon whom she is said to have married, and for whom she bore a male child. You began by citing the passage in Chapter Ten of the First Book of Kings in which the Queen of Sheba's visit to King Solomon is reported.

> "…when the Queen of Sheba heard of the fame of Solomon…she came to prove him with hard questions. And she came to Jerusalem with a very great train, with camels that bore spices, and very much gold and precious stones; and when she was come to Solomon she communed with him of all that was in her heart…And she gave the king an hundred and twenty talents of gold, and of spices, very great store, and precious stones; there

came no more such abundance of spices as those which the Queen of Sheba gave to King Solomon…And King Solomon gave into the Queen of Sheba all her desires, whatever she asked…So she turned and went to her own country, she and her servants." (*Pillars*, p. 36)

Professor Hansberry, you noted that the accounts concerning the life and activities of the Queen of Sheba were included in various traditions, including Islamic as well as the Ethiopian, each of which, in your view, rendered conflicting versions regarding the personal identity of the Queen and the geographical location of the Queen's kingdom. Moreover some sources hold that there is no compelling evidence that the Queen of Sheba ever existed personally, that she was only a legend. Among these sources, you held that the Ethiopian historical tradition is the most reliable. It says that the Queen of Sheba was an actual person of Ethiopian origin, and it identified the biblical Queen of Sheba with Queen Makeda, whose kingdom was in the boundaries of the present day Ethiopia. (*Pillars*, p. 37) You argued further that until the sixteenth century, Christendom in the West knew very little about the Ethiopian version of the Queen of Sheba story. You credited Father Francisco Alvarez as being among the first to provide Westerners with knowledge of Ethiopian history; he made this knowledge available to the Western world; "and in accordance with prevailing tradition, Alvarez regarded the ruined town of Axum—where he spent eight months—as having been the site of Sheba's capital city." (*Pillars*, p. 38) Further, one of the literary sources, according to you, in which the Ethiopian tradition is preserved and in which the identity and nationality of the Queen of Sheba can be documented is the *Kebra Nagast (The Glory of King)*. You noted that James Bruce, between 1769 and 1772, during his travel in Ethiopia, obtained copies of *Kebra Nagast* along with other manuscripts that he made available to Bodleian Library at Oxford University. (*Pillars*, pp. 39–40)

In your view, tradition avowed that the Makeda dynasty dates back to 1370 B.C. and was inaugurated by Za Besi Angabo whose dynasty maintained rule for 350 years. Queen Makeda's grandfather and father, respectively, were among the last two monarchs of the dynasty; the former was Za Sebado and is believed to have ruled between 1076 and 1026 B.C. He had a wife, Queen Ceres. Their daughter, the only child, named Ismenie, married Za Sebado's chief minister who became king when Za Sebado died. He reigned between 1026 and 1005 B.C., according to your research. The former chief minister and son-in-law who became king and his Queen Ismenie had two children, the first of whom was Prince Noural Rouz, who, while still an infant, was said to have been accidentally dropped into the fire "which caused his death." (*Pillars*, p. 42) The second child, the prince's sister, was Princess Makeda who came to be known as the Queen of Sheba. She was also injured during childhood. (*Pillars*, p. 42) Neither did this incident seem to affect her capacity to rule, nor mar in any significant way her beauty. She is said to have been one of the most beautiful women in the world. Her beauty nearly dumbfounded Solomon who went beyond the call of duty, as it were, to win her affection. Tradition attested of her intelligence, beauty, wealth and manners. Makeda assumed power after her father's death around 1005 B.C., and reigned until around 955 B. C. the year she is believed to have died. On the basis of your views, Queen Makeda became acquainted with King Solomon of Jerusalem through a famous merchant prince of Ethiopia named Tamrin who is reported to have owned hundreds of ships and camels. (*Pillars*, p. 43) King Solomon was one of Tamrin's customers to whom he often delivered products. Tamrin came to know King Solomon quite well and was deeply impressed with his style of administering justice and his wisdom. So impressed that upon returning from one of his trips, he conveyed to Queen Makeda his experience with King Solomon. This in turn aroused her imagination and interest and motivated her to visit King Solomon. Her journey to Jerusalem was coordinated and led by Tamrin. (*Pillars*, p. 44) The Queen is said to have spent six months with Solomon, drawing on his wisdom; after which time she was, reluctantly, ready to return to Ethiopia. Solomon, however, urged her to stay another season in order that he might continue to impart to her wisdom. (*Pillars*, p. 45) The virgin queen stayed another season, during which time she and Solomon enjoyed sexual intercourse with each other. During her stay Solomon also expressed the desire to marry the Queen and some sources indicate that the two did in fact marry. It is reported that when he first made a sexual advance to her, she resisted but later gave in. Solomon engaged in a risky act; if he were to approach her today, as he did back in his times, he no doubt would be accused of sexual harassment. [A similar charge has been brought against President Clinton, would you believe? Our President? And there is a big controversy surrounding President Clinton and Monica Lewinsky with whom he is said to have had sexual affairs; with this controversy the former Civil Rights leader Vernon Jordon, who is African American, is implicated]. Consider also what happened to Clarence Thomas who is now a Supreme Court Judge. A Black woman, Anita Hill, almost "destroyed" his reputation. The story was fully covered by CNN. A few months ago one of the Dallas Cowboys football players almost got in "trouble" on similar charges. In your research here discussed you described vividly how Solomon lured the Queen into having sex with him. (*Pillars*, p. 45) Immediately after Queen Makeda returned to her country she went into labor and gave birth to a son named Ebna Hakim. (*Pillars*, p. 45) When the son reached age twenty-two, Queen Makeda sent him to visit his father, Solomon, who had requested this. Solomon, along with the people of his kingdom, was delighted to see his son, whom he tried to persuade to remain in Jerusalem and become his successor. Solomon's only other male heir, Rheabom, apparently, due to a mental instability, was not suitable to become his father's successor. (*Pillars*, p. 47) What a sharp contrast between wise man Solomon and his other son, Rheabom. Ebna Hakim, contrary to his father's advice, returned to the land of his mother, Axum, (where she is believed to have been buried). It so happened that Ebna Hakim did not assume the throne of his father but the throne of his mother, following her death; on his ascendancy, he took the name Menelik I. You reported that during the course of imparting knowledge and wisdom to Queen Makeda, during her Jerusalem visit, Solomon was effective in convincing her of the significance of his Hebrew God and religion. She was also inspired by the learning and wisdom exhibited by Solomon. That she appreciated Solomon's religious instructions is evident by the fact that, as her son returned to Jerusalem, she urged him to deliver to Solomon on her behalf a message, according to which she wished to extend in her country the growth and development of the religion of Israel. (*Pillars*, p. 47) To this end Solomon selected from among his staff men learned in the law to fulfill this mission. These men, tradition avowed, migrated to Ethiopia, in keeping with the wishes of the queen and king, settled there and became teachers of the new religion, and thus stimulated the intellectual life of Ethiopia.

They and their descendants came to be known as the *Falashas*, the Black Jews of Ethiopia. Many believe that Haille Selassie's genealogy extends to the Queen of Sheba or the *Falashas*. The effect of their religious teachings also might in part explain why in the fourth century A.D., Christianity was, without a serious struggle, introduced into Ethiopia as the official religion. We must also keep in mind the view that organized religion such as that which was practiced by the Ethiopians in antiquity, predated most of the religions of the world.

Sincerely,

Dr. Johnny Washington, Ph.D. April 28, 1997

Christianity in Ethiopia

Professor William Leo Hansberry (1894–1965)   Destinicity Letter #34

Re: Frumentius, a Founder of Christianity in Ethiopia: Part IX

Dear Professor Hansberry:

This letter is concerned with the words and deeds of Frumentius and King Ezana, who played major roles in developing Christianity in Ethiopia. In your view, Frumentius and his brother Edesius were credited with having played a key role in introducing Christianity to Ethiopia in the fourth century A.D. Historical evidence and tradition also suggest that the two brothers converted King Azana to Christianity, and following his conversion, Christianity was made the official religion of Ethiopia. Scholars are uncertain as to the exact date when this occurred. Many believe that the conversion took place between 330 and 340 A.D. Here it will be instructive to consider briefly certain events that transpired before Frumentius Christianized Ethiopia. You, Joseph E. Harris and others, held that a number of Christians in Ethiopia already practiced the Christian religion before it became a state religion. A sizable number of Greek Christian traders, decades earlier, settled in Ethiopia as well as the surrounding areas, where they established prayer houses. Further it will be recalled that in the early history of Christianity, the Roman Emperors had begun to persecute many Christians. This practice resulted in many Christians fleeing their homeland under Roman rule and they settled in India, Asia, Ethiopia and in other countries that were tolerant of their beliefs. It has been reported that when Frumentius and Edesius first arrived in Ethiopia, they were startled to have found Ethiopians practicing Christianity but with no religious leadership among the people. (*Pillars,* p. 66) Some sources indicate that St. Matthew and Philip preached in Ethiopia. You observed that the Christians, who aroused Frumentius and Edesius' curiosity, were probably refugees and their descendants who had entered the country in order to escape religious persecutions. The story pertaining to the circumstances under which Frumentius and Edesius happened to have come to Ethiopia is noteworthy in itself. You related the story and its variations in the following way. A traveler and philosopher by the name Metrodore, of the city of Tyre, had visited Persia, India, and Ethiopia, and was delighted by what he enjoyed and observed in these respective countries. On the basis of your account, as he returned from one of his trips he described to another philosopher and merchant,

Meropius, of Tyre, the remarkable experience he enjoyed abroad. Accordingly, this prompted Meropius to set out on a similar journey, with whom Frumentius and Edesius went as fellow travelers, and some sources have it that the two brothers were Meropius' sons and students. According to your account, they sailed through the Red Sea, during the course of which a crisis erupted. You considered several versions of the crisis. According to one version, while sailing through the Red Sea, the boat on which Frumentius and his brother were sailing crashed as it hit rocks, and only Frumentius and Edesius survived the accident. Another version of the episode, which you took to be more credible, is that disagreement had erupted between the inhabitants of the Ethiopian port and the crewmembers of a Greek trading ship. Sometime before Meropius' ship had pulled into the Ethiopian port, a fight erupted between the local residents and the ship's crew. (*Pillars*, p. 65) Thus, as Meropius's Greco-Roman ship pulled into the port, it too was attacked by the local Ethiopians who killed everyone on the ship except Frumentius and Edesius. Apparently they were saved because of their youth. Following the massacre, on your view, King Ella Amida, Ezana's father, who held the throne at that time, requested that Frumentius and Edesius be brought back to his court where they were provided living quarters. Soon after, Frumentius, who also was well versed in Greek, became the king's secretary, and Edesius became one of the king's cupbearers. The two Tyrain brothers also served as tutors to the young Ezana and Shaiazana who were later enthroned as co-sovereigns of Ethiopia in 328 A.D.

You reminded us that, of the two brothers, Frumentius' achievements seemed to have been far greater than those of Edesius. During the early part of Frumentius' career he went to Jerusalem where he was said to have met Queen Helena, the mother of Emperor Constantine. You did not mention the exact purpose of Frumentius' trip to Jerusalem. You insisted that it can be inferred that it was a religious mission. In any event, while in Jerusalem, Frumentius conveyed to Helena the idea that there were great opportunities for the founding of a Christian church in Ethiopia. (*Pillars*, p. 69) She recommended to Frumentius that he was to relate this idea to Archbishop Althanasius in Alexandria. Frumentius was said to have accepted her recommendation, urging the Council of Bishops to allow Christianity to be developed in Ethiopia. Moreover, he recommended that a bishop and staff be sent immediately to Axum to facilitate in the development of the new religion. The Council accepted his report and recommendations. Frumentius was named Bishop at Axum. You noted that the date when Frumentius was consecrated as Archbishop of Ethiopia is uncertain but is believed to have been between 326 and 336 A.D. With the conversion and approval of the co-ruling kings, Ezana and his brother Shaiazana, who by this time had ascended the throne, Frumentius rapidly transformed Ethiopia into a Christian state, according to your account. To the degree that he did so he affected Ethiopia's immanent Destiny.

The Ethiopians readily accepted the new religion without much ado. In addition to serving in the capacity as bishop, he devoted much time to teaching and in building churches that also served as learning centers throughout Ethiopia. Frumentius' base of operation, however, is believed to have been around Axum where the famous learning center named "Fremona" was established. It is believed that that center was established in the honor of Frumentius, and the Ethiopians have immortalized him as Saint Frumentius.

On the basis of your views, during the early history of the Christian church and on through the Middle Ages, a number of bitter doctrinal disputes emerged among Christians throughout the Roman Empire, especially in the East. A case in point was the Council of Nicaea in 325 A.D.; during this time the Roman Emperor Constantine formulated a creed to which all Christians were to conform; its purpose was to put an end to all disputes and squabbles among Christians. However, this creed was ignored and rendered ineffective when a dispute erupted between Constantine's successor, Constantius II, who was influenced by Archbishop Arius and Patriarch Athanasius, the head of the Church of Egypt.

Sincerely,

Dr. Johnny Washington, Ph.D. April 29, 1997

Professor William Leo Hansberry (1894–1965)   Destinicity Letter #35

Re: The Nine Saints of Ethiopia: Part: X

Dear Professor Hansberry:

In your works on which I have been relying, you emphasized the fact that during the fifth century, theological squabbles and political conflicts often continued to erupt among Christians. Christians persecuted Christians. Such conflicts often affected Ethiopians and many in other quarters of the Christian world. They were divided on the metaphysical-theological question concerning the nature and identity of Jesus. At no other time in human history has such a metaphysical-theological question been so divisive. Because Christianity harbored many conflicts, brought on in part by inducing its followers to subscribe in a dogmatic manner to its creeds, Dr. Alain Locke and Dr. DuBois regarded Christianity and other forms of organized religion as objectionable. I am uncertain as to your religious beliefs, or your attitude towards organized religion generally. In recognizing the manner in which a religion such as Christianity, Islam or Judaism harbors conflicts, I attempted to offer an alternative mode of transcendence reflected in the Destiny model. As a philosopher, I am interested in seeking solutions to human problems. As a historian, your task was mainly to enable us to understand our past. You certainly did a good job in enabling us to understand the struggles and counter-struggles that occurred in the early Christian church. I will continue to recount below your interpretations of such struggles or debates. As you pointed out, one such debate focused on the nature and identity of Jesus. Thus, arose the question: was Jesus a man, or God, or both? A related question focused on the identity of Mary. Was she the mother of God or was she the mother of Jesus only, who was a man? Some preachers openly denied that Mary could be the mother of God. Because it was impossible for a human being to give birth to God, who is divine. In the same vein the above-mentioned priest urged that Jesus was not God, only an extraordinary human being who was without sin. In your view, most of the high religious officials of Constantinople, with the exception of Patriarch Nestorius, opposed the priest's view. As a consequence of supporting the position of the priest, in 431 the Council of Ephesus convened, where Nestorius was tried for heresy, was convicted and fled the country. During 451 there occurred the Council of Chalcedon where religious officials subscribed to the view that Christ's nature was twofold, human and divine, a view that both the Egyptians and Ethiopians opposed, on your analysis. The Egyptians and the Ethiopians shared the conviction that Christ was of one nature as suggested above, but here the terminology was a bit more specific. The term "consubstantialism" was introduced, and the people who employed the term came to be known as Monophysites. I cannot help but point out that the issue regarding the twofold or single nature of Jesus paralleled the ancient problem concerning the one and the many (Plato) or the finite-infinite problem in the views of Kant, Fichte and Hegel. This finite-infinite problem is also reflected in some debates among contemporary physicists as well as mathematicians. I have tried to simplify the finite-infinite problem by introducing immanent-transcendent Destiny; each is implicated in the other. Thus after the convening of the councils under discussion, the Roman Empire became much more intolerant of its Monophysites, many of whom were living in Syria, Egypt, Asia and other parts of the East. Some of the Monophysites whom the Church of Rome and the Church of Constantinople came to regard as heretics fled to Ethiopia where they sought security and protection so that they may practice their faith. Among the refugees, on your analysis, who came to Ethiopia to escape persecution were those who came to be known as the Nine Saints.

You indicated that there was a degree of uncertainty regarding the places of origin of the Saints. It is uncertain where they came from, the exact year they came to Ethiopia and whether they came as a group or individually. It is believed, however, that most of them came from Syria, except Abba Pantalewon who is believed to have come from Egypt. (*Pillars*, p. 95) You went on to remind us that scholars are even less certain about the hundreds of other refugees who came to Ethiopia during the early Christian era. What is known is that these refugees lived under extreme hardship, used rock caves for dwellings and lived off herbs and wild fruits. King Gabra Maskal I, who ruled between 545 to 580 A.D., is reported to have dedicated a church at Baraknaha to their memory, according to Harris and other scholars. Moreover, according to your research, in a small Ethiopian church there can be found mummified human bodies, and these are believed to be the remains of some of the above mentioned refugees. The Nine Saints were apparently refugees also, but we know more about them through their accomplishments. You, as well as Joseph E. Harris, reminded us that the Nine Saints made significant contributions. The names of the Nine Saint are: Abba Afse, Abba Alef, Abba Shema, Abba Guba, Abba Garima, Abba Yemata, Abba Aragawi (or Mikael or Michael), Abba Likanos, and Abba Pantalewon. (*Pillars*, p. 95) The Nine Saints were scholars, teachers, "preachers, builders of churches, and founders of monistic establishments." (*Pillars*, p. 96) As scholars they translated much biblical literature into Ge'ez, including selections from the Old and New Testaments.

The saints were credited with having built monasteries throughout Ethiopia, according to Harris. Besides serving as places of worship, these also served as learning centers. Today Christianity is still struggling with some difficult issues, including whether women should be ordained as

church pastors or priests; another issue has to do with whether or not abortion or euthanasia is morally permissible.
Sincerely,
Dr. Johnny Washington, Ph.D.   April 29, 1997

Professor William Leo Hansberry (1894–1965)   Destinicity Letter #36

Re: Prester John: Part XI

Dear Professor Hansberry:

Here I will elaborate on your discussion of Prester John, who just as well could have been called the Earthly redeemer of Medieval Europe. In my previous letter to you we considered the internal strife that the early Christian church suffered. On the basis of your research, Medieval Europe appealed to Prester John as a possible bearer of its political and spiritual Destiny, through which it was anticipated that the internal religious strife that it suffered could be resolved. During the early stages of Christendom, it will be remembered, it underwent a series of theological-political struggles that resulted in the persecution of thousands of Christians. Such internal struggles nearly led to Christendom's decline. Eventually, the tables were turned. It survived such struggles and reciprocal assaults. During the eleventh century Seljuk Turks invaded the Eastern quarters of the Christian world. The Muslims along with the Mongols engaged in aggressive actions against the Western quarters of the Christian world. During 1096, the Christian Crusaders came on the scene in an effort to route the Muslims from the Holy Land. Thus the future of Medieval Europe was in question. Fear and insecurity pervaded the land. In one of my previous pieces of correspondence to Dr. DuBois, I mentioned that the recent Bosnian military strife has its roots in this eleventh century conflict and others. Europeans at the time were looking for an Earthly redeemer with a powerful army who would assist them in fighting the Eastern and Western invaders. Thus arose among Europeans the rumor about a certain Prester John, a powerful Christian king who was willing to assist them in their struggles and counter-struggles against the Muslims. As you recounted the many rumors associated with Prester John, some are amusing and others induced inspirations. One rumor had it that Prester John ruled a kingdom located someplace in

Asia. Another rumor led people to believe that his kingdom was in Africa. You, yourself, seemed to have subscribed to the latter hypothesis.

Mysteries surrounded the identity of Prester John. Precisely who was he? Where was his kingdom? How many armies did he command? Was he an Asian? Was he an Ethiopian? Was he willing to establish an alliance with the Europeans? Was Prester John an actual person? Was he an embodiment of their Destiny? Did he represent the great white hope? Or, was Prester John a fictitious individual created by Europeans to inspire hope among themselves? Throughout the Middle Ages, in your view, such questions preoccupied the minds of European kings, queens and church officials, the educated ones and the masses alike. Even today scholars are still divided on this Prester John question. (*Pillars,* p. 113) You set out to put the rumor to rest by showing that the available evidence included mostly in letters between certain African king(s) and European leaders supported your hypothesis that there was an element of validity to the Prester John mystery. Thus in your view, a certain letter, along with a collection of other ones, is strong evidence in favor of the claim that Prester John was in fact an African king(s) with his kingdom located in Africa, or in and around what we now call Ethiopia. I believe you were perceptive in finding an additional clue to the Prester John mystery, in the name, "Prester John": "Prester John" designated an Ethiopian king or kings. In this connection you pointed out that many scholars are still uncertain about the historical Prester John problem, a problem analogous to establishing the validity concerning the claim about the historical Jesus, or Moses, for example. Others think that the royal letters were fraudulently written in perpetuating the myths associated with Prester John. However, you were of the opinion that the letters "possess a ring of truth, but which can be verified as being in substantial accord with the historical fact." (*Pillars,* p. 119) You suggested that the historic Ethiopian king, King Lalibela, about whom scholars know very little, easily fits the description of Prester John. He is believed to have reigned during the times in which the above-mentioned letter along with others was written. A similar letter, the one alluded to above, was received by Emperor Manuel around 1165. You suggested but did not state that Lalibela or some other Ethiopian king(s) could have been the author of some such letters. In your view throughout the Middle Ages sources indicated that there were many pieces of correspondence between Prester John and European kings, princes, popes and others. Prester John, and the leaders with whom he exchanged pieces of correspondence, discussed issues and problems of their respective kingdoms. In addition to seeking the support of Prester John in combating invaders of the East and West, it is reported that in some letters mention was made of the possibility of resolving the theological differences between European Christendom and the Ethiopian Christian community. Furthermore, certain sources indicated that Prester John, in his message of 1177, addressed to Pope Alexander III, desired to build a church or school in Rome. And in fact an Ethiopian church was acquired there, located near the aspe of St. Peter's Cathedral. However, no one seemed to know the exact date when the church was acquired and the condition under which it was acquired. (*Pillars*, p. 125) The fact that there were interactions between the Ethiopian kings and the religious or political leaders of Medieval Europe was evidence, you suggested, supporting your hypothesis that Prester John was the king (or kings of Ethiopia). Nevertheless, it must be kept in mind that when the rumors of Prester John initially arose, there was much confusion in the minds of Europeans as to the geographical location of Prester John's kingdom, you informed us. Some felt that it was in Asia or India. Marco Polo, Friar Oderic and others traveled throughout Asia in search of Prester John without success. In your view, it was Marco Polo who effectively convinced Europeans of the hypothesis that Prester John's kingdom(s) was situated in Ethiopia during which time Ethiopia was called Middle India. Before closing this letter, I would like to criticize your views or research approach. Besides Queen Sheba, you devoted, to my knowledge, little, if any, attention to the role of women in ancient or Medieval history. It is worth noting that during the times of Prester John, a number of Muslim women were active in the political arena in Egypt and in Yemen, a neighboring country of Ethiopia. Boydena Wilson, an Islamic scholar, has researched the lives of a number of such women. In Wilson's views, among these women was Umm Mustansir who ruled in eleventh century Egypt as a regent for her young son, al-Mustansir. The Jewish slaver, Abu SaCd al-Tustari was at one time her master and he made her sexually available to the Fatmid calipha al-Zahir (1021–1036). Through the sexual union of the two, al-Mustansir was born and in whose place she ruled. In Wilson's view, Umm Mustansir played a significant role in the theological debates and struggles that surfaced during that time. Umm Mustansir also engaged in intellectual correspondence with the Queen of Yemen, al-Malik al-Sayyida. It would be revealing to explore to what degree, if any, either of these queens was involved in the rumors surrounding Prester John. Professor Hansberry, you in effect concluded your Prester

164

John story by claiming that throughout the Middle Ages to the present day, rumors about Prester John had the effect of projecting Ethiopia on the international scene, a projection that enabled Ethiopia to enjoy presence, power, prestige, and respect. Today in view of all of the economic and social problems that the African American community and the people of African origins, the Africantude people, is suffering, indeed we need a certain Prester John as a role model who may inspire hope so that we can approximate our immanent or transcendent Destiny. Africa needs unity. Africa needs hope. Africa needs to embrace its noble Destiny. Where are the Piankhy the Greats? The Pharaoh Taharqas? The Prester Johns? And the Queen of Sheba? Has history deceived us? Will history surprise us by delivering a Queen of Sheba or a Prester John, "reincarnated"—the bearer of destinies? Are our great ancestors awaiting within the womb of time to come forth and snatch our Destiny away from the forces of oppression? What sorts of struggles are required to usher in the African Destiny Renaissance?

Sincerely,

Dr. Johnny Washington, Ph.D.   April 30, 1997

African Golden Age: Ghana, Mali, and Songhay

Professor William Leo Hansberry (1894–1965)   Destinicity Letter #37

Re: The World of Mansa Musa: Ghana, Mali, and Songhay: Part XII

Dear Professor Hansberry:

In the remaining letters in this portion of the series the focus will be on the ancient or Medieval kingdoms of Ghana, Mali, and Songhay. As the Muslim Arabs expanded into North Africa during the eighth century, they stimulated trade throughout North Africa as well as the economy of Ghana and the surrounding areas in West Africa. As historians noted, Ghana supplied the Muslims with salt, ivory, gold and other goods. Trade and commerce occurred between Ghana, Europe and other parts of the world. In the view of Joseph E. Harris, the Berbers and the Moors were the intermediaries through whom such commercial transactions were conducted. As you know, the kingdom of Ghana predates the rise of Islam, and the former enjoyed a thriving economy long before the rise of Islam and Christianity.

Ghana is believed to have emerged as a unified kingdom in the fourth century A.D. Although there is no direct causal connection between the rise of Ghana and the downfall of the Roman Empire, it so happened that as the Roman Empire fell, the political communities in West Africa became more organized. This is an illustration of the turning of tables in history. Do you have a theory to explain the occurrences and dynamics of historic events? Douthart, whom I mentioned in the Prologue of this work, explains turning points in the field of evolution by trading on the notion called "saturated complexity." When a given stage of evolution becomes saturated with complex entities, say the forms of matter that rely on the DNA molecules that harbor information by which living organisms are reproduced, such a stage of evolution gives rise to a new, and radically different stage, in which the brain provides a basic means by which information that is necessary for the perpetuating life is introduced and utilized. This shifting from one stage to the next is due to saturated complexity of information. Such a hypothesis may be extended to the field of history in explaining how historic transformations occur.

Are there any causal forces working in history? Do you think that God enters into human history, and guides it, as Judaic-Christian tradition would have us believe? Do you think there is merit to the cliché: "History repeats itself?" What is your opinion of Hegel or Marx's view of history? Do you think that Africa will ever reclaim the historic glory that it once enjoyed? What were the ancient Ethiopians' views of history? What were their views of Destiny? If they had a view of Destiny, how did it differ from that of the Egyptians? The Christian Ethiopians adopted Christianity's view of history or the Western view of history, I presume. Related to this is the issue of time and its relation to Destiny. In most ancient societies, people construed time as cyclical phenomenon, geared to the ever-recurring patterns in nature such as seasons or the phases of the moon or other heavenly bodies. Since the early Christians were awaiting the return of the resurrected Christ, they began to become conscious of time as involving a past, present and future. Time was construed as linear, similar to the time model scientists use today. For the earlier Christians the future assumed paramount importance. One's Destiny was tied to the future, an attitude that was adopted by Christian Ethiopians and Egyptians. Dr. Alain Locke has an outline of a theory of history which I discussed in my book, *A Journey,* (see pp. 106–107). Here I have gotten side-tracked by querying you on your theory of history. Let me now return to the issues at hand. Just as iron-smelting occurred early in

Ethiopia, so also it occurred early in Ghana, the development of which contributed to the rise of Ghana's commercial and political activities. Among the renowned kings of Medieval West Africa were Tenkamenin, Mansa Musa, Sunni Ali and Askia the Great. Already we mentioned the greatness of Prester John in the mid eleventh (or twelfth) century A.D. It was also around this time that Tenkamenin ruled Ghana during its peak, around 1067 A.D. This also was the year that the kingdom began to collapse.

Many historians venture to assert that the destruction of Ghana in the eleventh century A.D. was due to the invasion of the Black African Muslims, the Berbers, who had been defeated in North Africa by the Arabs.[7] These Berbers were called the Almoravids. With the destruction of Ghana, the kingdom of Mali arose under the leadership of Sundiata. Between the downfall of Ghana and the rise of Mali around 1230 there was much political struggle and unrest in the area. Sundiata, who ruled Mali from 1230 to about 1255, brought the old-kingdom under his rule and he gained control of the gold-producing lands of Wangara, Bambuk, and surrounding areas. Mali remained a powerful kingdom until around the fourteenth century. When Sundiata ruled Mali, its fame was largely limited to the Western Sudan region, as you know. Mansa Musa, who was Sundiata's grandson, changed all that, and more. Mansa Musa came to the throne in 1307, and seventeen years later, in 1324, he set out on his pilgrimage to Mecca. This pilgrimage became one of the most spectacular events in the history of the world. It was largely due to this pilgrimage that the fame of Mansa Musa and of Mali was echoed not only throughout Africa, but Europe and the Middle East as well. It is reported that Mansa Musa passed through Walata, later through Cairo, on his way to Mecca. He was accompanied by an entourage of some 80,000 people who carried with them pounds of gold and gold dust. Reports have it that he distributed this gold in displaying his wealth from Mali to Mecca, according to Harris. Among the scholars Mansa Musa brought back with him from Mecca was an Andalusian poet and architect named as-Sahili; he is credited with having designed the mosques and other buildings of Gao and Timbuktu, including the University of Sankore that was in Timbuktu. This illustrious university attracted outstanding scholars. Although learning and scholarship were already institutionalized in Timbuktu before the times of Mansa Musa, he also did much in promoting these and other cultural activities throughout the region. The University of Sankore stood as a monument to Mansa Musa's commitment in such an endeavor, only ruins remain today. The city of Timbuktu is believed to have been established around 1100 A.D. Felix DuBois, in his book, *Timbuctoo: The Mysterious,* credited the Moors as having been among the first to introduce scholarship into Timbuktu, following the founding of the city. These Moors were Black Muslims whose curricula included not only studies in the Koran, but other areas such as Philosophy, Law, Grammar, Mathematics, Medicine and Science. Many of these scholars were from the universities in Grenada, and Cordova, Spain, as well as from universities in Cairo and Mecca. Joseph E. Harris observed that such scholars and students of Timbuktu who were supported by kings and wealthy merchants devoted their lives exclusively to learning, teaching and research. Many served as advisers to the king. He usually granted these scholars sizable portions of land in exchange for their services. It would be interesting to compare the administrative style of certain Ethiopian kings and queens with the kings and queens during the Golden Age of West Africa. It would be equally interesting to compare and contrast the philosophic-theological views of the Islamic scholars in Medieval West Africa with the works of Christian scholars of Medieval Ethiopia, and Egypt.

Sincerely,

Dr. Johnny Washington, Ph.D.   April 30, 1997

Professor William Leo Hansberry (1894–1965)   Destinicity Letter #38

Re: The World of Sunni Ali: Ghana, Mail and Songhay: Part XIII

Dear Professor Hansberry:

One of your followers Joseph E. Harris as well as other historians maintains that shortly after Mansa Musa's death in 1332, the empire of Mali began to weaken. This was in part due to the lack of leadership exhibited by his successors, including Mansa Musa's son, Maghan, who reigned for four years and was unable to prevent the Mossi of Yatenja from raiding and burning the city of Timbuktu. A number of other factors contributed to the downfall of Mali. Most scholars attribute the downfall of Mali to the unification of the Songhay people under the leadership of Sunni Ali, also known as Ali Kolon. Sunni Ali began his career as a common soldier in the army of Mansa Musa who subjected the Songhay people to his rule. Sunni Ali, together with his brother Selmar Nar, and others, was inducted into Mansa

Musa's army. As a consequence, Sunni Ali was angered by the fact that he was forced to fight his own people who in his view were treated cruelly under Mansa Musa's rule. Thus Sunni Ali became resolute in his effort to liberate his people from what he perceived as cruel treatment.

It is reported by J.A. Rogers[8] and others that Sunni Ali and his brother devised plans by which to escape from Mansa Musa's army. This included mapping out the roads that led to Jenne. Then, whenever Mansa Musa would go on a military mission, they stole away and hid supplies of food, water and arms along the way. (*Great Men*, Vol. I p. 236) Sunni Ali rose to power in 1464, as the people of Gao rebelled because they no longer wished to pay taxes to the ruler of Mali. As for scholarship and learning, Sunni Ali did not seem to prize these to the same degree as Mansa Musa. He regarded scholars mainly as a threat to his rule, some historians observed. He seemed to have preferred conquest and political stability over anything else. It might be added that today the situation in Nigeria, and in many other apparently repressive governments where scholars are threatened, is similar to what it was like back during Sunni Ali's rule. Rogers held that the scholars at Timbuktu were threatened, intimidated or treated roughly by Sunni Ali. Many were forced to flee the city of Timbuktu to Walata, when he invaded Timbuktu in January 1468, during which time he executed a number of Timbuktu citizens. Rogers described the ways in which scholars were treated at Sankore thus: "The priests, the learned men, and all who made a living by religion, including the faculty of the University of Sankore, plotted against Sonni Ali, whereupon he put to death every one of his enemies within reach, and warned others to cease meddling in political affairs." (*Great Men*, Vol. I p. 237) On the other hand, the learned individuals who kept out of the political arena were treated well by Sunni Ali and were given land and money. He also attacked the practice of Islamic religion by prohibiting the practice of praying five times a day as the Koran enjoined. Instead he requested prayer time was to be restricted to evenings only, and rather than allowing five hours for prayer, he now held that now prayer was to be limited to only two minutes. (*Great Men*, Vol. I p. 237) Professor Hansberry, how would you evaluate the words and deeds of Sunni Ali? In comparison to many leaders in ancient or Medieval Ethiopia, how would you rate him? In your estimation, how would he rank in today's world? Rogers continued his story by focusing on Sunni Ali's capturing Timbuktu. Once Sunni Ali captured Timbuktu, his next move was to take Jenne, a town that had for centuries enjoyed political independence and a

thriving economy. Sunni Ali's last military mission involved the conquest of an area which included the country of Gomas. He died a tragic death. On his way home one day (during November 6, 1493), the horse on which he was riding fell into the Koni River, drowning Sunni Ali. "To preserve his body until Timbuctoo was reached, his son removed his intestines and placed it in honey." (*Great Men,* Vol. I p. 238) Many contemporary Africans who are scholars as well as scholars in African Studies find this area in Africa's history problematic. They regard Mansa Musa, Sunni Ali and other Islamic leaders as outsiders who imposed colonial rule on Africa similar to the colonial rule imposed by the European Christians a few centuries later. What is your assessment of the Muslim presence during the era here under discussion? A related question is this: do you think that the Christian's presence in Ethiopia as you described in your works was to the benefit of Ethiopians? I will end this letter with these questions to you.
Sincerely,
Dr. Johnny Washington, Ph.D.   May 1, 1997

Professor William Leo Hansberry (1894–1965)   Destinicity Letter #39

Re: The World of Askia the Great: Ghana, Mali, and Songhay: Part XIV

Dear Professor Hansberry:
    In this letter to you focusing on Askia the Great I will rely in part on J. A. Rogers' works that I have cited repeatedly in the course of my pieces of correspondence to you. Here I am basically offering a summary or a synoptic view of Rogers' commentary on what has been called the Golden Age of West Africa, of which Sunni Ali was one of the key leaders. The people over whom Sunni Ali ruled, including the scholars under his domain, were relieved from harsh treatment upon his death in 1493. Incidentally this was a year after Columbus's New World discovery. Sunni Ali was succeeded by his son, Sunni Boru who held the throne for only fourteen months. He lost the throne to one of his father's lieutenants, Askia the Great, also known was Mohammed I, who was recognized as one of the greatest kings of west Africa, greater than Sunni Ali. Others say that he was greater than Mansa Musa, who was the first African king to introduce Islam to West Africa. Like Mansa Musa (who himself was a Muslim), Askia the Great also em-

barked on a breath-taking pilgrimage to Mecca in 1495. Askia's pilgrimage to Mecca did not seem to capture the public's imagination to the degree that Mansa Musa's pilgrimage so captured it. Yet Askia's pilgrimage was noteworthy in itself. He took along with him large quantities of gold that his sizable entourage distributed along the way to Mecca. Askia the Great was motivated to reinvigorate the intellectual and cultural life of Timbuktu as Mansa Musa had done earlier. According to historians, Askia the Great, like Mansa Musa, persuaded scholars in Mecca to return with him to Timbuktu and Gao to assist in rebuilding the region. Harris went on to remind us that unlike Sunni Ali, Askia the Great treated the holy men and scholars with great reverence and respect, and, reciprocally, they held him in high esteem. Like Mansa Musa, Askia the Great was a Muslim whose creed he seemed to have devotedly held. Askia the Great, himself a scholar, included scholars in his court to interpret Islamic law; and he even visited the learning center at Baghdad, where he conversed with Caliph Motoweskel, to whom he is said to have granted large quantities of gold in supporting scholarship and cultural activities. However, a case would arise in which Askia the Great would set on a collision course with the scholars. (*Great Men*, Vol. I. p. 243)

Askia the Great ascended the throne when he was around 70 years of age, during which time he was aggressive in exercising and extending his rule beyond Lake Chad. (*Great Men*, p. Vol. I p. 242) In the latter part of his life, while still in office, he lost his eye sight, although his subjects were for the most part unaware of this affliction. Because he rarely, if ever, appeared before the public. Finally one of his sons discovered his father's weakness (blindness) and overthrew him on September 27, 1529. Scholars indicate that the *coup d'etat* was short-lived; the rebellious son died in office shortly thereafter. He was succeeded by his brother, Benkan, who was mean to his father. Aging and blind, nearly a hundred years old, Askia was determined to regain his power, and he finally did. It was through the loyalty and cooperation of one of Askia's other sons, Ismail, who one day paid his father a visit. In so doing, Askia dictated a battle plan to Ismail; he believed that if the plan was carried out, his son could be overthrown and that act would enable Askia to return to power. The plan worked, and Askia the Great did in fact return to power that he enjoyed until his death in 1538. Clearly he, along with Mansa Musa, and Sunni Ali, appeared to have been an embodiment of West Africa's immanent Destiny, or so it seems. I presume each regarded himself as a bearer of West Africa's Destiny. Whether the indigenous Africans held similar feelings I presume is a matter of debate.

Professor Hansberry, is it not the case that during your time many Africans regarded the presence of the Muslims in Africa during Medieval times as a foreign imposition comparable to the subsequent arrival of Europeans in Africa? Did the Ethiopians who witnessed the arrival of Christianity in their homeland hold a similar attitude towards the arrival of Christendom? In your assessment of the matter do you think that the Christians' economic-cultural contributions, if any, in Ethiopia out weigh the Muslims' economic-cultural contributions in West Africa generally? Some 50 years following Askia's death, the slave shipping depot on Gorée Island in Senegal opened (in 1588) and began shipping African slaves to the New World, although I doubt that there are any casual connections between the death of Askia and the opening of the Gorée slave depot.
Sincerely,
Dr. Johnny Washington, Ph.D.   May 2, 1997

Islamic God: Avicenna and Averroës' Proofs

Professor William Leo Hansberry (1894–1965)   Destinicity Letter #40

Re: The Muslim Philosophers: Avicenna and Averroës: Part XV

Dear Professor Hansberry:

I do not know your level of interest in academic Philosophy, but I assume you will appreciate my observations that follow inasmuch as they bear on the Islamic philosophy that apparently affected political events that played themselves out during the Golden Age of West Africa. You will recall that in my first few letters to you, I considered how the ancient Greek philosophic traditions were influenced by the African world-views. Here in this letter to you, the focus is on the early history of Islamic philosophy. Further, in some of my previous letters to you I highlighted your observations as to why the early Christian church became divided over the issues pertaining to the nature and identity of Jesus: a central theological-metaphysical question was, was Jesus God, or a man? As I have already suggested in some of my previous letters, this question parallels the age-old question concerning the relation between the infinite and finite and how the two can be united. Some centuries later the Protestant Reformation, with its many elements of strife, erupted. The history of Christianity is checkered with conflicts, before and after the Protestant Reformation. As you are well aware, such theological-metaphysical divisions are not unique to Christianity. All major religions are marred by conflicts similar to the ones mentioned above. Consider Buddhism as it arose in the sixth century B.C.; during its early history, the question also arose concerning the identity of Buddha (Siddhartha Gautama). This question resulted a few centuries later into the splitting of Buddhism into major two sects: the Mahayana sect, and the Theravada sect. Similarly, immediately after the death of Mohammed in 632, the Islamic movement became divided on the issue of Mohammed's successor. Eventually Islam resulted at the practical level into three major divisions, the Sunni, the Sufis, and the Shiite, although in the interest in brevity here, I will not elaborate on these. Rather, I wish to focus not so much on the practical conflicts that arose in Islam, but on some of the philosophic issues that shaped Islam by considering the views of Avicenna [Ibn Sina] (980–1037) and Averroës [ibn-Rushid] (1126–1198). Averroës influenced not only the development of Muslim philosophy but also the development of Jewish as well as Christian philosophic traditions. St. Thomas Aquinas (1225–1274) drew heavily on Averroës' views concerning, for example, the nature of God that in turn was incorporated into St. Thomas' celebrated cosmological proof for God's existence.

Mesopotamia, Syria and Egypt—these were some of the countries in which Christianity first took roots and in which the Greek philosophic and scientific endeavors thrived. As the Muslims came in contact with these countries through trade and conquests, its scholars developed an intense interest in the philosophic works of Plato and Aristotle, translated these into Arabic and interpreted them in such a way as to be made compatible with Islamic religion. Thus, arose Islamic philosophy and the doctrines of the Koran, the holy book of Islam. Professor Hansberry, Islamic philosophy never gained many followers in Ethiopia or for that matter in any East African countries. Why is this so? Islamic philosophy was institutionalized, as it were, by the Abassid Calip al-Ma'mum (813–833), who founded a school in Baghdad devoted exclusively to the study of Greek science and philosophy. During the early phases of the Middle Ages, the philosophic works of Plato and Aristotle and their predecessors were unknown in Europe. It was through the works of certain Muslim philosophers that Greek philosophy was reintroduced into Europe. Among such Muslim philosophers were Alfarabi (870–950), Avicenna and Averroës. Here we will briefly consider the views of the latter two. There are doubts in certain scholars' minds as to the ethnic identity of Averroës, yet, as previously noted, some authorities hold that substantial evidence indicates that Averroës was racially mixed between Black Africans and Arabs. He was born in Cordova, Spain and spent some time in Morocco. His views influenced not only many Islamic philosophers but Jewish and Christian philosophers as well. Some sources hold that he had more influence on Christians and Jews than on the Muslim world. This, of course, is a matter of debate.

What is more important is that certain scholars in Timbuktu undoubtedly drew on Averroës' works. As was mentioned, between the University of Sankore and the learning centers in Cordova and Seville, Spain there was much interaction and exchange. It was held that Averroës lived and worked as a scholar during the time in which Yakub Al-Mansur (1149–1199) defeated the Almoravids and their allies, the Arabs, and invaded Spain and Portugal in 1189. Yakub Al-Mansur is considered one of the greatest Moorish rulers of Medieval Spain, where he built hospitals, aqueducts and schools. As a commentator on Aristotle's works, Averroës regarded himself as dispelling certain mistakes his predecessors had made in interpreting Aristotle's works. Much of Averroës' works was directed against Avicenna, who in analyzing Aristotle's notion of substance, made a distinction between essence and existence. Avicenna insisted that the essence was ontologically prior to the existence: "essence" has to do with what a thing is; "existence" pertains to whether or not a thing exists. In Avicenna's view, essence and existence were distinct elements; they do not have identical meanings. The mind can think the essence of a thing without affirming or denying its existence. This distinction between essence and existence was a basis for his argument for the existence of God. The outline of this argument is as follows: things that now exist are contingent beings including trees, rocks, horses, houses, ice-fields, clouds, men and women. Contingent beings are mere possibilities, whose existence is not necessary. They are not compelled by their nature to exist. By contrast, God is a being whose essence implies, *necessarily*, His existence. God is not a mere possibility. God's existence is part of His nature. This is not the case of contingent beings. Averroës rejected Avicenna's essence-existence distinction. Avicenna had regarded essence (Platonic forms or transcendent Destinicity) as basic reality; existence, contingent beings (immanent Destinicity), as secondary. By contrast, Averroës, (in harmony with the teachings of Aristotle) held that basic reality pertained to contingent beings, i.e., individual substances, although the mind can distinguish between essence (transcendence) and existence (immanence). Individualized substance constitutes reality. Nothing more, nothing less. Contemporary philosophers, including the late Sartre, drew on such a distinction in developing his theory of existentialism. He held that human beings possess no essence as such or in and of itself; rather, their existence precedes their essence. By virtue of their freedom, the Destiny of human beings is open-ended with infinite possibilities exercised in the act of choosing. Avicenna was interested in the notion of Destiny that he related to the Koran.

> Someone asked the eminent *shaykh* Abū'Alī b. Sīnā…the meaning of the Sūfī saying, 'He who knows the secret of destiny is an atheist.' In reply he stated that this matter contains the most obscurity, and is one of those matters which may be set down only in enigmatic form and taught only in a hidden manner, on account of the corrupting effects its open declaration would have on the general public. The basic principle concerning it is found in a Tradition of the Prophet… 'Destiny is the secret of God; do not declare the secret of God.' In another Tradition, when a man questioned the Prince of the Believers, 'Alī (may God be pleased with him),' he replied, 'Destiny is a deep sea; do not sail out on it.' Being asked again he replied, 'It is a stony path; do walk on it.' Being asked once more he said, 'It is a hard ascent; do not undertake it.' The *shaykh* said: Know that the secret of destiny is based upon certain premisses, such as [1] the world order, [2] the report that there is Reward and Punishment, and [3] the affirmation of the resurrection of souls.[9]

Avicenna meant by the first premise that our Destiny is grounded in the world order that was created by God, who controls the world and everything in it, including human beings. There is a connection here between God, world order, human beings and Destiny. The second premise has to do with what may be called the free will problem. We are in harmony with Destiny to the degree that in exercising our free will, we seek to do good and avoid evil. When we do good and avoid evil, we approximate the ideal of perfection. When we do evil and neglect to do good, we enter into what may be called "sub-cendence," we abide in deficiency.

Avicenna made the connection between ethics and Destiny (God) in this manner: we are compelled to approximate our Destiny because in so doing we are to perform right actions for which God rewards us. God punishes us for performing wrongful actions that drive a wedge between us and our Destiny. In the Judaic-Christian ideal, there is a relation between the third premise and the afterlife: if we performed good deeds on Earth, we will be rewarded in heaven; if we performed bad deeds on Earth, we will be punished in hell. We are destined to go to heaven or hell. Avicenna also argued that when the resurrection occurs the soul returns to its point of origin which is in God, the point of singularity, as I have suggested in the beginning of this series of letters. This view also is related to the third premise mentioned above that Avicenna interpreted thus: this is why God the Exalted has said, "'O tranquil soul, return to your Lord satisfied and satisfactory.'" (*Medieval,* p. 210)

Muslim thinkers such as Avicenna and Averroës and their intellectual counterparts in Timbuktu and elsewhere created the science and culture that transformed the Middle Ages from ignorance and darkness to light, truth, beauty and goodness. Professor Hansberry, my criticism against Avicenna's view of Destiny is that his view is inadequate. He failed to make a distinction between transcendence and immanence modes of Destiny. However, such a distinction is implied in his view. He associated Destiny with the transcendent Allah (God), who in his views *controls* our Destiny on the basis of reward and punishment. However, Avicenna made no adequate room in his system for what I have called Destinicity that includes the Destiny as well as the ethnicity elements of a people. A problem with Avicenna's view is that he was too committed to the Platonic-Aristotelian world view that he attempted to reconcile with the Koran. While it is the case that Avicenna was prophetic in his thinking in connecting his notion of Destiny to the physical world order, to what I have called the unity of the world, that is, the Cosmic Destiny mode, the world order that he had in mind was based mainly on the physics and metaphysics of Aristotle that were widely accepted up to the times of Copernicus (1473–1543) who rejected Aristotle's views and offered a new model. He held that the sun was the center of the universe. It should be noted that he was not entirely original. Copernicus' paradigm was anticipated by philosopher Nicholas of Cusa (1401–1464) who held, "Just as the Earth is not the centre of the world, so the circumference of the world is not the sphere of the fixed stars, despite the fact that by comparison the Earth seems nearer the centre and the heavens nearer the circumference."[10] Modern physics informs us that even Copernicus' cosmology was also limited. Today the hypothesis is that the universe includes billions and billions of galaxies whose origin extends to the big bang singularity, the original black hole, that occurred between fifteen and twenty billion years ago. Moreover the order of the universe, that is, the Destiny of the universe is determined by the four fundamental forces of nature: the strong nuclear force, the weak nuclear force, electromagnetism and gravity. Further, I hold that the Destiny of the human race is connected to this cosmic Destiny, and that this cosmic Destiny is ultimately grounded in transcendent Destiny over which we have no control. Nor can we have any objective, scientific knowledge of our transcendent Destiny. We can, however, without depending solely on God (Allah), struggle against the elements of resistance in our immanent Destiny mode in approximating our transcendent Destiny. As Nietzsche maintained, our worldly struggles are difficult, and it requires strong souls to bear the heavy burdens through which our strengths increase:

> A full and powerful soul not only copes with painful, even terrible losses, deprivations, robberies, insults; it emerges from such hells with a greater fullness and powerfulness; and, most essential of all, with a new increase in the blissfulness of love. I believe that he who has divined something of the most basic conditions for this growth in love will understand what Dante meant when he wrote over the gate of his Inferno: 'I, too, was created by eternal love.'

—Friedrich Nietzsche, *The Will to Power*

Sincerely,
Dr. Johnny Washington, Ph.D.

May 3, 1997

Chapter 5

Destinicity and Struggles

Destiny: Immanence and Transcendence

Dr. Alain Locke (1886–1954)   Destinicity Letter #41

Re: Destinicity and Struggles: Part I

> Of what is great one must either be silent or speak with greatness. With greatness—that means cynically and with innocence.
>
> —Friedrich Nietzsche, *The Will to Power*

Dear Dr. Locke:

Medieval Islamic philosopher Avicenna's views of Destiny included in the set of letters to Professor Hansberry provided the transition to this set of letters that I shall now write to you. Moreover, I shall intermittently revert to some of the themes that surfaced in my letters previously addressed to you. Here I shall largely draw on the modern European philosophic tradition, with specific attention devoted to the views of Johann Gottlieb Fichte (1762–1814). First I shall repeat my definition of Destiny. In its broadest sense, it is an ideal of unity towards which a people strives in successive generations.

Moreover, Destiny includes immanence and transcendence, the nature of which is analogous to the actual and the ideal, the finite and infinite, modes of reality, respectively. The immanent mode of Destiny pertains to the natural and social environments, the empirical realm, and to the extent that this is the case, Destiny can be a basis for developing environmental ethics. This empirical realm is what Avicenna called the realm of contingent beings. What he called the realm of essence is on the par with the transcendent mode of Destiny, the Kantian noumena realm, within which, God, Freedom, and Immortality exist. It was Kant who informed us of our empirical knowledge. We must have faith that God, Freedom, and Immortality exist, without which life would be without meaning. This is where Kant's views diverged from the respective views of Avicenna and Averroës; both of whom held that one can come to know God (Allah). However, it will be recalled that Avicenna did indicate that we cannot know the Destiny of the human race.

In view of the distinction I make between transcendence and immanence, I hold that although we cannot obtain knowledge of our transcendent Destiny, we can obtain knowledge of our immanent Destiny. I have been suggesting all along that I accept Kant's view that we cannot have knowledge of God. This is true so long as we rely exclusively on the faculty of the intellect (Bergson). I am, however, tempted to go beyond Kant, and agree with Bergson who says that with the faculty of intuition that the moral hero employs, we can come to have knowledge of God. The moral hero may be construed as representative of the Cosmic Destinicity. Since most of us are not moral heroes or mystics, we are in no position to give an affirmative reply to the question: what constitutes the transcendent Destiny of African Americans, Africans, or any people? We must accept on faith that there is a transcendent Destiny, towards which we are all striving. We must posit Destiny. In addition to the recognition of this fact, the most that we can say about transcendent Destiny is that it is posited as an ideal of unity, the approximation of which enables us to exercise freedom. If Fichte held that the Ego posited the material world (the non-ego) so that our ethical duties can be fulfilled, I also hold that we are compelled to posit Destiny, so that we can fulfill our human potential generally. That is our Earthly vocation. We do not agree with Karl Marx that the transcendent world is merely the superstructure of the economic basis (immanent mode) of society. That the transcendent world is merely an illusion by which we become alienated from our authentic being, our genuine Destiny. Rather the transcendent world provides the grounds for the immanent world. The transcendent realm is that towards which we must strive, if we are to achieve authenticity, and genuine freedom. Although we are unable to obtain objective, scientific knowledge of the transcendent mode of Destiny found in the noumena realm, the field or discipline by which the immanent mode of Destiny is to be examined is Destinicity Studies to which I have already called attention.

Sincerely,

Dr. Johnny Washington, Ph.D.   May 7, 1997

Destinicity: A Whiteheadian Perspective

Dr. Alain Locke (1886–1954)   Destinicity Letter #42

Re: Destinicity and Struggles: Part II

> Man does *not* seek pleasure and does not avoid displeasure: one will realize
> which famous prejudice I am contradicting. Pleasure and displeasure are
> mere consequences, mere epiphenomena—what man wants, what every
> smallest part of a living organism wants, is an increase of power. Pleasure
> or displeasure follows from the striving after that; driven by that will it
> seeks resistance, it needs something that opposes it. Displeasure, as an ob-
> stacle to its will to power, is therefore a normal fact, the normal ingredient
> of every organic event; man does not avoid it, he is rather in continual
> need of it; every victory, every feeling of pleasure, every event, presup-
> poses a resistance to overcome.
>
> —Friedrich Nietzsche, *The Will to Power*

Dear Dr. Locke:

In the Prologue of this work I introduced the novel notion Destinicity.
Here I shall elaborate on such a notion. I can anticipate that some individuals will find my definition provided below unacceptable on the grounds that
it is too Euro-centric. Already I have been criticized by some Afrocentrists
as being too European in my "outlook." Do not most of these same critics
employ the English language, a European language? How many African
Americans speak Swahili or some other African language? Further, a number of Afrocentrists or any others who criticize my views for being too
Eurocentric, go on to praise Islam as the "Black man's religion," with its
roots in Africa. This view was echoed by the younger Malcolm X, for example. Some of these Afrocentric individuals who construe my works, as
well as yours, as being too European or too Eurocentric, will be surprised
to gather how deeply indebted Islam was to ancient Greek philosophy, including the views of Plato as well as Aristotle. We have already indicated the
influence these two had on Avicenna and Averroës, who in turn influenced
subsequent European and Jewish philosophers. The founder of Islam, Mohammed, was from Arabia. To what degree did Mohammed himself or any
of the other earlier Muslim leaders incorporate any African religious or cultural elements, in contrast to the Greek elements, into the teachings of Islam? By Destinicity I have in mind the primordial, active, creative, dynamic

180

impulse by which a people are lured, ultimately, toward the transcendent ideal of Destiny. Destinicity is comparable to the ancient Greeks' notion of reason by which the universe develops and consequently approximates unity and wholeness. Seen from another perspective, Destinicity bears some resemblance to the African notion *Nommo*, a universal, creative force that animates matter, becomes alive, or bears fruit through the word, through *Nommo:*

> Thus it is the earth, not God, that bears fruit. And it is not only manual labour that puts the fruits of the field at man's disposal. Sowing and reaping are only parts of human activity. Man must do much more than sow and reap, for the seed corn has of itself no activity of its own, it does nothing without the influence of man, it would not grow but would remain lying in the ground without his help, without the influence of *ubwenge*, active reason. How does man accomplish this? Through *Nommo*, the life force, which produces all life, which influences 'things' in the shape of the word. (*Mantu*, p. 124)

Like the notion of Spirit (*Zeitgeist*) in the views of Hegel (1770–1831) or Fichte, Destinicity, no matter what form it may assume—ethnic, national, world or cosmic—includes the passions, together with the rationality, of a people, their collective Eros, by means of which they actualize their potential within the context of community, through which their spirit is actualized and nurtured. Destinicity is also closely related to Bergson's notion of life impulse, the *élan vital*, the creative spirit within which all of life throbs, and by which life unfolds. The *élan vital* is the source of all reality, and it is on this basis that reality ascends or descends. According to Bergson, when reality ascends its freedom increases; when it descends its freedom decreases. The Spirit is constantly struggling to liberate itself from the resistance imposed by matter. The Spirit allows desires to ascend. Thus what Bergson call the Spirit, I called Destinicity, that creative force that pervades nature.

If "actual entity," along with its peculiar meaning, constitutes the centrality of the metaphysics of Alfred N. Whitehead (1861–1947), ethnic grouping(s) constitutes the cornerstone of Destinicity Studies. However, the starting point of Destinicity Studies is neither Whitehead's actual entity nor Leibniz's monad. Race or ethnicity is the launching pad. By juxtaposing "ethnic" with "Destiny," two notions that are apparently inherently in opposition to each other, the resulting synthesis is Destinicity. Its role is to unify experience. Destinicity makes it possible for us to approach the event horizon of transcendence. Destinicity Studies allows for the convergence of philosophy, on the one hand, and science, on the other. In asserting this claim, I am inspired by the works of the German philosopher Rudolph Hermann Lotze (1817–1881), whose works you, yourself, Dr. Locke, found interesting. In demonstrating the affinity between philosophy and science, Lotze reminded us that we tend to take certain concepts for granted, without seeking to determine their origin, significance, limits and validity. Philosophy can deepen our knowledge of the world by singling out the following concepts for special investigations and seeking to establish or determine the interconnection among them, by placing them within larger schemata to appreciate the fullness and richness of experience unfolding: causality, matter and energy, matter and spirit (Bergson), freedom and necessity (Spinoza) and means and ends (Dewey). Similarly, I maintain that in addition to the above-mentioned concepts, Destinicity Studies aims to investigate such concepts as: freedom and oppression, harmony and conflict, identity and difference, as well as origin and Destiny (and Destinicity). Transcendent Destiny is comparable to the Kantian regulative ideals in terms of which these concepts may be unified.

Some points Bergson makes regarding the nature of the self or personal identity I find useful in illuminating the notion of Destiny. Bergson held that memory constitutes the spirit of a person in that it enables experience to assume a qualitative tone and degrees of richness. I hold that memory plays a major role in the ways in which we ascribe values to items of experience. I believe you, Dr. Locke, did not emphasize the role of memory in your axiology. I want to maintain that our memory informs our attitude, that is, our values. The values that I placed on the little red fire truck when I was three years old on Christmas morning differ from the values I place on the same little red fire truck today. Thus, I believe you ignored or slighted an important faculty when you failed to assign memory a prominent role in your axiology. Memory adds vivacity, quality and tone to our sense meaning about which I spoke in the third letter in this work. Not only is memory important, but also the faculty of anticipation in the determination of values. The value of an item increases for me if it can satisfy me in the present as well as in the future; the faculty of anticipation, together with the faculty of memory, allows me to focus on the future. The notion of Destiny draws us toward the future which we can face either with hope or despair.

In this series of letters, I shall limit the discussion to examining the following concepts: conflict and harmony, and identity and difference. Science is concerned with discerning or inventing laws or concepts of nature whose range of validity is universal. Similarly, to the degree that Destinicity Studies is concerned with concepts such as those mentioned above, it provides the basis for a science of Destinicity (Destinology). Hegel construed his system of metaphysical categories as a science, insofar as it is concerned with the manner in which the categories are dialectically connected. Without emphasizing the dialectical nature of the concepts, the science of Destinicity also seeks to exhibit the interconnections among the above-mentioned concepts. What distinguishes Destinicity Studies (or Destinology) from, say, biology, physics, chemistry, anthropology or any other traditional scientific discipline is this: in the one, the stress is on the accumulation of quantitative data and the analysis of facts; in the other, the emphasis is on discerning the fundamental nature of experience as a unified whole within a broad framework, and from the standpoint of values. If mathematics is the cornerstone of traditional science, axiology, that is, value theory, is the cornerstone of Destinology. Destinicity Studies shall allow one to explore the Destiny landscape by enabling one to ascend to the highest points of the landscape of Destinicity to appreciate a panoramic view of the human condition, to discern its moments of despair and its moments of hope.

There are similarities between Whitehead's notion of God and the notion of Destinicity here under consideration. In describing God's relation to the world, Whitehead held that God was deficiently actual. In describing God in this manner, he was responding to a question concerning the existence of God. Does God exist or not? St. Thomas Aquinas (1225–1274) replied by offering the cosmological argument; St. Anselm (circa 1033–1109), by offering the ontological argument. (Neither of these two arguments shall be examined here). Whitehead broke with the religious-philosophic tradition by attributing to God a twofold nature, one of which is formal, the other is what he called "consequent." The former is comparable to Plato's forms found in the transcendent realm. In this sense, God's existence is merely formal, without substantial content. God is mainly the formal principle that provides gradation of order and value to the actualities and possibilities of existence. By contrast, the *élan vital*, the consequent nature of God, found in the immanent realm, is in effect a manifestation of the processes of the world, unfolding in accordance with the principle of creativity that precedes the formal as well as the consequent nature of God.

This principle is analogous to Hegel's notion of the Idea. Thus, this dimension of experience is where the mere possibilities are brought to fruition, made actual, where the spirit overcomes the resistance imposed by matter. Whitehead admitted that God exists but His existence is not altogether complete with respect to the world. In addition to the principle of creativity, the formal nature of God is the source through which the world, that is, the consequent nature of God, is determined. Just as God is deficiently actual with respect to His consequent nature, so also He is deficiently actual with respect to His formal nature, inasmuch as His formal nature is merely pure potentiality. Thus, from the standpoint of God's formal nature as well as His consequent nature, He exists in a deficient manner; yet, these two natures complement each other. Similarly, Destinicity includes the two above-mentioned modes of Destiny that parallel the twofold nature of God. Dr. Locke, it is interesting that I am here considering Whitehead's view to explain the origin of the universe.

The transcendence of Destiny is deficiently immanent; and the immanence of Destiny is deficiently transcendent. That is, the immanence of Destiny is imbedded in the transcendence, and the other way around. The ideal realm is striving to actualize itself; the actual realm is striving to idealize itself. Thus, the immanence and the transcendence are implicated in each other's modalities. The striving and inter-playing of modalities are what energize the Destinicity principle and what enables the whole system of Destiny modes to approximate immanent and transcendent unity. Parallels can be drawn between the notion of transcendent Consciousness or Ego, and *Osowo* (God) of the Bokis people of northern Nigeria, or of the Yourba's *Ori* (God).[1]

The Ego (universal consciousness) is compelled to posit the non-ego, the world, with all of its restrictions, or limits, without which the Ego would not be in a position to fulfill its potential. We need struggles in this world to render stronger men and women, to enlarge and enrich the human spirit whose purpose is to come to know itself, to enjoy freedom. Yes, our task on this Earth is to struggle, either in the defensive mode or offensive mode. Thus, a people are necessitated to posit Destiny, the conception of which includes the force of attraction that lures us onward beyond the immanent mode of Destiny to its transcendent ideal. Mention has already been made of this dialectical tension between these two Destiny modalities. The one limits us to this world, the other directs us beyond this world. For reasons different from the Afrocentrists, I believe you, Dr. Locke, would also criticize the Destiny model on the grounds that it makes an ontological commitment to a metaphysical entity that cannot withstand scientific scrutiny. My reply could take the following form: that we cannot have any objective knowledge of our transcendent Destiny. I hold that we are compelled to posit such a Destiny model to give our lives a sense of directionality and purpose. Additionally, I am also partly grounding Destiny in the realm of immanence, in culture whose institutions are investigated by sociology, anthropology, political science, economics and other disciplines. The elements of Destiny are accessible to scientific investigation in a manner in which science investigates many other cultural values, embedded in religions, art and other value-laden practices. Newton was not interested in determining the nature of gravity; rather he set out to examine its effect, its efficacy. Similarly we cannot know the nature of transcendent Destiny.

Sincerely,

Dr. Johnny Washington, Ph.D.   May 8, 1997

Manifest Destiny: Missouri Former Senator Benton

Dr. Alain Locke (1886–1954)   Destinicity Letter #43

Re: Destinicity and Struggles: Part III

> [A]ll expansion, incorporation, growth means striving against something that resists; motion is essentially tied up with states of displeasure; that which is here the driving force must in any event desire something else if it desires displeasure in this way and continually looks for it.—For what do the trees in a jungle fight each other? For 'happiness'?—*For power!*
>
> —Friedrich Nietzsche, *The Will to Power*

Dear Dr. Locke:

It is in this context I wish to dispel certain other possible views that might lead one to misconstrue the Destiny model. I have written these letters to you aiming to clarify my view of Destiny. Neither the Nazi German government's perverted usage of "Destiny" nor the Americans' usage of "Manifest Destiny" is compatible with the meaning I wish to assign to "Destiny." On the basis of the Nazi German government's usage of the notion of Destiny, it regarded the so-called Aryan race, of which the blond type members were paradigmatic, as genetically superior to other races, and thus, had a duty to assume the Destiny of the world. The erroneous belief held by many Germans was disseminated that the Aryans were justified in colonizing and ruling other ethnic groups, especially those of Asian or of African origins. Admittedly, the notion of Destiny was central to Hegel's philosophy of history, in which the Destinicity of a people is associated with the world Spirit or the principle of freedom. Construed as embodiments of the world Spirit, significant historic individuals, including Alexander the Great, Caesar and Napoleon, were regarded by Hegel as the means by which world history is transformed. Dr. Martin L. King, Jr. adopted Hegel's conception of history, as well as his notion of Destiny, of which King regarded himself as an expression. We would not be overstating the case, if we regarded Dr. King as an embodiment of the World Destinicity. Lerone Bennett, Jr., in his work, *What Manner of Man: Martin Luther King, Jr.*, indicated the ways in which Dr. King's views were related to the Hegelian notion of world history or spirit and its concomitant element of Destiny:

> Now, [during the late 1940s] as King unpacked his bags in the Crozer dormitory, the land was ablaze with controversy over the Progressive party and Harry Truman's up hill fight for a 'strong' civil rights bill. King watched, nourished, as the Negro ethos expanded. It was a sign, this, a sign of what King would later call the *Zeitgeist* stirring in the womb of time. In the hour when the Negro spirit changed from patient resignation to outreaching expectancy, King's *destiny* [emphasis added] began to take shape. Though King did not know it, the *Zeitgeist* was stalking him and preparing him for a later and more definitive engagement with a crumbling and outdated world view.[2]

In addition to the interest that Dr. Martin Luther King, Jr. had in certain views of European thinkers, Dr. W.E.B. DuBois, and other African or African American thinkers also adopted the views of many of the European thinkers. It is ironic that the very same German culture that produced Kant, Fichte, Hegel, Lotze and other philosophers, on whose ideas I have been drawing, produced also Adolph Hitler and Nazism to which he gave rise. (The racist position of Hitler and his movement, I outright reject, to say the least). Many of the theorists responsible for shaping the Nazi's racist ideology misrepresented or deliberately distorted the views of Hegel, Fichte and others. Dr. Locke, you never seemed to have gotten tired of reminding us of the fact that the U.S.'s own history is tainted with a notion of Destiny similar to the one once held by the Nazi Germans. You spent much of your time studying in Germany. Preceding the Nazi's racist ideology by a century, in the U.S. this superiority/inferiority myth was couched in what came to be called "Manifest Destiny." Newspaper editor John O' Sullivan as early as 1846 was among the first to use the term "Manifest Destiny" in describing the manner in which the white race in the United States was justified in its territorial expansion across the Mississippi River, toward the Pacific Ocean and beyond. Thus, the doctrine in question benefited many whites at the expense of such ethnic groups as African Americans, Asian Americans, and Hispanic Americans, on whose manual labor this country was largely built. Such an objective was echoed in an article by Missouri Senator Thomas Hart Benton (1782–1858), published in the May 28, 1846, issue of the *Congressional Globe:*

> Since the dispensation of man upon earth, I know of no human event, past or present, which promises a greater, a more beneficent change upon earth than the arrival of the van of the Caucasian race (the Celtic-Anglo-Saxon division) upon the border of the sea which washes the shore of Eastern Asia. The Mongolian, or Yellow race, is there, four hundred million in number, spreading almost to Europe; a race once the foremost of the human family in the arts of civilization, but torpid and stationary for thousands of years. It is a race far above the Ethiopian, or Black—above the Malay, or Brown (if we must admit five races)—and above the American Indian, or Red; it is a race far above all these, but still, far below the White; and, like all the rest, must receive an impression from the superior [white] race whenever they come in contact.[3]

Sincerely,
Dr. Johnny Washington, Ph.D.   May 9, 1997

Dr. Alain Locke (1886–1954)   Destinicity Letter #44

Re: Destinicity and Struggles: Part IV

> Values and their changes are related to increases in the power of those positing the values.
> The measure of *unbelief*, of permitted 'freedom of the spirit' as *an expression of an increase in power.*

> —Friedrich Nietzsche, *The Will to Power*

Dear Dr. Locke:

Here we are still on the trail of Senator Thomas H. Benton's Manifest Destiny doctrine that needs to be distinguished from the Destiny model. The Benton model points toward death and destruction, the Destiny model points toward hope. I suspect that the KKK that was founded a few years following the origin of the Manifest Destiny doctrine regarded itself as implementing the Manifest Doctrine. I should add that the founder of the Ku Klux Klan, Civil War General Forrest, was forced to surrender a contingent of his troops in my hometown, Gainesville, Alabama, where a monument is erected in his honor. I ended my last letter to you by citing a passage in which Senator Benton elaborated on the merits of Manifest Destiny. This passage that follows picks up the threads that we broke off in the previous letter. In this letter it shall become obvious that Senator Benton postulated the superiority of the white race; in opposition to this, he regarded as inferior Native Americans, Africans and other non-European groups: "It would seem that the White race alone received the divine command, to subdue and replenish the Earth! for it is the only race that has obeyed it—the only one that hunts out new and distant lands, and even a New World, to subdue and replenish..." (*American Issues*, p. 157) Senator Benton added:

> Civilization, or extinction, has been the fate of all people who have found themselves in the track of the advancing Whites, and civilization, always the preference of the Whites, has been pressed as an object, while extinction has followed as a consequence of its resistance. The Black and Red races have often felt their ameliorating influence. (*American Issues*, p. 158)

Some one hundred and fifty years have elapsed since Senator Benton provided the "official" rationale for Manifest Destiny. It seems as though my own personal Destiny sent me to Senator Benton's home state, Missouri, to remind the descendants of his constituency in the year 2001 and beyond, of the devastating effects that Manifest Destiny had on non-white people especially Native Americans. Apparently, Senator Benton was trapped in the immanent mode of Destiny where class, race and gender conflicts prevailed. It seems as though my mission here in Missouri is to walk a different path that is elevated beyond the one traveled by Senator Benton and to inform people about the Destiny model, one that enjoins global unity and harmony among all ethnic groups and races. The bad, warped Destiny that Senator Benton advocated needs to be routed or discarded. He accepted a bad Destiny for those adversely affected just as an auto mechanic may install in one's car a bad battery or taillight. The Destiny model inspires unity or harmony, both for the individual and the community. As in the teachings of Bergson, I define unity as involving concentration, relaxation and tension, as exhibited in the unity of life from a single celled organism to an ecological system. National Destiny pertains to the nation-state. The nation-state was a novel form of political arrangement that arose in sixteenth century Europe. In the U.S., "National Destiny" is expressed in the form of national interest, or national security. True, Hegel regarded the nation-state as bordering on divinity. He explicated world Destiny in terms of world history, made possible by significant individuals. Today, however, the nation-state seems to be on the verge of withering away, being replaced by what may be called centers of power in the world community. It is important to keep in mind that each of the Destiny modes is reciprocally related and the world Destiny is the dominant one, towards which the ethnic and the national modes of Destiny are converging, and by which unity is approximated.

I am cognitive of the fact that in describing Destiny, I have placed stress on the principle of unity as though that is its primary feature. In this regard, the Destiny model is comparable to that offered by the Medieval Muslim philosopher Avicenna who grounded Destiny in the order of the world, which is tantamount to the principle of unity on which the Destiny model is based. I doubt whether you had a great deal of appreciation for Medieval (Islamic-Christian) philosophy, since you regarded Christianity to be objectionable, in part because it induces dogmatic thinking.
Sincerely,
Dr. Johnny Washington, Ph.D.   May 12, 1997

Dr. Alain Locke (1886–1954)   Destinicity Letter #45

Re: Destinicity and Struggles: Part V

> It is not a matter of going ahead (—for then one is at best a herdsman, i.e., the herd's chief requirement), but of being able *to go it alone*, of being able *to be different*.

> —Friedrich Nietzsche, *The Will to Power*

Dear Dr. Locke:

Granted, the African American historic experiences are interwoven into the U.S. history, a history that is itself riddled with conflicts, wars, revolutions and other forms of social strife. Yet, due to slavery, segregation, the plantation and the urban life, among other factors, the African American experience exhibits certain oppressive features that are distinct from the history of other ethnic groups or of the larger society, generally. Moreover, the negative factors of slavery and racial segregation have rendered, intermittently, the African American experience fragmented or disjointed. Admittedly, the historic experiences of any people are somewhat fragmented. However, the racial oppression of African Americans, perhaps more so than that of any other ethnic group, has made evident the inharmonious features of the African American experience. Many of these features are constitutive elements in the social, juridical and political history of this country. Some authorities in African American Studies venture to maintain that the experiences of slavery and segregation were "negative merits" for African Americans. That is, due to the oppression brought on by the factors in question, African Americans were forced to create their own cultural forms. In the field of music, for example, such cultural forms include the following: the spirituals, the blues, jazz, gospels, rhythm and blues, soul, and rap. These forms of expressions ultimately became accepted or adopted by the larger society. Space does not permit one here to consider the various African American writers, painters, preachers and others who contributed to American culture. As was indicated, slavery and segregation were limiting factors imposed on African Americans who sought to overcome such limitations by the above-mentioned modes of cultural expressions, which are a distillation of Destinicity. Without devaluing the role of civil protest activities, riots, court actions and other measures, African Americans tend to deal with the limitations of freedom, that is, oppression, by engaging in creative, artistic expressions that are pervasive not only throughout this culture but worldwide. Whites or Europeans generally attempt to remove or deal with the limiting factors that characterize the immanent mode of Destiny. They tend to introduce greater laws and restrictions in the forms of treaties, legislative measures, backed by military force or other means of authority. This in part is an inadequate means of dealing with the limiting factors. How can one reduce restrictions by imposing greater restrictions? More laws? More police officers? More chain gangs? Are these appropriate ways to reduce "crime" and other social evils? Western history may be described as replete with laws (and treaties).

In the U.S., the Constitution, along with the Supreme Court, is virtually "sacred." In this context, one is also reminded of the previously existing Jim Crow laws and other measures aimed at controlling the freedom of Blacks. Prior to the 1960s, how many Blacks held public offices? Today many jails are being built. The incarceration rate is higher among Blacks than any other ethnic group in the U.S. Recent events in race relations in the U.S. reinforce the point that tensions and conflicts are also sources of creative transformation in U.S. history, a fact of which Dr. Martin Luther King, Jr. was well aware. Some recent events included the 1992 Los Angeles multi-ethnic riot following the Rodney King verdict; the intense racial tension that resulted during the O.J. Simpson trial and verdict; and the racial tension associated with the October 16, 1995 Million Man March. Dr. Locke, it is as though I am here preaching to the choir when I write to you describing the nature of social strife that the U.S. has recently undergone in struggling with its immanent Destiny and the vague possibility of unity. You understood the strife-ridden history of the U.S. quite well, and your novel axiological theory was aimed at enabling us to reduce racial and other forms of conflicts.

My criticism against your axiological views is that they could have been more powerful had you allowed for a transcendent principle(s) grounded in Destiny. Because of this peculiar sense of transcendence that is found in Plato's views and to a lesser extent Aristotle's views, the works of Avicenna and Averroës had such a wide appeal, influencing Jews, Christians and Muslims.

Sincerely,

Dr. Johnny Washington, Ph.D.   May 13, 1999

CC: Marcus Garvey

Dr. Martin L. King, Jr.

Dr Alain Locke (1886–1954)   Destinicity Letter #46

Re: Destinicity and Struggles: Part VI

The criterion of truth resides in the enhancement of the feeling of power.

—Friedrich Nietzsche, *The Will to Power*

Dear Dr. Locke:

This is a continuation of my previous discussion regarding the nature of conflicts affecting our immanent Destiny played out on the race or ethnic stage. Ethnic, national or world mode of Destiny—either mode of Destiny often undergoes internal as well as external conflicts. We witnessed this two-pronged strife in the early history of Christendom, where some such conflicts were played out in Egypt and Ethiopia. Similarly, there were internal as well as external conflicts that shaped the political (immanent) Destiny of Medieval West Africa. During more recent times, the ideological conflicts that occurred between Mr. Booker T. Washington (1856–1915) and Dr. W.E.B. DuBois (1868–1963) are familiar, so that a consideration of these need not be repeated here, for these two individuals were your contemporaries. However, you seemed to have had a greater appreciation for Mr. Washington than for Dr. DuBois. I read with pleasure your essay, "Booker T. Washington: Strategist of Southern Reconstruction." It should be added that a few years following your death in 1954, conflicts occurred between Malcolm X and Dr. Martin Luther King, Jr. If Frederick Douglass and later Booker T. Washington, during their respective times, were regarded as the bearers of African American Destiny, in the 1960s Dr. King, and to a lesser degree Malcolm X, were projected into that role. I elaborated on this point in my May 23, 1997 letter to Malcolm X; I plan also to write Dr. King, to bring him up to date on the plight of Black Americans. I am confident that he would be appalled in coming to know many of our youths have gone astray. Further, there is an effort to undermine Affirmative Action programs, and racial crimes are on the rise. Furthermore, many of the conflicts affecting the ethnic Destiny of a people may at first sight appear to have been generated, internally, by the group affected. However, it is also the case that on closer analysis the conflicts, or the source of the conflicts, may turn out to be connected, by way of causality, to an external source, generated by the pressures of the larger society, a point that you, yourself, described in many of your works. For example, in your essay, "The Negro in the Three Americas" (1944), you pointed out the ways in which the consequences of slavery and of colonialism, factors imposed from without, upon which the earliest stages of this capitalistic economy in part hinge, were the sources of many of the problems that Blacks in the U.S. and in the various Latin American countries, suffered. In a popular book *The Debt: What America Owes to Blacks* Randall Robinson demands reparations; that the one way the U.S. can do adequate justice to Blacks in addressing the wrongs that they suffered is to compensate them in terms of money for the sufferings that they and their ancestors underwent for hundreds of years. Later in this work I shall elaborate on the notion of reparations.

The U.S. Civil War is an illustration of a national Destiny that underwent internal conflicts, produced mainly by the practices of slavery and racism. The Northern states opposed the institution of slavery; the Southern states supported it. Both free and enslaved Blacks were caught between these opposing tendencies. Additionally, the Fugitive Slave Law of the 1850s, together with the U.S. Constitution generally, largely precipitated that conflict. On one side of this conflict was the Union army, led by General Ulysses S. Grant (1822–1885); on the other side was General Robert E. Lee (1807–1870). President Abraham Lincoln (1809–1865) represented the synthesis of the conflict. The Act of Reconstruction which offered former Black slaves "forty acres and a mule" was an attempt to restore equilibrium, in the form of racial reparations, into the African American community, and to the nation generally, that had been upset by centuries of slavery and the resulting Civil War. Yet many regarded the Act of Reconstruction as a failure. Later in this volume, I will explore more thoroughly the issue of racial reparations. Moreover, the collapse of the Soviet Union is another illustration of the disintegration of the national Destiny of a nation, the source of which was for the most part internal. Poverty, starvation, disease and political unrest are threatening the national Destiny of many societies on the continent of Africa, in Latin America, Asia and in the Caribbean. But the possibility of a true African Destiny still offers hope.

There can be conflicts between the ethnic, national or world Destiny modes; each can undergo internal strife, and each can clash with the other. Let us follow-up on the Lincoln illustration. During Lincoln's time, the national Destiny of the United States as a slave-holding nation was in conflict with the ethnic or collective Destiny of Black slaves. Here, then, is an illustration that exhibits a national Destiny mode on a collision course with the ethnic Destiny mode of a people. In an effort to avert such a collision, on September 22, 1862 Abraham Lincoln issued the Emancipation Proclamation, to become effective January 1, 1863, designed to abolish slavery in the United States. Some scholars hold that the Destiny of African Americans took a new twist on that day of January 1, 1863.

Moreover, many others observed that Lincoln was a reluctant Emancipator, who, in contemplating the emancipation of African American slaves, desired that once they were freed, they should be transported to another country, Haiti, Panama, or Liberia to establish a colony. Lincoln believed that due to the fundamental, biological and sociological differences between Blacks and whites, it was virtually impossible for the two races to live peacefully on the same soil. Frederick Douglass (1817–1895), a former slave who became an abolitionist and a representative spokesperson of the Destiny of African Americans, opposed Lincoln's colonization proposal. Among other issues, Douglass challenged Lincoln's accusation that Blacks were i) inferior to whites, and ii) that Blacks were the cause of the Civil War. Douglass subsequently inspired the works of Dr. W.E.B. DuBois, Booker T. Washington, yours, and others who attempted to discern the nature of the African Destiny.

Each of the leaders regarded himself as a means in removing or reducing the factors that restricted the freedom of African Americans. They were pathfinders, point-men, in the landscape of Destiny.

Sincerely,

Dr. Johnny Washington, Ph.D.   May 14, 1997

Ethnic Identity Revisited

Dr. Alain Locke (1886–1954)   Destinicity Letter #47

Re: Destinicity and Struggles: Part VII

> All events, all motion, all becoming, as a determination of degrees and relations of force, as a *struggle*.
>
> —Friedrich Nietzsche, *The Will to Power*

Dear Dr. Locke:

It goes without saying that the conflict and disunity that the whites caused Africans, and the people of African origin, to suffer induced in them an identity crisis. In attempting to address the identity crisis that African Americans suffered you introduced during the Harlem Renaissance, the appellation "New Negro." One wonders why this appellation did not catch on and acquire a sense meaning (C.I. Lewis) as did terms such "Black" and "African American." And since a people's identity is intimately related to their Destiny, when the one undergoes a crisis, the effect is manifest in the other, and vice versa. Destiny, in its immanent form, plays a major role in constituting identity. Let us now examine more closely the notion of identity (and difference) as pertaining primarily to African Americans in relation to the larger society. The history of the American people's preoccupation with racial identity extends some two hundred years and beyond. It was Thomas Jefferson who first devised racial categories for the census. Influenced by the tradition of the Enlightenment, the American so-called "Founding Fathers" established the framework in which the identity of this nation was shaped. They introduced the melting-pot doctrine. In regarding itself as a nation of immigrants, mostly those of European origins, the people of this nation were urged to ignore or reject racial or ethnic distinctions and to adopt the values and ideals of the nation-state. The Founding Fathers feared that if people allowed themselves to be differentiated into racial or ethnic groupings, this could result in factions within the nation. Ethnic distinctions, they felt, were incompatible with the fragile institutions of democracy. Rather than prizing the value of ethnic distinctions, the Framers of the Constitution stressed the rights of the individuals, to the exclusion of the rights of particular ethnic groups, a point stressed by many Affirmative Action opponents today. In spite of the efforts to dispel interest in ethnicity, racial identity or identity politics, ethnic neighborhoods were formed in most urban areas in the U.S., and many sought to define themselves by expressing pride in their European roots, including Irish-Americans, Greek-Americans, Polish-Americans and German-Americans. In one of your unpublished essays that I assigned the title, "The Quest for National Unity," you urged the U.S. to heighten its efforts in promoting ethnic diversity and racial integration, through which national unity could be approached. It is interesting that the U.S. "Founding Fathers," believed that they could establish political unity, while paying little, if any, attention to the factors of race or ethnicity. Granted, Jefferson devised racial categories, but he did so primarily for census purpose. I gather that in the 2000 census the category that allows for multi-ethnic heritage was included.

During the sixteenth century the notion of nation-state arose in Europe. It was John Locke and to a lesser degree Thomas Hobbes who later provided a philosophic foundation for the nation-state form of government. From the sixteenth century onward, the nation-state became the primary means by which people defined themselves. However, within the past few decades, this form of political organization has become uncertain. The future of the nation-state, whether of totalitarian or democratic form, is threatened throughout the world, a crisis fueled much by interlocking economies, the Internet, "fundamentalism" and ethnic discontent, or what may be called the "unhappy ethnicity"—this latter notion is a parody of Hegel's "unhappy consciousness." U.S. Senator Bob Dole, U.S. Representative Newt Gengrich and others hold that the moral character of the nation is being undermined by the entertainment industry giants, of which Time Warner is representative, that are saturating or polluting our minds and souls by producing and promoting grangsta rap artists such as Niggaz Wit' Attitude, 2 Live Crew, Geto Boys, as well as the late Tupac Shakur. These critics certainly desire to remove rap music from the media, an institution that plays a major role in shaping the identity of a people. Similar attacks were made against the blues, jazz, and R&B forms of music during the 1940s and 1950s, forms of culture into which the identity of African Americans and the nation is interwoven. Moreover, in this country some militia organizations, including the Michigan Militia, are reported to have begun to question the constitutional authority of the federal government, thus suggesting that the national government is no longer relevant in fulfilling the needs of the people.

Many hold that the Constitution itself, not the bureaucracy of the national government or other social agencies and institutions, is the primary identity-bestowing instrument of this nation. If the nation-state identity is a form of social construction, many authorities hold that the same is true of race or ethnicity. That is, racial differentiation that we make is an artificial construction, a point that you made in some of your essays.

Sincerely,

Dr. Johnny Washington, Ph.D.   May 15, 1997

CC: Marcus Garvey

Dr. W.E.B. DuBois

Dr. Alain Locke (1886–1954)   Destinicity Letter #48

Re: Destinicity and Struggles: Part VIII

> The will to power can manifest itself only against resistance; therefore it seeks that which resists it—this is the primeval tendency of the protoplasm when it extends pseudopodia and feels about. Appropriation and assimilation are above all a desire to overwhelm, a forming, shaping and

reshaping, until at length that which has been overwhelmed has entirely gone over into the power domain of the aggressor and has increased the same. If this incorporation is not successful, then the form probably falls to pieces...

—Friedrich Nietzsche, *The Will to Power*

Dear Dr. Locke:

Admittedly, during the 1960s African Americans devoted much attention to examining their identity. As was suggested a similar identity searching exercise occurred among Blacks in the 1920s. It was during this time that you introduced "New Negro." During this time Marcus Garvey often employed "Black," but "Black" did not become widely accepted until the 1960s, when many were torn between retaining the ethnic appellation "Negro," or rejecting it in favor of "Afro-American" or "Black." I believe your "New Negro" was a reaction to the identity associated with the Negro that Mr. Booker T. Washington knew, the Negro whose identity Mr. Washington helped to shape. The history of this ethnic appellation crisis extends to 1926 and beyond. It was George S. Grant who was among the first to urge African Americans to adopt "black" in reference to their identity. An eloquent discussion of this history is provided by Mary Frances Berry and the late John W. Blassingame in their book, *Long Memory*... In it they wrote, "Theoretical discussions of the use of the word 'black' were rare until the twentieth century. In 1926 George S. Grant insisted that it was the most logical term:" They added:

> The argument for it begins with the fundamental assertion that we are not Negroes (niggers) or colored people (cullud fellahs) but...BLACK AMERICANS;...By voluntarily choosing the logical mark which distinguishes our group from the group of White Americans, we endow both it and ourselves with a dignity, which...will operate to dispel the fallacious ideas of white purity, white beauty, and white superiority.[4]

In the 1980s, many found the appellation "African American" acceptable. Today various gangsta rap artists, including those previously mentioned, have re-appropriated "nigger" and changed its spelling to such a form as "Nigga," or "Niggaz" or some other form to distinguish it from the traditional "nigger." In adopting "nigger" or its derivation, one of the aims of rap artists is to reclaim that which was once used against African Americans; namely, "nigger," the word associated with noose-making, lynching and cross-burnings. In adopting nigger, the African American rap artists regard themselves as exercising vigorous self-assertion, an existential mode of freedom. Their using such an appellation enables them to reject or reverse the white man's meaning of "nigger," and ascribe a "positive" meaning to it. Their reversal of meanings associated with nigger is similar to the one introduced by Marcus Garvey, who attributed a positive value to blackness, any features having to do with Africa, and a negative value to whiteness, values often identified with white people or Europeans. In Garvey's value schemata, as in the system of values around which the Nation of Islam was once organized, white or European values were regarded as evil; by contrast, African values were elevated to the highest level. Some of the teachings of the Black Panther Party may also be traced to Garvey, whose inspiration was in part derived from Mr. Booker T. Washington. Thus the Black Panther Party, in seeking to embrace its Black identity, armed itself, and urged members of the Black community to do so as well, in order to overthrow the "system." The "system" in the Panthers' view, constituted the "other"; that is, in Hegelian dialectics, the "other" or "system," or "pigs" constituted the "difference" side of the identity equation. Blacks are defined in terms of whites and the other way around. If certain whites tend to employ "nigger" to demean Blacks, certain rap artists employ it to achieve the opposite effect. As was suggested, when rap artists employ "Nigga," clearly they are doing so to exercise their freedom to be different from the norms imposed by the larger society and to affirm the values of inner city youths alienated from mainstream America. Certain rap artists use "bitch" and "hoa" ("whore") in ways comparable to "Nigga." Here I may take the opportunity to criticize the rap artists whose lyrics are tarnished with such degrading words. Black women are not bitches or whores. Hence the switching of a derogatory term such as "nigger" into "Nigga" that comes to signify what may be called a negative merit in certain rap music may also be traced to the *Negritude* movement; yet, it is difficult to see how "bitch" or "hoa" or "whore" can be changed to have a positive meaning. In the 1930s, the Caribbean (Martinique) poet Aimé Césaire, together with the West African (Senegalese) poet Léopold Sédar Senghor, established "nigger" as the basis of *Negritude*, which became a cultural movement that inspired the decolonization of Africa. However, it will be recalled that during the Harlem Renaissance, you, Dr. Locke, coined the appellation "New Negro" and the roots of the *Negritude* movement can be traced to the works of Professors W.E.B. DuBois and William Leo Hansberry, as well as Marcus

198

Garvey, yours, and others. When I addressed a series of letters to Professor Hansberry I, in effect, thanked him for his contributing to the *Negritude* movement. As Berry and Blassingame observed: "The 1920's were the heyday of cultural nationalism. Howard University's William Leo Hansberry began his pioneering research into African history and Negritude..." (*Memory*, p. 412) On the basis of our considerations of the U.S. during the early phases of its history, it became evident that whites made a conscious effort to sort out and affirm their national identity, and in the process, degraded the identity of Blacks whose identity was defined in terms of the property of the slave master. As was indicated, Thomas Jefferson, no less than Abraham Lincoln, played a major role in this identity searching process. Neither quite knew what to do with African Americans. Both Jefferson and Lincoln seemed to have regarded African Americans as an anomaly, placed by Jefferson as well as Lincoln in the category of the "others." (It is interesting that Jefferson is reported to have been the father of several Black children whose mother was his slave).

Sincerely,

Dr. Johnny Washington, Ph.D. May 16, 1997

Dr. Alain Locke (1886–1954) Destinicity Letter #49

Re: Destinicity and Struggles: Part IX

> The degree of resistance that must be continually overcome in order to remain on top is the measure of freedom, whether for individuals or for societies—freedom understood, that is, as a positive power, as will to power.

—Nietzsche, *The Will to Power*

Dear Dr. Locke:

By highlighting the disunity or conflicts that pervade the immanent mode of Destiny, we are able to forestall a criticism that may unfold along the following lines. That because of the fact that Destiny places much emphasis on the transcendent realm, the other world, we are compelled to divert attention away from the problems and needs of this world. I believe that this is one of the reasons you did not subscribe to a notion of transcendence. I am clever enough to know that if one becomes too caught up, as it were, in the transcendent realm, one can lose contacts with the problems of this world. I am trying to have it both ways. Although we may posit transcendent Destiny, we nevertheless, whether consciously or unconsciously, are squarely rooted in this world inasmuch as we struggle against the limitations placed in our paths. As I have repeatedly indicated, Destiny as well as Destinicity possesses a twofold nature, one anchored in the transcendent world, the other, in this world, with all of its conflicts, paradoxes, and perplexities. I also assert with Hegel or with Whitehead that transcendent Destiny is implicated in immanence, and immanent Destiny is implicated in transcendence. The blues, jazz, the spirituals, and other forms of cultural expressions, for example, echo immanent suffering as well as transcendent lucidity. We need to come to the realization that the problems, that is, that the limitations that we face in this world are there for a specific reason: that they are thrown into our paths as obstacles to be overcome, by *working, fighting, gaming,* and, I might add, by *loving,* so that we can actualize our potential, our Destinicity as we journey along in approximating the transcendent mode of Destiny. During the 1960s, many community leaders perceived that the fragmentation of society was a grave social problem. Some dreamed of a society in which its many discordant elements were to be overcome through various forms of unity, including racial integration, political coalitions or workers' solidarity. Invariably the proponents of such ideals explored the notion Destiny without elucidating its meaning. For example, Martin Luther King, Jr. was a representative spokesman of the integration ideal, and he regarded himself as an embodiment of world Destiny (or Spirit). Yet, neither he nor Malcolm X defined Destiny.

Sincerely,

Dr. Johnny Washington, Ph.D.   May 19, 1997

Royce: Loyalty, Self-Esteem, and the Community

Dr. Alain Locke (1886–1954)   Destinicity Letter #50

Re: Destinicity and Struggles: Part X

> Once one has achieved a certain degree of independence, one wants
> more: people arrange themselves according to their degree of force: the

individual no longer simply supposes himself the equal of others, he seeks his equals—he distinguishes himself from others. Individualism is followed by the formation of groups and organs; related tendencies join together and become active as a power; between these centers of power friction, war, recognition of one another's forces, reciprocation, approaches, regulation of an exchange of services. Finally, an order of rank.

—Friedrich Nietzsche, *The Will to Power*

Dear Dr. Locke:

In considering the various modes of Destiny, I have done so as though the group is primary and the individual is secondary. That is, it may appear as though I have stressed various modes of collective Destiny including ethnic, national as well as world Destinies, to the exclusion of the individual. Many may criticize this approach. While I was writing my book, *A Journey,* I anticipated such criticism and introduced the views of one of your teachers, Josiah Royce. I focused on Royce's idea of "cause" and the related idea of loyalty. Like Hegel, Royce was regarded as an absolute idealist, the view that holds that spirit is primary and that matter is secondary. As systematic thinkers, both Hegel and Royce have been accused of developing a social-political philosophy that is so all-embracing that it stifles individuality or routes the individual. Royce, like Fichte, tried to find a place for the individual's autonomy by holding that what is important is the moral vocation which each individual is compelled to assume. Royce left undecided the question as to what was more important, the individual or the community. Instead, he wanted to reconcile the two by holding that one achieves individuality in and through the community of others, to whom the individual discloses his or her identity. There is the community of interpreters. The individual creates himself by fulfilling his/her duties in the community. This requires that the individual commits himself/herself to a cause, such as reducing world hunger, defending animal rights or building homes for the homeless. When one freely commits oneself to a cause, one is exercising one's freedom. The cause is that which brings to the fore the principle of inter-subjectivity and makes possible the community to emerge, so that one is interacting with others who are also fulfilling their respective duties.

Dr. Locke, you suggested in your essay, "Values and Imperatives," that you accepted Royce's loyalty principle. The loyalty-to-loyalty principle was intended to protect the individual or group from supporting an unethical cause. In some of your previous works you suggested that African Americans, as a group, in contrast to others, expressed great loyalty to this country. Loyalty also promotes group cohesion in which patriotism thrives. That is, one is not to be loyal to a particular cause that may degenerate into unethical activities. Such a commitment enjoins that one is to promote the harmony of humanity generally. This is a guiding ethical ideal. What is not conducive to universal harmony, made possible by the loyalty-to-loyalty principle, lacks moral worth. The commitment to a cause principle, along with the loyalty-to-loyalty principle, is the means by which one exercises freedom and individuality. This loyalty to loyalty to principle is a criterion of action for the individual as well as the community, a criterion comparable to Kant's categorical imperative, as you well know.

Many feminist thinkers incorporate the loyalty principle into their moral or ethical paradigm. By the same token, the caring principle to which Virginia Held and certain other feminist thinkers subscribe may be used to strengthen Royce's loyalty principle. These two principles combined reduce conflict within and among groups. Dr. Locke, the views expressed in your celebrated essay, "Values and Imperatives" illuminate the nature of value conflicts mainly by making clear the mechanism in terms of which such conflicts are generated or perpetuated. You reminded us that if we come to understand the workings of such a mechanism, this is a key step in the direction of reducing such a conflict.

Now the question arises: what can be done to enhance the individual's striving capacity? How might the individual enlarge his or her will to power (Nietzsche)? Individuals can form alliances, pool their resources and devote their collective efforts to a cause. When individuals devote themselves to a cause, their chances of overcoming the resistance, against which they are struggling, are increased many-fold. Here we have a basis for establishing a realm of action, a community in which freedom can be effectively exercised.

Our notion of freedom and individuality, as was suggested, is grounded in what may be called the primordial impulse that includes rational as well as irrational elements. This impulse is the source that enables us to strive towards the ideals associated with what I have called the transcendent Destiny.

According to Fichte, such striving is manifest in the form of irrationality or mere impulse when our aim exclusively is to satisfy our appetite whose objects impose limitations on our freedom. This is so because we become dependent on such material objects, found in the immanent Destinicity realm. By the same token, we can in the presence of the limitations imposed by the things of the world, envision other possibilities that point us in the direction of the transcendent world, where true freedom lies. We become cognitive of this through reflection on the limits with which we are confronted and against which we must struggle. We are condemned to working and fighting. We must never *retreat*. Never *surrender*. We must be prepared to attack on all fronts to overcome resistance imposed by the world. Our moral vocation on this Earth is to work and struggle against opposition by which we enter into spiritual ascension. We have no higher calling. A people must perpetually work and strive. This imperative is applicable to the masses as well to the elite. It was Marx who informed us with great clarity that our Destiny lies in our working on the world, in terms of which we not only produce objects that satisfy our needs but also we create our own humanity, our own Destiny.

There is no God to give us directions, according to Marx or Sartre. Moreover, nothing great in the world can be accomplished without a struggle (Hegel). In a similar vein, Frederick Douglass enjoined that power concedes nothing without demands. Whether we win or lose is beside the point. The goal is to strive and toil to the bitter end, for it is in and through this striving that we exercise our freedom. As the historians B.A.G. Fuller and Sterling M. McMurrin wrote:

> The eternal *struggle* to realize the ideal of absolute freedom and the equally eternal *failure* to do so, in which the life of the Ego consists, are seen in the situation of the finite egos, that is, of ourselves. To the finite egos, life seems cruel and harsh. The obstacles and the antagonists it opposes to our self-fulfillment appear as part of the non-ego [the world] and as an interference with our liberty, and our struggle with our environment and our fellow-men feels not like an exercise of freedom, but like a struggle against necessity. These limitations we cannot remove, neither can we *will* that they should cease to exist, for in so doing we should will our own destruction. We can, however, will to open our minds to the ideal at which the Absolute Ego is aiming. We can discipline ourselves to regard suffering and failure and misfortune and defeat, not as a thwarting of liberty by necessity, but as incidents in a *struggle* that we *welcome* and *freely* accept.[5]

We have an instinctive desire for freedom, informed by the primordial feeling, grounded in what a Kantian would call the transcendental ego, "the point of singularity" of the spirit. Thus, we can speak of conscience as resulting from our harmonizing the rational and irrational elements of our nature, an aim that many ethical or religious systems seek to achieve. Conviction that is informed by conscience is our moral lighthouse, our guide. There is a twofold test that may be employed to determine the validity of one's convictions. On the one hand, one can engage in dialogues with others, in an attempt to persuade them of the rightness of one's cause; on the other hand, one can ask one's self the question: Can I hold up my actions as a model for others to imitate through all of eternity? Fichte wrote:

> To listen to it [conscience], to obey it sincerely and unhesitatingly without fear and quibbling,—this is my only vocation, the whole purpose of my existence. My life ceases to be an empty game without truth and meaning. Something ought to happen simply because it ought to happen: that which my conscience requires of me, just of me who finds himself in this situation. That it may happen, for that, for that alone I exist. I have understanding so that I may recognize it, and strength so that I may accomplish it.[6]

Sincerely,
Dr. Johnny Washington, Ph.D.   May 21, 1997

Conscience as a Moral Guide: Fichte's Model

Dr. Alain Locke (1886–1954)   Destinicity Letter #51

Re: Destinicity and Struggles: Part XI

> What will become of the man who no longer has any reasons for defending himself or for attacking? What effects does he have left if he has lost those in which lie his weapons of defense and attack?

> —Friedrich Nietzsche, *The Will to Power*

Dear Dr. Locke:

I am especially interested in Fichte's notion of conscience, inasmuch as you, yourself, relied on this principle in your essay, "Values and Imperatives." While it is the case that you regarded it as a basis for guiding ethical conduct, you never provided a justification for its efficaciousness or validity. I think Fichte handled the notion much more effectively than you did. Moreover, I like the way he connected his notion of conscience with the elements of struggles, and counter-struggles, freedom, self-esteem and transcendent Destiny. Our striving is informed by conscience. The striving is a double-edged sword: on the one hand, it limits our freedom, inasmuch as the striving becomes attached to certain definite objects through which we seek enjoyment. To the degree we become dependent on objects of enjoyment, our freedom is reduced. A primary goal is to overcome through struggles and counter-struggles such dependence on things in this world, and to become independent, recognizing that our own being is grounded in transcendence, the singularity of Eternity, from which we have become alienated in our worldly attachments and to which we must strive to return. Buddha held that all suffering is due to our attachments to things of this world by which we are rendered unfree. Buddha's solution to this freedom problem was to urge us to become *detached* from things of this world. In a similar vein Bergson urged us to adopt an ascetic lifestyle so as to simplify life and to live within our economic means. "Keeping up with the Joneses," as it were, is a principle one should never adopt. Yet, competition is rooted in the human condition, and in nature generally. When Fichte pointed us in the direction of freedom, he maintained that the same impulse that harbors dependence and unfreedom also harbors freedom, the recognition of which is achieved when, through reflection, we are awakened to possibilities beyond the immediately given.

Since the time of Plato, philosophers have regarded the rational and irrational faculties as diametrically opposed. Fichte denied the opposition. In fact, he held that if our natural wants are placed within a broad context such that a particular want is regarded as constituting a larger plan by which one organizes one's life, the natural wants can be a means to allow one to approach spiritual freedom. In a word, the limitations imposed by the natural wants can be overcome, when such wants are placed in the context within which the various acts are regarded as constituting a series that can be inspected by oneself and others. There must be continuity to the plan by which one lives, in order to make life meaningful. An absurd, irrational life is one whose acts are fragmented, discontinuous or disjointed: duality— double-consciousness (W.E.B. DuBois). Is Sartre right in holding that we have no pre-existing plan by which to organize our lives; that we recreate ourselves through the choices we make? Certainly your axiology provides a coherent basis for organizing one's life. I believe your axiology could have been more effective if you had given sufficient attention to the notion of transcendence. When the rational and the irrational elements are unified in one's consciousness, we have self-esteem, according to Fichte. Self-esteem is an intrinsic value that protects the dignity of the individual. Self-esteem is not something one is born with. It is not a gift. Rather, it has to be earned through striving and struggling. Individual A has to be recognized by individual B and others, and the quest for recognition is a basis of self-consciousness. Because of the legacy of slavery, colonialism and segregation, Africans and the people of African origins have been described as people with low self-esteem. It is anticipated that some of the insights explored here in this letter shall address the self-esteem problem that the people of African origins and people generally face. I can also see how Nietzsche was inspired by Fichte who had much influence on Kant, Hegel, Schopenhauer and Marx. I trust that you are not bothered by my introducing Fichtean ideas into the African American philosophic perspective.

The opposite of striving or struggling is laziness or indolence. I suspect that such behavior is characteristic of the human population generally. Evilness arises when, due to laziness, and passivity, either in thought or action, we fail to act or strive in an effort to remove or embrace the resistance that is causing a particular problem in our lives. We have a tendency to seek the easy way out and avoid exerting the effort. In matters that require reflective thought or imagination, we often revert to clichés or conform to the prevailing "ways of doing things." Dr. Locke, this is similar to your own position regarding the nature of evil when you maintained that evil results from our allowing our lives to be guided by dogmatic creeds or attitudes, false absolutes. False absolutes are a source of inhibitions that can render an action without moral worth or distort us in a manner that we become inefficacious with low self-esteem. Whatever blocks our creative efforts is an evil. Boredom caused by repetitious or monotonous activities is a source of evilness (Whitehead). In matters of actions, we readily resort to a position of passivity or inaction. The framers of the U.S. Constitution were aware of the possibility of an entire nation under a democratic form of government becoming too passive, and as a remedy to this Jefferson made the following observation, it will be remembered. A little over a decade following the Revolutionary War of 1776, Thomas Jefferson raised the following questions: "What country can preserve its liberties if their rulers are not warned from time to time that their people preserve the spirit of *resistance*"[7] [Emphasis added]. He continued: the "tree of liberty must be refreshed from time to time with the blood of patriots & tyrants" (p. 15). What I have attempted to do here is to describe briefly, only in outline, some of the problems, themes and issues that researchers in the field of Destiny Studies may explore.

Destiny Studies offers an alternative to the approaches offered by many Black Studies programs, (including those grounded in Afrocentrism), Ethnic Studies, Gender Studies, Women's Studies, Environmental Studies and related academic disciplines. If many of these disciplines provide us only a fragmented, disjointed world-view, Destiny Studies allow for the diversity of perspectives, and seeks universal unity in knowledge, human consciousness, nature, culture and politics.

Sincerely,

Dr. Johnny Washington, Ph.D.   May 22, 1997

Chapter 6

Destinicity and Social Issues

Martin Luther King, Jr. and Malcolm X

Malcolm X (1925–1965)   Destinicity Letter #52

Re: Martin Luther King, Jr. and Malcolm X

> Terribleness is part of greatness: let us not deceive ourselves…Greatest elevation of the consciousness of strength in man, as he creates the over-man.

—Friedrich Nietzsche, *The Will to Power*

Dear Malcolm X:

In a series of previous pieces of correspondence, I directed letters to Professors W.E.B. DuBois, Alain Locke and William Leo Hansberry, the latter two of whom taught several decades at Howard University. I presume you are knowledgeable about Dr. DuBois. I feel obliged to mention to you the Pan-African movement inasmuch as it seemed to have inspired your own political views. As to the NAACP, it also appears that your own political views arose in part in reaction against it, along with reactions against organizations such as the National Urban League or the Southern Christian Leadership Conference (SCLC). During the 1960s in the minds of many, including yours, these organizations were perceived to be too moderate or conservative in dealing with the issues. It goes without saying that Dr. DuBois influenced a number of people, including one of your critics Dr. Martin Luther King, Jr. Nevertheless, the two of you helped to define the "Black Agenda" in the 1960s, and I believe without you there would not have been a Black Panther Party, about which I shall say more in my next two "letters" to you. I am confident that you would have been certainly interested in my series of letters directed to Professor Hansberry. In that portion of the series I focused much on African history, and I ended by considering the origins of Islamic philosophy, with specific attention devoted to the views of Avicenna and Averroës. You converted to Islam, and you inspired many others to do the same, including the then-boxing champion Muhammad Ali, who is now suffering from Parkinson's disease. You inspired others to adopt African names, or to affix African names to their children. One can find today t-shirts and caps with "X" inscribed on them, honoring your name. There are now a number of Internet sites devoted to your works. I suspect at some point in the future there will be a national Malcolm X holiday. A college in Chicago bears your name. Movies and videos have been made depicting your words and deeds. Parades have been held in your honor. In that set of letters, in echoing your thought, I considered the need to struggle, work, play, fight and love in order to approximate the African Destiny. To the degree I am interested in the African Destiny, I seem to have been a step ahead of thinkers such as Avicenna and Averroës who apparently were more interested in the teachings of Plato or Aristotle than the teachings of traditional African sages (philosophers). In fact Islam regarded African traditional religious values as objectionable. Could it be the case that I am more Afrocentric than was Avicenna or Averroës? I wonder to what degree that during the early years of your career, when you held an aggressive attitude toward all white people, you were aware as to how deeply indebted Islam was to the teachings of Plato and Aristotle? No, I am not claiming that the Prophet Mohammed himself consciously set out to incorporate Greek philosophy into his own teachings. What I am saying is that some of his followers such as Avicenna and Averroës did so. Believe me. Can we accuse them of being Euro-centric? I know this whole discussion regarding who influenced who gets convoluted, and it is often difficult to keep track of the threads of influence, in view of the fact that in my correspondence to Professor William Leo Hansberry, the point was made with compelling evidence that the earlier Greek philosophers were also highly influenced by the African philosophic traditions. In this connection we also pointed out how the Queen of Sheba, Frumentius, the Nine Saints of ancient or Medieval Ethiopia influenced Judaism as well as Christianity. Many hold that Judaism and Christianity had a direct bearing on the thinking of Mohammed, the founder of Islam. Thus, you can see that the threads of influence of the various traditions are interconnected and difficult to sort out. "By any means necessary!" is one of the slogans you coined. This is one of the slogans by which many of us remember you. When I visited Chicago during the Summer of 1964, I saw for the first time many of the members of your religion selling the newspaper *Muhammad Speaks,* on Chicago's west side streets. Often I would get off a city bus (the CTA) on the corner of Madison Street and Pulaski Road, and out of curiosity I would intermittently buy an issue of your paper, of which, I believe, you were the editor. During the times I read your paper in Chicago, your teachings were not popular in the parts of Alabama in which I grew-up. By contrast, in Alabama and in other parts of the south, "Martin Luther King, Jr." was virtually a household name, just as was the name Booker T. Washington a half century earlier.

I was moved, awe-struck, by the power of your rhetoric. However, some of your views I did not agree with. Yet, I admired you for having the knack to appeal to the masses of urban Black youths. Immediately following your death in 1965, Crane Community College on Chicago's west side changed its name to Malcolm X Community College in your honor. Many Black Panther Party members, including Mark Clark and Fred Hampton, who were killed in 1969 by the Chicago police, attended this college, and I often went to some of the meetings that they convened. [I am on the SMSU planning committee for the African American Heritage Month Celebration, and on February 2, 1999, at our invitation, a founder of the Black Panther Party, Bobby Seale, gave an exciting speech here at SMSU. He autographed for me a copy of his book, *Seize the Time* (1968)]. Moreover, Malcolm, when I was in Chicago, I also would on various occasions hear you speak on the local radio or T.V. You were a provocative, powerful speaker that seemed to me to have echoed many of the ideals of Marcus Garvey and Booker T. Washington. Here I wish to make a personal statement to you, Malcolm X. If I had to rank you and Dr. Martin L. King, Jr., on a scale of one to five, with five being the highest point, I would give Dr. Martin L. King, Jr. a higher rating than I would give you. The teachings of Dr. Martin Luther King, Jr. resonate with me better than your teachings. Why this is the case, I can offer no further explanation. I know perhaps you are thinking that I like the teachings of Dr. King better than your own teachings because of the fact that he taught nonviolence. Rather Dr. King gave me the impression that he was the bearer of the Destiny of the world. Ever since my meeting Dr. King or becoming knowledgeable about his works, I have been trying to understand the nature of Destiny. Below is a copy of an article[1] I published focusing my Dr. King experience.

I first heard about Dr. Martin Luther King, Jr. when I was approximately ten years old. This was in the mid-1950s, in Gainesville, Alabama, where I was born and reared. It will be recalled that the year 1954 was when Dr. King initiated his own career, and set the stage for what came to be known as the 1960s Civil Rights Movement. I recall vividly the first time I saw a photograph of Dr. King. He was featured in *Ebony* magazine. The story included one of his Montgomery struggles, in which he was attacked by police dogs, and sheriff's deputies whose faces were contorted by hate and hostility. These formidable images stand out in my mind today. One can imagine the emotional scar, the psyche crater, like the effects of the explosion of TNT in a virgin rainforest, these images left on the mind of a ten-year old such as myself. Televisions and certain other media sources were rare in the Black community back in the early and mid-1950s. There were no forms of sophisticated telecommunication, as we now have. Yet, Dr. King stayed in the public mind. He was the "talk of the town," especially in the Black community, which, then, was not racially integrated. Blacks or "Negroes" as they were then called, spoke quietly and approvingly of Dr. King. When I reached adulthood, Dr. King was still active. In 1964 I completed high school in segregated Alabama, and moved to Chicago that Summer in search of job opportunities. The Summer of 1964, was a turbulent one for Chicago; like other major U.S. cities, Chicago was undergoing social strife and intense racial tension that occasionally erupted in rioting and looting. I arrived in Chicago in time to witness riots, to be part of them. Those were exciting times, to say the least. For that was perhaps the first time I had witnessed such collective enthusiasm among the Black masses. This collective enthusiasm was distinct from the excitement I had experienced as a youth while attending religious revivals or attending homecoming rallies, and football games at the Black high school I attended in Livingston, Alabama. The energy associated with the Chicago riots reflects underlying political currents. This image is different from the above-mentioned one that included the police dogs attacking him. What I call my "Chicago image" of Dr. King is a pleasant one. I know it is odd to speak of an image of a mere "presence," but this seems to characterize my Chicago image of him. Indeed he had a commanding, magnetic presence. When he arrived in Chicago, it seemed as though the city itself became calmer, and peaceful. It seemed when he came on the scene that the city busses and trains came closer to keeping their schedules. As I watched him speak (I don't recall the title of his speech) I received the loving impression that he represented right and truth, goodness and love. My thoughts were at the time that if I make it to Heaven, I already have gotten a glimpse of Its representative by being in the presence of this heaven-sent prophet, whose path I was willing to follow. But as a teenager at the time, I did not become active in the Civil Rights struggle that Dr. King was leading. In fact, I went off to the Vietnam War. This was a moral conflict with which I was faced, whether to go to fight in Vietnam or follow Dr. King. A variation of such a conflict subsequently divided the nation. In any event, I fought with the U.S. First Infantry Division in Vietnam, between 1966 and 1967. (My own adult career started in Fort Leonard Wood, Missouri, where I began basic training on November 1, 1965, and additional training a few months later in Fort Polk, Louisiana). Dr. King taught us how to love. His ideals stood in contrast to those of the Framers of the U.S. Constitution. When they drafted the Constitution, it was made clear that the U.S. would raise and maintain an army, to defend itself.

Thus, the emphasis on violence and aggression is deeply rooted in the history of this nation. If the Framers spoke of violence and aggression, Dr. King by contrast spoke of nothing but love, simple love, combined with justice for all. That was one of the distinguishing features of Dr. King's teaching. Prior to Dr. King, Black Americans gripped by the consequences of slavery and segregation, were in the cellars and back rooms of history. In drawing attention to the cause to which he was dedicated, Blacks proudly ascended the highest peak of the landscape of history. Dr. Martin Luther King, Jr. took us to the mountaintop. Because of Dr. King, the Destiny of African Americans and the nation generally is now in sharper focus. His dream shall never fade away.

I know it might appear awkward in this context in which I am writing you, Malcolm X, a letter in which I am slightly criticizing you and praising Dr. King.
Sincerely,
Dr. Johnny Washington, Ph.D.   May 23, 1997

Letters to Malcolm X: The Black Panther Party

Malcolm X (1925–1965)   Destinicity Letter #53

Re: Elaine Brown, the Black Panther Party and the Black Agenda: Part I

To wait and to prepare oneself; to await the emergence of new sources; to prepare oneself in solitude for strange faces and voices; to wash one's soul ever cleaner from the marketplace dust and the noise of this age; to *overcome* everything.

—Friedrich Nietzsche, *The Will to Power*

212

Dear Malcolm X:

Although I ended my previous letter to you by referring to the teachings of Mr. Booker T. Washington and Dr. W.E.B. DuBois, now allow me to return to the issue at hand and focus on the Black Panther Party that was formed a year following your death and two years preceding the death of Dr. Martin Luther King, Jr. I am addressing this series of letters entitled, "Elaine Brown, the Black Panthers, and the Black Agenda" to you. Because in contrast to Dr. Martin Luther King, Jr., Mr. Booker T. Washington, Dr. W.E.B. DuBois, or Professor Hansberry, you projected a unique, uncompromising sense of militancy during your political career. You were bold. And in the 1980s there was a great deal of interest in your works, especially among African American youths alienated from the mainstream of society. Some scholars hold that the militant ideals you projected inspired today's rap music, a claim made not as a form of criticism of your views, but partly in an attitude of reverence towards your words and deeds. There is evidence that your works inspired the Black Panther Party that is now defunct. Elaine Brown gave a presentation here at Southwest Missouri State University, on February 22, 1996, during which time she also spoke approvingly of your words and deeds. In my African American Studies course, I have had the pleasure of introducing students to your teachings. I usually contrast your works to those of Dr. King. In what follows I will provide you a report of Elaine Brown's presentation.

"I have all the guns and all the money. I can withstand challenges from without and from within. Am I right comrades?" Those were the words that Elaine Brown (1943–) uttered as she assumed command of the Black Panther Party in 1974. She also allows these same revolutionary words to serve as an introduction to her book, *A Taste of Power*.[2] Her abovementioned presentation, whose title was the same as the book's, focused on its theme, which also reflects her own life and the history of the Black Panther Party. Here I will recount the highlights of her presentation. While connected to the Party, Elaine produced a series of musical recordings, *Seize the Time!* that inspired the Panthers and others to action. I shall here cite at length an extract from the cover of her musical recordings, *Seize the Time!*:

> In all societies, the way of life of the people, their culture, mores, customs, etc., evolve from the economic basis of that society. The United States is a capitalist society, the system of capitalism being one of exploitation of man by man, with by-products such as racism, religious chauvinism, sexual chauvinism, and unnatural divisions among the people. In other

words, it's a dog-cat-dog society. But it's not a dog-eat-dog world.

Men are not innately greedy, nor are they innately uncooperative with each other. Therefore, it is our goal, it is the goal of the Black Panther Party, and must be the goal of all men, to create conditions in which men can start being human, can begin to cooperate with each other, can live with each other, in fact, in peace. Men cannot do this without an arena in which to do so. In other words, in an exploitative system men are forced to exploit. In an unkind system, men are forced to be unkind. In a world of inhumanity, men will be inhuman. In a society that is warmongering, men are forced to be unkind in a word, inhumanity will be inhuman. In a society that is warmongering, men will war. These are the aspects or the way of life of a people who are part of a capitalist system.

And songs are a part of the culture of society. Art, in general, is that. Songs, like all art forms, are an expression of the feelings and thoughts, the desires and hopes, and so forth, of a people. They are no more than that. A song cannot change a situation, because a song does not live and breathe. People do.

Formed in 1966, the Black Panther Party, a paramilitary organization, sought to carry out your teachings of self- group-defense to its logical conclusion. Inspired by your slogan: "By any means necessary," Huey P. Newton, a founding member of the Party, urged African Americans and other oppressed people to take up arms to overthrow the "system," that is, the economic-political institutions under which they were oppressed. The Party regarded itself as the vanguard of the pending revolution. Elaine Brown opened her February 22, 1996, presentation by informing the audience that in keeping with Black History Month, she was compelled to place the history of the Party within the broader context of U.S. history. She added: "The Black Panther Party is an expression of the historic struggle of Blacks in this country." In reviewing that history, she began with the year 1861, when the U.S. Civil War erupted. In this context she reminded us of Abraham Lincoln's apparent motivation for issuing the Emancipation Proclamation of 1863.

He did so, Elaine reminded us, not on the basis of any benevolent motive, out of the love for Black people, but "because the North was losing the War, and Lincoln felt that if he freed the slaves, the slave holding states in the South would be weakened, militarily. Thus, he did so out of military necessity, 'to save the Union.'" Further, according to Elaine, following the conclusion of the Civil War, the Black Codes were enacted in the South, sinister laws aimed at protecting a form of Apartheid, disallowing the mixing of Blacks and whites, and degrading the humanity of the former.

214

The sharecropping system, another institution designed to rob Blacks of their humanity, was also organized around the Black Codes. Elaine did not leave the discussion without reverting to the Court case *Plessy vs. Ferguson* (1896) that permitted the "separate-but-equal" policy, whose centennial we recently (1996) witnessed. *Plessy vs. Ferguson* provided the constitutional basis for racist practices and institutions, until the 1954 *Brown vs. the Board of Education* decision in which the separate-but-equal policy was overturned. In seeking to determine whether or not African Americans had benefited from the *Brown vs. the Board of Education* ruling, Elaine Brown reverted to the opinion of African American sociologist, Dr. Kenneth B. Clark, the author of the celebrated book, *Dark Ghetto* (1967). Dr. Clark held that the nation's school system was more segregated in 1967 than it was in 1954. In tracing the "logic" of the U.S. history more closely, Elaine observed: "It wasn't until 1964 that Black people were told that they could enjoy public facilities." She added: "One needs to understand that Blacks have been free only within the past 30 years. The reluctance, she maintained, to allow Blacks to enjoy freedom has to do with the fact that "race as well as economic class is a big problem that this country does not want to face." On the basis of Elaine's view, the negative attitude that members of the larger society harbored regarding the racial identity, along with the various stereotypes, of African Americans, was readily communicated to many of the individuals of the various ethnic groups, including Asians, who immigrated to this country. According to Elaine, many such individuals associated with a particular ethnic group regard themselves as "better" than Blacks. "No group who comes to this country wants to be identified with Blacks," Elaine declared. Because of the history of slavery, its consequences and concomitant sufferings of African Americans, the larger society ought to express some form of atonement, according to which the [African] holocaust that Blacks suffered is publicly recognized. Elaine reminded us that in Washington, D.C., one finds monuments of George Washington, Thomas Jefferson and others, and the various war memorials. By contrast, there is no monument that attests to the sufferings that African Americans have undergone in this country. Nor have the cultural and economic contributions that they made to this country been fully appreciated, in the form of a memorial. At this point in her presentation, Elaine focused on certain African American individuals who were regarded by many as heroes or heroines. In her view, General Colin Powell was one of these whose place in history is problematic. She maintains that a problem with General Colin Powell is that he was the commander responsible for the invasion of a Third World nation, Iraq. She jokingly told the audience that Saddam Hussein never called her "nigger." Oprah Winfrey is another African American projected as a model representative not only of African Americans but American generally, according to Elaine.

However, Oprah, like Colin Powell, needed to become mindful of the fact that without the struggles of the previous generations of African Americans, she would not have been able to achieve the level of success that she now enjoys. At this point Elaine went on to speak briefly about the founding principles and strategies of the Black Panther Party. As we saw, many of the Party's principles were reflected in the cover of her musical recordings, *Seize the Time!* In her view, Huey P. Newton, during the aborning stage of the Party, said, "we are going to get guns to patrol the police, and protect our community from police brutality." Elaine stated that inasmuch as the Party regarded itself as the vanguard of the revolution, it was to set an agenda that included building coalitions among other oppressed groups. Among those with whom the Party established allies were: women, gays, Asian Americans, Hispanic Americans, and all other groups that were alienated from the American society. Other agenda items included providing: i) housing for poor people; ii) school breakfast programs; iii) and, testing or seeking a cure for sickle cell anemia. She said that the significance of the Party was that it did in fact accomplish those tasks and inspired others to continue to support such efforts. For example, the school breakfast program was subsequently adopted by the federal government. And since the times of the Panthers, more research has been conducted on sickle cell disease that mainly effects the people of African origins.

Sincerely,

Dr. Johnny Washington, Ph.D.   May 23, 1997

Malcolm X (1925–1965)   Destinicity Letter #54

Re: Elaine Brown, the Black Panther Party, and the Black Agenda: Part II

> In times of painful tension and vulnerability, choose war: it hardens, it produces muscles.
>
> —Friedrich Nietzsche, *The Will to Power*

Dear Malcolm X:

This is a continuation of my reporting on the February 22, 1996, presentation of Elaine Brown. Grounded in Marxism, Humanism, Pan-Africanism, the teachings of Marcus Garvey and your own teachings, the philosophy of the Panthers was to establish and implement a revolutionary agenda. Elaine Brown de-emphasized the role of violence that has been associated with the Panthers. She reminded the audience that she, as well as most of the other Panthers, was well aware of the fact that the few handguns and shotguns that they possessed were merely to project a "revolutionary image." They knew that they had no chance of seriously confronting the city police force or the National Guard. In this context, she described the significance of a dramatic event when the Panther Party "paraded" into the California Legislature with shotguns and handguns to protest the legislative effort to rescind the gun law. (At the time, the prevailing law permitted the carrying, openly, of handguns, a law designed to allow people to enjoy hunting. When the Panthers took advantage of that law and began patrolling the community with guns, the California Legislative body sought to repeal that law). Elaine informed the audience that following their "invading" the California Legislative body, a number of Black Panther Party chapters were established around the country, including Chicago and New York. During the questions and answers period, the first questioner asked Elaine to express her opinion about the movie, "Panthers." Elaine's reply was that it was adequate. She added that it could have been much better, had the producers and director not been so concerned with trying to please "Hollywood, and the demands of the marketplace." Another participant in the audience questioned her as to whether she thought that a Black Panther Party today would be needed or desirable? To which her reply was: "It is not so much that we need a Black Panther Party today. Rather what we need is a Black Agenda similar to the one established by the Panthers. The primary focus, however, of today's agenda should be on education, one of the most desirable, effective means to produce social transformation." In an effort to implement aspects of the Black Agenda, Elaine has embarked on an ambitious fund raising campaign. Her dream is to build a school that will enroll youths between the ages of two and fourteen. Its curriculum will include not only the traditional subject areas, but also business, music and art and serve as a sports center that focuses on ice hockey.
Sincerely,
Dr. Johnny Washington, Ph.D.   May 26, 1997

Racial Insanity

Dr. Alain Locke (1886–1954)   Destinicity Letter #55

Re: Racial Insanity

> *Not* to make men 'better,' *not* to preach morality to them in any form, as if 'morality in itself,' or any ideal kind of man, were given; but to *create conditions* that *require stronger men* who for their part need, and consequently will *have*, a morality (more clearly: a physical-spiritual discipline) *that makes them strong!* Not to allow oneself to be misled by blue eyes or heaving bosoms: *greatness of soul has nothing romantic about it.* And unfortunately nothing at all amiable.

> —Friedrich Nietzsche, *The Will to Power*

Dear Dr. Locke:

In this letter to you I am revisiting one of your unpublished papers entitled, "Insanity" included in your files at the Moorland-Spingarn Research Center at Howard University, where you taught between 1912 and 1953. I examined your essay, "Insanity" in my book, *A Journey*. In view of the recent acts of mass killing of human beings that occurred in the former Yugoslavia as well as Rwanda, Zaire and other parts of the world, it seems appropriate briefly to remind people of your views that may shed light on the situation.

In the U.S. the term "ethnic minority" is often used in describing such groups as: African Americans, Asian Americans, Hispanic Americans and Native Americans who suffered economic oppression. Invariably, in the past, and to a large degree today, such groups were treated as mere laborers, to be exploited to satisfy the needs of individuals and firms that controlled our capitalistic economy. The roots of such oppression in the Americas extend to 1492, when Columbus "discovered" the Americas, and Africans were subsequently brought to the "New World" as slaves. Thus, the ethnic minorities in question were denied adequate access to the institutions, offices, goods and services available to members of the larger society. They were—and to a large degree still are—excluded from the "norms" of the larger society, for it is the system of norms constituted by the larger society that bestows psyche health and wholeness, in the absence of which there lies what Dr. W.E.B. DuBois called a double-consciousness. In your view, the exclusion of Blacks and other ethnic minority groups from the mainstream of American life, from what may be called the "common good" can not only result into a double-consciousness, but also racial or ethnic insanity on the part of those who were excluded. If DuBois in his book, *The Souls*, described psyche disunity as it pertained to the isolated individual, you described it as if it pertained to certain social groups, of which African Americans are representative. You observed that the practice of disallowing one group from participating in the normal activities of a society often resulted in what may be called "minority group insanity," a form of "minority group neurosis." You described such neurosis as "a cycle divided between an historically conditioned 'depression phase' of submissiveness, sense of inferiority and helplessness, nervously exaggerated frustration and a manic phase of over-assertiveness, belligerence and sadistic seizures of power" (cited in my book, *A Journey*, p. 131). A more thorough analysis of this type of covert behavior may shed light on many of the social problems, including drive-by shootings, and other violent crimes and drug abuse practices, that are pervasive in the African American community today. If your analysis is correct, and I think that it is, the above-mentioned violent practices raise questions as to whether we have a right to blame the victims entirely. Who is to blame, society, or the perpetrators of the crimes? As I have already suggested your analysis also illuminates the "ethnic cleansing" problem in Eastern Europe and in Zaire.

What did you offer as remedies to this psychological problem brought on by ethnic oppression that disallows the members of the oppressed minority group from participating in the activities of the larger society? First, you suggested that the larger society needs to undergo structural rearrangement so as make its institutions open to all.

Is Affirmative Action the appropriate mechanism through which we are to achieve this inclusiveness? Once society becomes more inclusive, minority groups will be subjected to the same standard of "normalcy" as that which members of the larger society subscribe, and by which "objective" reality is constituted. Second, you reminded us that a given minority group itself may be the source of its own solution to the insanity problem: that among members of the oppressed minority, a "genius" or "social reformer" tends to emerge, who sets new "standards of normalcy" to which the entire society, minority as well as majority members, are urged to subscribe.

I maintain that Dr. Martin Luther King, Jr. and to a lesser degree Malcolm X, were such social reformers. Dr. King induced society to restructure itself, to open new channels, so as to allow sounder elements of sanity and sobriety to flow, having as their source the inexhaustible reservoir of love, grounded in the universal love.

You will observe that I began many of my letters in the series with passages extracted from the works of Nietzsche. In so doing, I believe I am in harmony with your own spirit, because you, yourself, also held him in the highest regard. You went so far as to regard Nietzsche as the father of modern value theory.

Traces of Nietzsche's principle of the elite, as well as the will to power, is found in your notion of the Talented Tenth. The same is true of your value theory generally. I suspect that Nietzsche had a similar influence on Dr. W.E.B. DuBois. Nietzsche's notion of the will to power can also be traced to Fichte, who influenced Kant, Hegel, Marx and others. Moreover, I should add that Nietzsche's will-to-power principle is comparable to my notion of Destinicity.

World Destiny, together with the principle of transcendent unity, makes it possible for all ethnic groups to approximate the ideal of transcendent harmony of harmonies through working and struggling in a concerted manner.

Sincerely,

Dr. Johnny Washington, Ph.D.   May 30, 1997

[PS. I recently gathered that a new Black Panther Party is now seeking to establish itself].

Ebonics vs. "Standard" English

Dr. Alain Locke (1886–1954)   Destinicity Letter #56

Re: Ethnic Identity

> You must know, then, that there are two kinds of fighting: one, with the laws; the other, with force. The first is proper to man, the second to beasts, but because the first is very often not sufficient, it is appropriate to have recourse to the second. Therefore, it is necessary for a prince to know how to make good use of the beast…
>
> —Niccolo Machiavelli, *The Prince*

Dear Dr. Locke:

Already in previous letters to you, or Dr. W.E.B. DuBois (1868–1963), I reminded readers that both you and Dr. DuBois were influenced by the views of Josiah Royce (1855–1916) and certain other Harvard University professors who had a great appreciation for German Idealism, especially the works of Hegel (1770–1831). This Hegelian influence in DuBois' thought is quite evident in his book, *The Souls*. In it he attributed to African Americans what he called a double-consciousness that has eclipsed their true Black identity. This double-consciousness is reflected in their dual striving, a principle that is prominent in both Fichte's and Hegel's respective views. African Americans are torn between striving on the one hand to adopt the American ideals, and on the other to adopt the norms in keeping with their African American identity, shaped largely by slavery and segregation.

During Fall 1996 the *Ebonics* debate was quite heated throughout this country, but it has subsided somewhat now. Some of the articles published in the *Palm Beach Gazette* explored the possible response Dr. DuBois or other African American scholars might have had towards *Ebonics*, but I did not see any article(s) seeking to determine the possible reactions you may have taken towards the *Ebonics* controversy. I suspect that you, like DuBois, would have found it acceptable, just as the two of you found acceptable the musical forms such as the blues, spirituals and other modes of artistic expressions in the African American folk tradition. Dr. Locke, it will be remembered that you and Dr. DuBois enjoyed similar social status and educational background. Both you and DuBois subscribed to the principle of elitism, which, it appears to me, was a variation of the teachings of Nietzsche's will to power. Both you and Dr. DuBois advocated the leadership of the Talented Tenth, a view informed by the teachings of Nietzsche. The Talented Tenth model, being a form of elitism, was at odds with the norms of the "average person's" out of which arose *Ebonics*. You, yourself, rarely, if ever, reverted to *Ebonics* in your writings. I certainly believe you would have accepted certain aspects of it. You often contrasted the folk tradition with the genteel tradition, and you prized the former as the main source of creativity and originality. Perhaps you, Dr. Locke, would have attempted to straddle the *Ebonics* debate. I believe you would have said that, on the one hand, as an assertion of the element of the African American identity or creativity it is important. On the other hand, inasmuch as African Americans as a group were members of the larger society with which they interact, the mastery of the so-called standard English is also important. Nevertheless many parents and educators are uncertain as to whether they should adopt curricula based on Afrocentrism, Eurocentrism or multi-culturalism. Opponents of *Ebonics* hold that it is a form of racial separation, which in itself is objectionable, they maintain. That if youths are not allowed to be educated in "standard" English, they will not be able to compete in the market place with their white counterparts who will have been taught the "correct" form of English, a position with which you would find agreeable. Those in favor of *Ebonics* hold that if teachers are allowed to be informed about *Ebonics,* they will be in a better position to impart knowledge to students. A theoretic basis of *Ebonics* seems to lie in Afrocentrism which is foreshadowed in the works of DuBois. The same is true with regards to your own works, especially your own book, *The New Negro*, and the views of Marcus Garvey and others. Aimé Césaire, one of the founders of the *Negritude* movement of the 1930s, acknowledged that he was indebted to your teachings that in part inspired the Harlem Renaissance movement of the 1920s. The compelling evidence indicates that the New Negro as well as the *Negritude* movement in part is a tradition out of which arose the *Ebonics*. Often when a certain cultural phenomenon arises within a subculture, it can initially appear threatening to the norms of the dominant culture. However, if such seemingly threatening phenomenon persists, it can become acceptable to the dominant group. This is true of many art forms. With the advent of the Black Theology of Liberation movement of the 1960s, Joseph R. Washington held that Black religion needs to reject its form of liberation theology and rather adopt the theology of mainstream religion. When the spirituals as a form of music were first sung in the primitive form, that is, in the folk music tradition out of which they arose as Blacks slaves struggling in the cotton fields, this form of music that in many ways conform to the *Ebonics* syntax, were disallowed by the slave masters, to whom the beating of drums were perceived as a threat. Ironically, today some of the greatest rap artists are white males who are skillfully imitating Blacks. I know of no white female rap artists. In many instances in the South, whites organized systematically to route such "race music," whose origins, I venture to claim, is similar to those of *Ebonics*. Such origins are anchored in the unique experience of the African American, shaped by oppression whose antecedent is slavery, sharecropping or segregation. When in the mid-1950s Elvis Presley "adopted" R&B, this new musical form became acceptable to white people generally. When the whites became interested in this music, its name was changed to "rock and roll," but its syntax and rhythmic structure re-

222

mained virtually the same. Now almost all contemporary music—secular as well as religious—is influenced by the African American field songs, work songs, and convict songs, blues, jazz and rhythm and blues. Dr. Locke you will recall that in the book, *The Souls,* DuBois suggested that the spiritual in part enabled African Americans to withstand, if not condone, the double-consciousness that they suffered. Such songs were one of the spiritual gifts, transcendent gifts, of hope and of freedom, offered to the nation and the world. DuBois wrote in the last paragraph of his book, *The Souls* the following: "Even so is hope that sang in the songs of my fathers well sung. If somewhere in this whirl and chaos of things there dwells Eternal Good, pitiful yet masterful, then anon in His good time America shall rend the Veil and the prisoned shall go free. Free, free as the sunshine trickling down the morning into these high widows of mine, free as yonder fresh young voices welling up to me from the caverns of brick and mortar below—swelling with song, instinct with life, tremulous treble and darkening bass. My children, my little children, are singing to the sunshine, and thus they sing…" (*The Souls,* p. 193)

Sincerely,

Dr. Johnny Washington, Ph.D.   February 15, 1998

The 1995 National Urban League Conference

Dr. Alain Locke (1886–1954)   Destinicity Letter #57

Re: Urban League's 1995 Annual Conference:

> A prudent lord, therefore, cannot and must not keep faith when this is to his disadvantage and the reasons that made him promise to do so no longer subsist.

> —Niccolo Machiavelli, *The Prince*

Dear Dr. Locke:

I will make this letter to you brief because I have accepted an invitation to attend the 11th Annual banquet of the Springfield, Missouri Chapter of the National Association of the Advancement of Colored People (NAACP). The banquet will be at 6:30 p.m. today, October 4, 1997; the tickets are $25.00 each. The banquet will be held a few blocks from Southwest Missouri State University campus at the University Plaza Convention Center. Thus, I am in a bit of a hurry; it is now 4:30 p.m. I need to begin walking towards my car. I have no idea as to whom the keynote speaker will be. I had intended to tell you earlier but it slipped my mind about a July 23–26, 1995, national conference of the National Urban League (NUL) that convened at the Miami Beach Convention Center, Miami Beach, Florida. I know you as well as Dr. W.E.B. DuBois would have been interested in such a conference, whose results I will now recount.

Dr. DuBois was one of the founding members of the NAACP. I believe your ties were greater to the Urban League than to the NAACP; nevertheless, you supported both. You published many articles in the NUL's magazine *Opportunity* that is now defunct. By contrast, the NAACP's magazine, *Crisis,* in which you also published articles, is still viable. I understand that the NAACP's membership a few years back reached as high as some 500,000 members. I am uncertain as to the exact membership of the National Urban League, but it, too, is doing well. Both organizations are instrumental in assisting African Americans in establishing a stronger cultural-political base in this civilization. Society's efforts to roll back the Affirmative Action initiatives have posed a major threat for both organizations and to the African American community generally. "Our Children = Our Destiny" was the theme of the annual 1995 Urban League Conference. When I first gathered that the Conference's theme was "Our Children = Our Destiny" I was especially intrigued by the "Our Destiny" part of the equation. Because for a number of years, I have been exploring the concept "Destiny." I was disappointed to determine on the basis of what I observed that no one at the NUL conference acknowledged the role I played in pioneering research in the Destiny model. Nevertheless, I anticipated with great pleasure attending the Conference because I was interested in seeing how the NUL might elucidate its Destiny model. When I arrived at the Conference on July 23, I was surprised to see how frequently "Destiny" was used with much enthusiasm by many of the presenters. However, no one provided a clear definition of "Destiny." Rather most of the presenters focused on the "Our Children" side of the equation. As was indicated in the Conference's program, most of the sessions and workshops dealt with issues associated with children: i) "Parents: the First Teachers"; ii) "Voices and Visions of Our Youths"; iii) "Impacting Youths: Program Models for Success"; and iv) "The Politics of Youth Programs"—these were the titles of the workshops that convened during July 24. There was no particular work-shop devoted to examining the concept of Destiny. I feel that was one of the shortcomings of the Conference. Dr. Locke, I believe Ossie Davis and Ruby Dee were on the scene during your times. In any event, in addition to Ossie Davis, among other illustrious participants of the Conference under discussion were Hugh B. Price, Myrlie B. Evers-Williams, Dr. Henry W. Foster, Jr., The Honorable Newt Gingrich, Patti LaBelle and Oprah Win-frey. The latter-mentioned is a prominent, wealthy TV talk-show hostess. You would be surprised to see the level of progress African American women have made during the past 40 years. Oprah is the paradigm of the successful African American female, just as General Colin Powell is pro-jected as a model of the successful African American male, although his parents are from the Caribbean. Many whites hold General Powell and Oprah in the highest regard, or so it seems. Many people seem to be am-bivalent towards Jesse Jackson, whose son is now a rising star in the politi-cal arena. I had my camera with me and I happened to have had the chance to get a few close-up pictures of Oprah during her presence at the NUL conference, although it was difficult to do so due to the size of the crowd of people.

Many vendors were present at the NUL conference with a variety of items on display for sale. The NUL had its own showcase and vendors. Among the NUL items for sale was a postcard, designed by artist Pat Richardson, depicting the "Our Children = Our Destiny" theme. I arrived at the Conference on July 23, in time to attend the Ruth Standish Baldwin Volunteer Recognition Luncheon, where Dr. Henry W. Foster, Jr. was the keynote speaker. His presentation, however, was preceded by Ossie Davis' presentation. He introduced Dr. Foster. Ossie began his remarks by raising a rhetorical question, "Not only are we to ask how are we to prepare our children for the future, but also how are we to prepare our future for the New World Order?" He went on to describe his New World Order to be one whose features are interconnected by the telecommunication super highway, that is, the Internet. Davis urged the conference also to focus on the issue of employment. He said that, owing to the rapidity of technologi-cal growth, we are to guard against creating a society in which the gap be-tween the rich and poor becomes greater. Dr. Foster held that he was con-cerned about the plight of youths. The environment of many is pervaded with violence. He held that as a physician, he wanted to approach the prob-lem by writing "prescriptions" for the Black community, which he regarded as the "patient." His first prescription enjoins that each Black citizen is to continue supporting the NUL.

That we must reclaim or re-appropriate the moral and cultural values of the Black community. That we no longer need to think of politics and politicians. Rather we need to think of policies and programs. Dr. Foster made reference to the negative impact that the reduction of Medicare and Medicaid will have on the Black community. His final "prescription" enjoined that youths be provided a sense of hope and self-esteem, an issue I have explored in some of my previous letters. Dr. Locke, when I heard Dr. Foster employ the "prescription" metaphor, I was reminded of some of your writings in which you introduced the prescription metaphor in approaching many of ills, including race problems that pervaded our publics schools.

On the evening of July 23, Hugh B. Price, the NUL president, gave his keynote address that was quite lengthy. At the outset he reminded the participants that the elimination of Affirmative Action was evident of the sea changes we are currently witnessing in U.S. politics. He went on to add that he stood before the audience to talk about the need to acquire a new vision and its concomitant values, in order to deal with the current political changes that are sweeping the U.S. political landscape. On the basis of Price's outlook, what is unique in the current situation is that in the past the political landscape attempted to be accountable in responding to the needs and aspirations of working people. But with the current changes, it had reduced its capacity to do so.

Rather, among the countries in the Western world, the economic gap between the rich and poor is the widest in the U.S. Thus the time has now arrived to make the U.S. live up to its economic ideals, so that all citizens, including women, African Americans and other minorities can fulfill their potential, the removal of Affirmative Action notwithstanding. He indicated that the starting point must be the inner cities, which we must rebuild. Educational opportunities needed to be enlarged. Youths now living in inner cities must come to know "that Affirmative Action, which means inclusion, is not about quotas, but about inclusion of the previously excluded." Price stressed the relation between "inclusion" and Affirmative Action, adding that "inclusion is morally right." He criticized California's then-Governor Pete Wilson for seeking to dismantle Affirmative Action practices.

On July 26, I attended the NUL workshop entitled, "The Future, Our Children and Technology." The moderator of this session was John W. Mack, president of the Los Angeles Urban League chapter. Jacquelyn B.

Gates, vice president of Ethics, NYNEX Corporation of New Jersey; Jose Montes de Oca, president and chief executive operating officer of LatinoNet, San Francisco; Anthony T. Riddle, executive director for Minneapolis Telecommunication Network, Inc., Minneapolis; and Gary Mendez, president of the National Trust for Development of African-American Men, Washington, D.C.—these were the panelists of that session. John Mack said that the information age represented the new revolution that will have an impact on various aspects of our lives, including jobs and education. One finds similar comments reflected in your own writings, Dr. Locke. Thus, I believe you would have appreciated this session, especially. Mack urged that the challenge facing the NUL is to make sure that African Americans are not hitchhikers on the technological "super highway." Rather, they should remain on the main path of the New Age.

Gates informed the audience about ways in which the super highway, the Internet, enabled students from different and diverse backgrounds, rich as well as poor, to enjoy parity among themselves and to play a greater role in educating themselves. She was optimistic about the voice mail service practices by which students can give and receive messages concerning homework assignments. Still she was optimistic that this new technology will also assist parents in monitoring the progress of the school performance of their children.

Riddle linked the information super highway experiences and potentialities to political realities, reminding us that, like our African American ancestors during the days of slavery, one has to fight for the capacity to communicate "to be free is to have the capacity to communicate," he said. To this I wish to add: "To be free is to have not only the capacity to communicate but also to act." What youths need is love, nurtured in close ties, and in interpersonal interactions. Technology tends to remove us from the interpersonal realm. That is a danger, he maintained. The crowning point of the conference was the appearance of Oprah Winfrey at the Whitney M. Young, Jr. conference dinner that I also had the pleasure of attending. I do not know how familiar you were with Miami Beach where the dinner was held in the Fontainebleau Hilton Hotel. "We are spiritual beings locked into our bodies to learn to love or fear," is the statement that Oprah expressed in opening her presentation. She added, "Our mission, our destiny is to learn how to allow the love element in our nature to prevail." Earlier in her presentation she attempted to satisfy the curiosity of her audience by, jokingly, stating that she had changed her hairstyle, to overcome the "problem" that many Black women suffer in dealing with their hair. On a more serious note, she focused her presentation on the issue of teenage murders: "kids are killing kids, because we do not have the time to look after them [children]," she reminded us.

The security guards protecting Oprah were many. The taking of photographs were discouraged. However, I did manage to get a few opportunities to snap her pictures. Her wealth is estimated to be over one billion dollars. The highlights of Oprah's presentation were reflected in this adage: "When you learn, teach; when you get, give." I trust that the some 4,000 individuals who attended the NUL conference benefited.

Sincerely,

Dr. Johnny Washington, Ph.D.   October 4, 1997

PS. I have some observations regarding a previous NUL event I attended. Let me share them with you. The Urban League of Palm Beach County convened its 17th Annual Equal Opportunity Day Banquet during April 4, 1992. It was held at the Palm Beach Gardens Marriott Hotel in Palm Beach Gardens, Florida. I was delighted to have had the opportunity to attend. The keynote speaker for the occasion was Attorney Willie Edward Gary, who gave an inspiring address that echoed the theme "Making a Difference in the 90s." The food was delicious.

The menu included: *Seasoned Fresh Fruit Coupe; Traditional Spinach Salad; Cornish Game Hen with Blended Wild Rice served with Honey Almond Sauce; Seasoned Fresh Vegetables; Chocolate Forte; Rolls and Butter; and Freshly Brewed Coffee Decaf, and Tea.*

Natural Disasters: The Boomerang Effect

Dr. Alain Locke (1886–1954)   Destinicity Letter #58

Re: Natural Disasters

> [I]n order to maintain the state, a prince is often forced not to be good, because, when the community—whether it be of the people,…that he judges to be in need of in order to maintain himself is corrupt, it is appropriate for him to follow its temperament in order to satisfy it, and then good works are inimical.

> —Niccolo Machiavelli, *The Prince*

Dear Dr. Locke:

Since time immemorial societies have struggled with issues of social justice which include providing assistance to those in need. The teachings of Buddha, Confucius, Mahavira, (Jainism), Zoroastrianism, Socrates, Jesus, Paul, Mohammed as well as Dr. Martin Luther King, Jr. reflected the social justice ideals. Such ideals have been a main focus of African American religions since the arrival of the first slaves. Inasmuch as the Destiny model initially arose out of my reflecting on the African American's quest for social justice, there is an inherent affinity between African American religion or liberation theology and the Destiny model. In contemporary society social justice issues include providing equal opportunity in the work world, subscribing to nondiscrimination principles on the basis of race, sex or religion, etc., and protecting a number of other civil rights—these are reflected in such ideals. As you were well aware, oppressed people such as African Americans are accustomed to monitoring the society's social justice practices. Some of the greatest moral reformers this country has known arose out of the African American community, including Sojourner Truth, Frederick Douglass, Dr. Martin Luther King, Jr. and Malcolm X. Among the institutions associated in the African American community that focus on justice and fairness, and in securing economic opportunities for the oppressed are the NAACP and the National Urban League (NUL) and the Black church. Many believed that in the Judaic-Christian tradition, as well as in those of others, there was a direct connection between social justice and catastrophic events. In ancient China it was believed that if the ruler were in a protracted poor mood, the people could anticipate a bad harvest or some other natural disaster. Similarly, the Hebrew prophets such as Amos and Hosea were of the conviction that when God's people tend to go astray, He punished them by causing catastrophic events to occur in their lives, floods, droughts, famine, or defeat in battle. Dr. Locke, you, like Dr. DuBois, did not appear to have been a great religious believer, so I do not know how far you wish to go with me in making this connection between catastrophic events and social justice. Yet, you spent most of your adult career advocating social justice on behalf of the oppressed. I personally believe that there is a ring of truth to the social justice/catastrophic hypothesis. I almost feel like apologizing to you for my subscribing to such beliefs, but I do not believe that that is necessary.

I was in the "event horizon" of Hurricane Andrew that terrified or devastated south Florida on August 24, 1992. It is not the case that I seek out natural disasters. In fact the very thought of an act of God such as a hurricane or a flood, or an earthquake evokes in me great fear and anxiety. Yet, it so happened that I already had planned to visit Springfield, Missouri and other parts of the Midwest before the 1993 flooding of the Missouri and Mississippi rivers. I spent much of July 1993 in Missouri. During July 10–12, I visited Kansas City, Missouri and Leavenworth, Kansas. While in Leavenworth, I had the opportunity to stand on the bank of the Missouri River and observe its intimidating force. As it flowed through Leavenworth, among its debris were dead trees, rotten fish, old cars, parts of houses, and garages and washing machines. It was an awesome scene to witness. I stood on the bank of the mighty Missouri, watching it swell. It formed vortexes and quietly devoured everything in its path and altered the landscape. During my brief pause on the Missouri River, I turned over in my mind the question: "Why would God permit such destruction?" [During early Spring 1998 a series of tornadoes ripped through the Orlando area; last week several tornadoes turned inside out a community in Birmingham, Alabama; and on April 16, 1998 several tornadoes made the city of Nashville reel and rock, killing some 11 people and ripped through that city on its way to the Carolinas and up the coast of the north Eastern parts of the U.S. On July 20, 1998, a huge tidal wave swept through Papua, New Guinea, killing some 1,200 people, and many thousands were missing. During mid-July 1998, in the U.S. a huge heat wave radiated through the U.S. South, where some 117 hundred people died due to heat related factors]. Some say that such bad weather patterns are the *El Nino* effect. Since time immemorial, floods, hurricanes, tornadoes and droughts have threatened the security of human beings. Certain authorities hold that the concept of God (or god) arose in prehistoric times as a result of people seeking ways to defend themselves from natural disasters. Our ancestors looked to the gods for protection. It seems that during the times of wars, social strife and natural disasters, people seem to develop a greater interest in religion.

The exact relation between God and natural disasters is uncertain, it must be admitted. In various aspects of the Judaic-Christian tradition, the point is made that when God gets angry with the people living sinful lives, he expresses it by causing or allowing bad things to happen to them. Hurricane Andrew, many people argued, was a case in point. This probably was a way in which god chastised the South Florida area. Others maintain that the Flood of 1993 or the mere consequences of tornadoes associated with the *El Nino* effect were the way God showed His dissatisfaction with the people affected, those whose homes were blown away or who had love ones killed or injured by the storms. It is difficult to reconcile these last two sentences with the view that God is all love. The attempt to reconcile the conflicting claims led to a philosophical-theological problem similar to one I raised at the outset: why would an all-powerful God permit or cause mass suffering (evil)? This is mere conjecture but it may be the case that God permits suffering such as the types described here to make us mindful of our Destiny. Besides religious sources, there is no way to confirm or deny that there is a relation between the will of God and natural disasters. Let us assume that the occurrence of a natural disaster such as a flood or a hurricane that a people suffer is a sign of their sinfulness. Does it follow that the extent, intensity or magnitude of such disasters is proportional to the sinfulness of the people in question? During prehistoric times and onward, natural disasters were a basis for myths such as the biblical Great Flood myth. In science we have the myth associated with the dying off of dinosaurs that is believed to have occurred about 60 million years ago, during which a great meteorite is said to have smashed in the area in and around Merida, Mexico. Like the flood against which Noah struggled, the Mississippi River flood of 1993 left its marks; the Noah Flood has religious significance. It is reported that the 1993 Mississippi River flood killed some 48 people and caused over 12 billion dollars in damages throughout the region. There are various interpretations of the moral qualities associated with the biblical God. As was suggested, one view is that He is the God of love. Another is that He is a Being who seeks revenge and shall punish or cause harm to those who are disobedient. A version of Hinduism appeals to the notion of the law of karma to explain suffering or what may appear as undeserved suffering. The law of karma holds that our words and deeds, both good and bad ones, are recorded in what may be called the cosmic registry. All of our actions are stored in a universal memory. Thus, the bad or good actions that one performed in this life as well as in a previous life, by virtue of the law of karma, will effect the manner in which one enjoys or suffers in this life. This is similar to what I have called the boomerang principle, according to which the good or bad deeds that one performs will return to the one who performed the deeds. Some people hold that the natural disasters that are being experienced on a regular basis worldwide are not due so much to our deeds, customarily regarded as sinful, that God deemed offensive. Rather, the natural disasters are caused by our abusing the natural resources; for example, our polluting the lakes and streams and destroying the ozone layer. The consequences that we let loose in nature are irreversible (Hannah Arendt).

Here is my comment on the nature of what may be called moral, in contrast to physical, evil. When a given group or individual's immanent Destiny fails to be compatible with transcendent Destiny, the outcome is strife within and among the human groupings. Dr. Locke, you were of the Baha'i faith, and one wonders what light would Baha'ism shed on the issue I am raising here of the connections between God, social justice and natural disasters. In certain instances, we may take only a few minutes to perform an action that adversely affects the environment; it could take hundreds or thousands of years to undo the negative effects of such an action.

Thus my claim is that the planet is undergoing a climatic imbalance because of the negative effect that we have produced. For example on August 9, 1993, in Guam an 8.1 earthquake caused the grounds to rumble and shake. On August 8, 1993, Tropical Storm Bret devastated the Caribbean and caused great floods and landslides in Caracas, Venezuela, killing some 100 people. On August 14, 1993 it was reported that a hurricane-like weather pattern was threatening to shatter the island of Puerto Rico. If these disasters are evident of God's anger, His anger seems to be worldwide. Environmental issues, during your times, were not on the forefront of our nation's conscience as they now are. However, I do believe that you would be sympathetic to many of the concerns I am here raising. You were a pacifist. Whether or not you were a vegetarian I am uncertain. There are principles in your axiology that are a basis for an environmental ethics. In fact in the introduction chapter of my book, *A Journey* I relied on your axiology to offer what I regarded as an outline for what I called "landscape ethics" in which I linked the notion of Destiny with the natural environment. My aim here in this present letter is to make a connection to the immanent/transcendent Destiny distinction, suggesting here that insofar as the realm of immanence pertains to the natural as well as the social environment, there is a need to link environmental studies (or ethics) to Destiny Studies.

Space will not permit me to dwell on it here, but it merits mentioning that considering these dreadful weather patterns or threatening natural events have great importance for the aborning science called Chaos Theory in which chaos or randomness that a tornado or a hurricane harbors is a main focus. Some researchers hold that similar chaos is found in the processes of evolution. Moreover, it maintains that if we abuse the natural envi-

232

ronment, this "negative karma" will be returned to us in the form of floods, droughts or hurricanes.

Sincerely,

Dr. Johnny Washington, Ph.D.   April 17, 1998

Religion *vs.* Evolution

Dr. Alain Locke (1886–1954)   Destinicity Letter #59

Re: Religion

> [T]he princes, not being able to fail to be hated by someone, must first endeavor not to be hated by the community, and when they cannot contrive this, they must strive with all their industry to avoid the hatred of those communities that are most powerful.

> —Niccolo Machiavelli, *The Prince*

Dear Dr. Locke:

In my latest letter to you, in which I dwelt on issues such as floods and hurricanes and other natural disasters, I probably did not leave you in good spirits. These are not pleasant topics. My last letter was not an inspiring one. It does, however, reflect the nature of reality, although the dark, negative side of reality. In that letter I suggested, if I did not explicitly state, that religion is important because it enables us to have the courage to face the negative moments of reality. Often when I attempt to give a public presentation on my views of Destiny, I am interrupted after I have spoken about 20 minutes, with the question: "How does your view of Destiny fit in with religion or God?" I have been asked such a question by Jews, Muslims, Christians, as well as by those in various other religious traditions. In response to this question I usually at the outset revert to my definition of Destiny, as an ideal towards which a people strives in successive generations.

The big bang theory, on which modern cosmology hinges, goes deeper than the theory of biological evolution offered by Charles Darwin. The big bang theory, like certain religious myths offered in attempting to explain the beginning or ending of the world, attempts to explain both the origin and

Destiny of the universe. That the universe originated in a point of singularity between 15 and 20 billion years ago, and that at some point in the distant future, the universe will return to the point of singularity, in the form of a big crunch. When it approaches its point of demise, it will begin to exhaust all of its energy and collapse onto itself, as in the case of the collapsing of a star, which, during its final phase, will be overcome through the exertion of gravity. Then the process will repeat itself: first, the big bang, followed by the emergence of the lighter chemical elements such as hydrogen and helium; next, simple and complex molecules; then galaxies, planets and possibly life. Some scholars hold that there is a conflict between the big bang myth and the creation myth found in, say, Genesis: Others deny such a conflict. A version of the Genesis myth is that God is eternal, that He always was, but the world, a contingent entity, was created by Him in space and time. Also that the creation occurred when God spoke and the world came into being. However, biblical sources are vague as to when God initially created the world. Creationists hold that the universe is the result of a single act initiated by the Creator. Certain creationists have based their calculations on biblical "evidence" that dates the initial creative act some 6,000 years ago. However, creationists are less interested in the initial act of creation than they are in an event such as the Great Flood for which Noah made great preparations. The importance of the Great Flood story is that it is believed to have left behind tangible evidence, deposited in rock sediments found in the remains of prehistoric plants and animals. Such evidence supports the hypothesis that the act of the Second Creation, following the Great Flood, did in fact occur—an observation, many claim, that is just as plausible as the evolution hypothesis. The biological evolution hypothesis holds that life is the result of millions of years of developmental processes, of saturated complexity, brought on in part by natural selection, conditioned by the environment. What is life? Samuel Butler in 1887 observed, "Life is two and two making five." To parody Butler, I hold that like Life, "Destiny is two and two making five." This "contradiction" in arithmetic points to the mystery of attempting to define life, whether artificial or "real" or attempting to define Destiny. Did the human race descend from apes, or were we all made in the image of God? Is the theory of evolution true? Or, is creationism true? Such questions today constitute the centrality of the evolution/creation controversy, into which, recently, an increasing number of parents, teachers, religious leaders and legislators have been drawn. Some raised questions as to whether evolution should be taught in public schools; others question whether prayer should be allowed in public schools. [I heard on an August 13, 1999 news report that certain school boards in the state of Kansas have ruled against allowing students to be taught theories of biological evolution. Preference is given to the creationist doctrine]. In some of your works you assigned both the school and the church the role of imparting values to youths, but to my knowledge you said nothing about whether prayer should be allowed in schools.

Dr. Locke, I suspect you have greater loyalty or sympathy with the evolutionist than with the creationist ideological tribe. Bergson tried to reconcile these conflicting hypotheses. Although I have no solution to the evolution/creation debate, I wish in this concluding remark to make an observation that, if investigated more fully, may illuminate the debated issue. It may well be the case that at a deeper level of reality, where matter and energy indistinguishably converge, the evolutionist and the creationist are struggling with similar, if not the same, problem. Contrary to what one may at the outset expect, it is not so much a problem of determining the truthfulness or falsity of religious claims. Rather, here the focus is on ascertaining the meaning or meaningfulness of a given claim of experience.

In the broadest sense both science and religion seek to uncover clues about the meaningfulness of the human condition. Science seeks to fulfill its task by "discovering" laws that govern the natural world, the understanding of which is made possible through examining such laws and the facts. By contrast, religion seeks to fulfill its task by enabling us to become inspired by the ideals found in the transcendent realm. Science is concerned with what may be called the immanent realm; religion, with the transcendent realm. Philosophy may be regarded as interested in both realms, the Hegelian Absolute. Indeed both science and religion seek to determine what C.I. Lewis called the "sense meaning" of experience, which, at the deepest level, is expressed through symbolic representations: "$E = M C^2$," "God," "Immortality," "Freedom," "Destiny,"—these are symbols with partly disclosed and partly concealed sense meanings. By "sense meaning" Lewis understands certain mental structures or schemata by which the mind imposes order on experience in determining its significance.

Dr. Locke, in explicating your axiology in the context of examining principles in evolution, I showed that your axiology involves schemata comparable to physicist Murray Gell-Mann's notion of schemata. The hypotheses that the evolutionist employs and the ones that the creationist employs seem to be at odds with one another. But in fact there is no opposition, if their respective hypotheses are regarded as symbolic representations of the same Ultimate Reality. In such an instance, symbolic representations, like a diverse stanza in a poem, do not contradict each other. Rather they are complementary, as in the Chinese yin/yang principle. The seemingly opposing perspectives can be reconciled in the Destiny model through which alone the meaningfulness of life can be fully appreciated. I do not regard religion as opposed to Destiny. What I try to avoid is getting into any debate about determining the nature of God, or "proving" His existence. In most instances, a given religion is intolerant of competing claims found in other religions.

Sincerely,

Dr. Johnny Washington, Ph.D.   February 20, 1998

Philosophy and Destinicity Studies

Dr. Alain Locke (1886–1954)   Destinicity Letter #60

Re: Re: Philosophy and Destinicity Studies

> There never has been, then, a new prince who disarmed his subjects: instead, when he has found them disarmed, he has always armed them, because by doing so their arms become yours, those of whom you are suspicious become faithful to you, and those who were faithful stay that way, and from being subjects become your partisans.
>
> —Niccolo Machiavelli, *The Prince*

Dear Dr. Locke:

In some of my works, I have emphasized that your own philosophic views as well as those of Drs. W.E.B. DuBois and Martin Luther King, Jr. were often keyed to addressing problems that African American suffered. In a similar vein, I developed an interest in academic philosophy as a career partly because I desired to make it applicable to problems in the African American community. First, I will consider ways in which philosophy generally may become more practical, and then go on to show how it may render a greater service to the African American community. In describing yourself as a Socratic mid-wife, this sufficiently indicates that you intended for philosophy to move beyond academia to the social or public arena. You were a "mid-wife" to many young African American artists during the Harlem Renaissance and beyond. In this midwifery role you stimulated public debate on issues pertaining to race relations, art, education and the like. One can attempt to make philosophy practical by establishing a method by which to engage in philosophic enquiry. In the history of philosophy, two such methods that have broad applications stand out; namely, Pragmatism and Phenomenology, of which Charles Sanders Peirce (1839–1914) and Edmund Husserl (1859–1938), respectively, were the founders. I know of no one who has effectively combined these philosophic approaches, but such a combination, if carried out, might be fruitful. I will not attempt such an ambitious task at this time. Yet we can find traces of such combinations in your axiology. Inspired by the teachings of Descartes, Kant, Hegel and others, Phenomenology provides a means by which to examine, descriptively, the essence of things, that is, the essence of phenomena as they appear, without making any commitment about the ontological status of things in question. It also seeks to focus on those dimensions of experience such as love, friendship, freedom, caring, anguish and anxiety, among others, that have been ignored by positive science. While your axiology does not specifically focus on these types of experiences, it is dealing with the normative experience generally. A few years back I queried a professor from the Sudan as to what role Phenomenology might play in the economic development of Africa. During the early 1990s I was quite active in the International Development Ethics Association (IDEA), which I have already mentioned in previous pieces correspondence. I wanted to know how a philosophic approach such as Existentialism or Phenomenology would be applicable in African economic development. His reply was that in his view, Phenomenology would be quite useful in conducting research on matters pertaining to African history and culture, etc. This is so because of its method that emphasizes describing the world as it appears. He was less certain as to what role Existentialism may play in development. Additionally, Phenomenology holds that one is to establish objectivity by relying on the principle of intersubjectivity. That is, by means of entering into a dialogue with others, one can achieve agreement and objectivity. Other than intersubjectivity, there are no objective rules or standards of social reality. Dr. Locke, you seemed to have moved in the direction of accepting the phenomenologist's criterion of reality as you formulated your axiological views. I have begun to incorporate some of the tenets of Phenomenology into the Destiny model. I am looking forward to the day when someone comes along and makes a greater effort in applying Phenomenology in studying the blues or jazz tradition within the context of the African American Destiny. Dr. Locke, indeed you were familiar with Phenomenology because you made reference to the views of Brentano, who was the forerunner of the Phenomenological method. The social sciences such as sociology, criminal justice, anthropology, history and economics are keyed to practical issues. Often perplexities and difficulties arise in these fields. Phenomenology offers a way out. It is applicable at this practical level in helping to shed light on such difficulties and examining the presuppositions that underlay the various disciplines. Although Husserl arrived at Phenomenology as a result of his reflecting on mathematics, he regarded Phenomenology as a means of unifying all of the disciplines. I will go so far as to say that Phenomenology may be useful in enabling us to examine our transcendent Destiny. Although Kant says that we cannot know the transcendent world, I believe Husserl would say that we can describe items, including the Destiny ideal, in such a world (see the works of Merleau-Ponty). Here I anticipate a possible criticism that may be brought against the Destiny model by some Afrocentrists who may attempt to say that I have not devoted adequate attention to traditional, pre-colonial African values; that I am urging people to adopt the Eurocentric value creed, inasmuch as I am embracing Phenomenology. My reply is that such a criticism is inapplicable. In his book, *Muntu,* Janheinz Jahn deals with the philosophic significance of "the word," that is, the African "Nommo" by tying it in with the blues musical expressions. In Jahn's view the word or linguistic expressions of a natural language is the creative force by which human culture is created and the human spirit is expressed. The blues is a bridge between the African cosmology and the African American cosmology, in view of the fact that the blues has its origins in Africa. The African stress on the family is another value that we may embrace in seeking the economic-culture development of the people of African origins. As I mentioned, I attended the World Conference of Philosophy (WCP) in Kenya, July 21–26, 1991, during which time I was approached by an African philosopher seeking ways to make Philosophy more applicable to problems in Africa. He had gathered from perusing my works that I had played a role in founding of the Institute for Applied Philosophy (IAP) in Fort Lauderdale, Florida (USA). This African philosopher under discussion said that he wished to learn more about the IAP on which to model his own institute. In August 1992, I attended a Pan-African conference in Nigeria to help establish an agenda for Africans and the people of

African origins worldwide. During the conference, I encouraged the organization to include as an agenda item the consideration of an African Destiny model.

Further I suggested the need for establishing a center on the continent of Africa (or in the U.S.) that would concentrate exclusively on examining the African Destiny, with the aim of inspiring what may be called the African Destiny Renaissance or the Destiny Movement. The recommendation was seriously entertained by a number of conference participants. Shortly I will spell out more explicitly what the mission of such a center (African/African American Destinicity Studies Center (A³DSC)) would include.
Sincerely,
Dr. Johnny Washington, Ph.D.   February 24, 1998

The 1996 Philosophy Born of Struggle Conference

Dr. Alain Locke (1886–1954)   Destinicity Letter #61

Re: Philosophy Born of Struggle Conference

> There is no doubt that princes become great when they overcome the difficulties and the opposition that they encounter, and thus fortune, especially when it wants to aggrandize a new prince who has more need to acquire a reputation than a hereditary one, arouses enemies for him …

> —Niccolo Machiavelli, *The Prince*

Dear Dr. Locke:

In my previous letter, I inadvertently failed to mention a forthcoming Rockefeller Institute focusing on Paul Robeson's works convening at Columbia College in Chicago during the 1998–1999 academic year. I should have indicated that since you made contributions to our understanding African American music and other forms of cultural expressions, it would be desirable if a similar conference were convened honoring your works. I also forgot to mention that a conference along this line already is held yearly, under the auspices of the Alain LeRoy Locke Society (ALLS). Professors Leonard Harris, Tommy Lott, Esme Bhan, J. Everet Green, myself and others are among the founding members of the ALLS that was founded a few years back. Professor Leonard Harris took the initiative in establishing the ALLS. He has also published a collection of some of your articles in a book entitled, *The Philosophy of Alain Locke: Harlem Renaissance and Beyond* (1989). "Alain Locke's 'Values and Imperatives': An Interpretation," is the title of one of my articles published in another work entitled, *Philosophy Born of Struggle* (1983), which Harris edited. The ALLS also publishes a well-organized newsletter, of which Professor Bhan is the editor. The ALLS, in connection with the Philosophy Born of Struggle III Conference, convened in your honor during October 17 and 18, 1996, at Rockland Community College, Suffern, N.Y.

The theme of the Conference was "The Harlem Renaissance and the Black Enlightenment." The title of the paper I presented on that occasion was, "Destiny: Alain Locke and the Black Enlightenment." A similar conference focusing on the views of yours was also held at The New School for Social Research, New York, N.Y., October 17–18, 1997. In the January 16, 1997 issue of the *Palm Beach Gazette*, p., 5A, under the title, "Philosophy Born of Struggle III: The Harlem Renaissance and the Black Enlightenment" the results of the October 17 and 18, 1996 conference were published. I shall herewith cite excerpts from remarks that were written by Professor J. Everet Green of Rockland Community College. I shall subsequently make comments on Professor Green's article.

> Professor Washington's paper focusing on African-American or German Idealism was vigorously debated by other speakers and members of the audience and was met with some opposition by most of the respondents primarily because he could not present any conceptual framework for the transcendental realm. Professor Washington confessed that he is a great admirer of Fichte and, like Fichte, he too seems to be committed to the principle of the unity of truth…So he wants to move beyond the multiplicity of truths to the Fichtean Absolute. Evidently this would be a rejection of Alain Locke's concept of cultural pluralism. Professor Washington believes that we are moving toward one economic order and consequently to bring about cultural and racial harmony we need to develop a transcultural, transhistorical world-view…
>
> What is not clear in Professor Washington's presentation is whether or not he endorses Fichte's idea that since there is one absolute and unconditional truth there can be only one absolute and unconditional philosophy which presents this truth. What is very evident is the fact that Professor Washington is committed to some principle or reality that is beyond all human diversity and beyond all historical particularity. This understanding some members of the audience found to be veering into un-

intelligibility. This is understandable in a modern philosophical climate that is hostile to any form of idealism. Professor Washington seems to be a promoter of enlightened cosmopolitanism that would transcend religious sectarianism.

My presentation generated much interest among the participants. However, in some instances Professor Green seemed to have misconstrued the points I was making. I wish here to clarify my position. First, Dr. Locke, it should be made clear that you did in fact have an affinity with the German Idealism tradition, Professor Green's view to the contrary notwithstanding. Both you and Dr. DuBois were highly interested in Hegel's views, who in turn borrowed many views from Fichte, especially his idea of the Absolute. In your essay, "Value Theory," when you considered your notion of a categorical imperative, you made specific reference to Kantian ethics, against which your views were formulated. At the Philosophy Born of Struggle III Conference under discussion, I divided my presentation into two parts: in part one, I focused almost exclusively on your axiology. In part two, I went on to outline my view of Destiny. I prefaced my presentation with the following comments: That although at the conference I would speak about my Destiny model, you, yourself, [Alain Locke] rarely used "Destiny." However, many of the issues you dealt with paralleled the Destiny model, especially regarding immanent Destiny. I went on to inform the audience that your own axiological views inspired my reflecting on Destiny, in the following manner. In your essay, "Values and Imperatives," you described what may be called the "four value fields": the religious, the ethical, the scientific and the artistic or aesthetic, each of which, you maintained, was related to the other. The unity of the four, in your view, lies in consciousness. The stress here is on the notion of unity on which the Destiny model hinges. I reminded the audience that you made it seem as though your value classification schema was complete. But in your essay, "Values and Imperatives," where you introduced the value schema, you seemed to have paid little, if any, attention to the political values; and you only mentioned in passing economic values, for your schema seems to have a limited place for economic values. What distinguishing psychological quality corresponds to the economic value model? How is wage labor or rent to be determined (Marx)? As to the realm of politics, what constitutes patriotism? Where do they fit into your value schema? I, moreover, informed the audience that I discerned what appeared to have been a weakness in your schema. The weakness largely lay in its incompleteness in that it did not include a major value field such as the economic or the political modality. Thus, it was in such a light that I searched for a unifying schema that would embrace each of your major value categories as well as those that you may have overlooked. My searching led me to the notion of Destiny that is both transcendent and immanent. Clearly your views focused mainly on the immanent realm; whereas I was also interested in both the immanent and the transcendent realms, and the last-mentioned is comparable to the Kantian noumena world. Although I mentioned Kant's views in this context to place things in perspective, I was, strictly speaking, interested in the views of Fichte. Professor Green held that I was interested in Fichte's views because he stressed unity to the exclusion of diversity. This is only partly correct. What I actually said was that I was interested in Fichte's views because he placed emphasis on the importance of struggles and counter-struggles, by which we attempt to overcome the resistance and contradictions in the world that characterize our immanent Destiny. To the degree that we struggle effectively, we are dealing with our immanent Destiny. Professor Green made it seem as though I was interested in Fichte's notion of the unifying, single truth, almost to the exclusion of everything else. This is not the case. However, I said that I was interested in his notion of action, his notion of work and struggle, and striving. These are the means by which the unifying principle of Destiny can be embraced. Admittedly I did make the claim that for the sake of my presentation, I wished to ignore any consideration dealing with religion or God. As previously indicated, it had been my experience in the past that each time I attempted to give a presentation focusing on Destiny, invariably once I progressed about fifteen minutes into my presentation, someone would interrupt me by introducing questions about God or religion. In some of your works you attributed such divisiveness to absolutist principles upon which most religions rest. This is why you seemed to have found a religion such as Christianity so objectionable, I informed the audience. The major religions, Christianity, Islam and Judaism, have been sources of strife among themselves, and at least two of them have adversely affected African Destiny. In my presentation under discussion my aim was not to deny the importance of religion in one's life. Professor Green made it seem as though I wanted to discard or route religion altogether. In this regard, he is mistaken.

Sincerely,

Dr. Johnny Washington, Ph.D.   February 26, 1998
CC: Dr. W. E. B. DuBois

Church Letters

Dr. Alain Locke (1886–1954)   Destinicity Letter #62

Re: The First Church Letter

> Princes, and particularly new ones, have always found more fidelity and more utility in those of whom they were suspicious at the beginning of their state than in those in whom they were confident.
>
> —Niccolo Machiavelli, *The Prince*

Dear Dr. Locke:

On January 26, 1998, I was inspired, by the teachings of St. Paul to write a letter to the churches in the African American community. My purpose was to inform its members of the practical relevance of the Destiny model. Heretofore I explained the Destiny model to audiences in such places as Nairobi, Kenya; Dakar, Senegal; Lagos, Nigeria; Tegucigalpa, Honduras; Managua, Nicaragua; San Jose, Costa Rica; and Ahmadabad, India, in addition to the many audiences in the U.S. I now feel the need to present my views to a specific audience. Initially this letter was intended mainly for the African American churches in the South Florida area, but it may be of interest to any religious or civic organization. Below is a copy of an open letter I addressed to churches, in anticipating that it would get media attention.

Sincerely,

Dr. Johnny Washington, Ph.D.   February 27, 1998

To: Churches in the South Florida Area and Beyond

Re: Exploring the African Destiny

Dear Church Members:

This letter is addressed to various churches in the South Florida area, especially those in Palm Beach County and beyond. The purpose of this letter is to inform the respective church members about the African Destiny model. I am at this time inviting you to examine closely this model.

Once you have done so, and if you find it adequate, or if it seems to address the needs of the community, consider adopting this model. I am confident that you will find it useful in approaching many of the problems the African American community, and the country generally, is facing today. Moreover, I wish to urge you also to consider establishing what may be called African Destinicity Day, analogous to African American Heritage Month. My rationale for such considerations is included in my discussion below. It is appropriate that I address this letter to you now. Because we are in between two significant events in the African American experience: the recent celebration of Dr. Martin Luther King, Jr's birthday, and African American Heritage Month. The African Destiny is in the spirit of both of these historic events. Moreover, it is especially imperative that we become more cognitive of our Destiny—we are now facing a new millennium. We need to acquire knowledge about ourselves, would not you agree? We need to determine who we are and what our goals are?

It merits repeating my definition of Destiny, according to which it is an ideal(s) of unity towards which a people strives in successive generations. "People" here could denote any people such as "Africans," "African Americans," "Hispanic Americans," "Native Americans," "Germans," "Kenyans," "East Indians" or humanity generally. Moreover I distinguish between transcendent and immanent Destiny modalities; the former is analogous to the Kantian noumena world; the latter to the world as it appears, i.e., the political realm and the market-place, as well as our natural environment. Transcendent Destiny relies on faith more so than knowledge. The Destiny model provides a schema in terms of which Destiny enquiry can occur in an effective manner. I anticipate that by now the question has arisen in your mind: What is the Destiny of Africans/African Americans? To which I shall respond shortly.

Admittedly a thinker such as Frantz Fanon in his book, *The Wretched of the Earth* (1961), devoted attention to the African struggle for independence from colonial rule. However, his model was inadequate in that he placed too much emphasis on violence, and he regarded the transcendent realm, i.e., a realm that is believed to be similar to, say, heaven, as objectionable. Many public and private schools, including colleges and universities, both white and predominantly African American, have adopted the Afrocentric model in organizing their curricula. The trouble with this model is that it focuses on the African American/African experience almost to the exclusion of the experiences of other ethnic or national groups. The African

Destiny is part of the world Destiny, a point that I highlight. Thus, inasmuch as the Destiny model is sufficiently general, it is the most appropriate one for exploring the African Destiny and human Destiny generally. In addition to the philosophers that I mentioned above, my philosophic views on Destiny are also informed by the African American intellectual tradition that includes the views of Sojourner Truth, Frederick Douglass, Booker T. Washington, Alain Locke, Ida B. Wells, Marcus, Garvey, W.E.B. DuBois, William Leo Hansberry, Martin Luther King, Jr., Malcolm X and others. I am also inspired by the teachings of the Bible, the Koran, the Vedic sources, the Bhagavad-Gita and other religious literature. Dr. Alain Locke insisted that society needed to subscribe to the principle of pluralism, in contrast to the melting-pot model that emphasized sameness of people at the expense of ethnic diversity. The pluralism principle enables each ethnic group to actualize its own potential. He held that the Destiny of African Americans lies largely in their contributing to the realms of art, religion and morals. Specificity or lucidity is not always a virtue. Destiny involves degrees of mystery. Certain phenomena or issues need not be subjected to scientific investigation. In response to the question, "What is the transcendent Destiny of African Americans or of Africans?" I arrived at negative results. Transcendent Destiny is a mystery, because it pertains to matters that are beyond our rational capacity or scientific capabilities. It involves matters of faith, comparable to our faith in God. However, I do in fact make a connection between the immanent Destiny of a people and their working to build or rebuild their community.

The framers of the U.S. Constitution, influenced by European Enlightenment, regarded ethnicity to be objectionable. In reversing this view, I seek to tie the immanent Destiny of a people to such particularities as one's ethnic grouping, precinct, or county region—that is, one's local community (Clifford Geertz). Destinicity also provides a means for exploring ethnic or religious diversity, tolerance and unity. Besides the Destiny model, there are few, if any, adequate models available in terms of which we can deepen our understanding of ethnic diversity and its concomitant unity. Indeed the Destiny model offers a schema for exploring ethnic issues at both the practical and theoretical levels. At the center of the Destiny model is the view that working and struggling, along with other burdens, derived perhaps from Original Sin associated with our eating the "forbidden fruit," are important means by which we deal with our Destiny. Life is a terrific paradox. One of the key aims to life is to come to know how to make use of one's burdens and problems, so as to enrich one's life spiritually. This might shed light on the question concerning the nature of the blues, jazz and other forms of African American music traditions. Such powerful modes of expressions grew out of the pain and suffering that African Americans had to bear, heroically. My view parallels the Christian tradition. On the basis of the Christian world view, by virtue of the fact that we are born into sin, our lives are filled with hardships and troubles, and that it is the Lord to whom we must turn in dealing with the trials and tribulations of our lives. The Lord is to assist us in dealing with our troubles. Similarly my view is that such burdens are merely obstacles placed in our paths as elements of resistance, out of which reality is constituted, to be overcome in this world, by working and struggling and fighting and loving. And to the degree that we work and struggle and fight and love, regardless of whether we are successful at any given task, we are dealing with our Destiny, both immanent and transcendent. Thus, our Destiny, partly submerged in pain and suffering, lies in working, struggling, fighting, gaming and loving, in seeking to overcome or accept the contradictions and perplexities of this world, in striving for the high water mark of existence. In approximating our Destiny in the manner in which I have described, we enhance our self-esteem. Social cooperation is also important in our worldly endeavors. Although this world that involves our immanent Destiny is characterized by disharmony and fleeting moments of joy, we can approach such problems more effectively if we reject the view that we, as isolated individuals, are important. My view focuses not so much on the isolated individual, but on our collective efforts and struggles in the community. The isolated individual needs to commit him/herself to a cause in which others are participating, by which loyalties and ties are established with others; in other words, we must subscribe to the principle of unity, which is ultimately grounded in the transcendence. God may be regarded as the Destiny of destinies. We must have faith in the efficacious nature of the Destiny ideal through which unity is possible. Thus, Africans on the continent of Africa, and the people of African origins throughout the Africantude community, are all unified in the one transcendent Destiny. In the spirit of Marcus Garvey, this is a call for unity among Black people worldwide. Just as the Bible enjoins that in Christ we are all unified, so also I hold that in transcendent Destiny we are all unified. The obstacles of the world, whichever problems facing individuals and groups, are largely negative merits, prompting us in overcoming or accepting such problems to seek unity, through which our immanent Destiny can be embraced and our transcendent Destiny approximated. I know many may be wondering whether the Destiny model is compatible with the Judaic-Christian religion. Yes, it is compatible not only with the Judaic-Christian religion, it is also in harmony with many other religious or ethical systems that emphasize universal harmony and peace. If God is our first most ultimate concern (Paul Tillich), Destiny is our second most ultimate concern. It is the commitment to these Ideals, from which radiate hope and peace, is what keep us from slipping into the sea of despair, absurdity and insanity. The ideals of God and Destiny allow us to live meaningful lives. Thus the self-knowledge path that the World Destinicity embarked upon at the outset is partly rescued from frustrations by making a greater commitment to the religious experience, God, the God of Whitehead, Fichte, Moses, and Saint Paul and transcendent Destiny, so that we may bear our burdens and struggle with greater determination.

Again, I at this time recommend that the first Wednesday of February, (that is, the first Wednesday during African American Heritage Month), to be established as African Destinicity Day, a day that we shall honor each Wednesday during the first week of February effective the year 2005 and henceforth. Such a day should be set aside as a day in which we reflect not only on our cultural contributions of the past but on our present and future possibilities as well. Such a day should involve activities of reverence in which we explore ways of approximating unity at all levels of existence, at the local, national, international and cosmic levels.
Sincerely,
Dr. Johnny Washington, Ph.D.

February 19, 1998

To: Churches in the South Florida Area and Beyond

Re: Exploring the African Destiny

Dear Church Members:

This is my second letter to you. I suspect that many are surprised to receive letters with such philosophic contents from me. I am doing so because many of the problems that the African American community is facing today are so complex and deep-seated that they merit moving the discussion to such a philosophic level to get to the root of the matter. Some of our own leaders, Drs. W.E.B. DuBois, Alain Locke and Martin Luther King, Jr. had a great appreciation for philosophy. It should be remembered that Dr. King effectively combined the Christian teaching with Western or Eastern (Hinduism) philosophy; at the center of his teachings was the ideal of social justice. In this light my philosophic letter writing activities are in harmony with an aspect of the African American intellectual tradition.

In an essay entitled, "Good Reading," Dr. Alain Locke described philosophy in terms of a mountain climbing metaphor. Like mountain climbing, philosophizing is both hazardous and exhilarating. The goal is to ascend the highest peak of the landscape so that one can appreciate a panoramic view of reality. From such a high and broad peak, one can get a better perspective of the landscape of Destiny. The juncture points of this landscape are the values, ideas and beliefs reflected in the great philosophic traditions pointing human beings toward their Destiny.

The views of Socrates, Plato, Aristotle, Avicenna, Averroës, Spinoza, Leibniz, Kant, Fichte, Hegel, Nietzsche, Bosanquet, Sojourner Truth, Frederick Douglass, Booker T. Washington, W.E.B. DuBois, Ida B. Wells, Marcus Garvey, along with Alain Locke—their views were prominent features in the landscape of the Western intellectual tradition. They were pathfinders in the wilderness of Destiny; the respective individuals were not devout Christians. Yet their views are in harmony with the Christian ideals. A good Christian is one who seeks to practice his or her religion. Similarly, one of the aims of this letter is to consider ways in which philosophy may become more practical in specific contexts. I invite you to join me in this endeavor. "In my Father's House there are many mansions": there are many ways to contribute to the African American cause, the African American Destiny. Some may travel the philosophic route; others may prefer the religious route. Still others may prefer to contribute to the cause by working or struggling by being involved in other professional, civic or religious endeavors. The attempt to render philosophy in the service of the community is not new. Such practices have been tried since Socrates' times and beyond. Socrates, Thoreau, Gandhi and Martin Luther King, Jr., among others, were interested in making philosophy practical as they dealt with social justice issues.

Charles S. Peirce (1839–1914), was the founder of the American version of philosophy called Pragmatism. As conceived and practiced by Peirce, Pragmatism was regarded as a method or a schema by which to ascertain the meaning of beliefs such as those pertaining to the nature of God, freedom, immortality, mass, energy, velocity or gravity. Meaning was to be established by determining the practical bearing that the consequences of the belief had on the environment, provided that one acted on one's beliefs. Peirce focused on scientific beliefs. Peirce also focused on certain specific metaphysical problems such as free-will, determinism and the nature of consciousness. He was also interested in the larger questions of the Destiny of the universe. He held that the principle of *agape*, (i.e., love) informed the Destiny of the universe. Other Pragmatists applied the method in examining problems in law, government, education, as well as the economy. Few, if any, attempted to do as Dr. DuBois, Locke or King did by employing philosophy in routing racial injustice. One wonders what would have been Dewey's or Peirce's response to today's problems of Affirmative Action, school desegregation, school violence or welfare reform. William James was a Harvard-trained physician who subsequently turned to the study of psychology and philosophy. To the degree that a number of public policy issues in this country and elsewhere in the world are grounded in Pragmatism, it provided a public forum to deal with a number of important issues. However, I discerned certain merits in the views of Kant, Fichte, Hegel, Nietzsche and other Idealists. Some described Idealism as a variation of the Christian tradition. I believe Martin L. King Jr. had a greater appreciation for German Idealism than for American Pragmatism.

Medical Ethics, Business Ethics, Environmental Ethics, and Economic Development Ethics—each, in turn, may be assumed under the rubric Applied Ethics that in turn may be assumed under the rubric Destinicity Studies. Within the past few decades, new areas of enquiry arose that made greater efforts to devote attention to the normative aspects of Medicine, Law and Business. Collectively these considerations came to be known as Professional Ethics, the aim of which lay largely in examining the normative principles that provided the basis for the respective disciplines. Professional Ethics was intended to address certain pressing dilemmas and perplexities that tend to arise within the various professions. Since the rise of modern science in the 17th century, the Western world has attempted to grapple with many value issues within the framework of secular ethics rather than from the traditional religious framework. Is it permissible to clone human beings? Is there anything wrong with manipulating the DNA material of animals or plants in attempting to increase their market value? How would such practices adversely effect bio-diversity? What's wrong with surrogate motherhood? When is euthanasia morally appropriate? When is abortion permissible from a Christian perspective? Is terrorism permissible as a means of inducing social change? What about the morality of war or revolution? Owing to the advent of contemporary technology and other factors, these and other questions have complicated moral discourse. The conceptual frameworks for Professional Ethics is customarily provided by the Kantian or utilitarian ethical tradition. William James explored issues that not only enabled us to understand Christian values, but religious experience generally. Pragmatism has much to offer in enabling us to investigate our immanent Destiny modes. In my previous letter to the church I considered the immanent/transcendent Destiny distinction. American Pragmatism, with the exception of William James' version of it, largely ignored the realm of transcendence that is analogous to biblical transcendence. Dr. Locke, as well John Dewey, was persistent in the effort to dispel the view that there are transcendent absolutes. I believe the two are misguided on this score. Many Pragmatists disliked the transcendence hypothesis because it was believed that it lends itself to absolute norms that could easily degenerate into a form of totalitarianism. Dr. Alain Locke often reminded us that Hitler's government in Nazi Germany was founded on absolute norms that he found objectionable. Dr. Locke criticized the absolutist principles found in Christianity; he did so on the basis that they encouraged cultural intolerance. I should add that some of the issues I am dealing with in this letter I previously considered in an article entitled, "Philosophy and the African-American People," where I proposed the founding of the A^{3}DSC. A main focus of the Center would be on economic issues, a major problem facing the people of African origins worldwide. Specific goals would include: i) establishing a plan to coordinate the economic-cultural development of the people of African origins; ii) and monitoring the radical macro phase-changes that are adversely effecting the people of African origins, and people generally, due largely to the technological advances that have occurred within the past few decades. As it stands, to my knowledge, there is no institution exclusively devoted to concerted and systematic efforts to coordinate the economic-cultural development at the local, national and international levels of the people of African origins. If Randall Robinson's reparations proposal is seriously considered in the future, part of the funds may be used to provide financial support for the A^{3}DSC. In my previous church letter, I recommended that during the year 2005, and henceforth, in the Month of February, African American Heritage Month, on the first

250

Wednesday of that month, we shall honor what I have called African Destinicity Day. This is not an attempt to replace any particular event that we already honor during African American Heritage Month. Rather African Destinicity Day is intended to draw attention to African Destiny in which we are reverently to reflect on our past, present and future possibilities. It is anticipated that by the year 2005, through the assistance of the African American churches, friends of the African American churches and other sources, plans will be well on the way in founding the A³DSC.
Sincerely,
Dr. Johnny Washington, Ph.D.

Reparations for African Americans

Dr. Alain Locke (1886–1954)   Destinicity Letter #63

Re: Reparations for African Americans

> And because not all your subjects can be armed, when you benefit those whom you arm, you can feel more secure about the others, and those who get preferential treatment are more obliged to you, while the others, judging it necessary that those who have more peril and obligations should also have more merit, excuse you.

> —Niccolo Machiavelli, *The Prince*

Dear Dr. Locke:

It will be remembered that the framers of the U.S. Constitution adopted from British democratic liberalism tradition the natural rights of life, liberty and the pursuit of happiness (property) principle. The practices of slavery or segregation denied Blacks the opportunity to exercise the last-two mentioned rights. Some lawyers maintain that this is another basis for bringing a lawsuit against the federal government. In addition to seeking a legal basis for the reparation lawsuit, some lawyers hold that we need to pursue the case within the broadest social context so as to establish allies among the various segments of the society, among the various religious, ethnic and civic groups. Your views are important in this context. In passing it will be remembered that you held that one of the most important ways to approximate the ideals of a racially integrated society is to urge the various groups to work together to this end. You felt that the ideals of a genuinely racially integrated society could be approximated only when parity between whites and African Americans in the area of education and other social institutions is achieved. Similarly, I suggest that the Destiny model provides a broad schema or forum for considering the reparations issue. In the event that African Americans are paid reparations, such payments might be regarded as a means by which they can effectively exercise the three above-mentioned natural rights, especially the rights of liberty and the pursuit of happiness. Moreover, a certain degree of economic security is required before such rights can be exercised. The reparation payments can be a way of approaching this economic security issue. It merits repeating that the Destiny model stresses immanent as well as transcendent unity, and regards the four Destiny modes as inherently related. To the degree that African Americans approximate their ethnic Destiny, facilitated by reparations, they may be culturally strengthening and enriching the national destiny mode.

Here at Southwest Missouri State University (SMSU) on November 14, 2000 I convened a public forum entitled, "The 1921 Tulsa Race Riots: The Call for Reparations." The Tulsa panelists included Dr. Vivian Clark-Adams, Chair, Liberal Arts Division, Tulsa Community College; Ms. Eddie Faye Gates, Tulsa Race Riot Commission; and Ms. Rita Duncan, Race Riot Survival Committee. First, we showed a 90-minute video focusing on the Tulsa riots, followed by a panel discussion. In addition to the larger Springfield community in general, students in my AAS 100 as well as my PHI 110 class attended the forum. Having done so, many wrote remarkable papers focusing on the Tulsa reparations theme. The students have begun to organize around the Destiny model, with reparations as one of its themes.

During the past twenty years in my ethics courses I have relied on the philosopher John Rawls' work, *A Theory of Justice* (1971). His theory is analogous to the social contract model of John Locke, Hobbes or Rousseau. When Rawls developed his theory of justice he had in mind an ideal, hypothetical situation, where rational adults, behind a veil of ignorance, deliberate about what sort of society to construct. According to Rawls, in establishing a social contract, the equality as well as the inequality principle would be acceptable to these individuals. The first principle holds that each individual is morally equal; the second principle holds that differences in social or political positions or practices among people are acceptable so long as the disadvantaged individuals benefit from the inequalities in society. The point may be made that while it is the case that racial reparations single out Blacks as different from their white counterparts or some other ethnic groups, society might be better off, insofar as it will be in a greater position to approximate the unity ideal, by allowing Blacks to receive reparations. Consequently society might experience fewer racial conflicts.

A work that has single-handedly fueled the current racial reparations debate is Randall Robinson's book, *The Debt: What America Owes to Blacks*. Also some prominent lawyers are beginning to work on the reparations case. They are: Willie E. Gary, Alexander J. Pires Jr., Richard F. Scruggs, and Dennis C. Sweet III. The November 2000 issue of *Harper's Magazine* includes a forum moderated by Jack Hitt that considers the views of these lawyers. Recently Michigan Representative John Conyers Jr. proposed a bill for the U.S. to apologize for slavery. Many of the people I have mentioned thus far working on the problem of reparations are lawyers or public officials, who are seeking legal provisions on which to base the reparations claim. Since one of my areas of specialization is philosophy, just as this is also one of your areas of specialization, you and I are in a position to illuminate the discussion from a philosophic perspective.

The question regarding racial reparations is comparable to the question that the country faced in the era of segregation, when the separate-but-equal policy was urged upon African Americans by whites. You and other African American political activists maintained that although the U.S. professed to adopt the separate-but-equal principle, in fact African Americans invariably received unequal treatment in areas such as housing, employment and education. Because of the practices of inequality that African Americans suffered, Booker T. Washington, W.E.B. DuBois, you and others held that the U.S. was morally compelled to provide funds so as to establish parity between African Americans and their white counterparts. Thus, it follows that the racial reparation debate is a continuation of the parity debate associated with the separate-but-equal doctrine that came to the forefront during the previous century. The current Affirmative Action controversy also bears upon the reparation issue. In the Prologue and in other works, I mentioned your extensive publications. Here I am especially interested in examining your published works in the area of education in which you advocated the parity principle to offset inadequate funding allocated to the predominantly African American schools, colleges and universities. Some of your specific works that bear directly or indirectly on the issues under discussion are indicated in the notes[3] section to chapter 6. Racial reparations pertain to economic justice effecting African Americans. Attention has already been called to Randall Robinson's book, *The Debt*. Thus the underlying economic justice issue of my proposal is significant in part inasmuch as it parallels a national debate or concern, and it illuminates the economic class and race problems that this country faces. As I indicated, you rarely if ever used the phrase "racial reparations," but you used the comparable notion, "parity." You used this in the context of attacking the separate-but-equal policy. In the passage below, you make the point that the demands for parity on the part of Blacks need to flow from the motive of public justice and not from the motive of mere charity.

> Since we cannot say that this dual system [of segregation] is on the wane, what ought, theoretically, to be an anomaly in our democracy exists as a definite and inescapably practical educational problem. Further, by reason of its being in ninety per cent of the instances a discriminatory separation, and only in ten per cent a voluntary group arrangement or a special effort to compensate the handicaps of socially disadvantaged group, the situation presents a problem of general public responsibility, and a clear issue of public justice and fair play. ("Negro Education," p. 567)

You go on to add in the same context: "Fortunately the last few years have seen a marked change of public attitude on the matter, not merely renewed effort to remove some of the most outstanding disparities, but more promising still, a shift of the appeal from motives of charity to motives of justice and 'square deal'" [emphasis added]. ("Negro Education," p. 567) Between the times you wrote this passage just cited and today, ones wonders whether the shifting in motives were successful. In the passage below you tied the reparation problem to the issue of taxes, an issue I will explore more closely shortly.

> A standardized public school education must become the standard in the education of the average Negro child. Otherwise, Negro education costs double and yields half. As an indirect, but heavily mulcted taxpayer, the Negro, under the present system in the South, either pays for someone else's education and himself goes without, or with the aid of the philanthropist pays twice, once through the public system, and once again through the special agency of the private school. ("Negro Education," p. 567)

The statement, "…Negro education costs double and yields half," is telling. It says, in effect, that through taxation Blacks carried a double burden: paid

254

double in supporting the education of white youths and their own; however, the quality of the education that the Black youths received was reduced to half of that of the white youths. Another way of interpreting the "costs double and yields half" principle is to paraphrase it, to be expressed in the form: "paid double, yielded half." Generally the conditions of slavery or segregation exacted a double burden on Blacks. As slaves or racially segregated laborers, they were forced to labor for the livelihood of their masters as well as the livelihood of themselves; under such conditions, they were also treated as less than human beings. Thus, the meaning of "paid double, yielded half," is now obvious. This principle in itself is a basis of filing a class-action lawsuit against the government demanding racial reparations. The tax exemption proposal that I state below in effect allows the "costs double and yields half" principle to be reversed. In the essay, "The Dilemma of Segregation," you explored some legal and political measures that can be taken in seeking social justice for Blacks. You wrote:

> I find the chief justification for legal procedure,—the appeal to the courts, to outflank this unwitting betrayal of the mass interests by the few and the equally lethargic and often over estimated prejudice and group resistance of the white community toward the admission of Negroes to the public school system on a basis of equality." ("The Dilemma," p. 407)

You continued:

> Negroes, therefore, should and must resort to the courts to secure any considerable or wholesale improvement of the situation. This becomes more imperative when you realize that separation and a parity of standards and facilities are naturally antagonistic and rarely if ever coexisted…Favorable decisions or pending appeals should be used by public-spirited and sanely balanced citizens' committees as bargaining points for progressive adjustment toward the equalitarian goals. ("The Dilemma," p. 407)

In the following passage, you introduced the political approach, which, at bottom, is also legal:

> The only other effective alternative of legal pressure would thus be political pressure, which in the nature of the case in the South is even more restricted than legal recourse. For, in the first instance, the legal system has through the machinery of appeal more reliable access to a wider circle of public opinion beyond the local community and a firmer tradition of impartiality. The political channels of effective pressure are purely local or

> primarily so in the American political system, and what little potentiality of appeal they have from the counter-pressure of biased local tradition and sentiment is itself legal. ("The Dilemma," p. 408)

Here again the issue of taxes arises:

> It will be noticed that the real cause of parity in the exceptional cases we are discussing is a legal one,—the charter agreement of the school system. Potentially the tax assessment acts of states and counties are just such legal contracts and guarantees when associated with compulsory school laws as now prevail in most states, even those who break this principle and obligation flagrantly in practice. No state would dare under modern conditions to legislate this qualitative differential openly. This proves, to my way of thinking, the value of legal recourse, granting even the probability of many adverse or equivocal decisions on technicalities. ("The Dilemma," p. 409)

You are stating that inasmuch as the South attempted to maintain a separate-but-equal, segregated system, supported by public taxes, coupled with mandatory education, the below-par education that Blacks received was a violation of contract on the part of the state. (It will also be remembered that, during the 1960s and onward, oftentimes the white parents resisting school integration convinced public officials to allow taxes to be used to fund "private academies" from which African American youths were excluded). This was another form of contract violation. This practice of contract violation was a basis for filing a claim for reparations or parity. A similar argument may be applicable to today's racial reparations debate as championed by such individuals as Dr. Vivian Clark-Adams, Mrs. Eddie Faye Gates, Ms. Rita Duncan, Mr. Randall Robinson and the lawyers mentioned in the November 2000 *Harper's Magazine* article. In an October 1994 newsletter entitled, "IRS Fact Sheet #94," the IRS entered into the debate. The IRS wrote:

> The basis of the claims date back to the post-Civil War period when Congress voted to provide former slaves 40 acres and a mule as a form of redress for their years in slavery. The bill was vetoed by President Andrew Johnson. Legislation has been proposed in Congress to study the reparations issue. However, there is no law currently, that allows for any reparation payable to African Americans for slavery.

The IRS warned Blacks against attempting to file a claim known as the "Black Tax" by which to seek reparations for slavery. The IRS statement was a reaction to the actions of some 20,000 Blacks who sought exemption from paying taxes on the grounds of racial reparations. Apparently these 20,000 Blacks were pursuing the reparation case on their own, filing as isolated individuals. To my knowledge there was no collective or concerted action involved. That is, they did not attempt to sue the federal government by filing a class-action lawsuit for reparations, along the lines that you proposed above. In our system of government, justice requires that "like cases are to be treated alike." I want to make a case for reparations by showing that the U.S. government has provided many safety nets for certain corporations not limited to bailing them out when they are on the brink of financial failures when they suffer loses. In bailing out them, the government often evokes a form of "reparations" in terms of interest free loans, or tax breaks, along with other remedies to their benefit. Chrysler Corporation is a celebrated case in point. In some cases, even our government gives "tax breaks" to some corporations that are not suffering but are thriving. It is reported that some of the corporations have received rebates. In an October 2000 report entitled, "Corporate Income Taxes in the 1990s" by Robert S. McIntyre and T.D. Coo Nguyen of the Institute on Taxation and Economic Policy, a clear scenario is presented regarding the tax paying patterns of 250 U.S. top corporations in the years 1996, 1997, and 1998.

The report indicates that of the 250 companies, forty paid less than zero in at least one year from 1996 through 1998. Here is a partial list of twenty-four companies that paid less than zero in taxes in 1998, while making profits: Lyondell Chemical, Texaco, Chevron, CSX, Tosco, PepsiCo, Owens & Minor, Pfizer, J.P. Morgan, Saks, Goodyear, Ryder, Enron, Colgate-Palmolive, MCI Worldcom, Eaton, Weyerhaeuser and General Motors. In many instances, some of these companies paid zero taxes, and yielded double tax breaks.

It merits mentioning that many of these companies have philanthropic organizations established through which they shelter their taxes and provide services to the society.

Such philanthropic organizations can take the lead in our reparations efforts, and the government and other agencies might follow. The authors of the tax report point out that:

> Had all 250 companies paid the full 35 percent statutory corporate tax rate on their $735 billion in pretax U.S. profits from 1996 to 1998, their federal income taxes over the three years would have totaled $257 billion. But

instead, they paid far less. In fact, tax breaks cut taxes for the 250 companies by $26.9 billion in 1996, $31.8 billion in 1997 and 39.3 billion in 1998, for a total of $98 billion in tax reductions over the three years (See page #7 of this report).

Thus it follows that if certain companies are allowed tax breaks, even in cases when they have not suffered losses, why is it difficult for the government at the federal, state, city and county levels, to restructure the tax laws in a manner that would allow Blacks to receive reparations? The federal government must take the initiative. If it does so, some of the companies that paid little or no taxes or received rebates could pass on some of their "savings" to Blacks in the form of reparations. After all, many of these companies' wealth resulted from a basis that has its roots in slavery or racially segregated labor.

This point merits repeating. The U.S. government should treat like cases alike. If some corporations receive tax benefits for "suffering" losses and setbacks, the same could be true of the situation of African Americans who suffered losses due to segregation and slavery. Why is it the case that they cannot get tax benefits in the form of reparations? It merits repeating: like cases are to be treated alike; this is basis for building a class-action lawsuit for reparations for Blacks demanding that, as in the case of certain corporations, African Americans shall be exempted from paying federal or state taxes. Moreover, that any and all Blacks who are not in the work force, such as retirees, children, the disabled, respectively, are to receive "refund" checks from the IRS; the exact amount they are to receive remains to be determined. Those who are not employed but are capable of finding a job may be motivated to do so in order to receive tax benefits. The same is true of many on welfare. Still, those who are in the low paying job sector of the economy may become motivated to develop skills that will place them in a position to seek better paying jobs. How long is such a tax exemption or any other reparation measures to last? Such measures could last as long as slavery in the U.S. lasted, (—it began in the U.S. in 1621 and lasted until 1863). Immediately before slavery ended, the U.S government embarked upon its Manifest Destiny policy that had a devastating effect on African Americans, Hispanic Americans, and especially, Native Americans. If Americans seriously consider such injustices, there is a legal and moral basis for reversing the above-mentioned principle "paid double, yielded half."

Presently many states already have lotteries established, and they use the revenues generated to assist in funding education. The lotteries are additional sources for generating reparation funds for African Americans. Some of the states that have lotteries are Arizona, California, Colorado, Connecticut, Delaware, District of Columbia, Georgia, Idaho, Illinois, Indiana, Iowa, Kansas, Kentucky, Louisiana, Maine, Maryland, Massachusetts, Michigan, Missouri, Montana, Nebraska, New Hampshire, New Jersey, New Mexico, New York, Ohio and Oregon. In New York, for the year 1999–2000, the lottery revenues were allocated in the following manner: Prize money, 51% or $1.9 billion; aids for education, 38% or $1.35 billion; retailer commissions, 6% or $222 million; contractor fees, 2% or $88.9 million; administrative costs, 3% or $101.5 million. California's distribution of lottery revenues is similar to New York's. California's state law has it that prize money is to be 52.5%; that the administrative costs are not to exceed 16%, and that 34% is to support public education. In California during the fiscal year 1999, sales in lottery ticket totaled $2.5 billion. Finally, consider the state of Missouri. During the last fiscal year the revenues generated in the sales of lottery tickets totaled $150.5. Some of the revenues generated by the lottery in the respective states can be used to pay racial reparations. It would be desirable for the states without lotteries to establish them to assist in paying their share of the reparations.

Let me suggest that certain government lands, whether federal, state or city, can be a basis for funding reparations. I proposed that some of these lands to be deeded to African Americans who will become the new "landlords." The various government agencies will in turn become the new tenants, who in terms shall be obligated to pay African Americans rents for the usage of these sources. This could be a way of satisfying the forty acres and a mule empty promise. The law could be formulated to protect the public so that the new landlords would be prohibited from selling the lands without government consent. Alternatively, the lands could be deeded to various sectors of the Black community throughout the U.S., and not to any particular individuals such that a trust fund could be established. Any and all revenues generated from the lands could be placed in the trust funds that in turn could be used to support community efforts as described below. Further each year, billions of dollars are generated by the sales of government surplus properties, e.g. military equipment, office items and confiscated criminal properties, etc. Some of the revenues generated by such sales could be used in support of the reparation efforts.

As I have repeatedly indicated, effective economic-cultural development plans cannot be exclusively imposed from the top, from such an organization as the World Bank or International Monetary Funds. These plans must arise from the various ethnic groups or local cultural groups seeking to affect their own respective immanent Destiny modes, comparable to the McGhee Plan I mentioned earlier that includes what is called the Community Renovation Center (CRC), where livelihood skills are imparted. Such a center(s) could be focused on not only imparting vocational skills, including instructions on how to use computers and other forms of technology, but also knowledge about finance, business and commerce. The curriculum could also focus on African American history and culture, with special emphasis on the history of slavery and segregation, highlighting the struggles and counter-struggles that Blacks have suffered.

In the *Harper's* article, the question arose as to who would be regarded as defendants, that is, who are to pay for the damages Blacks have suffered? In addition to the federal or state governments, many companies such as Aetna Inc., and the "Providence Bank, whose original wealth dates back to the family of John Brown, whose descendants underwrote Brown University enough to cover up his embarrassment where he made his money." (*Harper's*, p. 41) According to the lawyers, other possible defendants include a number of former large slave plantations of the south. A case could be made that an industry such as entertainment, especially music, owes Blacks reparations.

It is evident that many artists who contributed extensively to the development of the blues, jazz, R&B, and other artistic forms were exploited by the white record companies. Some artists never received royalties for their products. And some white artists illegally appropriated the Black artists' works without giving them due credit. Prior to the "integration" of baseball, many of the athletic clubs did not treat Blacks fairly. President Wilson made legal the segregation of all public facilities, and Blacks were systematically discriminated against in the civil service arena when it came to hiring, layoffs or promotions. The list of industries or organizations that treated Blacks unfairly is extensive. One other difficult question that the lawyers in the *Harper's* article raised focused on the issue of plaintiffs: which Blacks are to be included in the lawsuit? And how far back into history are we to take the case in establishing the class of plaintiffs? One suggestion was to go back as far as the 1940s to pick up claims, focusing on living victims. However, there are a number of individual Blacks who would oppose to being regarded as plaintiffs; a case in point would be someone such as Clarence Thomas, who may hold that it would only stigmatize Blacks as victims.

Perhaps, even someone such as Tiger Woods, who with mixed ancestors, may oppose being counted as a plaintiff (see the *Harper's* article). My approach to this problem is that that those living African Americans who wish to be a party in the lawsuit may be self-included. Those who oppose it may be self-excluded. The Destiny model provides a broad framework in terms of which those who are self-included and those who are self-excluded can be sorted out. In Part II of this volume, I shall focus on a different set of issues; namely, the origin and destiny of the universe, with special attention devoted to biological as well as robotic evolution that complements some of the themes dealing with cultural evolution in Part I of this volume. Thus, it is fitting and proper that I end Part I by considering racial reparations, in light of the horrible experiences that African Americans suffered brought on largely by segregation or slavery.

Sincerely,

Johnny Washington, Ph.D.   December 8, 2000

Part II
Evolution: Biological and Robotic

Chapter 7

Manohar A. Tilak's Theory

Assemblies in Evolution

Dr. Alain Locke (1886–1954)   Destinicity Letter #64

Re: Assemblies in Evolution

Dear Dr. Locke:

This letter is a precursor to a consideration of the collaborative works between Manohar A. Tilak and myself, where we examined the nature and destiny of human evolution. In this letter I will briefly consider an essay entitled, "Stages of Evolution and Their Messengers," by Edward Rubenstein.[1] He also seeks to shed light on the nature and destiny of human evolution, by examining what he called the basic logic by which nature makes so-called assemblies. By means of assemblies, cultural, physical or biological evolution occurs, without which entities such as galaxies, quarks, DNA molecules, figs, frogs, pigs, elephants and human beings would not have come to be. To this end the basic logic on which evolution hinges is the construction of assemblies. The question that Rubenstein explores pertains to the process by which the construction of such assemblies occurs. On his view, it does so by relying on intermediaries or what he calls messengers, "to transmit the needed information…in each step of the assembly along the way. Continuing sequences of assemblies are the veins of evolution" (p. 132). It seems that the role of messengers here is analogous to the role of the schemata in the views of Murray Gell-Mann. The template of a given item to be constructed at a site pertains to the information in the messenger's schemata. The messenger's role is to transmit information at what may be called a "construction site" so that assemblies can be constructed. At the outset there might appear to be little, if any, affinity between what Rubenstein has to say about evolution and your own views regarding the dynamic or processes of culture. In your philosophy of art you introduced the principle of the elite, or the Talented Tenth. When Rubenstein discussed cultural evolution, in contrast to physical or biological evolution, he held that the role of the messengers, including writers, artists, entertainers and researchers are to construct cultural items or practices and in turn impart knowledge of such cultural items or practices to society generally for the benefit of others. You would agree. Rubenstein began by describing the assembly's construction process at the level of elementary particles or the pre-biotic level:

> For instance, elementary subatomic particles, driven into motion by thermal agitation, may collide. If the attraction between them exceeds the momentum of their recoil, they may adhere. Their course as a union depends on the fitness of the combination, that is, on whether the strength of their mutual affinity exceeds the disruptive effects of their surroundings. If the attractive interaction is strong and the ambient conditions are mild, the composite is suited to its environment: if it is stable, it may have a long or a virtually infinite lifetime (p. 132).

In the evolution of the universe, such an elemental level of assemblies gave rise to a new continuum that included biological evolution; this in turn was followed by still another continuum within which is found cultural evolution: random interactions, natural selection and expression of symmetry are some of the features that physical evolution and biological evolution have in common, according to Rubenstein. According to Rubenstein, "the interplay of the forces relies on messenger particles to induce the assembly of structural particles—the fermions—into atoms and molecules." He informs us that the class of bosons is the subatomic particles that play the messenger role in physical evolution. Rubenstein describes in a bit more detail the role of the messengers in each stage of evolution.

> The messengers of physical evolution are intermediary subatomic particles belonging to the class of bosons. In the sense that any number can occupy a point in space-time, bosons are dimensionless. They transmit information in the language of the forces of nature, and their messages are simple imperatives: Move (p. 132).

In the biological arena, the role of messengers in constructing assemblies is quite pronounced in the RNA molecules. In cultural evolution the messengers are the symbols of acoustic, graphic or body language, by means of which meaning is imparted or shared in the culture so that assembly structures are perpetuated or new ones created. The next and final stage of evo-

lution that Rubenstein considers is what he calls "directed evolution." "The messengers of the directed evolution are human beings. Their messages, expressed in the language and methods of molecular biology, genetics and medicine and in moral precepts, express their awareness of human imperfections and reflect the values and aspirations of their species" (p. 132). It will be seen in Appendix A as well as Appendix B, that Tilak employs the notion "Direct Phase Evolution" that is comparable to "directed evolution" as used by Rubenstein. Through practices such as genetic engineering, which enables human beings to modify the course of human evolution, the journey of civilization has now entered the phase of directed evolution. It will be seen shortly that what Rubenstein is calling directed evolution, Tilak calls evolutionary-macro-phase-change. He holds that directed evolution has posed a threat to the survival of the human species. Some believe that directed evolution harbors negative consequences to life on the planet Earth; by contrast others hold that directed evolution can do much in protecting diversity, or reducing pollution in nature. Admittedly you rarely used the word "Destiny," and you seemed to have found objectionable the notion transcendence. Yet most of your views on culture and axiology enabled me to develop my notion of immanent Destiny, under whose rubric may be included the four types of evolutionary modes of which Rubenstein spoke: physical, biological, cultural and directed evolution.
Sincerely,
Dr. Johnny Washington, Ph.D.   March 5, 1998

Introducing Tilak's and Douthart's Ideas

Dr. Alain Locke (1886–1954)   Destinicity Letter #65

Re: Evolution, History and Destiny:

Dear Dr. Locke:

Some insights included in the materials under discussion we arrived at independently, without the assistance of Tilak or the other way round; other insights we arrived at as a result of our discussing and exchanging ideas. It merits repeating that Tilak invented the Evoluon Theory during the early 1970s, and I invented the Destiny model. Manohar Anant Tilak was born November 20, 1925 in Dhule City Maharashtra State, India and immigrated to the U.S. in 1962; he became a U.S. citizen in 1985. His academic background includes a B.S. degree from Benares Hindu University in India in 1947; during 1959 he earned his M.S. degree from Technical University, Berlin, West Germany. During 1961 he earned his Ph.D. degree from Technical University, Munich, West Germany. A research chemist with Eli Lilly & Company since 1961, Tilak invented the excess mixed anhydride method of peptide synthesis, and he holds many patents for other processes. He began his career in West Germany in 1961 at the Technical University, Munich and Max Planck Institute for Leather and Protein Research. Between 1963 and 1964 he was associated with the University of Pittsburgh, Pennsylvania here in the U.S. Between 1964 and 1966 he was employed at the Brookhaven National Laboratory, Long Island, New York, during which time he collaborated with Professor P. G. Katsoyannis in the world's first human protein hormone insulin research. As an evolutionist during the early 1970s, Tilak introduced a new model that he called the Evoluon Theory that delineates the various stages of evolution resulting from saturated complexity. It offers a description of the big bang singularity, the first stage of the evolution of the universe, which may be construed as Evl1 (Evoluon 1); the second stage, Evl2, (Evoluon 2) is characterized by duality where only matter and anti-matter, the electron and the positron, existed; and the third stage, Evl3, (Evoluon 3) characterized by the three dimensional configuration of the atom, evolved out of Evl2. Evl4 (Evoluon 4) is representative of the DNA molecule, the building block of life. We are now in Evl5, (Evoluon 5) which human beings with the faculty of the intellect are representative. He maintains that an appreciation of the Evoluon Theory may enable us to develop a broad perspective in terms of which we might be in a position to deepen our understanding of evolutionary-phase-change that the Earth is now undergoing. Deeply committed to the Vedic tradition, Tilak draws many parallels between the Evoluon Theory and the Vedic tradition. In this work I shall draw on two of his major publications, "New Postulate on the General Theory of Evolution" and "Evolvability, Direct Phase Evolution, Evoluon, and Rate of Evolutionary Change," that I have included in the Appendices of this work. The value of the Evoluon Theory is that it calls attention to the various fundamental changes, called phase-changes that occur in evolution.

Richard J. Douthart received his B.S. in chemistry and mathematics from the University of Arkansas at Little Rock in 1962. He then went on to earn a Ph.D. degree in biophysical chemistry at the University of Illinois at Champaign, Illinois. His thesis advisor was Prof. Victor Bloomfield, and his thesis research was on hydrodynamics of virus and DNA. In 1968 he took a job in industrial research at the Lilly Research Laboratories in Indianapolis, Indiana. While at Lilly he spent many lunch hours discussing Evoluon Theory with Tilak. His research at Lilly included interactions of nucleic acids with drugs. In a series of studies he characterized the biophysical properties of double-stranded RNA. Collaborating with Dr. Walter Kleinschmidt, Douthart did some research on the anti-viral and anti-tumor effects of interferon. In 1976 Dr. Douthart entered the emerging new area of recombinant DNA research. He was actively involved in the recombinant insulin program. He established one of the first DNA sequencing laboratories in industry and developed one of the earliest in house proprietary computer system for genome analysis. In 1982 Dr. Douthart took a position as Director of Biotechnology at Battelle Pacific Northwest Laboratory. He later became Scientific Advisor to Life Sciences. Douthart retired from Battelle in 1995 to become Vice President and Research Director of Biological Information Technologies. In 1997 he was contacted by Tilak and has actively been engaged in the resurrection of the Evoluon Theory using the Internet rather than noontime discussions at Lilly. Through Tilak, Douthart and I met in 1997, and since that time we have engaged in an active philosophical and scientific dialogue pertaining to evolution and related issues. Dr. Douthart is an adjunct professor of Biochemistry at Washington State University (WSU) and Scientist Emeritus at Pacific Northwest National Laboratory. He has served on the board of directors of the CADMS (Computer Aided Design in Molecular Sciences) Laboratory at WSU which he helped to found and served on the Biotechnology Advisory Board, Rochester Institute of Technology. In addition he has served on a number of boards and panels for the National Science Foundation and the Department of Energy. Among his various awards are the prestigious IR-100 Award for outstanding technology development and a Scientific Digest outstanding investigator award. He received the Federal Laboratory Consortium Award for technology transfer for his role in transferring the CAGE (Computer Assisted Genetic Engineering) software he developed for minority institutions. He has received three Battelle outstanding software development awards for the genome view software interface to the human genome. As adjunct professor, Dr. Douthart has been involved as research advisor to four masters and two Ph.D. students. He has co-authored with Professor Keith

Dunker of WSU a series of theoretical papers on protein secondary structure and folding. Douthart has over 50 publications in various aspects of Molecular Biology, Genetic Engineering, Human Genome, and Computer Science. He has given over 200 posters and presentations at various universities, companies, government symposia and scientific meetings. He has six patents ranging from anti-tumor agents to DNA sequencing technology. As was suggested, Douthart was introduced to the Evoluon Theory during its early stages of development. He is now strengthening the Evoluon Theory by formalizing its basic principles and aligning it with recent developments in AI, Physics, Biology and related fields.

Sincerely,

Dr. Johnny Washington, Ph.D.   March 6, 1998

Scientists' Reactions: Additional Letters

Dr. Alain Locke (1886–1954)   Destinicity Letter #66

Re: Letters from Scientists*

Dear Dr. Locke:

The letters that follow are a sample of some letters that Tilak received; some of them endorsed without reservations the Evoluon Theory. Others seemed to express a degree of uncertainty regarding the validity of the theory.

Sincerely,

Dr. Johnny Washington, Ph.D.   March 24, 1998

*See note 4 to Chapter 7

President, Academy of Science
Stockholm, Sweden
October 24, 1979

Dear Sir:

After considerable hesitation and self-debate, I am taking the liberty to send you a copy of my enclosed papers on the General Unified Theory of

Evolution. Peter Medawar in his book, *Art of the Soluble,* stated that the earlier attempts by Herbert Spencer, Pierre Teilhard de Chardin and Arthur Koestler, to formulate a general theory of evolution, were not successful. The Evoluon Theory that I have developed now cites factual details from a variety of scientific disciplines to support the idea of a gradation of the complexity of evolutionary phenomena. Exciting is the verifiable physical relationships between the number demarcating the stages of evolution, and the constituents entering the evolutionary processes. Evoluon Theory that I have developed presents a novel viewpoint about the evolutionary essence of human existence. Actual fact of the exponentiations of rates of change makes this point of view important enough to be represented in the social, economic and political thought processes, especially since the superpowers, in fact all the major political powers, seem busy "perfecting" weapons of "automatic" destruction, in which all "acting and analytical" abilities are being transferred to the exobiological functionalities…

Respectfully yours,

Manohar A. Tilak

Dr. P. K. Kelkar, Director
Indian Institute of Technology
Bombay
Powai, Bombay
16 February 1973

Dear Tilak:

I was pleased to have your letter. Your reactions are interesting. I think one of the basic factors in relation to what one might call the man's purpose of existence is to become a conscious party in his evolving towards an objective [Destiny].

The fact that in the ultimate analysis, the knower and the known merge and there is nothing further to be done is only true in the case of rare individuals. It does not at all mean that the vast majority of the people who are at various levels of evolution can give up all conscious efforts if any progress at all is to be achieved. If a person does not feel the need to know who really he is, then, I think, there is no point in expecting him to be interested in 'Adhyatmic' progress. All the leaders in this area have again and again emphasized that 'many are called but few are chosen.' It is in this context that I think the totality of the environment inclusive of the person concerned has to be taken into account before the final goal can be achieved…This is precisely the point. Optimization of our senses through material resources does not lead to significant knowledge of the 'Self.' Whatever, therefore, the process of evolution we may think of in terms of autonomous material transformation leading to in the last analysis to something analogous to life, it still will remain a part and parcel of the sensory phenomena. I am reminded of a story of Shankaracharya. He was telling about the world being 'Maya' and an elephant came at that time and charged the group where Shankaracharya was giving his discourse. Naturally, Shankaracharya, along with the rest of them tried to run away from the fury of the elephant.

One of the listeners of the discourse asked him while running, 'if everything is 'Maya' the elephant also must be 'Maya' and why you are running away?' Shankaracharya immediately replied, 'if elephant is 'Maya,' so also is 'palayana' i.e. running away also is 'Maya.' I think this story illustrates my point. Your intellectual efforts are certainly very commendable and I do hope it starts a thinking process in the minds of a large number of people…
Yours sincerely,
P. K. Kelkar

The New York Academy of Sciences
2 East 63rd Street
New York, N.Y. 10021
November 19, 1976

Dear Dr. Tilak:

On September 24, 1976, we mailed you an invitation for membership in the New York Academy of Sciences. Having had no reply from you to date, we are interested in learning of your decision. May we hear from you regarding this matter?

Should you decide to accept the invitation, it will be a pleasure to present your name to the Board of Governors at its next meeting.
Sincerely
Philip Siekevits, Ph.D., president

Professor Dr. H. Kuhn
Max-Planck-Institute for Biophysikalische Chemie
D-3400 Gottingen-Nikolausberg
Postfach 968
November 16, 1973

Dear Tilak:

Thank you very much for your reprint. I agree with you in the main points, and you may find somewhat related thoughts in my paper "Angew," in *Chem., Intern. Ed.* Vol. II, p. 838–862 (1972). (Unfortunately I have no reprints anymore). It seems to me that evolution was a succession of divergent and convergent phases which come to an end by a certain logic limit in the information capacity of the genetic apparatus and a fundamentally new phase to overcome this limit has now started. I tried to trace the different phases of evolution and found that a logic framework can be given for the general lines…
H. Kuhn

The University of Texas at Houston
School of Public Health
P. O. Box 20186
Houston, Texas 77025
December 4, 1973

Dear Dr. Tilak:

About a month ago you were kind enough to send me a reprint of your "New Postulation on the General Theory of Evolution." I have read the piece with considerable interest, particularly because the drift of your thinking is along the lines that have occurred to me. My own evaluation of human evolution suggests that man is transforming the natural ecosystem into the human ecosystem and that the climax of the process will be characterized by very heavy reliance upon sociogenetic transmission of information. I am more optimistic than you, I guess, for I think that even though man does now and will continue to depend upon what you label "direct phase evolution" for such informational transmission, he will maintain control. I have humanistic biases, perhaps, but I do not see a "being" emerging from the process you described. I should be interested to read other papers that come from your continued thinking about these matters.

Sincerely,

Frederick Sargent II, M.D.

The Continuum in Nature

Dr. Alain Locke (1886–1954)   Destinicity Letter #67

Re: Continuum

Dear Dr. Locke:

I believe I have already alluded to the scientific notion of continuum as Tilak used. In this letter I shall provide an explanation as to what he seemed to have had in mind. At the outset allow me to jar your recollection regarding the concept "field" which I believe you borrowed from physics. In physics a field is a context in which a certain form of energy, say, gravity, is manifest. Henri Bergson suggested that Life or Spirit is the ultimate continuum of the universe. The Life continuum has three distinguishing features: instinct, intelligence and intuition. It should be noted, to my knowledge, that although Bergson doesn't use "continuum," it is implicit in his views. Peirce held that the continuum of space-time includes fractures or breaks through which indeterminacy is manifest. Similarly Bosanquet suggested that the ultimate continuum of reality contains fractures or cracks by which elements from the realm of transcendence seep into our immanent world. In a similar vein some physicists hold that there are wormholes in the space-time continuum that can possibly allow us, in space travel, to exceed the speed of light; to travel to other universes, for example, and thereby circumvent the law of causality as we currently understand it.[2]

In your axiology, you described a value situation as constituting a field, which, one may say, is analogous to the electromagnetic or the gravity field. The value field is constituted by the interaction of the items that constitute the scene and the individual judging the scene. You reminded us that the religious value field is constituted by the feeling of exaltation; the ethical value field, by the feeling of tension; the scientific or logical value field by the feeling of acceptance or agreement; and the aesthetic value field by the feeling of repose or equilibrium. In your view the value field is dynamic, not static. And although it may appear that there are a number of value fields, at bottom there is only one value field. Shortly it will become evident that the notion continuum parallels your notion of field. If Tilak attributes to the realm of nature a continuum, you attributed to the realm of consciousness or feeling what is called the value field. That is, consciousness or feeling constitutes the field in which our valuing experience occurs. There are different kinds of continuum in the universe. Or, what may amount to the same point: that there is one ultimate continuum, with various distinguished features. The immanent Destiny continuum of the universe includes singularity, duality, chemical interaction, biologic interaction and cultural interaction. The limits of the initial continuum are imposed by the fact that once it accumulates a certain degree of information, that is, once it achieves saturated complexity, it triggers the rise of the subsequent continuum in the process of self-transcendence. Thus, in this sense each of the evoluon occurrences and its concomitant continuum involves self-transcendence, with Evl8 as the ultimate goal of the self-transcendence process.

The continuum of big bang, singularity, Evl1, is represented by a state of affairs in which at the outset, ordinary matter and energy were excluded. i) Following the big bang characterized as the initial singularity, the cosmos evolved. Dr. Locke I have introduced a term here "big bang" with which you may not be familiar. When you died in 1954, I do not believe that the big bang hypothesis had begun to receive widespread attention in the media and scientific journals, although it was first established as early as 1927. The widespread dissemination of the big bang hypothesis did not begin to occur until the early 1960s. The big bang hypothesis was intended to explain the origin of the universe, according to which it is the result of a large explosion that occurred between fifteen and twenty billion years ago. A fraction of a second following the big bang, the second continuum of duality, arose. ii) The continuum of duality, Evl2, is exhibited by particle-antiparticle interaction, the hydrogen-helium interaction in fusion or fission, comparable to the Chinese Yin/Yang dynamic. In this continuum the synthesis of elements occurred in the early universe under the range of the boundary conditions of temperature and concentration of particles similar to conditions within the interior of the active stars. When the information manifest as elemental particles or atoms achicved a certain level of complexity, this duality continuum was overcome; it gave rise to the third continuum of chemical interaction. The early universe continued to cool and expand, and matter as we know it today began to form. iii) As was mentioned, the third continuum includes chemical interaction. In this process the information involved achieves such a level of complexity that duality alone underwent self-transcendence, resulting in the distribution of the three-dimensional molecular structure. The continuum of chemical interaction can occur only on a planet such as the Earth that offers a rather favorable range of thermodynamic conditions involving temperature, pressure and concentration, as well as compositional boundary conditions required. The presence of all the necessary interactive elements constitutes the boundary conditions, including the thermodynamic range. The dynamics of this continuum occurred until the structure and path equilibrium of chemical cycles were established, thereby giving rise to a replicative system, directed by its own schemata of information. When this process reached its informational limits, it again exercised self-transcendence as though it were enjoying a Hegelian synthesis; iv) Chemical evolution or interaction gave rise to the fourth continuum of biologic evolution as described by Charles Darwin. The biologic continuum, Evl4, is based on permutative linear sequence of the four nucleotides, including the structural units in the DNA of living cells, given three at a time for genetic coding of each amino acid e.g., from A, B, C, D: ABC, CED, DCB, ABD…Dr. Locke, in discussing biologic evolution I believe I am now in the terrain with which you were familiar. You seemed to have accepted Darwin's model of evolution that you introduced as a basis for cultural evolution. Did it occur to you that a given cultural or ethnic group, insofar as it may be regarded as a complex system, possesses its own informational schemata that provide the basis by which the cultural or ethnic group's identity is constituted? I suspect that I am getting ahead of myself. Perhaps the comments I made here fit more appropriately in the context below in which I discuss the fifth continuum. You wrote much about cultural dynamics and evolution. You avoided adopting a biological theory of race that many such as the Nazis in the German government adopted during the Second World War to provide a rationale for the superiority of their racial group. There are infinite possibilities…and a very large number of life form variations are possible. Appearance or disappearance (demise) of the individuals or a given species does not matter in this continuum; it simply is a turnover. This turnover in fact is essential in order that the fifth continuum described below can arise and perpetuate itself. There is a continuum that is in the background of bio-evolution based on micro-events occurring in the cell. The macro form of this is the complexity of the "nerve system." Is this really a new continuum? To shed light on this question, we have to look at the most basic and general characteristics of micro life. It is replication, or replacement, which is a continuous process of cells other than the nerve cell. In fact replication is the very defining characteristic of life. All of a sudden the nerve cell does not replicate, nor is it replaced in the manner that other cells are replicated or replaced. The nerve cell is also in continuous active communication with the brain.[3] Due to its extroverted tendencies, the nerve cell reaches outwardly, in its endeavor to include the entire cosmic range of effects. That is, the phenomenon of life, by means of the nerve cell, along with the faculty of the human brain, has stretched its arms to embrace the being of the cosmos. This observation is attested by the Bhagavad-Gita in chapter 15, stanza 7, that describes in reference to Purushottama, the Being of the Cosmic realm, where attention is called to the process of transcendence from chemical interaction to the phenomenon of life. Dr. Locke, although I have mentioned five continuum manifestations, the one that is quite interesting is the fourth one whose defining feature is the DNA molecule; the fourth continuum is significant here in that it is the stage that ushered in human beings. Its significance lies in characterizing information based on linear sequence of the four nucleotides in the DNA. The fact that the nerve cell does not replicate is telling. The reason is that nerve cells represent a sort of phase change, a dramatic shift in the recurring patterns in nature resulting in a phase change characterized by discontinuity, that is, a break with the past, and the advent of a new organizing pattern which makes possible human intelligence. Apparently the DNA apparatus alone became inadequate for the perpetuation of life. Again, a mode of self-transcendence was required. Intelligence came on the scene and altered the planet dramatically.

Certainly animals possess the nerve cells on which hinge instinct which they found adequate to assure their survival. However, the lower animals were merely transitional entities that set the table for the appearance of human beings with larger informational schemata. v) The occurrence of the human beings started the continuum of post biologic transformation based on human activity involving the non-biologic material on the Earth. This is what may be called the fifth continuum, Evl5, The information system of such a continuum involves symbols and language made possible by the complex nerve system and the brain. This non-biologic material was not involved in any discrete structural or functional manner before the occurrence of human beings. Furthermore, the rate of the non-biologic-micro and macro structural and functional transformations has steadily increased in comparison to the rate of change in the pre-historic past. Since the Industrial Revolution the rate of change has increased exponentially and has become even more so now that the recent computer revolution is well on the way.

A certain ever-renewing of the population is needed for evolution to occur, a process that is in part provided by aggression. This may shed light on the fact that in the international community, there are invariably military or political strife activities, however small, occurring at various places. In the same vein since their origin, human beings have been acting on their environment by means of tools and weapons. Human beings, who are pre-occupied with the usage of tools and weapons (Bergson), are instinctive, the source of which lies deep within the brain. The strength or aggression of certain individuals or groups is essential for survival; however, it is counter-productive for each individual or group to attempt to exercise excessive strength or aggression. The net effect, however, inescapably enables the nonbiologic systems to play a major role in our survival.

The increase of activities on the non-biologic environment, functionally, is quite impressive. Sharp stones, spears, bows, arrows, catapults, guns, and more recently machines, engines, missiles, space vehicles, computers and robots were utilized by human beings in performing certain tasks. Human tribes, individuals, human organizations, nations, are all simply interactive intermediaries in this new continuum of post-biologic transformations. This will give rise to Evl6, where AI will be highly developed.

At this point I will revert to the notion field attributable to the natural environment. According to Tilak, the field in which human beings exercise their presence in the world is an indestructible continuum of actions, an observation that is attested by the Bhagavad-Gita, stanza #40. Elsewhere the Bhagavad-Gita says there is an unbreakable stream of beings whose activities constitute the eternal continuum or field. This is not the end of the evolution path. This mega-continuum is only the middle field of experience that constitutes the world in which we find ourselves. If the fourth continuum underwent a mode of self-transcendence, so did the fifth continuum of which human beings are representative.

In the post-biological continuum, the one that is on the horizon, it does not matter how many individual tribes, organizations or nations are affected or eliminated. Tilak maintains that we must keep in mind that nature expressed no preferences or favor toward human beings. What matters is human beings' capacity to transform perpetually their environment. Time is of no consequence. The essence of this continuum is the transfer of much human functionality to the nonbiologic material base. The continuum about which we speak has been named "Direct Phase Evolution." This is an appropriate term in this context because the mechanism of evolution is direct, in contrast to the indirect mechanism involving natural selections or mutations found in biological evolution.

Indeed this post-biologic continuum within which evolution is occurring stands in contrast to the phenomena of evolution understood in the traditional sense established by Darwin. As will be seen this post-biologic realm also shall give rise to a new ethical model for robots.
Sincerely,
Dr. Johnny Washington, Ph.D.   March 30, 1998

Chapter 8

An Outline of the Evoluon Theory

A Synopsis of the Evoluon Theory

Dr. Alain Locke (1886–1954)   Destinicity Letter #68

Re: Evoluons

Dear Dr. Locke:

It should be clear to you by now how Manohar A. Tilak arrived at results formulated in his work. He relied in part on the method of extrapolation. As he researched the nature of chemical interaction, he ventured to extrapolate the evolutionary process beyond human beings to manifestations representing post biological or universal material base and hierarchy. He claims that the gradation of the evolutionary complexity can be shown to be related at different stages of evolution to the physical details associated with the evolutionary processes. Further, Tilak's Evoluon Theory focuses on the following concepts or principles, some with which you are already familiar. In the interest of clarifying the issues, I also wish to engage in some repetitive comments intermittently throughout this letter. First, there is a gradation of evolutionary complexities based on the following eight logical fundamental attributes of the evolutionary processes: i) Presence or being (unity); ii) difference; iii) interaction; iv) replication; v) interrelation; vi) universalization; vii) transcendence; and viii) intermergence; the latter three mentioned are at this point in time beyond human existence. Second, the concepts of the evolutionary phase change and the direct phase evolution are reflected as new discrete items in the phase of post biologic evolution. The mechanism of evolution in the direct phase evolution is direct and abstract, in contrast to the indirect and accidental features that are closely linked to the mechanism of bio-evolution. Third, the capacity to mobilize and process bits of information is one of the most important and deciding hierarchical factors in evolution. Fourth, physical evidence indicates that the exponentiations of the rates of change are marked by the Industrial Revolution, where information processing of the exobiological nature resulted in a new hierarchical stage for the postbiological phase of evolution. Fifth, results regarding evolution included in the Evoluon Theory, in contrast to those found in many of the other theories, are telling, the concerning of which was arrived at by assuming a panoramic view of reality. Tilak believes that the results here under discussion are more adequate than the results of the other disciplines such as sociology, anthropology, political science or economics, etc. Dr. Locke, in your essay, "Good Reading" you urged that an advantage of our reading philosophy is that it enables us to adopt a panoramic view of reality to achieve breadth and depth in understanding it. Thus, this panoramic view is similar to Bergson's view of the generality of knowledge introduced by the faculty of intuition that enables us to overcome the boundary conditions of the intellect. In a similar vein you reminded us that such a view can be arrived at by studying science as well. This statement can be attested to by both Tilak (a scientist) and by myself (a philosopher). In scientific investigations, empirical observation is one of the most important means of conducting research. In the Evoluon Theory, we rely on the empirical approach, in addition to seeking to integrate all items or processes that interact within the boundary conditions. Further, Tilak rejects the view that there is a multiplicity of hierarchical processes reflecting a basic structure of the final process, as though evolution has a number of "ends" or "final goals." Because many scientists or humanists educated in an environment that subscribed to the norms of scientific relativism and cultural pluralism that rest on multi-approaches and diverse methodologies, they seem to feel that there is a multiplicity of answers concerning the evolutionary origin or destiny of life. Dr. Locke, you, yourself, subscribed to scientific relativism and cultural pluralism, and in today's society there is much talk about cultural or ethnic diversity. Considerations regarding cultural or ethnic diversity is often introduced as an alternative to racial integration or separation in the context of dealing with racial problems not only in the U.S. but in areas such as Ethiopia, the Sudan, the Middle East, Northern Ireland, Eastern Europe and other parts of the world, where racial or ethnic relations are unstable. Nevertheless, Tilak thinks that in the case of the self-organizing nature of matter, composition and thermodynamic parameters served as boundary conditions within which the summation of all interactions move toward a single unique outcome: the activity of replication, a determining factor of life. Implied in this notion of replication is the notion of unity, of which a living organism is paradigmatic. In order for replication to occur, the replicating item or organism must enjoy a high degree of unity. I shall add that insofar as replication hinges on the unity of the organism, Tilak's view of the goal of evolution is in harmony with my view of the Destiny model whose transcendent mode involves unity. True, the Destiny model also allows for diversity and conflict, manifest in the immanent realm of Destiny that includes the field of evolution.

Today there is much talk about ethnic or cultural diversity. I like the way that Tilak attempts to explain the nature of biological diversity, another area of contemporary interest. Here he comes at this issue from a different perspective. He says that accidental fits or the merely contingent modes in evolution, which allow for the possibility of diversity in nature, are not perpetuated indefinitely as hierarchical processes, or established as an alternative continuum, Eigen and Schuster's views to the contrary notwithstanding. Rather, Tilak is of the opinion that the entire spectrum of interactions moves along the minimum energy difference, maximum reaction rate, cyclic interaction path involving, on a global level, minimum diversity of chemical structural balance, from which replication emerges as the single and unique dynamically stabilized continuum of hierarchical system. Although diversity among plants and animals seems to be pervasive throughout nature, the principle of unity seems to outweigh the principle of diversity in nature. In any event nature seems to strike a balance between unity and diversity. This point regarding unity and diversity in nature might be of special interest to researchers today who are interested in understanding ethnic, racial, gender, political and religious diversity (and unity), an issue with which you, yourself, Dr. Locke were concerned. As I write this letter to you, I am reminded of your essay, "The Concept of Race as Applied to Social Culture,"[1] where you sought to explain the nature of ethnicity, that is, the difference as well as common threads among various racial groups. And in your essay "Races and Peoples" that relied on anthropologist Franklin Henry Giddings' main work you sought to explain the origin(s) of the human race and the manner in which racial diversity occurred. You were especially interested in the origin of the blond type found in the white race.

Let me get back to Tilak's views. He holds that besides the universal and unique nature of biological hierarchy on the Earth, we have another illustration of the emergence of a hierarchical singularity in bio-intelligence.

From among all the multitudes of species provided by nature, as interactive entities of a dynamically stabilized continuum based on a common replicative hierarchy i.e., the biosphere, there emerged over billions of years, along the path of blind, accidental mutations, only a single species with the bio-intelligence of the type represented by human beings. In his work, Eigen makes a statement that the exact path along which life originated cannot be ascertained by empirical means. Nevertheless, as was mentioned, in Tilak's view, the Evoluon Theory is capable of extrapolating to the highest level of generality, the end results of the total integration of all the interactions occurring within the boundary conditions, that is, the entire cosmic Destiny mode. Tilak seems to have been relying on a research method similar to the one offered by Husserl who invented Phenomenology, by which one is in a position to describe a particular scene or the broad field of human existence. You, yourself, seemed to have relied on phenomenology in formulating your axiology as well as developing your view of ethnicity. What I am trying to do in this work is to link the issues of axiology, ethnicity, Destinicity, Destiny, culture and history to the Evoluon Theory, that is, evolution generally, as described by Tilak. In evolution, since the hierarchy itself arises out of a basic logic that is intrinsic to a given, evolving system as a determining factor, of the evolutionary processes, all aspects of the phenomena necessarily effectively fit into the hierarchical constraints, and every process in the system of systems is interconnected. Otherwise one is simply observing an evolutionary phase change, based on the change of boundary conditions. Despite the differentiation in the untold number of species, past and present, all living things are based on a unique hierarchy common to the biosphere. Within the tolerable variations in the boundary conditions, including temperature, pressure, concentration and composition, the globally integrated evolution leads to the dynamic stabilization of interaction as a unique replicative process. That is, it leads to a hierarchy singularity of the biosphere. Outside of the tolerable boundary conditions, replicative stabilization will not occur. There can be no life as we know it on Venus or on Pluto; one is too hot, the other is too cold. Yet, AI can prevail in such seemingly hostile regions of the universe, and some believe that in the not too distant future we will be able to upload by means of lasers our personal identities into robots that we will have transported to such remote regions of space. Further, some believe that the boundary condition on Mars can be modified in such a manner to be conducive to life as we know it. Already the U.S. government is making plans to send human beings to Mars, a journey that would require a three-year non-stop flight, relying on today's technology. Dr. Locke, I suspect you would be pleasantly surprised to gather that the U.S. landed an unmanned space-craft on Mars on July 4, 1997 in search of life. Some scientists believe that they have found traces of Martian life and that there seems to have been a great Martian flood millions of years ago. Thus there are some similarities between the boundary conditions of Mars and those of Earth. Moreover, scientists draw attention to a number of similarities between boundary conditions of one of Jupiter's moons and boundary conditions of the Earth. [On September 4, 1998 a news report indicted that scientists have determined that billions of gallons of water lay frozen below the moon's surface]. Tilak holds that within the boundary conditions, the outcome at the hierarchical level will be unique. Biological evolution involves a nonlinear process. Tilak informs us that scientists whose thinking is conditioned by linear phenomena have to overcome certain obstacles in order to appreciate fully the evolutionary phenomena. There are some illustrations concerning how linear scientific thinking gives rise to erroneous results when the scientists seek to explain evolution. It will be remembered that Bergson informed us that the scientists had a difficult time understanding evolution and reality generally because, relying on concepts produced by the intellect, they had an inadequate model of time. According to Tilak, the overwhelming complexity of the living systems, as well as the entropy factors grounded in the laws of thermodynamics, a phenomena that many scientists did not completely understand, has led to some erroneous conclusions. Thus originated the erroneous notion of a single primordial cell that was assumed to have occurred accidentally only once and then "took on a life of its own" and subsequently gave rise to the diverse forms of life on Earth. Rather, Tilak is of the view that the whole field of the biochemical interaction moved towards replication, then endogenized and formed living cells that in turn differentiated and provided for the possibility of interaction through differentiation and aggression in the dynamically stabilized field of replication. Seen in this light, we are now in a position to offer a fresh definition of life whose distinguishing feature is replication. Moreover, one may say that the activity of human intelligence also needs a new definition. In the Evoluon Theory, the principle of life that arose at Evl4, as well as intelligence that arose at Evl5, is construed as constituting an evoluon, that is, a discrete stage of evolution. Thus the Evoluon Theory provides a description concerning chemical interaction and replication, each of which is internally determined or governed by its own dynamism. The essence of interaction is change and the distinguishing feature of replication is stabilization of the dynamics of the system. There is an evolutionary phase change between interaction and replication. Eigen and Schuster considered what may be called the "internal dynamics" of chemical interaction and replication phenomena. They also stated, emphatically, that the chemical interaction cycle at a minimum has three members, the number of elements that Tilak independently arrived at as characterizing the continuum of chemical interaction. They held that the process of replication involved in the RNA molecules is an unchanged cycle representing higher order of organization containing information of its own reproduction, a cycle with four members, that is, four nucleotides. Again this confirms Tilak's claim regarding the continuum of life that has four members which he assigned to Evl4. These evolutionary stages enable us to make certain claims, by means of extrapolation and analogy, regarding evolution beyond human or Earth phenomena. Projected at the stage of universalization is the occurrence of a Being based on the universal hierarchy, in contrast to the biological hierarchy that seems to be Earth-bounded. Such a Being has no birth or death. Yet it will be cognitive of its own existence but will not be limited by replication. It will effectively step outside of the boundary conditions that we described earlier. Yet, as in the case of any biological organism, it will be in a position to reconstitute its own kind. That is, mere replication will not be the determining attribute at the stage of universalization. Further, there will be a tremendous communication gap, on the one hand, between this Being and, on the other, human beings, just as there is a sizable communication gap between an amoeba and a person. In Tilak's view this is one of the reasons an explanation as to why human beings, heretofore, despite their efforts, encounter no communication that can be understood from creatures in interstellar space, assuming that such creatures exist. If we traveled at the speed of light, it would take over 250,000 years to escape the boundary conditions of our own galaxy. Both the evoluon of transcendence and evoluon of intermergence are beyond Evl6, the stage of universalization. Results published by Manfred Eigen and Peter Schuster indeed confirm the concepts of phase change expressed in the Evoluon Theory. Much of the foregoing observations have implications for environmental issues. Although all of life interacts with the environment, human beings' interaction is of a totally different order of magnitude, variety and complexity. Photosynthesis caused spectacular changes in the entire atmospheric composition of the primordial Earth, but the change still was limited and operated through the cellular mechanism. Every change in bioevolution must pass through a highly constrained biologic system. We have begun to exceed the boundary conditions of our planet whose conditions were established for us during billions of years of evolution. However, in doing so, we have given rise to some unhappy consequences that include air and water pollution and other unhappy environmental effects. If we damage one part of our environment, through, say, water pollution, we threaten the entire planet. Witness the many bad weather patterns, floods, tornadoes, droughts, mud slides, hurricanes, tidal waves and ice storms that the U.S. or other parts of the world has begun to experience with greater rapidity or intensity within the few past decades. Some of the views included in the Evoluon Theory provide the foundation for an environmental ethics. Dr. Locke, your axiology also provides the basis of an environmental ethics. It is imperative that any environmental ethics must include the principle of caring.

Sincerely,

Dr. Johnny Washington, Ph.D.   May 1, 1998

A New Evoluon Postulate

Dr. Alain Locke (1886–1954)   Destinicity Letter #69

Re: A New Postulate

Dear Dr. Locke:

As you were well aware, every academic discipline rests on a set of assumptions or postulates. Kant's postulates included the ideals of God, Freedom and Immortality. Similarly in both your axiology and theory of ethnicity or culture, you introduced certain postulates that you called functional equivalents that you construed also as operative in physics. Manohar A. Tilak has introduced a new postulate on which his views of evolution hinge formulated in his article, "News Postulate on the General Theory of Evolution…" You, Dr. Locke, were born a quarter of a century following the publication of Charles R. Darwin's book, *The Origin of Species*. It will be remembered that the publication caused a great controversy from the time it was first published up to the present. Many were threatened by the fact that Darwin had the audacity to inform us that we are not angelic creatures apart from nature. Rather, it is the case that apes, other primates, and human beings have a common ancestor; this is one of Darwin's assumptions or postulates. Bergson also tried to reconcile evolution and human freedom, maintaining that we are a part of nature, keyed to its principle of biological and physical necessity, a feature of which is expressed in the form of instinct, exercised during the course of performing one's moral duties in the closed society. Yet we exercise freedom, a degree of which is exercised by virtue of the intellect. When we employ the faculty of intuition which is much higher than the intellect—just as the intellect is higher than instinct—we enjoy the highest degree of freedom. The attempt to reconcile the two, freedom and necessity, is a difficult task. Bergson tried to solve the freedom problem by introducing the notion of "free causality." As you knew, Dr. Locke, Kant had a problem demonstrating how we can be both free and not free, members of the realm of dignity, on the one hand, and the realm of nature, of necessity, on the other.

During some of the earlier discussions I had with Tilak, I queried him as to how he approached the free will problem. In reply he seemed to have reverted to the Vedic tradition. He held that we possess free will so long as we do not become too attached to the affairs, activities or items of the world. Such attachment tends to rob us of our freedom. Already I have devoted attention describing the continuum of nature that I compared to your view of "field." I think such a comparison is limited, however.

Here it is anticipated that a more coherent picture of Tilak's view shall emerge, if we juxtapose the article that focuses on his idea of continuum with the one I shall now present. We recognized two major occurrences in evolution: i) the beginning of life, and ii) the emergence of human intelligence, the last mentioned of which I have elsewhere described as a point of "singularity" located in East Africa, in and around what is now Kenya or East Africa generally. This point about East Africa enables me to make the transition between my African history letters and those on which I am presently focusing, the Evoluon Theory. First, let us consider the development of life and intelligence. Since Darwin's time several major advances in cosmology, chemistry, physics, biology, AI and most of the other scientific fields have been made. Slowly a coherent picture of the evolution of our universe is emerging.

Sincerely,

Dr. Johnny Washington, Ph.D.   April 2, 1998

CC: Dr. W.E.B. DuBois

Life, Intelligence and Information

Dr. Alain Locke (1886–1954)   Destinicity Letter #70

Re: Destiny

Dear Dr. Locke:

The intellect is the faculty that distinguishes human beings from the lower animals, a point that is generally accepted. It will be remembered that Manohar A. Tilak held that items in our environment, whether animate or inanimate, by virtue of their structure and functionality, involve information that can be extracted or imparted by the organism in its struggle to survive. By interacting with the environment, the organism acquires information that is accumulated in what may be called its schemata (in the form or instinct or intellect). Such information was embedded, as it were, in the mere structure and functionality of the materials within our environment. In the message below, Tilak is maintaining that because matter contains certain information, and because matter is malleable and can be mobilized by the organism, it can be incorporated into the organism's immediate environment and thereby enhance or retard the organism's evolutionary process. I take it that, all things considered, this is what Tilak has in mind by the term "evolvability"; structures and features in the environment are malleable and possess the capacity to evolve along with us human beings. Witness Gell-Mann's view that complex systems, including computers, for example, tend to adopt schemata that take on a "life and career" of their own. Gell-Mann did an ample job in making a comparison between schemata that human beings employ and those that some other complex system such as a computer or an economy may employ, but he did not effectively demonstrate which features distinguish human beings from other complex systems. In passing, I shall mention the fact that a human being has the capacity to evaluate, or make judgments about, the information with which it is confronted; in so doing it regards some of the information as enhancing or inhibiting his or her self-efficaciousness as well as self-esteem. As Bergson would say, memory also plays a major role in this evaluative process; it adds tone or quality to one's experiences. Heretofore complex systems, other than human beings, have a quite limited, if any, capacity to make such value judgments regarding the information it has accumulated. Some researchers, including Tilak and Douthart, as well as myself, hold that in the not too distant future, in Evl6 and beyond, robots will have developed the capacity to make value judgments. Insects and other lower forms of life on the basis of chemical reactions or the nerve system, within which information is stored, react "instinctively" to their environment in a reflex manner. There is little, if any, room for indecision or hesitation in interacting with the environment. Notice how "instinctively" a roach or a mouse reacts in the presence of food. Whereas human beings, by virtue of their modes of perceiving their environment as well as by memory, often engage in behavior that involves indecision, or hesitation. Their perceptual apparatuses and memory that are products of the brain's capacity make possible freedom or consciousness, features that no other complex system at this point in time possesses. There are reciprocal interactions between the organism and its environment, by which the exchange of information occurs. We as human beings with a highly evolved intellect, have begun speeding up the evolutionary process.There is another interesting parallel between Bergson's view of energy and Tilak's view of information. When Bergson began to explore the possibility as to whether there is life in other regions of the universe, he gave a positive reply. His basic presupposition is that all of reality arose out of or is grounded in what he called *élan vital,* a universal spiritual force that is manifest in the form of Life. Life animates matter; in fact Life and matter interpenetrate each other, and the former seems to be pervasive. Matter is animated throughout the universe; and thus intelligence in varying degrees is animated throughout the universe. When certain forms of energy are accumulated, wherever such forms are located in the universe, they will burst forth as life, and no doubt undergo evolution and diversification in a manner similar to life on Earth. In *The Two Sources,* Bergson wrote: "…it is probable that life animates all the planets revolving around the stars. It doubtless takes, by reason of the diversity of conditions in which it exists, the most varied forms, some very remote from what we imagine them to be; but its essence is everywhere the same, a slow accumulation of potential energy to be spent suddenly in free action" (p. 255). Consider here the notion of information that was highlighted in Tilak's essay, "The Evoluon Theory: A General Unified Theory of Evolution…" (see Appendix A). Keep in mind that Tilak is employing "information" here in the broadest sense to include the structure and function of an object, along with chemical interactions at the molecular level. Our customary understanding of "information" includes data that runs computers and robots, and that which is found in books, magazines, face-to-face dialogue, radio, and TV, etc., the category to which Tilak occasionally shifts. In my usage of information in this context, I will be considering our customary usage of "information." I wish to begin by saying that we often employ "information" interchangeably with "knowledge," "data" or "facts." I believe we might need to distinguish between the first two concepts. It was Plato who held that knowledge is intrinsically good. I doubt whether he would hold that any and all information is intrinsically good. The insights embedded in Einstein's theory of general relativity may be called knowledge. The list of names and addresses included in the Manhattan phone book is mere information. Ignoring the distinction between empirical and *a priori* knowledge, generally knowledge involves information supported by evidence, and when it becomes organized it informs us about the world. One may be said to be knowledgeable about a state of affairs, provided that one sees the interconnections of the ideas he or she may entertain. If I "study" Einstein's theory of relativity to the extent that I can "see" the connections of its meaning, then I may be said to have knowledge of his views. Apart from the mere alphabetical order in which the names and numbers in the Manhattan phone book are organized, we would not be amiss if we hold that such a book contains no knowledge; it merely contains information or bits of data regarding a certain state of affairs. Some social critics hold that the mass media has begun to flood us with so much data that the bits of information are tantamount to the data found in a Manhattan phone book. We are "flooded" with information, ranging from stock prices to baseball scores to the price of hogs in Peoria or the price of cloth in Bombay. Some believe that the human brain is unable to "process" such information to sort out the useful from the useless. In some of my previous messages to you, Dr. Locke, I reminded you of my notion of Destiny that stresses unity as an antitoxin to the fragmented, disjointed experiences that we suffer in contemporary society. Knowledge in the Platonic sense may be intrinsically good; it does not follow that mere information is intrinsically or instrumentally good or useful. The accumulation of certain bits of information could easily result in what Spinoza (or Hegel) called inadequate knowledge. The consequences of acting on such inadequate knowledge can be fatal to the individual or society in general. Inadequate knowledge in many instances may be a source of moral evil. It was Leibniz who held that moral evil resulted largely from our limited perspectives in interacting within our environment. My claim is that much of the disconnected, fragmented information that we tend to accumulate plays a major role in limiting our perspectives. An advantage of the Destiny model is that it allows us as individuals or groups to overcome the narrow perspectives. In your value theory, you indicated that our dogmatic attitudes that often result in narrow perspectives and intolerance bring about moral evil. You disliked what you called value ultimates associated with dogmatism. Similarly Bosanquet held that moral evil was due to our subscribing to what he called "false absolutes," analogous to your value ultimates. I believe it would have been instructive had Tilak developed more thoroughly the connection between information (facts) and values. Reflective criticism as well as testing in experience is required to remove or route erroneous information, and conflicting values that do not square with experiences in establishing an adequate basis for our genuine knowledge through which we are liberated. Erroneous information can play an inhibiting role. Evil is whatever produces excessive inhibitions, whatever stifles creativity or growth. When Tilak discusses the role of information in the evolutionary process, he occasionally makes it seem as though most of the information is useful, true or good. As Gell-Mann tells us, although the schemata's role lies in part to provide the organism useful information about its environment, it does so my compressing the data stream of complexity of information into a common thread. Irregularities or random junk do arise, that is, useless or false information does enter into our schemata. If the organism acts on such false or erroneous information, it can result in the destruction of the organism.

Sincerely,

Dr. Johnny Washington, Ph.D.   April 17, 1998

CC: Dr. W.E.B. DuBois

Ethics and Scientific Research

Dr. Alain Locke (1886–1954)   Destinicity Letter #71

Re: History, Destiny and Evolution:

Dear Dr. Locke:

 Recently there has been much excitement within the scientific community as well as within the general public regarding the cloning of a sheep by some Scottish scientists who have begun to perfect the practices of genetic engineering. Most national governments have now enacted laws that prohibit the cloning of human beings. Scientists are now discussing the feasibility of such cloning in the future. Such an experiment raises many ethical questions, would you not agree? The foregoing observations lead us to consider the limitations of most traditional ethical views, including John Dewey's as well as your owns. As you were well aware, Dr. Locke, Dewey's ethics is linked to his epistemology. His point was that we have no *a priori* notion of right or wrong, good or bad. Reality is a dynamic, evolving process. Rules of right conduct are no more than hypotheses to be established in guiding experiments or human conduct. We do not know beforehand what the right course of action is. First we have to conduct the appropriate experiments. If we apply this scenario to the cloning illustration, we face a dilemma. Some hold that the issue itself is inherently morally reprehensible, or objectionable. No such experiment should be allowed. Are we up against a taboo? In Dewey's view, virtually any possible course of action or endeavor is available to be included in an experiment in which we examine its antecedent causal conditions, and future possibilities, in determining its possible or actual value. You, yourself, held that all values are merely claims that may or may not be valid. That the determination of the validity of a given value claim is achieved through testing in experience. We will not know whether it is right or wrong to clone human beings until we first do so by means of experiment. Your views as well as those of Dewey are open to the claim that any course of scientific investigation possesses moral worth. As I reflect on this line of thinking, a number of questions come to my mind. I shall mention a case that harbored several moral issues; namely, the syphilis experiment conducted at Tuskegee, Alabama's local Veteran Administration Hospital during the 1930s and lasted several decades. African American males were used as guinea pigs without their informed consent. A part of the experiment involved injecting the men with syphilis bacteria to ascertain the prolonged effects of the disease when allowed to go untreated. This experiment continued to occur long after they had discovered penicillin that was used as a cure for the disease. In fact some hold that the scientists outright deceived those participating in the experiment. Are those who approved of such an experiment blameworthy? I believe so. Any future experiments of that nature should be disallowed. This case seems to demonstrate that instances can arise where the search for Truth or Knowledge can—and often does—conflict with the Good and the Right. I myself

292

am not surprised by the fact that the Tuskegee experiment did occur. It is likely that it could have been prevented, or conducted in accordance with the Kantian categorical imperative, had the experimenters subscribed to the ethics of *caring* considered in an April 5, 1997 Destinicity letter addressed to Dr. DuBois. Tilak would say that such an experiment, along with the above-mentioned sheep cloning experiment, is largely a consequence of the phase change that the world is now undergoing. As Tilak points out, the phase change harbors many sorts of moral perplexities. In many instances our intellect is unable to comprehend the magnitude of the problem or its ethical ramifications. On the basis of what we have said so far regarding the point that each item of the universe involves information, it does not take much effort to embrace the claim that such bits of information can be used by an evolving complex system to acquire sophisticated schemata by means of which its evolving system takes on a life of its own, and pursue its own immanent Destiny. That is what Tilak seems to have in mind by "direct phase evolution," a view on which scientist Murray Gell-Mann would no doubt agree, I believe. Whether or not such a novel evolutionary being can ever acquire a sense of its transcendent Destiny in contrast to its immanent Destiny is a matter of debate. I do not think that many lay persons would doubt the fact that the direct phase evolution comparable to the sheep experiment described above is an actual occurrence; that the cloning of human beings is a real possibility; that we can in fact perform genetic engineering. However, many untutored minds might have a degree of difficulty grasping the meaning or ramification of direct phase evolution as intended by Tilak. I believe your mind would openly entertain the point that Tilak is trying to make. When you reviewed Roland Dixon's book, *The Racial History of Man*, you reminded us of the importance of genetic research on animals, including human beings. You were interested in understanding the nature of a hybrid race, of which, in your view, African Americans were representative. In examining the hybrid race you held that we can be in a greater position to understand the linkage between physical or biological features of race and the cultural features of race, a distinction that is often ignored or overlooked.

Sincerely,

Dr. Johnny Washington, Ph.D.   April 18, 1998

PS. [During the week of June 26, 2000 it was announced in the news that scientists have mapped the human gene pool. Some fear such knowledge by the rich and powerful might be used to the detriment of the weak and poor. Further it was announced February 10, 2001 that the scientists have determined that some thirty thousand genes are responsible for the formation of human beings, and that much of our makeup is determined by our environment].

Chapter 9

Evoluon Theory and Philosophy

Edmund Husserl: Phenomenology

Dr. Alain Locke (1886–1954)   Destinicity Letter #72

Re: Husserl's Contribution to Phenomenology

Dear Dr. Locke:

In this letter I will examine more thoroughly Edmund Husserl's contributions to phenomenology. Husserl (1859–1938) is indeed one of the most important philosophers of the 20th century. In addition to being one of the greatest contemporary philosophers of our time, along with Russell, Bergson, Dewey and Whitehead, yourself included, he had a great understanding of modern science. At the University of Leipzig, he studied Mathematics, Physics and Astronomy. You as well as Dr. DuBois had the opportunity, while in Europe, to study with some of Husserl's students. As you are well aware Husserl soon became disillusioned with Western science because in his views it subscribed to epistemological relativism, and it abandoned the search of absolute truth(s). He believed that the ancient Greek philosophers, Plato and Aristotle, had it right. They stressed contemplative knowledge, a kind of absolute knowledge, the importance of which allows one to achieve detachment from what he calls the "natural standpoint," the everyday world with which the average person as well as the scientist, is familiar. The problem with knowing this world is that our mind is filled with prejudice, presuppositions and assumptions that tend to distort our knowledge, a point with which you would agree.

Thus, the relativism of modern physics is problematic in that the scientific paradigm has been accepted by all the other sciences, including the social sciences, especially psychology, the science that ought to be concerned with the study of the spirit. According to Husserl, the fact that science abandons absolute certainty is one of the reasons for the major cultural crises of modern Europeans. You would find objectionable his quest for certainty. Husserl died in 1938, and before his death the Nazis had already come to power. He blamed their coming to power, and most of the other crises modern Europeans suffered, on relativism. By contrast, his contemporary, John Dewey, one of the founders of Pragmatism, attributed the crises of modernity to the quest for certainty; thus the title of his book, *The Quest for Certainty*. One of the consequences of subscribing to scientific relativism is that it sets up a dichotomy between the physical world, the natural standpoint, and the psyche process, the realm of the spirit.

Husserl reminded us that objective science has not adequately deepened our understanding of the subjective dimensions of human experience. Modern science lost interest in the emotional, spiritual side of human beings. Behavioral psychology, whose model is physics, was the dominant psychology during Husserl's times. Like physics, psychology also subscribed to the natural standpoint, with all of its prejudices, presuppositions and metaphysical baggage, including dualism, or the mechanistic model of human nature. The remedy he offered lay in what he called the Phenomenological Method, on which I will elaborate further shortly.

Three philosophers that had much influence on Husserl were Kant, Brentano and Descartes. In previous letters I mentioned your interest in the teachings of Brentano from whose works you drew in developing your axiology. Kant taught us that because of the manner in which the mind is structured, owing to the twelve categories of the mind, including space and time, we cannot engage in traditional metaphysical investigation such as attempting to know the nature of God, or whether or not the soul is immortal. We have to accept on faith the validity or meaningfulness of such ideals. The reason is that the categories of the mind are limited to the world of appearances, to color, shape, size, taste, smell, along with all of the so-called primary qualities, items of complexity, associated with things as they appear. We cannot know what he called a thing-in-itself, the reality of a thing. Kant assumed that things-in-themselves existed in the realm of what he called the noumena world. Thus Kant made a distinction between the realm of nature, the world of appearance (which is not an illusion) and the realm of dignity, within which reside our moral, religious and aesthetic values, the noumena world. Scientific knowledge is limited to this world. Husserl agreed with Kant that we can only know the phenomena world, the world of appearances, of phenomena. Hence the importance of the terms

"Phenomenology." Husserl's ultimate goal was to develop a science, called the Phenomenological Method, to study or *describe* the phenomena world. As I mentioned he denied that there is a noumena world. This is where he parted company with Kant.

In view of this, Husserl has inherited a problem with which people such as Locke, Hume and Kant grappled. Are the "items" of experience merely in consciousness or are they exclusively in the world? Or, are they in both consciousness and the world? To put this another way, when I perceive the red dot on my wall, where is the dot actually located? Is the dot on the wall as an objective entity that physics can investigate regarding its physical properties, or is the dot merely in my mind? It will be recalled that Bishop Berkeley, a British empiricist, held that "to be is to be perceived." Kant solved this problem by introducing what he called the *a priori* categories of the mind. (To elaborate on how Kant actually solved this problem will take us too far afield).

Like you, Husserl rejected Kant's approach to this problem, because Kant has fallen into the trap of establishing a dichotomy between mind and body. Kant accepted the Newtonian paradigm of reality. Husserl's solution to the dualism problem, the spiritual or mental world on the one hand and the physical world on the other, is found in the teachings of Brentano. Brentano's notion of intentionality is the key here. Consciousness is not a substance, as Descartes would have us to believe. Consciousness is "constituted," as it were, by its acts of intending, that is, by acts of always being directed toward something. Consciousness has an aim, a directionality. Each act of consciousness points towards an object. Consciousness is always either directed toward items in phenomena world, the world of houses, dogs, fish, stars, ice-field, quarks or clouds; or consciousness is directed "inwardly" to the acts of, say, remembering, imagining or dreaming. The latter category is just as "real" as the former category. To be conscious of my house is merely to "intend" my house. To be conscious of say, a black hole, is merely to intend such an object. What sort of ontological status my house or a black hole possesses, that is, what sort of reality such an object possesses for Husserl is without meaning. His Phenomenological Method is merely to describe such entities that are intended by consciousness. A main point to remember is that he does not want to make what may be called any ontological commitments, to attribute any underlying reality to a thing. That is, he does not want to say that only material things exist or are real or that only spiritual things exist or are real.

What do we actually describe when we describe objects in the phenomena world? Objects possess essences, features that they have in common. The Phenomenological Method is to assist us in describing the essences of things. Husserl's method was intended to provide an objective basis of the inner, subjective experience on which Nietzsche and Kierkegaard focused. Relying on the Phenomenological Method, the teachings of Heidegger and Sartre also concentrated on the subjective dimensions of human experience. The consideration of Husserl's method provides an ample bridge between the field of philosophy and field of science. The Destiny model, as well as the Evoluon Theory, draws on both science and philosophy. Dr. Manohar A. Tilak seeks to provide a description as well as an ontological analysis of the field of evolution.

Sincerely,

Dr. Johnny Washington, Ph.D.   March 4, 1998

Evolution of Consciousness: Fichte's Model

Dr. Alain Locke (1886–1954)   Destinicity Letter #73

Re: Fichte's Model

Dear Dr. Locke:

Heretofore in making commentaries on the views of Tilak, I have drawn on the views of Johann Gottlieb Fichte, David Hume, John Dewey and Henri Bergson. In this letter I will shift gears somewhat and briefly revert to Fichte's views that were largely overshadowed by those of Hegel. If the Evoluon Theory describes physical/biological complexity, it will be seen that Fichte's views describe spiritual complexity. Not only that but it will become clear, as we examine Fichte's views, World Destinicity and the Ego are analogous, and the former undergoes phase changes in a manner similar to the phase changes that occur in cosmic Destiny mode. Both Fichte and Hegel lived and worked prior to Darwin's 1859 publication, *The Origin of Species*. Like Hegel, Fichte's dialectical logic enabled him to subscribe to a dynamic view of history that unfolds in accordance with a principle of progression and developmental stages. In this regard his views of history paralleled Darwin's as well as Tilak's theory of evolution, it will become evident. There are many parallels between Fichte and Bergson as well as Nietzsche's views; each, for example, is in the voluntaristic philosophic tradition. Tilak does not seem to have much to say about human history. If Darwin were concerned with the evolution of life, Fichte was concerned with the "evolution" of Spirit, that is, the Ego, as it progressed from the level of pre-consciousness to the Absolute where the Spirit comes to know itself and enjoys absolute freedom and unity. Dr. Locke, I trust that you recalled that in one of my letters to you I discussed Tilak's view of the continuum and the concomitant schemata associated with each distinct feature, or periodicity, of the continuum. Similarly in a portion of his paper, "The Evoluon Theory: A General Unified Theory of Evolution" (Appendix B), Tilak reverted to what may be called the features of periodicity and other events associated with the evolutionary continuum. However, in this context, the accent is not so much on the continuum but on what he called an evoluon, periodicity, saturation and change of phase. I wish to bring to greater light some of Tilak's views expressed in his essay by drawing parallels between his views and those of Fichte, with whose views you were familiar. I will also briefly consider Hume as well as Kant's model of consciousness or mind. One of Fichte's key notions is Ego, which, elsewhere, I have compared to the point of singularity that resulted in the big bang associated with the origin of the universe. He began by describing the manner in which the Ego or Consciousness or Spirit seeks to approximate its immanent or transcendent Destiny mode. The ultimate aim of the Ego is to seek self-knowledge or freedom. Today some scientists are also placing emphasis on what may be called cosmic self-knowledge. As Eric Chaisson in his book *The Life Era* wrote:

> Only when complex organisms arrive at the dawn of the Life Era does *the Universe* acquire self-awareness, a knowing reality. In the words of biology's Nobel laureate George Wald, 'Matter has reached the point of beginning to know itself.' We are, he continues, 'a star's way of knowing about stars.' This, for me, is life's purpose and meaning, its raison d'être— to act as an animated conduit for the Universe's self-reflection. In short, we sentient humans are now among the purveyors of cosmic consciousness. Above all else, this is what grants us, not individually but as a species, a magnanimous worth and dignity among all creatures on planet Earth, indeed, among all known structures in the Universe.[1]

According to Fichte, in seeking self-knowledge, the Ego first postulated its own existence, followed by the postulation of something that is not itself, the non-ego, that is, the world. Here we have opposition between the Ego and non-ego. In its primordial stage the Ego (God or Brahma) enjoyed absolute unity, a unity that was disrupted by the postulation of the non-ego, the world whose existence was necessary for the enrichment of consciousness. This origin of the world hypothesis included in Fichte's model parallels in part the big bang hypothesis on which hinges contemporary cosmology. It may well have been the case that the initial postulation of the non-ego by the Ego may have triggered the big bang event, out of which arose the universe. In this context a question often arises, "What existed prior to the occurrence of the big bang, or before the beginning of space and time?"—the kind of question that a five year old child or a physics professor may ask. When our oldest daughter, Latanya, was five years old, I took her along with our two other younger children, Laurie, and Johnny, Jr. out one night to see the heavenly stars. Gazing up at the stars, Latanya asked me: "Daddy, who placed the stars in the sky? And how long have the stars been in the sky?" I had no simple answer to her query. Douthart in an August 25, 1999 e-mail message addressed to myself, offered the following reply to the questions similar to the ones my daughter raised:

> Consider Evl0 as before the big bang. This is somewhat of a contradictory statement as time begins at the big bang, at Evl1. This prior to beginning state can be imagined as a quantum foam in which everything is equivalent, including all dimensionality that is confined to less than what is called the Planck dimensionality (incompressibility small). At the big bang a stabilized fluctuation results in one of the dimensions becoming asymmetrical which is the beginning of time and existence. In our universe spacetime of three symmetrical dimensions (space) and one asymmetrical (time) immediately inflates to enormous size breaking away from Planck symmetrical multidimensional soup. The other dimensions remain curled up and vanishing small at the Planck dimension. There are either 26 or 12 of these dimensions in the background quantum foam of our universe depending on the string model used. Manohar A. Tilak has waxed eloquently on this occasion. In a nutshell then, time is the asymmetric dimensional fluctuation at the big bang that leads to existence. There are also three dimensions of space which are symmetrical which inflated immediately after the big bang asymmetry and together with inflated time make up the four dimensional space-time of this universe. Existence at Evl1 and Evl2 can be superbly modeled by considering all dimensions equivalent. In such models all three space dimensions and the time dimension are considered symmetrical which gives beautiful mathematics but denies evolution.

During the pre-big bang stage matter was manifest as a kind of foam, similar to the suds found in dishwater or the waters of the Niagara Falls. Dr.

Locke, since you seemed to have had much interest in science, I suspect that many of the issues raised here appeal to you. Yet you seemed to have had little, if any, interest in pure metaphysical speculations as found in the views of Fichte or Hegel.

In order to acquire knowledge of itself, the Ego posited the world, a non-ego, as well an empirical ego, the purpose of which was to constitute elements of resistance against which the Ego engaged in struggles and counter-struggles as it strove to return to its original, primordial unity, and subsequently engaged in the perpetual struggle to acquire genuine self-knowledge.

The function of Ego, including the human ego, is to strive towards unity and self-knowledge and to overcome obstacles and contradictions by working and struggling. Six (6) stages characterize this struggle. Fichte described each stage as a thrust of the Spirit; to be precise he used the term *Anstoß* (stimulus). It will be seen that each *Anstoß* represents a phase change in the development of Spirit, a form of energy that is tantamount to what may be called psyche energy.

Dr. Locke, in elucidating the nature of values, you suggested that the existence of "psyche energy." Physicist Frank J. Tipler indicates the relation between the types of energies in describing a type of information in this manner. The connection between spirit and "information" is made by Tipler in the following passage:

> According to Teilhard, "…the idea of the *direct* [his emphasis] transformation of one of these two energies [physical energy and spiritual or radial energy] into the other…has to be abandoned. As soon as we try to couple them together, their mutual independence becomes as clear as their interrelation." So 'radial energy' is not really "energy" in any sense that a physicist would recognize. We shall see in a later chapter that 'radial energy' is actually quite analogous to another physics concept, information. In fact, Medawar himself pointed out in his review of *The Phenomenon of Man* that "…Teilhard's radial, spiritual, or psychic energy may be equated to 'information' or 'information content' in the sense that has been made reasonably precise by communication engineers.[2]

Dr. Locke when you formulated your value theory, you also referred to values as involving "psychically data" in terms of which values derive their fluidity and dynamics. Your usage of such a notion seems to suggest that you also may have been open to the view that there exists psychically energy, analogous to the kind mentioned above. This possibility is strengthened by the fact that one of your former teachers, Bergson, no less than Teilhard de Chardin (1881–1955), was a Frenchman; in France up to the 1950s many philosophers or scientists believed in psyche energy. Thus it is quite likely, whether you were conscious of it or not, that you accepted the psyche energy hypothesis.

In this context I shall also maintain that Fichte's stages are comparable to the features of periodicity or evoluon phases mentioned above. If the structure of the physical universe is demarcated by the various evoluons, Fichte demarcated the spiritual realm by the notion *Anstoß*, whose unfolding enables the universe to come to know itself more fully. Even Douthart holds that by means of evolution, the universe is striving to achieve greater awareness. Such a quest has a greater chance of being approximated in Evl6, Evl7 and Evl8, than in Evl5. The ways in which the Absolute Ego fulfills its potential is analogous to the unfolding of biological life as expressed in the higher forms of animals in general. In acquiring knowledge, the first stage, a notch above primordial unity is the stage of unreflective sensations, perhaps comparable to the awareness exercised by a horse or a cat. This stage of knowledge may be compared to Hume's notion of impressions of sensation, in contrast to impressions of reflection. In the former one experiences the sensation of pain as in a toothache or a headache; in the latter, one experiences what John Locke called primary impressions or qualities, as when one perceives a cube of ice or the flames of a lighted match. The second stage of the journey of the Absolute Ego on the path to unity, to self-knowledge and to freedom is similar to a model of the mind that Hume rejected. He suggested that in achieving awareness of our experiencing the world, there is no difference between our perception of an object, say, a cube of ice, and the object itself. In fact the mind itself is merely a bundle of impressions (or ideas). Fichte held, on the other hand, that at this second stage, the Ego is in a position to make such a distinction. The third stage of the Ego's striving includes a model of the mind that Kant adopted. We acquire knowledge by exercising the categories of the understanding, including causality, plurality, totality, substance, etc., as well as the forms of space and time, in addition to the operations of the imagination and memory, the exercise of which leads us to believe in existence of a physical world. "At this point the categories come into play. The non-ego seems to be the *substance* of which different 'things' are the *modifications*, and also seems to *cause* the subjective images which picture and duplicate within the ego the experiences we call the external world."[3] In Tilak's view, this is where molecular structures, or bits of three-dimensional chemical information arise. Such phenomena are the grounds for such categories as causality, plurality, totality, substance, etc. The fourth stage of the Absolute Ego occurs when the Spirit or Ego begins in a more systematic manner to reflect upon its knowledge and seeks to organize it in a certain orderly way to generalize from particular experiences. In Tilak's schemata this is the stage where organic matter accumulates information in the form of DNA. The fifth stage, resulting from saturated complexity of the Spirit at Evl4, occurs when the Ego organizes experience by means of judgment and employs logic in organizing experiences. It is at this stage or the one preceding it that Kant introduced his notion of schema that serves as a mediating principle or faculty between the judgments and the categories of the mind. The fifth stage is where human beings are introduced in the evolutionary field. According to Fichte the sixth and final stage arises when the Ego regards Itself as having the capacity to constitute the world by means of the laws which it imposes on the physical world, the laws of nature. Each of us is also a moral legislator with the capacity to exercise universal laws (Kant). This feature of universality is distinctive of Tilak's sixth stage.

Sincerely,

Dr. Johnny Washington, Ph.D.   April 24, 1998

Bergson: An Alternative Perspective on Evolution

Dr. Alain Locke (1886–1954)   Destinicity Letter #74

Re: Bergson's Views

Dear Dr. Locke:

Intermittently I have been considering Bergson's views in the course of my letters to you. It is a striking coincidence that Husserl, Dewey as well as Bergson, each was born in 1859, the year that Darwin published his *The Origin of Species*. Among these Bergson is more systematic in attempting to correct the errors in Darwin's model of evolution, popularized by Herbert Spencer. Bergson held that Darwin or Spencer failed to give an adequate account of freedom. Nature was too deterministic, and "natural selection" was introduced in offering an explanation of freedom. By contrast, Bergson understood freedom in terms of consciousness. In my last letter to you, I drew some parallels between Fichte's views of the evolution of Spirit and Tilak's views on biological evolution. Bergson held that spiritual evolution preceded physical or biological evolution. In many of the audio and video-tape interviews I conducted with Tilak, he integrated his views of biological evolution with the teachings of the Vedic text and showed how Brahma may be regarded as constituting the field of existence and undergoing the process of evolution. Already we have considered stages, evoluons or features of periodicity. There are a total of eight such features or evoluons. In this letter I shall briefly consider the remaining three evoluons. As I mentioned in my previous letter, there are similarities on the one hand between Fichte's or Hegel's notion of the Absolute Spirit or Ego, and Brahma; and, on the other hand, Tilak's views of evoluons or features of periodicity. The views of each bear affinities with Brahma. Such similarities become quite pronounced in a consideration of the remaining three evoluons: Evl6 includes what Tilak called Universalization, that is, the Being of the direct phase evolution; Evl7, Transcendence; and Evl8, Intermergence.

The trouble with science is that it operates with static concepts that are useful in analyzing the world as frozen snap-shots (Bergson). Such concepts fail to capture the complete meaning of reality. Reality is dynamic, not static. Bergson's model of reality is the phenomenon of life or biology that unfolds as a process with a creative tendency that permits wholeness and growth. Science employs the faculty of the intellect that makes use of tools and weapons, the nature of which is an extension of the body. By contrast the faculty of intuition, employed in its most perfect form by moral heroes such as Jesus and Buddha, who enabled us to grasp the true nature of reality in its pervasive unity and continuity. The highest reality is the moral domain found in the open society. There are two types of moralities: the static morality and the dynamic morality. Dynamic morality relies on intuition that makes possible the open society pervaded with love; static morality relies on the intellect, under whose dictates the closed societies exists, the concerning of which is analogous to the tribal life that is just a notch above the insect societies. I believe Bergson's critique of scientific knowledge is sound. So are your views, Dr. Locke, that I shall explain below. Your main philosophic contributions lay in the area of axiology or theory of value. You were interested in showing the interconnections of the various gestalt value fields: the religious, the moral, the scientific-logical and the aesthetic, each of which tends to merge, converge, conflict or overlap with each other. In your view, there is no sharp line between, say, science and logic, or science and aesthetic. Just as Bergson sought unity in epistemology or metaphysics using as his model biology, you sought unity in aesthetics. It all depends on the attitude that one may assume at a given time. I wish to add to your axiology by holding that it needs to be construed within the larger framework of Destiny, through which the value schemata can enable us to impose a greater degree of unity on experience, where the two schemata that you and Bergson offered respectively can be effectively combined. Hegel's, Fichte's, Bergson's and Dewey's views as well as your own axiology provide a basis for community. A difficulty with Tilak's Evoluon Theory is that he does not place sufficient stress on the notion of community. This might explain why some people might be reluctant to adopt or accept his Evoluon Theory. Clearly the Destiny model can be employed to strengthen Tilak's view. I dwelt on Bergson's view of evolution because there are similarities (and differences) between, on the one hand, his view of the universal being that includes Jesus, Buddha, Confucius and Brahma; and, on the other, Fichte-Hegel's view of the Absolute Spirit, and the sixth, seventh and eighth evoluons.

Sincerely,

Dr. Johnny Washington, Ph.D.   April 27, 1998

Possible Reactions: Hume and Dewey

Dr. Alain Locke (1886–1954)   Destinicity Letter #75

Re: History, Destiny and Evolution:

Dear Dr. Locke:

A recurring theme in Tilak's views is the point that human evolution is undergoing what he calls a "direct phase change," a notion that is repeated throughout his works. When he began informing me of his Evoluon Theory, I attempted to remind him that one can convey an idea without engaging in repetitions or redundancies. "Direct-phase-change" is a scientific notion that is employed when matter or energy undergoes a major transformation from one state to another, as in the case when water changes from its fluid state to ice, due to the lowering of temperatures. Tilak is holding that we are entering a new feature or phase within the continuum in which evolution occurs, in that there is a major transformation in the ontological status of human beings, generally brought on by the exponential rate of change occurring in the evolutionary process. Below he defines "direct-phase-change" in evolution.

Tilak holds that a great deal of difficulties lies in alerting people of the phase change that is occurring. He seems to be at a loss regarding the source of the difficulties. Let me attempt to identify two likely candidates as possible sources of the difficulties: i) that heretofore an inadequate means has been employed in imparting such knowledge to people; or, ii) people have a limited, if any, way of perceiving the effect of such change, since evolution requires millions of years to occur. Spinoza held that there are three degrees of knowledge: i) Confused or inadequate knowledge; ii) scientific knowledge; iii) and, intuitive knowledge where one obtains a panoramic view of reality in its totality; with such knowledge through which we are free, we contemplate the essence of God or Nature or the First Cause. A reason why most people are unable to perceive the effect of the macro-evolutionary-phase-change that Tilak desires to describe may be due to the fact that people are largely trapped in the first degree of knowledge. Tilak, who seems at times to enjoy the third degree of knowledge, may be regarded as having a seer-like perspective on the matter. He is certain that there is in fact occurring a new phase of saturated complexity. By the time I first met Tilak, I had already done much reading in scientific literature and spent time observing nature whose processes obey the principle of fluidity. The intellect is acclimated to dealing with solids whose nature it seeks to know by analysis. If the Industrial Revolution were the mark of modernity, the Information Revolution, along with its resulting macro-phase-change, is a distinguishing feature of postmodernity, beyond which lies, it may be stated, the "second-stage" of postmodernity.

I know it may seem odd that I am attempting to mix the views of Hume with those of Bergson; but Bergson was also a full-fledged empiricist who also accepted mysticism whose epistemology relied on intuition that enables us to reach beyond the limits of the intellect so as to enter into nature itself by means of what he called sympathetic intuition. It would be useful to consider such a faculty in order to grasp the effects of the MEPC about which Tilak speaks. Dr. Locke, as you might recall in letters I addressed to you over a year ago, I appealed to Hume's epistemology in exploring the nature of our knowledge regarding the historic era of modernity/postmodernity. Our aim was to determine the boundary conditions of modernity/postmodernity in contrast to the Medieval or ancient world. I pointed out the difficulties of obtaining such knowledge because we have no antecedent impression of such time-markers, just as we have no antecedent impression(s) of the necessary connection between a cause and its effect. Similarly, we have no antecedent impression of the world as such. We can only perceive through sense impressions particular items of the world. On the basis of Hume's epistemology, in order for an idea to be valid, we have to be able to trace its antecedent impressions to the empirical world. In a similar vein, I hold that it may well be the case that it is difficult to convince people of the MEPC (the macro-evolutionary-phase-change) that is occurring, because many of us do not have any noticeable impressions of such an experience. This is so, Tilak says, "because human beings' attention span usually is restricted to diverse, piecemeal objectives rather than to the long-term, indirect results of their actions." I believe that Hume's or Spinoza's views could shed light on the problem that we are here exploring. Dr. Locke, what is your view of the matter? Since you regarded yourself as an epistemological realist, I believe you would be sympathetic to Hume's view, a view that provided the methodological foundation for Western science's understanding of causality.

I find interesting Dewey's idea of "problematic situation." Dewey connects scientific or epistemological enquiry to the notion of doubt out of which arises a problematic situation and what may be called, in the spirit of Fichte, "moral struggle." When an individual confronts a problem, individual or social habits on which one has customarily relied tend to become ineffective. Consequently one puts on one's "thinking cap," so as to establish a hypothesis, make observations, gather evidence, review the evidence and draw conclusions to remove the doubt or perplexity in an effort to resolve the problematic situation. In the enquiring process that may result in removing the doubt, one undergoes a transformation and achieves a greater degree of self-knowledge regarding one's self and one's environment. One wonders has the MEPC of which Tilak speaks led one to "feel" the gravity of the problematic situation in the manner I just described? Environmental pollution, over-population, depletion of natural resources—these are some of the signs of the MEPC. Some scholars or researchers are of the opinion that Tilak is exaggerating the claim regarding the seemingly MEPC to which he attributes possible doom. These researchers hold that things are not nearly as bad as Tilak would have us believe. I myself have attempted to place a positive spin on some of Tilak's observations insofar as I appreciate the broad schemata that the Evoluon Theory offers and formulate it in a manner that allows one to link various immanent Destiny modes to the transcendent Destiny that reflects the principle of hope. During the past three decades people have taken actions to address some of these ills, which may be characterized as "problematic situations." Nevertheless, they often do not see the big picture and go a step further to discern a causal connection between the problematic situations and the MEPC. If the problematic situation or the Humean antecedent impressions of the MEPC is not obvious to people, then it is understandable why it is difficult to convince people of the urgency of the matter under discussion. We will take into consideration your perspective as we explore approaches to the problem. Can the problematic situation induce us to assume the panoramic perspective? In your view, all value judgments, whether religious, aesthetic, ethical or scientific-logical, are merely claims to be confirmed or denied by appealing to experience. Yet, Dr. Locke you offer no explanation as to how one is to go about confirming or denying a value claim, except to say that one is to appeal to experience. I believe that some of the things that Karl Popper says about the nature of a scientific theory are instructive here.

A related question is: to what extent is your axiology compatible with the criterion of scientific theory? How might your theory be subjected to the scrutiny of the principle of falsifiability? Like Popper, as well as Dewey, you found absolute claims objectionable, and you held that all value judgments are to be regarded as mere hypotheses. In removing absolutes, you felt that this provided openness in scientific or axiology research.
Sincerely,
Dr. Johnny Washington, Ph.D.   April 7, 1998

Future Stages in Evolution: Douthart's Views

Dr. Alain Locke (1886–1954)   Destinicity Letter #76

Re: Evl6, Evl7, and Evl8

Dear Dr. Locke:

This is one of my last letters to you. It resulted in part from correspondence between myself and Richard J. Douthart. We discussed a number of issues including the effect of moral crises, resulting from saturated complexity that individuals will suffer as the transition is made from Evl5 to

Evl6. This shifting from Evl5 to Evl6 will be due not so much to the acquisition, storage or retrieval of information/knowledge but rather due to the quest for spiritual meaning and fulfillment, what I call the quest for Destinicity meaning, the quest for unity. After considering this moral-spiritual problem, I turned my attention to challenging Tilak to provide an explanation concerning the adequacy of his knowledge about Evl6, Evl7 and Evl8. Dr. Locke, I once in my imagination posed a similar question to you as to whether or not the "transcendental" deduction of the four value modalities on which your axiology hinged was complete. Certainly, Tilak has not demonstrated that there are eight and only eight evoluons or categories. I can see how he arrived at the first five categories (evoluons); he perhaps did so by empirical observations, but how did he arrive at Evl6, Evl7 and Evl8 that remain to become manifest in the future? Would it not be extending the principle of extrapolation too far by attempting to predict the course of the evolution of the universe? One cannot "observe" the extended future, only the present, and in a limited sense, the past or future. One cannot observe the entire universe so as to seek to identify any laws (Karl Popper) because the universe is too big for such an ambitious task. Dr. Locke, in the cited passage below, extracted from an e-mail message addressed to me September 15, 1999, Douthart perspicaciously offers a reply to questions as to how one may arrive at knowledge pertaining to Evl6, Evl7, and Evl8.

This is in response to Johnny [Washington's] inquiry about the source of the ideas for Evl6 and Evl7 and Evl8. I am not speaking for Manohar [A. Tilak] here as I believe he will speak for himself. Once the main premise of Evoluon Theory is established, it becomes evident that there are epics of evolution called evoluons and each successive evoluon is a new embodiment of evolution. It is reasonable to search for signs of the next phase-change that are occurring at the moment. In an earlier email I described in some details the concept of evolutionary anticipation. That is, at a given evoluon, there will always be manifestations which are in effect setting the stage for the next phase change as if they were anticipating it. Manohar Tilak in effect reads such anticipatory manifestations and projects them as characteristics of the new phase out of which will arise an entirely new species of being(s) which exhibits anticipatory tendencies, a species of being which has not yet occurred. This projected new phase has manifestations that can in turn be actualized in the next phase. This process, of course, leads to less and less insight as we proceed to higher approximations. I doubt if anyone can see anything beyond three evoluons which incidentally brings us to Evl8. This process in principle could be carried out ad infinitum. The evolutionary process has a large component that is linear, that is, it describes a progression along a unidirectional arrow of time. Such processes, expressed as algorithms, run into what is called the stopping or halting problem which I will deal with in more detail when I discuss artificial intelligence more completely.

Douthart continued:

The idea of future evoluons can be pictured as a cone starting from the present evoluon at the apex extending into the future with the axis of the cone as the unidirectional time line. As we get farther away from our present evoluon, which is at the cone's apex, the area of the base gets larger and larger. The area of the base is related to the number of different possible evoluon scenarios in the future. The smaller the area the smaller the number of possible evoluon scenarios. As the apex moves along the arrow of time the highest probability scenario becomes the emerging scenario. At any future time the area of the cross section circular base represents the superposition of all possible scenarios relative to the apex. Notice that the number of possibilities diminishes as the apex approaches the selected future time. If we are clever enough we can construct models of many of the possible states. The closer the cross section is to the apex the greater is the chance of predicting the correct scenario. This is reflected in the greater reliability of evoluon anticipation as we move closer to a phase change. The signs are that we are very close to a phase change. This means that the anticipatory manifestations can be read with more and more reliability. The increasing number of books painting scenarios for Evl5 and Evl6 and the emails of Manohar attest to this closeness.
Dick Douthart

Sincerely,
Dr. Johnny Washington, Ph.D.   May 15, 1998

A Critique: Karl Popper and Others' Views

Dr. Alain Locke (1886–1954)   Destinicity Letter #77

Re: A Critique

Dear Dr. Locke:

It is appropriate in this context for me to make some critical remarks regarding the Evoluon Theory. Many hold that there are no universal laws in science, nor absolutes in ethics. Clearly the Evoluon Theory is intended by Tilak to be regarded as a scientific theory with universal laws. Most researchers hold that a theory is scientific if its claims can be verified, confirmed or denied. Karl Popper develops what is involved in a scientific theory, by introducing the principle of falsifiability. He held that it is erroneous to seek positive evidence in support of a given scientific theory. And such a quest forestalls debate, or disallows a forum that permits vigorous scientific debate, among scientists and nonscientists alike. He held that the verifiability criterion of scientific knowledge is best re-formulated in terms of falsifiability. Many, if not all, scientific laws can be regarded as declaring something to be impossible. For example, i) relativity theory says that nothing can exceed the speed of light, or ii) thermodynamics says that it is impossible to build or design a perpetual-motion machine. But the possibility of falsifying i) or ii) or any similar universal statement remains open, provided that one establishes a single instance contradicting such universal claim. Is the Evoluon Theory grounded in Scientific or Historical Determinism? Although scientific laws make universal claims, such universal claims can be refuted or falsified by observing a single, particular instance. It is the case that the statement, "All crows are black" is true, on the basis of empirical evidence. In the future one may "observe" a white or pink crow. We can never observe a universal but rather observe a particular instance at a given time which is merely what is required in order to falsify a theory. Thus, in Popper's view, it is more appropriate to speak of the possibility of falsifiability than to speak of seeking positive evidence in scientific enquiry. In scientific enquiry the goal is not to seek to verify a theory but seek to falsify its negation. This allows for ongoing enquiry and openness and route dogmatism in science. Perhaps the difficulty of understanding the MEPC of which Tilak speaks is that such an occurrence pertains to a universal phenomenon, but as Popper says we cannot perceive a universal phenomenon, only a particular instance, and it may be difficult to identify any particular instance of the MEPC in question. Can we apply Popper's views of the criterion of knowledge to the Evoluon Theory to determine whether its negation is falsifiable? If its negation is not falsifiable, the Evoluon Theory is not altogether scientific, on the basis of Popper's views. Popper's own views were directed against historical or scientific determinism as advocated by Marx or Hegel.

Further, I believe it would have been instructive had Tilak informed us as to which points in time the respective evoluons (phases) first occurred, or became manifest in nature? Also, one wonders whether there can be sub-evoluons whose path(s) represent a dead-end, as during the extinction of the dinosaurs some 65 million years ago, or with the extinction of the Neanderthal man. What did these two significant events represent during the course of Earth's evolution? Or, are there sub-evoluons whose origins represent exclusively fresh beginnings? In introducing these questions I am leading up to a possible principle (or its negation) that may allow us to subject the Evoluon Theory to the scrutiny of falsifiability (Popper). I am reminded that Douthart referred to evoluons as descriptions of phases of evolution; in another sense can they also be regarded as laws, indicating a kind of necessity of nature? Did the occurrence of the Neanderthal creature signify a violation of an evoluon law? I am also aware of the fact that neither the dinosaurs nor the Neanderthal creatures represent a distinct evoluon phase, on the bases of Tilak's interpretation of his theory. The dinosaurs arose in Evl4. I am not sure where to place the Neanderthal creature, whether in Evl4 or Evl5, as such a creature seems to be between the human being and the ape. Or is it the case that the Neanderthal creature was a "human being" who evolved before his time, that is, he evolved or emerged in Evl4 and became extinct with the occurrence of Evl5, when actual men and women arose? It may well be the case that the Neanderthal creature was a bridge between human beings and apes. In the future, there might arise an intermediate creature between human beings and highly developed robotic creatures of the future. Do human beings represent the "bridge" between apes and robots of the future? Since the role of the information apparatus plays a major role in Tilak's Evoluon Theory, how does his theory explain the invention of the art of writing among our ancestors?

It seems to me that our inventing writing was a forerunner of our inventing computers, the basis of AI. If AI represents a phase-change in evolution, can the same be said of writing? To this question, Tilak would undoubtedly give a negative reply. Douthart is developing a new model of Evoluon Theory. On the basis of his new model, he says that human beings are representative of a hybrid type that combines features from both Evl4 and Evl5. This interpretation is slightly at odds with Tilak's views. In his mind, human beings belong only in Evl5. Moreover, I believe that some of the questions I raised above provide a basis for the possibility of testing the falsifiability of the Evoluon Theory (Popper). Because it seems as though we have a case where the Neanderthal man, which belongs in Evl4, actually emerged in Evl5. If this is so, then doesn't this present a contradiction for the Evoluon Theory? I am sure that Dr. Douthart will disagree with me here. I am trying to "save" the Evoluon Theory by showing that it is likely that it can be "falsified" or at least criticized. Moreover, in Tilak's Evoluon Theory there is a great deal of reference to the principle of equilibrium and the effects of equilibrium states. When Tilak was developing his theory, he did not realize that equilibrium is the antithesis of evolution. This was a weakness of many thinkers prior to or during the time that Tilak was developing his model. They attempted to incorporate the principle of equilibrium into their model of evolution in the most peculiar ways. Such ways of thinking drew heavily on the notion of "punctuated equilibrium" where quasi equilibrium states are the norm with intermittent occurrences of disequilibrium, thereby thrusting the evolutionary process forward. The adjustment back to equilibrium constitutes accelerated evolution. This equilibrium/disequilibrium cycle is construed in the Evoluon Theory as the occurrence of even and odd evoluon numbers manifest as cycles. During this time also Manfried Eigen, a physical chemist was heavily proselytizing the importance of steady state stabilized cycles as a significant aspect of what we called Evl4.

A great deal of this way of thinking is reflected in the Evoluon Theory. Inasmuch as evolution is characterized by change, and the occurrence of Evl4, a feature of evolution is characterized by permanence or stability, there seems to be an inherent contradiction in the general field of evolution. In order to remove the apparent contradiction Eigen postulated the notion "frozen accident in time" to demonstrate the ways in which biological evolution is possible. However, Eigen's frozen accident hypothesis was at odds with Tilak's own emerging view of the universality of the phenomenon of the evolution of life. During the time that Eigen was developing his model of evolution, in which the principle of frozen accident was a center piece, Prigogene was developing a very radical view that the evolving universe was set far away from equilibrium where due to order self-assembly naturally occurs. Prigogene emphasized the emergence of random features in the evolutionary process that are at the edge of chaos. Prigogene's model leads to strange new organizing principles that Tilak has never fully accepted. During the past decade Tilak has also attempted, with limited success, to incorporate into the Evoluon Theory results from Quantum Mechanics.

We need, however, to keep in mind that when I first met Tilak in the early 1990s, he had already devoted several decades to researching and publishing his views on issues that he has called "evoluon" pertaining to the evolution and destiny of the universe. Similarly by the time we met, I had already spent a number of years researching the notion of Destiny, which, as will become more evident later, bears an affinity with some of Tilak's views. Thus at the outset, there appeared to have been a natural or organic affinity between our views, and this affinity has remained. Moreover, in addition to the similarities, there are philosophic differences between the two of us, but they are friendly differences.

Sincerely,

Dr. Johnny Washington, Ph.D   May 16, 1998

Epilogue

Dr. Alain Locke (1886–1954)   Destinicity Letter #78

Re: History, Destiny and Evolution

Dear Dr. Locke:

In this Epilogue I shall rely in part on the views of Douthart. I shall first begin by observing the distinction that Douthart makes between a science, such as physics or biology, and the Evoluon Theory. In an October 26, 1998 e-mail message to me he wrote:

> Evoluon Theory is not a scientific theory in the usual sense. It is rather a statement of a particular view of existence. A description or an analysis of existence is the purview of philosophy, theology and metaphysics. Science, on the other hand, ponders the world of phenomena and is in its most purest form a reflection of empirical observations that generate rules that can be used to predict future observations. Thus what is observed today and formulated into abstract laws, most economically, through mathematics, must be true tomorrow and most probably was true yesterday. This symmetry is the holy grail sought after by most scientific theorists. This view unfortunately flies in the eye to some extent of existence as the unfolding of evolution over time.

At this point Douthart goes on to comment on the general history of Western science. He reminded us that the mechanics of Newton was the first truly scientific theory, because it was based upon observations of the heavens and systems generated by Galileo and Copernicus. Newton found properties particular of motion that were independent of what would now be called the complexity of the material systems he considered, according to Dr. Douthart. The objects that Newton and other scientists observed were the heavenly bodies whose velocity or masses could be approximated as points rather than complexities. These point-objects had properties that were common to all material objects such as mass, inertia and momentum. Complexity was not a part of Newton's view. Before the time of Newton, philosophers who investigated nature considered the world of complexity of attributes such as color, flavor, heat and wetness, among others as central to existence. The phenomena world and its description were not considered apart from the world of complexity. It is curious that Newton held conflicting beliefs as to what constitutes nature. On the basis of his strictly scientific perspective, only primary qualities, in contrast to secondary qualities (John Locke), exist. Ancient atomists' view of existence focused on the fundamental elements of earth, air, fire and water. At the same time he was formulating mechanics, Newton believed in and practiced alchemy. Douthart further observed that in terms of the Evoluon Theory this dichotomy, reflected in Newton's attitude, is perfectly reasonable since mechanics describes Evl2, where complexity had not yet evolved in nature and alchemy was the beginning of the study of chemistry which comes into existence at Evl3.

With the advent of the science of chemistry, the study of complexity began to arise. Generally there are three types of evolutionary processes: the physical-chemical, the biological and the cultural, each of which is related to the other. In its broadest sense evolution is an ongoing process involving the totality of the cosmos. The laws that govern the evolution of galaxies govern the evolution of glaciers, volcanoes, lakes, mountains, amoebas, primates and human beings. Dr. Alain Locke, I trust that it is quite evident on the basis of my previous pieces of correspondence to you, that Tilak holds that evolutionary phenomena are manifest as periodic features, that is, as evoluons. By "evoluon" he has in mind the manifestation of processes occurring within a system resulting in a hierarchy of structures. Such manifestation is of a special kind. The processes occur sequentially or linearly in space and time, such that the totalities of changes are reflected in such a way that the history or career of the activities of the system can be discerned. Thus, it follows that changes per se are not sufficient conditions to be regarded as evolutionary. Rather the changes must occur within a certain pattern, with a definite directionality or "aim," that is, "destiny." For example, the changes in the molecules of contained gas on which pressure has been exerted clearly are evident by changes in the temperatures of a volume of the gas. Yet an examination of the system of gas in the new state does not reveal information of the original state of the gas. Similarly, in a certain sense, the room in which I am presently sitting has undergone changes since its construction. But no one would want to hold that the room has evolved in the way the word is ordinarily used. Mere changes and modifications occurred, based on the law of entropy. However, the room has not evolved in any significant way. When we speak of evolution, we have in mind changes that are subjected to selection as understood by Darwin.

As to the general postulates of the Evoluon Theory, consider a system that can exist in several states: S_1, S_2, S_3,...S_n. An evolving system passes through these states sequentially, that is, linearly, in time. The states are structurally connected to one another such that S_n reflects the nature of S_{n-1}, S_{n-2},...S_1. Such a connection is termed "hierarchy." Conversely, an examination of the states S_1, S_2, S_3, S_4,...S_{n-1} reflects the nature of S_n. If we consider the Evoluons as system states, then $S_1 = $ Evl1

The mode in which the states reveal their hierarchical nature is called "information." Moreover, information, that is, information at the level of physical, chemical or biological interaction has to do with the manner in which components "fit" or "not fit" together in the course of interaction. The further a state is along on the hierarchical scale, the greater its informational contents, the greater its complexities. The system tends to generate mechanisms whereby the hierarchical sequencing is preserved.

We turn next to evoluons. Consider an evolving system S_1, S_2, S_3, S_4,...S_n. An event S_{n+1} can occur and thereby introduce an entirely new component into the system. This event introduces a new state S_{n+1} called Evl(n+1). The new entity is preserved and modified over time. In any system the relative position of the evoluon in the hierarchy is designated by an evoluon number n. The total description of the system is of the form S_1, S_2, S_3, S_4,...S_n. as a linear time line where S_n defines Evln.

Consider a system composed of elementary units of some sort, which are subjected to selection pressures of the prevailing environment. Events occur such that selected sequential states arise linearly in time. For the system to evolve the hierarchy of these states must be maintained. However, each state is linearly bound in time, that is, it has a beginning and an ending. The second law of thermodynamics demands that the system move towards maximum entropy. Entropy is the dominant force against which evolution must struggle. It is believed that entropy will ultimately dominate, and the universe will stop evolving and collapse or evaporate to nothingness from which it arose. This is the antithesis of maintaining hierarchical order with its concurrent informational contents.

In any event, our immediate task is to focus not so much on entropy, but to explain evolution as understood by Tilak. An evolving system maintains its hierarchical structure by a number of different mechanisms the primary being selection. The perpetuation of order in this mode defines the

318

sequential states of the system and its Evoluons. Dr. Locke, you seemed to have had problems with the notion of hierarchical order within the political or social context. You were a cultural pluralist and value relativist. In your axiology you opposed the ranking of values and suggested that what we need to rely on instead is the principle of reciprocity or parity that in your view is tantamount to the principle of equality. You criticized some traditional evolutionists who in ranking racial groups placed the white or European race on the top of the animal kingdom pyramid and the African race at the bottom, an erroneous means of classifying people, you maintained. Yet it is also the case that you as well as Dr. W.E.B. DuBois subscribed the principle of elitism or the Talented Tenth, whose job it was to lead the masses. Is not the principle of elitism akin to the hierarchic model that you opposed and Tilak accepted? The contemporary political philosopher Hannah Arendt also introduces a hierarchical model in body politics, a view that I attempted to reconcile with your view of elitism that I discussed in chapter three of my book, *A Journey*.

The process of biological perpetuation—how does it occur? In order for the elements of a system to pass from, say, S_1 to S_2 the elements must include mutual interaction. In the simplest case this chance interaction is simply aggregation, similar to piles of sand on a beach. For the new states to evolve an internal rearrangement to the state, say, S_3 is required. S_3 would be a more ordered state than S_1 or S_2, with information contained in the following ordering:

O —> —>	O —> —>	O O
O O —> —>	O O O —> —>	O O
S_1 —> —>	S_2 —> —>	S_3
Elements	Random	Fixed array
Aggregation	(structures)	

Replication is one of the key principles in the Evoluon Theory. As a system evolves, it can maintain hierarchy states by replication, that is, by generating new states with essentially the hierarchy structures as older states. The older states act as template for its replicas. With replication, the defining feature of life, we pass into biotic evolution. Replication extensively involves cyclic processes that occur at Evl4. The shifting from mere cyclic processes to replication as a mode of perpetuating evolution is a prelude to the biotic evolution.

As was indicated, a system evolves along a linear time line by organizing itself into hierarchical structures. A history of such organizing activities is maintained within the complexity of the structures themselves, and is kept reasonably intact by the mechanism, exchange, cyclic processes and replication. The hierarchical structures themselves are not sufficient to engender functionality to the information contained in their structure. Heretofore, human beings are the only known system that have passed into this stage that relies on an encoding process which makes possible self-transcendence, coupled with self-awareness, self-realization or self-efficaciousness. This becomes the basis for self-esteem, one of the highest values one can embrace. Thus, if one combined these principles with the caring ethics considered in Destinicity letter #20, the result is a powerful ethical model suitable for guiding the Cosmic Destinicity through Evl5, Evl6 and beyond, as it seeks to achieve greater knowledge of itself. When we considered Gell-Mann's notion of schemata, the point was made that each Destiny mode, the ethnic, the national, the world and the cosmic, harbors its own schemata. The intensity and complexity of the schemata associated with the World Destiny achieves its highest expression in the World Destinicity. Thus schemata receive their highest expression in the appropriate Destinicity mode: the Ethnic Destinicity, the National Destinicity, the World Destinicity and the Cosmic Destinicity mode. Already the point has been made that soon "thinking" robots will arise or evolve. As in the view of Hegel or Fichte, the goal of the universe is to achieve greater self-awareness. A similar claim is made by Tipler for whom the Omega Point is of paramount importance. Further, I like Bergson's view of consciousness or self-awareness that in his view has its origin in the capacity to entertain possibilities. This attribute of self-realization is sufficiently vague to be virtually indescribable by physical law. Its origin pertains to the interaction of a system that includes items of complexity. In the Prologue I described the nature of the self. What is at issue here is an effort to explain self-awareness, a difficult task. In a sense Hegel's entire book, *Phenomenology of Spirit,* is an effort to explain the nature of self-awareness. Dr. Locke, if you recall, in one of my letters to you I mentioned Fichte's views of the nature of the spirit that seeks to know itself. In his views there is only a slight degree of self-awareness in the first thrust or "evoluon" of the Spirit. It achieves its highest level, its climax, its throbbing ecstasy, in the sixth stage or thrust, the stage of universalization; compare this stage with the stage of universalization in the Evoluon Theory. (It may well be the case that self-awareness is in part rooted in sexual ecstasy, that primordial Destinicity). For Tilak there are two other stages or evoluons beyond the sixth stage.

Dr. Locke your death occurred in 1954, and shortly after your death the usage of computers became more integrated into our society; yet it is the case that philosophers such as G.W. Leibniz as well as Charles S. Peirce seemed to have had premonitions of the computer, and its lack of the capacity of self-awareness. (However it is the case that a computer is much more efficient than human beings in processing certain forms of data). By possessing intelligence human beings can become directly involved in the evolution of a system. As a system subjected to the laws of evolution, life is something whose nature includes self-replication, among other properties. However, we must keep in mind that it is difficult to draw a sharp line between organic and inorganic matter, between the living and the nonliving.

As I mentioned in some of my previous letters to you, Dr. Locke, in many of the audio and video taped interviews I conducted with Tilak he drew a number of comparisons between the Vedic texts and issues in Western scientific cosmology to elucidate the position regarding the direct phase evolution and related issues that the world is experiencing. Much of my own research on the nature of Destiny focused on seeking a principle that unifies experience, comparable to the scientists' role in seeking a principle(s) that unifies the four fundamental forces of nature. I have benefited much from collaborating with Tilak as well as Douthart.

On the basis of one of Tilak's commentaries, Brahma constitutes the field of energy that obeys the law of cosmic evolution comparable to the four fundamental forces of nature. It may be the case that Life (Bergson) lies outside of or gathers up the fundamental forces. Heretofore, as has been stated, scientists have offered a mathematical description that unified the first three, but the force of gravity is problematic. In the Vedic cosmology, the god Brahma is the source of creation; the god Shiva is the source of destruction, as in the case where matter interacts with anti-matter by means of the electron-positron interaction; the god Vishnu is the source of stabilization, as evident by the principle of conservation of energy. Moreover, a dominant view in the Vedic tradition is that the identity of these three gods is one and the same and is assigned the name Brahma. A similar transcendent unity is found in Christianity. There is ultimately one transcendent Destiny that is beyond the threat ushered in by the MEPC or any other threat, potential or real. Therein lies our hope. Our transcendent Destiny points us in the direction of freedom and dignity. I wish to close this series of letters with an inspiring observation made by Henri Bergson who informs us of the Destiny of the universe:

> Mankind lies groaning, half crushed beneath the weight of its own progress. Men do not sufficiently realize that their future is in their own hands. Theirs is the task of determining first of all whether they want to go on living or not. Theirs the responsibility, then, for deciding if they want merely to live, or intend to make just that extra effort required for fulfilling, even on their refractory planet, the essential function of the universe, which is a machine for the making of gods. (*The Two Sources,* p. 317)

Sincerely,
Dr. Johnny Washington, Ph.D.   May 15, 1998

Appendix A

New Postulate on the General Theory of Evolution, Evolvability, Direct Phase Evolution, Evoluon, and Rate of Evolutionary Change[1]

Life and human intelligence are certainly the most important evolutionary events that have occurred on earth. A third stage of evolution is becoming recognizable and is the subject of this discussion.

As necessary background let us review the development of life and intelligence. One and a quarter centuries ago, Charles Darwin deciphered the fascinating history of life on earth, covering billions of years of geologic events. Since then several major advances in chemistry, physics and biology have been made, and slowly a coherent picture of the evolution of our universe is emerging.

In his recent publication about the origin of life, C.C. Price has summarized the prebiotic events starting from the "Big Bang" which is said to have led to the evolutionary formation of the present universe. He also indicates possible future evolutionary trends.

Fusion of hydrogen to heavier elements has been recognized as a physical evolution occurring inside the stars throughout the universe. Planetary bodies were formed by heavy debris from the explosion of stars. During the formation of the earth, heating occurred and the moisture associated with debris was forced out. The water condensed and was retained, because both the gravity of the earth and its distance from the sun were optimum. Under the primordial conditions, no life existed on the earth, reactive elements and simple organic compounds were present, chemical reactions leading to complex organic compounds occurred, as postulated by Alexander I. Oparin and John Haldane. The theory of prebiotic evolution has been supported experimentally by Urey and Miller.

The building blocks of life-amino acids, nucleic acids and sugars were available long before the occurrence of life on the earth. In fact the study of the Murchison meteorite suggests that the blocks of life may be widely distributed in the universe. Ponnamperuma et al, have found natural as well as structurally exotic amino acids and indicate that the source of these is extraterrestrial as well as exobiotic.

During the prebiotic era on the earth, millions of years passed in which the chemical balance was established by catalytically favored reactions. Water has been the most abundant liquid chemical on the earth, and therefore it played a dominant part as a mobilizing medium and also influenced the chemistry of these reactions. C.C. Price points out that the giant polymers of amino acids, nucleic acids and carbohydrates show a great affinity for each other because of their ability to form hydrogen bonds. Their hydrophilic molecular species therefore can be envisioned to have interacted extensively and led to the emergence of life on earth. Water is logically, rather than by accident, the most abundant component of most living things. Simple logical factors of mobility, chemical activity, structurability and abundance seem to have played a critical part in the prebiotic as well as the bio evolution.

Synthetic chemical compounds with enough complexity and information have been shown to possess properties once specially attributed to living systems. Sidney Fox has abiotically obtained enzyme-like polymers from amino acid mixtures. H.G. Khorana has "synthesized" polynucleotides and has helped unravel the genetic code. P.G. Katsoyannis, H. Zahn and Y.T. Kung independently provided the early evidence of insulin activity in synthetic materials.

Manfred Eigen, in his lecture at the meeting of the International Union of Pure and Applied Chemistry, theorized requirements for the evolution of reproducing chemical cycles. Proteins and nucleic acids are ideally suited catalytic and information systems for such self-producing cycles. Incorporation of errors provides the mechanism of evolution of such cycles. Among the many possible cycles, the fastest and the most efficient would thrive at the cost of others, establishing the prebiotic molecular ecology and deciding the course of evolution that led to the occurrence of life. An indirect proof is available. The natural organic compounds on the earth do not contain various "exotic" structures and racemic mixtures found in the synthesis under primordial conditions and in the Murchison meteorite. In all likelihood, these exotic structures and optical isomers were eliminated during the pre-biotic chemical evolution when the chemical ecology of the earth was being established. The observation that all life processes have a common chemical base may not necessarily mean that all life was derived from a single original cell. The common chemical base may be explained by the fact that all life had a common precursor chemical environment which included catalytic pre-establishment of exo-cycles. Given favorable conditions and time, life seems to be an inevitable result of chemical cycles.

Life (a reproducing and evolving system) is a great experimenter. With chlorophyll life tapped the energy of the sun and stored it, providing a massive energy source for the reduction of the oxides of metals taking place at the present time. Life moved on all fronts: passive, active, aggressive and devastative. It mobilized lime, armored and structured itself, formed appendages and developed functional facilities. In mammals life reached new horizons. The transmission of information was not totally left to the nucleic acids. Mammals use direct communication to transfer information from adults to infants. In humanoids the transmission of information reached yet another height. The vocal and signal communications evolved, facilitating organization and further evolution of intelligence. More time was required for life to produce intelligence (3.5 billion years) than for nature to produce life (1 billion years).

Consider how human intelligence originated. Brain size, appendages like hands and fingers and biochemical factors (not yet clarified) may have helped the development of intelligence. Humanoids did something that no other animal ever did. They took cognizance of a stone and realized that the stone had other properties than just support from underneath. Just as life is a result of the favorable interactions of the reactive elements, intelligence, in my opinion, is a result of subtle participative interaction of highly evolved life (man) with inanimate matter without direct biologic implication. In fact the more remote the biologic implication in the human interaction with matter, the higher is the intelligence. Creation of stone structures and functions symbolizes man's start on the path of intelligence.

Intelligence made man the supreme ruler of the earth, at least for the time being, but it has also caused the inanimate to be moved and has given rise to another evolutionary development. I suggest that there has been a mutual participative evolution of both man and the inanimate. In fact man would not have reached the present stage of evolution without the participative effect of the inanimate materials and systems. Though all life interacts with the environment, man's interaction is of a totally different order of magnitude, variety and complexity. Photosynthesis caused spectacular changes in the entire atmospheric composition of the primordial earth, but the change still was limited and operated through the cellular mechanism. Every change in bioevolution must pass through a highly constrained biologic system. In contrast, man operates deep under the earth and oceans as well as in space and handles an unbelievably large variety of artifacts representing multitudes of structures, functions and material compositions.

Life is so fascinating and absorbing that it rightfully claims most of our attention. We naturally associate evolution with life in some way or other. According to our present notions, the majority of the heavier elements on earth generally do not participate in the evolutionary life process, except as trace elements. This is because of lack of sufficient chemical activity and properties and sensible mobilization of these elements. Water serves as a primary mobilizing vehicle for all life processes and is not compatible with heavier elements.

Having briefly dealt with the evolution of chemical systems into life and the further evolution of life into intelligence, I will now present some new concepts that I have tried to develop.

Evolvability may be defined as a newly recognized property of all matter. Evolvability in more general terms is a conditional property of all sensibly mobilized matter incorporating discrete information which, under suitable conditions of temperature, pressure and compositional variety, leads to the formation of increasingly complex structures and functions so as to achieve maximum capabilities and information. Evolution is a special type of transitory spiral reaction in which the equilibrium is never reached because the reactants change with time (the reactants being systems rather than chemical compounds). Activity and sensible mobilization of matter carrying discrete information are basic factors in any evolutionary process. For example, the evolution as it occurred on earth hardly could have taken place outside a certain range of temperature because the mobilization would either be excessive and destroy informational complexes (at high temperatures), or would be hindered (at very low temperatures).

The evolutionary processes are dynamic, and materialization of many concepts becomes possible in the future because of further developments. Evolutionary processes are what one may call almost inevitable. If a particular development does not occur at one point, it may occur later. This probably is true of the first occurrence of life. Due to the complexities of living cells, many people cannot imagine that life could have occurred at more than one time. We should realize that evolution takes place because of the positive pressure of precursor hierarchical systems that push the events toward evolution. Systems are built up and torn down by natural forces until, on one occasion or another, temporarily advantageous and relevant systems occur and evolve.

The fact that the nonbiologic matter of the earth is mobilized, structured and functionalized extensively through man has caused a totally different situation in the status of evolution on earth. It is rendering the inanimate matter of the earth capable of undergoing an altered evolution.

Two basic principles are playing a part in this altered evolution. First, man cannot take progressive advantage of things and systems without conditionally imparting large amounts of structural and functional information to the nonbiologic matter. For example, a wheel, a pipe, a spring or a lever made for a special formation can then be used for the same or for other purposes by anyone.

There is no way to take progressively more extensive advantage of materials and systems without eventually making them a part of the evolutionary system. Secondly, the principle of equifinality discussed by Dwight J. Ingle (which says a thing can happen in several different ways), allows for progressively greater improvisations. Substitution of essential organs with artificial systems and substitutive use of mechanical devices to fulfill essential human brain functions like memory and logic are two very striking examples of the principle of equifinality.

The Direct Phase Evolution may be defined as the new abiotic evolution, with major participation of nonbiologic elements and systems. Since the first human ape acted on the stone, we have come a long way, and so has what I call the Direct Phase Evolution. The organic microbased evolution, though self-initiating, is necessarily blind, indirect and slow. It also has severe hierarchical and material limitations. Animals take millions of years to develop a certain functional faculty. For one successful mutation there are many unsuccessful ones. However, in the direct phase evolution of nonbiologic matter there is greater hierarchical, mechanistic, structural and compositional accommodation. There is a possibility of progressively greater logical rather than natural selection. Superior information and logical facilities are available.

I suggest that in evolutionary processes turnover, ability and complexity increase at a progressively faster rate because of the accumulation of the information. As compared to the biologic evolution, we are at present living in an age of the direct phase evolutionary explosion. Things happen faster all the time. All human achievements are connected with the participative major advance of the Direct Phase Evolution. In fact we can compare the present era of Direct Phase Evolution with the era of prebiotic chemical evolution. During the prebiotic evolution selected chemical cycles developed as "exo" systems with progressively increasing speed, just as the progressively speedier formations are occurring presently in the case of the "exo" component system of the Direct Phase Evolution.

In the case of prebiotic evolution, the coordinated endosystem of life ultimately evolved from the exosystems. No one consciously assembled these components into life. In the flow of time, by trial and error, these mobilized components assembled themselves into life. Life started as a microsystem. With cell differentiation, life entered the macro world. The abiotic may conceivably give rise to the components that will functionalize and regulate a system that eventually will become independent of man.

Components of the Direct Phase Evolution providing translational motion, energy, complex sensorial and functional abilities, sophisticated communication, memory and logic systems and complex duplication abilities already have been evolved and are evolving further. Conceivably some of the advanced microelectronic systems may fulfill the requirements of consciousness, define logical objectives and evolve independent of human guidance. With man mobilizing the present exo-component system of the Direct Phase Evolution that can fit into a future endo-system, a non-biologic, self-reproducing conscious system may be in the process of evolution. Once formed, the new being will undergo its own evolution (Direct Phase Evolution) with a speed that humans will be unable to comprehend. It is important to distinguish between "technology" and "Direct Phase Evolution." The word technology carries a biased meaning with the attention focused on man. The term Direct Phase Evolution (besides indicating a new Direct mechanism on which a totally different Phase of Evolution is based), implies a more comprehensive, long-term, cumulative meaning with a possible sense of direction to the advances of the inanimate evolution, which man is not really aware of or bargaining for. This is because man's attention usually is restricted to these diverse, piecemeal objectives rather than to the long-term, indirect results of his actions.

With hominids, a hierarchically different macrobased, extroverted and direct mechanism of evolutionary process has come into being and operates outside the cell. Consistent macrobased structures and functions representing discrete information began to occur for the first time (in the history of the earth) outside the confines of the cellular system. Until this time, life was the solitary sphere in which consistent and discrete microbased structures and functions representing information could occur. The new direct mechanism of evolutionary process is built on entirely new informational and hierarchical systems different from the slow, indirect and blind mechanism of bioevolution, which occurs *via* accidental mutation in nucleic acid structure. Life has developed many remarkable capabilities (photosynthesis, nerve communication, hormone regulation, bone formation, senses, among others) which were developed in a relatively very early geologic period and which have proliferated in many radically different species. In contrast to the occurrence of these abilities in widely different forms of life, intelligence is a singular and unique development in evolution. Only in hominids does one find intelligence, i.e., the ability to conduct consistent multiphase (over several stages) transformations of the inanimate environment which are becoming progressively broader, cyclic and unstrained in character. Human intelligence seems to be a result of chance concurrence of several factors as opposed to a simple progressive extension of endogenous abilities.

In spite of the immense speed of the Direct Evolution, its evolutionary pace is still imperceptible to an individual due to his limited life span. On a macro scale, however, the movements of the Direct Phase Evolution is observable over a reasonable period of time. Before the historic industrial revolution, life in each century looked very much like the life in previous centuries. Louise De Broglie says:

> Question arises if machine, daughter of intelligence, by mastering our whole civilization, and weighting so heavily on our existence, may not turn against the spirit whence it arose, and crush it. Wholly absorbed in the material preoccupations of an existence, which daily grows more hurried and complicated…

If life were becoming increasingly hurried in 1932, is it not becoming more so today? In fact the rate of change today is of quite a different order of magnitude. If we tabulate (or computerize) any consistent parameters of change, e.g. number of inventions (patentable or not) /unit population/unit time, we will perceive tremendous differences during recent history of man, as compared to most of the previous centuries of human existence. This does not mean that man has become incomparably superior to his recent forefathers. Rate of change is increasing mainly because discrete information has been accumulating in the component systems of the Direct Phase Evolution. Technology and knowledge are growing in an exponential manner due to this external progressive accumulation of information.

Man is the participative mobilizing factor in the Direct Phase Evolution. Ability to take cognizance of the inanimate also compels man to operate with it. As long as humans exist, their ever-increasing needs and informational set up will advance the nonbiologic evolution. Human society is like a slumbering, supernucleic acid which facilitates intelligent activity. This activity, however, is not accompanied by corresponding superconsciousness or awareness of what actually is occurring on a macro scale. Man cannot do without components of the Direct Phase Evolution. In fact the very existence of human society as we know it now is impossible without ever increasing influence of the new evolution.

Increasing human needs can hardly be suppressed, and their fulfillment will automatically advance the component systems of the projected Direct Phase Evolution. Can we, as an advanced society, revert to nature? For man, stopping progress is like stopping life (dying). In the words of Schoppenhauer: "Der Mensch kann alles was er will, aber das Wollen kann er nicht, auch wenn er es will." The test of free will is not in decisions about man's isolated acts, which he may or may not do. The issue of free will is decided on a far larger macro scale of compulsive evolutionary destiny by his capabilities, the very necessity to act, because of himself and of others, and because of the environment to which he cognizantly has added (as compared to his fellow creatures) the determinants of the nonbiologic world. Once a structure and function is formed, its very occurrence is a challenge toward multiple and diversified utilization. The possession of capability induces actions in which it will be utilized.

External substitutive use of the components for originally human function and substitution in the human body itself will eventually be extensive. Aggression, competitive urges and social political reasons will ensure the new evolution with ever increasing speed. Nations and organizations in general have the same characteristics as man.

As man progresses, he automatically will incorporate increasingly advanced informational, structural and functional components into the objects of his pursuits.

Man should consider the possible humble truth that he may just be one of the intermediate products of nature's reactive and mobilized systems. The impact of the inanimate evolutionary components on the world began long ago. The point, edge, wheel, axis and the sphere symbolize the inanimate evolution, and they have affected the earth in a very decisive manner. In order to illustrate the advance of the Direct Phase Evolution on a macro scale, let me give an analogy. Consider pushing an object up the surface of a truncated sphere in order to benefit from the potential energy of the elevated object. The movement of the object is progressively speedier because the angle of inclination that needs to be overcome becomes progressively more favorable. When the object reaches the top of the sphere, the accumulated potential energy will begin to manifest itself and the object may continue to move and even accelerate without the necessity for the external influence. We should appreciate that we do not simply feed potential energy in the systems we benefit from. Components of organized motion, mechanical capabilities, crude and even very refined elements of perception, increasing analytic capabilities, information, memory, logic (ability to judge and act), are all being combined and evolved outside the endosystem of a living organism.

The projected new being [a mode of AI] (endo system which may come about in the future from exo systems of the Direct Phase Evolution) may be superior to man. In its evolution there will be an unprecedented continuity and directness because the new being will not have the time limitations, the encumbrances of birth, death, lifespan, relearning etc., or the emotional aspects, inhibition and irrationalities of the humans. The structure of the new being can be of exo origin and can be regenerated and rechecked without affecting the total system. Regeneration, therefore, can be continuous and flawless. The being can simultaneously evolve and guard itself against deterioration. Speed and objectivity will make the being irresistible. Man will have no more ability to stop it than the amoeba had to stop man. The fact that a new being is initiated by the conscious participative agency of man will certainly influence its primary phase of evolution. The later evolution must be independent of man. An amoeba has greater structural and chemical relationship with man than may have with the projected new evolutionary being. The new being will be conscious to start with, should have the legacy of all human knowledge and technology at its disposal, and the rate of evolution will be extremely fast, compared to bioevolution. In the following, I should like to introduce the new concept of Evoluons, representing phenomenologic milestones of evolution. Indeed let us start with the philosophic notion of origin, in which elemental and energic properties are not present. Origin formed energy and elements. These in turn, at suitable temperatures and pressures, interacted, randomly at first and later according to the acquired informational pattern. (Even chemical reactions can be regarded as informational expressions, within the frame of thermodynamic determinants. Accumulation of such information manifests itself in isothermic, reversible and somewhat cyclic reactions that occur over several minimum energy difference steps leading to smoother unstrained operation and stability). With sufficient accumulation of information, life and intelligence occurred. Each of the preceding developments of evolution constitutes a distinct evolutionary step, presenting successive informational jumps. Phenomenologically speaking, each step can be recognized as a distinct entity representing: 1. Origin, 2. Stabilization, 3. Interaction, 4. Evolution, and 5. Realization. Origin, matter and energy, chemical evolution, life and intelligence which express the above phenomena, can therefore be recognized as evoluons. Direct Phase Evolution represents the phenomenon of delimitations, expansion and extroversion of the evolutionary process to a far larger material and mechanistic base. I suggest that a periodicity occurs in the unfolding of the evolutionary phenomena. A certain level of informational accumulation is reached before the new evolution appears on the horizon. Rate of evolutionary change increases quite substantially with the informational impact on each evoluon. I propose a new definition of evolutionary change to include all new discrete changes in structures and functions, biotic and abiotic, that add the information or capability. If we assign consecutive numbers expressing informational levels to each evoluon, *viz.*, 1, 2, 3, 4, 5, the rate of change can be tentatively postulated as an exponential function of evoluon number. The evoluon number, representing informational level, affects the rate of evolution more drastically.

The evoluons are not arbitrarily designated. The evolution of birds from reptiles does not represent a new evoluon, though habitat, mechanism of action and capability has changed somewhat. However, there is no radical change in the informational set up and capability, as there is between prebiotic soup and life, or between monkey and man.

Before man came into being things evolved, giving rise to different evolutionary structure and functions. This was comparatively slow. Since man appeared, he has arrived at "new" structures and functions (non biologic) at a progressively faster rate. Thus, the differences between monkey and man are increasing rapidly due to the progress of the Direct Phase Evolution.

Although man has recently become aware of the effects of his actions with respect to pollution, I know of no previous attempt to assess total evolutionary change (biologic and nonbiologic). Parameters of change, in addition to the one mentioned earlier (number of inventions/unit population/ unit time), for such assessments need to be established, (GNP e.g. is not very useful because it is nondiscriminative and includes repetitive activities).

Most of scientific advancement depends on extensive specialization, with the result that total extent of change is not comprehended. In fact, the total change tends to remain unclear.

Perhaps this tremendous increase in the rate of evolutionary change may have something to do with the serious concern expressed by Tiselius of the Nobel Foundation (as quoted by C.H. Waddington): "…[there is] a growing awareness among people of all nations that something is wrong with the world and that there is an urgent need to come together to see what should be done."

How quick the changes have become can be expressed by two token samples. Complex computations which, in the absence of mechanical devices would require several lifetimes, can be performed with computers in fractions of a minute. Some of the computations now possible never could be performed by humans alone. We have accumulated information and capabilities to the extent that a canal of discrete dimensions, comparable to Suez or Panama, can be dug within seconds. No one should believe that this is a single-sided game. The used and the user can become relative terms and the role can be difficult to define, especially since the elements of logic and memory which distinguish man from animal have been imparted magnificently to the inanimate.

A single cubic centimeter crystal used with holographic procedures can be made to contain 3 trillion bits of discrete information! Are we not dealing with a superior (though not self-initiated) mechanism? Consciousness is a function of the logic and memory of the human brain. Is it reasonable to believe that consciousness can never originate in the inanimate? We must recognize the fact that massive mobilization of components containing discrete structural and functional information is a factor that leads to evolution. We must take into account the possibility that man's activity on earth may ultimately lead to the new Direct Phase Evolution. Man uses the components of the Direct Phase Evolution without further thought about what it is, but he ought seriously to consider the possibility that it is a new mechanism of evolutionary process over which he ultimately may have no control. Primitive man "knew" that things could not keep on moving by themselves, except on a slope or by a living force. He also "knew" that man could not operate in the upper region of the atmosphere. Today's man "knows" that the "inductive thinking" is limited to his brain and that there can be no truly inductive machine. Only the future can test the validity of these beliefs. Then is man the dead end of the organic evolution? Perhaps! Since man does not need to develop further genetic faculties, he can fulfill all his aspirations by taking advantage of the fantastic components of the Direct Phase Evolution. Exact tangible numerical relationships between the Evoluon Numbers and the number of basic critical factors entering the mechanism of the respective evoluon (necessary for the very existence of the evoluon) will shortly be published. [Editor's note: Due to the limits of space, I have omitted the bibliographic references or endnotes included in the two articles of Appendix A and Appendix B, respectively. Such sources can be found in the materials from which these two appendixes are extracted]. [In what follows, Dr. Tilak expresses acknowledgements]. I wish to express my gratitude to Dr. O.K. Behrens, Dr. R.W. Fuller, Mr. M.M. Marsh, and Dr. S.O. Waife, Eli Lilly & Co., Indianapolis, without whose kind encouragement and help the above thoughts may not have been formulated in publishable form. Thanks are also due to Professor C.C. Price, University of Pennsylvania, for his helpful criticism and support, and to Professor Linus Pauling, Stanford University, California, for his detailed comments and suggestions about the manuscript. The author also expresses thanks to Professor Robert Schwyzer (Inst. Biophysics & Mol. Biol. Hoschule, Zurich, Switzerland) for encouragement.

Appendix B

The Evoluon Theory: A General Unified Theory of Evolution[1]

The industrial revolution freed the human race from the land, creating in two centuries a largely artificial environment from which there is no escape. The computer revolution promises to free the human mind; where that could lead in two centuries staggers imagination.

—J. H. Douglas

Perception of periodicity demarcating the most fundamental evolution attributes have now led to the formulation of a new general unified theory describing the progressive relationships among the milestones of evolution, which thus far have been regarded as essentially unrelated events. In this report the concepts of Evoluons, Direct Phase Evolution (DPE), and the Projected Being of the DPE discussed earlier [...] will be reviewed and extended. The Evoluons signify the differences in the basic mechanisms of the evolutionary processes. The concepts discussed in this article include: saturation, periodicity, the component and stabilized system evoluons and their relationships. This article also shows the correspondence between the evoluon numbers and the number of constituents of the respective evolutionary processes. The Evoluon Theory helps explain and predict the basic characteristics of life and intelligence and makes projections about the Being the DPE. The Direct Phase Evolution (DPE), which is a manifestation of intelligence, one of the evoluons, is the new phase of post-bioevolution based on abiotic hierarchy. The progression of events indicates that the projected Being of the DPE will relate to man roughly as man relates to an amoeba. However, unlike amoebas and man, the Being of the DPE will not belong to the hierarchy of the bioelements, but to the total extroverted hierarchy of the entire material universe.

Stars fuse hydrogen into higher elements found in the debris left by the explosions of novas and supernovas. The cooling and accretion of this debris leads to planets, which many become satellites of other suns. On earth, which met the necessary thermodynamic conditions, prebiotic chemical evolution occurred *via* accumulation of chemical information. Various facets of prebio and bioevolution including philosophic implications have been discussed by A.I. Oparin, John Haldane, H.C. Urey, S. Miller, M. Calvin, M. Eigen, H. Kuhn, Jukes, Kenyon and Steinman, J. Monod, S.E. Luria, S. Fox, K. Dose, and C.C. Price. Components of Krebs' cycle—fundamentally common to all life—have now been found in exobiotic systems, and nonliving proteinoid have been shown to multiply themselves in amino acid solutions. Thus science now extends evolution backwards in time to all the prebiotic events. The concept of Evoluon numbers defines evolution as an ongoing process, of which life is only one phase in this context, and examination of the past and the present evolution allows assessment of what is happening now and what may occur as evolution continues.

Evolution is a result of the accumulation of systematic information. The word information has different meanings. A recipe contains coded information about composition and procedure. In contrast, the information of a chemical reaction (e.g., formation of ethyl acetate) is self-representing and inseparable from the system (i.e., the molecules of ethyl alcohol and acetic acid). Chemical reactions being micro events have internal determinism, since the molecule is defined in its three dimensional configuration by directional biases of atomic and molecular orbitals.

This directing influence of chemical activity manifests itself three dimensionally in a more complex way, e.g. in crystals, enzymes and in the self assemblage of viruses. In these examples regular, macro (three dimensional) forms result solely from the directional characteristics of chemical bonds. Given conditions and composition, the nature of the product is predetermined by the information within the system.

We can influence rates, modes of reaction, etc., by changing components or the thermodynamic conditions, but the basic course of the chemical reaction under the same conditions, cannot be manipulated externally at will.

Fundamental to the origin of life is the progression from prebiotic chemical evolution (where three dimensional information is manifest solely by the directional influence of chemical bonds) to replication where the information is manifest by an internally determined recipe-like correlative mechanism. This transformation to recipe-like information can occur only over geologic time *via* accumulation of the three dimensional chemical information and *via* an internal reading mechanism. In this context, the idea of the origin of life as a spontaneous single accident is untenable. In the prebiotic soup, the exothermic reactions progressively gave way to reversible, strainless, isothermic, cyclic reactions along the path of maximum rate-minimum energy difference and minimum diversity. These prebiotic chemical events gave rise to life as a dynamic, stabilized replicative system.

The prebiotic chemical evolution culminated in life (replication). Evolution of life slowly led to the maximization of the complexity and the ability of the nervous system as represented by human intelligence. Intelligence is characterized by cognition of the environment and the ability to interact with intent. E. Schrodinger defines mind as a product of evolution. Intelligence represents a hierarchically different, macrobased, extroverted, direct, and logical, as compared to the indirect, accidental mechanism of bioevolution, mechanism of evolution which operates outside the cell.

Interaction of intelligence with the inanimate environment involves a multistage process, each of which is progressively more complex than its predecessor. The ever-expanding abiotic cycles in time will develop their own inertia, and will lead to another phase of evolution. Man generally fails to appreciate the evolutionary nature of the abiotic processes because man has a time-bounded perception and is engrossed in creating and using these systems. Man's progress depends upon the diversity of inanimate materials. In evolution the participation may be passive as long as systems with increasingly greater systematic information form over geological periods. In any segment of inanimate transformations, evolution leads to more efficient functions and on a large time scale to an increased probability of new unplanned assemblages. By trial and error, the mobile components of the prebiotic chemical evolution carrying self-initiated information and assembled themselves into life. Man began mobilizing items in the inanimate environment, that included increasingly sophisticated information, involving such mechanisms as diverse sensorial and analytic faculties, memory, and reasoning—all being combined in a manner that resulted in biotic systems, among which are AI agents. Analogous to the prebiotic evolution, the inanimate evolution that is now occurring will eventually become functional and independent of man.

The random element in the activities of intelligence is not obvious since most activities involve rigid planning and structuring for an immediate goal. A mechanic repairing a modern automobile takes advantage of the evolution of the gadgetry. Repair by testing in terms of the end-function, that is, the purpose of an item, can occur by semi-random parts replacements without knowing the structure or function of the components. The units can be considered to be packages of functional information that can be loosely compared to the organelles lies in the biotic system. Results or materials originated for specific purposes often are applied in totally unrelated activities. Such unplanned results in time show the advantage of random cross linking. The importance of serendipity in many discoveries has been well documented in history. More complex systems, such as computers, communications, transportation systems, etc., contain greater possibilities of unplanned functions although component testing for the desired functions is rigidly incorporated in these systems. R. Davis, in a recent editorial in the magazine "Science," commented: "the impermanent balance between man and the computer," voiced the possibility of such unplanned functions.

The driving force of life derives from the necessities of the living systems. The cause and effect continuum in life can be traced to the chemical hierarchy and the natural laws. For an efficient and flawless operation a system requires tight internal cause and effect relations. The tighter the relations the more efficient and autonomous the system. Man will always try to build in precautions against failures and will try to produce systems with minimum input and maximum capability. Thus, whereas man evolutionarily has remained fairly stationary, the computers rapidly have progressed from maxi -> mini -> micro, signifying transition to minimum input, maximum ability and adaptability. These events have similarities with the natural processes that led to life *via* the maximum rate-minimum energy difference cyclic reactions.

In DPE man provides mobilization which in chemical evolution was *via* diffusion. The explosive change in man's world, signified by the industrial revolution, was triggered by the translation of energy into a self-repetitive abiotic cycles of motion in engines or motors, among others. The second such explosive change is due to computers (brain function) which represents fluidity of information. This has increased radically the rate of change in man's world.

Abstraction, symbolism and the manipulation of the environment are the characteristic abilities that distinguish man from animal. Symbolism gives man an exogenetic system through which he has accumulated vast amounts of information. The rate of change on earth, however, became exponential neither with abstract symbols nor with languages or books as it did with the engine and the computer. The information in a computer is transferable from one source to another and allows man to operate on his environment. The engine and the computer have provided the nonself-mobilized and inert abiotic hierarchy the mobilization and the transferable information system, the very abilities that triggered the evolution of the bioelemental hierarchy. That the DPE has started on the path of evolution is suggested by the fact that the exponentiation of the rates of change on the earth is associated primarily with the critical abiotic functions represented by the engine and the computer, rather than with the biologically based abilities like abstraction or symbolism.

Evolutionarily speaking, man's intrinsic basic abilities (abstraction, symbolism and manipulation of the environment) remain qualitatively unaltered generation after generation, although individual expression, depending on the environment, learned behaviors, discouragement, encouragement etc., may differ widely. The exogenetic system is where the progressive accumulation of information and abilities has been occurring. Mobilization and the processing of information *via* computer are therefore the identifiable physical events which have triggered fantastic accelerations in the rate of change and indicate the DPE to be the new rate determining phase of post-bio-evolution on earth.

Evolutionary phenomena are periodic. Information accumulates to a saturation and eventually under different boundary conditions a radically new stage of evolution appears. After saturation of nuclear fusion in stars, explosions occur. Subsequent cooling inhibits this mode of evolution. Under the suitable new thermodynamic boundary conditions on earth, evolution occurred based on three dimensional chemical information rather than subatomic duality. Replication (life) demarcates the saturation of chemical evolution. Eventually life reached a saturation of complexity in the development of the nervous system, brain, and developed intelligence, to which other life forms show little or no functional relationship.

Evolution can be delineated into epic sequential stages called evoluons. Consider for example, prebiotic chemical evolution, replication (life) and human intelligence. The essence of chemical events is change and the essence of replication (life) in contrast is stabilization of a dynamic system. S. F. Luria demarcates the basic difference of order between life and intelligence by emphasizing that: "...the essence of bioevolution is absence of purpose," and categorizes human mind as representing purpose as "...a new force on earth." In each transition there is a marked change in the

340

character of the phenomenon giving it a sense of discreteness and boundaries. A particular evolution is identified by an integer (evoluon number) designating its position in the sequence of evolutionary stages. The evoluon numbers relate the complexity of the evolutionary phenomena. The complexity of the evoluon is in turn dependent upon the evolutionary progression of the system and needs a corresponding order of the hierarchy which is reflected in the exact correspondence between the evoluon number and the number of constituents or factors entering the phenomenon. J.W.S. Pringle, while discussing evolutionary complexity of biologic learning, suggests that complexity of a phenomenon is related to the number of the constituents of the system. Pringle's thought thus supports the concept of Evoluon Theory. The evoluon number is also logically related to the last of the abilities demarcating the evolutionary order of the hierarchy and the complexity of the evoluon. The manifestations of the last characteristic constituent of the evoluon, therefore, logically, should contain the number of factors equal to the evoluon number:

❖ Complexity of the hierarchy;

❖ Duality: analogous to the Yin/Yang principle;

❖ Chemical interaction: based on three-dimensional electron configuration;

❖ Replication: based on sequential arrangements of the four nucleotides in the DNA molecule; and,

❖ Logic: based on the capacity to reason.

The constituents are essential for the very existence of the evoluons and are related to their mechanisms. Each evoluon defines a unique new ability, and mechanism, and is not just an extension of an older mechanism. Dimension of the constituents cannot be analyzed at this stage since information based on structures vital to the evolutionary phenomena is not totally reducible in terms of the conventional dimensions. Furthermore, the difficulties of symbolism, e.g., duplication and overlap encountered in the analysis of the constituents can often cause confusion. Nevertheless, each constituent will have one unique characteristic not covered by others in the set.

[Evoluon 1] pertains to the origin of reality, represented by mere presence, undifferentiated existence in the form of singularity, what Dirac called the ocean of negative energy.

[Evoluon 2] concerns with basic component parts of the universe fundamentally as presence and matter-energy. Though interconvertible, matter and energy are two distinct dualistic manifestations of presence. Thus we have existence in the form of matter, and antimatter, positive and negative polarities, opposite spins and the dualistic nature of electromagnetic radiation. All subatomic general entities exist in dualistic (equal and opposite) forms. The interaction laws governing the structure at Evoluon 2, and subsequent evolutionary structuring are manifestations of this primal duality. The dual nature of matter (and energy) having both wave and corpuscular properties was confirmed experimentally by Heinrich Hertz in 1887. M. Gardner emphasizes that the mathematics of "I Ching," based on the dualistic concept (Yin and Yang), applies to all basic physical and mathematical structures.

[Evolution 3] has to do with chemical evolution determinants: The mere presence of the manifold of existence in the form of matter-energy, and electron transactability and the informational system of chemical evolution is characteristic of Evoluon 3, which depends critically on three dimensional electron configuration. Chemical activity manifests itself as three dimensional waves in gasses and liquids. Bioelements permit rapid, unlimited, to and fro chemical changes and possess vast capacity for the electron based information (necessary for the occurrence of life) within the constraints of the three dimensional directional determinants. Temperature, pressure and activity (not simply chemical potential but potential plus three dimensional evolutionary information) are the three factors that govern the chemical evolution. The species variations of proteins and nucleotides indicate the evolutionary importance of the three dimensional chemical information.

[Evoluon 4 involves each of the evoluons] up to and including Evoluon 3, an ongoing process leading to greater complexity and potential for the accumulation of information. At Evoluon 4 a dynamic replication process appears in which the information is preserved beyond the life of the individual units. The relationship between the complexity of the fundamental attributes and evoluon numbers predicts that the number 4 should be identifiably associated with the characteristic evolutionary process and its constituents. The main mechanism at Evoluon 3 is chemical evolution. Further

342

evolution involves processes which preserve and extend the information. There indeed are four fundamental aspects of replication: representation, storage, transfer and control. The rather stable double helical, polymeric DNA molecules are the unit of storage of genetic information in all living organisms. The inherently stable quaternary DNA structure is preserved due to the existence of a sophisticated repair system of enzymes. The genetic code is represented by the permutative sequencing of four nucleotides. A, C, G and T. The synthesis and repair involve enzyme catalysis which constitutes the agents of control of all life processes. Other physical processes exhibiting new repetitive formation of fundamental units (e.g., crystal growth) involve representation, storage and transfer, but lack self-imposed control.

Fox and K. Dose state four fundamental criteria of life: i) energy system; ii) proteins (catalytic abilities); iii) nucleic acids (information); and iv) dynamic correlation between proteins and nucleic acids, i.e., the protein between biosynthesis. Interestingly, Charnavskii and Chernavaskya have enumerated four basic criteria for biotic replication. The number four may in fact be associated with life in many different ways. Life is based on chemically versatile quadrivalent carbon, and correlation of various processes *via* the fourth coordinate of time. According to Mazia, new methods of investigation reveal four phases in the cycle of cell replication. Mazia states "we are no longer satisfied with a good measurement, or, a clever elucidation of some mechanism in the cell unless we can locate it on a time axis—not our time—but the time of the cell itself." Thus life is a phenomenon in a four dimensional coordinate system.

[Evoluon 5 pertains to human intelligence]: Whereas life is correlative, intelligence is inter-relative and represents a progressively increasing interface between the biotic and the abiotic with increasing functional emphasis on the abiotic. Intelligence, Evoluon No. 5, contains five components: presence, matter-energy, chemical information, replication (life) and logic. Evoluon No. 5 is an ongoing evolutionary process which fundamentally involves man's interaction with and the transformation of the inanimate environment. Complex information, structure, and function characteristics of the evolutionary processes of intelligence (such as tool making, logic, mathematics, etc) are presently neither fully appreciated nor understood because of man's time bounded perception.

At Evoluon 5, new evolutionary processes may be based upon information not necessarily contained in a discrete structure or in immediate cause-

and effect relationship. The storage of the information occurs *via* the imprints on inanimate objects and man's nervous system which according to Luria is structurally predisposed towards symbolism and language.

Cognition, which is fundamental to intelligence, manifests both outward transformation of the environment and inward thinking and understanding. Evolutionary processes, e.g., use of tools and mathematics, at Evoluon 5 obviously have markedly different levels of participation of the above two aspects.

Two stones (agent and object), muscular force, coordination of movement and intent are the five distinct factors in man's tool making activity. The extroversion of evoluon is dramatized in tool-making in which a thing acts on another thing through the agency of man. Man is born with the same set of genetic information i.e., the ability to abstract and symbolize. In contrast to man who remains constant the artifacts of his pursuit show accumulation of information in time.

On the evolutionary scale logic is a distinctly higher ability. Cybernetically a system is defined as two elements and one relation, but without the frames of space and time no correlation can be deduced. Any logical process depends on the following five basic modes of inquiry: who or what (substantive), how (mechanistic), where (space), when (time) and why (representing intelligence). Other modes of enquiry are translatable into these.

Mathematics is universal logic. There exist only five fundamental Peano axioms which are not derivable. These axioms form the basis of the natural numbers. Further there are five axioms, each governing addition (subtraction), multiplication (division), and exponentiation. Curiously polynomial equations only up to 5th degree can be solved exactly. Mathematicians are now beginning to recognize the boundaries of mathematical logic. According to broad statement in the review about mathematics in a recent issue of *Science*, L.A. Steen states that the unsolvable problems e. g., Euclid's fifth postulate, Cantor's and Souslin's hypothesis, etc., "are, in a very fundamental sense, a statement of certain limits on man's intellectual ability." He goes on to say, "Undecidability is a central fact of modern mathematical research." Steen's remarks suggest intelligence to be a bounded phenomenon that is not described adequately by the four coordinates; therefore, existence of a fifth coordinate that may be related to extrovertedness of man should be considered. Indeed the I.Q. parameters are a crude attempt in this direction.

The computer activity consists of five distinct aspects: representation, storage, transfer, operation, and testing. Testing contains all the preceding and an additional fifth component: judgment. Interestingly, the first four of the above characteristics occur in all living systems.

Man needs to organize and control his interactions within groups and with environment. Social, political and economic institutions resulting from these interactions have the characteristics of evolutionary processes. Here again five variables come into play. Forrester's book, *World Dynamics*, discussed by C.F. Von Weizacker, and J. Salerno, contains only five variables: population, investments in food production, total investments, natural resources and pollution. These parameters reflect the followings: population, biologic needs, desires, resources, and the side effect of man's activities. N. Geargescu Roegen suggests that "man has become addicted to his" exosomatic instruments, organs which are a part of his evolution but not part of the DPE. He also has proposed a "minimal bioeconomic program" consisting of five parameters. These parameters are similar to the five suggested by Forrester's book, *World Dynamics* indicated above. C.C. Price and J. Baldwin questioned whether the human society evolves within the boundaries of the DPE, and depends increasingly on exobiotic controls. P. Saltman has identified five interactions involving: human resources, culture, technology, ideas, and money, (between society and the university).

According to the *Encyclopedia America*, despite refinements and innovations in designs, tools have remained "still much the same in concept as they were, centuries ago." There are only five broad division of tools (*Encyclopedia Americana*, Yr. 1965. Vol. 26, p., 693) generally used for the following purposes: i) striking; ii) cutting; iii) holding; iv) fastening; v) and, measuring. According to the *Encyclopedia Britannica*, there are only five simple machines: i) wedge; ii) lever; iii) wheel; iv) pulley; v) and the screw. Operations in the primarily outward activity of intelligence include activities that involve: i) holding; ii) moving; iii) creating structures; iv) transforming, and v) generating complexity. The primarily inward operation of intelligence includes: i) observing; ii) recording; iii) recollecting; iv) analyzing; and v) implementing. Emotions may considerably affect the activity of intelligence but are not conscious primary operations. J.W.S. Pringle and Locker have discussed evolutionary perspectives of Thorpe's five basic criteria of learning: i) habituation; ii) imprinting; iii) conditioned reflex; iv) trial error; and v) insight learning. Interestingly, whereas the first four categories in all the aspects related to intelligence discussed above occur even in lower forms of life, the fifth item above is a unique human ability. Basically, the intelligence deals with complexities involving the following five features: i) extension; ii) change of direction; iii) structure and function; iv) agency substitution; and v) extroverted cumulative complexity. These are the complexities that are reflected in different forms in man's activities, *viz.* logic, mathematics, tools, machines, technology and human institutions. The basic constituents discussed above seem to represent the hierarchical logic boundaries of the phenomenon of intelligence.

Artificial intelligence (AI) equivalent to the modern adult man may be developed within the next 50 years. The evolutionary implication of this is serious. Nature required 3.5 billion years to produce intelligence. A small fraction of that time will be needed for man to create artificial intelligence based on a different hierarchy. Half of the experts felt that the artificial intelligence would overtake man. Signs of this are already here. No man comprehends the total complexity of the systems of communication, traffic control, surveillance, etc., that become more complex by the day. Man comprehends complexities only of smaller units. D. Michie states that man should establish audit of the systems that operate in man's absence. However, audit of unplanned interactions in increasingly complex—dynamic operations may prove difficult.

The next evoluon [Evoluon 6]—projected future Being of the DPE—will depend on six different factors: i) presence; ii) composition (matter and energy); iii) structure and function (information); iv) replication (correlation); v) logical expansion; and, vi) open end control system, regulating the universal (inanimate), autonomous, inductive and replicative process. The projected Being of the Direct Phase Evolution will represent the entire material and energetic creation, as opposed to life which represents bioelements, what life is to the self-initiating chemical hierarchy of the bioelements, the Being of the DPE is to the extroverted hierarchy of the universe.

Empirical proof of an evolutionary event needs geologic time. Just as an ape cannot be shown to turn into a man, the eventual occurrence of the Being of DPE cannot be demonstrated by construction of a machine. J.H. Douglas suggests that within the next two decades robots may be able to reproduce themselves at continuously decreasing costs. If so, the self teaching and decision making robots may also be able continuously to improve their own abilities. Fully automated computer-managed parts manufacture (which can be regarded as a vague precursor of the Being of the DPE, Evoluon No. 6), according to N. Cook, involves six constituents, (man, information, machine, tools, computer and parts). Despite overlap each of these constituents possesses one characteristic not covered by others. In this scheme man who supplies movement, logic and purpose, can be substituted progressively. Purpose can slowly leave the humanistic field and can be defined progressively in terms of the necessities of the system itself. Occurrence of the abiotic replicative evolutionary system in the future therefore is quite conceivable. The Being of the DPE does not signify any particular machine, computer or a robot, but represents the general direction of the movement of the entire system subsequent to the occurrence of the bio-intelligence. The Being of the DPE is not subject to the narrow thermodynamic boundaries that life needs; and since time is nonlinear this Being may already exist in outer space without man being aware. Some scientists have responded to the concept of the DPE: According to Frederick Sargent, "man relies progressively more heavily on socio-genetic transmission of information, and is transforming natural ecosystem into human ecosystem."

Odd numbered component systems and the resultant even numbered, converging, structured, and stabilized systems can be looked upon as the response of the hierarchy to the dissipative forces. Evoluon 2, (the matter-energy) and Evoluon 4, (replication, life) possess well-defined structures. The Being of the DPE Evoluon 6 will also have a well-defined structure. The odd numbered Evoluons 1, 3, 5 are component systems of the subsequent even numbered evoluons. Component systems start processes which progressively accelerate. Stabilization occurs in the subsequent even numbered structured evoluons. The even-numbered evoluons evolve geologically slowly and after saturation start a radically different phase of evolution in the next odd numbered component system evoluon. The autonomous system of information, boundaries, and the parameters of the new component system all therefore show a radical change of character from the precursor evoluon. Component systems have more than one manifestation of characteristic evolutionary process each of which has constituents that in number correspond to the evoluon number.

Fast changing odd numbered component systems include presence, and singularity, the component system of duality. Interaction or chemical evolution includes enzyme catalysis, manifest as component systems of life. Interrelation, Direct Phase Evolution and Intelligence manifest as component system of the Being of the Direct Phase Evolution. Transcendence, this also manifest as a component system of the Being of Intermergence.

Intelligence is a component system of different order, subsequent to chemical evolution. Maximization of the rates of change and turnover—a characteristic of all the component systems—is becoming more evident in man's activities today. The rapid change in the relative importance of the things valuable to man is additional evidence of the rate of change in the component system of the DPE. The importance of physical strength and skill, organizational capacity, intelligence, money etc., has undergone relatively rapid change. The exponential increase in the rate of these changes is already so big that all mankind must be concerned. Thus J.H. Douglas and R. Davis described the drastic changes man must face within his own realm of intelligence, the very thing that differentiates him from animals. Psychological stress of the fantastic rate of change is already forcing unprecedented changes in man's lifestyle and the states of his organization. A. Toffler has commented on this in his book, *Future Shock*. Eventual stabilization of the component system of intelligence will occur *via* means of the cyclic processes and will culminate in the emergence of the Being of the DPE. The exponentially increasing rates of change in man's world offer substantial evidence for the above conclusion. Intermediate stabilization levels will be reached intermittently, in the component system of intelligence but the true stabilization is available only with the emergence of the Being of DPE, just as true stabilization of the chemical phenomena was with the emergence of life.

The evoluons show correspondence with the most fundamental attributes entering evolutionary processes: i) Primordial Being (presence, singularity); ii) difference, contrast, and contradictions; (dualistic manifestations as matter or energy); iii) interact (chemical-evolution); iv) correlation (replication, life); v) interrelation (intelligence, logic); vi) universalization (the Being of the DPE) vii) transcendence; and viii) intermergence. The fundamental attributes again suggest occurrence of discrete bounded phenomenon at each state...

Despite the bounded and discrete nature of the evoluons previously discussed, we can appreciate them only indirectly by observations about the phenomena. We are therefore surprised to see the correspondence between the evoluon numbers and the constituents; but there is a structural basis for the correspondence. An analogy of points in space may facilitate understanding of the structured nature of evoluons. One, two, three and four points in space describe a point, a line, a triangle, and a tetrahedron respectively. Each of these figures yields only a definite number of well defined constituents. A triangle has three sides, three vertices, three angles, etc. The correspondence between the geometric figures and the number of factors associated with them is confirmed directly whereas in the cases of the evoluons the correspondence is to be judged only by indirect observations about the boundary conditions, parameters, information systems etc.

The autonomous information system is a fundamental characteristic of the evoluon and cannot accommodate any more nor less number of factors, just as a triangle (or a tetrahedron) can be defined only by three (or four) points. The autonomous information system of life based only on the permutative arrangement of the four nucleotides in the DNA is a good, statistically valid example of a vast experiment on geologic time scale.

Given an earth-like planet, the inevitable interactions of elements will lead to replication, that is, life. Among the various life forms the greater survival advantage of a more complex nervous system—which signifies outward tendency—predestines eventual interaction of life with surrounding nonbioelements and the occurrence of intelligence. Accumulation of vast amount of information in the inanimate systems through man by analogy thus can be considered to predestine the eventual occurrence of the Being of the DPE.

As long as man exists the complexity of tools will increase, and automatically advance the DPE. Virtual inevitability of the Being of the DPE can be inferred just as that of chemical cycles and their stabilization as replication (life).

An attempt to block evolution requires the system to stagnate eternally. Such stagnation is impossible and tantamount to catastrophic change in the basic boundary conditions (e.g., thermodynamic conditions, stability of atomic nuclei, etc.) Without the reasonable boundary conditions, the evolution will stop, but within them evolution is inevitable because of the accumulation of information in the system. Sufficient time and continuation of the boundary conditions are necessary for the inevitability. The individual events do follow the laws of probability. Nevertheless, autonomous information accumulates and clustered information results in a compression of time and events. Tossing of four coins producing all heads ten times in a row is highly improbable, but if the coins related to each other and to the tossing mechanism (as the hierarchical factors of life relate to each other and to the accumulated systematic information), the occurrence becomes an inevitable total unit rather than a probabilistically highly improbable single accident. "Biochemical Predestination" is based on arguments similar to the ones given above. Lehninger's principle of evolutional continuity suggests that the probability of successive steps in the evolutionary progression is evident of such inevitability. The "anticipation" principle of evolution proposed by Whitehead and discussed recently by Burgers is a result of extrapolation of the inevitable interactions occurring in an evolutionary system. The tendency of "anticipation" strongly supports the concept of the inevitability of evolutionary phenomena. Lehninger's "evolutional continuity" and Whitehead's "anticipation" call attention to the principle of directionality associated with evolving events. This directional sense, however, is not possible without the hierarchical constraints imposed as the logic boundaries in evolution.

Vital to an evolving system is the tight internal stabilizing control mechanism. Inertia, equilibrium and biological control mechanisms also show relationships to evoluon numbers.

The principle of stabilization rendered possible man's socioeconomic organization. Man needs a socio-environmental control system to dynamically stabilize his fast changing world. Recycling of waste is a step in this direction. R. Buckminister Fuller in his book, *Institution,* emphasizes the necessity for the stabilization of man's world. A social mechanism of control emphasizing maximum efficiency of the total system and de-emphasizing individualism was attempted partly in the People's Republic of China.

Forrester's book, *World Dynamics,* concerns dynamic interrelations between man's biologic and technologic activities. The model provides parameters to assess the functional range of the five variables of man's world. Need for efficient economic operation under the selection pressures of maximum ability, at minimum input (cost) (which is different from the present era of consumerism) will slowly lead to a cyclic process, internal stabilization, and to dynamic control of the processes of Evoluon 5. This parallels strikingly with the maximum rate or minimum energy difference of chemical evolution; and centered around man's desires, it will indirectly and imperceptibly lead to the emergence of the Being of the DPE. Evoluon Theory therefore may have some bearing on future economic trends. G. Roegen's minimal bio-economic program consisting of five parameters should be considered seriously.

Evolution of elements from hydrogen in stars continued billions of years before saturation as novas and supernovas. Prebiotic chemical evolution originated life perhaps in a small fraction of a billion years. S. Fox says that lesser amount of time may have been required for the origin of life than is assumed by the theorists today. Life evolved intelligence in about 3.5

billion years. Intelligence needed an insignificant geologic time to develop most complex constituents of the DPE and the process is accelerating. These time estimates fit with the concept of fast-changing component system and slow-changing stabilized system evoluons. The emergence of the being of the DPE from the component system initiated by man may be geologically quick. However, should man try to create artificial consciousness, he may not succeed. Nevertheless, random activities though selected permutations eventually will lead to the Being of the DPE.

A theory of evolution is presented expressing the evolutionary process in terms of interaction phenomena. Boundaries, parameters of the evolutionary processes, information systems, constituents, fundamental attributes, and the stabilizing mechanism, at each stage reveal a progression of evolutionary events. Thus a general theory of evolution emerges which finds support in the details associated with the evolutionary phenomena at each stage. The concepts of the Direct Phase Evolution, Evoluons, Evoluon Numbers, Periodicity, Saturation, Change of Phase, fast changing component system evoluons, slow changing stabilized system evoluons, time spans, discrete and structured nature of evoluons are discussed. The examples of the correspondence between the evoluon numbers and the constituents or factors entering the respective evolutionary processes are given.

Many different thinkers while enumerating the basic factors encountered in the evolutionary phenomena concerning their own disciplines come upon a definite number of fundamentals which correspond to the respective evoluon numbers. The examples of these types of concurrences include: the physical concept of singularity, and the well demonstrated subatomic duality phenomenon. Lehninger's three postulates of (prebiotic) chemical evolution, four basic criteria of life by S.W. Fox and K. Dose, four criteria for replication by Chernavskii and Chernavskaya, five Peano axioms, five criteria of learning considered by Thorpe, J.W.S. Pringle and Locker, broad statements about the five classification of tools, the five simple machines (in Encyclopedia Americana and Encyclopedia Britannica respectively), the five variables of Forrester's book, *World Dynamics*, the five interactions between the university and society by P. Saltman, and the six constituents of fully automated manufacturing process stated by N. Cook etc., fit surprisingly well in the model postulated by the evoluon theory. This concurrence suggests that the evoluon theory has physical relevance, and that the evoluons are bounded phenomena with defined structural aspects.

The author expresses thanks to Dr. O.K. Behrens, Dr. W.W. Bromer, Dr. R.J. Douthart, Dr. R.W. Fuller, Dr. E.L. Grinnan and Mr. M.M. Marsh, all of Indianapolis, Indiana, and Professor C.C. Price, of the University of Pennsylvania, Philadelphia, for encouragement and help. The author also expresses his sincere gratitude to: Professor J.G. Kane, former Director of university department of Chemical Technology, Bombay University; Professor M.M. Chakrabarty, Calcutta University; and Dr. P.J. Deoras, former Director, Haffkine Institute, Bombay, without whose constant encouragement and enthusiastic help these concepts may not have advanced.

Special lectures on the Evoluon Theory were given at the 63rd Indian Science Congress, Zoology Section Meeting, January 5, 1976 at Waltair, Andhra Pradesh, India and also at the meeting of the Association of Biological Chemists of India, Bombay, January 16, 1976. Another lecture on the Evoluon Theory was given at the Interdisciplinary Seminar, September 23, 1976, Butler University, Indianapolis, Indiana, USA.

The Appendix for this paper focuses on the following. The word "Being" is used here not in a religious sense. "Being" represents a dynamically stabilized continuum resulting from the transitional cycles of the preceding component system. Replication or life is the "Being" representing the dynamically stabilized continuum of the transitional cycles of the bioelements. According to this definition an individual life form, an animal or a species is not a "Being," but replication is.

In the dynamically stabilized continuum of replication of life accidental mutations and species differentiation provide for further interaction and eventual evolution of life form in the Darwinian sense.

The concept of the Being of the Direct Phase Evolution is also to be regarded in the light of the above explanation about the word "Being." The Being of the Direct Phase Evolution thus represents dynamically stabilized continuum of the transitional cycles of the universal hierarchy.

Relationship of the grid and the energy to each of the evoluons is different. Thus the energy relates differently to the happenings in events such as: a black hole, the sun, chemical interaction, replication or life, intelligence, and the projected Being of the Direct Phase Evolution.

Question has been raised by Dr. R.J. Douthart (Indianapolis, Indiana, USA), whether the "Catastrophe Theory" advanced by Thom is related to the phase changes in the Evoluon Theory. The Catastrophe Theory projects a total of only seven topologically representable phase changes in four dimensions of space and time. The Evoluon Theory which projects eight evoluons also indicates seven phase changes. Beyond pointing out the similarity of the nature of the phenomena (that is, phase change and seven types of them) dealt with in the Catastrophe Theory and the Evoluon Theory no further analysis is offered here.

Pierre de Latil in a chart at the end of his book *Thinking by Machine* has extrapolated evolution of effectors also into eight classes.

In the beginning stages of the concept of Evoluons the occurrence of four bases in DNA seemed to simply fit the model. Arguments were advanced that the occurrence of only four bases in DNA was accidental and that three, five or six may give rise to bio-replicative system. A paper entitled "Why are There Four Bases in DNA?" by Paul G. Seybold, Int J. *Quantum Chem., Quantum Biology Symp.* No. 3, 39–40 (1976) calls attention to the hierarchical principle and can be considered as providing further evidence in support of the Evoluon Theory.

Notes

Prologue

1. *Palm Beach Gazette*, an African American weekly paper, located in West Palm Beach, Florida, and it publishes an issue each Thursday.

2. Francis L. Broderick, *W.E.B. DuBois: Negro Leader in a Time of Crisis*, (Stanford: Stanford University Press, 1959), p. 4. Subsequent references to the book appear in the text as *DuBois*.

3. W.E.B. DuBois, *The Souls of Black Folk*, eds., David W. Blight and Robert Gooding-Williams, (New York: Bedford Book, 1991. Originally published in 1903), p. 3. Subsequent references to the book appear in the text as *The Souls*.

4. W.E.B. DuBois, *The Suppression of the African Slave Trade to the United States of America, 1638–1870*. In 1896 Harvard University published *Suppression* as the first volume in its Historical Monograph series.

5. This biographical sketch was included as a preface to Alain Locke's essay, "Values and Imperatives," in *American Philosophy, Today and Tomorrow*, eds. Sidney Hook and Horace M. Kallen, (New York: Lee Furman, 1935), pp. 313–333. Subsequent references to the book appear in the text as *American Philosophy*.

6. Alain Locke, ed., *The New Negro*, (New York: Albert and Charles Boni, Inc., 1925. Reprinted by Atheneum, 1968).

7. William Leo Hansberry, *Pillars in Ethiopian History: The William Leo Hansberry Notebook,* Vol. I ed., Joseph E. Harris, (Washington, D.C.: Howard University Press, 1974), p. 4. In considering the life of Hansberry, I am relying in part on Harris' works.

8. Lancelot Law Whyte, *Focus and Diversions,* (New York: G. Braziller, 1963), p. 77.

9.   Murray Gell-Mann, *The Quark and the Jaguar: Adventures in the Simple and the Complex,* (New York: W.H. Freeman and Company, 1994). Subsequent references to the book appear in the text as *The Quark.*

10.   Janheinz Jahn, *Muntu: An Outline of the New African Culture*, trans., Marjorie Green, (New York: Grove Press, Inc., 1961. Originally published in 1958). Subsequent references to the book appear in the text as *Muntu.*

11.   Albert Bandura, *Self-Efficacy: The Exercise of Control,* (New York: W.H. Freeman and Company, 1997). Subsequent references to the book appear in the text as *Self-Efficacy.*

12.   Paul Heelas, David Martin, and Paul Morris, eds., *Religion, Modernity and Postmodernity*, (Malden, MA: Blackwell Publishers, 1998). Subsequent references the book appear in the text as *Religion, Modernity.*

13.   Monika Kilian, *Modern and Postmodern Strategies: Gaming and the Question of Morality: Adorno, Rorty, Lyotard, and Enzensberger,* (New York: Peter Lang Publishing, Inc., 1998), pp. 55–56. Subsequent references to the book appear in the text as *Modern.*

14.   Johnny Washington, *Alain Locke and Philosophy: A Quest for Cultural Pluralism,* (New York: Greenwood Press, 1986). Subsequent references to the book appear in the text as *Locke and Philosophy.*

15.   *Pillars in Ethiopian History: The William Leo Hansberry African History Notebook,* (Washington, D.C.: Howard University Press, 1974). Joseph E. Harris is the editor of the first-mentioned work. The following two works are by Frank M. Snowden, Jr.: i) *Blacks in Antiquity: Ethiopians in Greco-Roman Experience,* (Cambridge, MA: The Belknap Press of Harvard University, 1970); and ii) *Before Color Prejudice: The Ancient View of Blacks,* (Cambridge, MA: Harvard University Press, 1983).

16.   Christopher Ehret, *An African Classical Age: Eastern and Southern Africa in World History, 1000 B.C. to A.D. 400,* (Charlottesville: University Press of Virginia, 1998). Subsequent references to the book appear in the text as *An African.*

17.   John S. Mbiti, *African Religions and Philosophy,* (New Hampshire:

Heinemann Educational Books Inc., 1989. Originally published in 1969). Subsequent references to the book appear in the text as *African Religions*.

18. Sherwin B. Nuland, *How We Live*, (New York: Vintage Books, 1997), pp. 357–358.

19. E. Bolaji Idowu, *Olódùmarè: God in Yoruba Belief*, (Nigeria: Longman, 1962).

20. Similar interviews were recorded on audio tapes.

21. See especially his book, *Consilience: The Unity of Knowledge*, (New York: Knopf, 1998). Subsequent references to the book appear in the text as *Consilience*.

22. Gregory S. Paul and Earl Cox, *Beyond Humanity: CyberEvolution and Future Minds*, (Rockland, MA: Charles River, Inc., 1996). Subsequent references to the book appear in the text as *Beyond Humanity*.

23. Manohar A. Tilak, *Infinities to Eternities: The Cosmic Vision of Evolution*, (Nehru Chowk, Nasik [India]: Abhay & Associates, 1998). Subsequent references to the book appear in the text as *Infinities to Eternities*. Here many of the ideas are not completely developed, and some issues considered need further research so as to become more compatible with the findings of contemporary science or philosophy.

24. Henri Bergson, *Creative Evolution*, trans., Arthur Mitchell, (New York: The Modern Library, 1944).

Chapter 1

1. Eric J. Sundquist, ed., *The Oxford W.E.B. DuBois Reader*, (Oxford: Oxford University Press, 1996). Subsequent references to the book appear in the text as, *The Oxford*.

2. Dr. Goings received his B.S. degree in Education from Kent State University in 1972. In 1974 and 1977, he received his M.A. and Ph.D. degrees, respectively, from Princeton University. The latter two mentioned degrees are in History. The previous institutions

where he was employed before joining the Florida Atlantic University (FAU) faculty included the College of Wooster, Wooster, Ohio, 1976 through 1988. While at the College of Wooster, in addition to his teaching duties, he served in a variety of other capacities: between 1985 and 1986 and 1987 and 1988, he was the chairperson of the Department of History. His administrative skills were further exercised during his directorship of the Office of Off-Campus (International) Studies. Prior to that position, he was director of the Office of Black Affairs, 1980 through 1986. In between joining the FAU faculty and his employment at the College of Wooster, Professor Goings was an Associate Professor of history at Rhodes College in Memphis, Tennessee, from 1988 through June 1991. During his employment at Rhodes College, he was also chairperson of the Department of History. During his illustrious career, Professor Goings has taught a number of courses. Included among these are: i) "Intellectual History of African America;" ii) "History of U.S. Civil Rights Movement;" iii) "The U.S. Constitution and Civil Liberties: The Warren Court, 1953–1969;" iv) "History of American Presidential Assassinations;" v) "Fire in the Streets': American in the 1960s;" vi) "Comparative Slavery: Brazil, Cuba, Jamaica, and The United States;" vii) and, "Afro-American History Survey: West African Origins to the Present." The latter-mentioned course was offered by Dr. Goings during the 1991, Fall Semester; this course was among courses in the Ethnic Studies Program that I founded during my FAU employment between 1988 and 1993. Dr. Goings has been a recipient of various research grants, awards, and honors. In addition to the many articles which Professor Goings has published, his major publication includes his book, *The NAACP Comes of Age: The Defeat of Judge John J. Parker* (Bloomington: Indiana University Press, 1990).

Chapter 2

1. Aimé Césaire, *Discourse of Colonialism*, trans., Joan Pinkham, (New York: Monthly Review Press, 1972. Originally published in 1955). Subsequent references to the book appear in the text as *Discourse*.

2. See his book, *The Wretched of the Earth,* trans., Constance Farrington, (New York: Grove Press, 1968. Originally published in 1963). Subsequent references to the book appear in the text as *The Wretched*.

3.   William James, *The Varieties of Religious Experience*, (New York: The New American Library of World Literature, Inc., 1958). Subsequent references to the book appear in the text as *The Varieties*.

4.   John Dewey, *Theory of the Moral Life*, (New York: Holt, Rinehart, and Winston, 1960. Originally published in 1908). Subsequent references to the book appear in the text as *Theory*.

5.   Joy James, *Transcending the Talented Tenth: Black Leaders and American Intellectuals*, (New York: Routledge, 1997). Subsequent references appear in the text as *Transcending*.

6.   Virginia Held's work under discussion is included in the book, James White, ed., *Contemporary Moral Problems*, 4th ed., (New York: West Publishing Company, 1993). Subsequent references to Held's work appear in the text as *Contemporary*.

Chapter 3

1.   Abraham Pais, "Knowledge and Belief: The Impact of Einstein's Relativity Theory" in *American Scientist*, Vol., 76, March-April, 1988, p. 158. Subsequent references to this article appear in the text.

2.   The complete reference to the book is, Heinz R. Pagels, *The Cosmic Code: Quantum Physics as the Language of Nature*. (New York: A Bantam New Age Book, 1982). Subsequent references to the book appear in the text as *The Cosmic*.

3.   Donald J. Munro, *The Concept of Man in Early China*, (Stanford: Stanford University Press, 1969).

4.   This is an unpublished essay included in the Alain Locke Collection at the Moorland-Spingarn Research Center, Howard University. Subsequent references to this essay appear in the text as "Strategist." Locke himself never published this essay but it is included in my book, *A Journey*.

5.   Booker T. Washington, *Up From Slavery*, (New York: Dell Publishing Company, 1965. Originally published in 1900), p. 8. The introduction

to this 1965 edition was written by Louis Lomax.

6. Houston Baker, *Modernism and the Harlem Renaissance*, (Chicago: University of Chicago Press, 1987), p. 15. Subsequent references to the book appear in the text as *Modernism*.

7. Alain Locke, ed., *The Negro and His Music*, (Washington, D.C.: Associates in Negro Folk Education, 1936), p. 33. Subsequent references to the book appear in the text as *His Music*.

8. Albert Bandura, *Self-Efficacy: The Exercise of Control*, (New York: W.H. Freeman and Company, 1997). Subsequent references to the book appear in the text as *Self-Efficacy*.

Chapter 4

1. William Leo Hansberry, *Africana at Nsukka*, (New York: Viking Press, 1964). Subsequent references to the book appear in the text as *Africana*.

2. William Leo Hansberry, *Africa &Africans as Seen by Classical Writers*, Vol., II, ed., Joseph E. Harris, (Washington, D.C: Howard University Press, 1977). Subsequent references to the book appear in the text as *Africa & Africans*.

3. Henry Olela, *From Ancient Africa to Ancient Greece* (Atlanta, GA: Black Heritage Publication, 1981). Subsequent references to the book appear in the text as *Ancient Africa*.

4. Chiek Anta Diop, *The African Origin of Civilization: Myth or Reality*, trans., Mercer Cook, (New York: L. Hill, 1974), p. 232. Subsequent references to the book appear in the text as *African Origins*.

5. Henri Bergson, *The Two Sources of Morality and Religion*, trans., R. Ashley Audra and Cloudesley Brereton, (Notre Dame, IN: University of Notre Dame Press, 1935), p. 251. Subsequent references appear in the text as *The Two Sources*.

6. H. Odera Oruka, *Oginga Odinga: His Philosophy and Beliefs*, (Nairobi: Ini-

tiatives Publishers, 1992). Subsequent references to the book appear in the text as *Oginga*. See also Oruka's book, *Sage Philosophy: Indigenous Thinkers and Modern Debate on African Philosophy*, (Nairobi: Acts Press, African Center for Technology Studies, 1991). Subsequent references to this book appear in the text as *Sage*.

7. A thorough description of the political-social conditions of certain societies in Medieval Africa and others parts of the world during the turn of the last millennium is in the August 16–23, 1999 special issue of *U.S. News and World Report* magazine.

8. J.A. Rogers, *World's Great Men of Color*. Vol. 1. ed., John Henrik Clarke, (New York: Macmillan, 1972. Originally published in 1946); and J.A. Rogers, *World's Great Men of Color*. Vol 2. ed., John Henrik Clarke, (New York: Macmillan, 1972. Originally published in 1946). Subsequent references to the first-mentioned book appear in the text as *Great Men*, Vol. 1.

9. Cited in Forrest E. Baird and Walter Kaufmann, *Medieval Philosophy*, eds., Vol., II, 2nd ed., (New Jersey: Prentice Hall, 1997), pp. 209–210. Subsequent references to the book appear in the text as *Medieval*.

10. Cited in Albert B. Hakim, *Historical Introduction to Philosophy*, 3rd, ed. (New Jersey: Prentice Hall, 1997), p. 304.

Chapter 5

1. See my essay, "A Commentary on Oshita O. Oshita's Analysis of the Mind-Body Problem in an African World View," *Journal of Social Philosophy*, Vol. XXIV, No. 2, pp. 243–247, Fall 1993.

2. Lerone Bennett, Jr., *What Manner of Man: Martin Luther King, Jr.*, (New York: Pocket Books, 1965), p., 25.

3. Cited in *American Issues: A Documentary Reader*, eds., Charles M. Dollar and Gary W. Reichard (New York: Glencoe, 1994), p. 157. Subsequent references to the book appear in the text as *American Issues*.

4. Cited in Mary Frances Berry and John W. Blassingame, eds., *Long Memory: The Black Experience in America*, (New York: Oxford Univer-

360

sity Press, 1982), p. 394. Subsequent references to the book appear in the text as *Memory*.

5. B.A.G. Fuller and Sterling M. McMurrin, *A History of Philosophy*, 3rd., ed. (New York: Holt, Rinehart and Winston, 1955), Section II, p. 283.

6. Johann Gottlieb Fichte, *The Vocation of Man*, trans., Peter Preuss, (Indianapolis: Hackett Publishing Company, 1987), p. 75. Originally published in 1800. See also Fichte's book, *The Science of Knowledge*, eds., Peter Heath and John Lach, (New York: Meredith Corporation, 1970. Originally published in 1868).

7. Cited in George E. Connor's essay, "The Abuse of History and Theory in Pursuit of Citizenship and Public Affairs" included in *SMSU Journal of Public Affairs*, Vol. 3, 1999, p. 15.

Chapter 6

1. This article, "King Moved Lives from Cellar to Peak," was previously published in the Springfield *News-Leader*, February 21, 1994, p. 6A.

2. Elaine Brown, *A Taste of Power: A Black Woman's Story*, (New York: Pantheon Books, 1992).

3. "Negro Education Bids for Par." *Survey* 54 (September 1, 1925), pp. 567–570,— subsequent references to the article appear in the text as "Negro Education"; ii) "Adult Education for Negroes." *Handbook of Adult Education in the United States.* New York: American Association for Adult Education, 1926, pp. 121–131; iii) "The High Cost of Prejudice." *Forum* 78 (December 1927), pp. 500–510; iv) "Some Lessons from Negro Adult Education." *Proceedings of the Sixth Annual Conference of the American Association for Adult Education* (May 1934), n.p; v) "Reciprocity Instead of Regimentation: Lessons of Negro Adult Education." *Journal of Adult Education* 6 (October 1934), pp. 418–420; vi) "The Eleventh Hour of Nordicism: A Retrospective Review of the Literature of the Negro for 1934." *Opportunity* 13 (January-February 1935), pp. 8–12; vii); "Minorities and the Social Mind." *Progressive Education* 12 (March 1935), p. 28; viii) "The Di-

lemma of Segregation." *Journal of Negro Education* 4 (July 1935), pp. 406–411—subsequent references to the article appear in the text "Dilemma"; ix) "Values and Imperatives," *American Philosophy, Today and Tomorrow*. Ed., Sidney Hook and Horace M. Kallen. (New York: Lee Furman, 1935), pp. 313–333.

Chapter 7

1. Published in the June 1989 issue of *Scientific American*, p. 132. During the time the article was published, Edward Rubenstein was associate dean for postgraduate medical education at the Stanford University School of Medicine. All passages cited in this letter are extracts from page p. 132, mentioned above.

2. See Richard Morris' book, *The Edges of Science: Crossing the Boundary from Physics to Metaphysics*, (New York: Prentice Hall, 1990), pp. 201–208. Moreover, during November 16, 1998, it was reported in a newscast concerning scientific research that some scientists in California discovered that the speed of light is not *constant* as Einstein would have us believe. Rather, since the big bang, the speed of light has been decreasing at an alarming rate, and it is still doing so. How can it be that the speed of light is decreasing, while the expansion of the universe is increasing? Is not this a paradox? The light problem also seems to go against Tilak's view that, since modernity—an era he believes was ushered in by the Industrial Revolution of the 1800s—our activities are occurring exponentially.

3. John H. Bryant reminded me of a TV documentary entitled, "Bill Nye, the Science Guy" that described the nature and physiology of the brain. During the week of April 13, 1998, NASA launched a space ship into space that included small animals and insects to study the effect of a gravity-free environment on brain and nerve cell growth patterns.

4. The letters from scientists I have inserted here are included in the many papers or documents that Tilak has provided me, inasmuch as I am one of his collaborators. He has also provided me permission to include these letters and supporting materials in this volume. I should add that I have made a few minor Editor's stylistic changes in

some of these letters, in order to make them consistent with the style of this volume.

Chapter 8

1. Alain Locke, "Concept of Race as Applied to Social Culture" in *Howard Review*, 1 (June 1924), pp. 290–299.

Chapter 9

1. Eric J. Chaisson, *The Life Era: Cosmic Selection and Conscious Evolution*, (New York: The Atlantic Monthly Press, 1987), p. 229. I did not criticize the views expressed in Chaisson's book at this point in time, as that would have taken me too far afield. Further, I called attention to Chaisson's view in which stress is on cosmic self-knowledge, to highlight the fact that science and philosophy are converging on certain themes that were initially developed by the German Idealism tradition that includes Fichte, Kant and Hegel. Thus, I am confident that I am in good company when I draw heavily on the works of Fichte, whose works are eclipsed by contemporary analytic philosophy, phenomenology and existentialism, etc. Moreover, Chaisson placed much emphasis on ethics, whereas Tilak did not do so. If Chaisson placed emphasis on the positive aspect of cosmic evolution with galactic ethics at the center that offers hope, Tilak places emphasis on the negative aspect of evolution, the MPEC, without an adequate ethics. This is one of the weaknesses of the Evoluon Theory that makes it less appealing. I try to remedy this by introducing the caring ethics that is equally applicable to the galactic ethics. Tilak is unable to find an adequate ethics in the teachings of Husserl. Moreover, it is good for philosophers to know that a scientist such as Chaisson is interested in a cosmic ethics and metaphysics, and if they, as philosophers, do not go ahead and develop a unified view of the cosmos in which values and facts are integrated, scientists will do so instead. Analytic Philosophy has paralyzed contemporary philosophy that lacks the "nerves" to construct a grand view of the universe. Inasmuch as I am criticizing Analytic Philosophy and am an admirer of Hegel, Fichte and Bergson, I am an "outsider."

2. Frank J. Tipler, *The Physics of Immortality: Modern Cosmology, God and the Resurrection of the Dead*, (New York: Doubleday, 1994), pp. 112–113.

3.   B.A.B. Fuller and Sterling M. McMurrin, *A History of Philosophy,* (New York: Holt, Rinehart, and Winston, 3rd., edition, 1955, Contents—II,) p. 281.

Appendix A

1.   "New Postulate on the General Theory of Evolution, Evolvability, Direct Phase Evolution, Evoluon, and Rate of Evolutionary Change"was originally published in the September-October 1973 issue of the journal, *The Accelerator, Official Publication of the American Chemical Society.* The article is now included in Appendix A of Tilak's book, *Infinities to Eternities.* The version of the article on which I am relying is extracted from the latter-mentioned source. I have made a few Editor's changes in the article of Appendix A, so that so that its style and theme will conform to style or format of the present volume.

Appendix B

1.   "The Evoluon Theory: A General Unified Theory of Evolution" was originally published in the journal, *Science and Culture Supplement,* Vol., 45. No., 4, April, 1979. The article is now included in Appendix D of Tilak's book, *Infinities to Eternities.* The version of the article on which I am relying is extracted from the last-mentioned source. I have made a few Editor's changes in the article of Appendix B so that its style and theme will conform to style or format of the present volume.

Selected Bibliography

Alkalimat, Abdul. "Studies on Malcolm X." *Sage Race Relations Abstracts*. Vol. 17, Num. 4, Nov. 1992.

Angelou, Maya. *I Know Why the Caged Bird Sings*. New York: Random House, 1970. Originally published in 1969.

Ani, Marimba. *Yurugu: An African-Centered Critique of European Cultural Thought and Behavior*. Trenton, New Jersey: Africa World Press, 1994.

Appiah, Kwame Anthony. *In My Father's House: Africa in the Philosophy of Culture*. New York: Oxford University Press, 1993.

Arendt, Hannah. *Between Past and Future: Eight Exercises in Political Thought*. New York: The Viking Press, 1968.

———. *The Human Condition*. Chicago: The University of Chicago Press, 1958.

Asante, Molefi K. *Afrocentricity*. Trenton, New Jersey: Africa World, 1988.

Atkinson, Ronald. *The Roots of Ethnicity: The Origins of the Acholi of Uganda*. Philadelphia: University of Pennsylvania Press, 1994.

Baird, Forrest E., and Walter Kaufmann, eds., *Medieval Philosophy*. Vol. II, 2nd ed. New Jersey: Prentice Hall, 1997.

Baker, Houston. *Modernism and the Harlem Renaissance*. Chicago: University of Chicago Press, 1987.

Bandura, Albert. *Self-Efficacy: The Exercise of Control*. New York: W.H. Freeman and Company, 1997.

Bendix, Reinhard. *Max Weber: An Intellectual Portrait*. Garden City, New York: Doubleday & Company, Inc., 1962.

Bennett Jr., Lerone. *What Manner of Man: Martin Luther King, Jr.* New York: Pocket Books, 1965.

Bergson, Henri. *Creative Evolution.* Translated by Arthur Mitchell. New York: The Modern Library, 1944.

————. *The Two Sources of Morality and Religion.* Translated by R. Ashley Audra and Cloudesley Brereton. Notre Dame, IN: University of Notre Dame Press, 1935.

Berry, Mary Frances and John W. Blassingame, eds. *Long Memory: The Black Experience in America.* New York: Oxford University Press, 1982.

Bhagavad-Gita. Translated into English under the editorship of Christopher Isherwood and Swami Prabhavanaada. New York: New American Library, 1972.

Blight, David W., ed. *Narrative of the Life of Frederick Douglass: An American Slave Written by Himself.* Boston, MA: Bedford Books, 1993.

Bogdanov, Konstantin. *Biology in Physics: Is Life Matter?* London: Academic Press, 2000.

Bosanquet, Bernard. *The Value and Destiny of the Individual: The Gifford Lectures for 1912 delivered in Edinburgh University.* London: Macmillan, 1913.

————. *The Principle of Individuality and Value: The Gifford Lectures for 1911 delivered in Edinburgh University.* London: Macmillan, 1912.

Broderick, Francis L. *W.E.B. DuBois: Negro Leader in a Time of Crisis.* Stanford, CA: Stanford University Press, 1959.

Brown, Elaine. *A Taste of Power: A Black Woman's Story.* New York: Pantheon Books, 1992.

Buck, Christopher. "Alain Locke: Baha'i Philosopher," *Baha'i Studies Review* 10 (2001), pp. 7-50.

Byerman, Keith. *Seizing the Word: History, Art and the Self in the Work of W.E.B. DuBois.* Athens: University of Georgia Press, 1994.

Cairns-Smith, A.G. *Secrets of the Mind. A Tale of Discovery and Mistaken Identity.*

New York: Springer-Verlag, 1999.

Césaire, Aimé. *Discourse of Colonialism*. Translated by Joan Pinkham. New York: Monthly Review Press, 1972. Originally published in 1955.

Chaisson, Eric J. *The Life Era: Cosmic Selection and Conscious Evolution*. New York: The Atlantic Monthly Press, 1987.

Cone, James H. *Martin & Malcolm & America: A Dream Or A Nightmare*. Maryknoll, New York: Orbis Books, 1991.

Cox, Earl and Gregory S. Paul. *Beyond Humanity: CyberEvolution and Future Minds*. Rockland, MA: Charles River, Inc., 1996.

Crummell, Alexander. *Destiny and Race: Selected Writings 1848–1897*. Edited by W.J. Moses. Amherst, MA: University of Massachusetts Press, 1992.

Danielson, Peter. *Artificial Morality: Virtuous Robots for Virtual Games*. New York: Routledge, 1992.

Davis, Angela. *Women, Race and Class*. New York: Random House, 1981.

Davis, Lenwood. *Malcolm X: A Selected Bibliography*. Westport, CT: Greenwood Press, 1984.

Dawkins, Richard. *The Selfish Gene*. Oxford: Oxford University Press, 1989.

————. *The Blind Watchmaker: Why the Evidence of Evolution Reveals A Universe Without Design*. New York: W.W. Norton and Company, 1987.

Dear, Michael J. *The Postmodern Urban Condition*. Malden, MA: Blackwell Publishers Inc., 2000.

Dewey, John. *The Quest for Certainty: A Study of the Relation of Knowledge and Action*. New York: Putnam, 1960. Originally published in 1929.

————. *The Public and Its Problems*. Chicago: The Swallow Press, Inc., 1927.

————. *Theory of the Moral Life*. New York: Holt, Rinehart, and Winston, 1960. Originally published in 1908.

Diop, Chiekh Anta. *The African Origin of Civilization: Myth or Reality*. Trans-

368

lated by Mercer Cook. New York: Lawrence Hill Books, 1974.

Dodd, Nigel. *Social Theory and Modernity*. Malden, MA: Blackwell Publishers Inc., 1999.

Dollar, Charles M., and Gary W. Reichard, eds. *American Issues: A Documentary Reader*. New York: Glencoe, 1994.

DuBois, Felix. *Timbuctoo: The Mysterious*. Translated by Diana White. Westport, CT: Greenwood Press, 1970.

DuBois, W.E.B. *The Souls of Black Folk*. Edited by David W. Blight and Robert Gooding-Williams. New York: Bedford Books, 1991. Originally published in 1903.

———. *Dusk of Dawn*. New York: Harcourt Brace, 1940.

———. *The Suppression of the African Slave Trade to the United States of America, 1638–1870*. Cambridge: Harvard University Press, 1896.

Dyson, George B. *Darwin Among Machines: The Evolution of Global Intelligence*. New York: Addison-Wesley Publishing Company, Inc., 1997.

Ehret, Christopher. *An African Classical Age: Eastern and Southern Africa in World History, 1000 B.C. to A.D. 400*. Charlottesville: University Press of Virginia, 1998.

Eigen, Manfred and Ruthild Winkler. *Laws of the Game: How the Principles of Nature Govern Chance*. Princeton, NJ: Princeton University Press, 1993.

———. *Steps Towards Life: A Perspective on Evolution*. Oxford: Oxford University Press, 1992.

———. *Perspektiven der Wissenschaft*. Stuttgart: Deutsche Verlagsanstalt, 1988.

Eigen, Manfred and Peter Schuster. *The Hypercycle—A Principle of Natural Self-Organization*. Berlin: Springer-Verlag, 1979.

Elliott, Anthony, ed. *Contemporary Social Theory*. Malden, MA: Blackwell Publishers Inc., 1999.

Ellison, Ralph. *Invisible Man*. New York: Random House, 1952.

Fanon, Frantz. *The Wretched of the Earth*. Translated by Constance Farrington. New York: Grove Press, 1968. Originally published in 1963.

Farber, Marvin. *The Foundation of Phenomenology: Edmund Husserl and the Quest for a Rigorous Science of Philosophy*. 3rd ed., Albany, NY: State University of New York Press, 1943.

Ferguson, Eva Dreikurs. *Motivation: A Biosocial and Cognitive Integration of Motivation and Emotions*. Oxford: Oxford University Press, 2000.

Ferrari, Michel and Robert J., Sternberg, eds. *Self-Awareness: Its Nature and Development*. New York: The Guilford Press, 1998.

Fichte, Johann Gottlieb. *The Vocation of Man*. Translated by Peter Preuss. Indianapolis: Hackett Publishing Company, 1987. Originally published in 1800.

————. *The Science of Knowledge*. Edited by Peter Heath, and John Lach. New York: Meredith Corporation, 1970. Originally published in 1868.

Ford, Kenneth M., Clark Glymour and Patrick J. Hayes. *Android Epistemology*. Menlo Park, CA: AAAI Press/The MIT Press, 1995.

Franklin, John Hope, and Alfred A. Moss, Jr. *From Slavery to Freedom: History of African Americans*. 7th ed. New York : Alfred A. Knopf, 1994.

Franklin, John Hope, and August Meier, eds. *Black Leaders of the Twentieth Century*. Urbana, IL: University of Illinois Press, 1982.

Franklin, Stanley P. *Artificial Minds*. Cambridge, MA: The MIT Press, 1995.

Frazier, E. Franklin. *The Negro Family in the United States*. Chicago: The University of Chicago Press, 1968. Originally published in 1939.

Frederickson, George M. *The Black Image in the White Mind: The Debate on Afro-American Character and Destiny, 1817–1914*. New York: Harper and Row, 1971.

Fuller, B.A.G. and Sterling M. McMurrin. *A History of Philosophy,* 3rd., ed. New York: Holt, Rinehart and Winston, 1955.

Gates, Henry Louis. *The Signifying Monkey: A Theory of Afro-American Literary Criticism*. New York: Oxford University Press, 1988.

Geertz, Clifford. *Available Light: Anthropological Reflections on Philosophical Topics*. Princeton, NJ: Princeton University Press, 2000.

Gell-Mann, Murray. *The Quark and the Jaguar: Adventures in the Simple and the Complex*. New York: W.H. Freeman and Company, 1994.

Gleick, James. *Chaos: Making a New Science*. New York: Viking Press, 1987.

Goings, Kenneth. *The NAACP Comes of Age: The Defeat of Judge John J. Parker*. Bloomington, IN: Indiana University Press, 1990.

Grossman, James R. *The Land of Hope: Chicago, Black Southerners, and the Great Migration*. Chicago: The University of Chicago Press, 1991.

Hansberry, William Leo. *Africa & Africans: As Seen by Classical Writers: The William Leo Hansberry African History Notebook* Vol. II. Edited by Joseph E. Harris. Washington, D.C.: Howard University Press, 1977.

————. *Pillars in Ethiopian History: The William Leo Hansberry Notebook* Vol. I. Edited by Joseph E. Harris. Washington, D.C: Howard University Press, 1974.

————. *Africana at Nsukka*. New York: Viking Press, 1964.

Harris, Joseph E. *African-Americans Reaction to the War in Ethiopia, 1936–1941*. Baton Rouge, LA: Louisiana State University Press, 1994.

Harris, Leonard, ed. *The Philosophy of Alain Locke: Harlem Renaissance and Beyond*. Philadelphia: Temple University Press, 1989.

Haugeland, John. *Mind Design II: Philosophy, Psychology and Artificial Intelligence*. Cambridge, MA: The MIT Press, 1997.

Heelas, Paul, David Martin, and Paul Morris, eds. *Religion, Modernity and Postmodernity*. Malden, MA: Blackwell Publishers, 1998.

Hegel, G.W.F. *Phenomenology of Spirit*. Translated by A.V. Miller; with analysis of the text and foreword by J.N. Findlay. Oxford: Clarendon Press, 1977.

———. *Reason in History*. Translated, with an introduction, by Robert S. Hartman. New York: The Bobbs-Merrill Company, Inc., 1953. Originally published in 1837.

[h]ooks, [b]ell. *Ain't I A Woman: Black Women and Feminism*. Boston: South End Press, 1981.

Hopkins, Dwight N. *Introducing Black Theology of Liberation*. Maryknoll, New York: Orbis Books, 1999.

Howell, Stephen. *Afrocentrism: Mythical Pasts and Imagined Homes*. New York: Verso, 1998.

Hume, David. *A Treatise of Human Nature*. New York: E.P. Dutton & Company, 1911.

Husserl, Edmund. *Ideas: General Introduction to Pure Phenomenology*. Translated by W.R. Boyce Gibson. Atlantic Highlands, NJ: Humanities Press, 1931.

———. *Logical Investigations*. Translated by J.N. Findlay. New York: Humanities Press, 1970. 2 vols. (Based on revised Halle editions)

Idowu, E. Bolaji. *Olódùmarè: God in Yoruba Belief*. Nigeria: Longman, 1962.

Jacques-Garvey, Amy. *Philosophy and Opinions of Marcus Garvey*, ed. New York: Atheneum Press, 1977.

Jahn, Janheinz. *Muntu: An Outline of the New African Culture*. Translated by Marjorie Green. New York: Grove Press, Inc., 1961. Originally published in 1958.

James, Joy. *Transcending the Talented Tenth: Black Leaders and American Intellectuals*. New York: Routledge, 1997.

James, William. *The Varieties of Religious Experience*. New York: The New American Library of World Literature, Inc., 1958.

Jaspers, Karl. *Philosophy of Existence*. Translated by Richard F. Grabau. Philadelphia: University of Pennsylvania Press, 1971.

———. *Way to Wisdom*. Translated by Ralph Manheim. New Haven, CT: Yale University Press, 1954.

Johnson, Timothy V. *Malcolm X: A Comprehensive Annotated Bibliography.* New York: Garland Publishers, 1986.

Johnston, Victor S. *Why We Feel: The Science of Human Emotions.* Cambridge, MA: Perseus Books, 1999.

King, Jr., Martin Luther, *Why We Can't Wait: Martin Luther King, Jr.* New York: New American Library, 1964.

Kant, Immanuel. *Critique of Pure Reason.* Translated by Norman Kemp Smith. New York: St. Martin's Press, 1965. Originally published in 1787.

Lachs, John. *Fichte: Science of Knowledge. (Wissenschaftslehre),* New York: Cambridge University Press, 1982. Originally published in 1797.

Levinas, Emmanuel. *Discovering Existence with Husserl.* Edited and translated by Richard A. Cohen and Michael B. Smith. Evanston, IL: Northwestern University Press, 1998.

Levy, Steven. *Artificial Life: the Quest for a New Creation.* New York: Pantheon Books, 1992.

Lewis, David Levering. *W.E.B. DuBois: The Biography of a Race, 1869–1919.* New York: Henry Holt, 1993.

Lincoln, Eric and Mamiya Lawrence. *The Black Church and the African American Experience.* Durham: Duke University Press, 1993.

Locke, Alain, ed. *The Negro and His Music.* Washington, D.C.: Associates in Negro Folk Education, 1936.

———. *The New Negro.* New York: Albert and Charles Boni, Inc., 1925. Reprinted by Atheneum, 1968.

Lotze, Hermann. *Outlines of Psychology.* Translated, with a chapter on the anatomy of the brain, by C.L. Herrick. New York: Arno Press, 1973.

Lyotard, Jean-François. *Postmodern Fables.* Translated by Georges Van Den Abbeele. Minneapolis: University of Minnesota Press, 1997.

Machiavelli, Niccolo. *The Prince.* Translated by Paul Sonnino. New Jersey: Humanities Press, 1996.

Malcolm X. *Malcolm X Speaks*. New York: Grove Press, 1965.

Marable, Manning. *W.E.B. DuBois: Black Radical Democrat*. Boston: Twayne Publishers, 1986.

Mbiti, John S. *African Religions and Philosophy*. New Hampshire: Heinemann Educational Books Inc., 1989. Originally published in 1969.

McGary, Howard. "Alienation and the African-American Experience," *Philosophical Forum* 24 (1992), pp. 282–296.

Meier, August. *Negro Thought in America, 1880–1915: Racial Ideologies in the Age of Booker T. Washington*. Ann Arbor: University of Michigan Press, 1963.

Merleau-Ponty, Maurice. *Sense and Non-Sense*. Translated by Hubert L. Dreyfus and Patricia A. Dreyfus. Evanston: Northwestern University Press, 1964.

Minsky, Marvin. *The Society of Mind*. New York: Simon & Schuster, 1988.

Montmarquet, James A., and William H. Hardy. *Reflections: An Anthology of African American Philosophy*. Belmont: Wadsworth, 2000.

Moravec, Hans. *Mind Children: The Future of Robot and Human Intelligence*. Cambridge: Harvard University Press, 1988.

Murdock, G.P. *Africa: Its Peoples and Their Culture History*. New York: McGraw-Hill, 1959.

Monika, Kilian. *Modern and Postmodern Strategies: Gaming and the Question of Morality: Adorno, Rorty, Lyotard, and Enzensberger*. New York: Peter Lang Publishing, Inc., 1998.

Morris, Richard. *The Edges of Science: Crossing the Boundary from Physics to Metaphysics*. New York: Prentice Hall, 1990.

Munro, Donald J. *The Concept of Man in Early China*. Stanford: Stanford University Press, 1969.

Nietzsche, Friedrich. *The Will to Power*. Edited and translated by Walter Kaufmann and R.J. Hollingdale. New York: Random House, 1967.

Nuland, Sherwin B. *How We Live*. New York: Vintage Books, 1997.

Olela, Henry. *From Ancient Africa to Ancient Greece*. Atlanta: Black Heritage Publication, 1981.

Oruka, H. Odera. *Oginga Odinga: His Philosophy and Beliefs*. Nairobi: Initiatives Publishers, 1992.

————. *Sage Philosophy: Indigenous Thinkers and Modern Debate on African Philosophy*. Nairobi: Acts Press, African Center for Technology Studies, 1991.
Outlaw, Lucious. "African Philosophy," *The Journal of Ethics* 1 (1997), pp. 265–290.

Pagel, Heinz R. *The Cosmic Code: Quantum Physics as the Language of Nature*. New York: A Bantam New Age Book, 1982.

Palm Beach Gazette (Palm Beach, Florida), a weekly newspaper.

Popper, Karl. *The Logic of Scientific Discovery*, Routledge, 1977. First English ed., Hutchinson, 1959. First published as *Logik der Forschung* in Vienna: Springer, 1934.

Prigogine, Ilya and Isabelle Stengers. *Order Out of Chaos: Man's New Dialogue with Nature*. Boulder, CO: New Science Library, 1984.

Quarles, Benjamin. *The Negro in The Making*. New York. Collier Books, 1969.

Quine, Willard Van Orman. *From a Logical Point of View*. Cambridge: Harvard University Press, 1953.

Robinson, Randall. *The Debt: What America Owes to Blacks*. New York: Dutton/Plume, 1999.

Rogers, J.A. *World's Great Men of Color.*. Edited by John Henrik Clarke. 2 vols. New York: Macmillan, 1972. Originally published in 1946.

Rojek, Chris and Bryan S. Turner, eds. *The Politics of Jean-François Lyotard: Justice and Political Theory*. New York: Routledge, 1988.

Royce, Josiah. *The Philosophy of Loyalty.* New York, Hafner Publishing Company, 1971. Originally published in 1908.

Samuel, David. *Memory: How We Use It, Lose It and Can Improve It.* London: Weidenfeld & Nicolson, 1999.

Santayana, George. *The Sense of Beauty: Being the Outline of Aesthetic Theory.* New York: Dover Publications, 1955. Originally published in 1896.

Seale, Bobby. *Seize the Time: The Story of the Black Panther Party and Huey P. Newton.* Baltimore, MD: Black Classic Press, 1991.

Smolin, Lee. *The Life of the Universe.* New York: Oxford University Press, 1997.

Snowden, Jr., Frank M. *Blacks in Antiquity: Ethiopians in Greco-Roman Experience.* Cambridge, Massachusetts: The Belknap Press of Harvard University, 1970.

————. *Before Color Prejudice: The Ancient View of Blacks.* Cambridge, Massachusetts: Harvard University Press, 1983.

Spiegelberg, Herbert. *The Phenomenological Movement,* 2 vols. The Hague, Netherlands: Martinus Nijhoff, 1960.

Spinoza, Benedict De. *Ethics and On the Improvement of the Understanding.* Edited with an introduction by James Gutmann. New York: Hafner Publishing Company, 1949.

Steele, Shelby. *The Contents of Our Character: A New Vision of Race in America.* New York: Harper Perennial, 1990.

Sundquist, Eric J., ed. *The Oxford W.E.B. DuBois Reader.* Oxford: Oxford University Press, 1996.

Swimme, Brian. *The Universe is a Green Dragon: A Cosmic Creation Story.* Santa Fe, New Mexico: Bear & Company Publishing, 1984.

Thompson, Mildred I. *Ida B. Wells-Barnett: An Explanatory Study of an American Black Woman, 1893–1930.* Brooklyn, N.Y.: Carlson Publisher, 1990.

Tilak, Manohar A. *Infinities to Eternities: The Cosmic Vision of Evolution.* Nehru Chowk, Nasik [India]: Abhay & Associates, 1998.

Tipler, Frank J. *The Physics of Immortality: Modern Cosmology, God and the Resurrection of the Dead.* New York: Doubleday, 1994.

Tucker, Mary Evelyn, and John A. Grim. *Worldviews and Ecology.* Lewisburg, PA: Bucknell University Press, 1993.

Tulving, Endel, ed. *Memory, Consciousness, and the Brain.* Ann Arbor: Psychology Press, 2000.

The Upandishads: Breath of the Eternal. Translated into English, from original Sanskrit, under the editorship of Frederick Manchester and Swami Prabhavananda. New York: The New American Library of World Literature, 1957.

Washington, Booker T. *Up From Slavery.* Introduction by Louis Lomax. New York: Dell Publishing Company, 1965. Originally published in 1900.

Washington, Johnny. *Destiny: The Ideal of Unity* [A video recording]. Producers: David Dixon, J'Nell Jones, and Don Hendricks; Producer-Director, Tom Carter, 1999.

————. *Alain Locke and Philosophy: A Quest for Cultural Pluralism.* New York: Greenwood Press, 1986.

————. *A Journey into the Philosophy of Alain Locke.* New York: Greenwood Press, 1994.

West, Cornel. *The Cornel West Reader.* New York: Basic Civitas Books, 1999.

————. *Prophecy Deliverance! An Afro-American Revolutionary Christianity.* Philadelphia: Westminster Press, 1982.

White, James, ed. *Contemporary Moral Problems.* 4th ed. New York: West Publishing Company, 1993.

White, Michael. *Isaac Newton: The Last Sorcerer.* London: Fourth Estate, 1997.

Whitehead, Alfred North. *The Aims of Education and Other Essays.* Toronto, Ontario: The Macmillan Company, 1929.

————. *Process and Reality.* New York: Macmillan, 1929.

Whyte, Lancelot Law. *Focus and Diversions*. New York: G. Braziller, 1963.

Wilson, Edward O. *Consilience: The Unity of Knowledge*. New York: Knopf, 1998.

Wilson, William Julius. *When Work Disappears: The World of the New Urban Poor*. New York: Alfred A. Knopf, 1996.

Wiredu, Kwasi. *Philosophy and African Culture*. Cambridge: Cambridge University Press, 1980.

Zack, Naomi. *Mixed Race*. Philadelphia: Temple University Press, 1993.

Zamir, Shamoon. *Dark Voices: W.E.B. DuBois and American Thought, 1888–1903*. Chicago: University of Chicago Press, 1995.

Index

384

171; crises of modern Europe,
296; development, 55, 79, 89;
difference, 77–78, 125; diversity, 26, 134; evolution, 139,
264; expressions, 13, 37, 189,
199, 238; groups, 15, 27, 134;
identity, 19, 29, 41; intolerance,
249; pluralism, 280; saturation,
43; unity, 62; values, 184, 225

Dakar, Senegal, 2, 15, 74–75, 79–
82
Dark Ages, 55
Darwin, Charles, 232, 274, 298
Deficiently actual, God is, 182
Delhi, 83–84; *see also* Traveling in
Ahmadabad (India), 83–88
Descartes, René, 120
Destinicity: definition of, 13, 41,
128, 178, 189, 199; Studies,
13–14, 94, 181–182, 205, 248
Destiny: and Cosmic mode of, 12,
15, 28, 32, 175; definition of, 9,
177, 232, 243; and Ethnic
mode of, 127, 191, 251; Ideals,
41, 237, 245; and the immanent, 9–11, 15, 23–25, 27, 32,
35–36, 94, 134, 137, 152, 158,
170, 178, 190–191, 199, 240–
241, 244, 259, 265, 273, 292,
299, 308; modes, 10, 12, 16,
127, 135, 183, 198, 251; and
National mode of, 12, 127,
192; notion of, 2, 24, 174, 185,
199, 314; polarity, 10; "rotten,"
56; schemata, 12, 45; Studies,
13–14, 231; and the transcendent, 9–13, 16, 21, 25, 45, 135,
164, 184, 199, 243–244, 292,
299, 308, 320; and World
mode of, 12, 219, 319
Destiny model: and absolutism,
125; advocated, 12; and the
boundary conditions of modernity, 20; and a caring ethics,
115; and the Christian God, 9;
and conflict and diversity, 281;
and *Ebonics,* 37; and economic-
cultural development, 15, 89;
and ethnic experiences, 13; Ida
B. Wells and others, 38; and
Judaic-Christian religion, 246;
and Manohar A. Tilak's works,
28, 265, 298, 305; and the
McGhee's plan, 16; and narrow perspectives, 290; and National Urban League, 223; and
notion of struggle, 24, 26; and
notion of unity, 240; offers a
forum, 14; offers a ray of
hope, 32; and the Pan-African
model, 64; and Phenomenology, 236; and race or ethnicity,

396

STUDIES IN LITERARY CRITICISM & THEORY

Hans Rudnick, General Editor

The focus of this series is on studies of all literary genres that elucidate and interpret works of art in the context of criticism and theory. Theory and criticism are held to provide the hermeneutically most rewarding access to specific authors, works, and issues under consideration. Studies of a comparative nature with special reference to issues of literary history, criticism, and postmodern theory are the distinctive features of this monograph series. Emphasis is on subjects that may set trends, generate discussion, expand horizons beyond present perspectives, and/or redefine previously held notions about "major" and "minor" authors and their achievements within or outside the canon. Approaches may center on works, authors, or abstract notions of criticism and/or theory, including issues of a comparative nature concerning world literature.

For additional information about this series or for the submission of manuscripts, please contact:

> Peter Lang Publishing
> Acquisitions Department
> P.O. Box 1246
> Bel Air, Maryland 21014-1246

To order other books in this series, please contact our Customer Service Department:

> 800-770-LANG (within the U.S.)
> (212) 647-7706 (outside the U.S.)
> (212) 647-7707 FAX

or browse online by series at:

> www.peterlang.com